D1070673

HELM IDENTIFICATION GUIDES

NEW WORLD BLACKBIRDS

THE ICTERIDS

HELM IDENTIFICATION GUIDES

NEW WORLD BLACKBIRDS

THE ICTERIDS

Alvaro Jaramillo and Peter Burke

CHRISTOPHER HELM
A & C Black • London

DEDICATIONS

For Katja — Alvaro Jaramillo

For Dawn — Peter Burke

© 1999 Alvaro Jaramillo and Peter Burke

Christopher Helm (Publishers) Ltd, a subsidiary of A & C Black
(Publishers) Ltd, 35 Bedford Row, London WC1R 4JH

0-7136-4333-1

A CIP catalogue record for this book is available from the
British Library

All rights reserved. No part of this publication may be
reproduced or used in any form or by any means — photographic,
electronic or mechanical, including photocopying, recording,
taping or information storage and retrieval systems — without
permission of the publishers.

Typeset and designed by D & N Publishing, Membury Business Park,
Lambourn Woodlands, Hungerford, Berkshire, UK

Printed in Singapore

CONTENTS

INTRODUCTION **9**
ACKNOWLEDGEMENTS **10**
SYSTEMATICS AND TAXONOMY **12**
PLUMAGES AND TOPOGRAPHY **16**
BEHAVIOUR AND EVOLUTION **19**
NOTES ON THE SPECIES ACCOUNTS **22**
NOTES ON THE COLOUR PLATES, CAPTIONS AND MAPS **25**
GLOSSARY **26**
COLOUR PLATES AND DISTRIBUTION MAPS **27**
SPECIES ACCOUNTS **107**

1 **Casqued Oropendola** *Psarocolius oseryi* **107**
2 **Crested Oropendola** *Psarocolius decumanus* **108**
3 **Green Oropendola** *Psarocolius viridis* **112**
4 **Dusky-green Oropendola** *Psarocolius atrovirens* **114**
5 **Russet-backed Oropendola** *Psarocolius angustifrons* **117**
5X **'Yellow-billed Oropendola'** *Psarocolius (angustifrons) alfredi* **119**
6 **Chestnut-headed Oropendola** *Psarocolius wagleri* **123**
7 **Montezuma Oropendola** *Psarocolius montezuma* **125**
8 **Baudo Oropendola** *Psarocolius cassini* **128**
9 **Olive Oropendola** *Psarocolius (bifasciatus?) yuracares* **129**
10 **Para Oropendola** *Psarocolius bifasciatus* **131**
11 **Black Oropendola** *Psarocolius guatimozinus* **132**
12 **Band-tailed Oropendola** *Ocyalus latirostris* **134**
13 **Yellow-rumped Cacique** *Cacicus cela* **136**
14 **Red-rumped Cacique** *Cacicus haemorrhous* **140**
15 **Scarlet-rumped Cacique** *Cacicus microrhynchus* **142**
16 **Subtropical Cacique** *Cacicus uropygialis* **144**
17 **Golden-winged Cacique** *Cacicus chrysopterus* **145**
18 **Mountain Cacique** *Cacicus chrysonotus* **147**
19 **Selva Cacique** *Cacicus koepckeae* **149**
20 **Ecuadorian Cacique** *Cacicus sclateri* **150**
21 **Solitary Cacique** *Cacicus solitarius* **151**
22 **Yellow-winged Cacique** *Cacicus melanicterus* **153**
23 **Yellow-billed Cacique** *Amblycercus holosericeus* **155**
24 **Moriche Oriole** *Icterus chrysocephalus* **157**
25 **Epaulet Oriole** *Icterus cayanensis* **159**
26 **Yellow-backed Oriole** *Icterus chrysater* **162**
27 **Yellow Oriole** *Icterus nigrogularis* **164**
28 **Jamaican Oriole** *Icterus leucopteryx* **167**
29 **Orange Oriole** *Icterus auratus* **169**
30 **Yellow-tailed Oriole** *Icterus mesomelas* **172**
31 **Orange-crowned Oriole** *Icterus auricapillus* **174**
32 **White-edged Oriole** *Icterus graceannae* **176**
33 **Spot-breasted Oriole** *Icterus pectoralis* **178**
34 **Altamira Oriole** *Icterus gularis* **181**
35 **Streak-backed Oriole** *Icterus pustulatus* **183**
36 **Hooded Oriole** *Icterus cucullatus* **189**
37 **Troupial** *Icterus icterus* **193**

38 **Campo Troupial** *Icterus jamaicaii* **196**
39 **Orange-backed Troupial** *Icterus croconotus* **197**
40 **Baltimore Oriole** *Icterus galbula* **199**
40X **Hybrid Baltimore × Bullock's Orioles** *Icterus galbula × bullockii* **205**
41 **Bullock's Oriole** *Icterus bullockii* **206**
42 **Black-backed Oriole** *Icterus abeillei* **211**
43 **Orchard Oriole** *Icterus spurius* **213**
43X **'Fuerte's Oriole'** *Icterus spurius fuertesi* **217**
44 **Black-vented Oriole** *Icterus wagleri* **218**
45 **Black-cowled Oriole** *Icterus dominicensis* **221**
46 **Montserrat Oriole** *Icterus oberi* **226**
47 **Martinique Oriole** *Icterus bonana* **227**
48 **St Lucia Oriole** *Icterus laudabilis* **229**
49 **Audubon's Oriole** *Icterus graduacauda* **231**
50 **Bar-winged Oriole** *Icterus maculialatus* **233**
51 **Scott's Oriole** *Icterus parisorum* **235**
52 **Jamaican Blackbird** *Nesopsar nigerrimus* **238**
53 **Oriole Blackbird** *Gymnomystax mexicanus* **240**
54 **Yellow-headed Blackbird** *Xanthocephalus xanthocephalus* **242**
55 **Saffron-cowled Blackbird** *Xanthopsar flavus* **245**
56 **Pale-eyed Blackbird** *Agelaius xanthophthalmus* **248**
57 **Unicolored Blackbird** *Agelaius cyanopus* **249**
58 **Yellow-winged Blackbird** *Agelaius thilius* **251**
59 **Yellow-hooded Blackbird** *Agelaius icterocephalus* **253**
60 **Chestnut-capped Blackbird** *Agelaius ruficapillus* **256**
61 **Red-winged Blackbird** *Agelaius phoeniceus* **258**
61X **'Bicolored Blackbird'** *Agelaius phoeniceus gubernator* **266**
62 **Tricolored Blackbird** *Agelaius tricolor* **269**
63 **Tawny-shouldered Blackbird** *Agelaius humeralis* **273**
64 **Red-shouldered Blackbird** *Agelaius assimilis* **275**
65 **Yellow-shouldered Blackbird** *Agelaius xanthomus* **277**
66 **Red-breasted Blackbird** *Sturnella militaris* **281**
67 **White-browed Blackbird** *Sturnella superciliaris* **284**
68 **Peruvian Meadowlark** *Sturnella bellicosa* **285**
69 **Pampas Meadowlark** *Sturnella defilippi* **287**
70 **Long-tailed Meadowlark** *Sturnella loyca* **290**
71 **Eastern Meadowlark** *Sturnella magna* **292**
71X **'Lilian's Meadowlark'** *Sturnella magna lilianae* **300**
72 **Western Meadowlark** *Sturnella neglecta* **302**
73 **Yellow-rumped Marshbird** *Pseudoleistes guirahuro* **306**
74 **Brown-and-yellow Marshbird** *Pseudoleistes virescens* **307**
75 **Scarlet-headed Blackbird** *Amblyramphus holosericeus* **309**
76 **Red-bellied Grackle** *Hypopyrrhus pyrohypogaster* **311**
77 **Austral Blackbird** *Curaeus curaeus* **312**
78 **Forbes's Blackbird** *Curaeus forbesi* **315**
79 **Chopi Blackbird** *Gnorimopsar chopi* **317**
80 **Bolivian Blackbird** *Oreopsar bolivianus* **320**
81 **Velvet-fronted Grackle** *Lampropsar tanagrinus* **322**
82 **Tepui Mountain-Grackle** *Macroagelaius imthurni* **323**
83 **Colombian Mountain-Grackle** *Macroagelaius subularis* **325**
84 **Cuban Blackbird** *Dives atroviolacea* **327**
85 **Melodious Blackbird** *Dives dives* **329**
86 **Scrub Blackbird** *Dives warszewiczi* **331**
87 **Great-tailed Grackle** *Quiscalus mexicanus* **333**

88 **Boat-tailed Grackle** *Quiscalus major* **338**
89 **Slender-billed Grackle** *Quiscalus palustris* **342**
90 **Nicaraguan Grackle** *Quiscalus nicaraguensis* **344**
91 **Common Grackle** *Quiscalus quiscula* **346**
92 **Greater Antillean Grackle** *Quiscalus niger* **350**
93 **Carib Grackle** *Quiscalus lugubris* **353**
94 **Rusty Blackbird** *Euphagus carolinus* **357**
95 **Brewer's Blackbird** *Euphagus cyanocephalus* **360**
96 **Baywing (Cowbird)** *Molothrus (Agelaioides) badius* **364**
96X **'Pale Baywing'** *Agelaioides (Molothrus) badius fringillarius* **367**
97 **Screaming Cowbird** *Molothrus rufoaxillaris* **368**
98 **Giant Cowbird** *Scaphidura* **(***Molothrus) oryzivora* **371**
99 **Bronzed Cowbird** *Molothrus aeneus* **374**
100 **Bronze-brown Cowbird** *Molothrus (aeneus) armenti* **378**
101 **Brown-headed Cowbird** *Molothrus ater* **380**
102 **Shiny Cowbird** *Molothrus bonariensis* **385**
103 **Bobolink** *Dolichonyx oryzivorus* **390**

BIBLIOGRAPHY 396
INDEX 427

INTRODUCTION

This book deals with the family Icteridae, known as the blackbirds, troupials or New World Blackbirds. The name blackbird comes originally from the Blackbird (*Turdus merula*) which is an Old World thrush. When British settlers arrived in North America, they gave names to the local birds based on what they knew from back home. Thus, the North American members of the Icteridae, many of which are largely black, became 'the blackbirds' even though they are entirely unrelated to the European thrush of that name. This confusion in names seldom causes much of a problem, but perhaps it can be avoided by using the colloquial name 'the icterids' for this most interesting group of birds. Ironically, the family name, Icteridae, refers to the yellow colour of the genus *Icterus*, the New World orioles.

The icterids are a diverse family that ranges throughout the Americas, from Alaska to Cape Horn, including a sizable component in the Caribbean. They are truly an American family. Icterids are medium-sized to large passerines, which usually sport a long straight bill, long tails, strong legs and well-developed facial musculature which allows them to 'gape' while foraging. The action of gaping, involves the insertion of the bill tips either into the ground or any gap or nook and then forcibly and powerfully opening the bill so that the gap can be expanded. It has been suggested that the ability to gape in this group may be the single most significant reason why this is such a successful and widespread family of birds. Icterids share some similarities to several Old World groups of birds. Some icterids outwardly resemble the Old World Orioles (Oriolidae), Starlings (Sturnidae) and Weavers (Ploceidae). Like weavers, the orioles, caciques and oropendolas create intricately woven hanging nests, and most of the caciques and oropendolas are colonial, like the weavers.

In North America, where the pastime of birdwatching is immensely popular, I would hazard to guess that for many the abundant Red-winged Blackbird is the first bird which birders learn to recognize easily by sight and sound. Perhaps it is even true to say that for many, the antics of the Red-winged Blackbird were the seed which began an interest in birds for many birders. Sadly, as experience increases and the search for less common birds grips the interest of the birder, the common Red-winged Blackbird is one of the first birds to be ignored, or to be put in the category of 'dirt bird'. A species that is not interesting because it is so common and easy to see. One of the aims of this book is to excite a new interest in these common birds, even from these nature enthusiasts who have unknowingly learned to ignore these amazing birds. In contrast, scientists have not ignored the icterids. In fact, different species of icterids have been involved in some rather important and influential ornithological work. We hope that both by summarizing some of these scientific findings for the lay person, and by outlining the curious behaviour of these species, an interest in both science and icterids will be promoted.

The aim of this book is to summarize the natural history of the icterids. A significant amount of space is devoted to species' description and identification. However, this book is not intended to be a field guide. Its size alone precludes that it would achieve that role effectively. However, we hope that the book will serve as a resource to which the reader can turn when the field work is done. Neither is this book a handbook to the icterids as the detail included is less than that which would be expected for a tome of that type. Some of the icterids, such as Red-winged Blackbird and Brown-headed Cowbird, surely rate among the best known birds in the world. Red-winged Blackbird alone is the subject of several previous ornithological books. Our book is intended to be a starting point for anyone interested in the icterids. The natural history summaries are written such that an overview of what is known about each species can be gained by reading them, and the bibliography at the end of the book aims to lead the reader to more detailed works on topics of interest. The colour plates have been produced so that an accurate representation of plumage variation due to sex, age or geography of a species is illustrated. Some of these plumages have not, to our knowledge, been illustrated before. We have attempted to be thorough in our examination of the literature, but so much has been written about the icterids that we are conscious of the fact that there are gaps in our research; some of these are intentional. In the Neotropics, only a few species have been researched to any degree, while others are almost unknown in life. Many may be common and widespread but details of their natural history are little known. We have tried to identify true gaps in the knowledge of these species, and we hope that this will encourage detailed observations, tape recordings or research to be conducted in order to close these gaps. If we can encourage a greater interest in this most fascinating group of birds, particularly with respect to finding out what is not yet known about them, then our book has served its purpose.

The author would be grateful to receive any new information, corrections, or updates to what is written here for use in future editions. You may contact the primary author at alvaro@sirius.com or c/o the publishers, A & C Black.

ACKNOWLEDGEMENTS

This work would not have been possible without the use of several specimen collections. We thank the present staff, previous curators, collectors and preparators of this invaluable resource for us. Specifically we would like to thank the following people and institutions (listed alphabetically by institution): Louis Bevier, Leo Joseph, Dave Agro and Robert Ridgely (Academy of Natural Sciences, Philadelphia); Mary LeCroy, Paul Sweet and Richard Sloss (American Museum of Natural History, New York); Art Clark and Wayne Gall (Buffalo Museum of Science, Buffalo, New York), Karen Cebra (California Academy of Sciences); Michel Gosselin (Canadian Museum of Nature, Ottawa); Charles Dardia (Cornell Ornithology Collection, Ithaca, New York); Dick Cannings (Cowan Vertebrate Museum, University of British Columbia, Vancouver); Peter Lowther and David Willard (Field Museum of Natural History, Chicago); Luis Coloma (Museo de Zoología, Universidad Catolica, Quito, Ecuador); Juan Carlos Torres-Mura and Jose Yañez (Museo Nacional de Historia Natural, Santiago, Chile); Carla Cicero (Museum of Vertebraze Zoology, Berkeley, California); James Dean and Phil Angle (National Museum of Natural History, Smithsonian Institution, Washington, D.C.); Jim Dick, Glen Murphy, Brad Millen and Mark Peck (Royal Ontario Museum, Toronto); Janet Hinshaw (University of Michigan Museum of Zoology, Ann Arbor, Michigan), Dennis Paulson (University of Puget Sound and Burke Museum, Seattle). Finally thank you to the creators and administrators of the Biodiversity and Biological Collections Web Server (http://www.keil.ukans.edu/). A big thanks to Lucie Metras, of the Bird Banding Office, National Wildlife Research Centre, Canadian Wildlife Service for providing us with records of blackbird banding recoveries.

This book was greatly aided by expert consultation from the following people: John Arvin (orioles, oropendolas), Marcos Babarskas (Argentina), Michael Baltz (Bahamas), Jon Barlow (Lilian's Meadowlark), Dusty Becker (W Ecuador), William Belton (SE Brazil), Chris Benesh (oriole songs), Ethan D. Clotfelter (Bronzed Cowbird), Mario Cohn-Haft (Brazil), Sam Droege (Rusty Blackbird), Jon Dunn (meadowlarks), Nancy Flood (orioles), Brush Freeman (Texas orioles), Kimball Garrett (orioles), Sharon Goldwasser (oriole songs), Hector Gomez de Silva Garza (Mexico), Michel Gosselin (orioles and cowbirds), Paul Greenfield (Ecuador), W. J. Hamilton (Tricolored Blackbird), A. Bennett Hennessey (Bolivian species), Steve Hilty (Colombia and Venezuela), Jocelyn Hudon (eye colour), Kevin P. Johnson (grackle systematics), Kenn Kaufman (SW meadowlarks), Lloyd Kiff (nests), Niels Krabbe (Ecuador), Dan Lane (Band-tailed Oropendola), Scott Lanyon (systematics, and copies of unpublished manuscripts), Wesley Lanyon (North American *Sturnella*), Greg Lasley (Texas orioles), Paul Lehman (ID of North American orioles), David Lemmon (Carib Grackle), Juan Mazar Barnett (Argentina), Adolfo Navarro S. (Mexico), Kevin Omland (*Icterus*), John P. O'Neill (Selva Cacique), Roberto Otoch (Pale Baywing), Kenneth C. Parkes (moult), Van Remsen (Velvet-fronted Grackle), Jim Rising (orioles), Scott Robinson (oropendolas, caciques), Adriana Rodríguez-Ferraro for sharing her unpublished manuscripts on Olive Oropendola, Sievert Rohwer (meadowlarks), Paul Salaman (Colombia), Charles Sibley (systematics), David Sibley (oriole ID), F Gary Stiles (Costa Rica), David Wege (Neotropics), Andrew Whittaker (Brazil), Tony White (Bahamas) and Robin W. Woods (Falkland Islands).

The following kindly read and commented upon many of the species accounts: John Arvin, Robert Behrstock, William Belton, Ethan D. Clotfelter, Jon Dunn, W.J. Hamilton, A. Bennett Hennessey, Sebastian Herzog, Steve N.G. Howell, Andrew Kratter, David Lemmon, Sjoerd Mayer, John P. O'Neill, Kenneth C. Parkes, Jim Rising, Scott Robinson, Paul Salaman, P.William Smith, F. Gary Stiles, Richard Webster and Robin W. Woods.

We obtained many of the bird records from American Birds and its continuation Audubon Field Notes. A great resource for Canadian records was the publication Birders Journal. Perhaps this is one of the first bird books which made an extensive use of the Internet in order to gather information. We would like to thank all of the participants and listowners of BIRDCHAT, NEOORN, Frontiers of Bird Identification, Calbird-l, South-Bay Birds, Tweeters, OBOL and rec.birds for helping out with my many requests. The following fine folks sent me notes or information on distribution: David F. Abbott, Kenneth P. Able, Jon Anderson, Lyn Atherton, Bruce Baer, Tom Bailey, Jim Barton, Dave Bergman, Rob Bierregaard, Bill Bousman, Jack Bowling, Karen Bridgers, Ned Brinkley, Hank Brodkin, Wayne Campbell, Dick Cannings, Rick Castetter, Chuck Carlson, Duane Carmony, Don Cecile, David Christie, Drew Clausen, Glenn Coady, Mario Cohn-Haft, Charlie Collins, Alan Contreras, Mort Cooper, Tom Crabtree, Brian Dalzell, Phil Davis, Paul DeBenedictis, Tom de Boor, Rich Ditch, Rob Dobos, Cameron Eckert, Tom Edell, Jim Eidel, Vic Fazio, Steven Feldstein, Gary Felton, Dick Ferren, Jim Flynn, Kurt Fox, Jim Fuller, Steve Ganley, Mark Gawn, Bill Glenn, Lex Glover, Dale Goble, Troy Gordon, Jim Grandlund, D. H. Grey, Gary A. Griffith, Frank Haas, Charity Hagen, George Hall, Steve Hampton, Robin Harding, Jim Hengeveld, John Henly, Sebastian Herzog, Paul Hess, Bill Hill, Rob Hilton, John Hirth, Tim Janzen, Bob Jennings, David Keating, Dave Keller, Steve Konings, Ray Korpi, Thomas E. Labedz, Laurie Larson, Harry LeGrand, Denis Lepage, Paul Lehman, Tony Leukering, Catherine Levy, Tom Love, Gail Mackiernan, Roger Malone, Blake Maybank, Steve McConnell, Brad McKinney, Ian McLaren, Craig McLauchlan, Roger McNeill, Martin Meyers, Marty Michener, Paul Millen, Burt Monroe Jr., Joe Morlan, Jim Mountjoy, Harry Nehls, Nancy Newfield, Hal Opperman, William Otto, Noreen Palazzo, Helen Parker, Rob Parsons, Dennis Paulson, Aevar Petersen, Mark Phinney, Michael Price, Bill Principe, Nick Pulcinella, Dan Purrington, H. Raney, Luis Miguel Renjifo, Ron Ridout, Jim

Rising, Russell Rogers, Steve Rottenborn, Bill Rowe, Mauricio Rumboll, Lawrence Rubey, Tom Ryan, Dan Sandee, Eric Scheuering, Sy Schiff, Carol Schumacher, David F. Suggs, Paul Salaman, Leanna Shaberly, John Shipman, Rick Simpson, Michael R. Smith, P.William Smith, John Sterling, Brad Stovall, Bob Stymeist, Dan Tallman, Bill Taylor, Peter Taylor, Max Thompson, Stuart Tingley, Vivek Tiwari, Francis Toldi, John L. Trapp, Declan Troy, Derek Turner, John van der Woude, Willem-Pier Vellinga, Alan Versaw, Noel Wamer, Tony White, Sheri Williamson, Jeff R. Wilson, Bill Wimley, Terry Witt, David Wright and David Yee. I hope I did not forget anyone, I apologize if I did.

Thanks to the following individuals for providing us with photographic material, information and comments valuable in the production of the plates: David Beadle, Bob Behrstock, Sean Blaney, Tony Blunden, Michael Bradstreet, David Brewer, Kim Caldwell, Art Clark, Coyote Creek Riparian Station, David Christie, Don DesJardins, Jim Dick, Jon Dunn, Tim Dyson, Wayne Gall, Tom Goodier, A. Bennett Hennessey, Matt Holder, Phill Holder, Colin Jones, Diane Kodama, Greg Lasley, Mary LeCroy, Paul Lehman, Jon McCracken, Bert McKee, Ian McLaren, Craig McLauchlan, Doug McRae (aka Julio Iglesias), Steve Metz, Borja Milá, Brad Millen, Glenn Murphy, Mark Peck, Ron Pittaway, Mike Runtz, Doug Sadler, Brett Sandercock, Richard Sloss, Don Sutherland, Paul Sweet, Doug Tate, Guy Tudor, George E. Wallace, Alan Wormington and Dale Zimmerman. Thanks to Louis Bevier and Doug Wechsler at VIREO for sending us photos and allowing access to the collection.

A big thank you to Terry Taylor and J. W. Hardy for making available to us recordings included in their forthcoming ARA cassette tape on the icterids (Voices of the Troupials, Blackbirds and Allies. Family Icteridae By J. W. Hardy, G. B. Reynard, A. J. Begazo and Terry Taylor). We also want to thank Sjoerd Mayer for making available to us his Birds Sounds of Bolivia CD ROM (http://ourworld.compuserve.com/homepages/bird_songs_international/homepage.htm). Bob Behrstock, A. B. Hennessey, Dan Lane and Lawrence Rubey were kind enough to provide us with copies of their own personal recordings.

The help of the following individuals is greatly appreciated: Bill Bouton for lending me his recorder during a trip to southern California; Juan Mazar Barnett for providing difficult to access Argentine literature; Barry Kent McKay for offering leads on Montserrat Oriole information; Bert McKee for his detailed observations on Guatemalan blackbirds; Dave Powell for translating papers from German; David Wege for providing obscure references and other assistance. Gracias to Fernando Ortiz Crespo who helped me locate Gary Stiles. Sam Droege was kind enough to send an advance copy of his paper on Rusty Blackbird authored by Russell Greenberg and himself.

A big thanks to Sherman Suter, Louis and Catherine Bevier, Paul O'Brien, Jonathan Alderfer, Leo Joseph, James Dean and Ottavio Jenni for making my visit to Washington D.C. and Philadelphia a heck of a good time, even if we did miss the Salt Marsh Sharp-tailed Sparrow.

Some personal thanks from Alvaro. None of this would have been possible without the continuing love, encouragement and support from Katja Rimmi. I would like to thank my parents (all four of them) and my brothers and sister for their encouragement and love. Aun que mis parientes en Chile estan seguro que soy loco, ellos siempre me dieron toda la ayuda, amor, y comodidad durante mis viajes a mi tierra – gracias. Thanks too to the Berkeley Posse, and any of Data's friends, particularly los primos Pinti and Poncho as well as Mona and Horacio for making my new home just that much more fun. My interest in birds would not have flourished without the help of my teenage birding buddies, Karl Konze and Ian Richards – thanks for making those years such fun guys. As well, I thank Hugh Currie, Luc Fazio and Jim Fairchild for the many birding trips before I could drive; I still appreciate the time you spent hanging around with a young birder who asked way too many questions. Jim Rising, Mary Gartshore and John Reynolds have taught me countless important lessons which have helped me as a scientist and naturalist. My co-workers at the Coyote Creek Riparian Station, particularly Chris Otahal and Diane Kodama have been most patient with me during the time I was writing the book, thank you for the support. I am grateful to the people who have helped me while in the field in South America, particularly Esteban and Patricia Bremer of the Estacion Biologica Punta Rasa, Argentina; Manfredo Fritz of the Instituto Entomologico de Salta, and Alejandro Suarez and Bonnie Bochan at Jatun Sacha in Ecuador, and the folks at Fundacion Maquipucuna in Ecuador. A big thanks to Vic Smith and Eagle-Eye Tours (http://www.eagle-eye.com/) for giving us the opportunity to travel throughout Latin America. Both of us would like to thank Sammy Maudlin and Bobby Vitman. Thanks to DJ Spinz, Carlos Vives, Bosstones, Pennywise, Veruca Salt, Rage against the Machine and others for creating the tunes that kept me going. Speaking of energy, thank god there is Yerba Mate.

Peter would like to especially thank his wife Dawn Burke for her endless supply of encouragement, time, love and effort in helping complete the plates. Acting as a go-between, she sent a huge number of faxes of drawings to Alvaro for me which was a critical part of their preparation. My parents and brother have supported my desire to paint with love and generosity from my earliest childhood memories. The Brenner family has been a constant source of encouragement and I thank them all deeply. I would also like to say thanks to all the friends I have made that have given me the support to paint these plates. Special mention to Doug McRae, who has been an endless source of compliments, advice and a great lifetime friend. Finally to my late brother David, who will not get to see my work, but who gave me the inspiration to become an illustrator.

We would both like to thank Robert Kirk and A & C Black for their initial interest in our project and the gentle coaxing we have received during the production of the book.

SYSTEMATICS AND TAXONOMY

The Icteridae are a family which until recently was considered a subfamily (Icterinae) of the Emberizidae. Emberizids are part of the large group of birds that are incorporated into the nine-primaried oscines, a group of recently-derived songbirds which have nine instead of ten primary feathers. Within the nine-primaried oscines, the icterids appear to be most closely related to the Parulids, New World Warblers (see Curson *et al.* 1994 for a summary of that group) and more distantly related to the Emberizidae (sparrows and buntings) Thraupidae (Tanagers) and Cardinalidae (Cardinals, Grosbeaks) (Bledsoe 1988). The Icteridae is a diverse family, but its members share many morphological, behavioural, and vocal characters which reveal their relationship. Membership of the group has been argued for Dickcissel (*Spiza americana*) in the past; however, Dickcissel does not appear to be closely related to the icterids (S. Lanyon pers. comm.). The Brazilian Scarlet-throated Tanager (*Compsothraupis loricata*) is said to be vocally, behaviourally and visually reminiscent of an icterid (Juan Mazar Barnett, Andrew Whittaker pers. comm.); however, no research has been conducted into this question. The bill morphology of Scarlet-throated Tanager is very similar to that of several other tanagers, rather than to the icterids. Until more data is available it is safest to assume that it is indeed a tanager. An illustration of this tanager is included in Plate 31.

There is some discrepancy regarding how many species of icterids should be recognized. Peter's Check-list of Birds of the World (Blake 1968), a standard reference list for the birds of the world, lists a total of 91 species. Its publication prior to the discovery of Pale-eyed Blackbird (*Agelaius xanthopthalmus*) (Short 1969). Even so, the treatment by Blake (1968) was generally conservative and many of the forms lumped there are now widely regarded as separate species, for example Peruvian Meadowlark (*Sturnella bellicosa*) and Pampas Meadowlark (*Sturnella defillippi*). More recently Sibley and Monroe (1990) recognized a total of 97 species. We are recognizing a total of 103 species in the Icteridae. None of these additional species which we are including are new species. We have not conducted systematic work to determine species limits in the blackbirds, but all of the species we have included have been recognized in other published works in the recent past. Once more research is carried out on this group, the number of species might rise even further.

Relationships within the Icteridae

Even though the icterids have received a great deal of research attention with respect to their ecology and behaviour, until recently very little research on the systematics within the group had been conducted. The AOU (1983) recognized three tribes within the Icteridae: the Icterini (orioles, caciques and oropendolas); the Dolichonychini (Bobolink); and the Agelaiini (all remaining genera and species). Based on the pattern or restriction sites of mitochondrial DNA of 47 species of icterids, Freeman and Zink (1995) concluded that there was no evidence for the division of Bobolink into its own tribe. Their data did show evidence for a fundamental separation between the Icterini and the rest of the icterid species. Lanyon, S. M. (1994) studied the relationships within the genus *Agelaius*, the marsh blackbirds. He discovered that contrary to expectations, all *Agelaius* blackbirds were not each other's closest relatives. In fact the *Agelaius* genus splits into two or maybe three groups. Thus *Agelaius* is said to be polyphyletic, that is the genus has more than one common ancestor, so more than one lineage is involved. This suggests that a taxonomic change is needed, where some *Agelaius* blackbirds have to be moved to a new genus or to an existing genus other than *Agelaius*. On the other hand, S. M. Lanyon's (1994) data supports the assertion that the *Quiscalus* Grackles, Orioles, Meadowlarks and *Psarocolius* Oropendolas groups are each monophyletic. This means members of each one of these groups are each other's closest relatives and all share one common ancestor. Similarly, all of the parasitic cowbirds in the genera *Molothrus* and *Scaphidura* were found to be monophyletic (Lanyon 1992). Thus, brood parasitism appeared once in the blackbirds and all brood parasitic cowbirds are descendants of this original brood parasitic cowbird. The non-parasitic Baywing (also known as the Bay-winged Cowbird) does not belong in this group and therefore it should be moved from the genus *Molothrus*. On the other hand, since the Giant Cowbird (*Scaphidura oryzivorus*) was found to belong in a monophyletic group with the other cowbirds, it should be moved to the genus *Molothrus* and the monospecific *Scaphidura* should be dropped. S. M. Lanyon's (1994) data show that within the *Agelaius* and allies there are at least two major lineages, one of primarily South American species (South American *Agelaius*, Scarlet-headed Blackbird, the South American all-black Blackbirds, Marshbirds, Oriole Blackbird, Mountain-Grackles and Velvet-fronted Grackle), the other of primarily Caribbean and North American species (Grackles, *Dives* Blackbirds, *Euphagus* Blackbirds, brood parasitic Cowbirds, North American and Caribbean *Agelaius*, and Jamaican Blackbird). The Yellow-headed Blackbird, meadowlarks and Bobolink are more distantly related to the *Agelaius* and allies. Ongoing research at the Lanyon lab at the University of Minnesota is studying the relationships within the grackle group and the oriole group. The Grackle group is monophyletic, as are the orioles, including the Troupials (K. Johnson and K. Omland pers. comm.). Within the oriole group there are two well-defined lineages, one of which includes Epaulet, Black-cowled, St Lucia, Montserrat, Martinique, Orchard, Hooded and Black-vented orioles. The other lineage includes Baltimore, Black-backed, Bullock's, Orange,

Altamira, Yellow, Streak-backed and Jamaican orioles, and perhaps Yellow-backed, Audubon's and Scott's orioles (K. Omland, S. Lanyon unpub data). It appears that oriole plumage patterns are not necessarily due to close evolutionary relationships (K. Omland, S. Lanyon unnpub data).

Species Limits

Of all of the concepts used in ornithology, perhaps no other has been so widely discussed and debated as what constitutes a species. The answer to this question has important ramifications to biologists studying ecological or behavioural aspects of a bird's life, as much as it has ramifications to the birder who wants to know how many bird species they have seen in a lifetime. Part of the problem is due to the fact that the processes of evolution and speciation are ongoing, and we sometimes encounter populations which are in the middle of this process and a clear answer is not possible. Another aspect to the species debate involves science and philosophy; there are different, competing species concepts. These concepts are competing in the sense that biologists are assessing which one of the concepts gives the best overall solution to sorting out the complexities and dynamics of populations. There are many different species concepts, but the two most prominent ones are the Biological Species Concept (BSC) and the Phylogenetic Species Concept (PSC). Other lesser known and accepted concepts include the Evolutionary Species Concept (Wiley 1981, Frost and Hillis 1990), the Recognition Species Concept (Paterson 1985) and the Cohesion Species Concept (Templeton 1989). The most important characteristic of a biological species (BSC) is that it is reproductively isolated from all other biological species. Mayr (1969) defined a biological species as groups of interbreeding natural populations that are reproductively isolated from other such groups. Thus if two populations of birds, lets say Baltimore Orioles and Bullock's Orioles, hybridize extensively where the ranges of the two meet, they do not fulfill the prerequisites of being full biological species and under a strict reading of the BSC they should be lumped as one species. In fact, this has occurred, as for many years these two species were considered one under the name Northern Oriole. A very big and chronic problem of the BSC is that in ornithology it has never been applied systematically and evenly, even though the BSC has been the most widely accepted modern species concept. One may ask, how about if two species interbreed but exceedingly infrequently? The two species of North American Meadowlarks do interbreed at times (Lanyon, W. E. 1994, 1995) but it is very rare. In this case, the BSC ignores these cases since over the vast area where these two species are found together (sympatrically) they do not interbreed. Furthermore the young of hybrid crossings by these two meadowlarks are infertile; this means that they cannot contribute to the next generation and allow genes from one species to infiltrate the gene pool of the other. Thus, in situations where interbreeding (hybridization) occurs but the young are infertile the two species still remain reproductively isolated and are treated as separate biological species. The level of hybridization which proponents of the BSC allow between two species varies, and when a hybrid zone exists between two species (such as in Baltimore and Bullock's Orioles) it adds another level of complexity. Basically, the opinion of some proponents of the BSC is that if the hybrid zone is stable and therefore neither of the species is becoming genetically swamped (introgression) then they should remain as two species (Mayr 1982). Given that interbreeding, or the lack thereof, is of such importance in the application of the BSC, biogeography throws in a great big problematic wrench which affects the concept's usefulness. What happens when two very similar looking populations are not found together (this is termed allopatry) so that one cannot 'test', through their ability to interbreed, whether the two populations are best treated as part of one biological species or two? Basically the answer to this problem has been in the hands of the taxonomist conducting the research and in many cases the determination of whether one or two species are involved has been entirely subjective.

The PSC defines a species as the smallest diagnosable cluster of individual organisms within which there is a parental pattern of ancestry and descent (Cracraft 1983). A phylogenetic species is a population which is on its own independent evolutionary trajectory. Members of a phylogenetic species have one common ancestor (monophyletic), they are part of a single evolutionary lineage. One of the most serious problems with the BSC is that there are situations when two populations hybridize extensively and have to be classed as one species, but in fact they are not each other's closest relatives. There is good evidence that Baltimore and Bullock's Orioles are not each other's closest relatives (Freeman and Zink 1995). Thus, when the two species were lumped under the name Northern Oriole the available evidence suggests that two separate lineages were involved. These species composed of separate lineages are termed paraphyletic; they misrepresent the patterns of evolution and can be very misleading when trying to understand topics such as biogeography, phylogeny and comparative biology (Zink and McKitrick 1995). The PSC does not have this problem as all phylogenetic species are monophyletic; however there is a good deal of controversy on what actually constitutes 'a smallest diagnosable cluster'. Critics of the PSC have suggested that if one has enough resolution one can find consistent differences between any two populations, and thus almost all populations will have to be split off as different species. In one clever study, Barrowclough (1993) and two other well-known avian systematists determined that there are 1.97 phylogenetic species per biological species, thus if a PSC was used there would be between 15845 and 20470 species of birds in the world.

The debate over which species concept best helps biologists understand nature still continues. However, now that new molecular techniques which allow ornithologists to pry into the very code of an bird's DNA are becoming more common, a new methodology has emerged which greatly aids in answering questions such as: is there gene flow (hybridization) between two populations? How long have these two populations been separate? Are these groups monophyletic? These techniques have gone a long way towards giving us a new set of data which ornithologists can use to more objectively address questions of speciation and evolution. As more of this type of work is conducted, more situations are being encountered where it is shown that the established dogma of the past is not correct, and as a consequence many new species are being 'split' from old ones. This is not only due to the availability of these molecular techniques but also to a general shift towards an acceptance of the PSC as a less subjective manner of classifying birds. However, it should also be pointed out that species concepts are merely tools that are used to better understand nature, and that different species concepts may be applicable in different circumstances. The concept of the species is human made, not a biological reality, and evolution does not work at the level of the species, but at the level of the individual, so there are always going to be grey areas where opinion is the only way to come to the resolution of a problem.

Nomenclature (Taxonomy)

Species are named using the binomial system of nomenclature, first introduced by Linnaeus, the father of modern taxonomy. The binomial is composed of two words, the first which is always capitalized is the genus, and the second which is never capitalized is the specific name. As an example, Red-winged Blackbird (*Agelaius phoeniceus*) belongs in the genus *Agelaius*, which includes several other species, and within that genus it belongs to the species *phoeniceus*. In print these binomials, more properly termed scientific names, are either italicized or underlined. In some cases a species may be found to vary geographically to such an extent that a subspecies is named. Subspecific names are added as a third term after the full scientific name. Thus Red-winged Blackbird of the Pacific Northwest is given the full name *Agelaius phoeniceus caurinus*. Note that when a subspecies is named, the subspecies that the original species was named from (the Type) takes the species name as the subspecific epithet. So in this example, *Agelaius phoeniceus phoeniceus* is the subspecies of northeastern North America, where the first Red-winged Blackbird was named from. This original subspecies is called the 'nominate' subspecies. The purpose of this scientific name system is partially to provide an international standard for naming organisms, regardless of the native language of the scientist or birder.

The naming of species and subspecies is governed by a complex set of rules, contained in a document called the International Code of Zoological Nomenclature. Its purpose is to make sure that species are named consistently and unambiguously, so that no two animals will share the same scientific name. Note however that botanists use a different code and some plants and animals share the same Genus name! In order to name a new species, the biologist has to provide a specimen of the new species which is curated in a museum. This original specimen which is associated with the description of the new species is termed the 'type specimen', or more properly the holotype. If any ambiguity arises over the validity of a named species, the type specimen must be checked in order to make sure that the newly named species is not the same as another one which was named earlier. If the two named species are determined to belong to the same species, then the more recently described species is termed to be synonimized with the older species. The code makes it clear that the earliest description of a species is the one which is valid. This is the concept of priority and it can be quite complex in some situations. It's worth mentioning that many changes in the names of blackbirds have occurred due to priority, or names that were 'occupied' by another species. Many of these cases are noted in the text. Sometimes a biologist might choose to name a species based on a series of specimens rather than on a single specimen. This series is termed the 'type series' and specimens belonging to the series are called 'sintypes' or 'cotypes'. Sometimes, particularly in older literature, the scientific name would be written followed by a person's name, for example *Agelaius phoeniceus* (Linnaeus). The person named is the scientist who first described the species, his or her name is not italicized but is always capitalized. In this example the name is in parenthesis; this signifies that since the original description was documented the species *phoeniceus* has been moved to a different genus from the one it was originally ascribed to. Linnaeus originally named Red-winged Blackbird as *Oriolus phoeniceus*, thinking it best belonged in the genus of Old World orioles. Since that time, Red-winged Blackbird was deemed not to be an Old World oriole, but an *Agelaius* blackbird, so the name of the describer Linnaeus is put within parentheses. This shuffling of a species from one genus to another is relatively common, and reflects a better understanding of species relationships, as in theory birds within a genus are each other's closest relatives. When this is not the case, the genus either has to be expanded, in order to include all of the descendants of one lineage or broken up. In the blackbirds, new genetic evidence has been published (Lanyon, 1992, 1994) that shows that several genera will have to be changed since they either include unrelated species or do not include all of the descendants from one common ancestor. These potential changes are noted in the text.

Subspecies and Geographic Variation

Many species vary geographically, in terms of their morphology, ecology and behaviour. The main way of acknowledging geographic variation has been to name subspecies or races of a species. A subspecies is a group of similar populations of a species which inhabits a geographic subdivision of the range of a species, and which differs taxonomically from other populations of the species (Mayr 1969). This means that subspecies are defined both on differences in appearance as well as distribution. Subspecies are subpopulations of a species which differ morphologically, usually in size or aspects of plumage colouration, from all other populations of the species. Furthermore these populations cannot breed in the same region; different subspecies always breed separately from each other except where two subspecies' ranges abut. During the non-breeding season more than one subspecies may be found in the same region. The description of subspecies has suffered from an inconsistent application of what is different enough to be deemed a new subspecies. Thus, some subspecies are almost impossible to identify unless a good series of specimens is available, while others are strikingly and unambiguously different. Geographic variation is in fact quite complex and it is far from being two dimensional; it is clear that the subspecies system is not close to being adequate in terms of describing geographic variation (see Power 1970a for a good example pertaining to blackbirds). Partially due to this realization, the naming of new subspecies has fallen out of favour. Nevertheless, geographic variation sometimes does follow simple perceptible patterns. For example, in dry desert areas birds tend to be paler in colour than those from moist areas. Several of these patterns have been proposed to follow certain rules of geographic variation. For example, Gloger's Rule states that animals inhabiting cool and dry areas are paler than those from warm and moist areas. Bergmann's rule states that animals that live in cooler areas are larger than those which inhabit warmer sites. There are several other rules which have been proposed. The naming of these rules and the observations of general patterns does make it clear that animals vary according to the ecology of the area in which they live. In some cases the variation is clinal, which means that as one proceeds in one geographic direction a trait or a series of traits change in a slight but directional manner. As an example, bill depth (thickness) in Red-winged Blackbirds is clinal in Central North America with birds from Canada having the greatest bill depth which becomes progressively smaller as one proceeds toward Texas (Power 1970a). Clinal variation does not have an obvious break, it is gradual and steady. This can mean that individuals from the extreme ends of a cline are readily separable but those in the middle are not. Clinal extremes cannot be named as different subspecies since the intermediate birds will not clearly fall into one or other subspecies. Nevertheless many subspecies are parts of clines, and this has further decreased the usefulness and applicability of the subspecies concept.

In the species accounts we have described the differences between named subspecies for any icterid species which is not monotypic (has no subspecies). There is no up-to-date reference on the recognised subspecies of birds of the world. Therefore, the subspecies which we have included have been those recognised by Peter's Check-list of Birds of the World (Blake 1968), those in the fifth edition of the American Ornithologists Union Check-list of North American Birds (AOU 1957) and those subsequently published in peer-reviewed journals. In some cases, we have lumped subspecies if this has been suggested in the recent literature. We caution that due to the inconsistent manner of the naming of new subspecies, many of the subspecies we list are little more than points on clines or are so poorly differentiated that they are largely not diagnosable, particularly in the field. Other subspecies are consistently different and diagnosable, and may be identified in the field. We have made an additional effort to suggest when field identification is possible. We caution the reader that field identification to the subspecific level is often not possible. In fact identification to the subspecies is only certain when the bird is on the breeding grounds. We encourage that difficult subspecific identifications be confirmed with a visit to a museum collection.

PLUMAGES AND TOPOGRAPHY

Plumage Pattern Terms

Bib – A black patch on the throat and anterior part of the face.
Epaulet – a brightly-coloured patch at the bend of the wing (shoulder), usually including the lesser and median coverts.
Handkerchief – a square patch at the base of the folded primaries.

Plumages

In North America the accepted method of naming plumages and moults follows the Humphrey-Parkes system (Humphrey and Parkes 1959). However, outside of North America this system is not well known. The following is a comparison of what these two systems call different plumages. The first plumage name given is the Humphrey-Parkes name, the alternate name is in parentheses.

Juvenal (Juvenile) – The first plumage obtained after the downy nestling plumage. Note that juvenal is an adjective, referring to the plumage while juvenile is a noun. Thus a specific bird may be a juvenile, and it wears a juvenal plumage.
***1st Basic* (1st winter or 1st non-breeding)** – This is the first plumage after the juvenal. In species with one post-juvenal plumage type per cycle during their first year this is the plumage that will be retained for the entire year.
***1st Alternate* (1st summer or 1st breeding)** – In species with two plumage types their first year, after the juvenal, this is the plumage gained after the 1st basic is lost.
***Definitive Basic* (Adult winter or Adult)** – Definitive refers to the adult stage, this is the plumage type after no more age-related plumage changes occur. It is the adult plumage gained after the complete moult. In species with one plumage per year, the definitive basic is the only plumage obtained.
***Definitive Alternate* (Breeding adult)** – In species with two plumages per year, this is the plumage gained after the basic plumage, usually through a partial moult.

Moult

In the Humphrey-Parkes system the moults are named for the plumage that comes after the moult. Thus the moult which changes the juvenal plumage to the 1st-basic is the 1st-pre-basic moult. These are the equivalent mopult sequences of the two systems, with the Humphrey-Parkes terminology on the left side of the equation:

First pre-basic	=	Post-juvenile
First pre-alternate	=	First pre-breeding
Definitive pre-basic	=	Adult post-breeding
Definitive pre-alternate	=	Adult pre-breeding

A bright first-basic (first-winter) male Brewer's Blackbird showing pale feather tips, which create a pale-headed appearance. As these tips wear off, by the late winter, the bird will appear entirely glossy black.

Icterid topography

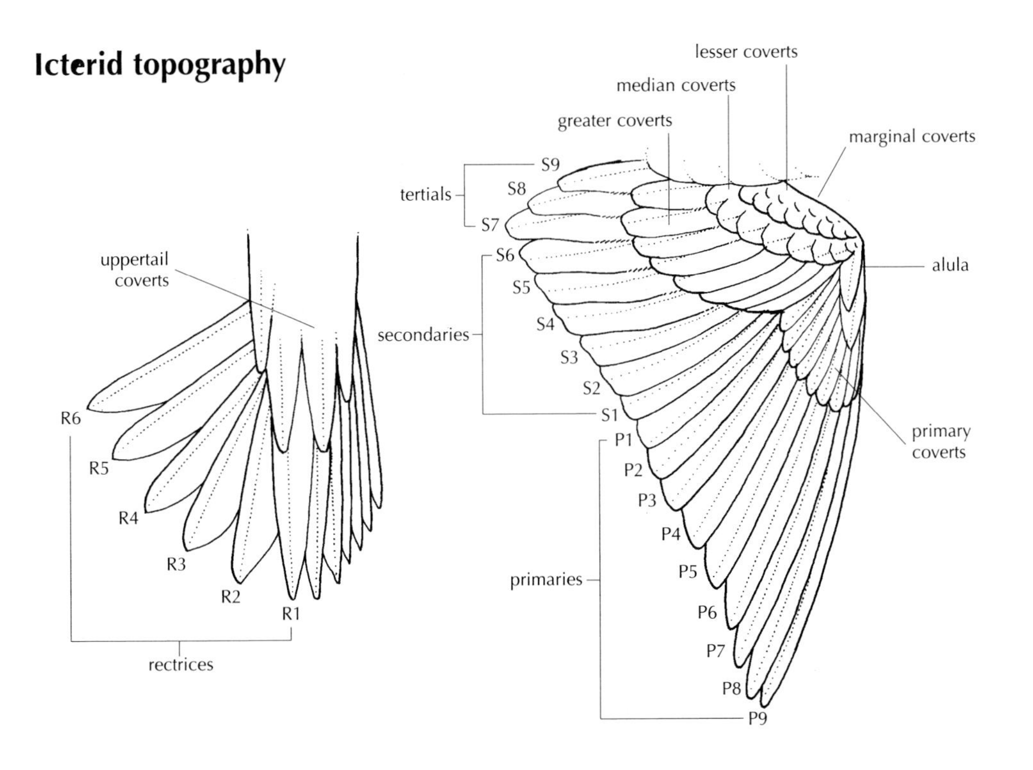

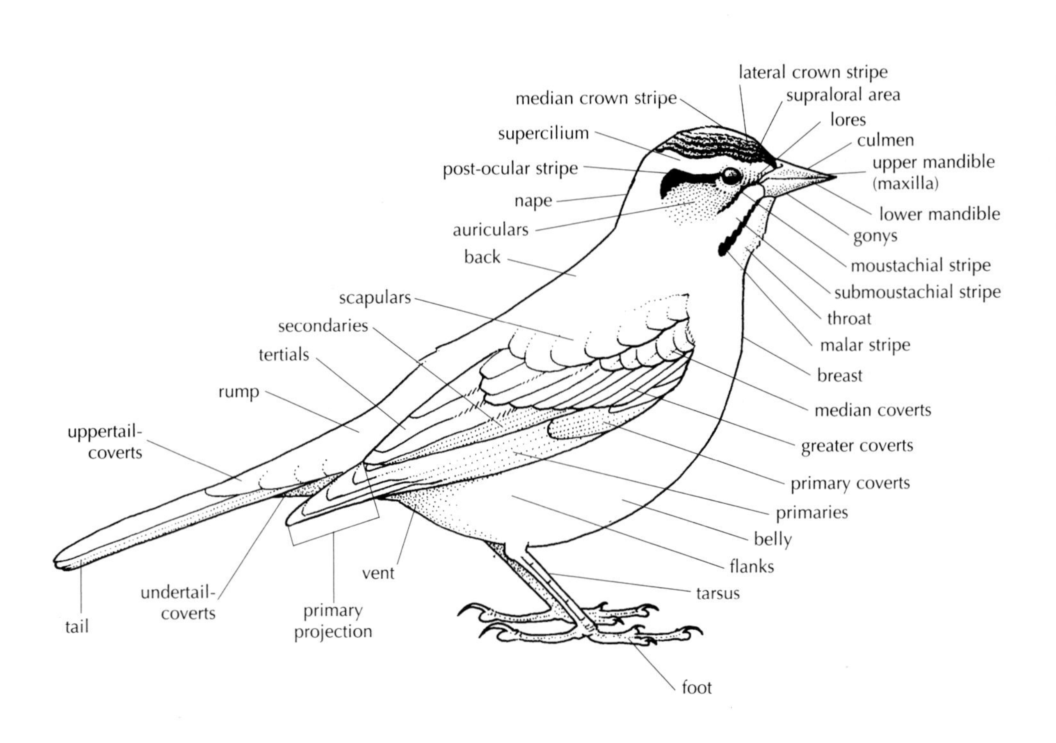

Icterids fall into two patterns of moult as adults. Either they have one complete moult (the definitive pre-basic) moult per year and therefore only one plumage per year or they undergo two (definitive pre-basic and pre-alternate) moults a year, in which case they have a breeding (alternate) and non-breeding (basic) plumage during a year. Almost all of the Icterids fall into the first pattern; Bobolink is the only icterid that follows the second pattern. Nevertheless, many blackbirds undergo visible plumage changes without undergoing a moult. This is achieved by feather wear and fading. Many species of blackbirds in fresh plumage have coloured feather tips which give the bird a different appearance than when these tips have worn off. Perhaps the most obvious example is Rusty Blackbird, which in fall and winter is black with a largely rusty back, face and breast due to rusty feather tips. As the winter progresses these rusty tips wear off, revealing an entirely black plumage. Thus, Rusty Blackbirds effectively have both a breeding and a non-breeding plumage, but this is achieved without a moult. Since no moult occurs, both the rusty plumage and the black plumage are different states of the basic plumage, and are not given separate names under the Humphrey-Parkes system.

Tail Moult

Note that some icterid genera moult their tail centripetally (outermost rectrices first, inner ones last); this is rare in passerines. All of the blackbirds except the orioles, caciques and oropendolas show centripetal moult (Parkes 1997). The orioles, caciques and oropendolas moult their tail centrifugally, as is typical of passerine species. This separation fits neatly in the major separation between these two major icterid lineages suggested by the molecular data (Freeman and Zink 1995). However, Bobolink does follow a different pattern moulting the tail centrifugally but twice a year rather than once (Parkes 1997). Centrifugal and nearly simultaneous tail moult is also known in the group, being particularly common in *Quiscalus* grackles (Parkes 1972).

BEHAVIOUR AND EVOLUTION

Displays

Icterids perform some very elaborate and exciting displays to watch. Some species, such as Red-winged Blackbird, display so obviously that even people not familiar with birds will ask "what is the black bird with the red wings that is always singing and puffing himself out in the marsh?" The reasons why a bird displays are countless, ranging from aggressive to amorous interactions. Perhaps a few displays common to many of the icterids should be outlined here. Many icterids perform a display called the 'bill tilt'. When icterids are 'bill-tilting' they sleek their body plumage and hold the bills up, pointing towards the sky. Usually one bird begins the 'bill-tilt' aimed at another bird, and the second bird reciprocates. 'Bill-tilting' is a sign of aggression and occurs either between or within sexes, between different species of blackbirds, and it most often occurs during feeding aggregations. The 'bill-tilt' does not appear to be present in the caciques, oropendolas and orioles. The 'song-spread' (or 'ruff out') display is typically associated with male Red-winged Blackbirds while on their marsh territories. The 'song-spread' display appears to serve the purpose of communicating to other birds that a territory is taken and that the owner is willing to fight to reserve that privilege, as well as serving a mate attraction function. 'Song-spread' displays are always accompanied by the song. While it is most common for males to perform this display, females of some species perform their own version as well. 'Song-spreads' are seen in *Agelaius* blackbirds, cowbirds, Yellow-headed Blackbird, grackles and other icterids. They are lacking in many species, and in particular the caciques, oropendolas and orioles do not perform them. The oropendolas replace the 'song-spread' of their relatives with the 'bow' display. Only males perform the 'bow' display, and in most cases only the males which are dominant at the colony. 'Bow' displays are again always given with song. In this display the oropendola bows down, with the bill pointing downwards as the wings are raised and the tail is cocked. While the song is uttered, the oropendola brings the tail forward and lowers the head until the bird is nearly upside down, hanging from the branch. The different species vary in the manner they give the display, but all are somewhat similar. Some oropendolas, such as Crested and Green beat their wings after the display creating an audible flapping noise. Several of the meadowlarks and the Bobolink give a flight (or aerial) display which functions like the 'song-spread' and 'bow' displays to secure a territory or attract a mate. These are only given by males. Again, song is always performed with these displays. Flight displays vary depending on the species giving them, but they tend to be most common in species which inhabit open grasslands. The general tendency is for the male to fly up, with an odd fluttery wing-beat and then parachute flight to the ground singing at the same time; some species sing both on ascent and descent, while others sing as they perform a level, fluttery flight.

The Role of Blackbirds in Science

Icterids have played a major role in biology, mainly because they are common, easy to study and have a varied set of behaviours, mating systems, and territorial and non-territorial breeding systems. Observations of how male Red-winged Blackbirds choose their territories and how females choose to settle within those territories spurned the idea of the Polygyny Threshold Theorem, which suggests when females are better off mating with a male that already has a mate than finding a male that does not have a mate (Verner and Willson 1966, Orians 1969). The study of icterid mating systems has focused on polygynous species such as Yellow-headed Blackbird and Red-winged Blackbird. Most recently, advanced molecular techniques have allowed biologists to assess the paternity of offspring within a nest. Gibbs *et al.* (1990) determined that not only were males polygynous in Red-winged Blackbird, but that females would go off and mate surreptitiously with neighbouring males so that a territorial male didn't always father all of the chicks in the nests of females nesting in his territory. Populations of colour-banded Red-winged Blackbirds had been studied for many years and biologists had no evidence that females were mating with other males at times. This paper showed that one cannot necessarily assume that observed behaviour is a good guide to determining who fathers the young at any one nest. More recently, Webster (1994a) has clearly shown that the mating system of Montezuma Oropendola is harem-defense polygyny, a system unknown in birds, but common in mammals (e.g. Elephant Seals (*Mirounga* sp.), and Red Deer (*Cervus elaphus*)).

The evolution of plumages and sexual dimorphism is closely tied to the evolution of mating systems. Thus, there is a clear relationship between size dimorphism and the level of polygyny in animals. Polygyny is where one male has more than one female as a mate, and the level of polygyny affects the strength of sexual selection on males. Sexual selection is a subset of natural selection, and in its most basic sense is the selection for qualities which give an individual advantages in mating, and thus an increase in the number of young produced. Sexual selection is the driving force in the evolution of ornate displays, as well as fighting and display apparati such as antlers in deer and fighting spurs in pheasants for example. The relationship between size and polygyny is clearly present in icterids, where the oropendolas and caciques, which experience the highest level of

Representative icterid displays. Green Oropendola 'bow display' sequence (left); 'bill tilting' in Carib Grackle (top right); 'flight display' in White-browed Blackbird (lower right).

polygyny, show the greatest amount of size dimorphism, and monogamous species (such as the Jamaican Blackbird) show very little size dimorphism (Webster 1992). Dimorphism in colour (dichromatism) is also generally regarded as an effect of sexual selection. The elaborate plumage of the peacock is thought to have evolved through an intense amount of competition between males for access to mating females, as well as very choosy females. In most birds which show a great deal of sexual size dichromatism the pattern is for the male to be bright and the female to be dull; when the two sexes are not dichromatic the tendency is for both of them to be dull (North American emberizid sparrows are a good example). When the ecological situation creates competition between females for access to males, the males tend to become choosy and often female plumage evolves to the bright end of the spectrum (the Phalaropes are the classic example). Thus, it has been a boon to researchers that icterids show a variety of different states of sexual dichromatism, from Baywing where the sexes are inseparable based on their plumage to Bullock's Oriole where the male is significantly brighter than the female. However, icterids defy the generalities of sexual dichromatism in one significant way. In species where the two sexes share the same plumage, they usually share a bright plumage not a dull plumage, examples of this abound in the tropical orioles. Also, within certain groups, such as the grackles, the males are black and shiny whereas the females of some species are brown and dull; in the more monomorphic (both sexes alike) species the females are black and sometimes even shiny, albeit not as accentuated as in the males. Even within a species the same pattern is shown. The northern populations of Streak-backed Oriole are clearly dichromatic, with bright males and dull females. However, towards the south of the ranges the females become more male-like in their plumage so that in the southern extreme of the range the two sexes are almost identical in plumage (monochromatic). This pattern suggests that in the icterids sexual selection is acting strongly on female plumage, such that dichromatism and monochromatism evolve through selection on females not males as has been assumed (Irwin 1994). It is when general patterns such as these are not followed that a deeper understanding of the underlying biology can be understood, thus it seems likely that the evolution of plumage brightness is more complex than had been previously thought. A related line of study has focused on trying to answer the question of 'Why do some immature birds delay obtaining the full, bright, plumage pattern of adults until their second year of life?' The study of this phenomenon on Baltimore Orioles shows that young males, in their dull immature plumage, may trick adult males into believing they are actually females. This 'female mimicry' allows the young males to gain close access to females and sometimes to mate with them (Flood 1984). Another competing hypothesis suggested that the evolution of delayed plumage maturation was due to the fact that young males had very little chance to breed in their first year, and thus a dull plumage would minimize their conspicuousness and thus the chance that they would be predated. The fact that some orioles do not show delayed plumage maturation and that others do, including some females, will allow scientists to further test these theories and gain a better understanding of plumage evolution in the process.

These are just a sample of what the study of icterids has provided to the furthering of biological understanding. There are many other examples, and many are summarized in the species accounts. It is perhaps not an overstatement to say, that more than any other group of American birds, the icterids have offered scientists new insights, new hypotheses and new methods of testing these hypotheses.

NOTES ON THE SPECIES ACCOUNTS

The sequence of species largely follows that of Sibley & Monroe (1990), with additional 'splits' inserted in an appropriate sequence. The sequence used in Sibley & Monroe (1990) is already somewhat out of date as new research into the phylogeny of the blackbirds has been published or is in process (Lanyon, S. M. 1994, Kevin Omland pers. comm., Kevin Johnson pers. comm.). As some of these studies are ongoing it does not seem appropriate to entirely re-arrange the checklist sequence of the icterids in our book until these are completed. However, we chose to put several species pairs which are now thought to be sister species and were formerly distant in the sequence, next to each other. The cowbirds were re-arranged to follow Lanyon (1992).

All accounts are numbered; these numbers do not pertain to any official numbering scheme. They are numbered solely for easy reference. Note that whenever another species of icterid is mentioned within an account, its reference number will be listed immediately after the English name, but only for its first appearance in any given account. The accounts are headed with the most readily accepted English Name for the species. All North American bird names are based on those accepted by the American Ornithologists Union (AOU 1983, plus supplements). If an alternate name exists, it is listed in the Notes section. Some well-marked subspecies, or subspecies groups, are given their own sub-accounts after the full account of the species under which they are classified. The names of these well-marked forms are given in quotation marks, and their number is the same as that of the full species but is followed by an 'X'. In many cases, further research may reveal that these well-marked forms may be best treated as full species. The plate number where the colour illustration of the species is to be found follows the English Name. On a second line is the scientific name of the species, along with the author who first described it and the year when it was described.

Identification

Unlike other books in this series, we have chosen not to give an overall length measurement here. These length measurements are very inaccurate and do not take into account the geographic variation in body size that many icterids show. We suggest that readers look at the measurement section in order to assess body sizes of icterids. The purpose of this section is to highlight how to identify the species in question. We have made an effort to note which species pose the greatest identification problems, and we make direct comparisons to those species in terms of the identification criteria. The extent of this section depends upon the degree of complexity of the identification problem. In some cases, vocal characteristics which aid identification are mentioned, but more details will be found in the Voice section.

Voice

This section is divided into two parts, song and calls. Songs are the more complex vocalisations usually used by icterids to defend a territory or to attract a mate. Calls tend to be simpler in structure and used in a variety of situations. The determinations of what comprises a song and what comprises a call are not always clear cut, especially when the complete repertoire of vocalisations is not known, as is the case with most of the birds in this book. Where appropriate we have noted when there is a substantial amount of geographic variation in the vocalisations of a species. The information incorporated in this section has been garnered from both the literature as well as from recordings and field experience. Attempting to describe a sound in words is exceedingly difficult. We have used descriptions from the literature as well as our own descriptions, and hope that these serve their purpose. The subjectivity and interpretational nature of verbal voice descriptions can be overcome by the use of printed sonograms (visual plots of sound frequency against time) but we have found that these are nearly meaningless to those not intimately associated with their use. Thus, we chose not to use sonograms for this book.

Description

This section gives a detailed description of the plumage and structural features characteristic of the species. First a general description of the structure is given, with particular reference to bill morphology. In addition the wing formula is described. The primaries are named P1 (innermost) through P9 (outermost). Note that in the UK, the primaries are numbered in the opposite manner, with P1 being the outermost and P9 the innermost primary. The symbols > (greater than), > = (greater than or equal to), < (less than), < = (less than or equal to), = (equal to) and ≈ (approximately equal to) are used to characterize the lengths of adjoining primaries. Wing formulas are divided into three parts separated by semicolons; the first outlines the relative primary lengths for the wingtip, the second states which inner primary is most similar in length to the outermost primary (P9) and the third describes which primaries show obvious emarginations. Primary emarginations are to be looked for on the outer vane of the primary; if this part of the primary narrows abruptly towards the tip, it is said to be emarginate. In some cases, primary emarginations are not listed. If plumages differ due to age or sex, all plumage types are described. The adult male is described first followed by the adult female, the immature male (first basic), immature female (first basic) and the juvenile. If there are significant differences between the fresh and worn plumage these are listed separately. The nominate subspecies is described unless otherwise noted. All descriptions were taken from museum specimens, except that of Selva Cacique.

Geographic Variation

This section describes the known geographic variation for each species. For the most part this entails the description of named subspecies. The breeding range of the subspecies is noted here. However, if no subspecies is named for a species this does not mean that geographic variation within the species does not exist. If published works have mentioned possible clinal variation within a monotypic form then this is stated. Those species for which there is no published subspecies are listed as monotypic. More information on subspecies and geographic variation is given above under Systematics and Taxonomy. If substantial differences exist between subspecies, a detailed plumage description may be given here that can be compared to the description of the nominate subspecies under Description. More often, a short account is given of any distinguishing features shown by the subspecies. In some situations, subspecies may be grouped into subspecies groups if they share certain characteristics with some, but not all, subspecies. All measurements of subspecies are listed under Measurements below. If there are distinct geographic differences in vocalisations, these may be mentioned here but are described in more detail under Voice.

Habitat

An attempt is made to give an accurate representation of the habitat characteristics needed by each species. In some cases, breeding habitats and non-breeding habitats are different and these are noted in the text. If habitat choice varies depending on geography this is also stated.

Behaviour

Perhaps it is in their elaborate behaviour that the icterids shine. However, many aspects of the behaviour of the icterids, particularly the Neotropical species, are not known. This account attempts to summarize the known displays as well as other aspects of behaviour such as mating systems and foraging. In many cases, these accounts are scant as there is little available data; in others there is so much data that a very general summary of the species' behavioural repertoire is given. In these cases, we refer readers to the reference section and the bibliography for leads on where to obtain more information.

Nesting

This section describes the breeding biology of the species. This entails a description of the nest, nest site, nest construction, breeding period, presence of coloniality, eggs, clutch size, incubation period, nestling period, fledging period, extent of biparental (both sexes) care and known details of brood parasitism. For many species, this information is just not known or has not been published, and the varying lengths of this section between species reflects these gaps in knowledge. There is a certain amount of overlap between Nesting and Behaviour; these are separated as clearly as possible, only dealing with behaviours that have direct relevance to nesting within Nesting. The timing of breeding varies extensively, due to geography, particularly in the tropics. Whenever possible we have listed different published nesting dates; in most cases these pertain to birds observed tending a nest, but sometimes refer to birds caught or collected which showed physiological evidence of nesting such as enlarged gonads.

Distribution and Status

This section begins with a general remark about the relative abundance of the species. These statements of relative abundance are not quantified but are a subjective attempt to describe how common a species is. These relative terms in decreasing order of abundance are: Abundant, Common, Uncommon, Rare and Extremely Rare. The term 'local' is used to identify when a specific state of abundance only occurs in a patchy manner. If known, the elevational range in which the species is found is listed. This is primarily of importance in the Neotropics, as in North America the elevational distribution of the more widespread species does not appear to be so restricted. If the breeding distribution differs greatly from the non-breeding distribution, these are given separately. If the species is divided into separate subspecies, a more in-depth understanding of the distribution will be gained by reading Geographic Variation in conjunction with Distribution and Status.

Movements

This section highlights the migratory movements for each species, including the approximate routes of migration. If the species is sedentary then this is noted here. As is the case with most of the Neotropical species very little is known about their patterns of movements, whether truly migratory or not. We openly expect that many of the species listed here as sedentary to perform some type of seasonal movement which has not yet been determined. This appears to be the case with many of the oropendolas and caciques as they are present in some areas during part of the year and absent during others. For many North American species, we have tried to summarize what has been learned from banding (ringing) programs; however, the number of recoveries for some species such as Common Grackle, Red-winged Blackbird and Brown-headed Cowbird are so immense that their analysis was beyond the scope of this book. Wherever known, maximum life spans (longevity records) are listed. This section also summarizes any recent range expansion that the species has undergone, as well as the history and patterns of vagrancy. Information on vagrants was garnered from a variety of sources including Field Notes, and its predecessor American Birds, official state books, Birders Journal which lists Canadian rarities, and personal communications with birders and state or provincial bird records committees. We were careful to exclude any reports which were obviously erroneous, but we also want to make it clear that many of the reports we have included are not necessarily officially accepted ones. Many states and provinces do not have an official body which deals with their rare bird records, and many historical records have not been adequately assessed by modern committees even in places where the latter exist. People conducting research on vagrants listed here should attempt to inspect the original notes or evidence available for each record. The word 'record' is here used loosely to mean a vagrant bird which has achieved some level of scrutiny, all officially accepted vagrants are called 'records'. The term 'report' expresses that this vagrant may not have an official standing; many of these observations are from areas where there is no existing body that examines reports of vagrant birds.

Moult

The Humphrey and Parkes (1959) system of moult and plumage nomenclature is used here. A complete moult is one which replaces all of the bird's feathers, while an incomplete moult replaces all of the body plumage, as well as some of the primaries, secondaries and rectrices, but not all of them. The final moult type, the partial moult replaces part or all of the body plumage, including the secondary coverts, but no flight feathers or rectrices are replaced. Not all individuals of a species fall into these categories, so sometimes further discussion describes variations on the general pattern. If known, the timing of the moults is given, but as this is related to the breeding period there is often a great deal of variation in moult timing in tropical species. There is almost no literature available regarding the moults of the Neotropical species of blackbirds. What is listed here is a summary of what has been published as well as details observed through the study of museum skins. This section should be considered a basic approximation of what the real moult patterns are for each species, and surely some of our assertions will be found to be in error once more detailed work is conducted. Much of our moult data for North American birds comes from Pyle *et al.* (1987), Pyle (1997a), and Pyle (1997b).

Measurements

Linear measurements are given for both sexes as well as almost all described subspecies for polytypic species. The following measurements are given: wing, tail, culmen and tarsus. Wing refers to the unflattened wing chord, which is the length of the wing from its bend to the tip of the longest primary. The tail is measured from the base of the rectrices, where they meet the body, to the tip of the longest rectrix. The culmen is the length of the culmen from its base (even if covered by feathers) to the tip of the bill. The tarsus is the length of the lower section of the leg, from the intertarsal joint to the last large scale before the toes diverge and the leg scales break up into several smaller adjoining scales. For all measurements, a mean (average) is given as well as a range from the minimum to the maximum value for that measurement. Given that we did not expect most readers to be statisticians, we chose not to provide the standard deviation as a measure of variance, over the more easily understood range. Before the listing of each measurement the sample size (n) is given in parentheses; if none is noted then the sample size for the previous measurement, or for the entire set of measurements (given at the start of the list) is the corresponding sample size. The higher the sample size, the more reliable the measurement means and ranges.

Measurements were obtained both from the literature and from specimens measured by the authors. Every researcher measures slightly differently, and there is an added error when one lumps measurements taken by more than one person. However, larger sample sizes provide better approximations of the true mean and range of the population. Where measurements were taken from a published source, this source is listed in References.

Notes

This section is reserved for the communication of any miscellaneous note pertaining to the species. In many cases, these include taxonomic notes, or nomenclatural considerations. Known hybrids are listed here. In addition, any information relating to changes in the population status and conservation are noted here.

References

The references listed are not necessarily all which were used in writing the species account, but include the most important or specific ones. These references should provide ample leads in finding out where to obtain more detailed information on the species.

Abbreviations Used

The abbreviation NP stands for National Park, while NWR is a National Wildlife Refuge (United States). The term Co. is a short form for County.

NOTES ON THE COLOUR PLATES, CAPTIONS AND MAPS

Plates

All species of Icteridae are illustrated in colour. Where there are significant differences in the plumages of the different sexes or subspecies these are illustrated as well. Some of these plumages have not been adequately illustrated before. However, there are examples where no specimen could be found for a specific age class or subspecies which we surmised was slightly different, and these could not be illustrated. Nevertheless, the plates illustrate the majority of plumage variation within the family. The watercolour plates were prepared using a variety of reference materials including photographs, museum specimens, field notes, sketches and personal experience. All of the perched birds are illustrated to scale, except where noted. Illustrations of flying birds are drawn to a smaller scale.

Captions

These give a brief summary of the distribution and main features that help to identify the illustrated plumage. Where different subspecies are involved, their range and characteristic features are given. These descriptions are short and are to be used as a guide; the reader is asked to refer to the text for more details.

Maps

Each species, and in some cases subspecies, groups or identifiable forms have a distribution map opposite the plate. Where the species is a year-round resident, the range is depicted in green. This does not necessarily mean that there is no migration within this range, but that the species may be found there at any time of the year. Regions where the species is only a breeding (usually summer) visitor are depicted in yellow. Areas where the species is found only in the non-breeding (usually winter) season are depicted in blue. Where the winter distribution's southern or northern limit fluctuates annually, a dashed line is used to denote this fluctuation. Green, yellow or blue arrows are used to highlight a small and isolated area of occurrence, which may not be obvious given the scale of the map.

GLOSSARY

Allopatric When two or more taxa are not found in the same area.

Cocha An Amazonian oxbow lake.

Emergent Used to describe trees in forests which are noticeably taller than the average height of the forest canopy.

Monotypic Has only one subpecies if referring to a species, or only one species if referring to a genus.

Polyandrous A mating system where females mate with more than one male.

Polygamous A mating system where both sexes mate with more than one individual of the opposite sex.

Polygynous A mating system where males mate with more than one female.

Sympatric When two or more taxa are found in the same area.

Terra Firme Upland forest in the Amazon basin, these forests remain above the water line during the wet season.

Várzea Flood plain forest in the Amazon basin, usually referring to white water areas. These forests usually flood at least once a year, during the wet season. The term igapó is the equivalent in black water sites.

Xeric Dry.

PLATES 1–39

2 Crested Oropendola *Psarocolius decumanus* Text page 108

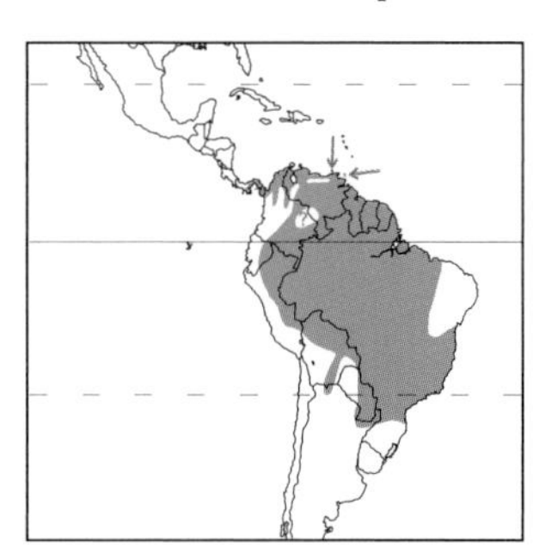

2a Adult male (nominate, N. South America): Black with bright blue eye, pale bill, and chestnut crissum and rump.
2b Adult female (nominate): Smaller and slightly paler than the male, may show some chestnut edging on back.
2c Adult male in flight (nominate): Black with a chestnut rump. The tail is largely yellow, showing two black central rectrices.
2d Adult male (*insularis*, Trinidad and Tobago): Shows chestnut edging on wings and back.

6 Chestnut-headed Oropendola *Psarocolius wagleri* Text page 123

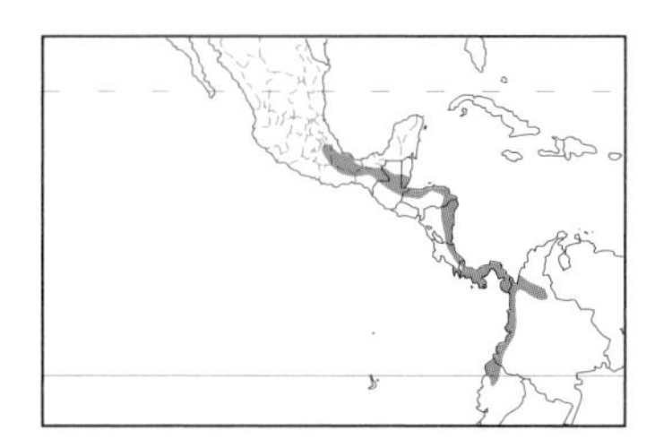

6a Adult male (nominate): Pale bill with a noticeable casque. Largely black with a dark chestnut head, rump and crissum, and bright blue eyes. Very long-winged.
6b Adult female (nominate): Similar to the male, but smaller and shows a duller iridescence, relatively shorter wings, and a smaller bill casque.
6c Adult male in flight (nominate): Looks black with a chestnut rump and long, pointed wings. The tail is largely yellow, with a dark central pair of rectrices and dark edges to the outer rectrices.

12 Band-tailed Oropendola *Ocyalus latirostris* Text page 134

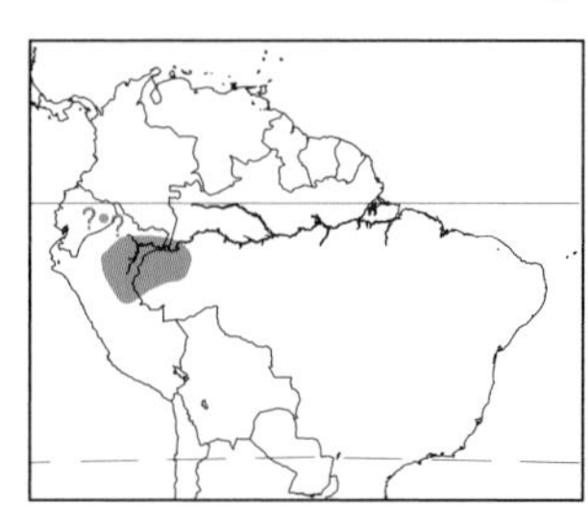

12a Adult male: A small black oropendola with a dark chestnut nape, blue eyes and a distinctive tail pattern, which is often hidden when perched.
12b Adult female: Similar to the adult male, but smaller and showing a duller iridescence. From below the tail looks yellow with a black terminal band. (In both sexes.)
12c Adult male in flight: Looks black but with a distinctive tail pattern. The tail black with yellow edges, but has a black outer edge to the outer rectrices and dark tips to all yellow rectrices.

2a
2d
2b
2c
6a
6b
6c
12b
12a
12c

1 Casqued Oropendola *Psarocolius oseryi* Text page 107

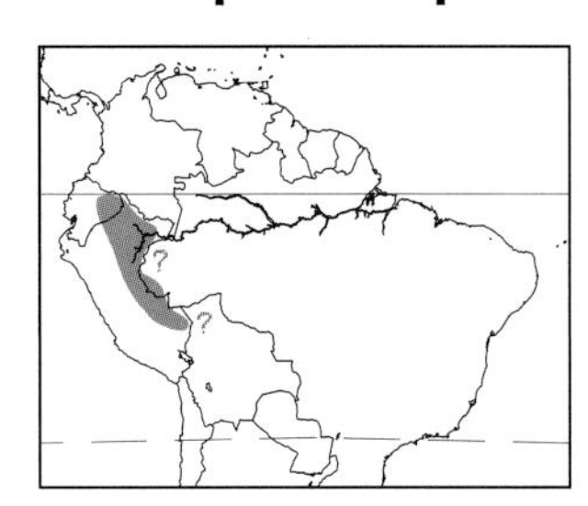

1a Adult male: A small oropendola with a prominent bill casque. The body is almost entirely chestnut, except for the greyish throat, and olive-yellow breast and neck sides. The primaries and secondaries are prominently edged with yellow.
1b Adult female: Similar to the male, but smaller in size and slightly paler.
1c Adult male in flight: Looks mainly chestnut. The tail is black with yellow edges, the outer rectrices are edged dark.

3 Green Oropendola *Psarocolius viridis* Text page 112

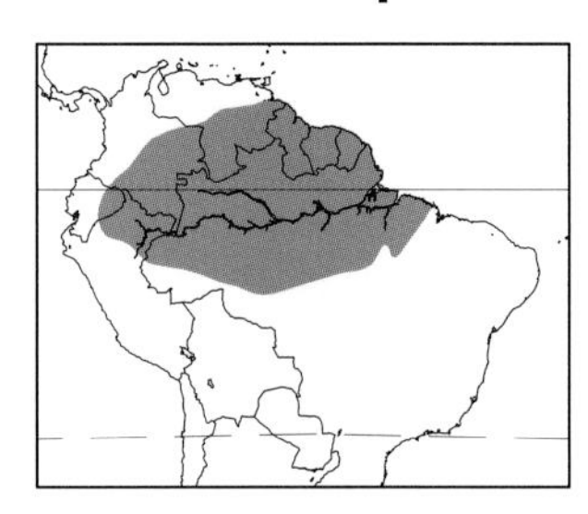

3a Adult male. A large green oropendola with a chestnut posterior body colouration. The bill is pea-green with a red tip. The eyes are sky-blue, and usually there is a small patch of pink bare skin around the eye. The wings are dark with olive edging.
3b Adult female: Similar to the male, but substantially smaller. Some may show chestnut edges to the lower mantle.
3c Adult male in flight: Appears green with a chestnut rump and dark olive wings. The tail is black with yellow sides, and a dark outer fringe to the terminal portion of the outer rectrices.

9 Olive Oropendola *Psarocolius (bifasciatus?) yuracares* Text page 129

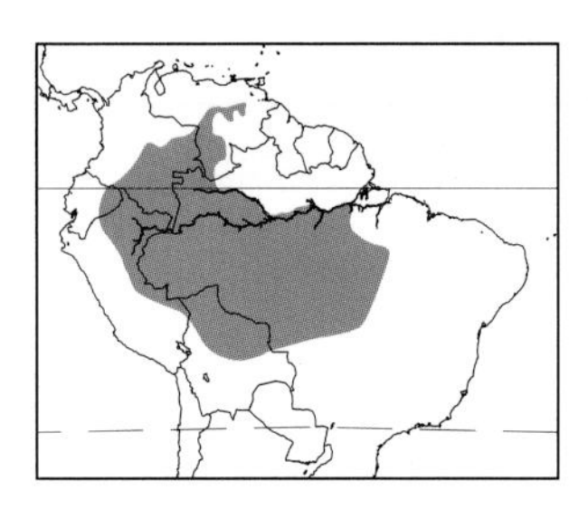

May be conspecific with Para Oropendola (10) Plate 3
9a Adult male (*yuracares*, Amazon basin, west of the Rio Tapajós): Appears chestnut with an olive head, neck, breast and upper back. The wings are chestnut. The bill is black with an orange-red tip, while the face shows a vivid bare cheek patch and a small bare patch at the base of the culmen; the eyes are dark.
9b Adult female (*yuracares*): Similar to the adult male, but smaller and lacking the male's long head plumes.
9c Adult male in flight: Looks largely chestnut, including the wings, but with an olive head, neck and breast. The tail is yellow with an olive pair of central rectrices.
9d Adult male (*naivae*, south of the Amazon river between the Rio Tapajós to the Rio Xingu): Probably best considered a hybrid swarm between Olive and Para Oropendolas. Similar to Olive Oropendola, but the head, neck, breast and upper back are largely black, showing a variable amount of olive colouration. (Drawn to a smaller scale.)

1c
1a
1b
3a
3b
3c
9a
9d
9b
9c

7 Montezuma Oropendola *Psarocolius montezuma* Text page 125

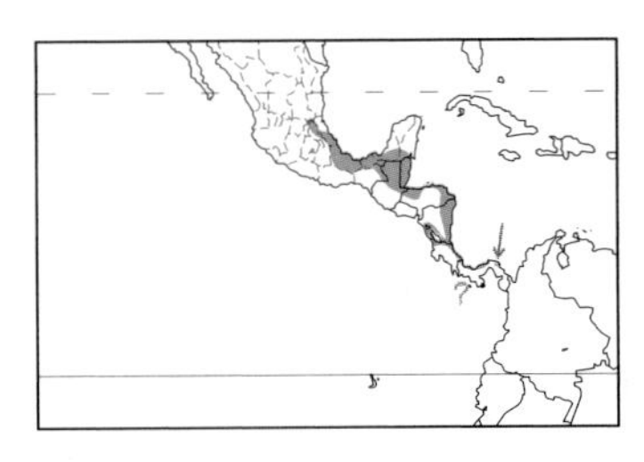

7a Adult male: A large chestnut oropendola with a black head. The face is adorned with a bare blue cheek patch and a pink malar wattle, as well as a small pink patch at the base of the culmen. The bill is black with a large orange-red tip which extends to approximately one third to one half the length of the bill.
7b Adult female: Similar to the male but noticeably smaller and with a proportionately smaller malar wattle.
7c Adult male in flight: Appears largely chestnut, with blackish primary tips and a black head. The facial patches are apparent in flight. The tail is yellow with the central pair of rectrices black and only slightly shorter than the longest yellow rectrices.

11 Black Oropendola *Psarocolius guatimozinus* Text page 132

11a Adult male: A large, shiny black oropendola with chestnut restricted to the wing-coverts, scapulars, lower back, upper- and undertail-coverts. The face is adorned with a bare blue cheek patch and a pink malar wattle, as well as a small pink patch at the base of the culmen. The bill is black with a small orange-red tip.
11b Adult female: Similar to the male, but noticeably smaller and less glossy. Note that on both sexes the tail is largely yellow with the central pair of rectrices black; unlike in the other species in this group, the central rectrices are noticeably shorter than the yellow rectrices. Usually the black central rectrices are only 2/3 the length of the yellow rectrices; this is visible in flight.

10 Para Oropendola *Psarocolius bifasciatus*  Text page 131

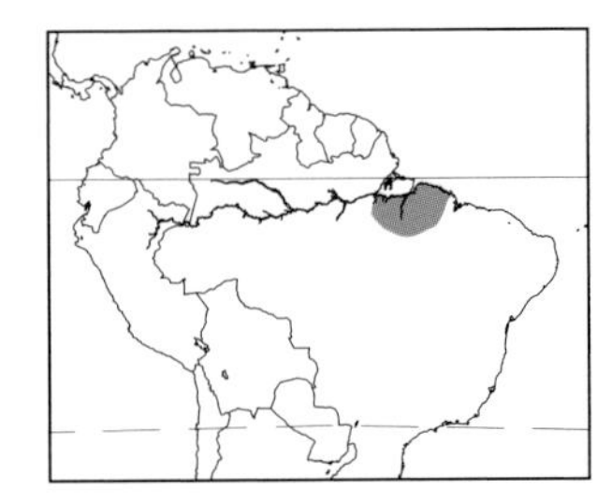

10a Adult male: A large chestnut oropendola with a black head. The face is adorned with a bare pink cheek patch. The bill is black with a large orange-red tip, roughly 1/3 the length of the bill.
10b Adult female: Similar to the male, but smaller and not as richly coloured. The crest plumes are shorter.

8 Baudo Oropendola *Psarocolius cassini*  Text page 128

8a Adult male: A large chestnut oropendola with a black head, neck, breast and mid-line on the belly. The face is adorned by a bare pink face patch and wattle, as well as a small pink patch at the base of the culmen. The bill is black with a small orange tip to the bill, covering roughly the terminal third of the bill.
8b Adult female: Similar to the adult male, but smaller and slightly duller in colouration.

7a
7b
7c
11b
11a
10a
8b
8a
10b

4 Dusky-green Oropendola *Psarocolius atrovirens* Text page 114

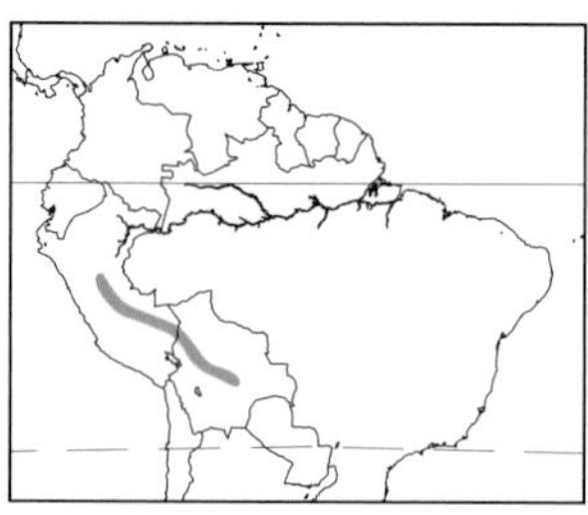

4a Adult male: A dark olive oropendola with a pale bill. The head is dark olive with a paler throat and cheek. The body is dark olive, becoming russet near the tail. Note that the eye is not easily visible on the dark face.
4b Adult male, variant: Some adults show a yellow forehead and pale blue-grey eyes. Note that juveniles always show a yellow forehead.
4c Adult female: Similar to the adult male but smaller and slightly duller in colour. Note that from below, the tail appears largely olive due to the entirely olive outer rectrices and the olive tips to the yellow rectrices.
4d Adult male in flight: Appears dark, with a paler (russet) rump. The tail shows litlle yellow. The outer rectrices are olive as are the central two pairs; thus there are only three pairs of yellow rectrices and these are all tipped olive.

5X 'Yellow-billed Oropendola' part of Russet-backed Oropendola *Psarocolius (angustifrons) alfredi* Text page 119

5Xa Adult male (*salmoni,* both slopes of western cordillera and west slope of central cordillera in Colombia): Dark brown, almost blackish head with a noticeable yellow forehead extending back beyond the eyes. The back is chestnut and the rump and lower back russet. The underparts are largely chestnut, becoming russet on the undertail. The bill is pale.
5Xb Adult male (*alfredi,* SE Ecuador south to Bolivia, ranges to lowlands): Head pale olive (yellowish-olive) and paler than the rest of the body. There is a small yellow patch on the forehead that does not extend past the eyes. The eyes stand out on the pale face, sometimes the eyes are pale blue-grey. The body is pale chestnut, becoming russet near the tail. The bill is pale.
5Xc Adult male (*oleagineus,* coastal range of N Venezuela): The most olive form in this group. The head, neck, breast and belly are olive, becoming russet-olive on the back and flanks and pure russet near the tail. The face shows a yellow forehead patch that does not extend past the eyes; this is most extensive on juveniles and often missing in adults. The bill is pale greenish-white. Note that the outer rectrices are olive.
5Xd Adult male (*neglectus,* E slope of the E cordillera in Colombia, as well as the Venezuelan Andes): Dark olive head and upper breast, becoming chestnut-olive on the body and paler russet near the tail. The yellow forehead extends past the eyes as a short supercilium. The bill is pale. This form is somewhat intermediate between the more olive *oleagineus* to the east and the more chestnut forms to the west.
5Xe Adult male in flight (*alfredi*): Appears dark with a paler head and a russet rump. The tail is extensively yellow, with brownish tips to the outer rectrices and brown central rectrices.

5 Russet-backed Oropendola *Psarocolius angustifrons* Text page 117

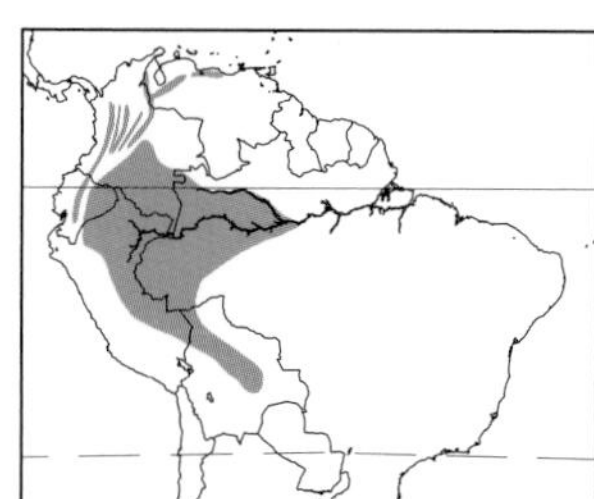

5a Adult male (*angustifrons* – lowlands of E Ecuador, W Brazil and NE Peru): A rich brown oropendola with a black bill. The back and underparts show a distinctive russet wash, becoming pale russet near the tail. The head is pale brown, with a beady dark eye.
5b Adult female (*angustifrons*): Similar to the adult male but smaller and duller, sometimes showing a paler throat. Note that from below the tail looks largely yellow, with brown restricted to the terminal portion of the outer rectrices.
5c Adult male in flight (*angustifrons*): Appears dark brown in flight, with a paler russet rump. The black bill is visible from a distance. The tail is largely yellow, with brown central rectrices and limited brown tips to the yellow rectrices.

4b
4c
4a
4d
5Xa
5Xb
5Xe
5Xc
5Xd
5c
5b
5a

PLATE 5 CACIQUES 1

14 Red-rumped Cacique *Cacicus haemorrhous* Text page 140

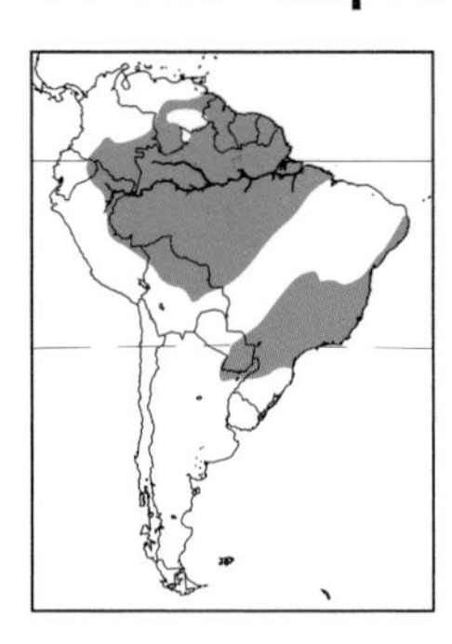

14a Adult male (*haemorrhous*, N South America, north of the Amazon basin): A black cacique with an extensive scarlet rump, reaching up to the lower edge of the mantle. The bill is pale and the eyes blue. Note the long wings and relatively short tail. This subspecies shows a strong blue iridescence.
14b Adult male (*pachyrhynchus*, Amazon basin): Only the head is illustrated. Similar to *haemorrhous* but shows an expansion at the base of the culmen. In addition, this subspecies is not so strongly glossed and tends to show slightly more extensive scarlet on the rump and back.
14c Immature (1st-basic) male (*haemorrhous*): Like the adult male, but smaller, showing a dark eye and a reduced and duller-coloured rump patch. The rump patch appears dark barred towards the mantle. The body plumage is not as iridescent as on the adult.
14d Adult male (*affinis*, SE Brazil, N Argentina, E Paraguay): Not as iridescent as *haemorrhous* or *pachyrhynchus* and the bill is more slender than either of those forms. Iridescence is restricted to the upperparts on this form.
14e Adult female (*affinis*): Dark brown body colouration, not black. The female of *affinis* is only weakly iridescent on the back. The rump patch is scarlet-orange.
14f Adult male (*haemorrhous*) in flight: Appears black with an extensive scarlet rump and lower back. Note the long and pointed wings.

16 Subtropical Cacique *Cacicus uropygialis* Text page 144

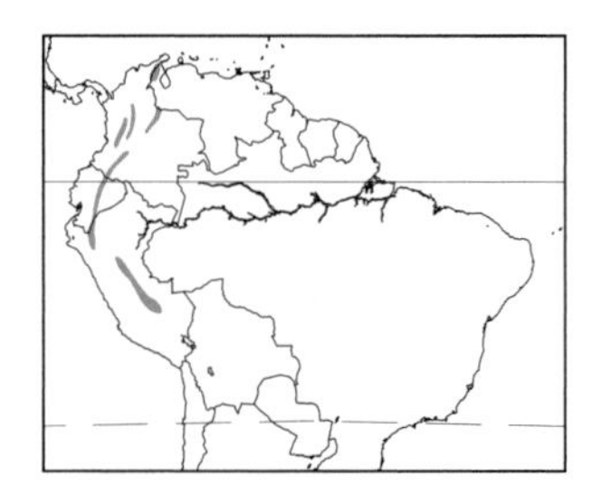

16a Adult male: A black cacique with a restricted red rump patch. The patch does not include the lower rump and only extends to the upper edge of the lower back. The black plumage is only slightly iridescent. The wings are not exceedingly long, but the tail is relatively long. The eyes are blue and the bill is pale.
16b Adult male in flight: Appears black with a restricted red rump. The wings are not as long and pointed as on Red-rumped Cacique. In addition, the tail is long and slightly graduated.

15 Scarlet-rumped Cacique *Cacicus microrhynchus* Text page 142

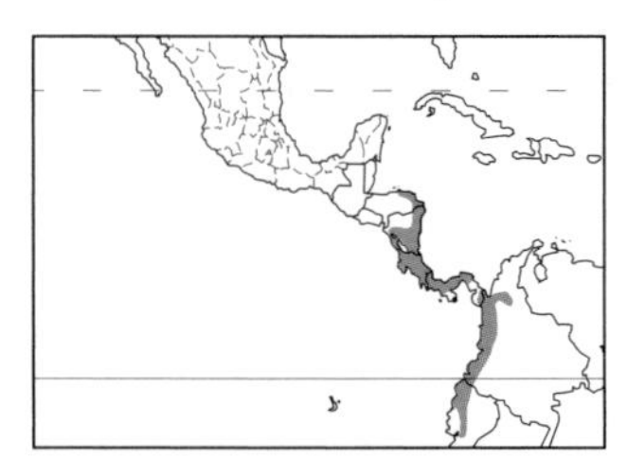

15a Adult male (*microrhynchus*, Honduras to Panama): A small black cacique with a restricted scarlet rump. The rump patch extends from the lower back to the upper rump. The black body is only slightly iridescent. This is a short-tailed and relatively long winged species.
15b Immature (1st-basic) male (*microrhynchus*): Similar to the adult but slightly smaller, brownish-black rather than black, dark-eyed and showing a pale orange rump which may appear barred.
15c Adult male (*pacificus*, Pacific lowlands from E Panama to Ecuador): Only the head is illustrated. Similar to the nominate form but shows a slightly deeper bill with a more expanded culmen, and deeper base to the lower mandible.
15d Adult male in flight (*microrhynchus*): A small black cacique with relatively long wings and a short tail. The scarlet rump patch is small, not reaching the lower rump and uppertail-coverts.

14a
14b
14c
14d
14e
14f
16a
16b
15a
15b
15c
15d

PLATE 6 CACIQUES 2 AND GIANT COWBIRD

98 Giant Cowbird *Scaphidura (Molothrus) oryzivora* Text page 371

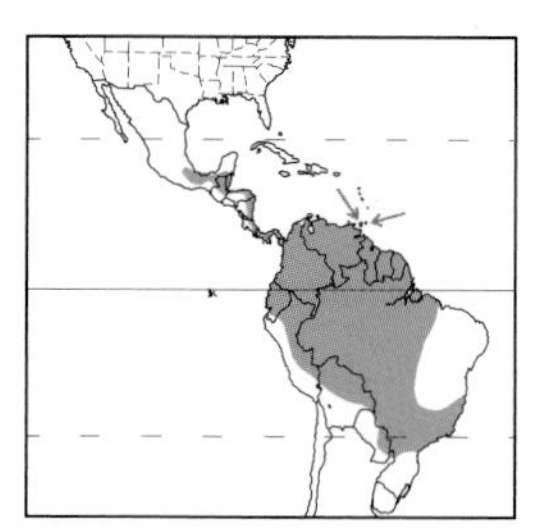

98a Adult male (*oryzivora*, South America, E Panama, Trinidad): A large iridescent black icterid with long wings and tail. The head appears peculiarly small, partially due to the expandable neck-ruff. The black bill is approximately as long as the head, showing a slight curve to the culmen. The eyes are often dark red.
98b Adult male (*oryzivora*): Pale-eyed individual. The eyes range from whitish-yellow to deep yellow on some individuals. Their frequency varies geographically.
98c Adult female (*oryzivora*): Similar to the male, but smaller and less iridescent. Always lacks the neck-ruff, it does not appear as small-headed as the adult male.
98d Juvenile male (*oryzivora*): Similar to the adult male but blackish-brown, with brown eyes and a pale bill.
98e Adult male in flight (*oryzivora*): A long-winged, large black icterid, showing a small-headed and deep-chested appearance. The wings make an audible noise during flight.
98f Adult female in flight (*oryzivora*): Similar to the adult male, but smaller, shorter-winged and not as small-headed in appearance. The wings do not make an audible noise in flight.

13 Yellow-rumped Cacique *Cacicus cela* Text page 136

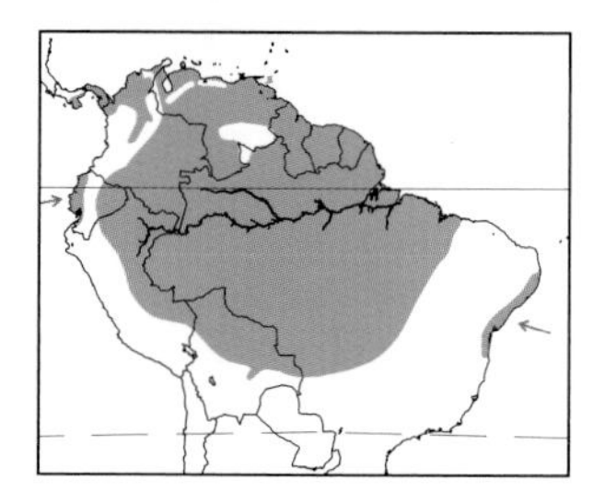

13a Adult male (*cela* South America east of Andes): A large black cacique with a yellow rump, base of tail, wing patch, crissum and belly. The bill is pale, with a noticeable curve to the culmen and shows sky-blue eyes.
13b Adult female (*cela*): Similar to the adult male, but noticeably smaller and duller black. Females show relatively shorter wings than adult males.
13c Immature (1st-basic) female (*cela*): Similar to the adult female, but greyish-black and dark-eyed. The bill shows a dark base.
13d Adult male (*vitellinus*, Panama to W Colombia): Similar to *cela*, but the yellow is slightly more orange in colour. In addition, the wing patch is restricted to the inner greater coverts and the yellow at the base of the rectrices is restricted. Note also that the base of the bill is slightly greyish.
13e Adult male in flight (*cela*): A large black cacique with the posterior half of the bird appearing almost entirely yellow. The tail is black with an extensive amount of yellow at its base. The yellow wing patches are noticeable in flight.

22 Yellow-winged Cacique *Cacicus melanicterus* Text page 153

22a Adult male: A large black cacique with a long crest and yellow patches on the rump, lower back, crissum, wing-coverts and sides of tail. The bill is pale and the eyes are dark. Note the slim proportions of the body and bill of this cacique.
22b Adult female: Similar to the male in pattern but noticeably smaller, and dark grey instead of black. The grey belly is often washed yellow. Note that from below the tail appears almost entirely yellow.
22c Adult male in flight: A black cacique with large yellow wing patches and a yellow posterior half to the body. The tail is black with yellow outer rectrices, the outermost showing a thin greyish outer edge.

98b
98f
98e
98a
98d
98c
13e
13d
13a
13c
13b
22a
22b
22c

18 Mountain Cacique *Cacicus chrysonotus* Text page 147

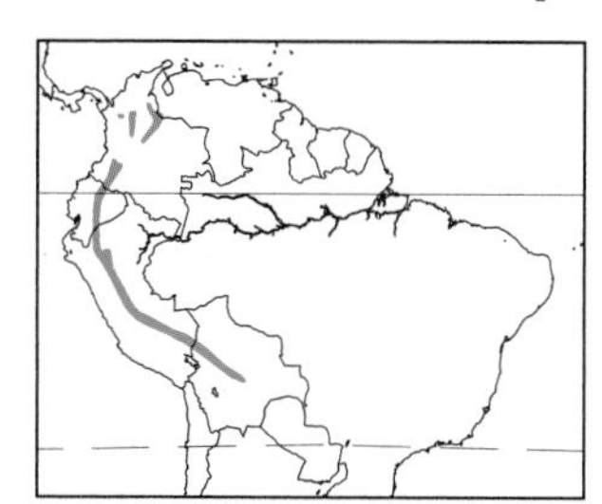

18a Adult male (*leucoramphus*, Andes of Venezuela to Ecuador): A medium-sized black cacique with a large yellow rump patch and a noticeable yellow wing patch on the median and inner greater coverts. The eyes are blue and the bill is pale with a darker base. Note that the wings are relatively short but the tail is long.
18b Juvenile (*leucoramphus*): Similar to the adult, but with a dark eye and darker bill colour. In addition, instead of being black, the body plumage is blackish-brown. (Only head illustrated.)
18c Adult male in flight (*leucoramphus*): A black cacique with noticeable yellow wing and rump patches. Note that the rump patch extends up to the lower back.
18d Adult foraging (*leucoramphus*): An acrobatic species, often hanging while foraging. Note that it can sometimes hide part of the yellow on the wings.
18e Adult male (*chrysonotus*, S Peru to Bolivia): Similar to the nominate form but lacks a yellow wing patch. Some individuals may show one or two yellow feathers on the wing-coverts. Note that the rump is often hidden. The eye colour ranges from blue to white.
18f Adult male in flight (*chrysonotus*) . A black cacique with no yellow on the wings or tail, but an obvious yellow rump patch extending to the lower back.

17 Golden-winged Cacique *Cacicus chrysopterus* Text page 145

17a Adult male: A small black cacique with a small yellow rump patch, not reaching to the lower back, and a yellow wing patch comprised of yellow median and inner greater coverts. The bill is pale blue-grey and the eyes vary from orange through yellow to pale blue.
17b Adult male: Head illustrated only. A male showing bluish-white eyes.
17c Juvenile: Similar to the adult but duller, being blackish-brown rather than black. The eyes are always dark and the bill colour averages darker than on adults.
17d Adult male in flight: A small black cacique with a small yellow rump patch and noticeable yellow wing patch. The wings are somewhat rounded.

19 Selva Cacique *Cacicus koepckeae* Text page 149

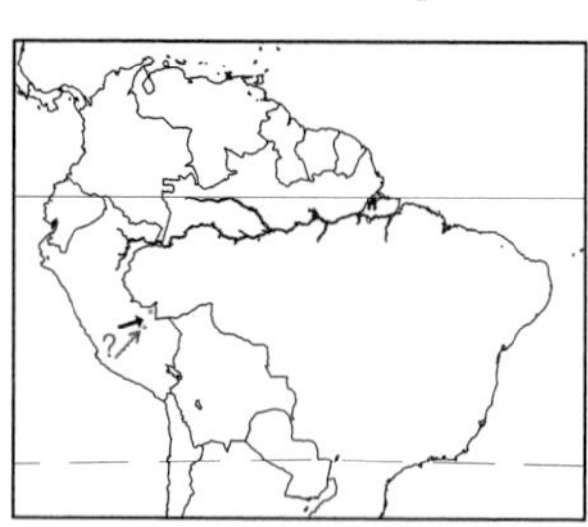

19a Adult male: This species is exceedingly rare, and it is crucial to eliminate all other identification contenders. This is a medium-sized cacique that is entirely black, showing yellow only on the rump. The rump patch averages larger than on Golden-winged Cacique. Note that the bill is whitish-blue and that the eyes are blue. The bill is longer and slimmer than on *chrysonotus* Mountain Cacique.
19b Adult male in flight: A black cacique with yellow restricted to the rump. The tail is relatively long.

18a
18c
18f
18b
18d
18e
17b
17a
17d
17c
19b
19a

21 Solitary Cacique *Cacicus solitarius* Text page 151

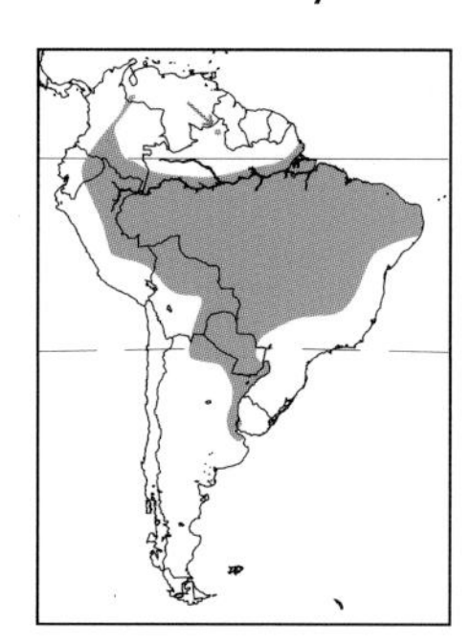

21a A large black cacique with dark eyes and a pale-coloured bill. The wings are short and somewhat rounded but the tail is long. The bill is deep at the base, unlike in the similar Ecuadorian Cacique.
21b Juvenile: Similar to the adult but the plumage is brownish-black. In addition, the bill tends to be paler than on adults.

20 Ecuadorian Cacique *Cacicus sclateri* Text page 150

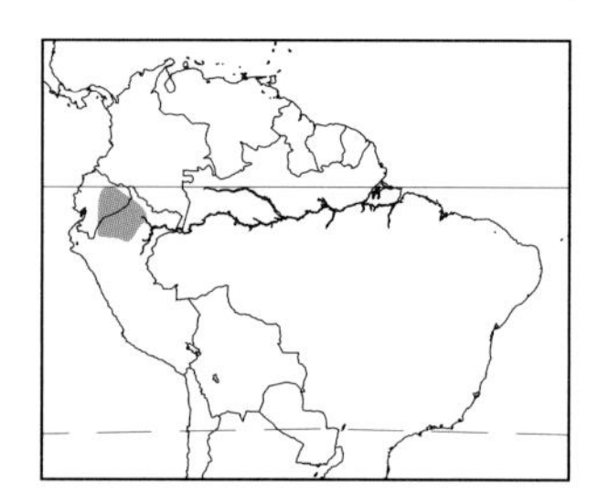

20 A slim medium-sized black cacique with bright blue eyes. The slender, shallow-based bill is whitish-blue in colour. The black plumage is only slightly iridescent. This species is very similar to Selva Cacique in proportions and structure.

23 Yellow-billed Cacique *Amblycercus holosericeus* Text page 155

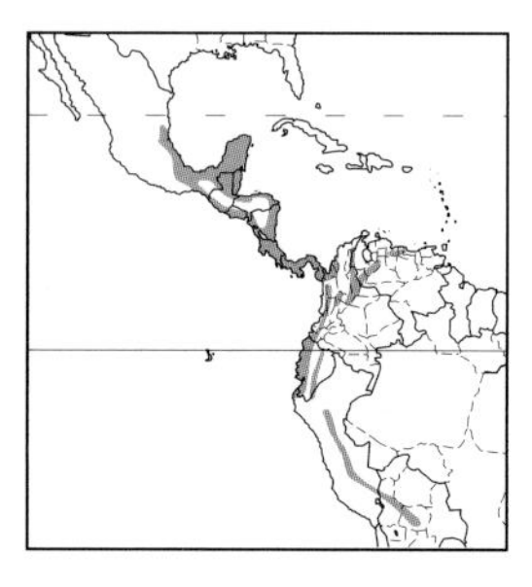

23a Adult male (*holosericeus*, Mexico to NW Colombia): A dull black cacique with short rounded wings and a broad tail. The eyes are bright yellow, and the bill is whitish-yellow, thick at the base and chisel-like due to a flattened culmen. At close range it may be possible to see that the nostril is covered by a flap (operculate nostril) unlike in the true caciques.
23b Adult male (*australis*, Andes of Venezuela to Bolivia): Similar to the nominate form but slightly smaller and the bill shows a grey base.
23c Juvenile (*holosericeus*): Similar to the adult, but shows a browner plumage. In addition, the eyes are dark.
23d Adult male in flight (*holosericeus*): A smallish black cacique with very short and rounded wings. Seldom observed in flight out in the open, keeps to deep cover. The tail tends to be held above the horizontal while in flight.

21a
21b
20
23b
23d
23c
23a

PLATE 9 ORIOLES 1

43 Orchard Oriole/43X 'Fuerte's Oriole' *Icterus spurius/Icterus s. fuerteri* **Text page 213**

43a Adult male (*spurius*, breeds in USA and Canada, winters in Central and South America): A small black and chestnut oriole; the head, back, tail and wings are black while the underparts, epaulet and rump are chestnut. The wing feathers are neatly edged white, with some chestnut edging on the tertials and inner greater coverts. The bill is black and slightly decurved, showing a small amount of blue-grey at the base of the lower mandible.
43b 1st-summer immature (1st-alternate) Male (*spurius*): Like the adult female but has a black face and throat bib. A variable number of chestnut feathers are present on the underparts.
43c Adult male 'Ochre Oriole' (*fuertesi*, breeds in S Tamaulipas to S Veracruz in Mexico, winters from Guerrero to Chiapas): Similar to *spurius* in all respects but the chestnut of the plumage is replaced by ochre.
43d Adult female (*spurius*): A slender and petite oriole with olive-green upperparts and more yellowish underparts. The wings are dark olive, showing two distinct white wingbars. A slightly paler supercilum is visible. Note the small bill with a slight downcurve.
43e Juvenile: Similar to the female, but duller. The underparts are paler yellow and the upperparts are often washed with buff or brown. The bold wingbars are buffy-white. Note that young juveniles show a pinkish bill which gradually becomes darker.

25 Epaulet Oriole *Icterus cayanensis* **Text page 159**

25a Adult male (*cayanensis*, NE Brazil (Parà) and the Guianas; Amazonia to E Peru and N Bolivia): A slender, long-tailed black oriole with yellow epaulets. The bill is long and decurved.
25b Adult male (*tibialis*, E Brazil from Maranhão, Piauí and Ceará south to Rio de Janeiro): Similar to *cayanensis* but has yellow thighs.
25c Adult male (*pyrrhopterus*, C Bolivia, Paraguay and SW Brazil south to Uruguay and N Argentina): Like the nominate but with a dark chestnut epaulet. In addition this form is shorter billed.
25d Adult male (*valenciobuenoi*, S Goiás, W Minas Gerais, W São Paulo, and SE Mato Grosso in Brazil): This form is intermediate between the chestnut-shouldered and yellow-shouldered forms, showing a tawny-yellow epaulet colour. (Drawn to smaller scale.)
25e Hybrid Epaulet x Moriche Oriole: Intermediate between the two, always showing a yellow epaulet but the extent of the yellow rump, thighs and crown is variable.

24 Moriche Oriole *Icterus chrysocephalus* **Text page 157**

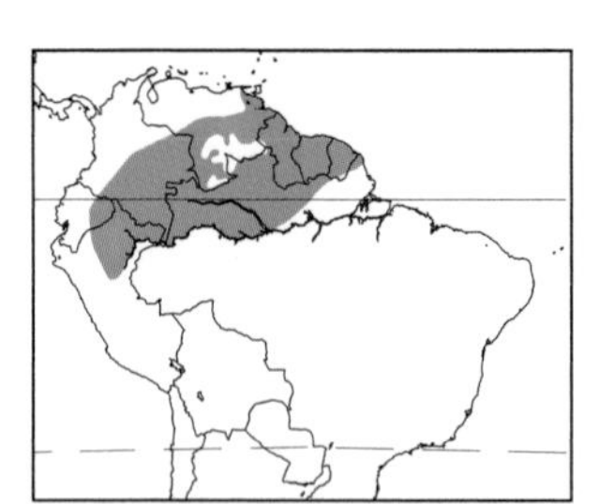

24a Adult male: A slim and long-tailed oriole with a thin and decurved bill. The body plumage is black, showing bold yellow patches on the crown, epaulets, rump and thighs.
24b Juvenile. Similar to the adult but the yellow patches are dull in colour and the black body plumage is replaced by blackish-brown. The greater coverts are thinly edged yellow, forming an indistinct lower wingbar. (Illustrated at smaller scale.)

43a
43e
43d
43b
43c
25c
25a
25b
25d
25e
24a
24b

40 Baltimore Oriole *Icterus galbula* Text page 199

40a Adult male: Black hood and back. The underparts, rump, and lower back are bright orange. The wings are black with an orange epaulet, wide white lower wingbar and noticeable white fringes to the primaries, secondaries and tertials. The tail is largely orange, having black central rectrices and black bases to the outer orange rectrices. **40b 1st-summer immature (1st-basic) male**: Variable. Similar to the adult female, but usually shows more extensive black on the face. Compared to the adult male, shows orange-olive mottling on the back, an incomplete black hood and a dull coloured tail. **40c Adult female**: Variable, some are almost indistinguishable from the adult male, while most are much duller. The individual illustrated is relatively male-like. Note that the orange of the underparts is duller than on the adult male, while the head and back are rarely entirely black. The most male-like individuals show a dull coloured tail, not the black and orange tail of the adult male. Compared to the first alternate male, the wings are not as worn. **40d Adult female**: Variable, this individual is not very male-like. Shows dull orange underparts, and sometimes a dusky streak on the throat. The crown and back are olive-grey with an orange wash and streaked with black. Note that the face is variable; may show an orange suffusion and or an olive colour with little orange. Shows two bold white wingbars. **40e Juvenile**: The head, breast and underparts are washed golden-orange. The crown, nape and back are olive with an orange wash. Two white wingbars are present and the flanks are washed with olive. Note that younger individuals show a pinkish or orangish bill colour. **40f pale 1st-fall (1st-basic) individual**: Some individuals (young females??) in the fall and winter are rather similar to Bullock's Oriole, showing a grey belly and almost unstreaked upperparts. Note that Baltimore Orioles tend to show a dull face pattern, lacking a dark eyeline and an orange supercilium. The face may or may not be washed orange. The undertail-coverts are always yellowish. Given a good look most of these individuals show some darker streaking on the upperparts.

41 Bullock's Oriole *Icterus bullockii* Text page 206

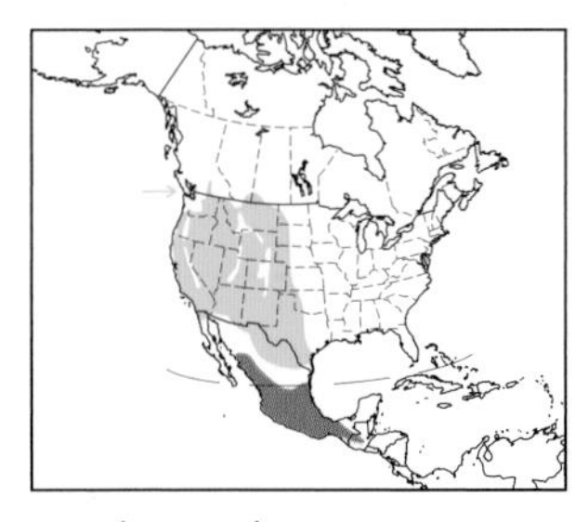

41a Adult male: An orange oriole with a black throat stripe, black eyeline, crown and back. The orange supercilium is prominent and extends well behind the eyes. The wings are black with a large white wing panel comprised of white lesser, median and greater coverts. The remiges are crisply fringed white. Note that the orange outer rectrices are tipped black. **41b 1st-summer immature (1st-alternate) male**: Similar to the adult female, but shows a complete black throat stripe and black lores. In addition, the wings are relatively more worn than on the female. **41c Adult female**: Orange-yellow on the face, including a prominent supercilium, throat and breast. Otherwise the underparts are whitish-grey, sometimes showing yellow on the crissum but not always. The upperparts are grey, with indistinct or no streaking on the back, and a grey rump. Note that on the face, not only does the darker eyeline contrast with the orange-yellow supercilium but the darker crown does as well. **41d Adult female, variant**: The head is shown only. Some females, perhaps older ones, show a black throat stripe and appear much like 1st-alternate males. Note that they often lack black on the lores. **41e Juvenile**: Similar to the adult female, but the yellow on the face and throat is duller in colour and more restricted in extent. In addition, the grey upperparts are sometimes washed saffron, the wing colour is duller and the wingbars are less extensive. Younger individuals lack the blue-grey of the bill, showing a pinkish or orange colour.

42 Black-backed Oriole *Icterus abeillei* Text page 211

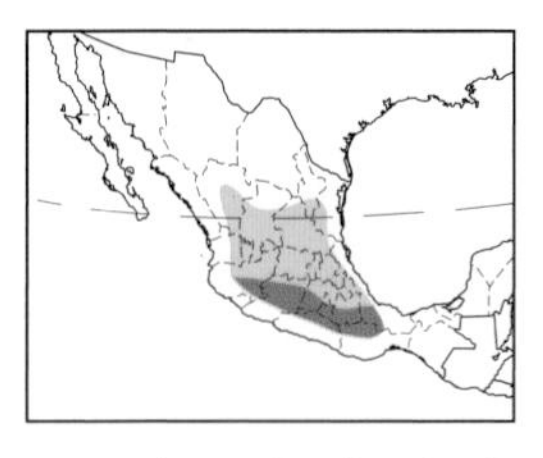

42a Adult male: Shows black auriculars, crown, nape, back, rump and flanks. The throat shows a black central stripe; other than the flanks the rest of the underparts are orange. Note that there is a short orange supraloral stripe and two small orange eye crescents which create a spectacled appearance. The tail is black with orange outer rectrices which are tipped black. The wings are black with an extensive white wing panel as on Bullock's Oriole; in addition, the remiges are crisply fringed white. **42b Adult female**: Shows a 'ghost pattern' of the male's colouration. However the black is replaced with olive-grey. On the underparts the orange is restricted to the throat and breast, as well as the crissum. The belly is grey, showing darker flanks. **42c Immature (1st-basic) male**: Similar to the female, but shows blackish lores and a black throat stripe. The back feathers have noticeable dark centers. In addition, the orange of the underparts is brighter.

40c
40f
40d
40e
40a
40b
41a
41e
41c
41b
42b
41d
42a
42c

36 Hooded Oriole *Icterus cucullatus* Text page 189

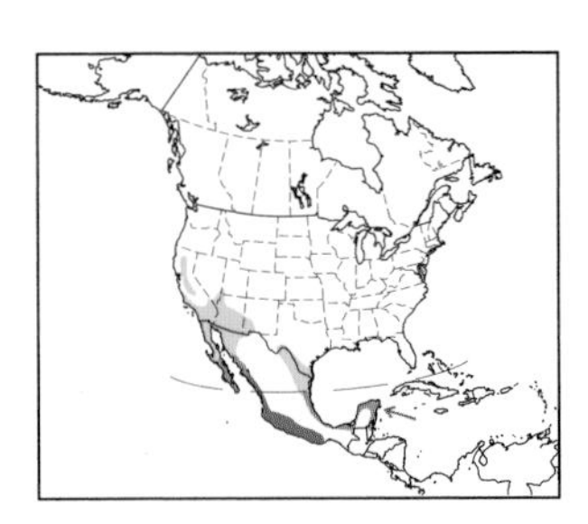

36a Adult male (*nelsoni*, breeds from California to SW New Mexico and adjoining parts of Mexico): A small, slim and dull orange oriole with a black face, bib, black back and tail. The wings are black with a bold white epaulet and a thin white lower wingbar. The remiges are crisply fringed white. Note that the bill is long, thin and decurved.
36b Adult female (*nelsoni*): Olive above, and dirty yellow below. The wings show two white wingbars. The long tail is dark olive.
36c Immature (1st-basic) male (*nelsoni*): Similar to the adult female but has black lores and a throat bib. The underparts are often brighter yellow than on the female.
36d Juvenile (*nelsoni*): Similar to the adult female, but the underparts are duller olive-buff, becoming creamy on the belly. The upperparts are darker olive. The bill is often pink-based, and younger individuals show a noticeably shorter bill than adults.
36e Adult male, fresh plumage (*nelsoni*): Like the worn male, but the nape is often tipped olive and the black back feathers are tipped orange-olive, partially obscuring the black back. (Shown at a smaller scale.)
36f Adult male (*igneus*, Yucatán region of Mexico and Belize): This is the brightest orange race. Note that the eastern, orange races of Hooded Oriole are shorter-billed than the western forms.
36g Adult female (*igneus*): Similar to female *nelsoni* but much brighter. The underparts and face are dull orange and the upperparts are olive with a noticeable orange wash.
36h Immature (1st-basic) male (*igneus*): Similar to the 1st-basic male *nelsoni* but brighter orange on the head and underparts. In addition, the back is darker, creating a greater contrast with the dull orange head than on *nelsoni*.

29 Orange Oriole *Icterus auratus* Text page 169

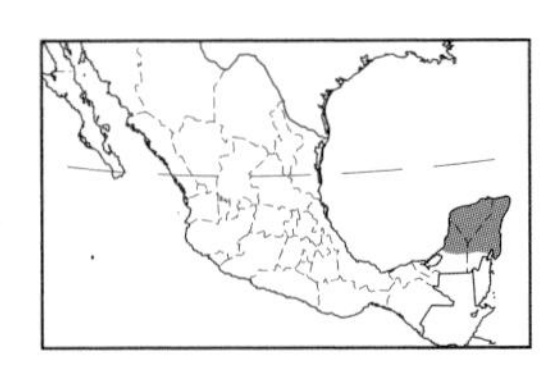

29a Adult male: A small and slim, brilliant orange oriole with a black face and throat stripe. The tail is entirely black, as are some of the lower scapulars. The tail is black with a white epaulet, and an extremly indistinct white lower wingbar. The primaries are edged with white at their bases, creating a white 'handkerchief'; The secondaries and tertials are crisply fringed white. Note that unlike Hooded Oriole, the slender bill of this species is straight.
29b Adult female: Similar to the male, but the orange is duller, washed with green on the mantle. In addition, the lesser and median coverts are largely black.
29c Immature (1st-basic) male: Similar to the adult female but the tail is olive-brown. The olive wash on the back extends up to the nape.

36a
36b
36c
36d
36e
36f
36g
36h
29a
29b
29c

35 Streak-backed Oriole *Icterus pustulatus* **Text page 183**

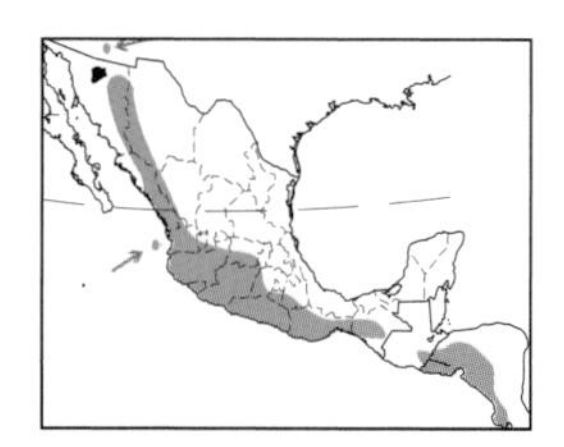

35a Adult male 'Scarlet-headed Oriole' (*microstictus,* breeds from Sonora and Chihuahua south to Jalisco): A bright orange oriole with a reddish-orange head. The back is noticeably streaked with small black streaks. The face and throat bib are black. Note that the black wings are heavily edged with white, forming two bold wingbars connected by white edging to the greater coverts as well as white fringes to the remiges, which are particularly heavy at the base of the primaries forming a white 'handkerchief'. In all forms note the straight bill. **35b Adult female 'Scarlet-headed Oriole'** (*microstictus*): Similar to the adult male in pattern but much duller. The back, nape and crown are olive-green and the underparts are pale orange-yellow, brightest on the face. **35c Juvenile 'Scarlet-headed Oriole'** (*microstictus*): Similar to the adult female, but the back is only faintly streaked. The underparts are white on the belly and the wings are weakly edged white, the wingbars are not prominent. The bird illustrated has already begun moulting into 1st-basic plumage and has thus gained black lores and bib. **35d Adult male 'Tres Marias Oriole'** (*graysoni,* Tres Marías Islands, Mexico): Similar to 'Scarlet-headed Oriole' in pattern but lacks most of the streaking on the upperparts. The plumage is more orange-yellow as well. **35e Adult female 'Tres Marias Oriole'** (*graysonii*): Drawn to a smaller scale. Similar to the adult male but the back is greenish and the underparts are a paler orange-yellow. **35f Adult male 'Streak-backed Oriole'** (*sclateri,* El Salvador to Costa Rica): Similar to 'Scarlet-headed Oriole' but the back streaks are wider, such that the back is almost evenly black and orange. The orange body colouration is not as reddish as that of 'Scarlet-headed Oriole'. Both sexes similar. **35g Adult male 'Streak-backed Oriole'** (*alticola,* Guatemala and Honduras): Similar to *sclateri,* but when worn shows an almost entirely black back as the black streaks average wider in this form. (Drawn to a smaller scale.)

34 Altamira Oriole *Icterus gularis* **Text page 181**

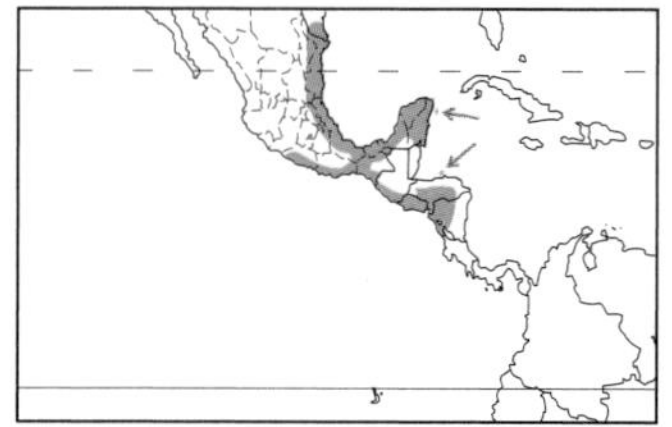

34a Adult male (*tamaulipensis,* S Texas to Campeche, Mexico): A large, stocky oriole with a thick-based bill. A bright orange oriole, brightest on the head, with a black face and bib as well as a black back. The wings are black, with an orange epaulet, wide white lower wingbar, and a distinct white 'handkerchief' at the base of the primaries. The secondaries and tertials are thinly fringed white. **34b Immature (1st-basic)** (*tamaulipensis*): Similar to the adult but the back is olive and the orange body plumage is not as bright. Note also that the wings are blackish and lack the bold white patterning of the adult. **34c Juvenile** (*tamaulipensis*): Like the 1st-basic plumage but lacks black on the face and throat. **34d Streak-backed variant** (*tamaulipensis*): Drawn to a smaller scale. Very rarely some (2nd-basic??) individuals show orange edging to the back feathers and may look deceivingly like Streak-backed Oriole. Note Altamira Oriole's thick-based bill. **34e Adult male** (*flavescens,* Guerrero, Mexico): Drawn to a smaller scale. This is the palest form. It is like *tamaulipensis* but the orange is replaced with bright yellow or yellow-orange.

33 Spot-breasted Oriole *Icterus pectoralis* **Text page 178**

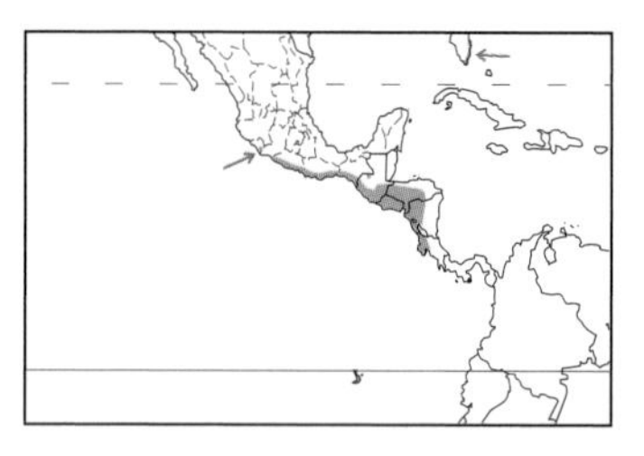

33a Adult male (*guttulatus,* S Mexico to Nicaragua): In all forms, the bill is slender and noticeably decurved. This plumage shows an orange body with a black face and throat bib surrounded by black spots on the breast. The back is black. The black wings show an orange upaulet, as well as a small white 'handkerchief' at the base of the primaries and a bold white wedge created by wide edges to the tertials. **33b Immature (1st-basic)** (*guttulatus*): Patterned like the adult, but the back is olive, the body plumage is orange-yellow and in many cases breast spotting is missing. Otherwise only a few breast spots are present. The wings are dull, but show the distinctive white tertial wedge. **33c Juvenile** (*guttulatus*): Similar to 1st-basic plumage but lacks black on the face and throat, note the beady-eyed appearance. The wings show the white tertial wedge as well as two thin buffy wingbars. The bill is pinkish at the base of the lower mandible. The breast is often washed olive. **33d Adult male** (*carolynae,* Guerrero and Oaxaca, Mexico): Similar to the other races, but the black spots on the breast coalesce into a nearly solid patch. It is paler orange-yellow than *guttulatus.* (Drawn to a smaller scale.)

35f
35g
35d
35e
35a
35c
35b
34c
34d
34b
34e
34a
33c
33d
33b
33a

32 White-edged Oriole *Icterus graceannae* **Text page 176**

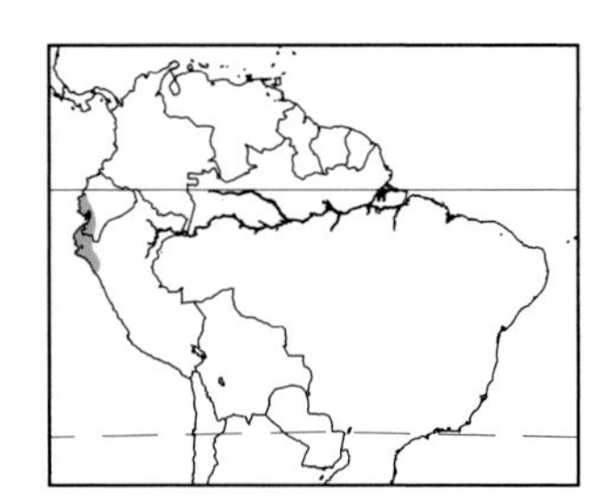

32a Adult male: An orange-yellow oriole with a largely black tail; the outer rectrices being greyish-white. The face and throat bib are black as is the back. The wings are black, with a yellow epaulet and a large white wedge on the tertials. Note that the bill is straight, particularly the culmen.
32b Immature (1st-basic): Similar to the adult in pattern but the underparts are paler, more yellow. The back is olive and the wings are blackish rather than black. Note that as on adults the tail looks largely white from below. The white tertial wedge is noticeable.
32c Adult male in flight: The tail appears entirely black; also the yellow shoulder is separated from the white of the tertials by a black band.

30 Yellow-tailed Oriole *Icterus mesomelas* **Text page 172**

30a Adult male (*taczanowskii*, Ecuador to NW Peru): A golden yellow oriole with a black face and bib, as well as a black back. The black tail shows three pairs of yellow outer rectrices, these are obvious when the tail is observed from below. Note that the black wings have a yellow epaulet which connects with yellow inner greater coverts and crisp white fringes to the tertials. All forms show a bill which is slightly downcurved.
30b Immature (1st-basic) male (*mesomelas*, Mexico to Honduras): Similar to the adult, but the back is olive instead of black, the black comes in patchily in older immatures. The wings are blackish, and the yellow of the greater coverts is often not solid. The tail is olive with yellower outer rectrices. Immatures of the other forms similar.
30c Juvenile (*mesomelas*): Similar to the 1st-basic plumage but lacks black on the face and throat. The greater coverts are entirely dark, but tipped with olive-yellow creating a thin olive-yellow lower wingbar. The yellow on the lesser and median coverts is reduced.
30d Adult male (*carrikeri*, Panama, N Colombia and NW Venezuela): Similar to *taczanowskii* but lacking white fringes on the tertials.
30e Adult male in flight (*taczanowskii*): The yellow edges to the tail are noticeable. Also, the yellow of the epaulet is connected to the white of the tertials by the yellow inner greater coverts.

31 Orange-crowned Oriole *Icterus auricapillus* **Text page 174**

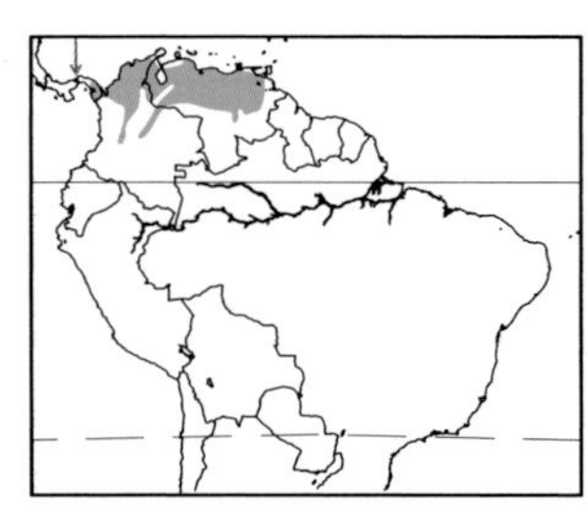

31a Adult male: A slim and long-tailed oriole with a small and nearly straight black bill. The black on the face and bib is extensive. The rump and underparts are orange-yellow and contrast with the vivid orange of the crown, nape and neck sides. The back and tail are black. The black wings show an orange-yellow epaulet but no pale edging on the remiges.
31b Juvenile: Shows a beady dark eye on an orange-yellow face, which contrasts with the olive crown, nape and back. The underparts are pale yellow, becoming white on the belly and washed with olive on the breast. The wings are olive-grey, with a nearly unoticeable yellowish-green upper wingbar and a more noticeable yellowish lower wingbar. The tail is olive.
31c Adult male in flight: The black wings are set off by an orange-yel low epaulet, otherwise the bright orange nape is a feature that is visible in flight. The tail is entirely black.

32a
32b
32c
30a
30e
30c
30b
30d
31b
31a
31c

26 Yellow-backed Oriole *Icterus chrysater* Text page 162

26a Adult male (*chrysater*, S Mexico to NW Nicaragua): A yellow-backed oriole with an extensive black face and throat bib. The crown, underparts and rump are also yellow. The tail and the wings are entirely black without paler edging. The bill is long and straight.
26b Adult female (*chrysater*): Similar to the adult male, but slightly duller in colour, the back and crown are washed olive.
26c Immature (1st-basic) (*chrysater*): Drawn to a smaller scale. Similar in pattern to the adults, but even duller than the female: The back and crown are olive-yellow, while the wings are brown with contrasting blackish coverts. The tail is olive.
26d Juvenile (*chrysater*): Like the 1st-basic plumage but lacks black on the face. The face is olive, with a golden supercilium and dusky lores. The lower mandible is pinkish-white.
26e Adult male in flight (*chrysater*): A rather rounded-winged oriole, showing a yellow back contrasting with black wings and tail.

27 Yellow Oriole *Icterus nigrogularis* Text page 164

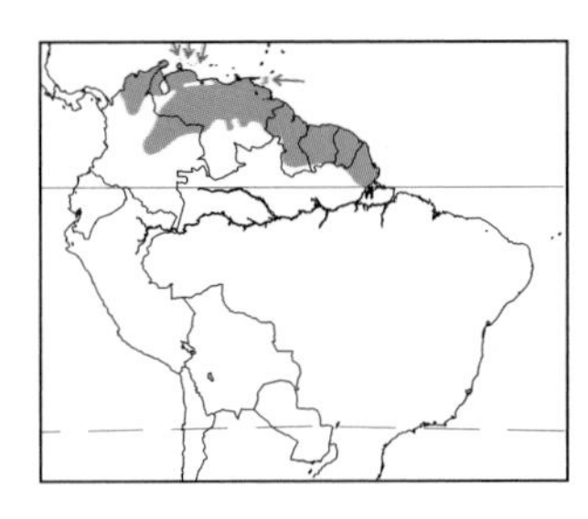

27a Adult male (*nigrogularis*, Colombia to N Brazil): A largely yellow oriole, with a yellow back. The lores are black and the black throat bib is thin. The wings are black with a yellow epaulet. There is a thin white lower wingbar, and white edging at the base of the primaries creates a white 'handkerchief', the secondaries and tertials are also fringed white. The tail is black.
27b Adult female (*nigrogularis*): Similar to the adult male but duller, showing an olive wash to the back and nape.
27c Juvenile (*nigrogularis*): Lacks black on the face. The upperparts are olive-yellow, brighter on the face. The underparts are yellow, palest on the belly and washed with olive on the throat and breast. The wings are brownish-black, showing a yellowish upper wingbar and a buffy-white lower wingbar.
27d Adult male in flight (*nigrogularis*): A rounded-winged oriole with a largely yellow body. The wings are black with a yellow epaulet, while the tail is entirely black.

26a
26b
26c
26d
26e
27d
27a
27b
27c

28 Jamaican Oriole *Icterus leucopteryx* Text page 167

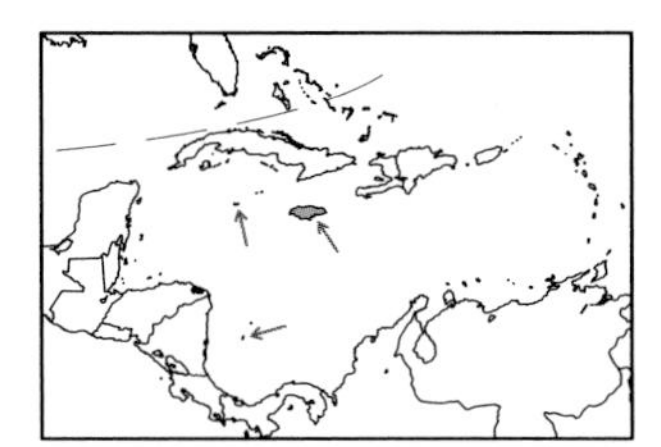

28a Adult male (*leucopteryx*, Jamaica): The most olive-coloured of the orioles. The body, including the back, is olive-yellow, becoming yellow on the belly. The face and throat bib are black. The wings are black with a large white patch on the greater and median coverts, the lesser coverts are olive-yellow. In addition, the secondaries and tertials are fringed white. The tail is black. In all forms the bill is thick at the base and has a slightly downcurved culmen.
28b Immature (1st-basic) (*leucopteryx*): Similar to the adult but with an even more strongly olive body plumage, particularly on the back and flanks. The greater coverts are not entirely white, showing blackish bases, particularly the outer set.
28c Adult male (*bairdi*, Cayman Islands): Extinct. Similar to the nominate form but bright yellow instead of olive-yellow.
28d Adult male in flight (*leucopteryx*): An olive-yellow oriole with large white wing patches.

48 St Lucia Oriole *Icterus laudabilis* Text page 229

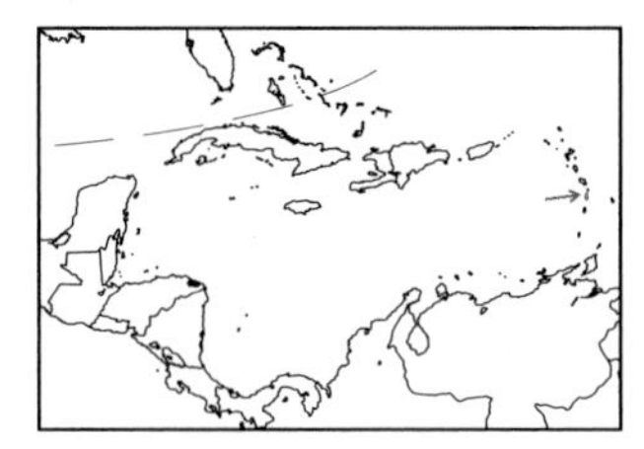

48a Adult male: A slim black oriole with russet-orange epaulet, rump, lower back, belly, thighs and crissum. The tail is relatively long.
48b Adult female: Similar to the adult male but the orange is paler, more orange-yellow.
48c Immature male (1st-basic): Similar to the adult male, but orange replaced by golden-olive. The nape and lower breast is chestnut-brown. The lower wingbar is thin and golden-olive. The tail is largely olive, sometimes with newer black central rectrices.
48d Immature female (1st-basic): Similar to the immature male but the chestnut-brown is more extensive on the upperparts and lower breast. The epaulet averages duller in colour.
48e Adult male in flight: Largely black, with a russet-orange epaulet and posterior half of the body.

47 Martinique Oriole *Icterus bonana* Text page 227

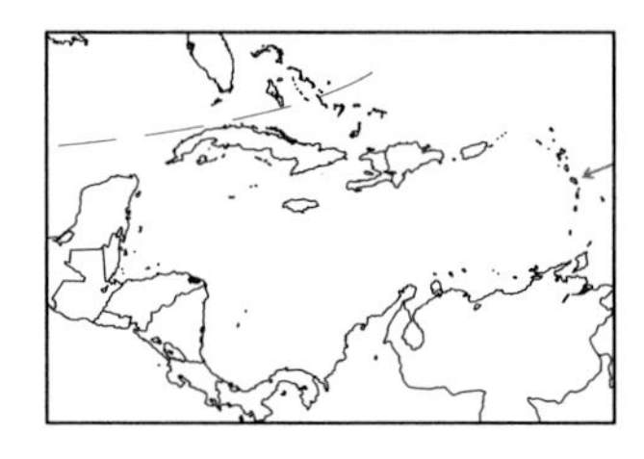

47a Adult male: The head is a deep chestnut, becoming black on the back and scapulars. The chestnut of the head extends down to the chest where it blends in with the tawny underparts, which are palest on the belly. The tawny extends to the lower back and rump. The wings are black with a tawny epaulet. The tail is entirely black.
47b Adult female: Similar to the male but slightly duller in plumage.
47c Adult male in flight: Largely black anteriorly, and tawny on the posterior half. The black wings show a tawny epaulet.

46 Montserrat Oriole *Icterus oberi* Text page 226

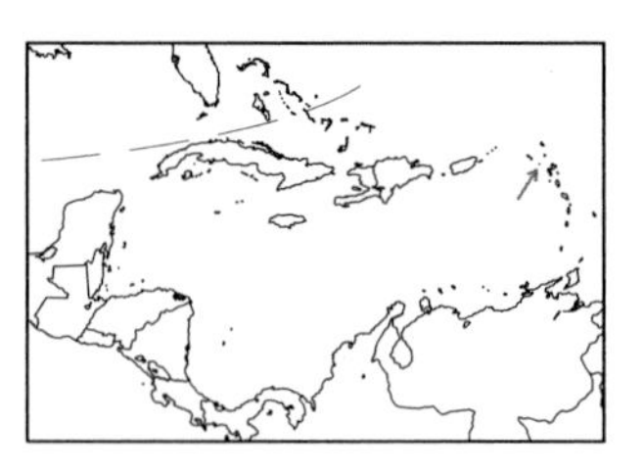

46a Adult male: Largely black, ending on the upperparts at the breast. The rest of the underparts are ochraceous-tawny, extending to the rump and lower back. The wings and tail are entirely black.
46b Adult female: Quite unlike the male. The female is olive above and olive-yellow below, extending to the face. The lores are dusky. The olive-brown wings show two dull olive wingbars. The tail is olive.
46c Adult male in flight: Looks largely black with an ochraceous-tawny rear half of the body, the wings and tail are entirely black.

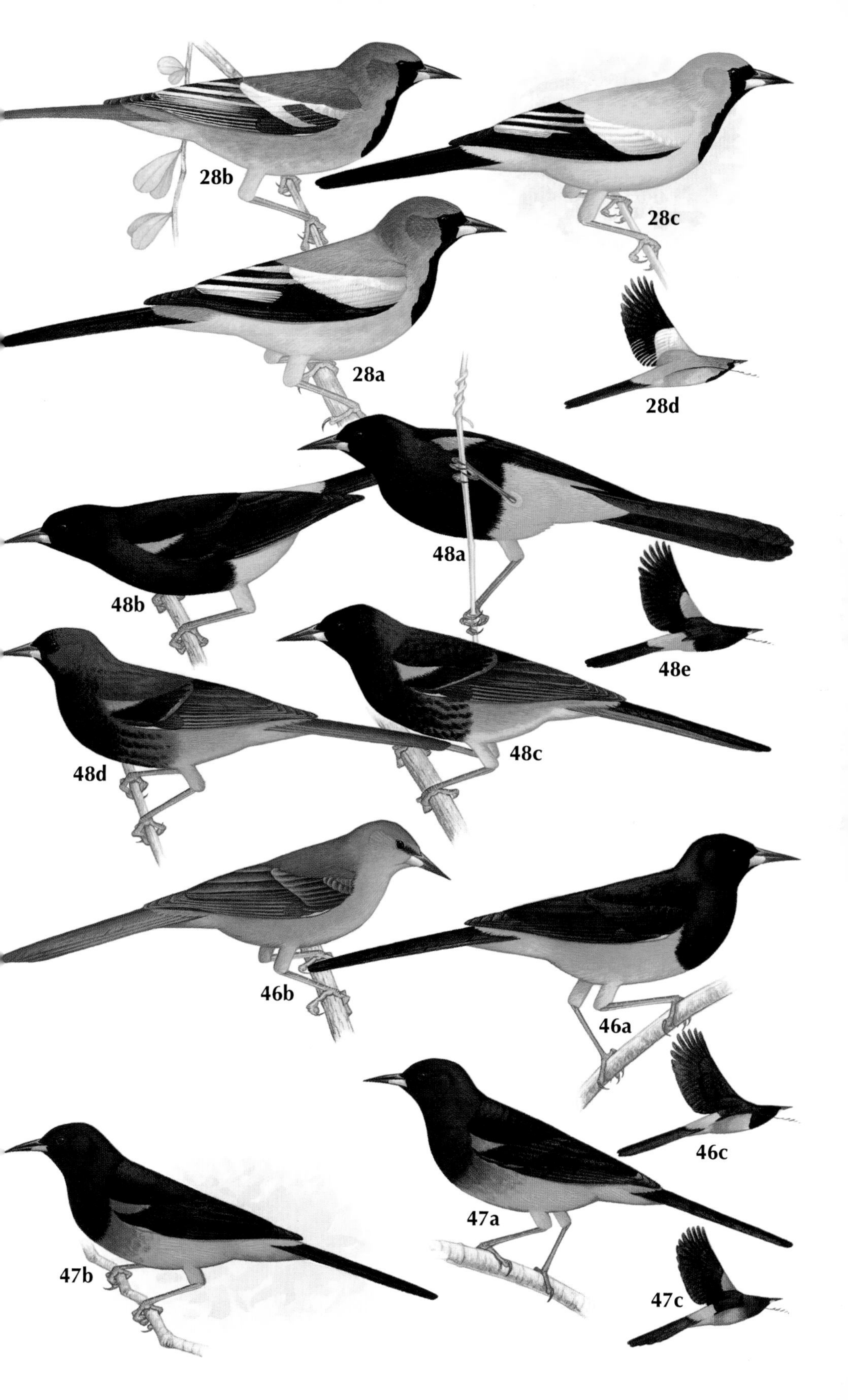
28b
28c
28a
28d
48a
48b
48e
48d
48c
46b
46a
46c
47a
47b
47c

44 Black-vented Oriole *Icterus wagleri* Text page 218

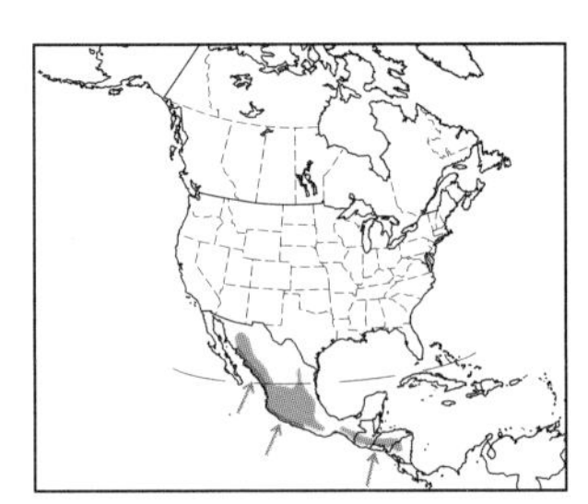

44a Adult male (*wagleri*, S Sinaloa, Mexico to Nicaragua): A black hooded, and black-backed oriole. The underparts, rump and epaulet are orange. Where the black and orange meet on the breast there is often a variable chestnut wash, most noticeable in the northern subspecies *castaneopectus*. The black vent and uppertail-coverts are not very obvious. The long bill is decidedly downcurved.
44b Immature (1st-basic): Variable. There is a variable amount of black on the face and breast, which is not neatly defined. This is an individual showing a large amount of black. Otherwise the head is olive-orange, turning olive on the back. The underparts are pale orange. The wings are blackish-brown with olive-orange lesser coverts and two thin greyish-white wingbars.
44c Immature (1st-basic): Variable. This is a rather dull individual, showing almost no black on the face.
44d Juvenile: Similar to the 1st-basic plumage but all black on the face is lacking. As well, the wings are brownish and show two cinnamon-buff wingbars. The underparts are a duller orange-yellow than those of the 1st-basic.

45 Black-cowled Oriole *Icterus dominicensis* Text page 221

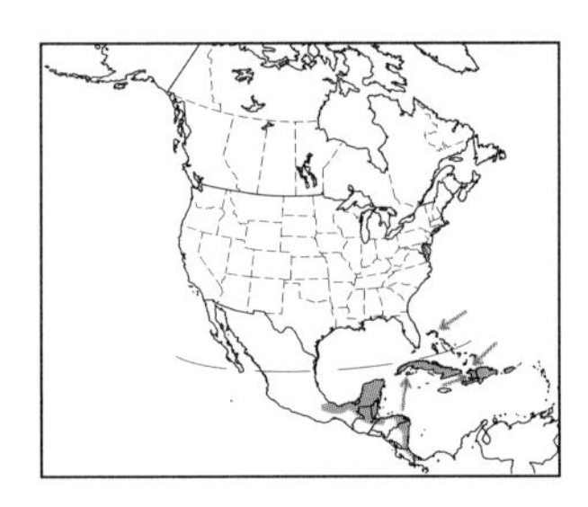

More than one species may be involved.
45a Adult male (*prosthemelas*, Mexico to Nicaragua): A largely black oriole with yellow belly, thighs, crissum, rump, lower back and epaulets. Otherwise the wings and tail are entirely black. There is often a chestnut band where the black of the breast meets the yellow of the belly. The bill is long and downcurved.
45b Adult female (*prosthemelas*): Shows a neatly set-off black forehead, face and wide throat bib. The crown and upperparts are olive, while the underparts are yellow. The wings are black with yellow lesser coverts and an entirely black tail.
45c Immature (1st-basic) male (*prosthemelas*): Similar to the adult female but the tail is olive. As well, the black on the forehead is more extensive.
45d Adult male (*portoricensis*, Puerto Rico): Similar to *prosthemelas* but the black is much more extensive on the underparts such that the yellow is restricted to the thighs and crissum. Note the black upper- and undertail-coverts which are tipped yellow.
45e Immature (1st-basic) (*dominicensis*, Hispaniola): Largely olive, with a chestnut olive head and breast and a black face and throat, the colours all subtly blending into each other. The olive wings show yellow median coverts. Most of the Caribbean forms have immatures similar to that of *dominicensis*.
45f Adult male (*northropi*, Bahamas): This is the most yellow of the Caribbean forms, with the black only extending down as far as the breast. Note that there is no chestnut wash where the black breast ends as in *prosthemelas*.
45g Immature (1st-basic) (*prosthemelas*): This is the most divergent immature plumage of the Caribbean forms. It is almost entirely olive, becoming yellow on the rump and belly and shows a yellow throat patch. The blackish-brown wings show two creamy-yellow wingbars. The lores are black.

44c
44a
44d
44b
45a
45b
45c
45g
45e
45f
45d

51 Scott's Oriole *Icterus parisorum* Text page 235

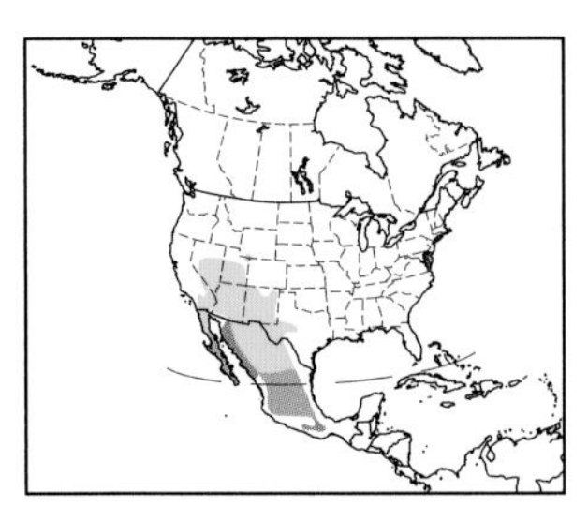

51a Adult male: A black hooded, black-backed oriole with bright yellow underparts and rump. The wings are black, and have a yellow epaulet bordered in white, the lower wingbar is white and prominent. The remiges are thinly fringed white. The tail is black with the bases of the outer several rectrices yellow.
51b 1st-summer immature (1st-alternate) male: Similar to the adult, but the back is olive, streaked with black. The underparts are olive-yellow, rather than yellow and the wings and tail are olive rather than black. The lesser coverts are olive and the wingbars are not nearly as prominent.
51c 1st-fall immature (1st-basic) male: Similar to the 1st-alternate male, but the black on the head and back is more restricted.
51d Adult female: Only the head is illustrated. Largely olive above and yellow below, with a variable amount of black on the face and black streaking on the back.
51e Immature (1st-basic) female: Similar to the adult female, but usually lacking black on the face. The underparts are duller and the streaking on the upperparts is not so noticeable.
51f Juvenile: Similar to the 1st-basic female, but even duller. The underparts are light yellow, almost white on the belly. The bill is pinkish at the base, and shorter than in adults.
51g Adult male in flight: A strikingly black and yellow oriole. Note the yellow rump and underparts which contrast with the black head, breast and back. The black tail shows yellow bases to the outer rectrices.

49 Audubon's Oriole *Icterus graduacauda* Text page 231

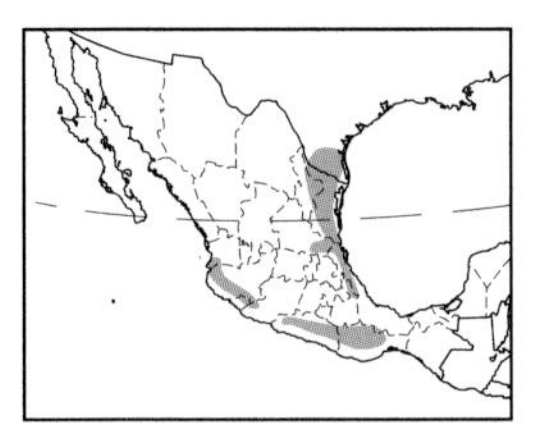

49a Adult male (*audubonii*, S Texas, NE Mexico): A yellow-bodied oriole with a black head and breast. Note that the lower edge of the black breast is jagged, not smooth. The back is washed olive. The black wings have yellow epaulets, a white lower wingbar and white fringes to the remiges. The tail is black.
49b Adult female (*audubonii*): Only the head is illustrated. Similar to the adult male, but the back and nape are more strongly olive in colour.
49c Adult male in flight (*audubonii*): A largely yellow oriole with black head and tail. The black wings show yellow epaulets.
49d Immature (1st-basic) (*audubonii*): Similar to the adult female but even more extensively olive on the upperparts. The wings are brownish, with contrasting blackish coverts and tertials. The tail is also olive.
49e Adult male (*dickeyae*, Guerrero and Oaxaca, Mexico): Like *audubonii* but the back is pure yellow, not washed with olive. The edge of the black hood on the nape is irregular, not smooth. In addition, the median coverts are black and only the tertials are fringed white.
49f Juvenile (*dickeyae*): Like the 1st-basic plumage but lacking black on the head. Instead the head is greyish-olive. The wings are brownish, lacking wingbars.

50 Bar-winged Oriole *Icterus maculialatus* Text page 233

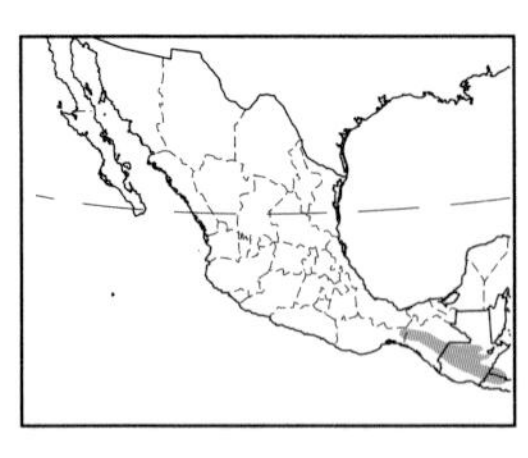

50a Adult male: The head, breast, back and tail are black; while the underparts and rump are lemon-yellow. The wings are black with a yellow epaulet, and a wide white lower wingbar. The lower wingbar is wide and the edges are parallel.
50b Adult female: Unlike the male. The entire upperparts are olive while the underparts are lemon-yellow. The forehad, face and a large, wide throat bib are black and crisply delineated from the rest of the plumage. The olive-brown wings show a yellowish upper wingbar and a noticeable white lower wingbar. The tail is olive.
50c Immature (1st-basic) male: Similar to the adult female, but the sides of the head and crown are speckled black.
50d Immature (1st-basic) female: Similar to the adult female but the auriculars are olive, not black.
50e Adult male in flight: Shows black wings, tail, back and head. The rump, underparts and epaulets are yellow.

51a
51b
51c
51d
51e
51f
51g
49e
49a
49b
49d
49f
49c
50a
50c
50b
50d
50e

37 Troupial *Icterus icterus* Text page 193

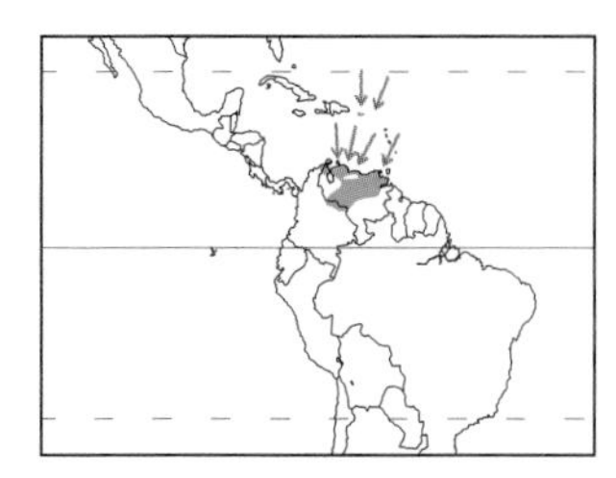

37a Adult male (*icterus*, Llanos of Colombia and Venezuela): A large oriole with a black hood, showing a ragged lower edge. The underparts, neck collar and rump are orange. The back and tail are black. The black wings show a large longitudinal (on the closed wing) white wing patch composed of an orange epaulet, fringed in white, white inner greater coverts and white edges on the tertials and inner secondaries. The eye is yellow and surrounded by a bare patch of blue skin.
37b Immature (1st-basic) (*icterus*): Similar to the adult, but orange-yellow rather than orange. The black of the plumage is blackish-brown, rather than flat black. Note that the eye patch is smaller.
37c Adult male (*metae*, Venezuela and Colombia near the Meta river): Similar to *icterus* but the orange collar extends to the nape. As well, some orange may be present on the lower back. The white wing patch is divided by black greater coverts.
37d Adult male in flight: (*icterus*) A large oriole, which looks orange with a black hood. The white wing patch is conspicuous in flight.

38 Campo Troupial *Icterus jamacaii* Text page 196

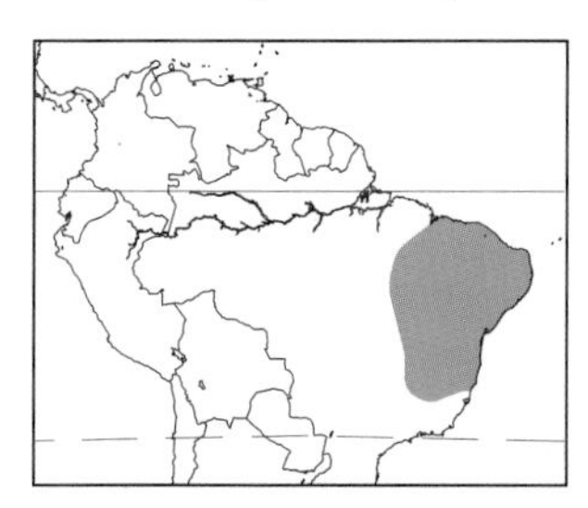

38a Adult male: A large oriole with a black hood, showing a ragged lower edge. The underparts, neck collar and rump are orange. The back and tail are black. The black wings show orange epaulets and a large wedge-shaped white patch on the tertials and inner secondaries. The eye is yellow and surrounded by a tiny bare patch of blue skin.
38b Adult male in flight: A large oriole, which looks orange with a black hood. The black wings have orange epaulets and limited white on the secondaries.

39 Orange-backed Troupial *Icterus croconotus* Text page 197

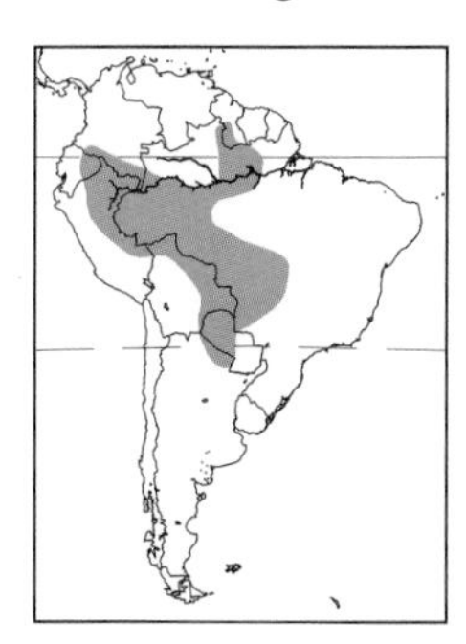

39a Adult male (*croconotus*, Amazonia): A troupial with an orange crown, nape and back; however the lower scapulars are black. The face and throat bib are black, showing a smooth edge, unlike the other troupial species. The tail and wings are also black. The wings show an orange epaulet and a small white wedge-shaped patch on the folded tertials and inner secondaries. The yellow eyes are surrounded by a small bare patch of blue skin.
39b Immature (1st-basic) (*croconotus*): Similar to the adult but orange replaced by orange yellow, and black replaced by brownish-black. The eyes are dark.
39c Adult male in flight: Looks largely orange, with a black face and throat. The wings are black with an orange epaulet and limited white on the secondaries.

53 Oriole Blackbird *Gymnomystax mexicanus* Text page 240

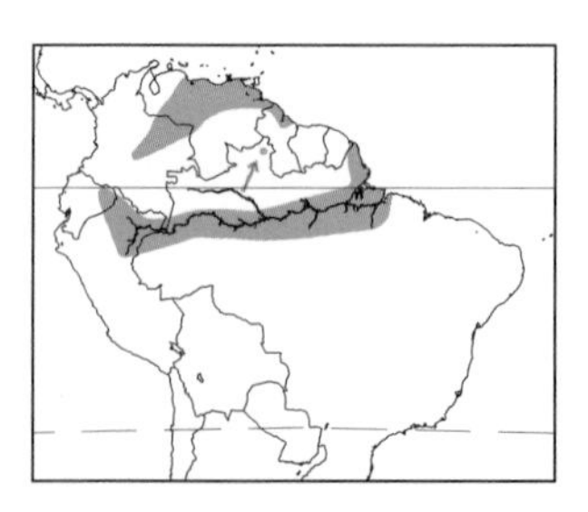

53a Adult male: A large icterid with a yellow head and underparts. The back, rump, tail and wings are black, except for a yellow epaulet. There are two blackish bare patches of skin on the face, one surrounding the eyes and the lores and the other along the malar line.
53b Juvenile: Similar to the adult but slightly paler yellow. A black cap is also present. Younger individuals show pinker facial skin.
53c Adult male in flight: Shows a yellow head and underparts, contrasting with black wings, tail and upperparts. The yellow epaulets are noticeable.

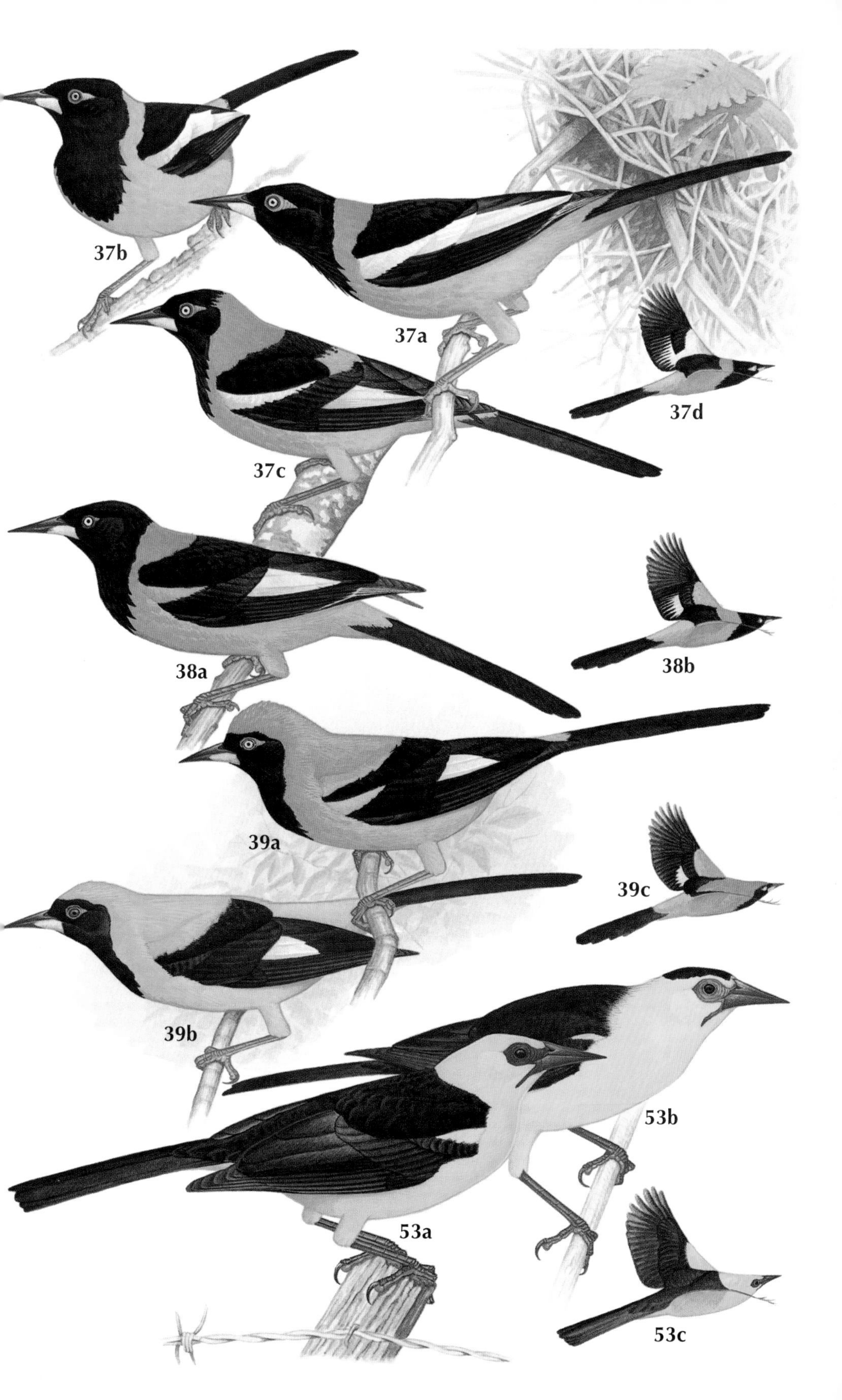

37b
37a
37d
37c
38a
38b
39a
39c
39b
53b
53a
53c

59 Yellow-hooded Blackbird *Agelaius icterocephalus* Text page 253

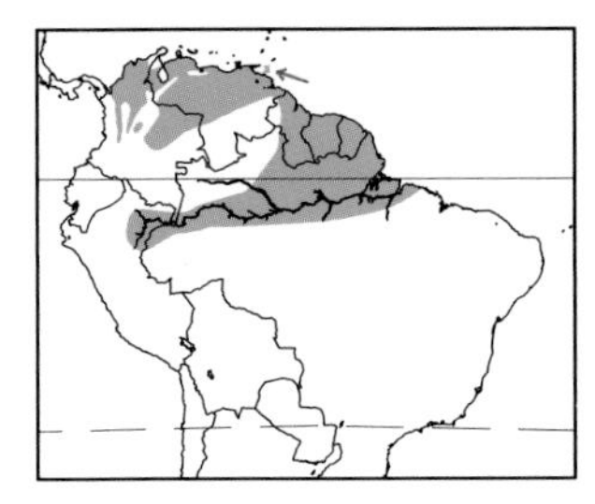

59a Adult male (*icterocephalus,* lowlands of N South America): A small blackbird with a yellow head contrasting with a black body. A small black mask is also present. Note the short-tailed appearance of this species.
59b Adult female (*icterocephalus*): Greyish-olive above and greyish below, obscurely streaked both on the upper and underparts. It shows a yellow supercilium and throat.
59c Adult male fresh plumage (*icterocephalus*): Like the worn male, but with olive tips to the nape feathers and olive tips to the back and underpart feathers.
59d Immature (1st-basic) male (*icterocephalus*): Similar to the female, but the yellow on the throat is brighter and more extensive. The young male also shows black feathers on the underparts and blackish wings.

56 Pale-eyed Blackbird *Agelaius xanthophthalmus* Text page 248

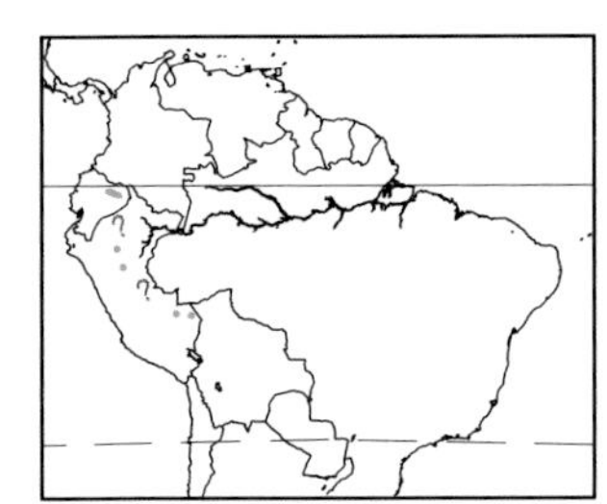

56a Adult male: An entirely black icterid showing a slight blue iridescence. The bill is slender and the wings are short. The eye is brilliant yellow.
56b 1st-basic female: Similar to the male but blackish-brown instead of black. The throat is weakly streaked with ochre brown. Some are yellowish on the throat and breast, streaked with brown (younger individuals?):

57 Unicolored Blackbird *Agelaius cyanopus* Text page 249

57a Adult male (*cyanopus,* E Bolivia, S Brazil, Paraguay, N Argentina): Entirely black, slightly glossy. Note the slim shape, long tail and thin, straight bill.
57b Adult female (*cyanopus*): Yellow below, obscurely streaked with grey. The upperparts are dark olive, streaked blackish and edged rusty on the scapulars. The ear-coverts are black. Note that the blackish wings are widely edged rusty. The tail is blackish.
57c Immature (1st-basic) male (*cyanopus*): Similar to the adult female but black on the throat and spotted with black on the underparts and back.
57d Adult female (*xenicus,* NE Brazil): Blackish with olive edging to the wing-coverts and tertials. The belly is olive-yellow.
57e Adult female (*atroolivaceus,* coastal SE Brazil): Dark olive-grey, with a yellowish wash on the underparts, palest on the throat. The coverts and tertials are thinly edged rich brown.

59d
59a
59c
59b
56b
56a
57c
57a
57b
57d
57e

58 Yellow-winged Blackbird *Agelaius thilius* Text page 251

58a Adult male (*petersi*, Argentina, S Paraguay, S Brazil and Uruguay): A small slim blackbird with a sharply pointed, straight bill. Entirely black with yellow epaulets.
58b Adult male, fresh plumage (*petersi*): Like the worn male but the upperparts are tipped brown, creating a streaked appearance. There is a pale brown supercilium, the underparts are tipped olive and the wing feathers are crisply edged chestnut.
58c Immature (1st-basic) male (*petersi*): Intermediate between the adult female and male. Overall, like the female but with black on the throat, and extensive black streaking on the underparts.
58d Adult male in flight: Entirely black with yellow linings and epaulets.
58e Adult female (*petersi*): Buffy below with brown streaks. The upperparts are a mix of brown and rusty streaks. The face shows a crisp, buffy supercilium on a brown face. The wings are edged rusty-brown.
58f Adult female (*alticola*, altiplano of Bolivia and Peru): Similar to female *petersi*, but larger and more heavily streaked blackish. The underparts are greyish rather than buff. In addition, females of this form sport a large yellow epaulet.

60 Chestnut-capped Blackbird *Agelaius ruficapillus* Text page 256

60a Adult male: Black with a chestnut crown and throat patch. Somewhat short tailed, and straight-billed.
60b Immature (1st-basic) male: Similar to the adult male but the plumage is heavily veiled with grey-buff tips, concealing the underlying pattern of black and chestnut.
60c Adult male in flight: Looks largely black, only in good views are the chestnut throat and crown patches apparent.
60d Adult female: Dull and nondescript, somewhat resembles female Shiny Cowbird. Olive throughout, obscurely streaked darker and showing a yellowish throat. The supercilium is indistinct.
60e Juvenile: Similar to the adult female, but buffy-brown rather than olive. More heavily streaked on the underparts. The wing-coverts are edged warm brown.

58a
58c
58b
58f
58e
58d
60c
60b
60e
60a
60d

65 Yellow-shouldered Blackbird *Agelaius xanthomus* Text page 277

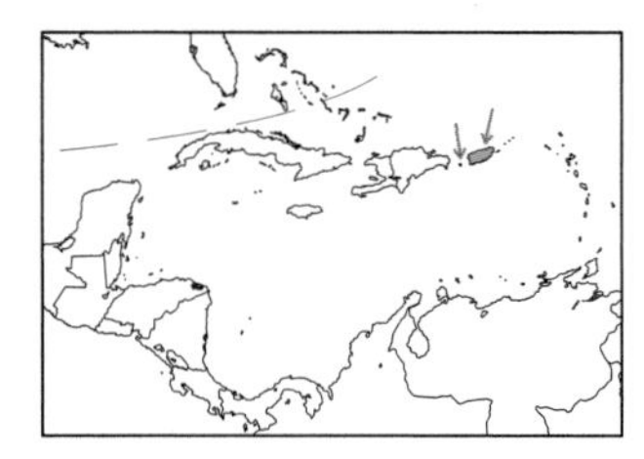

65a Adult male (*xanthomus*, Puerto Rico): A small slim blackbird with a strait black bill. The body is entirely black, save for the yellow epaulet.
65b Immature (1st-basic) (*xanthomus*): Like the adult but blackish-brown. The median coverts are black, edged creamy-white.
65c Adult male in flight (*xanthomus*): Entirely black with a yellow epaulet.
65d Adult male (*monensis*, Mona Island): Similar to *xanthomus* but the median coverts are whitish, paler than the rest of the epaulet.

63 Tawny-shouldered Blackbird *Agelaius humeralis* Text page 273

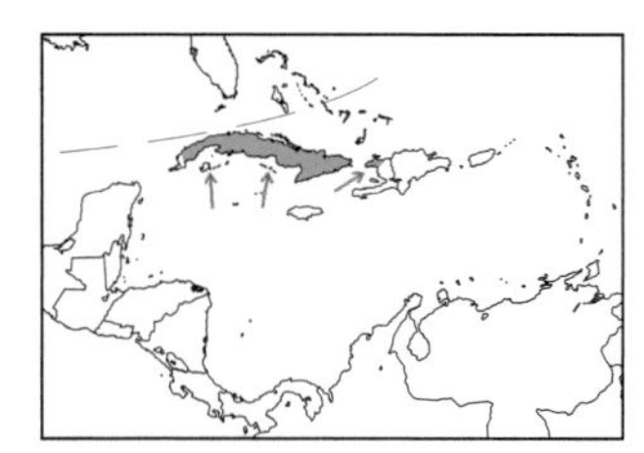

63a Adult male: A small icterid with a short spiky bill. Adult male. Entirely black with a tawny epaulet with a buff border.
63b Immature male: Like the adult but the body is brownish-black. The median coverts are black, edged buff, and the rest of the epaulet is not as deeply coloured as the adult's.
63c Adult male in flight: Entirely black with a tawny epaulet.

64 Red-shouldered Blackbird *Agelaius assimilis* Text page 275

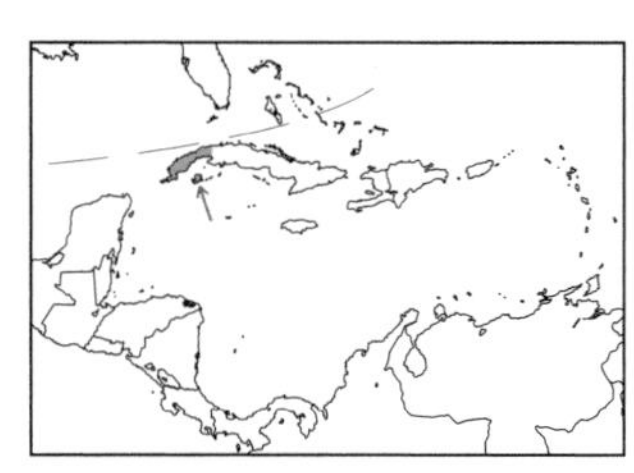

64a Adult male: A smallish blackbird, with a sharply pointed black bill. The plumage is entirely black except for the white-bordered red epaulet. Extremely similar to Red-winged Blackbird.
64b Adult female: Similar to the adult male, but entirely black without iridescence. Lacks the epaulets.
64c Adult male in flight: Appears black, with a red epaulet which is bordered white.
NB Map should show distribution extending further east to cover western half of Cuba.

65a
65d
65c
65b
63a
63b
63c
64b
64a
64c

61 Red-winged Blackbird *Agelaius phoeniceus* Text page 258

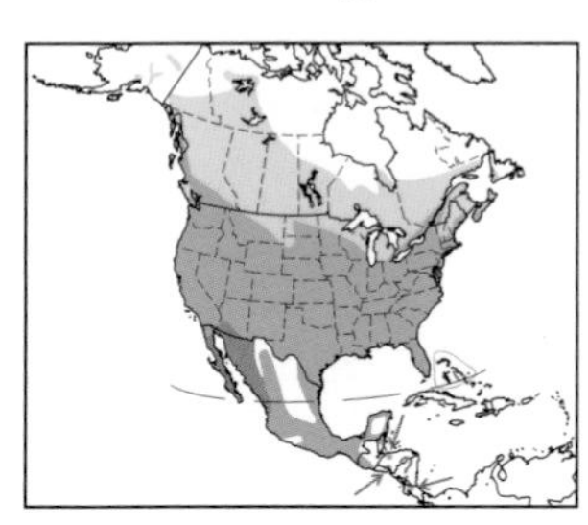

61a Adult male, summer (*phoeniceus*, NE North America): A striking black icterid with a red epaulet bordered in pale buff. The bill is somewhat deep at the base and finely pointed.
61b Adult male, early fall, fresh plumage (*phoeniceus*): Similar to the summer male but the upperparts are tipped brown and rusty, thus appearing streaked above. Note that the greater coverts and tertials are thinly edged pale brown.
61c Adult male in flight: Appears largely black, but the vivid buff-bordered red epaulet is visible.
61d Immature male, summer (1st-basic/alternate) (*phoeniceus*): This plumage is variable, some individuals are more heavily marked than others. Similar to the adult male in summer but shows black in the epaulet. In addition, the upperparts and wings are variably edged or tipped brown.
61e Immature male, fall (1st-basic) (*phoeniceus*): Like the fresh adult male but the brown and rust tipping is substantial, such that the upper and underparts appear streaked. The epaulet is marked with black.
61f Juvenile male (*phoeniceus*): Similar to the female but the back is richly edged chestnut. Appears larger than the female in direct comparison. The ear-coverts are dark, forming an ear patch and contrasting with a paler lower eye crescent. The face is washed with buffy-ochraceous.
61g Adult female, fall, fresh plumage (*phoeniceus*): Boldly striped blackish on the white underparts. The back is brown streaked with chestnut and buff. The face shows a whitish supercilium, becoming peach near the bill; the throat and malar stripe are suffused with peach as well.
61h Adult female, summer, worn plumage (*phoeniceus*): Much of the paler edging wears off, particularly on the upperparts. Thus, the upperparts begin looking blackish-brown with a reduced amount of brown and buff streaking.
61i Immature female, summer (worn 1st-basic) (*phoeniceus*) . Only the head is illustrated. Similar to the adult female but lacks the peach, or pinkish suffusion on the face.
61j Adult female in flight: No obvious field marks. A small, streaky brown icterid. Some show a substantial amount of red on the epaulets.
61k Juvenile female (*phoeniceus*): Only the head is illustrated. Similar to the juvenile male but smaller and paler. The ear-coverts are not as solidly dark. Compared to the adult female it lacks the pink on the face and has a horn-coloured bill .
61l Adult female (*mearnsi*, N Florida): Similar to *phoeniceus* but buffier below and browner above. The bill is longer and more slender.
61m Adult female (*aciculatus*, Kern Valley, California): Similar to *phoeniceus* but darker on the belly, browner on the back and showing a long and slender bill.

61a
61b
61d
61c
61e
61f
61l
61k
61m
61g
61i
61h
61j

62 Tricolored Blackbird *Agelaius tricolor* Text page 269

62a Adult male, summer, worn plumage: Glossy black with a dark red epaulet bordered in white. Note that in all plumages the bill is long and slender.
62b Adult male in flight: Black, with a white-bordered red epaulet. Note the more pointed wings than 'Bicolored Blackbird'.
62c Adult male, fall, fresh plumage: Only the head is illustrated. Similar to the summer male but copiously tipped with cold grey-buff on the upper and underparts. The epaulet border is buffy-white.
62d Immature male, fall (1st-basic) fresh plumage: Similar to the fall adult male, but even more heavily tipped grey-buff. The epaulets shows some black markings.
62e Immature male, summer (1st-basic): Only the head is drawn. Similar to the summer adult male, but showing black in the epaulet. Variably edged or tipped grey-buff on the upperparts.
62f Adult female summer,, worn plumage: Blackish-brown, noticeably streaked only on the breast. The throat is whitish. A whitish supercilium is present, but may be worn-off on some individuals. Tends to show a cold grey-brown colour. Often indistinguishable from female 'Bicolored Blackbird'.
62g Adult female in flight: Appears blackish, has slightly more pointed wings than 'Bicolored Blackbird'.
62h Adult female, fall, fresh plumage: More noticeably streaked than in the summer. Note the cold grey-buff colours on the upperparts and neck.
62i Juvenile: Similar to the adult female, but more obviously streaked below. The supercilium is more pronounced.

61X 'Bicolored Blackbird' *Agelaius phoeniceus gubernator* Text page 266

61Xa Adult male, summer, worn plumage (*mailliardorum*, coastal C California): Entirely black with a red epaulet showing no obvious border. Often hides the epaulet and appears entirely black.
61Xb Adult male in flight (*mailliardorum*): Appears entirely black except for the red epaulet. Note that the wings are more rounded than on Tricolored Blackbird.
61Xc Adult male, fall, fresh plumage (*mailliardorum*): Similar to the summer male, but the upperparts are edged with brown and chestnut.
61Xd Immature male, fall (1st-basic) fresh plumage (*mailliardorum*): Similar to the fall adult male but even more heavily edged with brown, buff and chestnut. The epaulets show black tips or bases to the feathers and the wing feathers are edged brown.
61Xe Immature male, summer (1st-basic/alternate) worn plumage (*mailliardorum*): Similar to the summer adult male but shows black in the epaulets. It may also show a variable amount of brown tipping on the upperparts.
61Xf Adult female, summer, worn plumage (*mailliardorum*): Blackish-brown, noticeably streaked only on the breast. The throat is whitish, with a peach or pinkish wash. A whitish supercilium is present, but may be worn-off on some individuals. Tends to show a warm blackish-brown colour. Often indistinguishable from female Tricolored Blackbird.
61Xg Adult female in flight (*mailliardorum*): Appears blackish, has slightly more rounded wings than the Tricolored Blackbird.
61Xh Adult female, fall, fresh plumage (*mailliardorum*): More noticeably streaked than in the summer. Note the chestnut, cinnamon or warm brown edging on the upperparts and neck. The underparts show a buffy wash.

62b
62a
62c
62e
62d
62i
62h
62g
62f
61xg
61xh
61xb
61xc
61xf
61xa
61xd
61xe

54 Yellow-headed Blackbird *Xanthocephalus xanthocephalus*

Text page 242

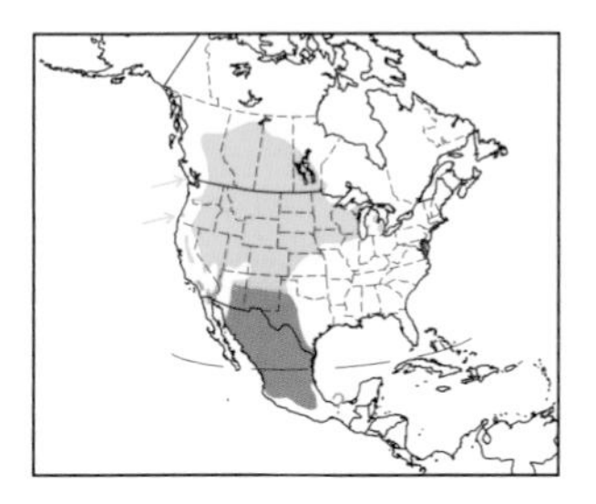

54a Adult male, summer: A large blackbird with a vivid yellow head and black face mask. The wings are black with a large white wing patch on the greater primary coverts. Note that the feathers around the vent are yellow, but rarely observed in the field.

54b Adult male, early spring, in flight: Appears large, black and yellow-headed. The white wing patches are obvious in flight. Note that this individual shows some golden-brown tipping on the nape, this wears off completely by summer.

54c Adult male, fall, fresh plumage: Only the head is illustrated. Similar to the adult male but the nape and crown are heavily tipped golden-brown.

54d Immature male, summer, (1st-basic/alternate), worn plumage: Similar to the adult female but the body is darker, showing some black feathers. Additionally, the yellow on the head is more extensive and immature males are noticeably larger than females. Note the small white wing patch.

54e Immature male, fall (1st-basic): Like the summer immature but the yellow on the face is golden yellow, rather than pure yellow. The body is brownish-black, not as dark as that of the summer male. Note that the yellow on the breast extends down as yellow streaking.

54f Immature (1st-basic) male in flight: Similar to the adult female, but noticeably larger and shows a limited amount of white on the greater primary coverts.

54g Adult female, fall, fresh plumage: A blackish-brown icterid with a bold yellow supercilium, malar stripe, throat and breast. The lower breast is streaked with white. The wings are entirely dark.

54h Adult female, summer, worn plumage: Only the head is illustrated. Similar to the fall female but paler yellow on the face and breast. The ear-coverts become yellow-brown, not solid brown as when fresh.

54i Adult female in flight: A dark icterid with no wing patches and a yellow throat and breast.

54j Juvenile: Quite unlike the adults. Blackish-brown above, showing a cinnamon to cinnamon-buff head and a dark face mask. The underparts are cinnamon-buff, palest on the belly. Note that the blackish-brown wings show two bold cinnamon wingbars.

54a
54b
54c
54d
54e
54f
54g
54h
54i
54j

67 White-browed Blackbird *Sturnella superciliaris* Text page 284

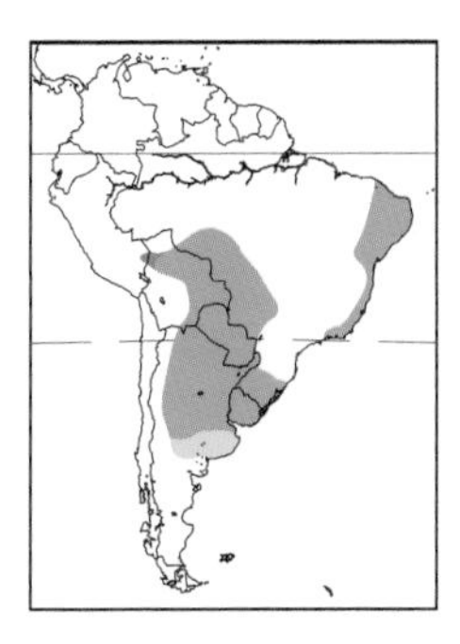

67a Adult male, summer, worn plumage: A small black icterid with a noticeable white supercilium. The throat and breast are red. Note the dumpy and short-tailed appearance of this icterid.
67b Adult male, fall, fresh plumage: Only the head is illustrated. Similar to the summer male but the upperparts are heavily tipped brown. The red chest is tipped buff, making it appear less vivid.
67c Adult female, summer, worn plumage: A small streaky icterid with a pale supercilium. Note that the breast is streaked but the belly is not. The belly often shows some red colouration. Basically indistinguishable from female Red-breasted Blackbird but has a shorter bill.
67d Adult female, fall, fresh plumage: Only the head is illustrated. Similar to the summer female but more richly coloured.

66 Red-breasted Blackbird *Sturnella militaris* Text page 281

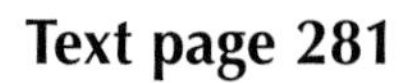

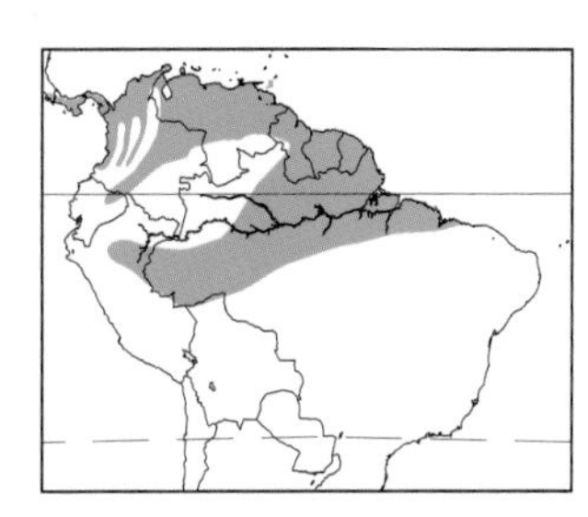

66a Adult male, summer, worn plumage: A small black icterid with a red throat and breast. Like White-browed Blackbird, this species is short-tailed and dumpy-looking.
66b Adult male in flight: Appears small and short-tailed with a vivid red breast and epaulets. The wing-linings are dark.
66c Adult male, fall, fresh plumage: Similar to the summer male but the upperparts are tipped with brown, making them appear streaked. The red breast is tipped buff.
66d Adult female, summer, worn plumage: A small streaky icterid with a pale supercilium. Note that the breast is streaked but the belly is not. The belly often shows some red colouration. Basically indistinguishable from female White-browed Blackbird but has a longer bill.
66e Adult female, fall, fresh plumage: Only the head is illustrated. Similar to the summer female but more richly coloured.
66f Adult female in flight: Appears small, brown and streaky. Note the short tail and rather pointed wings. The epaulets may show a red wash. Streaking on the vent and breast separates it from Bobolink.
66g Juvenile: Similar to the adult female but more extensively streaked on the underparts. Crisp pale fringes to the upperpart feathers gives it a scaly appearance.

75 Scarlet-headed Blackbird *Amblyramphus holosericeus* Text page 309

75a Adult male: A striking species. It is large and slender with a straight and sharply pointed bill and a long tail. The body is black, showing a rich scarlet head, neck, breast and thighs. The lores are dusky.
75b Immature (1st-basic/alternate?): An older individual showing a pattern similar to the adult's. However, the head is duller in colour, more orange-red and dusky patches may be present on the nape and breast.
75c Immature (1st-basic): A younger individual showing only a restricted amount of orange-red on the throat. The juvenile appears to lack red altogether, but soon after fledging, red is moulted in.
75d Adult male in flight: A large, long and slim black species with a vivid scarlet head.

67d
67c
67a
67b
66b
66f
66e
66d
66a
66c
66g
75a
75d
75c
75b

70 Long-tailed Meadowlark *Sturnella loyca* Text page 290

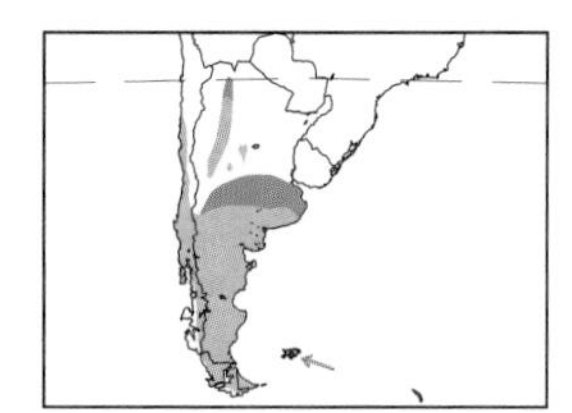

70a Adult male, summer, worn plumage (*loyca*, Chile and S Argentina): A large meadowlark with a long tail and long bill. The body is blackish-brown streaked throughout with brown. The supercilum is red before the eye and white behind the eye. The throat, breast and upper belly are vivid red.
70b Adult male in flight (*loyca*): Similar to the other red-breasted meadowlarks but shows a noticeably long tail and pale underwings.
70c Adult female, summer, worn plumage (*loyca*): Similar to the male but browner. The red on the underparts is restricted to the belly, while the breast is streaked brown and the throat is white. The supercilium is white, sometimes becoming pinkish before the eye.
70d Juvenile male (*loyca*): Similar to the adult female but the underparts are warmer buff, and the crissum and belly are brown and streaked, not black. The throat and supralores show a reddish wash.

68 Peruvian Meadowlark *Sturnella bellicosa* Text page 285

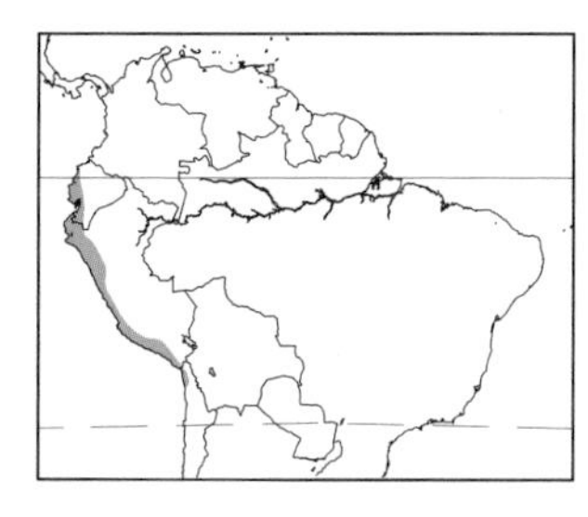

68a Adult male: A medium-sized meadowlark with a short tail. The upperparts are brownish and streaked. The supercilium is red before the eye and white behind the eye. The throat and breast are red; the entire red patch is surrounded by black, on the face, sides of the breast and upper belly. The flanks and undertail-coverts are buffy with darker streaking.
68b Adult male in flight: Similar to the other red-breasted meadowlarks but shows a short tail and pale underwings.
68c Adult female: Similar to the male, but lacks solid black except on the cheeks and crown. The throat is buff, sometimes showing a hint of red. Otherwise the red on the underparts is restricted to a small patch on the lower breast; the upper breast is streaked.

69 Pampas Meadowlark *Sturnella defillippi* Text page 287

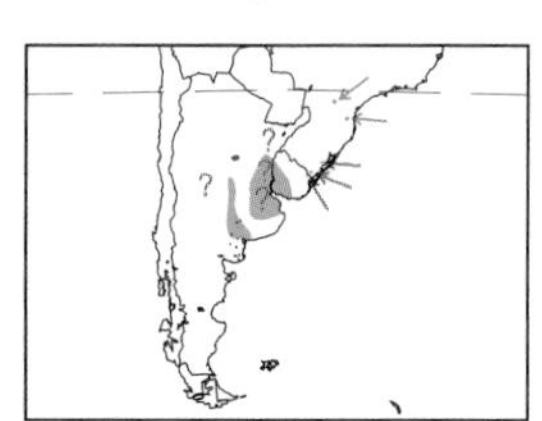

69a Adult male: Smaller and blacker-looking than Long-tailed Meadowlark, also proportionately shorter-tailed. The upperparts are blackish with restricted brown streaking. The throat, breast and upper belly are vivid red, contrasting with the blackish breast sides, face, belly and undertail-coverts.
69b Adult male in flight: Similar to the other red-breasted meadowlarks but shows a short tail and blackish underwings.
69c Adult female: Similar to the male, but the throat is buffy, the breast is streaked and the red on the underparts is restricted to the lower breast. The body plumage is not as blackish as the male's, and is more obviously streaked. Very similar to female Long-tailed Meadowlark but smaller and proportionately shorter-tailed.

70b
70a
70c
70d
68b
68a
68c
69b
69a
69c

71 Eastern Meadowlark *Sturnella magna* Text page 292

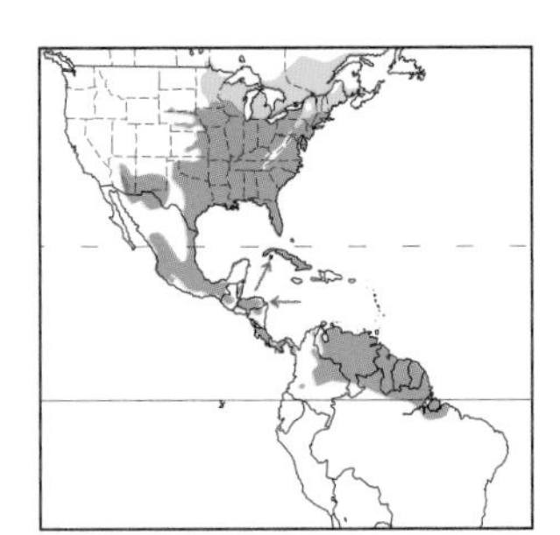

71a Adult male, summer, worn plumage (*magna*, NE US, E Canada): A plump and short-tailed meadowlark. Largely brown and streaked, with a bright yellow throat, breast and upper belly which is divided by a black 'V' on the breast. Note that the yellow on the throat does not extend to the malar region. Compared to Western and 'Lilian's' Meadowlarks, this species is darker, showing buffier flanks, and richer brown colouration on the coverts and tertials. The dark bars on the tertials, greater coverts, and central rectrices blend together at the midline of the feather. The song is a series of sweet descending whistles. See the text for more information.
71b Adult male, fall, fresh plumage (*magna*): Like the summer adult but the underparts are tipped with buff, the black 'V' is often indistinct. The upperparts are more crisply fringed, often appearing scaly.
71c Tail, dorsal side: Note that the two outer rectrices (R5 and R6) are almost entirely white and the third one from the outside (R4) may be entirely white or may show some dark on the inner edge. The fourth rectrix from the outside (R3) is dark, but shows a variable amount of white.
71d Juvenile: Similar to the adult, but the yellow underparts are much duller. The 'V' on the beast is replaced by a set of dark streaks in the shape of a 'V'. The upperparts are crisply edged paler, appearing scaly. Younger individuals may show an obviously short bill with a pinkish base.
71e Adult male (*alticola*, Chiapas, Mexico to Costa Rica): Like *magna* but paler brown on the upperparts. The yellow of the throat sometimes invades the malar area. The ear-coverts are pale, contrasting to a greater degree with the dark eyeline. Shows more white on the tail than *magna*. This taxon may in fact be more closely allied to 'Lilian's Meadowlark'.
71f Adult male (*meridionalis*, Andes of Colombia and NW Venezuela): Like *magna* but larger and longer-billed. The streaking on the breast sides and flanks is broad and blackish. The malar area is yellow, sometimes marked with fine dark smudges. The wings show a chestnut wash.
71g Adult male (*inexspectata*, E lowlands from Honduras to Nicaragua): Like *magna*, but very small in size and bill proportions. Slightly warmer brown on the upperparts than *magna*.

71X 'Lilian's Meadowlark' *Sturnella magna lilianae* Text page 300

71Xa Adult male, summer, worn plumage: A plump and short-tailed meadowlark. Largely brown and streaked, with a bright yellow throat, breast and upper belly which is divided by a black 'V' on the breast. Note that the yellow on the throat does not extend to the malar region. Compared to Western and Eastern Meadowlarks, this species shows a paler face with contrasts to a greater degree with the blackish eyeline. This is a pale form, similar to Western Meadowlark in overall colour, and paler than Eastern Meadowlark. The dark bars on the tertials, greater coverts, and central rectrices do not blend together at the midline of the feather. Song similar to Eastern Meadowlark's.
71Xb Adult male, fall, fresh plumage: Similar to the summer adult but buffier overall, and the underparts are veiled by small buffy feather tips. The upperparts are fresh and scaly-looking.
71Xc Tail, dorsal side: Shows an extensive amount of white. The outer four rectrices (R3–R6) are entirely white. The next rectrix in, R2, sometimes shows substantial white.
71Xd Juvenile: Similar to the adults but the breast 'V' is replaced by a pattern of black streaks. The underpart colours are duller and the upperparts are neatly fringed pale, giving a scaly appearance.

71b
71a
71c
71e
71d
71g
71f
7
71xa
71xb
71xc

PLATE 28 MEADOWLARKS 3 AND BOBOLINK

72 Western Meadowlark *Sturnella neglecta* Text page 302

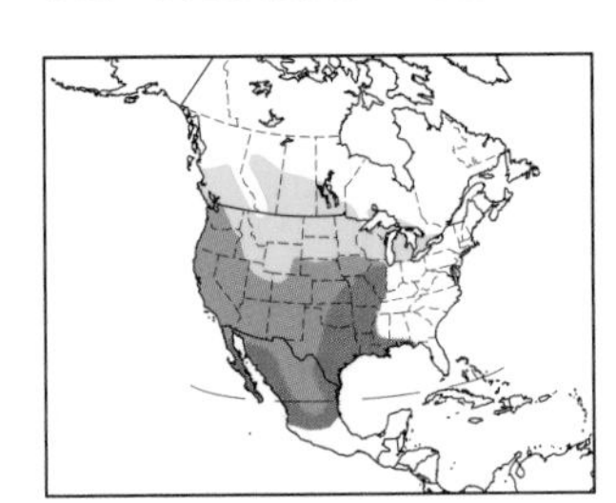

72a Adult male, summer, worn plumage: A plump and short-tailed meadowlark. Largely brown and streaked, with a bright yellow throat, breast and upper belly which is divided by a black 'V' on the breast. Note that the yellow on the throat extends to the malar region, more so on males. Like Eastern and unlike 'Lilian's' Meadowlarks, this species shows a darker face with contrasts to a lesser degree with the blackish eyeline. This is a pale species, similar to 'Lilian's Meadowlark' in overall colour, and paler than Eastern Meadowlark. The dark bars on the tertials, greater coverts and central rectrices do not blend together at the midline of the feather. The song is a series of pleasant, flute-like notes unlike Eastern or 'Lilian's' Meadowlarks. See the text for more details.

72b Adult male in flight: A dumpy, short-tailed meadowlark with white sides to the tail. Note that the extent of white on the tail is less than on either Eastern or 'Lilian's' Meadowlarks.

72c Adult male, fall, fresh plumage: Similar to the summer adult but the underparts are heavily veiled with buff tips, nearly obscuring the black 'V' on the chest and the yellow colour. The upperparts are crisply fringed, giving a scaly appearance.

72d Juvenile: Similar to the adult, but the underparts are duller yellow, nearly buffy. In addition, the black 'V' on the chest is replaced by a series of dark streaks. The bill shows a pinkish base, and is noticeably shorter on young individuals.

103 Bobolink *Dolichonyx oryzivorus* Text page 390

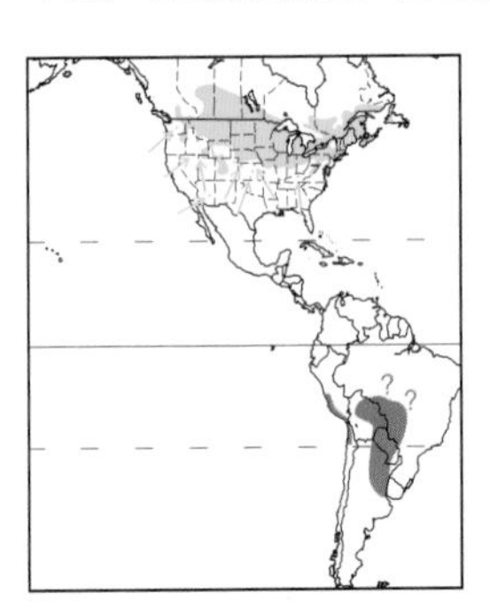

103a Adult male, summer (worn definitive alternate): It is a small, short-tailed, and short-billed icterid with pointed tips to the rectrices. A spectacular-looking species with a black face and underparts, a buffy-white nape, buff streaking on a black back and a greyish-white rump and lower back. The wings are black with pale edging on the remiges and a large white patch created by the largely white scapulars.

103b Adult male, summer (worn definitive alternate) in flight: Flies with a peculiar fluttery wingbeat while on the breeding grounds, often performing song flights. The patchy mix of black and white on the upperparts is distinctive. Note the long and pointed wings.

103c Adult male, early spring (fresh definitive alternate): The white and buff of the upperparts is largely hidden beneath brownish feather tips. The underparts are similarly brown-tipped, hiding most of the black of the plumage. However, a ghost pattern of the summer male's plumage is apparent. The wings are heavily edged warm brown.

103d Adult male, late summer (moulting from definitive alternate to basic): Patchy-looking, the black of the underparts being replaced by buffy-brown. The bill becomes pink.

103e Adult male, winter (definitive basic): Similar to the adult female, but larger. A brown and streaky species with a noticeable dark post-ocular line, and two well-defined buff 'suspenders' on the back. Shows a white throat.

103f Adult female, summer (definitive alternate): Brown and streaky, with a short horn-coloured bill. Note that the buffy-brown underparts are not streaked on the breast, the streaking is confined to the flanks. The adult female shows a whitish throat.

103g Immature, fall (1st-basic): Similar to the adult summer female but more richly coloured. The breast is often obscurely streaked. The throat is yellowish-buff, not whitish.

103h Immature in flight (1st-basic): A small, brown, streaky species with long and pointed wings. Listen for a distinctive 'pink' flight call.

103i Juvenile: Similar to the immature but streaking is confined to the breast sides. The upperparts appear scaly when fresh, due to complete pale fringes to the feathers.

72d
72a
72c
72b
103b
103d
103a
103c
103e
103h
103g
103f
103i

74 Brown-and-yellow Marshbird *Pseudoleistes virescens* Text page 307

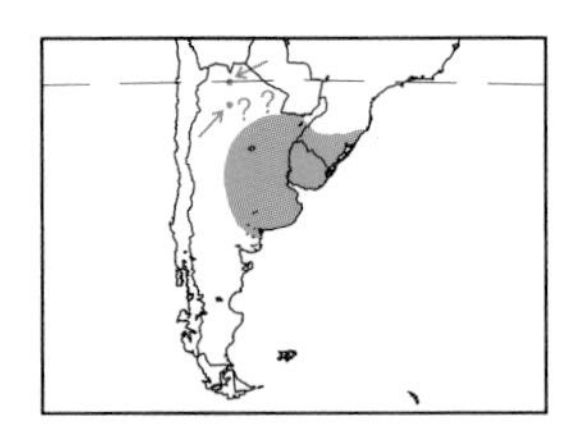

74a Adult. A large and stocky blackbird with a long, thin black bill. The body is largely olive-brown, darkest on the head and extending to the flanks. The lower breast and belly are bright yellow. The wings are brown, except for a smallish yellow epaulet.
74b Adult in flight: Appears largely brown with yellow underparts and yellow epaulets.
74c Juvenile: Similar to the adult, but the throat is pale yellow and the breast is pale yellow, streaked with brown. The rest of the underparts are pale yellow. The bill shows a horn-coloured base to the lower mandible.

73 Yellow-rumped Marshbird *Pseudoleistes guirahuro* Text page 306

73a Adult. A large and stocky blackbird with a long, thin black bill. The body is dark brown, darkest on the head. The rump and underparts, up to the chest and including the flanks, are yellow.
73b Adult in flight: Large and stocky. Appears brown with yellow underparts, rump and epaulets.
73c Juvenile: Similar to the adult, but the throat is pale yellow and the breast is pale yellow, streaked with brown. The rest of the underparts are pale yellow. The bill shows a horn-coloured base to the lower mandible. As in the adult, the rump and the flanks are yellowish, separating it from juvenile Brown-and-yellow Marshbird.

55 Saffron-cowled Blackbird *Xanthopsar flavus* Text page 245

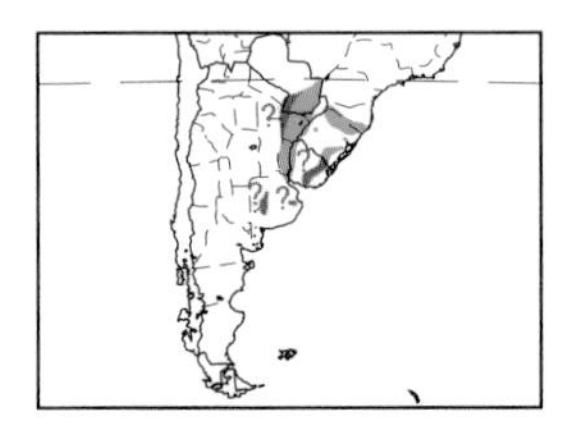

55a Adult male: A beautiful, small and slender species with a fine spike-like bill. The head, underparts and rump are yellow. The black extends from the nape to the rest of the upperparts. The face is highlighted by black lores. The tail and wings are black, the latter sporting yellow epaulets.
55b Adult female: Similar to the male, but the upperparts are greyish-brown and obscurely streaked. The yellow face shows a dark eyeline which creates a bold yellow supercilum which contrasts with the olive crown. The rump is yellowish and the belly white.
55c Adult male in flight: Yellow head and underparts, black on the upperparts contrasting with a yellow rump. The black wings show yellow epaulets.

74b
74a
74c
73a
73b
73c
55a
55b
55c

81 Velvet-fronted Grackle *Lampropsar tanagrinus* Text page 322

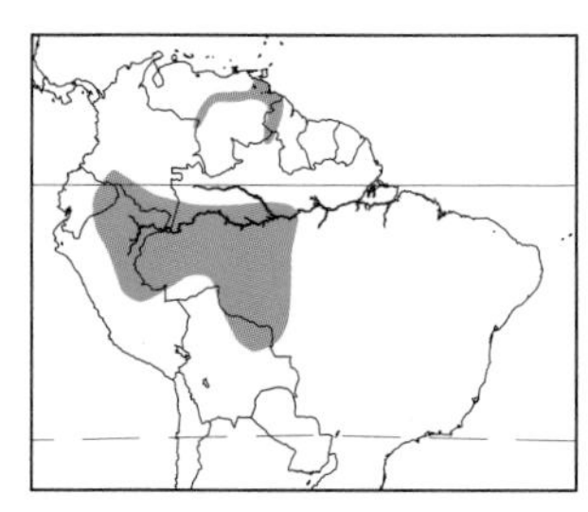

81a Adult male (*tanagrinus* Amazon basin from E Ecuador and NE Peru east to Brazil): A slim, entirely black icterid with a short bill. Note the flat crown, and the long fan-shaped tail. The plumage shows a slight blue iridescence.
81b Adult male (*tanagrinus*): Only the head is illustrated. If observed well, the plush feathers on the forehead may appear black, contrasting with the glossy blue of the head.
81c Adult male in flight (*tanagrinus*): Note the rather rounded wings and long tail, otherwise looks entirely black.
81d Immature (1st-basic) (*tanagrinus*): Only the head is illustrated. Similar to the adult, but largely lacking iridescence. The plumage is brownish-black rather than black.
81e Adult male (*violaceus,* NW Mato Grosso in W Brazil): Similar to *tanagrinus,* but has a vivid violet gloss.

83 Colombian Mountain-Grackle *Macroagelaius subularis* Text page 325

83a Adult male: A slim and very long-tailed icterid. Entirely black with a purple-blue iridescence. The axillaries and epaulets are dark chestnut, often difficult to differentiate from the black of the body.
83b Adult in flight: Slim and long-tailed, appears entirely black. The chestnut epaulets and axillaries may not be readily visible in flight.

82 Tepui Mountain-Grackle *Macroagelaius imthurni* Text page 323

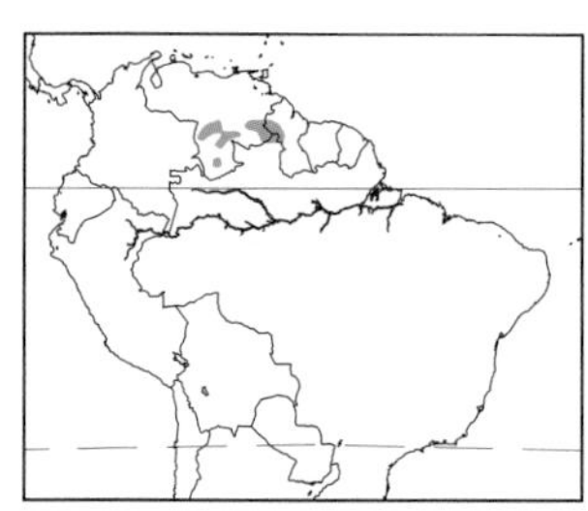

82a Adult male: A slim and very long-tailed icterid. Entirely black with a blue iridescence. The axillaries are yellow, or tawny yellow, some may be pale chestnut at their bases. Note that the epaulets are black, like the rest of the wing. The bill is longer and thicker than that of Colombian Mountain-Grackle.
82b Adult in flight: Slim and long-tailed, appears largely black. The golden axillaries are usually visible in flight.

76 Red-bellied Grackle *Hypopyrrhus pyrohypogaster* Text page 311

76a Adult male: A large and stunning icterid. It is stocky, thick-billed and long-tailed. The body plumage is largely black, showing shiny shaft streaks on the head feathers. The lower breast and belly are scarlet as are the vent and undertail-coverts. The eye is pale yellow.
76b Adult male in flight: A large stocky blackbird with a long and square-tipped tail. Appears largely black, with red on the belly.
76c Immature (1st-basic): Only the head is illustrated. Similar to the adult but blackish-brown rather than black. The eye is greyish-yellow. The head lacks the pointed, feathers with shiny shafts of the adult.

81a
81b
81e
81d
81c
83a
83b
82b
82a
76a
76b
76c

PLATE 31 'BLACKBIRDS' 8 AND SCARLET-THROATED TANAGER

77 Austral Blackbird *Curaeus curaeus* — Text page 312

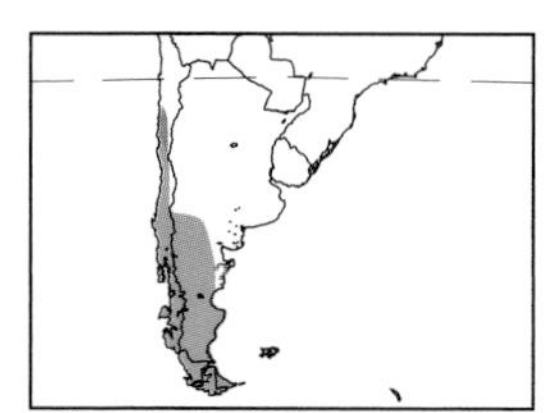

77a Adult male: A large, long-tailed icterid with a long, straight and sharply pointed bill. The body plumage is entirely black lightly glossed blue. The feathers of the crown and nape are pointed and glossy.
77b Adult in flight: Appears large, black and square-tailed. Note that the tail is long.
77c Immature (1st-basic): Similar to the adult male but brownish-black and lacking the pointed head feathers on the head.

79 Chopi Blackbird *Gnorimopsar chopi* — Text page 317

79a Adult male: A medium-sized black icterid showing a light blue iridescence. The bill is short and has a slightly downcurved culmen. When observed at close quarters, a diagnostic series of ridges and grooves may be visible at the base of the lower mandible. The crown and nape feathers are pointed and glossy.
79b Adult in flight: Medium-sized and all black, the wings are somewhat pointed and the tail is of moderate length and square-tipped.
79c Immature (1st-basic): Similar to the adult but blackish-brown instead of black and almost lacking iridescence. The bill grooves can be missing early in life. Also lacks the pointed head feathers of the adult.

78 Forbes's Blackbird *Curaeus forbesi* — Text page 315

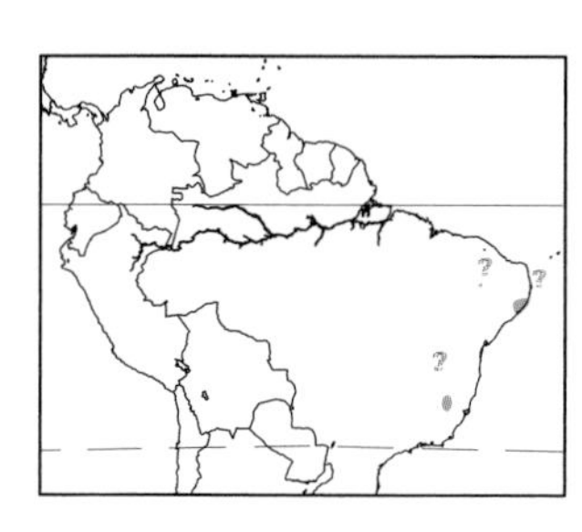

78a Adult male: A small to medium-sized, black icterid. It is slim, long-tailed and sports a long, straight and pointed bill. The body is entirely black with a light blue gloss. The nape and crown feathers are pointed and glossy. Note the different bill shape from Chopi Blackbird, and the smaller size. The vocalisations differ greatly.
78b Adult in flight: A small to medium-sized, black icterid with a long tail and moderately pointed wings.

Scarlet-throated Tanager *Compsothraupis loricata* — No species account

See pages 12 and 318

Xa Adult female: An all-black, icterid-like tanager. It has been suggested that this species is in fact an icterid based on plumage colour and vocalisations. It is illustrated here for completeness, but this species is mentioned only in the Introduction. Note that the bill is tanager-like, not icterid-like.
Xb Adult male: Similar to the female but with a scarlet throat patch.

77b
77a
77c
79c
79b
79a
Xb
Xa
78b
78a

85 Melodious Blackbird *Dives dives* Text page 329

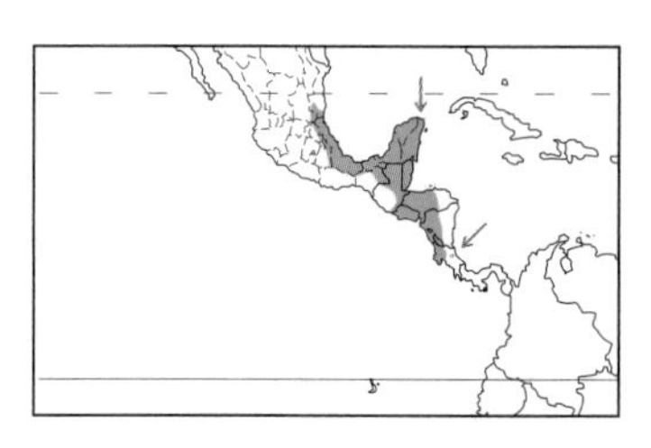

85a Adult male: A medium-sized black icterid that shows a slight blue iridescence. The bill is slightly shorter than the length of the head and the culmen shows a slight downcurve. The tail is relatively long.
85b Adult in flight: A medium-sized black icterid with a long tail and somewhat rounded wings.
85c Immature (1st-basic): Similar to the adult but the plumage is brownish-black, lacking a noticeable gloss.

86 Scrub Blackbird *Dives warszewiczi* Text page 331

86a Adult male (*warszewiczi* W Ecuador to La Libertad in W Peru): A medium-sized black icterid that shows a slight blue iridescence. The bill is slightly shorter than the length of the head and the culmen shows a slight downcurve. The tail is relatively long. Visually, this race is nearly indistinguishable from Melodious Blackbird; however, their vocalisations differ, and they are allopatric.
86b Adult male in flight (*warszewiczi*): Medium-sized black icterid with rounded wings and a moderate length tail. The tail is relatively shorter than that of the allopatric Melodious Blackbird.
86c Immature (1st-basic) (*warszewiczi*): Similar to the adult, but the body is brownish-black, and lacks a noticeable gloss, particularly on the underparts.
86d Adult male (*kalinowskii,* La Libertad to Ica in W Peru): Similar to *warszewiczi* but much larger, with a proportionately longer and deeper bill. The body iridescence is slightly more violet.
86e Pair duetting: They sit close to each other and bounce up and down as both sing in unison.

84 Cuban Blackbird *Dives atroviolacea* Text page 327

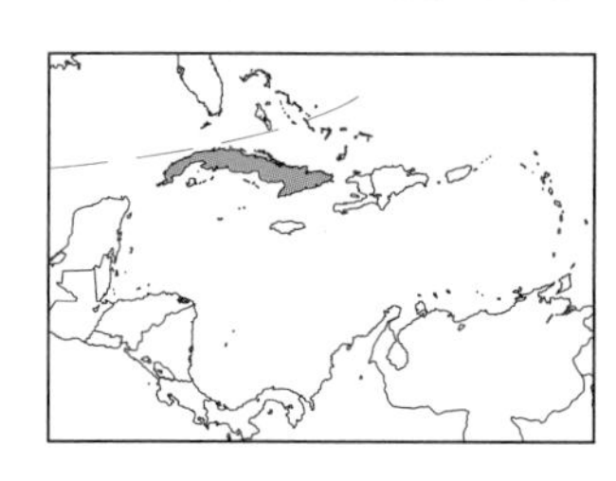

84a Adult male: A medium-sized black icterid with a vivid violet iridescence. Note that the bill is short, and the culmen is slightly curved.
84b Adult in flight: A medium-sized black icterid with moderately pointed wings and a tail of moderate length.
84c Immature (1st-basic): Similar to the adult male but the body plumage is brownish-black, only barely glossy on the upperparts.

85a
85c
85b
6e
86a
86c
86b
86d
84c
84a
84b

52 Jamaican Blackbird *Nesopsar nigerrimus* Text page 238

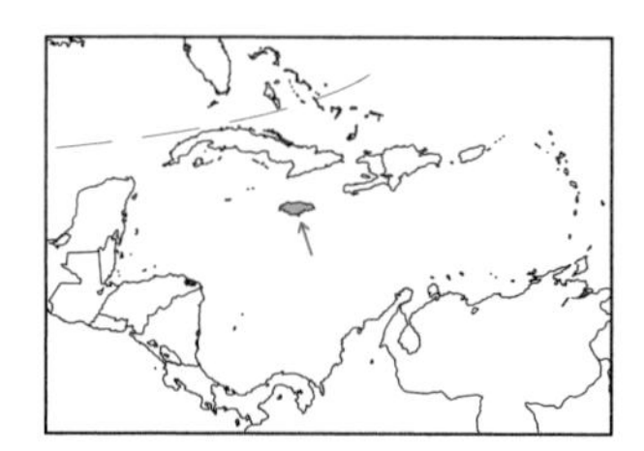

52a Adult male: A petite black icterid with a finely pointed, slim, straight bill. The tail is relatively short. The plumage is glossed greenish-turquoise on the head, wings and tail, and blue on the body.
52b Adult male in flight: Appears small, black and relatively short-tailed.

80 Bolivian Blackbird *Oreopsar bolivianus* Text page 320

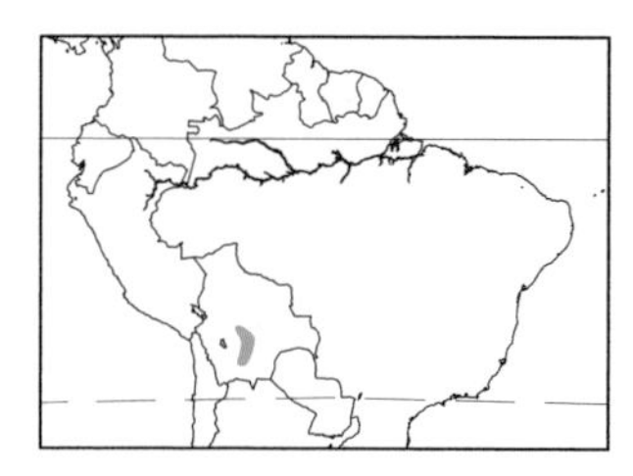

80a Adult male: Entirely black with a dull gloss, shows contrasting brownish primaries which are not obvious when perched. The bill is approximately the length of the head and shows a strong downcurve to the culmen.
80b Adult in flight: Black with contrastingly browner primaries.
80c Immature (1st-basic): Similar to the adult but has a brown cast to the black body plumage and is even less glossy.

93 Carib Grackle *Quiscalus lugubris* Text page 353

93a Adult male (*luminosus*, Grenada, Grenadines and Los Testigos Island, Venezuela): Entirely black with a vivid purple gloss, greenish-blue on the wings. The eyes are pale yellow. Note the long legs, and bantam cock-like posture. All Carib Grackles show a short tail for a grackle, but it is typically graduated and may be folded into a wedge-shape. **93b Adult male in flight** (*luminosus*): Appears black, mid-sized and somewhat pointed-winged. The tail is obviously graduated, and somewhat short for a grackle. **93c Adult female** (*construsus*, St Vincent Island): The darkest subspecies. These females are dark brown, somewhat paler on the underparts particularly the throat. Lacks an obvious supercilium. The eyes are yellow. **93d Adult female** (*guadeloupensis*, Montserrat, Guadeloupe, Marie-Galante, Dominica and Martinique): A lighter coloured subspecies. The head and underparts are grey-brown and the chest is obscurely streaked. The upperparts and vent are darker. Note the dark eyeline which contrasts with a pale supercilium. The eyes are yellow. **93e Adult female in flight** (*guadeloupensis*): Appears brown, with a paler head and underparts and a graduated tail shape. **93f Juvenile male in moult to 1st-basic** (*guadeloupensis*): Only the head is illustrated. Similar to the adult female but patchily marked with black feathers. The eye is becoming paler, as full juveniles have dark eyes.

92 Greater Antillean Grackle *Quiscalus niger* Text page 350

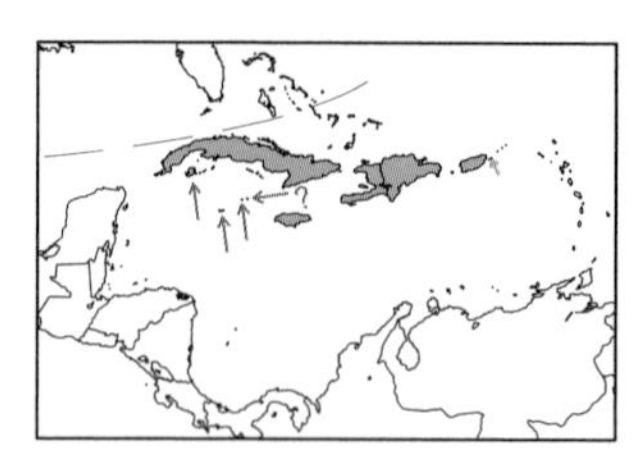

92a Adult male (*niger*, Hispaniola): Plumage entirely black, glossed purple on the head, back and rump while the tail is glossed blue-green. The eyes are yellow. Like Carib Grackle but larger, deeper-billed, and longer-tailed.
92b Adult male in flight (*niger*): Appears black, mid-sized and somewhat pointed-winged. The tail is obviously graduated.
92c Adult male (*caymanensis*, Grand Cayman Island): Similar to *niger* but more blue-glossed rather than purple. The eyes are yellow.
92d Adult female (*caymanensis*): Similar to the male but smaller, not as highly iridescent and with a less keel-shaped tail. The eyes are yellow.
92e Immature (1st-basic) male (*caymanensis*): Only the head is illustrated. Similar to the adult male but patches of brownish-black are retained from the juvenal plumage.

52b
52a
80a
80c
80b
93f
93c
93d
93e
93a
93b
92b
92a
92e
92d
92c

87 Great-tailed Grackle *Quiscalus mexicanus* Text page 333

87a Adult male (*prosopidicola*, SC US and NE Mexico): A large grackle with a enormously long and keel-shaped tail. The body is glossed purple on the head, breast and back becoming blue on the lower back and belly. The bill is thick, and the forehead is flat, giving an aggressive expression exacerbated by the yellow eyes.
87b Adult male in flight (*prosopidicola*): Large and black with a large, keel-shaped, graduated tail.
87c Adult female (*prosopidicola*): Dark brown, paler on the head and underparts, the throat being the palest part of the body. Shows a paler supercilium contrasting with a dark crown and cheek patch. Often a dark malar stripe is present, the eyes are always yellow.
87d Adult female in flight (*prosopidicola*): Large and brown, with a noticeably graduated tail.
87e Immature (1st-basic) male (*prosopidicola*): Similar to the adult male but smaller and shorter-tailed. the body is dull black, only weakly iridescent. On younger birds the eye may still be dark or dark yellow rather than the bright yellow of adults. Perhaps it is not possible to separate Boat-tailed and Great-tailed Grackles in this plumage unless heard calling.
87f Immature (1st-basic) female (*prosopidicola*): Similar to the adult female but the underparts are richer in colour, and the upperparts are less iridescent. The supercilium is broader and more extensive.
87g Juvenile (*prosopidicola*): Similar to the adult female but noticeably streaked below. The back is greyish-brown and the underparts are buffy-grey. Often shows bare grey facial skin around the dark eyes.
87h Adult female (*obscurus*, coastal Nayarit to Guerrero, Mexico): The darkest subspecies. In this form the underparts are only slightly paler than the upperparts. The underparts are a richer chestnut-brown than *prosopidicola*, no supercilium is present. The eyes are always yellow.
87i Adult female (*nelsoni*, S California and SW Arizona to NW Mexico): This is the smallest and palest form. The upperparts are grey-brown, paler than *prosopidicola* and the underparts are a pale grey-buff, sometimes subtly barred. The supercilium is distinct and a darker malar stripe is usually visible. The bird illustrated is a particularly grey individual.
87j Adult male display: The wings do not come up above the horizontal during the display.

88 Boat-tailed Grackle *Quiscalus major* Text page 338

88a Adult male (*major*, coastal Texas and Louisiana): A large grackle with a enormously long and keel-shaped tail. The body is glossed purple on the head and neck, becoming blue on the mid-back and lower breast. Does not appear as aggressive as the Great-tailed Grackle due to the steeper forehead, shallower bill and dark eyes.
88b Adult male (*torreyi*, Atlantic coast from New York to N Florida): Only the head is illustrated. Similar to major but shows pale yellow eyes.
88c Adult female (*major*): Upperparts, wings and tail dark brown, slightly glossed turquoise. The head and underparts are buff-brown, lightest on the throat and supercilium. Tends to show little or no malar stripe. The eyes are brown. Compared to female Great-tailed Grackle, the plumage is warmer and paler, the eyes are dark and the forehead is steeper and the bill thinner.
88d Immature (1st-basic) female (*major*): Like the adult female but the plumage is more colourful, noticeably washed with cinnamon. The ear-coverts are darker, contrasting with the paler supercilium. Dark eyes.
88e Adult male display: The wings are 'flipped' well above the horizontal while this species is displaying.

87g
87i
87c
87a
87j
87h
87f
87d
87e
87b
88d
88c
88a
88b
88e

91 Common Grackle *Quiscalus quiscula* Text page 346

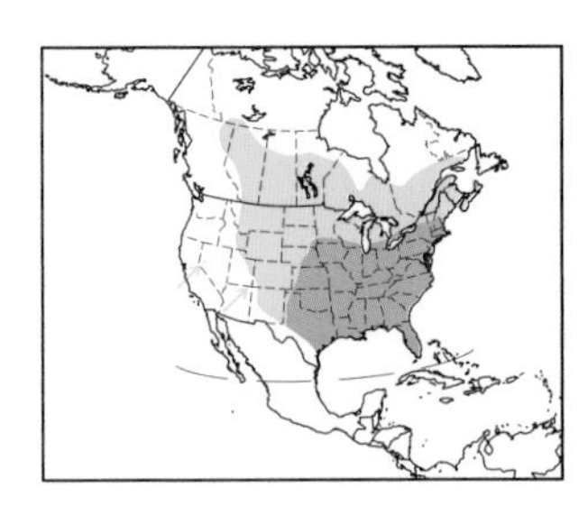

91a Adult male 'Bronzed Grackle' (*versicolor*, the northern and western portion of the range): A mid-sized grackle with a noticeably keeled tail. The body is entirely black, glossed with blue on the head, while the back and belly are glossed bronzy-olive. The blue hood comes to a distinct border with the body gloss colouration. The greater coverts are always glossed purple. The eyes are yellow.
91b Adult male in flight (*versicolor*): Medium-sized, black with a long, keeled tail. Does not undulate like Red-winged Blackbird.
91c Adult female (*versicolor*): Similar to the male but the iridescence is duller, particularly on the underparts. In size, it is noticeably smaller and the tail is not held in a keel shape.
91d Adult female in flight (*versicolor*): Similar to the adult male but the tail is not held in a keel shape.
91e Juvenile (*versicolor*): Entirely dark brown, with blackish lores. The tail is long and graduated, but not keeled. The eyes are dark.
91f Adult male 'Purple Grackle' (*stonei*, largely in the region of the Appalachians): Basically an intergrade between *versicolor* and *quiscula*, and quite variable in appearance. The nape shows a purplish-gold gloss, mixed with blue. The back mixes a bronze and purple gloss, while greater coverts mix a blue and purple gloss.
91g Adult male 'Florida Grackle' (*quiscula*, Florida north to coastal Louisiana and South Carolina): Black with a purple-blue iridescence on the face, purple-gold on the nape and breast sides. The back and underparts are glossed olive-green, while the wings are glossed blue; the greater coverts are always blue never purple. The bill is slightly longer and sleeker than on *versicolor*.

89 Slender-billed Grackle *Quiscalus palustris* Text page 342

89a Adult male: Extinct. A very slim, medium-sized grackle with a long, keeled tail. The bill was particularly thin, and the legs particularly long and thin. The body was black, glossed with purple on the head and breast, becoming blue posteriorly. It is presumed that the eyes were pale.
89b Adult female: The wings, tail, vent, undertail-coverts and thighs were dark brown. The rest of the body plumage was cinnamon-buff, palest on the belly, throat and supercilium. The lores were dusky. The thin black bill showed an orange-red base to the lower mandible. Dark eyes?

90 Nicaraguan Grackle *Quiscalus nicaraguensis* Text page 344

90a Adult male: A smallish grackle with a thick bill for its size, and long legs. The body is entirely black with a turquoise gloss on the head, upper back and upper breast becoming purple on the lower back, belly and rump. The wings are glossed greenish. The eyes are yellow.
90b Adult male in flight: Small, black and showing a keel-shaped, graduated tail.
90c Adult female: Noticeably darker on the upperparts than the underparts. The upperparts are brown, darkest on the wings and tail. The face shows a distinct pale supercilium. Pale below, whitish on the throat, becoming buff on the breast and paler again on the belly, contrasting with the dark brown flanks, thighs and undertail-coverts. The eyes are yellow.
90d Adult female in flight: A brownish icterid with a long, graduated tail and obviously paler underparts.

91b
91a
91e
91d
91f
91c
91g
89a
89b
90b
90d
90a
90c

95 Brewer's Blackbird *Euphagus cyanocephalus* Text page 360

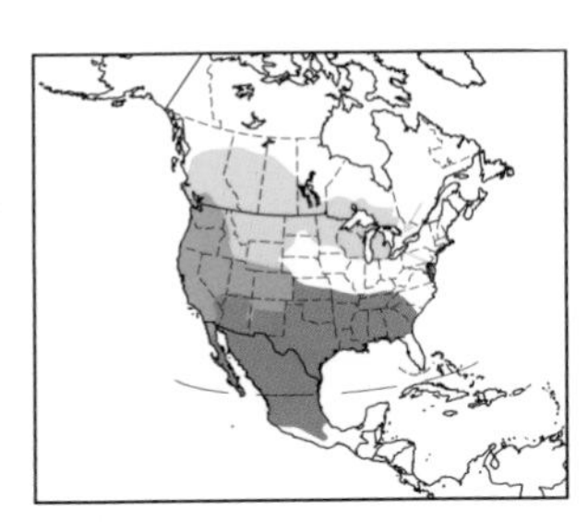

95a Adult male summer, worn plumage: A medium-sized black icterid with a purple gloss on the head, becoming green on the body, wings and tail. The long tail is square-tipped and not keeled as in the grackles. The eyes are bright yellow.
95b Immature (1st-basic) male in fresh plumage: Similar to the adult male, but some individuals show noticeable cold-buff feather tips on the head, back and underparts. Note that the tertials and greater coverts are entirely black. Most 1st-basic males are indistinguishable from the adults, only a minority show pale feather tipping.
95c Adult female: Entirely grey-brown, only slightly paler on the underparts. A slightly paler supercilium is usually noticeable. The wings and upperparts show a dull greenish gloss. The eye is almost always dark (see text):
95d Juvenile male: This individual is undergoing the first pre-basic moult. The dull brown juvenile feathers are replaced by the glossy black of the adult male. As this moult takes place the eye changes from dark to pale.

94 Rusty Blackbird *Euphagus carolinus* Text page 357

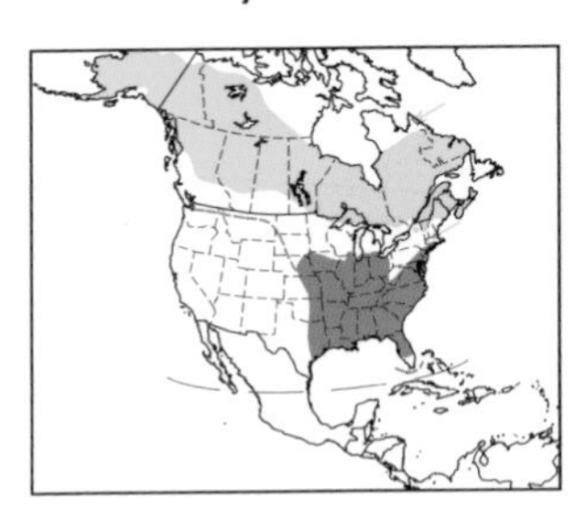

94a Adult male summer, worn plumage: A medium-sized blackbird with a square-tipped tail and thin bill. The body is entirely black, faintly glossed greenish (sometimes purplish) on the body. The eyes are yellow.
94b Immature (1st-basic) male, fall: Black plumage is hidden by an extensive amount of rusty, buff and brown feather tips. The crown, auriculars, nape and back are rufous. The supercilium, throat and belly are buff-brown, barred with black on the underparts. The rump, tail, wings and lores are black. The fresh adult (definitive basic) male is similar but more narrowly edged rufous and buff, thus appearing blacker.
94c Adult female, summer, worn plumage: Dark grey, almost black and unglossed. May show a faint greenish gloss on the wings.
94d Immature (1st-basic) female, fall: Similar to the fall immature male but the underparts are softly barred with grey. The wings, rump, and tail are medium grey, not black. Note that the black on the lores is less extensive on the female:
94e Juvenile moulting into 1st-basic male: The brownish juvenal plumage is briefly held. Patches of black feathers tipped buff or rusty increase as the moult continues. The eyes are dark at first, and become paler as the moult nears completion.

95b
95a
95d
95c
94e
94a
94c
94d
94b

97 Screaming Cowbird *Molothrus rufoaxillaris* Text page 368

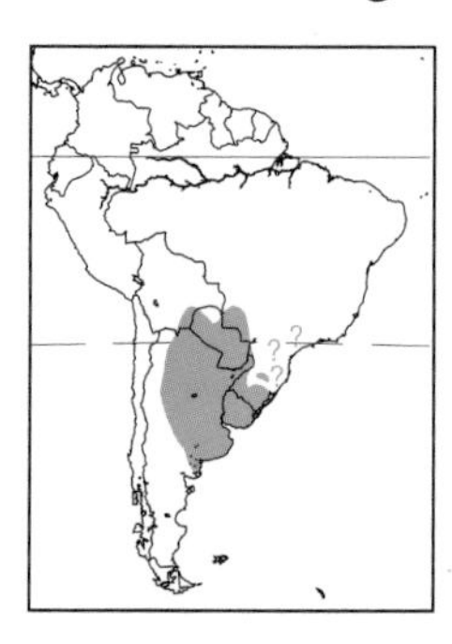

97a Adult male: A stocky black cowbird with a short and thick bill. Entirely black with a dull blue iridescence, much less glossy than Shiny Cowbird. The eye is deep red, this is difficult to see in the field. Note the long primary extension.
97b Adult male: Only the head is illustrated. Note the dark red eye, and the bulky, square-headed appearance of Screaming Cowbird.
97c Adult male: In the hand one may see the chestnut axillary patch, this is only rarely visible in the field.
97d Juvenile: Mimics the pattern of Baywing. Grey with rufous wings. The tail is contrastingly dark. Only separable from juvenile Baywing by the thick bill, and long primary extension. Juvenile male Screaming Cowbirds are noticeably larger than Baywings.
97e Juvenile in first pre-basic moult in flight: Note the blackish wing-linings lacking in Baywing.

96 Baywing *Agelaioides (Molothrus) badius* Text page 364

96a Adult male: A stocky icterid with short wings and a short, finch-like bill. Entirely olive-grey with black lores and mask. The blackish wings are widely edged rufous, such that they appear solidly rufous. The tail is blackish and contrasts with the paler grey body.
96b Adult female: Basically indistinguishable from the adult male.
96c Juvenile: Much like the adult in pattern, but obscurely streaked. Also shows a yellowish gape mark.
96d Juvenile in flight: Has grey wing-linings, unlike the black wing-linings of juvenile Screaming Cowbird.
96e Male guarding a nest from a pair of Screaming Cowbirds: This nest being an old nest of a Firewood Gatherer.

96X 'Pale Baywing' *Agelaioides (Molothrus) badius fringillarius* Text page 367

96X Adult male: Similar to *badius* Baywing but paler buff in colouration. In addition the dark mask is more extensive on the face. The tail is sandy-brown, not blackish.

97a
97c
97b
97d
97e
96a
96b
96x
96e
96c
96d

99 Bronzed Cowbird *Molothrus aeneus* Text page 374

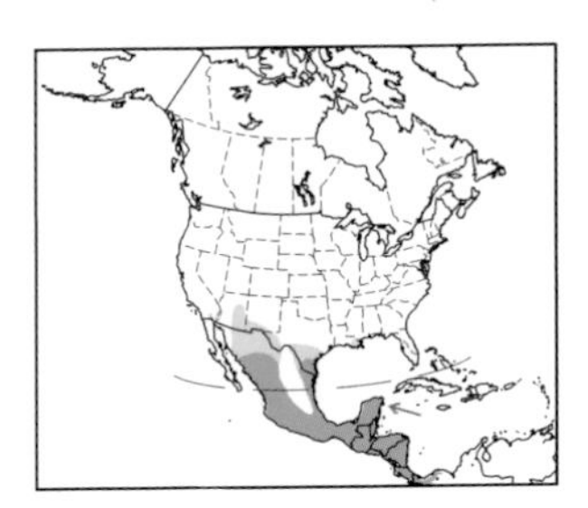

99a Adult male (*aeneus*, SC Texas to Panama): A medium-sized cowbird with bright red eyes and an erectable nape ruff. Entirely black with a greenish-bronze gloss to the head, breast and back. The median coverts and uppertail-coverts are glossed purple, while the greater coverts and tertials are glossed turquoise.
99b Adult male in flight (*aeneus*): A stocky black cowbird, the wings are moderately pointed.
99c Immature (1st-basic) male (*aeneus*): Similar to the adult but not as iridescent. Larger than the adult female.
99d Adult female (*aeneus*): Similar to the male, but blackish instead of black, and lacking noticeable iridescence.
99e Adult female in flight (*aeneus*): Not as black as the male, perhaps slightly more rounded-winged.
99f Juvenile (*aeneus*): Dark brown, almost blackish, with obscure streaking on the underparts. The lower mandible is horn-coloured and the eyes are dark.
99g Adult female (*loyei*, SE Arizona and SW New Mexico to the western Mexican lowlands to Nayarit): Unlike female *aeneus*, this female is grey-brown, noticeably paler than *aeneus*. The pale greyish underparts are subtly streaked. The eyes are red, not dark as in Brown-headed Cowbird.
99h Juvenile (*loyei*): Similar to the adult female but paler below, light grey-buff and more noticeably streaked. The bill shows a horn-coloured base to the lower mandible and the eyes are dark.

100 Bronze-brown Cowbird *Molothrus (aeneus) armenti* Text page 378

100a Adult male: The smallest cowbird. Entirely black with a brownish-bronze gloss on the head, back and underparts. The lesser, median and greater coverts are glossed purple, while the rest of the wings are glossed turquoise. The eyes are red. A small nape ruff is present, but not nearly as well developed as on Bronzed Cowbird.
100b Adult male in flight: Looks small, black and chunky.
100c Adult female: Similar to the male but paler black and only weakly glossed turquoise on the upperparts.
100d Juvenile: Entirely dark brown, subtly streaked. The eyes are dark.

99d
99b
99a
99f
99e
99c
99g
99h
100a
100d
100b
100c

101 Brown-headed Cowbird *Molothrus ater* Text page 380

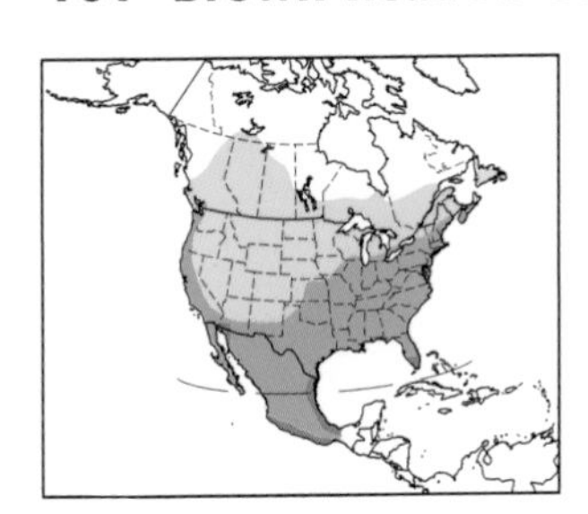

101a Adult male (*ater*, E North America): A black cowbird with a brown head. The body is mainly greenish-glossed, except for a purple collar.
101b Adult male in flight (*ater*): Appears entirely black, not the long and rather pointed wings.
101c Adult female (*ater*): Dull brown, with a paler throat, darker malar stripe and pale supralores. Lacks a noticeable face pattern. The black bill often shows a horn base to the lower mandible.
101d Adult female in flight (*ater*): Looks brown and small, but with rather pointed wings.
101e Adult female, summer, worn plumage (*obscurus*, Pacific coast to northern Mexico): Like *ater* but tends to show a paler head when worn. Slightly smaller in size.
101f Juvenile (*ater*): Similar to the adult female but noticeably streaked on the underparts and pale edged on the upperparts.
101g Juvenile (*artemisiae*, Great Basin and E Great Plains): In the two western subspecies, *artemisiae* and *obscurus*, some juveniles are very light and strikingly scaly-looking on the upperparts and boldly streaked on the underparts. It is not clear if this variant plumage is sex-related.

102 Shiny Cowbird *Molothrus bonariensis* Text page 385

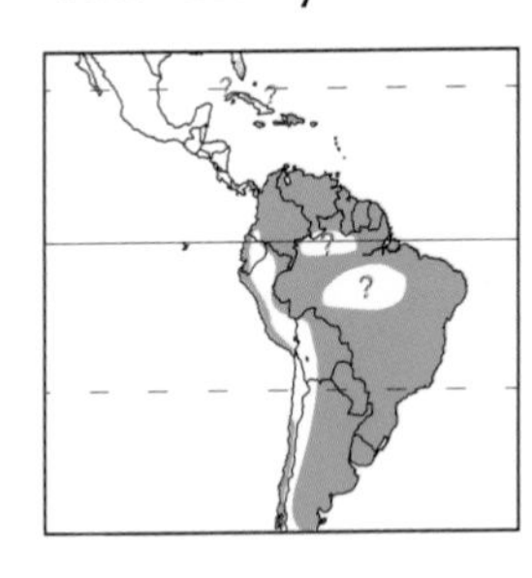

102a Adult male (*bonariensis*, S of Amazonia to S Argentina and Chile): An entirely black cowbird showing a strong purple-blue gloss on the head, nape, back and rump, while the breast is more pure purple glossed. The wings and tail show a greenish iridescence. Compared to Screaming Cowbird this species is glossier, longer-billed and not as square-headed.
102b Adult male in flight (*bonariensis*): Looks black with relatively pointed wings.
102c Adult female (*bonariensis*): Entirely dark grey-brown and not glossed. The underparts are slightly paler, palest on the throat and supercilium. Although pale, the throat does not appear whitish. The bill is more long and slender than on Brown-headed Cowbird, and is entirely black. In addition, the face pattern of Shiny Cowbird is more striking, with a noticeable supercilium and darker cheek.
102d Adult female in flight (*bonariensis*): Appears dark brown, small and relatively pointed-winged.
102e Adult female *'melanogyna'*, a dark morph of *bonariensis*. Black like the male but not heavily iridescent. In direct comparison, it is noticeably smaller than a male and gives the typical female vocalizations.
102f Adult female (*cabanisii*, E Panama to N Colombia): Similar to *bonariensis* but more ochre-brown, in addition the throat is paler. Also averages larger in size.
102g Juvenile (*bonariensis*): Like the adult female but noticeably streaked on the underparts.

101a
101f
101c
101g
101d
101e
101b
102b
102a
102c
102g
102d
102e
102f

1 CASQUED OROPENDOLA *Psarocolius oseryi* (Deville) 1849 — Plate 2

A small oropendola of the forest interior, it is restricted to western Amazonia and is not well known.

IDENTIFICATION Casqued Oropendola is chestnut on the body and wings, showing a grey throat as well as a lime-green breast and neck sides. It is sympatric with several species of oropendola but it is only likely to be confused with Green (3) and Olive (9) Oropendolas. The three species share a greenish-yellow and chestnut plumage. Casqued Oropendola shows much more extensive chestnut on the body, particularly on the upperparts. Green and Olive Oropendolas have entirely green heads, while Casqued Oropendola has a chestnut crown, nape and face. Note also that the throat of Casqued is grey, turning lime-green on the sides of the neck and breast. On Green and Olive Oropendolas these areas are more evenly coloured green. In addition, Olive Oropendola has a noticeable pink bare patch on the face. The bill shape of Casqued Oropendola is characteristic; it is short and stout, and bulges noticeably at the base of the culmen, forming a distinct casque. The expansion at the base of the culmen on the bills of Olive and Green Oropendolas is not nearly as extreme. Casqued Oropendola's bill is pale yellow horn with a darker tip, unlike the black and red bill of Olive Oropendola and the red-tipped greenish bill of Green Oropendola. Also note that the bright yellow edges to the chestnut primaries and secondaries are diagnostic of Casqued Oropendola. In flight, this oropendola undulates noticeably, unlike the much more level flight of other oropendolas.

VOICE Colonies have been described as very noisy.

Song: A ringing *squaaaaaa-oook* is given during display (Ridgely and Tudor 1989). The song may also be represented as an *OOP-Koooheee*, lasting for almost 2 seconds. The first and second notes are low in pitch, roughly 1 kHz in frequency, with the first note lower than the second. What is striking is that a high-pitched note (the *heee* part of the call) is given simultaneously with the second low note; this high note meters 6.3 kHz in frequency. Since two very different notes are given at the same time, it makes it difficult to adequately describe the song of this species.

Calls: The typical contact note is a loud *chak* similar to that of many other oropendolas. A loud and piercing whistle that descends evenly, lasting roughly one second may be an alarm call. A low, hollow *whook* is also given. Other notes include a buzzy *whEEoo*, a muffled *fffttt-fffttt*, and a *chhhhhppp*.

DESCRIPTION A medium-size oropendola with a bulbous base to the culmen. The bill is short and stocky. The culmen is slightly curved. The primaries of the male are noticeably long and pointed, less so in the female. Wing formula: P9 < P8 < P7 ≈ P6 > P5.

Adult male (Definitive Basic): The bill is yellow, the eyes whitish or palest blue. The lores are black, while the rest of the face, crown and nape are chestnut with a thin band of olive at the base of the casque (forehead). The chestnut of the crown and nape is tinged with olive. The chestnut crown continues on to the neck, back, scapulars, rump and uppertail-coverts. There are some visible olive bases to the feathers on the sides of the neck; these may be more obvious in worn plumage or when ruffled by the wind. The throat is grey with an olive wash that becomes progressively brighter towards the breast. The upper breast is bright lime-yellow, turning chestnut on the lower breast, flanks, belly, vent and undertail-coverts. The area immediately around the vent is surrounded by olive feathers. The wings are black and chestnut. The coverts and scapulars are chestnut, while the tertials are black on their inner half and chestnut on the more visible outer half. The blackish outer secondaries and the inner primaries have bold yellow outer edges. The underwings are black. The tail's central four rectrices (R1–2) are dark olive, the next three (R3–5) are yellow while the outer rectrix (R6) is yellow on its inner half and olive on the outer half. The central (olive) rectrices are slightly shorter than the longest yellow feathers, but the tail is generally graduated, with the outermost rectrices being the shortest. The legs and feet are black.

Adult female (Definitive Basic): Plumage exactly as in adult male, but the female is noticeably smaller. Females also have a proportionately smaller casque and shorter primaries.

Juvenile (Juvenal): Similar to adults, but with a more loosely-textured plumage and slightly less vivid colour. The eye is dark.

GEOGRAPHIC VARIATION Monotypic.

HABITAT A forest specialist, Casqued Oropendola avoids edge and open habitats. In E Peru, it nests in várzea forests ranging from the younger transitional (mid-successional) to old (mature) forests. No nests were found in terra firme forest, which is above the level of the floodplain (Leak and Robinson 1989). In contrast, Ridgely and Tudor (1989) state that this species is typical of terra firme, not várzea forests. Leak and Robinson (1989) did find that Casqued Oropendolas used terra firme forest to forage in, but not for nesting.

BEHAVIOUR Early in the nesting period, females are aggressive toward each other, perhaps due to squabbles over nest placement. During this time, females may steal nesting material from other females. Within a colony, the nesting schedule of the females is highly synchronised. This high level of female synchronisation may preclude the ability of any one male to monopolise access to multiple females. Males are observed to be aggressive, supplanting males displaying to females. Severe fights where males grapple with each other and fall to the ground, as in Montezuma (7) and Chestnut-headed (6) Oropendolas, are rare. Observations conclude that most males display and consort with only one female. However, it is not known how long any one particular consortship lasts. Copulations are not observed at the colony, and probably occur in nearby forest. (Leak and Robinson 1989). Koepcke (1972) states that this species is very aggressive in the defense of the nesting colony, more so than Band-tailed Oropendolas (12) with which it sometimes nests.

NESTING In E Peru, breeds September– November. Casqued Oropendolas are colonial; five colonies in Peru ranged from 15–25 nests (Leak and Robinson 1989). The nests are hanging baskets as is the rule in the oropendolas, although Koepcke (1972) illustrates the nest as being rather wide and short, reminiscent of that of Red-rumped Cacique (14), rather than of a *Psarocolius* oropendola. Cecropia (*Cecropia* sp.) trees are invariably chosen as the nesting tree, particularly those that are isolated from the canopy (Koepcke 1972, Leak and Robinson 1989). The nests are secured to the stalks of cecropia leaves or to the leaves themselves. These isolated trees make it more difficult for raiding monkeys to reach the nests; Brown Capuchins (*Cebus apella*) are known nest predators. Cuvier's Toucans (*Ramphastos cuvieri*) also raid nests. Both of these species are harassed and mobbed if they come close to the nest tree. In C Peru, the species of Cecropia chosen for nesting is attended by a fierce stinging ant, which presumably affords this oropendola some nest defense; colonies are not clustered around wasp nests (Koepcke 1972). Piratic Flycatcher (*Legatus leucophaius*), which is known to pirate nests of other oropendola species, harasses females and nests in Casqued Oropendola nests. It is not known if they only use old nests of this species, or actually take over active nests (Leak and Robinson 1989). Colonies of Casqued Oropendolas sometimes have small numbers of Band-tailed Oropendola nests mixed within (Koepcke 1972, A. Whittaker pers. comm.). Colonies are not clumped together and the nests of the two species of oropendolas are mixed together rather than being separate on the nest tree. There are no known cases of Giant Cowbird (98) parasitism in this species. However, given the small number of colonies which have been studied this is not unexpected and should not be construed as evidence that the Giant Cowbird never parasitises this species.

DISTRIBUTION AND STATUS Uncommon and local, mainly in the lowlands (below 400 m). Has been recorded to 750 m in Peru (Ridgely and Tudor 1989). In E Ecuador it is found west to the confluence of the Rio Coca and Rio Napo. In E Peru (below 400 m on the east slope of the Andes) it is found from Loreto in the north to Puno in the south. Recently, Casqued Oropendola has been found in Brazil: in Eirunepé, Rio Juruá in 1992, and a colony has been described from the upper Rio Juruá, in Acre during 1994.

MOVEMENTS Not well known, but considered a permanent resident and entirely non-migratory.

MOULT It appears that adults undergo one moult per year, the complete definitive pre-basic moult. Presumably this moult occurs after the nesting season. The extent and timing of the first pre-basic moult is not known.

MEASUREMENTS Highly sexually size dimorphic, but not for an oropendola. Males: (n=4) wing 216 (202–225); tail 145 (137–150); culmen 50 (49–52); tarsus 40 (36–43). Females: (n=2) wing 163, 179; tail 115, 149; culmen 40, 48; tarsus 28, 39.

NOTES This species differs substantially in shape from other oropendolas, being shorter and stockier, in addition to the casque. It was formerly classified in the monotypic genus *Clypticterus*. Based on Chapman (1931b) and Schäfer (1957), the word 'Oropendola' refers to the Spanish name for the old world 'Oriole'.

REFERENCES Hardy *et al.* 1998 [voice], Koepcke 1972 [behaviour, nesting], Leak and Robinson 1989 [habitat, behaviour, nesting, sexual dimorphism], Moore 1993 [voice], Moore 1994a [voice], Pearman 1994 [distribution], Ridgely and Tudor 1989 [identification, voice, habitat, distribution], Williams 1995 [distribution].

2 CRESTED OROPENDOLA *Psarocolius decumanus* (Pallas) 1769 — Plate 1

The common oropendola in most of South America, it is not as strictly tied to forest as some of the others and can be common in rural areas.

IDENTIFICATION A large black oropendola with restricted russet on the rump and undertail-coverts. It has a bright blue eye and pale, whitish bill. This largely black oropendola may be identified even at some distance by its pale bill. The only other oropendola that is similar in plumage is Black Oropendola (11); however, this species has a orange-tipped black bill as well as a bare blue face patch and pink wattle. Para (10) and Baudo (8) Oropendolas have more extensive chestnut on the wings and back. In addition, they possess orange-tipped black bills and colourful bare face patches like Black Oropendola. Several cacique species are black with pale bills and blue eyes, similar to Crested Oropendola but these are smaller and lack the largely yellow tail of the oropendola. Giant Cowbird (98) is frequently observed near Crested Oropendola nests and could be confused for this species due to its large size and black plumage. Adult Giant Cowbirds have black bills, and a diagnostic small-headed appearance. The bill is blunter and shorter than that of Crested Oropendola. However, note that juvenile Giant Cowbird has a pale-coloured bill. Giant Cowbird always has an entirely black tail, unlike the largely yellow tail of Crested Oropendola. Crested Oropendola is a strong flier, often seen flying swiftly high over the canopy of the forest. It is certainly one of the strongest fliers of the oropendolas. It does not visibly undulate.

VOICE The frequency of singing is variable depending on the stage of nesting. It only sings during the breeding season, and is very quiet during the non-breeding season (Schäfer 1957).

Songs: As the male gives his display, he will perform an incredible, liquid, descending or ascending *Cr-crreeeEEEooooooooooooo*. This call sometimes ends with one or two low notes, for example *CreeeeeeeEEEE-wooo-poooo*. The song has a distinctive vibrating quality reminiscent of a finger being run across the top of a plastic comb. The display includes the audible sound of the beating wings which may last for 3 or 4 seconds after the vocal element is finished. The song is structurally complex, being composed of two different sounds given concurrently. The main sound is the high-pitched rattling (the comb sound) which may rise or fall, the other sound is a low growling or popping sound. During the territorial display the male will perform a gurgling rattle followed by liquid plopping, *crrrrrrr-OOOOO-whooop-whooop*; sometimes the rattle is replaced by a hoarse *tsreee-kleee*. Snyder (1966) describes the song in Guyana as a musical hodgepodge of bubbling plus a falling piano-like sequence ending with a loud *how-hoo*. It is not known if variation in the songs of this species falls into geographic patterns, or is strictly individual in nature.

Calls: The common call is a loud *clack* or *wak* given by both sexes, which often sounds to be closer than it actually is. When alarmed, the calls are quicker and sharper, *wak-wak-wak-wak*. Sometimes a more melodious *kaueek* is heard. Females also give a whining scold, particularly when chasing off Giant Cowbirds, or if in confrontations with other females. A high-pitched alarm note is given when predators are present. Males sometimes snap their beaks audibly as a sign of aggression or during the pre-copulatory display (Schäfer 1957).

DESCRIPTION Stout, chunky oropendola with a deep-based, spike-like bill. Old males have longer, more pointed wings than either young males or females. Wing Formula: P9 < P8 < P7 > P6 > P5; P9 ≈ P5; P8–P5 emarginate.

Adult male (Definitive Basic): Bill pale whitish-yellow or ivory, without any dark markings. Eyes turquoise-blue. Body largely black with little or no gloss. The rump, uppertail-coverts and undertail-coverts are chestnut. Some of the scapulars may be edged with chestnut, but see Geographic Variation. The thigh feathers are black, washed with chestnut. A thin black crest is often visible in the field. The tail is yellow with two black central rectrices. The dark central rectrices are shorter than the rest creating a complete yellow fringe around the black center of the tail. The legs and feet are black.

Adult female (Definitive Basic): Similar to male, but smaller and duller, tending to brownish-black rather than the jet black of the male. The crest of females is much reduced, and not visible in the field under usual conditions. The wings are shorter and more rounded than those of the adult male; this shows up as a visibly shorter primary extension.

Immature male (First Basic): Like adult male, but the wing is shorter and more rounded, and the tail is shorter. The difference in length of the outer primary (p9) and the third and fourth from the outside (p7 and p6) is much greater in adult males. Immature males are substantially smaller, in linear dimensions and mass, than adult males.

Juvenile (Juvenal): Similar to female, but even duller throughout. The bill has a brownish tint, and the eyes are brownish. In addition, there may be thin brown or chestnut-brown fringes present on the scapulars, back and underparts.

GEOGRAPHIC VARIATION Four subspecies are recognised. Populations from the north are larger than those from further south, but there is a great deal of variation in size within populations, particularly in males.

P. d. decumanus occurs in northern South America, east of the Andes, from Colombia south to N Peru and the Amazon river in Brazil. It is described above.

P. d. melanterus is the northern form, found from Panama and the west slope of the eastern Colombian Andes and the east slope of the central Andes range in Colombia. This form is said to be blacker than the other three, with little chestnut tipping to the feathers. Apparently, many specimens cannot be identified to race questioning the appropriateness of maintaining this subspecies.

P. d. insularis is restricted to the islands of Trinidad and Tobago. This race is the most chestnut of the forms, having many of the body feathers, particularly on the lower back, edged with chestnut. This subspecies is significantly smaller than the nominate. Birds from the Paria peninsula, Venezuela, are similar to *insularis*, in that they have an abundance of chestnut edges to the upper- and underparts and may best be regarded as that subspecies.

P .d. maculosus is the southernmost race, found south of the Amazon river, south to NW and NE Argentina. It differs from the other subspecies in possessing a variable amount of yellow, rarely white, feathers scattered throughout the body feathers and wing-coverts. In addition, this form averages browner than *decumanus*. It is slightly smaller than *decumanus* and has a shorter bill. However, there is variation in this form, as northern birds from the Rio Purus, Brazil, show very little, if any, yellow feathering on the body (Gyldenstolpe 1951). Specimens from Goiás, Brazil also lack yellow spotting. On the other hand, individuals from Bolivia and NW Argentina are heavily spotted (Gyldenstolpe 1945a). Individuals from Goiás, Brazil were previously set apart as the subspecies *australis*. These birds are said to have a more vivid bottle-green gloss than other *maculosus* (Gyldenstolpe 1945a).

HABITAT A species of lowland tropical forest edge and clearings. It lives in habitats ranging from disturbed agricultural areas and plantations to the edge of mature tropical forests. In the Amazonian lowlands, it tends to prefer more humid várzea forests as well as cochas and the shores of rivers; it is quick to colonise openings in the forest. It is also present along the lower reaches of montane forest and even more open deciduous forests in parts of Brazil. Its habitat choice is quite variable and the main prerequisite is a tall nesting tree, particularly if well isolated from the rest of the forest. Often the roosts are in dense bamboo thickets. In more open savanna habitats it has a preference for gallery forests. Crested Oropendola avoids arid areas. This species commonly forages in plantations (coffee, banana, cacao etc.) and is not at all shy of inhabiting urbanized areas.

BEHAVIOUR This oropendola is to be found singly or in small foraging flocks. In forests it may join up with *Cyanocorax* jays, forming mixed species flocks. Sometimes it may descend to the ground and forage there, particularly on lawns at the edge of forest. However, almost all foraging occurs in the mid or upper levels of the canopy. Feeds on large insects and fruit, but may nectar at large flowers (*Erythrina* trees for example). A rather acrobatic forager, given its size, often probing and searching the undersides of leaves. At night, Crested Oropendolas roost communally, often in flocks of hundreds. Females which are incubating remain at the nest, but males return to the roosts even during the breeding season. The oropendola roosts may be joined by other species including Giant Cowbird.

This species is highly polygynous and females greatly outnumber males. In Venezuela, at an average-sized colony, 15–30 females may be present in contrast to only 3–4 males (Schäfer 1957). One male is dominant at a nesting colony and he presumably fathers most of the young, but subordinate males may linger nearby hoping to achieve some copulations. As in other oropendolas, the male gives an exaggerated and elaborated bow display. The singing male bows forward so that the head is below the level of the perch; at this point the wings are extended and brought over the back where they are vibrated rapidly. The tail is brought forward, over the back, and the male bows even more noticeably, sometimes nearly hanging upside down. He begins to recover from the bow as the tail is lowered and the wings are flapped more slowly making a noticeable sloshing sound, as the song ends. The bird then stops flapping and resumes a normal posture. The crest is noticeably elevated throughout the display. Displays given to females are the most intense. In a version of the display given to other males, likely with a dominance or territorial significance, the tail is not cocked so far over the back, the bow is not as deep and the vocalisations are different, basically a shorter version of the song. Displays are given roughly 10–20 times an hour, (Schäfer 1957); not a high rate of display in comparison to 'Yellow-billed' Russet-mantled Oropendola (5X). Young males rarely give a full intensity display. While at the nest tree, the male makes noisy flights between perches. These conspicuous movements may be considered a display by the male. The noticeable noise is caused by the pointed, highly emarginate primaries of the adult male.

Copulation takes place at, or very close to, the nest tree. Receptive females will approach males with feathers pressed close to the body, giving the body a sleek appearance. They take on the stereotypical passerine pre-copulatory pose with head slightly raised, tail cocked and a fluttering of the wings, which may be almost not noticeable at times. The male may respond by fluffing up his neck feathers, displaying and sometimes pecking at the cloaca of the female (Tashian 1957). The male's pre-copulatory display may include audible bill snapping, in addition to wing fluttering (Schäfer 1957). Copulation is short and afterward both birds may preen or bill wipe for a short time, before flying off without any post-copulatory display.

NESTING Crested Oropendola is strictly colonial, with colony size ranging from two to 30 or more nests. Nests are commonly clumped within the colony. Smaller colonies are less clumped than the larger ones. The nests are long, hanging baskets with an entrance on top and tend to be over 125 cm long (Schäfer 1957). The preferred nest trees tend to be those that are tall (25–35 m) and isolated, so that their canopy does not touch that of other trees. This protects the colony against access by predators such as snakes and monkeys. The same colony trees are used year after year (Schäfer 1957). Nests are placed an average of 20 m from the ground, ranging from 10–35 m (Schäfer 1957). Along the north slope of the coastal mountains of Venezuela the highest nesting Crested Oropendolas may uncommonly nest in mixed colonies with *oleagineus* 'Yellow-billed' Russet-backed Oropendolas.

In Trinidad, nesting takes place January–May, during the start of the dry season. In eastern Venezuela, Crested Oropendolas are well into the nesting season in mid March while in N Venezuela, nesting takes place February–July. In Surinam, Crested Oropendolas nest December–January, some into May; the earliest fledglings are observed in the first week of February (Haverschmidt 1968). In eastern Ecuador, birds appear to be on eggs during December. Crested Oropendolas in the Magdalena valley, Colombia, were nest building in mid April (Boggs 1961). In the subtropical zone in Cochabamba, Bolivia, they are nesting during October (A. B. Hennessey pers. comm.). The breeding cycles of different females at the colony are not synchronous, and it is uncertain if this is also the case within 'clumps' of nests.

The nests are woven from a variety of materials ranging from fibres stripped from bromeliads, heliconias, *Tillslandia* ('Spanish Moss'), bark strips, orchid roots and coconut fibres. Commonly, clutch size is of 2 or even only 1 egg. Eggs bluish-green, or grey with dark spots, blotches or scrawls. Incubation lasts 15 days in Trinidad (Tashian 1957) and 17–19 days in N Venezuela (Schäfer 1957). It begins after the first egg is laid (Schäfer 1957), thus in two egg clutches there will be a size hierarchy in the nestlings as one will be older than the other. Fledging occurs at 31–36 days in Trinidad (Tashian 1957) or 28–34 days in N Venezuela (Schäfer 1957). Giant Cowbird (98) regularly parasitises nests of this oropendola. In addition, Shiny Cowbird (102) has been known to parasitise nests of this oropendola, but no young have been observed to fledge (Friedmann and Kiff 1985).

DISTRIBUTION AND STATUS Common and conspicuous, becoming more local in the south of its range. A bird of the lowlands, most typical of the tropical zone usually being found below 1000 m, but sometimes as high as 2600 m (Hilty and Brown 1986). In Panama, west to W Chiriquí on the Pacific Slope and W Colon and N Coclé (Ridgely and Gwynne 1989). In Colombia, distributed from the NW lowlands west of the Andes, but absent from the Guajira Peninsula, it is present in the Magdalena and Cauca river valleys but is now very rare there. It also ranges to the lowlands east of the East Andes in Colombia but is absent from the Llanos. East from Colombia it is found through most of Venezuela. In Venezuela, it is absent from the Andes and higher regions of the coastal mountains, but is present throughout the lowlands both north and south of the coastal mountains. Also found on the islands of Trinidad and Tobago. From Venezuela, it ranges east to the Guianas and into Brazil; south through E Ecuador, E Peru, and N Bolivia (all east of the Andes); extending south in the Yungas forest along the Andean foothills to Jujuy, Argentina; E Paraguay and N Misiones in Argentina and Santa Catarina in Brazil. Not present in the caatinga region of E Brazil.

MOVEMENTS Appears to be non-migratory in the strict sense, but highly mobile, performing some regular seasonal movements. In northern Venezuela, Crested Oropendola breeds mainly at the edge of the tropical and subtropical zone, moving to the lowlands of the tropical zone during the non-breeding season (Schäfer 1957). During the non-breeding season much wandering occurs.

MOULT The complete definitive pre-basic moult occurs after the young have become independent. The breeding season varies depending on the locality, and therefore the timing of this moult also differs. For example, specimens from Putumayo, Colombia, are moulting in early June while birds from Santander, Colombia, are moulting in early November. Adults from Trinidad moult in July, while Surinam individuals moult in mid October, and in Goiás, Brazil, moult occurs during May. The moult in N Venezuela occurs July–November (Schäfer 1957). The available evidence suggests that pre-alternate moults are lacking. The timing and extent of the first pre-basic moult is unknown.

MEASUREMENTS Males are much larger than females.

P. d. decumanus. Males: (n=18) wing 226.3 (201–268); tail 194.1 (167–225); culmen 63.2 (58–71); (n=9) tarsus 46.4 (45–58). Females: (n=12) wing 174.6 (160–203); tail 162.2 (139–200); culmen 49.4 (41–66); (n=10) tarsus 41.1 (37–45).

P. d. melanterus. Males: (n=11) wing 233.3 (215.9–250.2); tail 198.8 (184–221); culmen 63.6 (55.9–66.5); (n=10) tarsus 51.9 (48.3–55.4). Females: (n=10) wing 168.6 (160–180); tail 151.6 (137.1–159); culmen 47.7 (46.0–49.3); tarsus 41.3 (39.0–43.3).

P. d. insularis. Males: (n=6) wing 215.5 (190–250); tail 193.0 (160–218); culmen 60.8 (58–65); (n=3) tarsus 50.7 (50–52). Females: (n=10) wing 166.5 (145–188); tail 161.1 (148–180); (n=9) culmen 47.6 (45–48), (n=6) tarsus 39.7 (38–50).

P. d. maculosus. Males: (n=21) wing 225.6 (200–247); tail 179.8 (156–204); (n=20) culmen 58.9 (54–66); (n=9) tarsus 48.8 (44–53). Females: (n=12) wing 181.6 (158.8–255); tail 157.0 (134.6–203); culmen 43.8 (37.1–66); (n=10) tarsus 38.8 (37–40).

NOTES Previously placed in the genus *Ostinops*. The plumage of this oropendola has a distinctive musky odour (Wetmore 1939), which is caused by the strong smell of the oil produced by the uropygial gland (Schäfer 1957).

REFERENCES Belcher and Smooker 1937 [nesting], Chapman 1920 [geographic variation, measurements], Dalmas 1900 [description, measurements], Drury 1962 [voice, behaviour, nesting], Gyldenstolpe 1945a [geographic variation, measurements], 1945b [measurements], 1951 [geographic variation], Hardy *et al.* 1998 [voice], Hilty and Brown 1986 [distribution], Moore 1993 [voice], Ridgway 1902 [measurements], Snyder 1966 [voice], Tashian 1957 [behaviour, nesting], Todd 1917 [geographic variation].

3 GREEN OROPENDOLA *Psarocolius viridis* (Müller) 1776 — Plate 2

This oropendola is more likely to be found within forest rather than around openings. It is behaviourally and vocally similar to Crested Oropendola, but visually it is more like the otherwise quite different Olive Oropendola.

IDENTIFICATION Green Oropendola appears mainly green in the field. With russet towards the posterior section of the body. The bill is pea-green with a reddish tip, and the tail is yellow with dark central rectrices. Green Oropendola is most likely to be confused with the broadly sympatric Olive Oropendola (9). Both of these species appear largely green in the field. Olive Oropendolas are larger than Green Oropendolas, but sexual size dimorphism creates a great deal of overlap between the two species. Olive Oropendola has a bare pink face patch, and upon close inspection also shows a bare pink patch at the base of the culmen; Green Oropendolas lack both of these face patches. Green Oropendola has a pale green bill with an orange or reddish tip, while Olive Oropendola sports a black bill with an reddish tip. Male Green Oropendolas have blue eyes, while Olive Oropendolas always show brownish eyes. The distribution of chestnut on the plumage differs between these two species. Its extent is greater on Olive Oropendolas than Green Oropendolas. The russet is restricted to the crissum, lower belly and rump on Green Oropendola. Olive Oropendolas have chestnut extending to the upper belly and back as well as on the wings. In addition, the chestnut colour of Olive Oropendola is darker, contrasting more strongly with the olive head and anterior body than on Green Oropendola; this colour is a paler chestnut, or russet as opposed to chestnut on the latter species. Green Oropendolas do not show an obvious interface between the olive-green and russet parts of the body, but there is a more subtle blend between the two colours. Finally, there is a difference in the tail pattern. Green Oropendolas have an olive outer web to the outermost rectrix that is lacking on Olive Oropendolas. When observed from below, the tail of Green Oropendola appears yellow with two lateral olive stripes that reach three-quarters of the way to the tail tip, since the outer rectrices are shorter than the inner ones; Olive Oropendolas show an entirely yellow tail from below. In addition, the central dark rectrices are olive in Olive Oropendola and nearly black in Green Oropendola. Differences in vocalisations may also aid in identification.

VOICE Song: A typical song begins with a screechy, high-pitched note and ends in low, hollow popping, or more liquid sounds. For example, *zweeeEEE-whopwheerupwhopwheeerup* or *zeeeeeEEEE-papaparaparapUUUp*, or *zeeeeee EEWHaruuuup*. The initial high screech is distinctive for this species, lacking the trilled nature that sounds like the running of a finger along a comb, of several other oropendola species. This note also undulates in its frequency range, giving it an almost warbled nature. However, there appears to be at least two song types. The second is an explosive, descending *trrreeeeeeeeeeeeeppppp*, finishing with the sounds of flapping wings; this song can sound similar to that of Crested Oropendola (2).

Calls: A loud *chak-chak*, each note similar to the *chak* notes of other oropendolas, but invariably they are doubled in this species. Also gives an odd mewing, reminiscent of calls of North American sapsuckers (*Sphyrapicus* sp.).

DESCRIPTION A medium-sized to large oropendola with a thick bill, showing an expanded culmen, that forms a slight casque. The tail is moderately graduated, and the central tail feathers are almost as long as the longest yellow rectrices. Wing formula: P9 < P8< P7 ≈ P6 > P5; P9 ≈ P4; P8–P5 emarginate.

Adult Male (Definitive Basic): The bill is pale green (pea-green) or bluish-green with a red or orange-red tip. Eye of adult male brilliant sky-blue. On some individuals bright pink bare skin is visible on the lores, and sometimes around the eyes. On other individuals this bare skin may be dull in colour and not noticeable; it is not known if the degree of brightness of the facial skin is correlated with the breeding season. The head, nape, back, throat and breast are olive with a lime-green tone. The body plumage varies considerably, some individuals showing a greater infusion of this lime-green (yellowish) colour, see Geographic Variation. The lower back and rump are chestnut, while the uppertail-coverts are brownish-olive. On the underparts, the hind flanks and crissum are chestnut. The thighs are dull chestnut, with olive tips to the feathers. The wings are blackish with olive edges and a small number of chestnut feather tips. The coverts are broadly edged dull olive, such that they appear largely olive from a distance. The underwing-coverts are blackish with olive tips. The tertials and closed secondaries look solidly olive due to their olive outer halves; the browner inner halves are hidden when the wing is folded. The tail is yellow with the central pair of rectrices (R1) a contrasting dark brown, appearing blackish in the field. The tips of the outer vanes of the outermost rectrices (R6) are olive-brown. The legs and feet are black.

Adult Female (Definitive Basic): Similar to male, but noticeably smaller in size. Females tend to show brown, not blue, eyes.

Juvenile (Juvenal): Similar to adults, but with a duller coloured and more loosely textured plumage. Unlike adults, juveniles show a grey wash on the breast and the throat is pale grey. The thighs are dull olive, not chestnut as on adults. Juveniles and immatures may be sexed by size differences, but note that young males may be substantially smaller than adult males. Young males lack blue eye of mature males.

GEOGRAPHIC VARIATION Currently considered monotypic. In the past, birds from Peru and W Brazil were separated as a separate subspecies, *P. v. flavescens.* This form was described as yellower and brighter than the nominate, showing less grey-olive in the feather centers (Bangs and Penard 1918, Gyldenstolpe 1945a). This subspecies is now synonymized with the nominate, but may reflect a cline in the plumage colour of this species.

HABITAT A forest oropendola, this species is particularly fond of terra firme (dry) forest. Green Oropendola tends to keep to the canopy and avoids forest openings or settled areas. It appears to be more tolerant of disturbed and open forests in Venezuela than in western Amazonia (pers. obs.).

BEHAVIOUR This species is most likely to be observed at the breeding colonies. Otherwise it is often seen flying over the forest or forest openings during early morning or late afternoon. One male is usually present at the colony and defends his status from other males. He spends most of the day near the colony, often displaying and calling at the top of the nesting tree or a nearby tree. Females announce their arrival at the colony by giving *chak chak* calls. Males perform a bow display, as in all oropendolas. The display is accompanied by song. The sequence of events during the display is as follows: first the male partially opens the wings and vibrates them before holding them stiffly locked over the back in a 'V'; he then begins to sing, and with the body plumage raised and the tail cocked he bows down, bringing the head slightly below the level of the body in a weak display. He recovers, and rights himself as the song finishes, but terminates the display by shaking the wings audibly. In the most extreme displays, the male may hang upside down as he beats his wings. During very weak displays, the bow is not as extreme and the wings are not beaten. The display is most similar to that of Crested Oropendola. Mating occurs away from the colony. The pre-copulatory display involves the male jumping with spread wings onto a larger branch in the forest canopy, often bowing the head and bill and cocking the tail upwards as it vocalises towards the female which sits roughly one meter away. She then cocks her tail upwards and the male mounts; several mounting attempts may occur (Jaramillo pers. obs.).

NESTING An obligate colonially-nesting species. There is one record of a joint colony of Green Oropendolas and Red-rumped Caciques (14) from Surinam (Haverschmidt 1972). The nests are long, hanging baskets of woven plant fibres, measuring roughly one meter in length. Nesting colonies vary in size, with 5–10 nests being common, but they may be as small as only two nests. Most nests are aggregated in a clump on the nesting tree, usually with a few nests more distant. The success rates of clumped versus unclumped nests has not been studied in this species. The trees chosen for the colony site tend to have an open mid story, usually as a consequence of the trees being tall emergents. In Manaus, Brazil, only the Mulateiro tree (*Peltogyne* sp.) is used for nesting, probably due to its slick, smooth trunks which deter climbing predators (Oniki and Willis 1983). Otherwise nest trees are situated in areas where the forest is more open or lower in stature such that access to the nests is not impeded by nearby vegetation. Green Oropendolas do not choose to nest in solitary trees in cleared areas as Russet-backed (5) and Crested Oropendolas often do. Nest trees are used year after year. Nesting begins in March in eastern Venezuela (Caura forest, Bolívar), and in June in eastern Ecuador (Jatun Sacha, Napo). In Surinam, this species nests December–April (Haverschmidt 1968, 1972). The nesting period in Manaus, Brazil, ranges from September–March (Oniki and Willis 1983). Hilty and Brown (1986) cite the nesting period in Meta, Colombia, as March–April. The nests are built by the female alone. Eggs white with reddish spotting (Haverschmidt 1968). No information is available on the incubation period. All care of the eggs and young is conducted by the female, with no help from males. This oropendola is a common host of Giant Cowbird (98). Young cowbirds successfully fledge from Green Oropendola nests (Jaramillo pers. obs.).

DISTRIBUTION AND STATUS Uncommon to locally common to 500 m. Found in northern South America east of the Andes. In Colombia, ranges from E Vichada south to Vaupés and west to Meta. It is likely also found south to Amazonas (Hilty and Brown 1986). In Venezuela, it is found mainly south of the Orinoco river from SE Sucre and NW Monagas including the Delta Amacuro, south through Bolívar and Amazonas (Meyer de Schauensee and Phelps 1978). Inhabits most of the forested areas of Guyana, Surinam and French Guiana. In Ecuador, restricted to the eastern lowlands of Napo and Pastaza provinces. Found in northern Peru mainly in the northern department of Loreto. Widespread in northern Brazil, in Amazonas, Roraima, Pará, Amapá, N Maranhão, N Rondônia and N Mato Grosso. Absent from Bolivia, but it should be looked for in the northern department of Beni.

MOVEMENTS A resident species that does not appear to undergo any noticeable migratory movements. Its abundance changes depending on the time of year, implying that a certain amount of local movement occurs, possibly tied to local differences in fruit abundance. Although there is no banding record of a bird in the wild, a female Green Oropendola survived in a zoo for 18 years (Crandall 1937).

MOULT There is one complete moult a year in adults, the definitive pre-basic moult. This occurs subsequent to nesting, and therefore its timing varies according to the breeding schedule in a given area. In French Guiana, adults are in moult in May, while in Guyana specimens in moult were examined from early July. In Amapá, Brazil, adults are moulting during November while in E Pará, Brazil, the moult is ongoing during April. No evidence was found to suggest a pre-alternate moult. The extent and timing of the first pre-basic moult of juveniles is not well known.

MEASUREMENTS Males: (n=12) wing 235.4 (221–267); tail 172.8 (158–205); (n=10) culmen 65.1 (57–72); (n=12) tarsus 52.3 (43–57). Females: (n=12) wing 190.0 (180–200); tail 148.0 (135–160); (n=10) culmen 51.5 (49–55); (n=12) tarsus 43.8 (38–48).

NOTES Green Oropendola resembles Olive Oropendola in plumage, but the two species do not appear to be closely related. Behavioural and morphological aspects are similar to Crested Oropendola, but it is unknown if this is due to a close genetic relationship.

REFERENCES Bangs and Penard 1910 [Geographic Variation], Crandall 1937 [behaviour], Gyldenstolpe 1945a [geographic variation], Haverschmidt 1972 [nesting], Oniki and Willis 1983 [nesting], Robert Behrstock personal recordings [voice].

4 DUSKY-GREEN OROPENDOLA *Psarocolius atrovirens* Plate 4
(Lafresnaye and d'Orbigny) 1838

Dusky-green is a good description of the colour of this dull oropendola. It is not well known and is the only oropendola species restricted to highlands.

IDENTIFICATION Dusky-green Oropendola is a dull olive oropendola of medium size with a pale bill. The tail is brown and yellow, but the yellow is more restricted in this species than in most other oropendolas. This species can easily be mistaken for the sympatric 'Yellow-billed' Russet-backed Oropendola (5X) of the race *alfredi*. However, Dusky-green Oropendola is more extensively olive, lacking a russet back or russet wash to the breast. The olive colour of the head is darker on Dusky-green Oropendola than on 'Yellow-billed' Oropendola, such that the head of Dusky-green is close to being the darkest part of the plumage while the head of *alfredi* 'Yellow-billed' is paler than the body plumage. In addition, Dusky-green Oropendola has less yellow on the tail and paler olive central rectrices. The pattern of dark colouration on the tail differs from 'Yellow-billed' in the following way: R2 (second from inside) is yellow on its outer half on *alfredi* and the outermost tail feather (R6) shows yellow on the base of the inner vane. On Dusky-green Oropendola, R2 and R6 are entirely dark olive (Figure 4.1). Also note that Dusky-green Oropendola is typically a species of higher elevations than *alfredi* 'Yellow-billed Oropendola', which tends to be absent above 1500 m. The more olive subspecies of 'Yellow-billed Oropendola' from Venezuela (*oleagineus* and *neglectus*) resemble Dusky-green more closely than *alfredi*; however, these are easily separated on geographical grounds alone as they are not migratory. Nevertheless even these races of 'Yellow-billed Oropendola' show russet on the back, unlike Dusky-green Oropendola.

Crested Oropendola (2) is sympatric with Dusky-green Oropendola, but the two species are unlikely to be confused. In the field, Crested Oropendola appears largely black, with an obvious pale ivory-coloured bill and bright blue eyes. It has no olive, green or brown colours, only black and russet. Dusky-green Oropendola undulates slightly in flight, unlike Crested Oropendola.

VOICE Song: The song lasts just over a second in duration. The songs are extremely variable and there is no single typical pattern. Many songs have an introductory section of several low notes, followed by a dry rattle and a final longer, low note with a hollow quality. The dry rattle resembles the sound of running a finger across the teeth of a plastic comb. Other songs lack this rattle, but are composed of a repeating series of notes, which perhaps could be classified as a slower version of the rattle. This repetitive series of notes is perhaps the most distinctive part of Dusky-green Oropendola's song. Overall, the song has a liquid or plopping quality that resembles that of the Russet-backed Oropendola complex. Note that the preliminary *whoop* or *poop* notes separate the songs from those of Russet-backed Oropendola which tend to begin with a rattle, although this is variable. Song examples are: *whoop-poop-poop-krrRREEEoooo, who-hoo-hoo-ttrrrrrrrr-phwaaaa, whoop-poop-kweeooo, whoo-kwEEoooo*. Another song type has a whinny and more muffled character and can be described as: *wrrok wrok whoooOOOO* or *whao-whOOOeee*. Males have a large repertoire of song types. A male may sing one song type continuously for several minutes and then change to a new song type; this often occurs when he flies to another tree or changes position in the same tree (A. B. Hennessey pers. comm.). The full song repertoire of males is not known, but it appears to be sizeable. Dusky-green Oropendola is an active and prolific singer, often commencing singing before other species in the morning. It sings throughout the day at the colony, but less frequently while in foraging flocks. The songs given by males accompanying foraging flocks of females appear to be weak and uninspired. Limited playback experiments suggest that males respond to playback by changing their songs or by matching the song type offered during playback. The intensity of the display accompanying the song increases in response to playback (A. B. Hennessey pers. comm.).

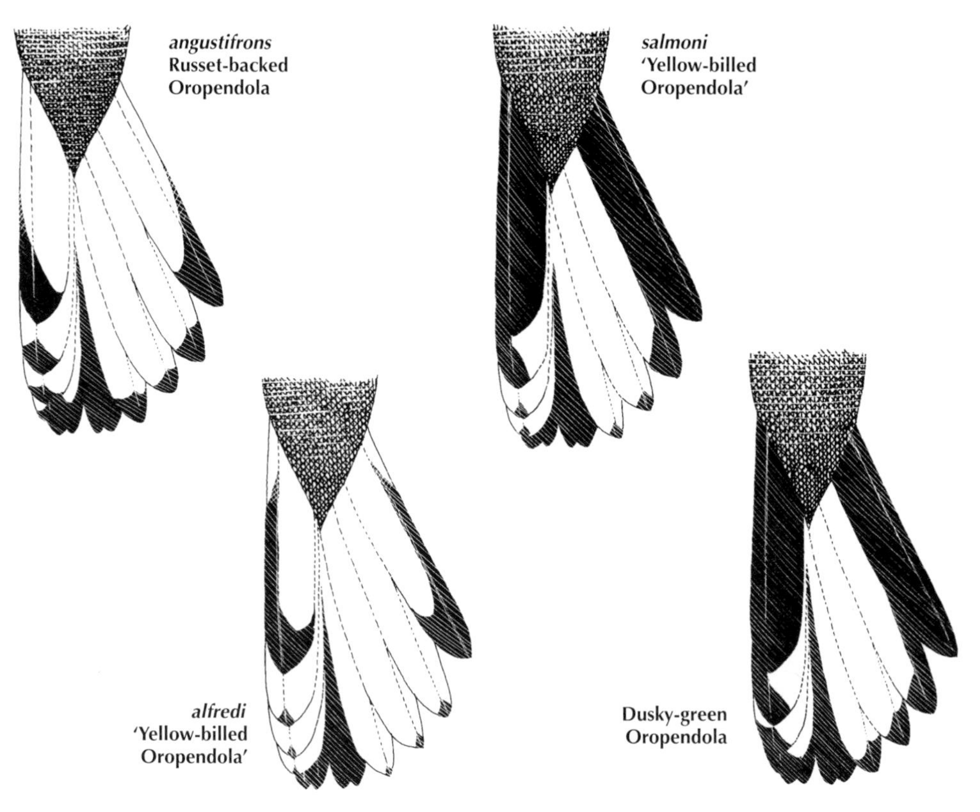

Figure 4.1 Tail pattern in Russet-backed and Dusky-green Oropendolas. From left: angustifrons *Russet-backed Oropendola;* alfredi *'Yellow-billed Oropendola';* salmoni *'Yellow-billed Oropendola'; Dusky-green Oropendola.*

Calls: The contact call is a deep *chog* or *chuk*; a loud *tschuik* has also been described.

DESCRIPTION A medium-sized oropendola with a long, deep bill. There is no obvious expansion (casque) at the base of the culmen. Wing formula: P9 < P8 < P7 > P6 > P5; P5 > P9 > P4; P8–P5 emarginate.

Adult male (Definitive Basic): The bill is pale, either a greenish-white colour or a brighter pea-green. The eyes are brown, dark blue, light grey or pale blue; the blue eye may be associated with age. Grey eyes appear dark at a distance (A. B. Hennessey pers. comm.). The face, crown and nape are olive, having contrasting dark lores. Some individuals show yellow on the forehead, sometimes to the eye. Typically, Dusky-green Oropendolas show a pale area on the side of the face or neck, sometimes appearing as yellowish patches on the neck sides or back of the auriculars. The mantle is olive. The lower back quickly changes to russet, including the rump and uppertail-coverts. Underparts are olive, with a whiter throat, the vent and undertail-coverts are russet. The thighs are olive. The flanks are olive, but there is a russet wash to the hind flanks. The wings are blackish with olive coverts and wide olive outer edges to the secondaries and tertials, making the wings appear largely olive when folded. The underwings are blackish. The tail shows a mixture of yellow and olive: the central four tail feathers (R1 and R2) are olive; while the next two pairs from the inside (R3 and R4) are yellow with small olive tips; the next pair (R5) are olive on the outer half, but yellow on the inner half; the outermost rectrix (R6) is entirely olive. Since the outer tail feathers are olive and the others are tipped olive, there is a substantial amount of olive, as opposed to yellow, visible on the underside of the tail. The legs and feet are black.

Adult female (Definitive Basic): Extremely similar to male, differing mainly in its smaller size. Like the male, females may have brown or pale blue eyes.

Juvenile: Similar to adult, but duller and with an obviously loosely textured plumage, particularly on the flanks, vent and rump. Juveniles have pale yellow-green bills, yellower than those of adults. In addition, juveniles sport a yellow forehead which is usually lost later on in life. Unlike adults, the underparts are greyish-green, not olive. The throat is white with a yellow wash, not solidly olive-green. The crown is dusky, not olive as on adults. The wings, back and rump are similar colours to those of the adult. The eyes are always brown. The legs and feet are black.

GEOGRAPHIC VARIATION Monotypic. No described geographic variation.

HABITAT Most frequently found in edge situations in humid subtropical forest. The majority of this type of habitat is only accessible along roads, which are usually found in flatter and more open areas, possibly causing a bias in the perceived habitat choice of this oropendola (A. B. Hennessey pers. comm.). A great deal of foraging occurs in the deeper parts of subtropical forests. The colonies are located along stream courses or the similar habitat created by roads. Streams and their associated ravines are often frequented. In fact, the presence of stream valleys, particularly where a sufficient flat floodplain is available, appears to be a prerequisite for the presence of this species. Dusky-green Oropendola is rarely sympatric with the *alfredi* race of 'Yellow-billed' (Russet-backed) Oropendola in Bolivia. More commonly, Dusky-green Oropendola is sympatric with Crested Oropendola, but mainly at the lower elevational range of the former species. 'Yellow-billed' Oropendola occurs at even lower elevations, usually 800–1000 m particularly where there is agriculture. Dusky-green Oropendolas appear to avoid agricultural areas (A. B. Hennessey pers. comm.).

BEHAVIOUR Dusky-green Oropendola uses a variety of foraging techniques. On trees, it moves by hopping with wings closed and side-stepping along branches; flap-hops are used on longer jumps. It continues to probe for insects as it moves. The species commonly hangs while foraging, even hanging upside down if this will aid it in reaching a far-flung morsel. Rather than manipulating things with the bill, or gaping, Dusky-green Oropendola pokes rapidly, moving the head so that the angle of the probe is constantly changing. Like other oropendolas, Dusky-green tends to forage in the canopy or subcanopy of the forest. Most commonly it is observed nimbly hanging and probing into dead leaf clusters before quickly moving on to the next cluster. It also forages on the forest floor, which is rare in oropendolas. When foraging on the ground, the entire feeding flock descends. They noisily search for insects in the forest floor detritus. Floor-foraging flocks are easily detected by their loud *chuck* calls and the noise they make while moving leaf material (A. B. Hennessey pers. comm.). They may also forage on cliffs and in emergent aquatic vegetation. It appears that insects are usually the main quarry, but fruits also make up a sizeable proportion of the diet. This species' resourcefulness is illustrated by the following observation at a Bolivian site: at a police road-check in the forest, a light is kept on below a shed-like roof throughout the night, attracting large numbers of nocturnal insects. The local flock of Dusky-green Oropendolas returns daily to the shed after daybreak, and quickly feeds on the large concentration of insects. The oropendolas do not then tend to return to the area until the following morning (A. B. Hennessey pers. comm.).

Like other oropendolas, this species is polygynous. Due to the smaller size of colonies, the level of polygyny is not as high as in other species. Males are always at attendance in the colony. Females forage in small flocks, usually of less than 10 individuals. Female foraging flocks are typically followed by a male who sings sporadically as he follows them. Mating probably takes place while foraging away from the nesting colony, but this needs verification. During many observations at nesting colonies, mating was never observed (A. B. Hennessey pers. comm.). Mating away from the colony creates a situation where a single male cannot monopolise all of the matings at a single colony.

NESTING Forms small colonies, building long, hanging basket-like nests as is typical of the genus. Nests have been recorded from Bolivia July–November. Differences in moult timing suggest that populations from Puno and Cuzco, Peru, breed later than those from Cochabamba, Bolivia. The colonies are almost always located along clearings created by roads or streams cutting through the montane forest. Of 30 colonies, 75% were placed over a river, 20% over a road and 5% in other clearings (A. B. Hennessey pers. comm.). In many cases, colonies exist near buildings or settlements, and this species does not appear averse to nesting in areas where people are present.

The colonies tend to comprise of less than 10 nests placed in one tree. The smallest colonies have three nests, and solitary nesting is a possibility but is as yet unknown. The nesting trees may be large or small, densely leaved or nearly dead but all share the attribute of having a substantial amount of open airspace around them, by being isolated from the forest. The use of introduced conifers as nest trees is known in Bolivia. One colony in Bolivia had the peculiarity of being secured to roots emerging horizontally from a cliff edge. The three nests were placed 1–3 m above the surface of a river. Cliff nesting actually seems to be somewhat regular in this species (A. B. Hennessey pers. comm.).

The tightly-woven nests are of rootlets or grass stems, with the construction material dependent upon the site. Three nests at Cotapata, Bolivia (1300 m), were lined with green leaves of a broad-leaved bamboo species that is found at higher altitudes, suggesting that the acquisition of nesting materials may span a significant elevational range (A. B. Hennessey pers. comm.). Eggs off-white, with brown splotching which is concentrated at the broader end of the egg. Clutch size not known. One nest examined in Bolivia had only one egg (A. B. Hennessey pers. comm.). Giant Cowbird (98) parasitism is not known in this species; in fact the cowbird is rarely observed in the altitudinal range frequented by Dusky-green Oropendola.

DISTRIBUTION AND STATUS Uncommon to common between 800 and 2600 m; usually above 1000 m. One sight report from 450 m, at Campamento Macunucu, Parque Nacional Amboro, Santa Cruz, Bolivia (S. Herzog pers. comm.). Only found in Peru and Bolivia along a thin strip on the east slope of the Andes. In Peru, it ranges from Huánuco department south to Puno department. In NW Bolivia, it occurs from La Paz department through Cochabamba to westernmost Santa Cruz.

MOVEMENTS Sedentary, no evidence of long or medium distance movements. Likely undergoes altitudinal movements during different seasons, but this has not been confirmed. In the Zongo valley, Bolivia, Dusky-green Oropendola is at times common and at other times entirely absent, suggesting that some movements do occur.

MOULT Adults undergo a complete definitive pre-basic moult after the breeding season. Adults from Machupicchu, Cuzco, and Oconeque, Puno, Peru, undergo pre-basic moult May–June. Birds from Cochabamba, Bolivia, moult during April. The extent of the first pre-basic moult is not known. Pre-alternate moults appear to be lacking. Juveniles are known from Junín, Peru, from December (Fjeldsa and Krabbe 1990).

MEASUREMENTS Males: (n=10) wing 213.3 (185–244); tail 166.7 (146–188); culmen 51.7 (45–55); tarsus 49.7 (48–51). Females: (n=10) wing 182.5 (171–204); tail 145.9 (138–159); (n=9) culmen 44.0 (42–48); (n=10) tarsus 40.8 (39–42).

NOTES It has been reported that Dusky-green Oropendola may hybridise with the *alfredi* subspecies of 'Yellow-billed' (Russet-backed) Oropendola in SE Peru (Ridgely and Tudor 1989). In many ways Dusky-green Oropendola resembles the highland forms of 'Yellow-billed Oropendola' closely, including the tail pattern, behaviour, habitat choice and colouration. A review of the entire 'Russet-mantled' and Dusky-green Oropendola complex needs to be conducted in order to assess the significance of these similarities. If Dusky-green Oropendola was not sympatric with a form of the 'Russet-mantled' Oropendola group, it would likely be considered conspecific with it.

REFERENCES Fjeldsa and Krabbe 1990 [voice, habitat, nesting], Hardy *et al.* 1998 [voice], A. Bennett Hennessey personal recordings and notes [voice, behaviour, nesting], Ridgely and Tudor 1989 [distribution, notes].

5 RUSSET-BACKED OROPENDOLA *Psarocolius angustifrons* Plate 4 (Spix) 1824

A common oropendola of the western Amazon basin. The phylogenetic relationships of the complex that includes this form and the 'Yellow-billed' Oropendola (5X) are not well understood.

IDENTIFICATION The identification of the yellow-billed forms of this species is treated below in a separate account. Russet-backed Oropendola is a dull, largely brownish-russet oropendola with a black bill and an entirely dark forehead. This large oropendola is only likely to be confused with other members of the genus, particularly Green (3), Crested (2), Casqued (1) and Olive (9) Oropendolas. Russet-backed is a dull-coloured, brownish oropendola with a dark bill. Green and Olive Oropendolas are both more obviously green on the body, with contrasting chestnut plumage on the rear halves of their bodies. Green Oropendola has a pale green bill with an orange tip, while Olive Oropendola has a black bill with an orange tip, and extensive pink bare facial skin. Casqued Oropendola is smaller than Russet-backed and has an obvious bulge at the base of the culmen, as well as a paler bill with a dark tip and culmen. Casqued Oropendolas also have grey throats, yellow-green breasts and dark chestnut bodies. Crested Oropendola is quite different, as it is mainly black and has a noticeable ivory-coloured bill. Eye colour is not reliable in helping to identify Russet-backed Oropendola as it varies from brown to pale blue in this species.

VOICE Song: The song has a liquid or a plopping quality. Songs always appear to begin with a rattle, which then changes into the plopping sound, and they often end in a low-pitched, drawn-out note. Songs tend to have an explosive, whip-like quality. Song examples are: *prrrr-WHOOOOpeeeuuuuu*, or *whrrroooOOOkpee OO*, or *wrrrrrrrkpeeeeeeOOOOO*. Also a shorter *prrr-WHOP-up*. The initial rattle differs from the songs of Dusky-green Oropendola which usually begin with a series of low liquid notes, or a repetitive series of notes too slow to be called a rattle.

Calls: The most common call is a low, resonant *chugh chugh chugh* or *quip quip quip*, often delivered in pairs of notes. The exact tone and emphasis of this note changes depending on the state of excitement of the individual.

DESCRIPTION A medium-sized to large oropendola with a short crest which is not usually visible in the field. The culmen and gonys are both straight, and the bill is roughly as long as the head. The bill is deep at its base. Wing Formula: P9 < P8 < P7 > P6; P9 ≈ P4; P8–P5 emarginate.

Adult Male (Definitive Basic): The bill is black and the eyes are brown or occasionally pale greyish-blue. The head is olive merging into the warm chestnut-brown of the body; the crest feathers are darker than the rest of the crown. The upper- and underparts are chestnut-brown, brightest (tending to russet) on the rump and vent. The chestnut underparts usually show an olive wash and the thighs are entirely olive. The throat is paler green than the head and is detectably greener than the rest of the underparts. On some, the olive of the head extends down to the breast. The wings are blackish-brown, darker than the body and the coverts are tipped chestnut. The tertials are usually edged chestnut. The tail is yellow with the central tail feathers (R1) blackish-brown, while the next pair (R2) are blackish-brown on their inner half and yellow on their outer half, the outer pair of tail feathers (R6) show a dusky outer edge and tip. The yellow tail feathers (R3–R5) have small dusky tips (Figure 4.1). The legs and feet are black.

Adult Female (Definitive Basic): Like male, but very much smaller, it tends to show a paler throat.

Juvenile (Juvenal): Like adults, but with a stronger yellow wash on the head and the yellow on the tail may look duller, more olive-yellow. The eyes are always dark.

GEOGRAPHIC VARIATION The black-billed form is monotypic, see 'Yellow-billed Oropendola' (5X) which is currently considered part of this species for the other subspecies. It appears clear that the Russet-mantled Oropendola is composed of more than one variable species.

HABITAT This species is attracted to forest edge, particularly clearings and openings in the forest, or river islands and shores of lakes (cochas) and rivers. Also common in the seasonally-flooded (várzea) forest near rivers as well as swamp forests. Typically this is a bird of the lowlands, and most of the 'Yellow-billed Oropendola' complex is found in highlands. However, at least in Ecuador, the *angustifrons* Russet-backed Oropendolas range up to the subtropical zone around the vicinity of Baeza, Cascada San Rafaél and Cordillera de Guacamayos. In subtropical montane forests, Russet-backed Oropendola is found both in open forested situations and more commonly along the forest edge. It is not unlikely that the cutting of these highland forests has created more habitat for this species and has allowed it to range higher than in the past.

BEHAVIOUR Forages in small flocks, usually at canopy level. It may mix with other oropendolas, Yellow-rumped Caciques (13) and mixed foraging flocks led by Violaceous Jays (*Cyanocorax violaceus*). It gleans from leaves, and forages on fruit as well as probing into bark and dead leaves for insects. At all times of the year, this species roosts in large groups, often in dense vegetation on river islands. The mating system is polygynous with one or several males at attendance in a colony. Dominant males have the ability to monopolise most of the matings at the colony and there is a high degree of competition between males to achieve the role of colony master. When several males are present at a colony, typically only one, the dominant bird, will perch on the nest tree. Subordinate males perch on trees farther away. Males display as in other oropendolas, giving the explosive song as the tail is raised, the wings are partially spread and the body falls forward so that the bird may almost be upside down. The wings are not flapped at the end of the display. Less intense displays are given where the male does not topple forward. Courtship behaviour does not occur away from the nest tree. Mating also occurs at the nest tree. The females often forage in small flocks and these may be accompanied by a male; it is likely that these attendant males are subordinates attempting to mate with the females. It is unknown if this is a successful strategy, or indeed if this is what is occurring, particularly since courtship does not occur away from the nest tree.

NESTING Nesting occurs between November–January in eastern Ecuador. This species is colonial, and colony sizes tend to be small, usuallyof less than 10 nests but sometimes of many more. Some colonies are spread out over several trees with several males in attendance. It is not unusual for this species to nest in the same tree as a colony of Yellow-rumped Caciques; the two species usually coexist quite peacefully. However, in some cases the oropendola will attack nearby caciques and drive off the females from the nests nearest its own. These scuffles can become quite violent, with the oropendola even grabbing adult caciques in its claws and killing nestlings (Robinson 1985a). The empty nests presumably afford the oropendola some protection from nest predators, some of which give up after only searching a few nests in a colony. The nests are hanging baskets that range in length from 76 to 140 cm. Females conduct all of the nest building and incubation; nests may take up to 51 days to build, however, building is not continuous. Clutch 1–2 eggs which take 19–20 days to hatch. Nestlings fledge after 25 to 30 days. This species is regularly parasitised by Giant Cowbird (98).

DISTRIBUTION AND STATUS Common, largely a bird of the lowlands but ranges to 2000 m in E Ecuador (Baeza). Inhabits the lowlands east of the Andes in SE Colombia (Meta, Vaupés, Caqueta and Amazonas); E Ecuador (Napo and Pastaza) except the very far SE; NE Peru (Loreto); and W Brazil (W Amazonas) as far east as the junction of the Rio Purús and the Amazon River, where it is relatively rare.

MOVEMENTS Apparently sedentary, but local movements may occur in E Ecuador.

MOULT The complete definitive pre-basic moult occurs after breeding in adults. In the lowlands of E Colombia, birds are in moult June–July. There is no evidence that pre-alternate moults exist. The extent and timing of the first pre-basic moult is not known.

MEASUREMENTS Males: (n=10) wing 235.2 (208–257); tail 204.8 (185–225); culmen 59.2 (56–63); (n=9) tarsus 54.8 (50–60). Females: (n=10) wing 189.6 (176–200); tail 164.6 (156–176); culmen 49.0 (46–54); (n=9) tarsus 44.7 (42–50).

NOTES The 'Yellow-billed' form *neglectus* of Colombia apparently hybridises with *angustifrons* (Russet-backed); however, the extent of this hybrid zone (if any) is not known. A hybrid specimen has been described as being like *angustifrons*, but with a dark-based pale bill; another is similar but has an entirely yellow bill. However, there are specimens of *angustifrons* from Ecuador which are not hybrids and which show pale on the bill. The frequency and relevance of the existence of yellow-billed *angustifrons* needs to be studied, and may perhaps be age-related. In addition, some birds from Ecuador (Macas) appear to be intermediate, appearing most like *alfredi* 'Yellow-billed' Oropendola but with reduced yellow on the forehead, a mix of dark and yellow on the bill, and yellow outer rectrices. 'Yellow-billed Oropendolas' are birds of the Andean subtropical zone, quite unlike typical *angustifrons* which are only locally found in the subtropical zone. However, note that in the south, *alfredi* 'Yellow-billed Oropendola' ranges into the lowlands. Some vocalisations are similar between the two forms and others differ. Curiously, *alfredi* and *angustifrons* share a similar tail pattern while all of the highland forms share a different tail pattern. Full species status for 'Yellow-billed Oropendola' is probably warranted, but there is a need for more study, particularly where the two forms are reported to interbreed. It is probably quite likely that *alfredi* and *angustifrons* are more closely related to each other than either is to the highland forms.

REFERENCES Drury 1962 [behaviour, nesting], Fraga 1989 [nesting], Gyldenstolpe 1951 [measurements, geographic variation], Moore 1993 [voice], Parker 1985 [voice], Ridgely and Tudor 1989 [habitat, notes], Robinson 1985a [nesting], Zimmer 1930 [notes, hybrids].

5X 'YELLOW-BILLED OROPENDOLA' *Psarocolius (angustifrons?) alfredi* (DesMurs) 1856

Plate 4

This oropendola is the only one regularly observed in the Andean subtropical zone, but it ranges into the lowlands in southern Peru and Bolivia.

IDENTIFICATION This species and its close (conspecific?) relative, Russet-backed Oropendola (5), are very similar. 'Yellow-billed Oropendolas' from the Andes of Colombia and Ecuador are easily identified from the allopatric Russet-backed Oropendola by their yellow bills and foreheads. In addition, 'Yellow-billed Oropendolas' tend to be deeper chestnut on the upperparts and may have almost black heads depending on the subspecies. Russet-backed Oropendolas have more extensive yellow on the tail than all except *alfredi* 'Yellow-billed Oropendolas', the outermost two rectrices are largely yellow. In addition, their black bills are shorter and deeper than those of 'Yellow-billed Oropendolas'. 'Yellow-billed Oropendolas' show long, and finely-pointed bills which are not so deeply based as those of Black-billed Oropendola.

For the most part, this is the only oropendola found in the Andean highlands, except in Peru and Bolivia where Dusky-green Oropendola (4) is found. These two species are quite similar. However, Dusky-green Oropendola is more extensively olive, not as rufous as *alfredi* 'Yellow-billed'. Dusky-green has a greenish, rather than yellow bill. In addition, Dusky-green's head is the same colour as the back and there is little yellow on the forehead, in *alfredi* 'Yellow-billed' the head is noticeably greener and paler than the chestnut back and has noticeable yellow on the forehead. The tail of *alfredi* shows more yellow than on Dusky-green Oropendola, and the dark feathers are blackish rather than olive (see Dusky-green Oropendola for more details). In the Amazonian lowlands of Peru 'Yellow-billed Oropendola' overlaps with Casqued Oropendola (1). This species is smaller, shows an obvious bulge at the base of the culmen and lacks yellow on the forehead. In addition, Casqued Oropendolas have grey throats, olive-yellow breasts and largely chestnut bodies.

The Venezuelan *oleagineus* form of 'Yellow-billed Oropendola' is allopatric with similar oropendolas, probably overlapping only with the largely black Crested Oropendola (2). However, the plumage of *oleagineus* 'Yellow-billed Oropendola' is strikingly similar to that of Dusky-green Oropendola. Both of these species are largely olive-green with greenish-yellow bills and usually lack yellow on the forehead, even though it is sometimes present on Dusky-green Oropendola. Both forms also show reduced yellow on the tail and olive central rectrices. Range alone is sufficient to distinguish between these two forms, but their close visual similarity is noteworthy.

Note that montane forms, at least, have a weak flight, and glide frequently. Their comparatively short and broad wings give them a slow and heavy flight style.

VOICE The song varies geographically and more work is need to adequately assess which subspecies of this group have distinctive songs. In addition, individuals appear to have a repertoire of different song types. Further work may find that these song types are as variable as those of Dusky-green Oropendola (A. B. Hennessey pers. comm.). It appears that the song of lowland *alfredi* is similar to that of the lowland, black-billed form (species?) *angustifrons* (5) from the north, while songs of highland populations differ from those of the lowlands. The frequency of singing differs depending on the nesting stage. In *oleagineus*, the song is heard from January–September, peaking in April–May but this subspecies does not sing between October–January (Schäfer 1957).

Song: Like the nominate form of this complex, and Dusky-green Oropendola, the songs have a whip-like quality as well as liquid plopping sounds. Songs of Dusky-green and the Russet-backed/'Yellow-billed' Oropendola complex can be quite similar. The songs of *alfredi* from Peru and Bolivia begin with a rattle, as is typical of *angustifrons*. Songs of *alfredi* can be described as: *wwr-rrrrwwwoopeeeOOOO, rrrrrrpoopkeeeea, rrrrrr pooPKKwaaaa*, or *rrrrrrprrRRRRR-pOOaaa*. The songs last just over one second. In Bolivia, *alfredi* shows a considerable amount of variation in its songs, perhaps more than has been appreciated in the past (A. B. Hennessey pers. comm.). In Venezuela, *neglectus* gives a liquid *wwww-poookeeea*, lacking the rattle at the beginning of the song. Northern Venezuelan *oleagineus* has a variable song. The song can be described as a melodious, chime-like series of 3–6 notes which gain in volume and finish in an explosive sound. In addition, female *oleagineus* sing during the nest-building period, giving a song similar to that of the male (Schäfer 1957). There is little information on the songs of the Andean highland forms, but it appears that there are different song types involved. Hilty and Brown (1986) describe songs west of the Colombian Andes, belonging to the form *salmoni,* as a *Whoop-KE-chot!*. The form *atrocastaneus* from Ecuador sings a loud *BONK* and then a weaker, higher-pitched trill at the end (P. Greenfield pers. comm.)

Calls: A harsh *chugh* is commonly given; this call is almost exactly like that of *angustifrons* Russet-backed Oropendola, however several of the oropendolas have calls that resemble each other. Another call given by *alfredi* is a softer, muffled *phuup* or *whuup* and a sharp, hollow *chak.. chak..chak.* Venezuelan *oleagineus* gives a doubled *chak-chak,* as well as a more liquid *whoop* and a *queek-queek.* A loud *BRZZT!*, reminiscent of the call of Blue Grosbeak (*Guiraca caerulea*) has been heard from *alfredi* in Peru (J. Arvin pers. comm.).

DESCRIPTION A medium-sized to large oropendola with an obviously yellowish bill. The wing formula is: P9 < P8 < P7 < P6 > P5; P9 ≈ P3; P8–P5 emarginate. The bills of this form are not as deep as those of Russet-backed Oropendola, and therefore tend to look longer. The form *alfredi* is described here.

Adult Male (Definitive Basic): The bill is yellow, while the eye colour ranges from brown to grey-blue. The yellow forehead does not extend beyond the eye. The head is greenish-olive, becoming russet on the nape and towards the back. The back is russet with greenish feather tips. The rump and uppertail-coverts are chestnut. The underparts are greenish-russet, becoming pure russet on the crissum. The thighs are olive. The throat is pale, showing a yellow wash as does the face. The wings are blackish with olive tips to the lesser coverts. The median and greater coverts, on the other hand, have russet tips. The primaries, secondaries and tertials are blackish with the secondaries edged olive on the outer edge, while the tertials show russet edges. The tail is largely yellow, with the central two feathers (R1–2) brownish-olive; the outer vane and tip of the outer tail feather (R6) is brownish-olive, as are the extreme tips of the other yellow feathers (R3–5) (Figure 4.1).

Adult Female (Definitive Basic): Similar to the male, but much smaller. The eyes are brown or pale blue.

Juvenile (Juvenal): Similar to adults, but with duller plumage and brown eyes. In addition, the plumage texture is quite loose and this may be noticeable in the field, particularly on the flanks, given a good view. The duller plumage is most noticeable on the belly as juveniles may show greyish-brown rather than olive underparts, and browner napes, and duller russet or chestnut on the back. The bill is vivid yellow, brighter than on adults, and the yellow forehead patch is extensive, also more so than on adults.

GEOGRAPHIC VARIATION Six subspecies are recognised, differing in the extent of yellow on the forehead, and the colour saturation of the back and head. More than one species may be involved.

P. a. alfredi ranges from SE Ecuador (Morona-Santiago) south along the east slope of the Andes and adjoining lowlands through E Peru, except the far NE (Loreto), into Bolivia as far south as Cochabamba and Santa Cruz. The location of the contact zone between the yellow-billed *alfredi* and the black-billed *angustifrons* is not well understood but it appears to be in SE Ecuador. The head is grey-brown with a yellowish wash, the body is olive-russet, becoming rufous on the rump and vent. The head is paler in colour than the body. The tail is more extensively yellow than in other forms. The outer two pairs of tail feathers are largely yellow, not olive as in some of the other subspecies. This form is described above. The birds from Santa Cruz, Bolivia, were previously separated as a different race, *australis.* This form was said to be darker than typical *alfredi,* but individual variation in *alfredi* appears to explain *australis.*

P. a. salmoni can be found in Colombia, on both slopes of the Western Andes and on the west slope of the Central Andes, south to Nariño. It is found mainly in the subtropical zone, but does range to the tropical zone. This is a dark form with a distinctly dusky-olive head. The throat is paler. On *alfredi,* the back is darker than the head; the opposite is the case in *salmoni.* The darker head allows the yellow forehead to stand out much more prominently than on *alfredi.* In addition, the forehead patch reaches farther back past the eyes. The back is also a darker colour, almost chestnut. The underparts are correspondingly darker, with the thighs being dark olive. The eyes appear to usually (always?) be brown. The tail pattern (Figure 4.1) differs from *alfredi,* with the outer pair (R6) of tail feathers being completely olive-brown. The inner two (R1) are olive-brown, yet the next two (R2) have olive-brown inner vanes or may be entirely olive-brown. The tips of all others (R3–R5) are olive, while the rest of the feathers are yellow. The juvenile is like the adult, but shows paler cheeks. In addition, the back is browner and the underparts are grey-brown, lacking an olive wash. The throat is yellowish and the undertail-coverts are saffron, paler than the tawny undertail-coverts of the adults. The yellow forehead is extensive and reaches to behind the eye. Note that *salmoni* is reported to intergrade with *sincipitalis* in Huila, and with *atrocastaneus* in southern Nariño (Meyer de Schauensee 1946). Wing Formula: P9 < P8 < P7 > P6; P9 ≈ P4; P8–P6 emarginate.

P. a. atrocastaneus This subtropical zone form is an inhabitant of the west slope of the Andes in Ecuador. Like *salmoni,* this is a dark, rufescent form with a striking yellow bill and yellow on the forehead. However, it does not reach the intense colour saturation found in *salmoni.* The back is not as chestnut and the head not as deep olive. The outer tail feathers (R6) of this form are olive. The eye colour is brown, but some individuals may show a blue ring outlining the pupil which is encircled by the rest of the brown eye. Wing Formula: P9 < P8 < P7 > P6; P9 ≈ P4; P8–P5 emarginate.

P. a. sincipitalis Inhabits the west slope of the E Andes in Colombia from Santander to the head of the Magdalena valley. Now rare in Cundinamarca and may be extirpated (Olivares 1969). This form is similar to *salmoni* but paler, like *atrocastaneus*, and has more extensive yellow on the forehead and supercilium. The outer tail feathers (R6) are olive. Juveniles have grey-brown bellies, lacking the olive or green colour of adults. The nape of the juvenile is brown, not greenish as on the adult. The pale bill and yellow forehead is well-developed on juveniles. Juveniles have chestnut fringes to the tertials. Wing Formula: P9 < P8 < P7 > P6; P9 ≈ P4; P8–P5 emarginate.

P. a. neglectus This form is found in the subtropical zone of the Venezuelan Andes from Tachira to Lara, as well as in the Sierra de Perija on the border between Colombia and Venezuela, and along the east slope of the E Andes in Colombia from Cundinamarca south to Caqueta. This yellow-billed form is similar to *sincipitalis*, but the back is less chestnut, more olivaceous. In addition, the yellow on the forehead is more restricted and does not extend backward as a yellow supercilium. The throat and breast are concolourous; on *sincipitalis* the throat tends to be more yellow. The outer tail feathers (R6) are olive. Apparent hybrids of this and *angustifrons* Russet-mantled Oropendola are known from the locality of Florencia, Colombia, at 750 m. These birds lack yellow on the forehead but have yellow bills, and the body colouration is similar to *angustifrons*, but with more olive on the underparts.

P. a. oleagineus is found in the coastal mountains of Venezuela, west of the Jaracuy depression including Distrito Federal, east to Carabobo and the interior mountains of Cerro Golfo Triste along the Miranda–Aragua border. This is the most olive form, as well as the smallest in the group. Structurally it has rather weak feet (Schäfer 1957). Usually it lacks yellow on the forehead, this is the norm for females. The bill may look dull and greyish, usually it is pale pea-green with a dusky base. The eyes are always brown. In this subspecies the back is olive with a slight tawny tint, becoming tawny on the lower back and rump. The coverts are olive (some show tawny inner coverts), and the tertials have olive outer halves. The secondaries and primaries have yellow-olive edges. Like Dusky-green Oropendola, the tail is largely olive. The outer two (R5 and R6) tail feathers are olive, as is the central pair (R1); the rest are largely yellow with olive tips except for R2 which is yellow on its outer half and olive on the inner half. This form is noticeably variable in its plumage colour and has a tendency towards leucism, showing pale patches anywhere on the body (Schäfer 1957). Juveniles of this race are similar to adults in plumage colour, but have the entire bill yellowish and always sport a noticeable yellow forehead. In fact it is likely that a yellow forehead is a sign of immaturity in this form and that adults do not maintain a yellow forehead. Immatures keep the yellow forehead until almost a year old.

HABITAT Lives in openings in subtropical montane forest, cloud forest or lowland tropical forest. The types of openings used most commonly include the edges of settlements, towns or agriculture. Note, that it is only the form *alfredi* which is found in the tropical zone (lowland) forests edge, but it does range to the subtropical zone in Bolivia and Peru. The forms *neglectus, oleagineus, salmoni, atrocastaneus* and *sincipitalis* are almost entirely restricted to the subtropical zone. The subtropical races tend to be found most commonly near the edge of the forest, or where it has been thinned, but there does appear to be a tendency for this oropendola to be found near rivers. However, when foraging and in the non-breeding season, the highland forms are much more likely to be found inside the forest; only while breeding are they largely confined to openings.

BEHAVIOUR Forages in flocks of usually less than 10 individuals. Most of the foraging occurs at or near the canopy. It is especially fond of probing into large epiphytic bromeliads for food as well as flaking bark from tree trunks. The slightly more pointed bill than that of the *angustifrons* race of Russet-backed Oropendola may be an adaptation to foraging in bromeliad clumps. Apart from insect prey (their major food source), these oropendolas also take fruit. It is not uncommon for them to forage for nectar at large arboreal flowers. Venezuelan *oleagineus* rarely flocks with other species, but are indifferent to birds which may temporarily join them while feeding (Schäfer 1957). In Bolivia, *alfredi* roosts in large groups.

The breeding system is polygynous. As in other oropendolas, the sex ratio is highly skewed towards females, as high as 6:1 female to male ratio. Although more than one male is present at the colony, only one appears to be dominant. Males spend a great deal of time displaying and sitting at the nest tree. They appear to be attracted by the sight of nests, whether of their own species or of another sympatric oropendola species. The typical pattern is for males to establish themselves and begin displaying at the colony before the females arrive to build the nests. All of the work associated with nesting and rearing of young is performed by the female.

The bow display is exaggerated and dramatic. The male bends down and leans forward, bowing the head deeply and actually falling over the perch so that he is nearly upside down as he utters the song, while keeping the body feathers fully erect. The tail is elevated and thrown forward during the bow. The crest is raised, particularly during the final half of the display. Sometimes the male does not fall over, but the head is bowed forward so that the bill is between the legs. Males display almost constantly, sometimes every 5–10 seconds! They may perform a total of 80–100 displays per hour, a much higher display rate than in Crested Oropendola (Schäfer 1957). Before mating, males chase the female into the forest. Copulations occur near the nest tree, but not at the tree itself.

NESTING As expected this oropendola is colonial. The colonies are small, rarely as large as 25 nests and sometimes as small as only two or three nests. Lowland *alfredi* has larger colonies than the highland forms. In *oleagineus*, the colony is typically spread out over several trees (up to 10, extending along an area of 200 m) and the colony site is not necessarily the same year after year. Nests in these colonies tend to be placed 5–8 m above the ground, rarely as high as 15 m (Schäfer 1957). The nests of the highland forms are often positioned directly over roads, or over streams or torrents. Mixed colonies with Yellow-rumped Caciques (13) are known, for *alfredi* and *oleagineus*. At certain elevations where the two species are found together, this oropendola may uncommonly nest in the same tree as Crested Oropendola, at least in N Venezuela (Schäfer 1957).

The timing of nesting is variable depending on local differences in rainfall. In Colombia Hilty and Brown (1989) mention juveniles being fed by adults in late June in the W Andes above Cali (*salmoni*); in April there was an active colony in the Upper Anchicaya valley, Valle (*salmoni*). At Milligalli, Colombia, Goodfellow (1901) saw an active colony in September. In N Venezuela, *oleagineus* nests February–August; nest building begins in February and nests are being lined with leaves in mid March. Fledging peaks in May. (Schäfer 1957)

Schäfer (1953, 1957) best summarises the nesting behaviour of this species, with reference to the race *oleagineus*. The nests are long, hanging baskets with an entrance near the top. Nest length 76–146 cm. Nest construction takes 114–153 days. The later nests are typically those of young females. It takes a female an average of 21 days to build a nest, with 13 days to build replacement nest. Nests are lined with leaves of heliconias and bromeliads, and placed in the lower branches of trees rather than near the crown. However, where 'Yellow-billed Oropendolas' nest with Crested Oropendolas they may nest at the top of the nest tree. Lays 2 eggs, although sometimes the clutch is only 1, over 4–6 days and incubate for at least 18–19 days; incubation begins after the first egg is laid. Curiously, it has been noted that of the two eggs, one is consistently and noticeably smaller than the other. Rarely the clutch size may be as high as three (recorded in two of 150 nests). The young may fledge in 30 days; however there is a high mortality of male nestlings while in the nest. It is quite unusual for more than one young to fledge from a nest.

The eggs of *salmoni* are salmon-coloured with dark spots and blotches on the wide end (Sclater and Salvin 1879). The eggs of *oleagineus* are variable in colour, being white to pale green with purplish or reddish-brown spots and scrawls, mainly clustered nearer to the wide end (Schäfer 1957). Giant Cowbird (98) is known to parasitise nests of *alfredi*, but it is largely allopatric with the montane forms and there are no records of it parasitising those subspecies.

DISTRIBUTION AND STATUS Uncommon to locally common. Ranges along the Andes usually above 700 m, but *alfredi* ranges lower than this in Peru and Bolivia. In addition, *oleagineus* is found as low as 200 m in the humid forests of N Venezuela (only on the north slope of the coastal range, higher on the south slope) while further south in Venezuela the species is not usually found below 700 m. Usually observed below 2000 m but not unusual to 2500 m. Ranges from Sucre in Venezuela west and south along the coastal mountains and Venezuelan Andes to the Perija mountains on the border between Colombia and Venezuela. Continues south along both slopes of the central and western Andes of Colombia as well as the W slope of the east Andes to the upper Magdalena valley, south along the west slope of the Andes in Ecuador; continues south from SE Ecuador on the east slope of the Andes along E Peru to the east slope of the Bolivian Andes to N Santa Cruz.

MOVEMENTS Usually considered sedentary, but may undergo minor altitudinal movements in the coastal range in N Venezuela (Schäfer 1957). In the Andean foothills of Bolivia, at the edge of the lowlands in Rurrenabaque, there is a noticeable seasonal pattern of occurrence. This oropendola is exceedingly common during the breeding season (May–November) but is almost absent January–early April (A. B. Hennessey pers. comm.). This is strong evidence of either migration or a local movement, possibly to the lower elevations.

MOULT The definitive pre-basic moult is complete, taking place after breeding is finished. Pre-alternate moults appear to be lacking. The extent of the first pre-basic moult is unclear, but it is known that as a minimum the body feathers are replaced during this moult. Adults from Santander and Huila, Colombia (*sincipitalis*), undergo the definitive pre-basic moult July–September; however, *salmoni* individuals from Colombia are undergoing their definitive pre-basic moult late November–January. *Atrocastaneus* from Ecuador moult in August. *Alfredi* in SE Ecuador moult April–July; similarly in the highlands of S Peru this species moults June–July. *Oleagineus* undergoes the definitive pre-basic moult in November (Wetmore 1939), but sometimes as early as mid-September. This is in contrast to Schäfer (1957) who notes that the extreme dates for completion of the moult are July 20 and October 5, and that most adults have finished by the start of September, late moulting individuals being younger ones (first pre-basic). The timing of the moult varies depending on the colony.

MEASUREMENTS ***P. a. alfredi***. Males: (n=8) wing 235.6 (209–252); tail 199.1 (172–219); culmen 61.1 (59–65); tarsus 50.6 (46–53). Females: (n=10) wing 185.3 (175–194); tail 162.2 (142–176); culmen 46.7 (43–49); tarsus 42.7 (39–45).

P. a. salmoni. Males: (n=10) wing 232.8 (204–265); tail 216.5 (181–249); culmen 59.8 (54–62); tarsus 52.2 (47–58). Females: (n=10) wing 184.2 (173–198); tail 176.1 (165–190); culmen 48.5 (47–51); tarsus 41.5 (39–43).

P. a. atrocastaneus. Males: (n=4) wing 236.8 (233–240); tail 211.5 (205–220); culmen 59.8 (58–62); tarsus 49.5 (46–55). Females: (n=7) wing 187.7 (180–195); tail 167.7 (157–183); culmen 50.3 (49–52); tarsus 42.1 (38–49).

P. a. sincipitalis. Males: (n=6) wing 227.8 (215–236); tail 196.5 (187–227); (n=4) culmen 54.5 (52–57); (n=6) tarsus 47.3 (44–50). Females: (n=10) wing 180.4 (172–190); tail 160.3 (152–175); culmen 45.8 (43–47); tarsus 39.6 (36–44).

P. a. oleagineus. Males: (n=3) wing 229.3 (210–265); tail 188.7 (182–195); culmen 57.0 (51–66); tarsus 49.3 (46–53). Females: (n=2) wing 174, 204; tail 166, 177; culmen 44, 50; tarsus 40, 46.

NOTES The form *neglectus* of Colombia apparently hybridises with *angustifrons*, however the extent of this hybrid zone (if any) is not known. In SE Ecuador, birds appearing intermediate between *alfredi* and *angustifrons* have been recorded. For more discussion see Notes under Russet-backed Oropendola.

The extent of geographic variation in this form is extreme, with some forms blending into others and some maintaining themselves as discrete entities near the contact zone. The northern form, *oleagineus*, is quite different from the others but appears to be linked to them by the intermediate *neglectus*. The understanding of the systematics of this complex is far from resolved. It is not inconceivable that even within 'Yellow-billed Oropendola' more than one species may be present! The division suggested here, based on bill colour, may be a superficial one in nature and entirely incorrect. In many ways *alfredi* and *angustifrons* are more similar to each other than they are to the highland subspecies. The phylogenetic break, if any, may be between the highland and lowland forms. Schäfer (1957) notes that the uropygial glands of this species produce a rather smelly oil which gives members of *oleagineus* (and presumably the entire group) a musky odour in life.

REFERENCES Bob Behrstock personal recordings [song], Chapman 1914 [geographic variation] 1917 [geographic variation], Fjeldsa and Krabbe 1990 [geographic variation], A. Bennett Hennessey recordings [song, calls], Ridgely and Tudor 1989 [geographic variation], Schäfer 1953 [nesting, behaviour, *oleagineus*], Sclater and Salvin 1879 [nesting], Zimmer 1930 [notes, hybrids].

6 CHESTNUT-HEADED OROPENDOLA *Psarocolius wagleri* Plate 1
(Gray) 1845

A small, long-winged oropendola largely confined to Central America.

IDENTIFICATION Chestnut-headed Oropendola is a very long-winged, small oropendola with a noticeable expansion (casque) at the base of the culmen. It shows a largely black plumage with dark chestnut on the head, rump and undertail-coverts. Eye colour is bright sky-blue. Chestnut-headed Oropendola is broadly sympatric with Montezuma Oropendola (7) in Mexico and Central America. Montezuma Oropendola is much larger, but due to the high degree of sexual size dimorphism this feature can be difficult to use in identification. Chestnut-headed Oropendola is a black bird with a brownish or chestnut head; Montezuma Oropendola is largely chestnut with a black head. However, at a distance it is difficult to see the actual colour of these two species and they may appear entirely blackish. The bill of Chestnut-headed Oropendola is pale ivory-coloured, quite unlike the reddish-tipped black bill of Montezuma Oropendola. The face of Montezuma Oropendola has bare patch of pale blue skin while the face of Chestnut-headed Oropendola is fully feathered. These differences in bill and facial pattern are also useful in separating Chestnut-headed from Black (11) and Baudo (8) Oropendolas, with which it may be sympatric. Note that Chestnut-headed Oropendola, particularly the male, has very long and tapered primaries. In flight, it appears much more pointed-winged than the rounded-winged appearance of Montezuma Oropendola. In addition it has a quicker wingbeat.

In Colombia, the similar Crested Oropendola (2) poses another identification problem. Like Chestnut-headed Oropendola this is a dark oropendola with a pale bill; note however, that Crested Oropendola has an entirely black head. Crested Oropendola is also larger, but again beware of size dimorphism. Chestnut-headed Oropendola is a chunkier-looking bird, not as long and elegant as Crested Oropendola. Crested Oropendolas do not show the broadly-expanded casque at the base of the culmen that Chestnut-headed Oropendola shows. Otherwise, Chestnut-headed Oropendola is only likely to overlap with yellow-billed forms of Russet-backed Oropendola (5X). These have largely russet, chestnut and olive-coloured bodies, with yellow foreheads, quite unlike Chestnut-headed Oropendolas. The two species are not likely to be confused.

VOICE Song: The song is recognisable as that of an oropendola in its explosive, crashing quality. It begins with a soft note *guaa*, sometimes given more than once, followed by a longer, crackling, crashing note. For example: *guu-guu-PHRRRRTTTT*. Males from different colonies have song dialects specific to the colony.

Calls: Several calls are shared by both sexes. The *chuck* call is the common contact call, and a *cack-cack* is given when alarmed. A whining call is thought to be a threat call, given when males are competing for a female, or when females are competing for a nest site. Females give a gurgled *wee-chuck-chuck-chuck*. Males give a liquid *guaa* or *wauu*, often interspersed with *chuck* calls. The male's deeply-emarginated primaries are noisy in flight, particularly during displays. Females lack such pointed primaries and make no noticeable noise when flying. Slud (1964) also noted that this oropendola gives an oriole-like chatter.

DESCRIPTION The long wings are more pointed, and the outer primaries more incised, than in other oropendolas. Wing Formula: P9 < P8 < P7 > P6; P9 ≈ P5; P8–P5 emarginate; P9–P7 notched. The wingbeats are quick, unlike those of larger oropendolas. The dark central tail feathers (R1) are obviously shorter (up to 20 mm) than the longest yellow rectrices (R2–R3).

Adult male (Definitive Basic): Bill pale greenish ivory, sometimes with a dusky tip, eyes whitish-blue. The bill is extremely wide and deep at the culmen base, forming an obvious casque. The head, neck, nape and breast are dark chestnut-brown. There is a thin, wispy crest of four feathers. The back and scapulars are black with an obvious bluish gloss. This colour extends to the lower breast and upper belly on the underparts. The flanks and crissum are chestnut, while the thighs are blackish-brown. The rump and lower back are also chestnut. The wings are black and very long. The black lesser and median coverts are glossed blue, like the back, while the greater coverts and primaries are not nearly as glossy. The tail is yellow, with blackish central tail feathers which are somewhat shorter in length than the longest yellow feathers. The outer rectrix (R6) has a dull blackish outer edge. The legs and feet are black.

Adult female (Definitive Basic): Like the adult male, but smaller and slightly duller in colour, particularly with respect to the body gloss. The crest is lacking, the frontal shield (casque) is less developed and the primaries are not nearly as long and tapered as in the male. The soft part colours are as in the male.

Immature Male (First Basic): Very similar to the adult male, but smaller. In addition, the primaries are not so long and pointed.

Juvenile (Juvenal): The bill is noticeably duller in colour than in adults and the eyes are brownish. Juveniles resemble the adults but are duller with the black of the body largely replaced by sooty-chestnut, without any iridescence. They may show some yellow on the supraloral area. Juvenile plumage is not held for long.

GEOGRAPHIC VARIATION Two subspecies are commonly recognised.

P. w. wagleri is found from Mexico south to Guatemala, Belize, N Honduras and NE Nicaragua. It is described above.

P. w. ridgwayi occurs from SE Honduras and NC Nicaragua south to Ecuador. It resembles *wagleri*, but is blacker below and the head is paler brown. The culmen is broader at the base, creating a more noticeable casque, and is appreciably bulged in profile. However Wetmore *et al.* (1984) did not notice any geographic variation in a large series of birds ranging from Mexico to Panama and doubted the existence of *ridgwayi*, unless it is restricted to South America. They did not have enough South American material to address this question.

HABITAT Inhabits humid lowland tropical forest, keeping to the canopy in contiguous forest and venturing to the foothills and subtropical forest (at least in Costa Rica) (Slud 1964). It is also to be found in forest edge situations and openings in the forest, both natural and man-made. It shows a preference for rivers cutting through contiguous forest. Chestnut-headed Oropendola also lives in older patches of second growth and plantations. In addition it is found in clearings where large trees have been left standing. This species is not shy and will nest in isolated trees near buildings, settlements or roads. Colonies in 'unnatural' situations such as these are quite common. This oropendola is absent from dry tropical forests (Slud 1964, Wetmore *et al.* 1984).

BEHAVIOUR Foraging occurs mainly in single sex flocks. This species includes a high fruit content in its diet. Birds are most commonly observed high in the canopy. Stomach samples show that this oropendola forages largely on large arboreal insects (for example, grasshoppers and katydids but also scorpions!) as well as fruits and berries (Hallinan 1924, Wetmore *et al.* 1984). They also nectar at flowers (Leck 1974). Apparently, there is little interchange between individuals of different colonies. Mating is highly polygynous, and is likely polyandrous as well (polygamy). The sex ratio is highly skewed in the direction of females; in colonies in Panama there are five times the number of females than males. Copulation does not occur at the colony site. Males spend most of their time displaying to females. They perch just above the female, while she is perched or in the nest. He ruffles his plumage, particularly the lower back and rump, and spreads the tail as he bulges out his face feathers, making the bluish eyes prominent, and elevating the crest. He will then give the primary song as he fluffs the full body feathers and rises on his toes. Males do not bow forward or tip over like many other oropendolas. Aerial chases are often observed early in the season, often with more than one male chasing the female through the forest. Several males may be at attendance at a colony tree. Each will focus his attentions on a particular female while she is laying eggs. His attention turns to another female as soon as the first female begins to incubate.

NESTING The nesting period commences synchronously with the dry season in Panama (December) and continues until May or June. A very small number may also breed in the wet season (June–July); it is not known if these are merely second clutches, late birds or individuals following a completely different strategy. In Costa Rica, the season is similar, ranging from January to June. The beginning of the season is heralded by males returning and calling from traditional nesting trees. The females begin to congregate a few days later and start the nest-building process (Chapman 1931b). Colonies may be as small as four nests, or as large as 132 nests, but average 30–40 nests. The nest is a long hanging basket 0.6–1 m in length, woven of thin vegetable fibres with an entrance at the top. Nests are not infrequently placed near wasp nests. Clutch typically 2 eggs, with an average fledging rate of 0.40 chicks per nest (Smith 1983, based on 10 years of data). The eggs are pale blue with blackish spots, more heavily concentrated at the broader end. Incubation lasts approximately 17 days. Nestlings take approximately 35 days to fledge. Parasitism by botflies (*Philornis*), which grow as larvae under the skin of the victim, are the main source of nestling mortality. This oropendola is known to have been parasitised by both Giant (98) and Bronzed (99) cowbirds; the latter unsuccessfully (Friedmann and Kiff 1985).

DISTRIBUTION AND STATUS Uncommon to locally common. A bird of the lowlands, rarely reaching 1200 m in elevation. Ranges from S Mexico (E Puebla, C Veracruz) east along the wet forest zone on the Caribbean slope of Tabasco and Chiapas through N Guatemala, S Belize, N Honduras, N Nicaragua, E Costa Rica; also along the Pacific slope of southernmost Costa Rica (Golfo Dulce region), and along both slopes of Panama into the Pacific lowlands of Colombia and northernmost Ecuador. In Colombia, the range extends inland from Chocó into Antioquia.

MOVEMENTS Sedentary, with no evidence of migration. However, during the non-breeding season flocks may wander widely. Probably most oropendolas are long-lived, but so few have been banded that this has not been confirmed; however, a banded Chestnut-headed Oropendola was recorded to have lived for at least 26 years! This is the oldest known icterid.

MOULT A complete definitive pre-basic moult occurs after the breeding season. Timing varies due to the local timing of breeding. In western Panama, the timing of the pre-basic moult is early November–mid December. In W Colombia, the definitive pre-basic moult occurs during October. No information exists on the extent of first pre-basic moults. No evidence exists for the presence of pre-alternate moults.

MEASUREMENTS Males: (n=10) wing 214.1 (187.0–227.0); tail 133.5 (124.8–144.0); culmen 66.1 (62.4–72.4). Females: (n=10) wing 152.1 (145.4–161.9); tail 110.0 (100.0–117.3); culmen 49.9 (46.6–52.0).

NOTES Often classified in its own genus, *Zarynchus*, based on its small size, large frontal shield and extremely elongated primaries. Also known as Wagler's Oropendola. The bow display of Chestnut-headed Oropendola is quite unlike that of most other oropendolas, suggesting that it is indeed more distantly related to the rest.

REFERENCES Chapman 1931b [description, behaviour, nesting], Fleischer and Smith 1992 [nesting, parasitism], Howell and Webb 1995 [description] Smith 1983 [behaviour, nesting], van Rossem 1934 [geographic variation], Wetmore *et al.* 1984 [geographic variation, measurements].

7 MONTEZUMA OROPENDOLA *Psarocolius montezuma* (Lesson) 1830 Plate 3

The common oropendola of Central America, it is not a bird that is overlooked by even the most casual observer.

IDENTIFICATION A large chestnut oropendola with a black head. The bill is black with a reddish tip, and the head is adorned with a pale blue naked patch of skin and a pink wattle. Throughout most of its range, the only other oropendola present is Chestnut-headed Oropendola (6); it is unlikely to be confused with any other species. Both are large black and chestnut birds, but note that Montezuma Oropendola is largely chestnut with a blackish head while Chestnut-headed Oropendola is mainly blackish with a chestnut head. In addition, Chestnut-headed Oropendola has a pale ivory-coloured bill and has a fully feathered face, unlike Montezuma's black and red bill, and blue and pink bare face patch. In flight, many of these differences are visible, but Chestnut-headed Oropendola also has relatively longer and more pointed wings than Montezuma Oropendola as well as a quicker wingbeat. In the Panama Canal Zone, Montezuma and Crested Oropendolas (2) overlap. Crested Oropendola is almost entirely black with an ivory-coloured bill and no bare face patches. Montezuma Oropendola has dark eyes, while Crested and Chestnut-headed Oropendolas have pale blue eyes which are visible from a distance.

VOICE **Song**: A strange-sounding, two-part song is given, accompanied by the bow display. The first part of the song is a series of bubbly, conversational, tinkling notes overlaid by metallic sounds that ascend the scale as they increase in volume. The second part is a loud, descending, liquid gurgle that carries for a long distance. The complete song, which lasts under 3 seconds, can be interpreted as: *tic-tic-glic-glac...gluuuuuuluuuuu* or *ticki-ticki-ticki-ticki-WHAAAoooo*. The conversational sounds which begin the song set it apart from songs of other oropendolas except those of Black Oropendola (11) which are quite similar. Display songs are loud and are heard without much trouble at 150 m distance.

Calls: A common call is a whining *waaaah*. The alarm call is a sharp *cack* and females give a *cluck*. A frog-like *crrrrk* is sometimes uttered. An odd sound is given by males which sounds like the ripping of cloth.

DESCRIPTION A large oropendola with a deep bill and culmen, showing a tendency to form a slight casque. The tail is graduated, but the central rectrices are slightly shorter than the others. Wing Formula: P9 < P8 ≈ P7 ≈ P6 > P5; P9 ≈ P4; p8–p5 emarginate. Emargination and notching are both more developed in old males.

Adult Male (Definitive Basic): The bill is black with the terminal half or third reddish or orange. The bare facial patch is pale blue on the cheek, greyish-blue around the eyes and pinkish on the lower edge of the malar area. This pink 'wattle' is very swollen on adult males. There is a small bare, pink, patch at the base of the culmen. The eye is brown. The head, neck and upper breast are black while the rest of the body is chestnut, darkest on the underparts. The black of the head and the chestnut of the body blend into each other, not showing a distinct break where the two colours come together. The wings are chestnut with the primary tips blackish. The underwings are blackish-brown. The graduated tail is largely lemon-yellow with the middle pair of rectrices (R1) blackish. The strong legs and feet are black.

Adult Female (Definitive Basic): Similar to male, but noticeably smaller. There are some slight plumage differences like the colour tone of the black on the head, which may look duller and browner on the female. Similarly, the middle tail feathers are browner on the female than the male. Often the chestnut of the belly and thighs can be darker on females, sometimes approaching black. These differences are not absolute, females are best identified by their size and behaviour at the colony.

Immature Male (First Basic): Similar to adult male, but immatures are smaller. In addition, they possess a smaller pink wattle, below the blue face patch, which is not swollen as on adult males.

Juvenile: Similar to adults but duller, with black colours replaced by dull sooty-black. The bill may be paler black with a duller tip and show less of a discrete demarcation where these two colours meet.

GEOGRAPHIC VARIATION Monotypic. No described geographic variation.

HABITAT Inhabits humid forest, secondary forest, forest edge and corridors of gallery (riparian) forest. It is rarely observed deep in forest, except near clearings, and appears to be primarily an edge species. In cleared areas it is associated with banana, cacao plantations or shade-grown coffee plantations; this type of farming maintains enough trees for Montezuma Oropendola to feed in. Isolated trees are preferred for nesting and these may be well away from the forest edge. It has a preference for roosting in dense bamboo thickets, particularly those by water.

BEHAVIOUR The mating system of Montezuma's Oropendola is highly polygynous. Only a small proportion of the males perform most of the copulations, based on the observed copulations at colonies. Dominant males defend nesting aggregations of females from other males, and compete with each other for status and access to females. This social system (female defense polygyny) is rare in birds, and appears to be restricted to the Icteridae, but is common in many species of mammals, for example: Red Deer/Elk (*Cervus elaphus*) and Elephant seals (*Mirounga* sp.). As in those species, male Montezuma Oropendolas are much larger than the females. Males can achieve this level of female monopolisation partly because within a colony nesting is staggered, not synchronous, thus 'alpha' males can shift their focus to the females or group of females that are coming into receptivity at a particular time. A rigid dominance hierarchy is established, with the 'alpha' male (most dominant) excluding all other males from the colony. The body size (weight) of males appears to play an important role in competition for mates, as high-ranking males are heavier than low-ranking males. The extreme sexual-size dimorphism in this species may have evolved due to this competition between males. Fighting is rare, but when it does happen it is severe, with males grappling in mid-air and falling to the ground. When the 'alpha' male leaves to feed, the 'beta' (second in the line of dominance) male takes over and excludes all those under him on the hierarchy, until the 'alpha' male returns. 'Alpha' males spend most of the day at the colony, particularly when the greatest proportion of females are sexually receptive. The number of males at a colony is correlated with the number of receptive females present. Colonies that are very large may have more than one male defending a group of females. Females appear to be sexually receptive when lining their nests with leaves. Copulation by dominant males take place at the nesting colony. However, in reality, subordinate males do achieve some copulations. These subordinate males court females away from the colony and are rarely observed copulating. A recent study (Webster 1995) determined the paternity patterns of nestlings based on their DNA 'fingerprint'. Of the 21 nestlings sampled, seven were sired by the 'alpha' male, four by the 'beta' male and the remaining 10 were sired by lower ranking males. This contrasts with the observation that 90–100% of the copulations at a colony are performed by the alpha male. Presumably lower ranking males mate with females away from the colony. While the actual mating success of the 'alpha' male is substantially lower than calculated from the number of observed copulations, he probably sires many more young than any subordinate male. There is circumstantial evidence that in large colonies, some low ranking males may consort with only one female.

During the pre-copulatory display, the male approaches the female and bows to an angle 45° below the horizontal, once in front of the female. The male is silent and does not ruffle the plumage. He will then move around and rapidly peck at the outer yellow tail feathers of the female. A more intense display follows, with the male extending the neck and pointing the bill downward, while ruffling the nape feathers, spreading the tail and drooping the wings. Often the 'shredding cloth' call is given at this point. If the female does not leave, the male will mount and mate, without uttering vocalisations. The female does not display to the male, other than by lifting her tail and bill, and there are no post-copulatory displays or vocalisations. In larger colonies, low-ranking males often disrupt the copulation attempts of higher ranking males. The full song display ('bow display') in this species resembles that of other oropendolas. The male bows forward, cocks the tail upwards and spreads the wings as he sings. The neck does not tend to be ruffled, nor does the male flap the wings during the display. During a high intensity display the male will flip almost upside down while pointing the yellow tail upwards.

Females forage away from the colony in small groups, averaging five females, and stay hidden in the canopy. These female groups are often attended by one or more males which actively display, and presumably mate with some of the females. Males forage alone, unless they have joined a female foraging party. Because they are smaller than males, females are better able to forage at the tips of branches, probing into dead leaves and gleaning from live leaves. Males, on the other hand, stay closer to the trunk and larger branches while feeding, as they are heavier and less agile. Fruits and large insects appear to be the mainstay of this species. There is one report of a Montezuma Oropendola which killed a juvenile Black-faced Grosbeak (*Caryothraustes poliogaster*), presumably to eat it (Wolf 1971).

NESTING In Costa Rica, breeding occurs in the dry season (January–May, with lesser numbers breeding into September), often beginning just as the rains stop. Colonial nesting occurs as a rule, with up to 172 nests in a colony tree; thirty or fewer nests are more common, and there may be as few as three. Within a colony tree, nests tend to be clumped rather than evenly spread. Up to 89% of nests in a tree are within 'clumps' (Webster 1994b). Colonies tend to be located in the same general area year after year, often, but not always, in the same tree. The structure of the nesting tree is stereotypic. It is usually an isolated, umbrella-shaped tree, often in cleared land or near the forest edge. However, there has to be forest nearby for the oropendolas to forage. On rare occasions, a large colony will overflow to nearby trees. The nests are long and basket-like, as in other oropendolas, and are built in two to three weeks by the female alone. Nests are up to 120 cm in length. Incubation lasts for approximately 15 days and the nestling fledges after 30 days. Clutch size 2 eggs; however, more than one nestling rarely fledges per nest. The eggs are white or pinkish, with large dark spots. Within a colony nesting is not synchronous, although the timing of nesting is synchronous in nest clusters within the colony. Clustering of nests probably helps to prevent nest predation, and may also help decrease brood parasitism by Giant Cowbird (98). Cowbirds are chased by nesting oropendolas, particularly when they approach the female owner's nest. Therefore, cowbirds have less opportunities to approach a clump of nests, than isolated ones, as more females will be available to chase them off.

DISTRIBUTION AND STATUS Common to uncommon in tropical lowlands, rarely to 1500 m. Range extends from the Atlantic slope of Mexico as far north as S San Luis Potosí, Querétaro and Hidalgo southeast along Veracruz and N Oaxaca to Tabasco, N Chiapas, S Campeche and S Quintana Roo but not found in the Yucatán Peninsula proper. In Guatemala, found only in the northern half (Petén and Atlantic lowlands), absent from El Salvador, and found only in the Atlantic lowlands of Honduras. It is found in both the Atlantic and Pacific lowlands of Nicaragua. However, in Costa Rica it may be found most commonly throughout the Caribbean lowlands and central valley, while it is local in the extreme NW (Guanacaste), in the Pacific lowlands. In Panama, this oropendola may be found primarily in the Caribbean lowlands of Bocas del Toro and to a lesser extent east to the Canal Zone; it is rare on the Pacific slope having been found in Veraguas and the Canal Zone, and there is also one record far to the east from Chepo, Panama province (Ridgely and Gwynne 1989). Note that north of Nicaragua, Montezuma Oropendola is absent from the Pacific slope, and south of here it inhabits the Pacific slope only in NW Costa Rica with scattered records from Panama.

MOVEMENTS Sedentary, but wanders in flocks during the non-breeding season. A specimen from Chepo (Panama province), Panama, is well east of the regular range and may represent a stray record or perhaps the species is resident east of where presently known, but in low numbers.

MOULT. The complete definitive pre-basic moult occurs after breeding, the timing of which depends on location and the start of the dry season. In southern Mexico, the pre-basic moult occurs in mid October. Pre-alternate moults appear to be lacking. The extent of the first pre-basic moult is not known, particularly with respect to whether any of the flight feathers are moulted.

MEASUREMENTS Extremely sexually dimorphic, males may weigh more than twice as much as females. Young males are smaller than adult males. Males: (n=10) wing 256.0 (230–273.1); tail 195.5 (176–215); culmen 76.0 (73–78.7); tarsus 57.1 (54.6–61). Females: (n=10) wing 196.6 (188.0–205.7); tail 154.3 (144.8–162); culmen 57.7 (55–61.0); tarsus 46.0 (44–49).

NOTES This and all the bare-faced oropendolas were previously separated into the genus *Gymnostinops*, and are sometimes still placed in this genus (Sibley and Monroe 1990 for example). It has also been suggested that this group should all be lumped as one species, Great Oropendola (name proposed by Ridgely and Tudor 1989), which in our opinion is a rather extreme view given the differences in morphology, behaviour, voice and distribution patterns.

REFERENCES Davis 1972 [voice], Fraga 1989 [nesting], Hardy *et al.* 1998 [voice], Howell 1964 [behaviour], Howell and Webb 1995 [behaviour, nesting, distribution], Moore 1994b [voice], Skutch 1954 [behaviour, nesting], Webster 1994a and 1994b [behaviour, nesting] 1994c [nesting, behaviour, parasitism], 1995 [behaviour, mating system], Wolf 1971 [predatory behaviour].

8 BAUDO OROPENDOLA *Psarocolius cassini* (Richmond) 1898 — Plate 3

This rare Colombian endemic has only been encountered three times in the wild. Also known as Chestnut-mantled Oropendola.

IDENTIFICATION A large black and chestnut oropendola with a largely black bill. It has a pink face patch. The range of this rare species should only overlap with Chestnut-headed Oropendola (6), however, both Black (11) and Crested Oropendolas (2) are found nearby. Black Oropendola is most similar, but it is marginally smaller than Baudo Oropendola and with a relatively smaller bill. The back, wings and flanks are chestnut on Baudo Oropendola, while Black Oropendola shows a black upper mantle, black flight feathers and tertials. Note that both species show chestnut on the lower mantle and rump. However, on Black Oropendola there is a contrast between the chestnut wing-coverts and the black primaries, secondaries and tertials. As noted above, these areas are all evenly chestnut on Baudo Oropendola. Note that the black regions of Black Oropendola are glossy, not matt as in Baudo Oropendola. In addition, the bare face patch of Baudo Oropendola is pinkish, while it is blue, with a pink wattle and culmen base in Black Oropendola. The Black Oropendola's dark central tail feathers are strikingly shorter than the longest yellow rectrices, while on Baudo Oropendola the central tail feathers are nearly the same length as the yellow rectrices. Crested and Chestnut-headed Oropendola lack a bare face patch and have entirely pale bills. Furthermore these two species are noticeably smaller than Baudo Oropendola and are nearly entirely black in colour.

VOICE There are no known recordings, or published descriptions of the song or calls of this species.

DESCRIPTION A large, long-billed oropendola. The bill is proportionately wider at the base than in its relatives. The crest in this species is longer and more pronounced than in Montezuma Oropendola (7). Wing Formula: P9 < P8 < P7 > P6; P9 ≈ P4; P8–P6 emarginate.

Adult Male (Definitive Basic): Bill black with an orange tip, restricted to less than one-third of the bill length. The base of the frontal shield also shows some orange colour, due to a bare patch of skin there. The bare facial patch is pinkish. There is a small, pinkish, malar wattle present. Head and underparts black, except for the undertail-coverts. The chestnut is restricted mainly to the upperparts, from the back to the uppertail-coverts. The undertail-coverts and flanks are also chestnut. The wings are chestnut with the tips of the primaries blackish, but the wings appear completely chestnut when closed. The tail is lemon-yellow with the central two tail feathers blackish; these are slightly shorter (20 mm) than the longest yellow feathers. The legs are black and the eyes dark.

Adult Female (Definitive Basic): similar to the male, but much smaller and the black colours are duller in colour.

Juvenile (Juvenal): Unknown.

GEOGRAPHIC VARIATION Monotypic.

HABITAT Lowland forest and forest edge. The only recorded observation of a flock took place in riverside forest on sandy ground.

BEHAVIOUR One flock of 10 birds was observed high up in the canopy. This flock was observed at the shore of the Río Dubasa, a tributary of the Río Baudo (von Sneidern 1954). The collector of the type specimen noted this species as being very shy, and he only saw one individual.

NESTING Unknown, but assumed to be colonial nester like all of the other oropendolas. Assuming that the definitive pre-basic moult occurs just after breeding, it can be calculated that the breeding season occurs a month or two before July (see Moult).

DISTRIBUTION AND STATUS Extremely endangered, it has not been observed since 1945. However, there is a recent report from Utria Sound NP (P. Salaman pers. comm.). The species is known from four specimens. Specimens have been taken in the Serranias de Baudó and de los Saltos in Chocó Department, Colombia, between 100–365 m.

MOVEMENTS Unknown.

MOULT Largely unknown. Presumably adults undergo one complete moult per year, the definitive pre-basic. Two of the specimens, a male and female from the Rio Baudo, upper Chocó, Colombia, were undergoing their definitive pre-basic moult during late July. No information available on the first pre-basic moult or on the presence of pre-alternate moults.

MEASUREMENTS Male: (n=1) length 533.5; wing 270.5; tail 203.2; culmen 85.9; tarsus 61.5. The pale bill tip is 25.4 mm in length in this bird. Female: (n=1) wing 205; tail 167; culmen 65; tarsus 37.

NOTES Also known as Chestnut-mantled Oropendola. A specimen reported to have been a hybrid between this and Black Oropendola has recently been re-examined and found to be a perfectly good Black Oropendola (Ridgely and Tudor 1989). This species belongs to the 'Great Oropendola' group previously classified as the genus *Gymnostinops* and which has been suggested should be considered a variably polytypic species (See Notes under Montezuma Oropendola). The red data book lists this species as indeterminate. More information is needed to properly assess the size of the remaining population and if it has declined in numbers or if it has always been rare.

REFERENCES Collar *et al.* 1992 [habitat, status], Richmond 1898 [description, measurements], Ridgley and Tudor 1989 [description, notes], Ridgway 1902 [description], von Sneidern 1954 [distribution].

9 OLIVE OROPENDOLA (Amazonian Oropendola)

Psarocolius (bifasciatus?) yuracares **Plate 2**

(Lafresnaye and d'Orbigny) 1838

A large, green and chestnut oropendola of Amazonia. It is invariably a bird of the canopy. Hybridises with Para Oropendola (10) and is often considered conspecific with that form.

IDENTIFICATION Olive Oropendolas are characterised by a black bill with an orange-red tip, bare pink facial skin and a vivid yellow-olive head and front half of the body, which contrasts with chestnut wings, belly, lower back and rump. All of its closest relatives, the other large oropendolas with bare face patches (the *Gymnostinops* subgenus), are chestnut and black, lacking all olive colouration. Green Oropendola (3) closely resembles Olive Oropendola, both in the olive and chestnut plumage and in the red-tipped bill. However, Green Oropendola lacks the bare pink face patch on the cheek, and the bill base is pale greenish or blue-green rather than black. The extent of chestnut is greater on Olive Oropendola, with the chestnut restricted to the crissum, lower belly and rump on Green Oropendola. The wings of Olive Oropendola are chestnut while those of Green Oropendola are blackish with green feather edges, making the folded wings appear green, particularly on the coverts. Additionally, adult Green Oropendolas show pale blue eyes, while those of Olive Oropendola are always dark. The vocalisations also help in clinching an identification. The song of Olive Oropendola most closely resembles that of Crested Oropendola (2), not the more similar Green Oropendola nor that of the closely related Montezuma Oropendola (7) (see Voice).

The rare Casqued Oropendola (1) could also be confused with Olive Oropendola since Casqued Oropendola is largely chestnut and yellowish-olive in colour. Unlike Olive Oropendola, Casqued lacks a bare face patch of pink skin. Casqued Oropendolas have a yellowish-grey or pale yellow bill with a darker tip, not a black bill with an orange-red tip. As the name implies, Casqued Oropendola has a noticeable expansion at the base of the culmen which forms a casque; the casque is not nearly so well developed on Olive Oropendola. The head of Olive Oropendola is entirely yellowish-olive. In contrast, Casqued Oropendola has a largely chestnut head, with a grey chin and throat, and greenish-yellow sides of the neck which continue to the lower breast.

VOICE Song: A two part, loud gurgling *tek-tek-ek-ek-ek-ek-oo-guhloop!* or *cc-rr-rr-rr-rr-whh-heeeeeoooooppp*, lasting approximately 1.5 seconds. The song has also been described as *psooEE-OH,o,o,o,o,o,o,o*. The first part is descending in frequency and sounds grating or crackling, and has a metallic overtone (when heard up close). The explosive, whip-like, terminal portion of the song has the typical liquid quality of oropendola songs. Sometimes several audible wingbeats are given after the song. Note that songs from localities as geographically diverse as Venezuela, Peru and Ecuador are quite similar. This species has complex song, composed of two different sounds given simultaneously. The first sound is the descending, screechy, metallic note; its dominant frequency begins at 8 kHz and quickly descends to 1 kHz, becoming more grating as it does so. At the same time, the lower-pitched gurgling notes are given, with power centred at approximately 1 kHz, and ending with a complex mix of high and low notes, the sharpness of the sound giving a whip-like quality. Songs most resemble those of Crested Oropendola. However, Crested Oropendola's song usually includes a clearer and more striking, descending series, before the final flourish; in addition, Crested Oropendolas habitually beat their wings after a song, often for several seconds.

Calls: the common call is a loud *tac* or *chak*, which also functions as an alarm vocalisation, being given more frequently when the bird is alarmed. Also utters a *drrOT*, the flight call is a softer *dwot*, also a mewing *nhye*.

DESCRIPTION A large oropendola with an obvious bare face patch. The wings are not noticeably pointed, and the primaries are truncate, not heavily tapered. The form *yuracares* is described. Wing Formula: P9 < P8 < P7 < P6 > P5; P9 ≈ P4; P8–P5 emarginate.

Adult Male (Definitive Basic): The bill is black with an orange-red terminal third; the culmen spreads to an obvious frontal shield (casque). Eye colour is brown. It shows a bare face patch, and also a bare patch at the base of the culmen, both of which are pink. The head, upper back, neck, breast and upper belly are vivid yellow-olive. The green of the face and base of culmen is a darker olive than that of the neck or breast. A wispy crest of four feathers is present, but is not always visible in the field. The crest feathers are approximately 7 cm long. There is an abrupt transition between the green of the head and the dark chestnut of the rest of the body. The lower back, rump, belly, flanks, thighs, vent, crissum and wings are chestnut. The primaries are blackish, with chestnut outer edges, thus appearing chestnut when the wing is folded. The underwings are blackish with chestnut linings. Tail yellow with olive central rectrices (R1); these central tail feathers are slightly shorter than the longest yellow rectrices. The legs and feet are black.

Adult Female (Definitive Basic): Like the male but smaller, with less developed crest and frontal shield.

Juvenile (Juvenal): Like adults, but duller and lacking vivid colour on the naked face patch.

GEOGRAPHIC VARIATION Para Oropendola (10) may be conspecific (Haffer 1974), but we treat it as a separate species. Excluding Para Oropendola, there are two forms recognised within Olive Oropendola. The nominate *yuracares* occurs throughout the major portion of this species range. This subspecies is found east to the lower Rio Tapajós region of Pará, Brazil (Haffer 1974). The colour of the foreparts of nominate Olive Oropendola varies to a noticeable extent, with some birds being much more lime-green than others, which have less of a yellow undertone and more of an olive one. It is not clear if this variation is related to geography or not. Birds from the Río Caura in Venezuela were previously classified as a different race, *caurensis,* due to an apparently smaller bill and darker plumage than the nominate form (Todd 1913).

The eastern subspecies *neivae* is found from the Rio Tapajós (Gyldenstolpe 1945a) to the Rio Xingu region and as far east as the Rio Tocantins (Haffer 1974), Pará, Brazil. It is characterised by its dark olive (olive mixed with black) head and breast. In fact it is better described as intermediate in appearance between Olive Oropendola and Para Oropendola and appears to represent a hybrid zone between these two forms (Haffer 1974). If this is the case, and we suspect that it is, it would be incorrect to have a name (*neivae*) assigned to birds fitting this description as they would not fit the criteria for a subspecies. The hybrid zone, based on a limited number of specimens, appears to form a smooth cline between Olive on the west and Para on the east (Haffer 1974). According to Haffer (1974) birds from the west bank of the Rio Tapajós and the upper Rio Xingu resemble Olive Oropendola but are a bit darker on the head; birds from the mouth and east bank of the Rio Tapajós are darker still; specimens from the middle of the Rio Xingu are roughly intermediate between Olive and Para Oropendolas (this is the type location for *neivae*); at the mouth of the Rio Tocantins, specimens are practically indistinguishable from typical Para Oropendolas.

HABITAT Inhabits lowland rainforest and forest edge, usually feeding close to the canopy. It will come into cleared areas if enough large trees are left standing. It avoids seasonally flooded (várzea) forest as it is a bird of the dry terra firme forest, particularly unbroken forest tracts.

BEHAVIOUR Usually observed alone or in small groups foraging high in the canopy. It may join mixed species canopy flocks, sometimes with other oropendolas, for example Green Oropendolas. In Bolivia, evening flights of Olive Oropendolas of 100 or more flying in groups of *c.* 20 birds, have been observed (A.B. Hennessey pers. comm.). Presumably these birds were headed to a communal roost.

A highly polygynous species, based on observations at the colony. The sex ratio is severely skewed towards females with ratios as high as seven females to one male. The little that is known about this species suggests that like Montezuma Oropendola, this species is harem polygynous. Early in the breeding season, males spend most of their day resting, displaying or preening from a favourite perch at the colony tree. As the lining of nests begins and females prepare to lay eggs, the male's rate of display increases dramatically and tends to take place at the nest site rather than from the preferred perch.

The male bow display is similar to that of other oropendolas. He bends down and bows his head deeply, while drooping the rapidly vibrating wings. He culminates the display by bending forward more dramatically and cocking the tail upwards as the rump feathers are raised; the final liquid gurgle part of the song is given at this point. In 'less extreme' displays the wings are not flapped. There is no audible wing noise while displaying.

Colony activity is high during the nest-building and chick-feeding stage, but declines noticeably during the incubation stage. At times an oropendola colony may appear almost deserted while incubation is underway. Males are always present at the colony, and observations suggest that one of the males is dominant and spends most of his time at the colony. During the receptive period of the females, the dominant male is seldom away from the colony. Other males will display near the colony tree, but these are driven off by this male if they approach too closely. Attendance by other males appears to be highest during the nest-building stage. Copulations do not occur at the colony; this may allow males other than the dominant male access to mate with the females. Females forage singly, or in groups, and often in the company of lone males. Males do not consort with other males during the breeding season and are usually encountered foraging alone if they are not accompanying females. (largely from Rodríguez-Ferraro unpub. ms).

NESTING A colonial nesting species which builds long hanging basket nests like all other oropendolas. Colonies are not large; two colonies in Venezuela contained 14 and 17 nests. Nests are clumped within a colony, usually in groups of two to seven. In Venezuela, nesting starts in January coinciding with the dry season (January–April). The colony nest trees are similar in structure to those chosen by other species of oropendola. These trees are tall, isolated from the continuous canopy, and umbrella-shaped, tending to lack branches close to the ground. Nest trees are used year after year. Colonies may be shared with other icterids, which include Red-rumped Caciques (14) in Venezuela. While nesting of the two species occurs on the same tree, the caciques nest away from the oropendolas. In fact, male oropendolas are aggressive towards the caciques and even though they are much smaller the caciques may chase away male oropendolas (largely from Rodríguez-Ferraro unpub. ms).

DISTRIBUTION AND STATUS Uncommon to fairly common to 500 m. One sight record from 1200 m from the Serrania Bellavista, La Paz, Bolivia (S. Herzog pers. comm.). Ranges throughout the Amazonian lowlands, from SE Colombia and S Venezuela south through E Ecuador and S Peru to C Bolivia, northeast though Mato Grosso, Brazil to the Rio Xingu. In Brazil, the distribution lies largely south of the Amazon river, except west of Manaus (C Amazonas), where it is also found north of the river.

MOVEMENTS Apparently sedentary, with no evidence of migration.

MOULT Adults change their entire plumage once a year through the definitive pre-basic moult. In Amazonas, Peru, birds are moulting in July, while in Santa Cruz, Bolivia, the moult takes place in April. The definitive pre-basic moult takes place after the breeding season, therefore it is safe to say that in E Venezuela this moult takes place after April. The timing and extent of the first pre-basic moult is not known. There is no evidence for pre-alternate moults.

MEASUREMENTS ***P. y. yuracares***. Males: (n=10) wing 252.5 (226–292); tail 200.1 (182–246); culmen 78.7 (73–83); tarsus 57.5 (55–60). Females: (n=8) wing 207.7 (192–219); tail 167.4 (154–183); (n=7) culmen 60.9 (58–63); (n=8) 46.8 (42–50).

P .y. neivae. Male: (n=1) wing 245; culmen 65; tarsus 52. Females: (n=2) wing 194, 197; culmen 52, 54; tarsus 41, 42 (Snethlage 1925).

NOTES This species and Para Oropendola are usually considered conspecific in most recent literature (Ridgely and Tudor 1989, Sibley and Monroe 1990). The noticeable differences in plumage and size of these two species, and the fact that the hybrid zone has not been thoroughly studied, lead us to recommend that they are regarded as separate species until more research is conducted to clarify this situation. Given that other icterids (for example, Baltimore and Bullock's Orioles) hybridise freely where they come into contact, and after a great deal of research they are considered to be separate species and perhaps not closely related (sister) species, it does seem premature to lump these two oropendolas. The *neivae* form of Olive Oropendola is darker and appears to form the link between these two species. In fact, it may be more appropriate to drop the subspecific name for that form and consider it a hybrid swarm between the two oropendolas (see Geographic Variation). Should it be concluded that these two species best be lumped as one species, the common name Amazonian Oropendola seems appropriate.

REFERENCES Bob Behrstock personal recordings [song], Chapman 1917 [geographic variation], Haffer 1974 [hybridization, geographic variation], Moore 1993 [voice], Ridgely and Tudor 1989 [voice], Rodríguez-Ferraro Unpublished Manuscript [behaviour, nesting], Sick 1993 [voice, behaviour], Snethlage 1925 [geographic variation, measurements], Todd 1913 [geographic variation].

10 PARA OROPENDOLA *Psarocolius bifasciatus* (Spix) 1824 — Plate 3

A dark oropendola, endemic to Brazil, which is found near the lower reaches of the Amazon River and which hybridises with the very different Olive Oropendola (9). Its plumage pattern is much more similar to that of the geographically distant Montezuma Oropendola (7).

IDENTIFICATION Para Oropendola is a large oropendola with chestnut body and wings, becoming blackish on the head and breast. The bare face patch is pink, and no wattle is present. It has a black bill with an orange-red tip. Para Oropendola is sympatric with Crested Oropendola (2) and Green Oropendola (3); it overlaps and hybridises with Olive Oropendola. None of these other oropendolas closely resemble Para Oropendola in plumage, the closest is Crested Oropendola which is largely black. Para Oropendola has chestnut wings and back; these areas are black on Crested Oropendola. Crested Oropendola has an ivory-coloured bill and pale blue eyes, in contrast with Para Oropendola which has a black bill with an orange-red tip, dark eyes and a bare patch of pink facial skin. Green Oropendola is largely green in colour. Para Oropendolas lacks green or olive in its plumage. The bill of Green Oropendola is pale green with a red tip, not black-based as in Para Oropendola. Also note Green Orpendola's pale blue eye and lack of bare facial skin on the cheeks. Most problematic is the identification problem with Olive Oropendola. Olive and Para Oropendolas are visually very different, but have a similar pink bare patch on the face and black bills with orange-red tips, both characters typical of the *Gymnostinops* subgenus of the oropendolas. Olive Oropendola is chestnut with a greenish or yellow-olive head while Para Oropendola is chestnut with a black head and breast. The identification problem arises because the two forms hybridise within a region in the lower Amazon basin. The *neivae* subspecies of Olive Oropendola is probably best considered a hybrid swarm between the two species (see Olive Oropendola Geographic Distribution). Hybrids (*neivae*) show a mixture of black and olive on the head and breast, but they are variable. As a rule of thumb Para Oropendolas should not show any olive on the plumage while Olive Oropendolas should not show any solid black on the head and breast. There are several other differences between Olive and Para Oropendolas. Firstly Olive Oropendola averages larger, but considering variation in size due to age and sex this is of limited use in the field. Olive Oropendolas have olive-coloured central tail feathers, these are blackish-brown in Para Oropendola. In addition, the chestnut body plumage of Olive Oropendola is brighter, having more of a red or orange component than the more brownish-chestnut of Para Oropendola. Para Oropendola is visually most similar to its allopatric relatives,

Montezuma, Black (11) and Baudo (8) Oropendolas. It differs from these three species, in lacking a wattle along the malar area and also from Black and Montezuma Oropendolas in having pink as opposed to blue facial skin. Note also that Baudo and Black Oropendolas have more extensive black on the plumage, particularly on the underparts.

VOICE There are no known descriptions or recordings of the song of this form.

DESCRIPTION A large oropendola with a stout bill which is slightly decurved towards the tip. The casque is noticeably expanded, but not bulbous. This oropendola has a long thin crest and like its congeners, a moderately graduated tail. Wing formula: P9 < P8 < P7 > P6 > P5; P9 ≈ P4; P8–P5 emarginate.

Adult Male (Definitive Basic): The bill is black with a red tip that extends to nearly half the length of the bill. It has a naked, pink face patch. The face patch extends back level with the hind edge of the eye and there is no wattle on the malar area. The face patch continues as a small, bare pink border to the bottom and rear of the eye. The head, neck and upper breast are blackish-brown. It has three, long (90 mm) crest feathers, which may be visible in the field. The mantle, rump and rest of the underparts are chestnut, blending into the black of the head and neck. The thighs are also chestnut. The wings are black with chestnut outer edges to the primaries and secondaries, such that when folded they appear mainly chestnut. The wing-linings are chestnut-black. The tail is yellow with the central pair (R1) entirely dark brown. These black feathers are slightly (20 mm) shorter than the longest yellow feathers. The strong legs and feet are black.

Adult Female (Definitive Basic): It is similar to the male but substantially smaller.

Juvenile (Juvenal): Unknown.

GEOGRAPHIC VARIATION Monotypic. Hybridises with Olive Oropendola producing a hybrid swarm which has been named *Psarocolius yuracares neivae* (see Olive Oropendola Geographic Variation for more information).

HABITAT A forest species living both in terra firme and seasonally flooded forest in the Amazon lowlands.

BEHAVIOUR No information available.

NESTING No information available, presumably colonial as is typical of the genus.

DISTRIBUTION AND STATUS Uncommon, restricted to Pará and Maranhão states, Brazil. The precise western boundary of this species' range is complicated by hybridisation with Olive Oropendola. Individuals that appear to be pure Para Oropendolas are observed to the mouth of the Rio Tocantins, Pará. Ranges east to NC Maranhão and as far south as the confluence of the Rio Araguaia and Rio Tocantins.

MOVEMENTS No information available, presumably sedentary.

MOULT The definitive pre-basic moult is complete, and may be the only moult adults undergo. No other information is available, but probably similar to others in the *Gymnostynops* subgenus.

MEASUREMENTS Male: (n=1) wing 262; tail 210; culmen 72; tarsus 47. Female: (n=3) wing 202 (193–208); tail 155 (140–170); culmen 54.0 (53–55); tarsus 43.7 (40–46).

NOTES Hybridises and forms a hybrid swarm with Olive Oropendola. See Notes and Geographic Variation sections for Olive Oropendola.

REFERENCES Haffer 1974 [geographic variation, distribution].

11 BLACK OROPENDOLA *Psarocolius guatimozinus* (Bonaparte) 1853 — Plate 3

Black Oropendola is not really an appropriate name for this species as there are other oropendolas which are more black than this one. What is distinctive about this species is that it is restricted to the wet forests of the Chocó region of E Panama and W Colombia, extending to the Magdalena Valley in the latter country.

IDENTIFICATION A large, dark oropendola, with bare facial skin. The entire body is black except for the lower back, secondary coverts and rump, which are chestnut. Black Oropendola is the glossiest, most iridescent of the *Gymnostinops* ('Great Oropendola') subgenus and the one with the longest crest feathers. The bill is black with a small orange tip. The bare facial skin is as in Montezuma Oropendola (7); the area around the face is sky-blue, while the wattle along the jaw line is pink, as is the small amount of bare skin at the base of the culmen. The Black Oropendola does not come into contact with Montezuma Oropendola; nevertheless note that the latter has black restricted to the head, and the body and wings are entirely chestnut. The extremely rare Baudo Oropendola (8) is similar again. Like Montezuma Oropendola, Baudo Oropendolas show less black on the plumage than Black Oropendola. The black is restricted to the head, breast, mid-line of the belly and thighs on Baudo Oropendolas, and the mantle, wing-coverts and primaries are chestnut. Black Oropendolas have chestnut restricted on the wings to the secondary coverts; only the rump to lower mantle and some scapulars are chestnut on the upperparts. Note that the tertials of the Black Oropendola are almost entirely black and the primaries are entirely black, the same set of feathers are chestnut on Baudo Oropendola. Black Oropendolas show a contrast between the black primaries, secondaries and tertials and the chestnut wing-coverts, all of these feathers are evenly chestnut on Baudo Oropendola. Curiously, the central dark tail feathers are noticeably shorter than the longest yellow feathers on Black Oropendola, a condition that is not as extreme in any of the 'great oropendolas' (subgenus *Gymnostinops*). Baudo Oropendolas have a restricted amount of bare skin on the face, and only the facial area and the base of the culmen are pink; as noted above, Black Oropendolas have a blue face and pink wattle along the jaw line.

Two other largely black oropendolas overlap in distribution with Black Oropendola. These are Crested (2) and Chestnut-headed (6) Oropendolas. The features which place Black Oropendola into the *Gymnostinops* ('Great Oropendolas') genus, the bare facial skin and bicolored bill are alone enough to separate it from Crested and Chestnut-headed Oropendolas. Both Crested and Chestnut-headed Oropendolas have pale, ivory-coloured bills which can be seen from a distance; Black Oropendola's bill looks black at a distance. Furthermore, the bill of Chestnut-headed Oropendola shows a noticeable bulge (casque) at the base which gives it a distinctive look, even at a distance. Closer views reveal that Crested and Chestnut-headed Oropendolas usually have sky-blue eyes, unlike the dark eyes of Black Oropendola. The wings of both of these species are entirely black, save for chestnut covert tips in some individuals; Black Oropendolas have chestnut wing-coverts. In good light, note the chestnut head of Chestnut-headed Oropendola, unlike the black head of Black Oropendola. Vocalisations may also aid in the identification of these species. Finally, the *salmoni* subspecies of the yellow-billed form (species?) of Russet-backed Oropendola (5X) may range overlap with Black Oropendola. This population of 'Yellow-billed Oropendola' has a pale bill, and the body plumage lacks black altogether, being a mix of chestnut and olive, with an obvious yellow forehead. Largely black species of Caciques have pale bills, are smaller in size, and may have red or yellow rumps; most species lack yellow on the tail.

VOICE Song: A loud and resonant *skol-l-l-l-wool!*, similar in nature to that of other oropendolas (Hilty and Brown 1986). Wetmore *et al.* (1984) describe the song as 'hollow sounding', and state that sometimes the hollow note was followed by a series of high-pitched notes *kwee, kwee, kwee* or *keea, kee-a, kee-a*. These final notes were often given as the male took off in flight. Recordings show that the song resembles that of Montezuma Oropendola, being composed of both a low liquid gurgling series of notes which are given simultaneously with a series of quick, high-pitched metallic sounds (Hardy *et al.* 1998).

Call: A low *cruk* (Hilty and Brown 1986).

DESCRIPTION A large oropendola with a long, thin crest of four feathers. The stout bill is moderately expanded at the base of the culmen. The wings have a short outer primary. Wing formula: P9 < P8 < P7 >= P6 > P5; P9 = P3. The tail is moderately graduated. The central rectrices (R1) are distinctly shorter than the rest of the tail (see below)

Adult Male: Bill black with an orange-red tip which is approximately one-third of the length of the bill. Note that Ridgely and Gwynne (1989) and Hilty and Brown (1986) state that the bill is tipped yellow; however, Rodriguez (1982), Wetmore *et al.* (1984) and Ridgely and Tudor (1989) mention that the bill is tipped with orange. It is not clear if this difference in the description of bill tip colour is due to a biological cause, either geographic variation or age, or to perceived colour differences. However, the information from Wetmore *et al.* (1984) is from notes made from a bird collected in the field confirming that the bill tip colour is indeed orange on some, if not all, Black Oropendolas. The eyes are chestnut-brown (Rodriguez 1982, Wetmore *et al.* 1984), appearing dark at a distance. In contrast, Sclater and Salvin (1879) reported that the eyes are blue, which appears to be erroneous. The bare face patch is blue and extends to the level of the rear edge of the eye, although a short spur extends back along the rear edge of the eye itself, curling up around the eye. There is a bare, pink, malar wattle, separated from the blue face patch by a strip of black feathers. A small patch of bare skin at the base of the culmen is pink, like the malar wattle. Four long (80–100 mm) black crest feathers are present, giving this species the longest, thinnest crest of any of the *Gymnostinops* subgenus oropendolas. The head and nape, upper back and most of the underparts are black with a faint blue gloss. The lower back, rump and uppertail-coverts are dark chestnut, but the longest uppertail-coverts are black. On the underparts, chestnut is restricted to the crissum. The thighs are black. The wings are black with chestnut lesser and median coverts, as well as chestnut tips to the greater coverts; however, the tertials are entirely black. Thus, the primaries, secondaries, alula and primary coverts are entirely black. The underwings are black. The tail is yellow with black central tail feathers; however, the black feathers are exceedingly short. The central rectrices (R1) are as much as 70 mm shorter than the next pair of rectrices (R2); they appear to be half to two-thirds the length of the tail on a live bird. The thick legs and feet are black.

Adult Female (Definitive Basic): Similar to male, but significantly smaller and with less intense black colouration and not as iridescent. The crest is shorter, averaging only 25–30 mm in length. The bare parts are as in the adult male.

Juvenile (Juvenal): Unknown. If this species fits the pattern of its close relatives the juvenile is similar to the adults but duller in colour.

GEOGRAPHIC VARIATION Monotypic, no described geographic variation.

HABITAT Restricted to humid forests, often found at the forest edge. It appears that closed forests are not preferred and that edge is a vital characteristic of its chosen habitat. In Panama, it is stated to prefer the edge of humid forests, as well as clearings with taller trees (Ridgely and Gwynne 1989). Also found in less humid forests, in its Colombian range, where it is not as numerous (Hilty and

Brown 1986). At the Parque Naciónal Los Katios, Chocó, Colombia, Black Oropendola prefers the highland forest of the Darién highlands, rarely being found in the lowland forest (Rodriguez 1982).

BEHAVIOUR Usually seen foraging solitarily, or in small groups, seldom in large groups. Tends to forage at or near the canopy, in the manner of most other oropendolas. Wetmore *et al.* (1984) describe observing a Black Oropendola feeding at flowers by probing them and gaping while the beak was inside the flower to open the corolla, presumably to drink the nectar. The mating system is unknown; however, the great degree of sexual size dimorphism suggests that this species is polygynous like other members of the group. Males display at or near nests, and several males are at attendance at the colony (Wetmore *et al.* 1984). The bow display is similar to that of Montezuma Oropendola. The male sings as he falls forward, hanging upside down for an instant, while keeping a tight grip on the branch with his strong legs.

NESTING Colonial as in other oropendolas. The nesting colonies are often along rivers (Ridgely and Tudor 1989). Birds in breeding condition have been reported from S Cordoba, Colombia, in April and a breeding colony has been observed in late June from the upper Río Verde de Sinú in Colombia (Hilty and Brown 1986). Wetmore *et al.* (1984) describe a colony in Pucro, Darién, Panama, that was in the early stages of nesting in late January. They also note that females collected during February in Pucro were at the laying stage. The Panama colony was comprised of 20 nests, secured to a single tree at the edge of a banana plantation. The nests were lined with leaves. The eggs are pale pink, with a few large red-brown spots (Sclater and Salvin 1879).

DISTRIBUTION AND STATUS Uncommon or rare to locally common (Antioquia, Río Verde del Sinú and SW Cordoba according to Hilty and Brown 1986). Uncommon in Parque Nacional Natural Los Katios, Chocó (Rodriguez 1982). Ranges to 800 m. Restricted to Colombia and Panama. In Colombia it is found in N Chocó, north to the lower Atrato River valley (Rodriguez 1982), upper Río Salaquí and the upper reaches of the Río Sucio; also in W Antioquia (Río Verde del Sinú) and southernmost Cordoba. The distribution of this oropendola continues inland curling around the northern base of the western and central Andes of Cordoba, reaching as far as Remedios, Antioquia, which lies between the Serrania de San Lucas and the northern extension of the central Andes. To the south, the range is reported to extend to N Caldas, based on a sight record there (Hilty and Brown 1986). In Panama, it is found in the lowlands of the Darién, in particular the lower Tuira and Chucunaque river valleys, west to Santa Fé in W Darién. Also reported from E Panama, in the upper Bayano River, based on sight records (Ridgely and Gwynne 1989).

MOVEMENTS Sedentary, no reported seasonal migrations.

MOULT Largely unknown. It seems safe to assume that its moults are like those of other large oropendolas, mainly that there is one complete moult per year, the definitive pre-basic moult. Birds undergo this moult during April in Panama (Marraganti, 150 km east of Panama City). The extent and timing of the first pre-basic moults is unknown. It is not known if pre-alternate moults exist in this species.

MEASUREMENTS Males: (n=10) wing 252.1 (227–270); tail 204.2 (182–212); culmen 66.6 (64–70); tarsus 58.0 (54–60). Females: (n=9) wing 202.7 (195–207); tail 167.6 (153–182); culmen 53.7 (52–56); tarsus 47.0 (45–50).

NOTES A specimen reported to have been a hybrid between Black and Para Oropendolas has recently been re-examined and found to be a Black Oropendola (Ridgley and Tudor 1989). Therefore, previous statements that these two species hybridise are unfounded.

REFERENCES Hilty and Brown 1986 [voice, behaviour, nesting, distribution], Ridgely and Gwynne 1989 [habitat, distribution], Ridgely and Tudor 1989 [description, nesting], Rodriguez 1982 [habitat, distribution], Sclater and Salvin 1879 [nesting], Wetmore *et al.* 1984 [voice, behaviour, nesting, measurements].

12 BAND-TAILED OROPENDOLA *Ocyalus latirostris* (Swainson) 1838 — Plate 1

This little known icterid appears to form a link between the caciques and the true oropendolas, as based on outward morphologically.

IDENTIFICATION This small black oropendola is more likely to be mistaken for a cacique than an oropendola. Its distinctive tail pattern, a crisp black terminal band to the yellow tail is not shared by any other cacique or oropendola. From below, the folded tail looks yellow with a black tip; note also that unlike other oropendolas the tail is more square-tipped rather than graduated (rounded). When perched, Band-tailed Oropendola often appears entirely black, as the yellow of the tail is hidden on the folded tail, particularly when the bird is observed from above or from the side. At such times, it may recall Solitary Cacique (21) but note the blue eyes and thicker bill, with a moderate casque, on Band-tailed Oropendola. In addition, the bill of the oropendola is dark near the forehead, but pale on the lower mandible and the tip of the upper mandible; Solitary Cacique has an entirely pale bill. The bill shape and colour differences can also be used to separate this species from the rare and local Ecuadorian Cacique (20), which shares its blue eye colour with Band-tailed Oropendola. Note that given a close view, Band-tailed Oropendola shows a chestnut cast to the nape. Band-tailed Oropendolas may flock with Yellow-rumped Caciques (13) which are largely black, but which have obvious yellow rumps and yellow wing patches. The other sympatric oropendolas are rather different in colouration. Crested Oropendolas (2) are also largely black with blue eyes, but consider Crested Oropendola's large size, tail pattern, ivory-coloured bill, and chestnut rump and undertail-coverts.

VOICE **Song** (from D. Lane recordings): Unlike the true oropendolas, this species appears to give several different notes which may be regarded as part of the song. Probably homologous to the song of true oropendolas is a hollow *chh-tooUUUUU* or *chh-eUUUUU*. Apart from this, males also give several different notes, most of which sound rather squeaky and metallic. Other notes include a *chetow, ch-zzp, zeeoo,* as well as a squeaky warbling reminiscent of the song of Giant Cowbird (98). These calls are difficult to characterise and have a rather large (700–8000 Hz) frequency range. Most peculiar is a low, mechanical rattle given by males.

Calls: A loud *chuk* or *chuff* (D.Lane recordings). The notes are given singly, not paired. They are almost indistinguishable from the call notes of Russet-backed (5) or Casqued (1) Oropendolas. Sometime a more musical *chuue* is given and also a squeaky *ch-whooo.*

DESCRIPTION A small oropendola with a thick bill. The bill is noticeably broad, with an obviously enlarged frontal shield (casque). The wings are long, showing a noticeable primary extension, particularly on males. Wing formula: P9 < P8 ≈ P7 ≈ P6 > P5 > P4; P5 > P9 > P4; P8–P5 emarginate. The plumage has a curious velvety texture. A short crest of feathers immediately behind the base of the culmen is raised when excited (D. Lane pers. comm.).

Adult Male (Definitive Basic): The bill is pale yellow, becoming dusky at the tip. The eyes are pale blue. This species is largely black with the mantle, rump and lower back black with a dark violet iridescence. The head, crown, nape and upper back are dark chestnut, appearing black at a distance. This chestnut colouration extends down the sides of the neck as a half-collar. The face and underparts are entirely black, with a strong violet gloss on the breast, but less so on the face, throat or belly. The wings are black with a dull green gloss; the underwings are black. Most noticeable is the black and yellow tail. The central four rectrices (R1 and R2) are black while the next four pairs (R3–R6) are yellow with a clean-cut black tip; the outermost tail feather (R6) also has a black outer edge. The legs and feet are black.

Adult Female (Definitive Basic): Similar to the male, but smaller. The plumage pattern is like the males, but the iridescence is not as strong on the female. The chestnut crown, nape and collar are paler and less extensive. In addition, the throat is a dull brownish-black rather than jet-black.

Juveniles (Juvenal): Unknown, presumably similar to the adults but not as bright or iridescent.

GEOGRAPHIC VARIATION Monotypic.

HABITAT Found in seasonally-flooded (várzea) forest, and permanently flooded swamp forest, in Amazonia, as well as on river islands. On river islands, it inhabits those which are older and less disturbed (Rosenberg 1990). In Peru, a nesting colony was active in terra firme forest distant from a river (D.Lane, A. Kratter unpub. data), suggesting that habitat choice is variable.

BEHAVIOUR Forages in the canopy, subcanopy and mid-story, as is typical of oropendolas. Most commonly searches the undersides of leaves for insects, although fruit has been reported from the stomachs of specimens (D. Lane pers. comm.). Other stomach contents have included caterpillars (specimen label data). Band-tailed Oropendola occurs in small flocks, from 10–30 individuals in size. It is also known to flock with Russet-backed Oropendolas, and Yellow-rumped Caciques (Ridgely and Tudor 1989). However, at the colony Band-tailed Oropendolas may act aggressively towards Yellow-rumped Caciques (D. Lane pers. comm.). Male displays are not as extreme and exaggerated as in *Psarocolius* oropendolas. Males puff up their throats and lean forward slightly as they sing. Two to three males were observed singing in one Peruvian colony, greatly outnumbered by females. Males sing at the nest tree, apparently only during the morning, at least during the nestling stage. Some males produce an audible wing noise in flight around the nest tree, but this is not constant and not all males appear to produce this sound (D. Lane, A. Kratter unpub. data).

NESTING This species has a variable nesting pattern and appears to be the only oropendola that nests solitarily, albeit in a colony of another oropendola species. Koepcke (1972) describes Band-tailed Oropendolas from east-central Peru (S of Pucallpa) as commonly nesting within colonies of Casqued Oropendolas. A pure colony of Band-tailed Oropendolas has also been found in Peru (D. Lane, A. Kratter unpub. data). The known solitary nest was with Casqued Oropendolas in Brazil (A. Whittaker unpub. data). Most known nests of Band-tailed Oropendola have occurred with Casqued Oropendola even though other oropendolas were present in the same area. This nesting association is peculiar and may be more than just a random ocurrence. Nests of Band-tailed and Casqued Oropendolas are similar, as are some of their behaviour patterns (Koepcke 1972).

The Peruvian colony was active in mid August, at which time at least some of the nests contained young (D.Lane pers. comm.). Breeding condition specimens have been collected in mid June from Leticia, Colombia (Hilty and Brown 1986). The Peruvian colony was located on a 30 m high Cecropia tree in a gap in the forest. The tree was not an emergent, but it was not touching, or shaded by, neighbouring trees. Most of the 15–20 nests were situated both near the top and centre of the tree. The canopy of the nest tree was dense, such that the nests were not clearly in the open as is the case with most oropendolas (D. Lane, A. Kratter unpub. data). All nesting colonies found by Koepcke (1972) were also in Cecropia trees, in a tree species which is guarded by an aggressive stinging ant. Nests are secured to leaves, so that when these leaves are shed the nests fall with them. The nests are wide, and shorter than most oropendola nests, resembling those of Red-rumped (14) and Yellow-rumped Caciques (Koepcke 1972).

DISTRIBUTION AND STATUS Rare, found to 300 m. Largely a Peruvian species, found in Loreto, and N Ucayali south to Pucallpa. Also in the Leticia region of southernmost Colombia, and adjacent Acre (upper Rio Juruá) and Amazonas in Brazil (Gyldenstolpe 1945a). There are two historical records from Ecuador, one from Archidona and the other from Sarayacu (Ridgely and Tudor 1989). However the Archidona specimen actually refers to a Chestnut-headed Oropendola and the locality is likely incorrect (Gyldenstolpe 1945a).

MOVEMENTS No known migratory or seasonal movements, but this species is very poorly known.

MOULT Little known. Adults likely undergo only one moult a year, the complete definitive pre-basic. Adult specimens nearing completion of this moult exist from mid June from Amazonian Colombia. The extent and timing of the first pre-basic, or presence of a first pre-alternate moult are not known.

MEASUREMENTS Males: (n=5) wing 217.6 (200–235); tail 118.2 (108–132); culmen 37.8 (35–40); tarsus 35.6 (33–38). Females: (n=3) wing 178.7 (137–212); (n=2) tail 89, 106; (n=3) culmen 36.3 (32–39); tarsus 33.7 (30–36).

NOTES The systematic relationship of this rare oropendola is not well understood. Some taxonamists place it in the genus *Psarocolius*; however vocal, morphological and nesting behaviour suggest that this oropendola is intermediate between *Psarocolius* and the caciques. At this stage it is appropriate to maintain the monospecific genus *Ocyalus* for this species.

REFERENCES D. Lane and A. Kratter unpub. data [habitat, behaviour, nesting, voice], Hardy *et al.* 1998 [voice], Koepcke 1972 [nesting], Remsen 1977 [distribution], Ridgely and Tudor 1989 [behaviour, distribution], Rosenberg 1990 [habitat].

13 YELLOW-RUMPED CACIQUE *Cacicus cela* Plate 6

(Linnaeus) 1758

The most common and widespread cacique in Amazonia, this black and yellow bird is a familiar sight in the American tropics.

IDENTIFICATION A large black and yellow cacique with obvious yellow on the tail, wings, rump and vent. The bright blue eyes are noticeable from a distance. The yellow of this cacique is always visible, thus it should not be confused with the all-black caciques or the 'red-rumped' caciques which show no yellow. It is probably sympatric with the smaller Selva (19) and Golden-winged (17) Caciques both of which are black and yellow. The latter has not been found sympatrically with Yellow-rumped Cacique, but it is possible that the two coexist in central Bolivia; Mountain Cacique (18) is found at higher elevations than Yellow-rumped Cacique. Selva Cacique is extremely rare and any possible sighting of this species should be documented with great caution. Selva Caciques resemble Yellow-rumped Cacique in having a yellow rump, albeit more restricted. Unlike Yellow-rumped Cacique, it lacks yellow on the wing, underparts and tail. The bill of Selva Cacique is straight, lacking the distinctive curve to the culmen that Yellow-rumped Cacique shows. Golden-winged Cacique lacks the yellow on the tail and crissum that Yellow-rumped Cacique shows, although, it does have the yellow rump and wing patch. Golden-winged Caciques are smaller, tending to show a shaggy crest and often have yellowish eyes. The rump patch of Golden-winged Cacique is significantly smaller than that of Yellow-rumped Cacique; this can be seen in flight as can the lack of yellow on the tail. The combination of yellow on the tail, rump and crissum is diagnostic for Yellow-rumped Cacique. The larger oropendolas, particularly the black-bodied species, may be confused with the cacique due to their yellow-edged tails. However, all oropendolas lack yellow on the rump, wing and underparts.

The smaller black and yellow Moriche Oriole (24) has yellow on the nape, unlike Yellow-rumped Cacique. Yellow-tailed Oriole (30) is much smaller and is largely yellow, with entirely yellow underparts, unlike the cacique.

VOICE The repertoire of vocalisations in this species is huge, making a summary of their vocalisations somewhat futile. In addition, *C.c.cela* is an adept mimic.

Song: The primary song ('cela song') is usually made up of four syllables, the first of which is a screech typical of Yellow-rumped Caciques. The following notes are variable. It lasts under two seconds in duration. Each colony has a song (or several songs) specific to its colony and all males conform to these songs while at the colony. The colony song changes from year to year, and sometimes within a season. While in full display another song is given ('harsh song'), which is more complex and continuous, not discrete. This longer song can last up to 20 minutes and includes many loud whistles, and it is here that this cacique (form *C.c.cela*) will incorporate mimicry of local birds, frogs and other assorted noises. The harsh song is extremely variable, but includes the *tchak* notes typical of this cacique. The loud volume and varied nature of the song may be its most distinctive features. Females rarely sing the 'cela song' but never sing the 'harsh song'. It appears that the 'harsh song' is aimed mainly at other males and has some aggressive or threatening significance, while 'cela song' is correlated with slight alarm and may function to advertise the presence of the male to females (Feekes 1981).

Vitellinus is not known to mimic or to produce mechanical-sounding noises. The vocalisations of *vitellinus* are said to be a conversational and musical series of notes: *wick-a-weo char-che-ar, chut-chu-chu.* Also gives a more muffled song, *Aaghh-a-whaaghee*, or *cluuk-cluuk-whaaagooo.* Males of *vitellinus* sing five to eight song types; all males in a colony share the same song types. The number of shared songs between colonies is directly proportional to the distance between colonies, with nearby colonies sharing more song types. Dialects are stable during a breeding season, but some songs change gradually over the months. Less than half of the song types are present in a recognisable form during the next breeding season. Some new songs (17%) are introduced by dispersing males from other colonies, but for the most part new songs are derived from old songs (45%), the rest are of unknown origin (Trainer 1989). An analysis of song types of *vitellinus* in Panama discovered that of the seven song types, specific songs were associated with certain behaviours. One song was more likely to be sung by males in groups, while another was sung by lone males. A third was sung by flying males, and yet another when supplanting another male, while the remainder were not so closely associated with specific contexts (Trainer 1987). There appear to be consistent rules that govern the sequence of song types given by males at the colony. Social interactions as well as social context appear to influence (or determine?) the organisation of song types at the colony (Trainer 1988). The form *flavicrissus* of the Pacific coast is also not known to be a mimic.

Calls: The common call of *cela* is a repeated harsh *tchak* or *chaaak*, uttered more loudly and frequently when alarmed. Females utter a rough *rrrrrrr*, particularly when behaving aggressively. The subspecies *vitellinus* gives a standard icterid *chuk*.

DESCRIPTION A chunky cacique with a noticeably curved culmen. The bill is slightly shorter than the length of the head. The primaries of the adult male are long and greatly tapered, those of females and young males are not so tapered. Wing Formula: P9 < P8 < P7 <≈ P6 > P5 > P4; P4 < P9 < P5; P8–P5 emarginate and notched, more so on the male. The nominate form is described here.

Adult Male (Definitive Basic): This plumage is obtained during the third year (third basic plumage). The bill is ivory-coloured and the eyes are sky-blue. Largely a black bird with a slight blue gloss. The lower back, rump, uppertail-coverts, crissum, vent, hind flanks and base of the tail are bright lemon-yellow. On the tail, the central four feathers (R1–R2) are yellow on the extreme base only, appearing largely black as the yellow bases are covered by the yellow uppertail-coverts. The outer feathers are more than half yellow, such that the tail looks yellow with a black inverted 'T' pattern from above. From below, the tail looks largely yellow. There is also a lemon-yellow wing patch made up of most of the greater coverts except the outermost, and some of the inner median coverts. The rest of the wing is black, including the wing-linings. The legs and feet are blackish.

Adult Female (Definitive Basic): Similar to male, but noticeably smaller in direct comparison, and sooty-grey (blackish) rather than black. The distribution of yellow is as on the male. Frequently, females have grey eyes (perhaps only younger females?).

Second Year Immature (Second Basic): Females as adult females. Males are duller than adult males, showing olive edges to the belly feathers but have blue eyes with dark flecks, and yellow bills. Their greater size should distinguish them from females; however, they are slightly smaller than mature males.

First Year Immature (First Basic): The bills have patches of brown toward the base and the eyes are purplish-brown, appearing dark at a distance. They are substantially duller, greyer-plumaged than adults. Young males are larger than females, but smaller than adult males.

Juveniles (Juvenal): Similar to 1st-year immatures, but with a loose plumage texture.

GEOGRAPHIC VARIATION Three subspecies are recognised which fall into two groups: *vitellinus* (Saffron-rumped or Lawrence's Cacique) and *cela* (Yellow-rumped Cacique). It is clear that there are consistent morphological, behavioural and vocal differences between these two groups and it may be determined that they are best separated as two different species in the future. The two groups almost come into sympatry in N Colombia, but no intermediate specimens are known.

Yellow-rumped Cacique ***C. c. cela***, is found east of Andes. This is the most widespread subspecies, found throughout the tropical lowlands of South America east of the Andes as far south as C Bolivia and E Brazil. In Colombia, *cela* is found along the coast of Santa Marta, and comes quite close to the range of *vitellinus*. This population shows no evidence of intergradation towards *vitellinus* (Todd and Carriker 1922). *Cela* is described above.

Saffron-rumped Cacique is found west of Andes and is comprised of two subspecies. The race ***C. c. vitellinus*** is found from the Canal zone in Panama east into W Colombia (south to the Rio Salaqui) to the western and southern base of the Santa Marta mountains, and south to the middle of the Magdalena valley. It differs from *cela* in being larger, with a bill showing a more strongly curved and broadly-based culmen. The yellow of the body is a deeper colour, more orange, than on *cela*. In addition, the yellow on the coverts is much less extensive, being restricted to the innermost greater coverts. The yellow on the base of the tail is much less extensive, only covering the basal third or less on the outer rectrices. Wing Formula: P9 < P8 < P7 > P6; P9 ≈ P5; P8–P5 emarginate.

The other race in this subspecies group, ***C. c. flavicrissus*** occurs in the Pacific lowlands of Ecuador from Esmeraldas south to NW Peru (Tumbes). There is a gap in range between *flavicrissus* and *vitellinus* along the Pacific Coast of S Colombia. This form is like *vitellinus* in having less yellow on the tail than *cela*, but it has a larger yellow wing patch, like *cela*. It is smaller than *vitellinus*, has less of an orange colour on the rump, and a smaller bill which has a plumbeous-coloured base. Wing Formula: P9 < P8 < P7 > P6; P4 < P9 < P5; P8–P5 emarginate.

HABITAT This cacique is partial to open forest and forest edge. It has benefited from the creation of roads and roadside clearings, perfect habitat for this species. It avoids dense forest, but will venture into it to feed. Nesting occurs in more open areas. In much of Amazonia it was likely originally confined to river courses, seasonally flooded várzea forest, and the edge of cochas and swamps. In other parts of South America it can be found in open woodlands as well as the more open savanna and swamp forest along the coast of the Guianas. In the basic sense this species is a generalist in its habitat choice, providing that some large trees or forest patches remain in the area. In addition, a suitable nest tree needs to be present in the area (see Nesting).

BEHAVIOUR Largely from Robinson (1986c, 1986d, 1986e, 1988). At the colony, females outnumber males, but the number of males fluctuates throughout the season. The maximum number of males is present during nest-building, when females are sexually receptive. The mating system is polygamous and males may be highly polygynous if successful. Mating does not occur at the nest tree, but in the nearby forest. Males at the colony spend a good deal of time displaying to a specific receptive female. Consortships last a few days and males will then change their attention to another female. Males form a dominance hierarchy, with only the males high in the hierarchy achieving consortships with females. The subordinate males spend most of their time acting as sentinels, warning of incoming danger. Rather than being altruists these sentinels are actually 'satellite' males, waiting for any opportunity to mate with a female temporary left unattended by one of the breeding males. Dominant males use a noisy flight, while subordinates do not. When the dominant males are absent, subordinate males may fly to the display perches, but rarely display. Several males are present at the colony at any one time, and tolerate each other because the dominance hierarchy is strictly followed, decreasing the amount of conflict at the colony. Fights break out when a subordinate male tries to become a dominant male. The dominance hierarchy is determined by the weight of the males. The heaviest males become dominant and this occurs typically in the first year or two after maturity. The constant stress created by having to stay at the nest tree guarding females, with little time for feeding, makes males lose weight. As they lose weight some males lose their status as the season progresses, only to return the next year heavier and able to defend their status once again. Older males usually cannot defend their status and are relegated to the 'satellite strategy'. At night all individuals except incubating females leave the colony to spend the night at a communal roost.

Immature males (less than three years old) live a different existence from the adults. First-year males do not visit the colonies and presumably spend their time in the forest. Second-year males do visit the colonies; they do not sing or display, but instead perch quietly near the colony. They are subordinate to all adult males. These young males frequently chase adult females and harass them throughout the nesting cycle. They will peck at females in the nest, and sometimes grapple and fight with them. Young males attack females feeding young and will even enter the nest and remove or kill nestlings. Once the young fledge they still do not escape harassment by 2nd-year males. The reasons for this behaviour are not understood. Adult males do not offer any assistance to the females against the young male's attacks.

Nesting females also set up a dominance hierarchy. Groups of females form nesting associations with females that have nested nearby for several seasons; new females are forcibly excluded and have to fight to obtain a nesting spot. The choice sites, near wasp nests, are taken by the oldest and most established group of females. Established females will also rob newcomers of their nesting material, not allowing them to finish their nest. Only the heaviest young females are able to establish themselves in a colony. By the time they are five years old, most females belong to an established group and interestingly weigh approximately 10% less than when they were younger and trying to become part of a nesting aggregation. First-year females are not present in colonies. During their second year, young females begin to attempt nesting at the colony. Very few complete their nests, partially due to the stealing of material and harassment by older, established females. Small 2nd-year females which do not complete a nest generally leave the area. Young females that are successful in nest construction lay fertile eggs, but these are almost always abandoned before the young fledge. Nevertheless, the ones that are successful in building a nest are more likely to complete a nest the subsequent season and to join a nesting cluster.

The display of the male *cela* is striking. He sits almost horizontally and fluffs out his feathers, particularly the yellow rump, and flutters his wings as the bill is pointed downwards, he bows forward and sings. The 'wing-fluttering' begins above the level of the body with the wings brought slowly down as the tail is raised and the rump is prominently displayed.

The 'harsh song' is given during this display which is aimed at females or other males. When soliciting copulations, the male sleeks his plumage, maintaining the body in a horizontal position and keeping the head withdrawn; he keeps the wings tightly folded high on the body. The only feathers which are raised are the yellow wing patch and the undertail-coverts. He bows his head up and down until the female gives the soliciting posture, with wings quivering and tail raised. At this point she will fly off with the male in pursuit, presumably to mate in the forest.

Yellow-rumped Caciques typically forage in pairs or small groups, usually comprising several females and one male. Males also often feed solitarily. They may end up feeding with other species of caciques or oropendolas at a productive fruit tree. Yellow-rumped Caciques often join mixed canopy flocks of larger birds such as Nunbirds (*Monasa*), Aracaris (*Pteroglossus*) and Jays (*Cyanocorax*). Fruits make up a great proportion of their food, particularly early in the nesting season, as do insects, primarily large ones such as cockroaches (*Blattidae*) and Katydids or other arboreal *Tettigoniidae* (crickets and katydids). The foraging behaviour of females varies, depending on the nesting stage. In general, they forage on the outer foliage of the canopy taking 75.2% of prey from live leaves and 22.1% from dead leaves (Robinson 1986d). When not nesting, females forage at a leisurely rate, carefully probing and searching for caterpillars in the canopy. As energy requirements increase, during laying, incubating and when feeding fledglings, they forage at a faster rate, probing and hanging to reach food, in the mid- or low canopy, mainly catching spiders, caterpillars and some *Tettigonidae*. When they are feeding nestlings or nest-building, energy demand is at its highest, and the foraging rate is fastest, with birds often foraging in the understory where a large proportion of *Tettigonidae* (Katydids and Crickets) are pursued.

NESTING In Surinam, breeding occurs at all months of the year, but there is an obvious increase as the dry season commences (December to May). In Colombia, nests are known from October to April. Breeding occurs between July and February in eastern Peru and from January–June in Trinidad. Nests in October in Pará, Brazil. The form *vitellinus* begins nesting in April–early May in Panama (Wetmore *et al.* 1984), in contrast to October–November in the Magdalena valley, Colombia (Boggs 1961). Yellow-rumped Caciques always nest in colonies, but colony-size is variable. In smaller colonies, the nests are built at the same time and nesting is synchronous; however, in larger colonies several waves of birds will build and lay eggs, making them asynchronous as a whole. The nest-building, incubation and care of the young is performed exclusively by the female. Nests are clumped together, often around or near an active nest of paper wasps (*Polistinae*). Both the clumping of nests and building near wasp colonies has the effect of protecting the nests from predators. Nests on the periphery of the colony seldom fledge young. The caciques also prefer nesting on islands, where there are few predators. In Panama, *vitellinus* also seeks out nesting sites near wasps or in Cecropias protected by ants (*Azteca*), often those over water.

The nest is a hanging basket, averaging 43 cm in length, noticeably smaller than oropendola nests. Cacique nests are made from palm strips almost exclusively, and are thus quite uniform in appearance. Nest-building lasts from five to 22 days, averaging approximately 10 days. Yellow-rumped Caciques often nest side by side with several species of Oropendola and Red-rumped Caciques (14). The clutch size is normally two and the eggs are bluish with brown or chestnut spots; the egg-laying period lasts two to five days. Incubation begins after the second egg is laid and lasts approximately 13–14 days; the young take 24–30 days to fledge. Successful nests commonly fledge only one young, the fledgling period lasting 35–65 days.

This species is parasitised by Giant Cowbird (98). However, in Peru at least, such parasitism is rare as Yellow-rumped Caciques seldom leave the colony unattended giving little opportunity for the cowbirds to lay. The form *flavicrissus* has been observed to host eggs of Shiny Cowbird (102), but fledging has not been recorded from nests of this cacique (Friedmann and Kiff 1985). Cacique nests are pirated by Piratic Flycatchers (*Legatus leucophaius*) as well as Orange-backed Troupials (39), and Russet-backed Oropendolas (5) may drive off caciques nesting near its nest. Nest predators commonly raid cacique colonies; these include Capuchin monkeys (*Cebus* spp.), Cuvier's Toucan (*Ramphastos cuiveri*), Black Caracara (*Daptrius ater*) and Great Black-Hawks (*Buteogallus urubitinga*). The toucans and caracaras are the most common nest predators, but they are usually successfully mobbed by the caciques.

DISTRIBUTION AND STATUS Common and widespread in the lowlands to 900 m; a species almost restricted to the tropical zone but sometimes found in the upper tropical zone. One sight report from 1200 m in the Serrania Bellavista, La Paz, Bolivia (S. Herzog pers. comm.). The distribution is disjunct. In Panama, found from the Panama Canal Zone on the Caribbean slope and from Veraguas province on the Pacific slope south and east to Colombia. In Colombia, from Chocó north to the Santa Marta region but absent from Guajira, ranges south to mid-way up the Magdalena Valley (Tolima). Another population is found in the Pacific lowlands of Ecuador, almost reaching the border with Colombia and extending south to Tumbes in N Peru. East of the Andes it ranges throughout the lowlands from E Colombia (César to Putumayo and east), throughout Venezuela, Trinidad but not Tobago, the three Guianas, to Ceará, Brazil; south through E Ecuador (all provinces east of the Andes), E Peru (all departments east of the Andes), to NE Bolivia (Pando, Beni, La Paz, Cochabamba and Santa Cruz); in Brazil occurs in all states north of the Amazon and south to Mato Grosso, Goiás and Piauí. Another isolated population occurs in coastal NE Brazil from Pernambuco south to SE

Bahia. No records exist from Paraguay but it should be looked for in the north.

MOVEMENTS Sedentary, staying in the general area near the nest tree throughout the year.

MOULT Adults undergo a complete moult (definitive pre-basic) after the breeding season. Since the timing of breeding varies between populations and weather regimes, moult timing also varies. Early June specimens from Trinidad are well into their moult; while mid to late June birds from E Colombia are beginning to moult. A specimen from Goias, Brazil, was just finishing its pre-basic moult in June. The extent of the first pre-basic moult is unclear, but its timing is similar to that of the definitive pre-basic. It is unknown if pre-alternate moults are lacking, but this is suspected.

MEASUREMENTS ***C. c. cela***. Males: (n=12) wing 154.4 (135–168); tail 106.7 (95–123); (n=11) culmen 35.6 (34–41); (n=10) tarsus 32.9 (31–35). Females: (n=12) wing 127.8 (118–135); tail 91.6 (87–98); culmen 30.4 (28–33); (n=10) tarsus 28.5 (27–30).

C. c. vitellinus. Male (n=10) wing 172.5 (162.6–183.0); tail 112.4 (106.7–118.5); culmen 38.5 (36.0–40.6); tarsus 33.1 (31.0–35.1). Female (n=10) wing 133.5 (120.0–140.1); tail 93.7 (88.4–103); culmen 32.2 (31–34.2); tarsus 28.6 (27.0–32.2).

C. c. flavicrissus. Male (n=11) wing 152.7 (140–161); tail 102.3 (96–107); (n=8) culmen 32.7 (31–34); (n=10) tarsus 31.7 (30–33). Female (n=3) wing 116.5 (112–119.4); tail 84.1 (83.3–85); (n=2) culmen 29.2, 30; tarsus 26, 27.9.

NOTES The three forms were originally described as different species. They are morphologically and vocally different, and further work is needed to address the question of species status for these forms. In the Santa Marta region of Colombia, *cela* and *vitellinus* occur in close proximity to each other, but show no evidence of intergradation. It is reasonable to suggest that at least two species should be recognised. The plumage of this cacique has a distinctive musky odour (Wetmore 1939).

REFERENCES Belcher and Smooker 1937 [nesting]; Feekes 1981 [voice, habitat, behaviour, nesting], 1982 [voice, behaviour]; Hilty and Brown 1986 [nesting]; Lawrence 1864 [geographic variation, measurements]; Robinson 1985a [nesting, predators], 1985b [nesting], 1986a [description, measurements], 1986c [behaviour], 1986d [foraging, behaviour, nesting], 1986e [behaviour], 1988 [behaviour, immatures]; Sclater 1860 [measurements]; Trainer 1987 [voice *vitellinus*, behaviour], 1988 [*vitellinus*, voice, behaviour], 1989 [voice *vitellinus*]; Wetmore *et al.* 1984 [voice, nesting, measurements, *vitellinus*].

14 RED-RUMPED CACIQUE *Cacicus haemorrhous* Plate 5
(Linnaeus) 1766

This widespread species is the only cacique with a red rump in most of lowland South America.

IDENTIFICATION A black cacique with a large, red rump patch, blue eyes, and a thick bill with a slight casque. This is the only 'red-rumped' cacique in most of its range; however, it may come into contact with Subtropical Cacique (16) along the base of the eastern Andes of Colombia. Subtropical and Red-rumped Caciques are similar in size, however Red-rumped Cacique has a more extensive (to the lower edge of the mantle) and brighter red rump patch. The plumage has a noticeable blue gloss on Red-rumped Cacique which is lacking on Subtropical Cacique. In addition, Red-rumped Cacique has a straighter culmen and a deeper bill. Note also that the Red-rumped Cacique has a more noticeably notched tail than either Subtropical or Scarlet-rumped (15) Caciques. Scarlet-rumped Cacique is found west of the Andes so it is entirely allopatric with Red-rumped Cacique. In addition, it is much smaller, has a more restricted extent of red on the rump, is duller black, and has a thinner bill and noticeably curved culmen (form *pacificus*). When perched, Red-rumped Caciques hide their colourful rump and look all black with a blue eye. In these situations it may be confused for Solitary Cacique (21) which is roughly the same size. However, Solitary Cacique prefers moister habitats, forages lower down in the forest and has a dark eye. Ecuadorian Cacique (20) is all black with a blue eye, but it is much smaller and slimmer than Red-rumped Cacique and does not show a blue gloss.

VOICE Utters a variety of sounds, its variable repertoire making it difficult to state which sounds are distinctive for the species.

Song: The song is made up of three notes *dang-da-dang* and is uttered particularly when a female lands at the colony tree. This song is not colony specific. While displaying a soft series of *ke ke ke ke* notes is given, followed by a harsh grating call or a bell like *klang-klang* or *weuu-kleuu*, the full series may be considered its display song. The harsh grating call resembles the sound made by running a finger over a hair comb; however, it ascends the scale and appears to go 'over the top' at the end.

Calls: A common call given at the colony is a loud *tjew*. This loud call, which is frequently given, is perhaps the most distinctive sound of this cacique. Another common call is a harsh *gwash*, or *chwak* which is repeated often. Females give a series of loud shrieks. There are a variety of other sounds given, from sweet whistles to mechanical grating sounds and bugle-like calls.

DESCRIPTION A long-winged, short-tailed cacique, showing a noticeably notched tail. On males, the primaries are long and tapered, less so on females. The bill is long and shows a decurved tip to the culmen. The gonys is largely straight. Wing Formula: P9 < P8 < P7 > P6 > P5; P5 > P9 > P4; P8 to P5 emarginate. The nominate race is described.

Adult Male (Definitive Basic): The long bill ranges from greenish-ivory to pale yellow with a greener tip and base. Yellowish bills are perhaps more common in the southern race, *affinis*. The eyes are sky-blue. The body is almost entirely black, with a slight blue gloss. The rump patch is salmon-red, and extends up to the lower edge of the mantle. In addition, most of the uppertail-coverts, except the longest ones, are red. The legs and feet are blackish.

Adult Female (Definitive Basic): Like the male, but obviously smaller and slightly duller in colour, particularly on the rump and underparts. Some have dark eyes (young females?).

Immature (First Basic): Similar to the adults, but slightly duller in colour and tend to be smaller.

Juvenile (Juvenal): Duller still than immatures and always with dark eyes. The rump patch at this age is dull red and often barred dusky. In the field it may be the more loosely-textured plumage of juveniles may be noticeable, particularly on the flanks.

GEOGRAPHIC VARIATION Three subspecies are recognised.

C. h. haemorrhous inhabits the Guianas, northern Brazil, SE Venezuela and E Colombia. This race is characterised by having an extensive blue gloss on the upper and underparts, and the culmen is not swollen at the base but has a noticeable ridge. It is described above.

C. h. pachyrhynchus is found in the Amazon basin and along its southern tributaries, in Peru and Brazil. It is identified by the obvious swelling at the base of the culmen. It is not as glossy as *haemorrhous* and the gloss is more purple than blue. On average, the rump patch extends forward more than in the other two. North of the Amazon river, this form likely intergrades with *haemorrhous*, but few specimens exist from this zone.

C. h. affinis lives in forests south of the dry chaco-caatinga belt, in SE Brazil, N Argentina and E Paraguay. It is less glossy than *haemorrhous* and *pachyrhynchus*; the upperparts show a slight blue gloss yet the underparts are not glossed. This is most obvious in females which may show an obvious brownish colouration to the underparts. The culmen is not ridged or swollen at the base. The rump patch is on average smaller in this form and tends to be more orange in colour, particularly in females.

HABITAT Lives in a variety of habitats, but is much more of a forest bird than Yellow-rumped Cacique (13). It avoids wet várzea forest and rivers, preferring the upland terra firma forest. However, in eastern Venezuela, colonies are often located in swamps, or at least over water (Jaramillo pers. obs.). Colony trees tend to be along the forest edge or in more open areas within the forest. This species will nest in trees that are nearly covered by the forest canopy, unlike oropendolas and some other caciques. At times Red-rumped Caciques may be seen feeding in clearings and more open areas, particularly *affinis*. Habitat choice may be less restricted for *affinis*, which is not sympatric with Yellow-rumped Cacique.

BEHAVIOUR The mating system is polygynous in this species. However more males are present at the colony than in Yellow-rumped Cacique. Males do not have display perches, but will display throughout the nest tree, while following specific females. In display, the male positions his body horizontally often on fully extended legs. At the same time the body plumage is fluffed and the tail is raised. The wings are shaken and vibrated, and eventually raised well above the body while the red-rump is flashed. A short set of vocalisations is given at this point. The wings stop vibrating for an instant and the male begins to bob the head and body forward while giving a harsh note or a bell-like sound. Precopulatory displays differ in that the body plumage is not fluffed out and the bill is pointed upwards, with the head kept withdrawn close to the body, the wings are drooped, the tail cocked and the undertail-coverts raised. Males follow the females into the forest where copulation takes place.

NESTING Strictly a colonial nester. In Surinam, the breeding period is restricted to the dry season (September–May, peaking in January–February). No breeding occurs during the wet season as is the case in Yellow-rumped Cacique (Feekes 1981). However, Haverschmidt (1968) noted that this species breeds between June and November, and also sometimes in January. In eastern Venezuela, Red-rumped Caciques are actively nesting in mid March. Nesting peaks in March in Colombia. In the Manaus area, Brazil, Red-rumped Cacique nests from late December to late May, which is late in the wet season, and quite unlike other sympatric caciques and oropendolas. In Belém, Brazil, Red-rumped Cacique also nests late in the wet season after the sympatric Yellow-rumped Cacique has finished nesting (Oniki and Willis 1983). In northern Argentina, the breeding season begins in September to October and lasts until January. The birds nest in colonies which typically use the same tree year after year. Nest trees may be of any height, but they are usually in more enclosed habitat than those of Yellow-rumped Cacique. However, Red-rumped Cacique will nest in the same tree as Yellow-rumped Cacique and Crested Oropendolas (2). The race *affinis* is partial to nesting in palms at the forest edge. In the Manaus area, Brazil, only the Mulateiro tree (*Peltogyne* sp.) is used for nesting, probably due to its slick, smooth trunk which deters predators (Oniki and Willis 1983).

Colony size varies, being as small as 10 nests or as large as 40 or more. Nest-building and colony formation may be relatively quick. A colony with three nests under construction was discovered in late December, by late January 24 nests were present and by early February 35 nests were present (Oniki and Willis 1983). The nests of this species are not clumped to the degree observed in other caciques and oropendolas and may be quite evenly distributed throughout the tree. In some cases, clumping may be as extreme as in Yellow-rumped Cacique, and colonies are known to form around nests of wasps as is commonly the case in the latter species. However, wasp nests are not a prerequisite for colony formation. The nest is a typical hanging basket with an opening at the top. Nests average 38 cm in length and are constructed from a variety of vegetable materials. The nests of Red-rumped Cacique are shorter and bulkier than those of Yellow-rumped Cacique. In a situation where the two species are nesting side by side, these two species use different nesting materials from which to construct their nests (Loat 1898). All nests are built by the female and this takes an average of a week. Clutch size usually 2 eggs, white with purple or reddish spots and blotches, clustered at the broader end. Incubation lasts approximately 17 days and the young take 25–28 days to fledge. Most of the nestling care is performed by the female, but occasionally a male will feed the young. Females will accompany and feed fledglings for some time, probably for several months. Giant Cowbird (98) parasitises this species.

DISTRIBUTION AND STATUS The status of Red-rumped Cacique is variable, being common in the south of its range and in the Guianas, scarce in western Amazonia, but common again in the south (*affinis*). It is found to 1000 m. Ranges from E Ecuador through SE Colombia, SE Venezuela and the Guianas south along E Peru and N Bolivia to N Mato Grosso and E Belém, Brazil. The southern race is found from N Argentina (E Formosa, N Corrientes, Misiones) north along the E half of Paraguay into S Brazil south of the dry Caatinga to Pernambuco; south along the Atlantic Coast to Rio Grande do Sul and west to N Corrientes, Argentina.

MOVEMENTS Largely sedentary. However, this species is not always present in parts of Rio Grande do Sul, Brazil (Belton 1985), suggesting that seasonal movements may occur.

MOULT Adults undergo one complete moult a year, the definitive pre-basic, after the breeding season. The timing is dependent on when breeding occurs and this varies depending on the population. Birds from Pará, Brazil (*haemorrhous*), are undergoing moult in April. In N. Argentina (*affinis*), the moult period appears to be between January and April. No evidence exists for the presence of pre-alternate moults. The juvenal plumage is lost through the first pre-basic moult which has a timing similar to that of the definitive pre-basic moult. The first pre-basic moult is incomplete, involving the body plumage and coverts, and most or all of the primaries, secondaries and tertials.

MEASUREMENTS ***C. h. haemorrhous***. Males: (n=28) wing 175.8 (168–187.5); tail 105.9 (100–115); culmen 38.2 (34–41). Females: (n=7) wing 138.9 (132–143); tail 89.2 (83–94); culmen 32.9 (30–35); (n=4) tarsus 28 (25–30).

C. h. pachyrhynchus. Males: (n=13) wing 193.5 (183–210); tail 116.5 (107–125.5); culmen 40.4 (38–42). Females: (n=6) wing 146.6 (139.5–149); tail 95.2 (89–98); culmen 34.4 (33–35.5).

C. h. affinis. Males: (n=20) wing 175.2 (165–185); tail 115.9 (109.5–122); culmen 36.3 (34–39). Females: (n=11) wing 139.8 (137–142); tail 99.9 (97.5–101); culmen 30.9 (29.5–32).

NOTES The disjunct nature of the distribution of this species is interesting. Similar patterns are seen in many other neotropical species, having a population on either side of the dry chaco-caatinga belt. The genetic differentiation of birds showing this distribution pattern has not been assessed.

REFERENCES Belton 1985 [habitat, movements], Feekes 1981 [voice, habitat, behaviour, nesting], Hilty and Brown 1986 [nesting], Oniki and Willis 1983 [nesting], Parkes 1970 [geographic variation, measurements], Straneck no date [voice].

15 SCARLET-RUMPED CACIQUE *Cacicus microrhynchus* Plate 5 (Sclater and Salvin) 1864

A small red-rumped cacique of the tropical zone of Central America and the Pacific Coast of northern South America.

IDENTIFICATION A small black cacique with a restricted red rump patch. It has pale blue eyes, as well as an obvious curve to the culmen. Throughout its range, this is the only red-rumped cacique. The allopatric Subtropical (16) and Red-rumped Caciques (14) are much larger than Scarlet-rumped Cacique. Subtropical Cacique is also proportionately longer-tailed. Red-rumped Cacique has a noticeable bluish gloss to the plumage, as well as the rump patch being brighter scarlet and extending further up the back, to the edges of the mantle. Note the thinner bill of Scarlet-rumped Cacique (but also note geographic variation). A perched Scarlet-rumped Cacique may appear all black, similar to a Yellow-billed Cacique (23). However, Scarlet-rumped Cacique has blue, not yellow eyes, and is a different shape. Scarlet-rumped Caciques are smaller, shorter-tailed and thinner-billed than Yellow-billed Caciques. Yellow-rumped Cacique's (13) yellow wings and tail patches are always visible, and easily separate that species from Scarlet-rumped Cacique. Scarlet-rumped Tanager (*Ramphocelos passerinii*) of Central America and Flame-rumped Tanager (*Ramphocelus flammigerus flammigerus*-group) of W Colombia are also black birds with a red rump patch. They may be differentiated from the cacique based on their dark eyes, more extensive red on the rump, thick blue-grey bill with a slight hook, and different vocalisations.

VOICE Song: This cacique has a pleasant, musical song. The song is made up of clear whistles, for example *wheew-whee-whee-whee-wheet*. Often these notes are ascending, followed by a similar set of notes that descend in pitch. Some of these songs are fast and have a bubbling nature to them. While displaying, males give a screeching version of the song. There is a good deal of variability in songs, but not nearly as much as in the song of the sympatric Yellow-rumped Cacique. Females sing a similar, but slightly higher-pitched song. The *pacificus* race in Ecuador sings a melancholy *whip, wheeo wheeo wheeo*, each progressive *wheeo* descending the scale. It also gives a series of closely spaced *wheeo* notes.

Calls: One of the calls is a burry *reeo-reeo* or *shreew-shreew*. The corresponding call of the race *pacificus* is similar but sweeter, lacking the burry quality: *teeo* or *keeo*. The form *pacificus* also gives a drawn out *shweeeeee*, *shweee-ooooo*, or *sheeeoooo*, as well as a guttural *kraaaaa*.

DESCRIPTION A small cacique with a thin bill and a slightly curved culmen (see Geographic Variation). Wing Formula: P9 < P8 ≈ P7 ≈ P6 > P5; P9 ≈ P4; P7–P5 weakly emarginate. The nominate form is described.

Adult Male (Definitive Basic): The bill is pale yellow at the base, becoming greenish toward the tip. The eyes are sky-blue. Most of the body plumage is sooty-black, except for the rump patch. The rump patch is scarlet (with an orange tone) and includes the extreme lower back and rump. The uppertail-coverts are black. Usually the rump patch is covered by the black wings, making perching birds appear all black. Some of the feathers on the neck have pale bases; this is not visible in the field. The legs and feet are black.

Adult Female (Definitive Basic): As male, but slightly smaller and not as pure black. Females are blackish-grey.

Juvenile (Juvenal): Similar to adults, but the rump is duller, a brownish-orange rather than scarlet, often with some black tips to the feathers. The pale bill may show a brown suffusion, unlike in adults. The eyes are dark brownish-grey.

GEOGRAPHIC VARIATION Two subspecies are recognised. Pacific Cacique (*pacificus*) may be best considered a different species, but not enough data is available to confirm this. Note that vocalisations and morphology differ. The two forms are entirely allopatric. Subtropical Cacique has until recently been lumped with this species.

C. m. microrhynchus is found from E Honduras south to E Panama in E Panama and E San Blas provinces. It is described above.

C. m. pacificus (Pacific Cacique) is found from E Darién in Panama to El Oro in Ecuador. It is similar to *microrhynchus* but has a larger bill with a swollen base to the lower mandible. The concealed patch of white on the neck and dorsal area is more extensive, and is formed by white bases to these feathers. Differences in vocalisations appear to be consistent.

HABITAT This species lives in undisturbed, moist lowland forest or older second growth. It avoids cleared areas and has thus suffered a population decline as forests have been felled. In Costa Rica, it is an inhabitant of the upper and middle strata in heavily forested areas, but it also ventures out to more open areas particularly fruit plantations (Slud 1964). In southwest Costa Rica, it may be found in tall mangroves (Slud 1964). It is absent from arid areas in Central America, for example the Guanacaste region of Costa Rica. The subspecies *pacificus* is found in humid, lowland tropical forests, including edge and second growth (Hilty and Brown 1986). Uses dense stands of trees, particularly near rivers or wetlands for roosting.

BEHAVIOUR Travels and forages in pairs or small groups, often with a mixed-species flock, particularly canopy flocks and those including Black-faced Grosbeaks (*Caryothraustes poliogaster*) or larger flocking species (Nunbirds, Fruitcrows, Oropendolas etc.). Pairs break off from the flock during the breeding season. This species forages at all levels, but frequently remains relatively high in the canopy. When feeding, tends to probe and glean, particularly from epiphytes, often hanging upside down or performing other acrobatics to search a difficult place. The diet appears to be largely insectivorous, but does eat some fruit and nectar from flowers. A curious and active bird, but not as obvious as some of its larger relatives. While foraging, whistles or makes soft calls, unlike the raucous calls of many other caciques. At night, roosts in dense, vine-covered trees near water, often numbering many hundreds of birds. The mating system is monogamous, at least in solitary nesting individuals and the mating system in colonies is not known. Females and males often take part in duets; sometimes the female responds to the male's song while she is in the nest. The male's courtship display is similar to that of other caciques. He bows forward while raising the tail, spreading and slowly moving the wings while giving a screeching version of the song. Male *microrhynchus* will feed the young in the nest (Wetmore *et al.* 1984), unlike polygynous caciques. This cacique is quite curious, and may tamely approach observers while scolding.

NESTING Nest-building occurs between March and late June in Panama (Wetmore *et al.* 1984). Costa Rican birds nest early, many beginning before February and some continuing until May. In Costa Rica and Panama, this species nests solitarily for the most part. However, in eastern Panama province, coloniality is common with few individuals nesting solitarily (Wetmore *et al.* 1984). The form *pacificus* is also colonial in Colombia, but the colonies are small. The nest is the typical, long, basket-shaped nest of a cacique, roughly 60 cm in length or even as long as 70 cm. Nest placement varies from under 10 m to over 30 m off the ground, preferably near a wasp's nest. The female builds the nest, the male may accompany her on her searches for nest material but he does not help. Nest-building takes from one to over two weeks

(Skutch 1996). Sometimes two nests are observed side by side, one old and one new. This nesting pattern has been attributed to a bird returning to the same nest tree as used in the previous year (Skutch 1996). Clutch is of two eggs, which are white with blackish and brown spots as well as dark scrawls, particularly around the larger end of the egg. Both parents help to feed the nestlings.

DISTRIBUTION AND STATUS Common to uncommon. *Pacificus* is found to 800 m along the western Andes, the nominate form is found as high as 1100 m. Ranges from E Honduras (Olancho) south and east through Nicaragua, Costa Rica and Panama, and being found in the lowlands of both slopes of these countries. Also found in South America, west of the Andes in the Pacific lowlands, from W Colombia (lower Cauca valley, Antioquia) south to El Oro, Ecuador.

MOVEMENTS Sedentary, but forms larger flocks during the non-breeding season which wander about widely.

MOULT Not well known. Adults appear to undergo only one complete moult in the year (definitive pre-basic) after nesting has finished. Adult specimens from late July and early August from Costa Rica (Heredía, Limón provinces) were undergoing pre-basic moult. Specimens from W Panama were in extremely fresh plumage in early November. The extent of the first pre-basic moult is unclear, but it appears that a body moult is involved and at least some (all?) juveniles moult the tail and perhaps the wings as well. Juvenile specimens are known from June, suggesting that the first pre-basic may be timed similarly to the definitive pre-basic, largely in July. There is no evidence that pre-alternate moults occur in this species.

MEASUREMENTS ***C. m. microrhynchus***. Males: (n=10) wing 129.0 (122.0–136.5); tail 89.7 (83.3–96.5); culmen 29.6 (28.8–30.2); tarsus 28.8 (27.4–28.8). Females: (n=10) wing 113.9 (109.7–117.3); tail 80.4 (77.5–87.3); culmen 27.5 (25.7–29.2); tarsus 26.9 (25.7–27.8).

C. m. pacificus. Males: (n=10) wing 132.2 (122.2–137.2); tail 91.1 (87.8–94.9); culmen 29.6 (28.2–31.0); tarsus 29.5 (28.2–31.5). Females: (n=10) wing 119.4 (113.1–124); tail 83.7 (76.0–99); culmen 27.5 (25–28.9); tarsus 26.4 (24.5–28.5).

NOTES The two subspecies of Scarlet-rumped Cacique are allopatric, differing in morphology and vocalisations as well as nesting behaviour. They may be best considered separate species, in that case the name Pacific Cacique has been proposed for *pacificus* (Ridgely and Tudor 1989)

REFERENCES Bob Behrstock personal recordings [voice of *pacificus*], Ridgely and Tudor 1989 [distribution], Rodriguez 1982 [nesting], Skutch 1972 [voice, behaviour, nesting], Wetmore *et al.* 1984 [voice, geographic variation, habitat, behaviour, nesting, measurements].

16 SUBTROPICAL CACIQUE *Cacicus uropygialis* (Lafresnaye) 1843 — Plate 5

This is the only red-rumped cacique found in the subtropical zone of the Andes.

IDENTIFICATION A large black cacique with a restricted red rump patch and blue eyes. This species has a small, scarlet-orange coloured rump patch, unlike Red-rumped Cacique's (14) larger and brighter pure scarlet-red rump and lower back. The rump patch of Red-rumped Cacique extends to the lower edge of the mantle, not the extreme lower edge of the lower back as in Subtropical Cacique. The culmen of Subtropical Cacique is more curved than that of Red-rumped Cacique and the body plumage is not as glossy as that species. The two species also differ in their wing formula, with Red-rumped Caciques having noticeably narrower outer primaries, and with P9 longer than P4 and shorter than P5; P8 or P7 is the longest. Subtropical Caciques have P9 longer than P3 but shorter than P4, and P6 or P5 are the longest, thus giving a much more rounded wing. It can be surprisingly difficult to see the red rump of this species under normal viewing conditions. This can cause confusion with the sympatric, but all black, Yellow-billed Cacique (23). Eye colour sets these species apart. Subtropical Cacique has blue eyes and Yellow-billed Cacique has yellow eyes. Furthermore, Yellow-billed Caciques seldom venture from heavy cover, usually *Chusquea* bamboo, unlike Subtropical Caciques which may be observed high in the canopy, in edge habitats.

VOICE Song: The song appears to be a series of whistled *wheeop* notes followed by a *wheep-wheep-wheep-wheep-wheep*. Also gives a series of conversational, warbled whistles.

Calls: The common call is a *wheeop* reminiscent of the call of a Yellow-bellied Sapsucker (*Sphyrapicus varius*). A raspy *ckrr-ckrr-ckrr* also given along with an odd whinny.

DESCRIPTION A large cacique with a relatively long tail, which is moderately graduated. It has a short, shaggy crest. The bill is thick and has a noticeably curved culmen. Wing Formula: P9 < P8 < P7 < P6 > P5; P9 ≈ P3; P8–P6 emarginate.

Adult Male (Definitive Basic): The bill is pale ivory-coloured. The eyes are pale blue. Its entire body plumage, except for the rump patch, is black with a slight blue gloss. This gloss is not noticeable under most viewing conditions. The scarlet-orange rump patch extends from the extreme lower back to the lower rump. The uppertail-coverts are black. The legs and feet are black.

Adult Female (Definitive Basic): Similar to the adult male, but noticeably smaller. On good views, the black of the female can be seen to have a dull brownish wash. This is perhaps not possible to assess without direct comparison to a male bird.

Juvenile (Juvenal): Similar to the adult, but the plumage is blackish-brown rather than black. Iridescence is almost entirely absent from the body plumage except for the tips of the feathers of the upper back. In addition, the rump patch is orange, not scarlet.

GEOGRAPHIC VARIATION Monotypic. No evidence of geographic variation.

HABITAT Open montane forests in the subtropical zone and the upper tropical zone. It prefers forests on slopes, where it keeps to the top of the irregular canopy. It will venture into open areas that retain a few large, bromeliad-laden or flowering trees, but tends to keep to the forest and forest edge. The nesting habitat is not known.

BEHAVIOUR Most regularly observed foraging in pairs or small flocks of less than six birds. Flocks will search for food both in the understory as well as in the canopy and intermediate heights. The canopy appears to be the preferred foraging strata for this cacique. While foraging they are quite acrobatic, hanging upside down to gape into rolled, dry leaf clumps. Subtropical Caciques also probes tree bark for food, but does not hammer (like Yellow-billed Cacique). As well as insect prey, Subtropical Caciques probe and gape into flowers, often destroying them in the process, presumably to sip nectar. Subtropical Caciques join mixed-species foraging flocks, particularly those with large species such as *Cyanolica* and *Cyanocorax* jays and larger tanagers. Nothing is known of the mating or social system of this species.

NESTING Almost nothing known. Most reports of *'uropygialis'* nests refer to those of Scarlet-rumped Cacique which was formerly lumped with this species. An active nest (or colony?) was observed between March and May at Cueva de los Guacharos NP, Colombia (Hilty and Brown 1986). It is not clear if this species is colonial or not.

DISTRIBUTION AND STATUS Uncommon to locally common. Only found in the subtropical zone and upper tropical zone, from 1300 m (rarely 1000 m) to 2300 m. Found in the Andes from Venezuela to Peru. In Venezuela, restricted to the Sierra de Perijá in Zulia and Tachira, basically along the western border of the country. In Colombia, it has a disjunct range. It is found in the West Andes, from Cordoba south to Valle; in the Central Andes from Antioquia south to Quindio; on the western slope of the East Andes from Cundinamarca south to S Huila (head of Magdalena valley); the east slope of the East Andes in S Norte de Santander; also in W Putumayo (Hilty and Brown 1986). In Ecuador, only found in the subtropical zone on the east slope of the Andes. Similarly, it is restricted to the east Andean slopes in Peru, from the border with Ecuador south to Cajamarca. Appears not to be present in the environs of the North Peru Low, a gap in the Andean highlands through which the Río Marañón flows. Also found south of the gap, from S San Martin, west of the Río Huallaga southeast to Cuzco.

MOVEMENTS Appears to be sedentary. However, with any Andean species such as this one it is not unlikely that seasonal movements occur due to local changes in food supply.

MOULT One complete moult occurs a year in adults, the definitive pre-basic moult July–August in Ecuador. The juvenal plumage is replaced by the first pre-basic moult. This moult involves the body plumage. It appears that the juvenal primaries, secondaries and rectrices are not changed. The first pre-basic moult takes place during the same time as the definitive pre-basic moults. There is no evidence for pre-alternate moults.

MEASUREMENTS Male: (n=8) wing 156.5 (145–165); (n=9) tail 129.6 (107–142); culmen 30.8 (29–33); tarsus 33.0 (32–36). Female: (n=9) wing 138.4 (132–153); tail 117.9 (110–134); culmen 30.4 (28–32); tarsus 31.0 (28–33).

NOTES Up until recently, this species was considered a subspecies of Scarlet-rumped Cacique (15). Both Hilty and Brown (1986) and Ridgely and Tudor (1989) suggested that the differences in vocalisations, morphology and altitudinal distribution probably warranted the separation of these two forms as separate species. More recently, the two have been considered different species, i.e. Ortiz *et al.* 1990.

REFERENCES Hardy *et al.* 1998 [voice], Zimmer 1930 [identification].

17 GOLDEN-WINGED CACIQUE *Cacicus chrysopterus* (Vigors) 1825 — Plate 7

A small and slender cacique of southern South America, the only yellow and black cacique in the far south.

IDENTIFICATION A small black, shaggy-crested, cacique with a yellow rump and wing patch. It has yellow or whitish eyes, somtimes pale blue. The other yellow and black caciques either have more or less yellow in their plumage. Golden-winged Cacique comes to marginal contact with Yellow-rumped Cacique (13) and Mountain Cacique (18). Yellow-rumped Cacique is much larger and has the yellow of the rump extending to the base of the tail as well as onto the undertail-coverts. Its bill is much more curved. Mountain Cacique is variable in appearance, with the population that is likely sympatric with the Golden-winged lacking yellow on the wing-coverts. In addition, Mountain Cacique has the yellow of the rump extending to the lower back, unlike on Golden-winged Cacique. The little-known Selva Cacique (19) also lacks yellow on the wings. Note that the eye colour of Golden-winged Cacique tends to be yellow or whitish, only rarely blue as in the above-mentioned species. Also, particulary when elevated, the shaggy crest of Golden-winged Cacique is characteristic. The other sympatric caciques are either all black or have red on the rump. Other black and yellow blackbirds like the marshbirds (73 and 74), Yellow-winged Blackbird (58) and Saffron-cowled Blackbirds (55) are birds of open country and marshes, not forest or forest edge. The different habitat and behaviour of these birds is enough to separate them from this species, but beware that the small size of this cacique suggests an oriole or *Agelaius* blackbird.

VOICE Song: A loud series of notes delivered slowly with a final, loud burst. There are usually two hollow, low-pitched introductory notes, the first slightly lower than the second. An intermediate element is variable, sometimes a single buzz or purr, at other times a doubled *whick whick* for example. The final section is a loud, whistled *WHEEOO* or *WHOOP-WHEEO*. From a distance only the loud, final part is likely to be heard. The entire song may then sound like: *cuuk-kaa-Prrrr-WHIIP-WHEEEO* or *cuuk-kaa-whick whick-WHEEEO* to give two examples. Individual songs are variable, but the general pattern appears to be followed throughout the range. The notes and pitches are varied. The overall impression is of a pleasant, albeit loud song. The loudness and odd hollow nature of some of the sounds are typically cacique-like. Pairs appear to duet, with each individual singing a slightly different song in response to the notes given by its mate. It is known to mimic.

Calls: The common call is a mewing *charr* or *wreyur*.

DESCRIPTION A small, slender, black and yellow cacique. The bill shows a straight culmen and gonys. There is no frontal shield and the culmen tapers to a point at the forehead. A short crest is present which can be made quite conspicuous when raised. The wings are somewhat rounded and the primaries are not conspicuously emarginate as in some of the larger caciques. The tail is not obviously long, and is truncate, not noticeably graduated. Wing Formula: P9 < P8 < P7 ≈ P6 > P5; P9 ≈ P2; P8–P6 weakly emarginate.

Adult Male (Definitive Basic): The bill is pale greyish-blue. However, Wetmore (1926) noted that the bill colour varies, some being darker than others. The eye colour is also variable, ranging from orange to yellow to bluish-white and white. The body is largely black (slightly glossy) with a contrasting yellow wing patch and yellow rump band. The yellow wing patch extends along the lesser and median coverts, and the innermost greater coverts, the outermost greater and median coverts are black, this patch can be hidden by the scapulars at times. The yellow on the rump is restricted to a band that does not extend to the lower back or uppertail-coverts.

Adult Female (Definitive Basic): Like the male, but somewhat smaller.

Immature (First Basic): Similar to the adults but the primaries and secondaries appear duller and browner, contrasting with blacker coverts and tertials. The yellow rump feathers may be tipped with black.

Juvenile (Juvenal): Like the immature but the plumage is blackish-grey rather than pure black. The yellow rump and wing patches are dull and obscured by blackish feather tips. The eyes are greyish-brown.

GEOGRAPHIC VARIATION Monotypic, no described geographic variation.

HABITAT Forests and forest edge, from moist yungas forest along the base of the Andes to drier gallery woodlands on the edge of the chaco as well as dry woodlands (for example, cerrado and chaco woodlands). It is fond of second growth and other open forests bordering either rivers or roads. Golden-winged Cacique avoids lowland rainforest.

BEHAVIOUR Tends to move about in pairs or family groups rather than in flocks, and it is usually seen alone. Feeds in trees on fruit and insects, by diligently probing into epiphytes, dry leaves and flaking bark. It keeps to the middle strata, but does feed in the canopy. It is an active and curious cacique, resembling an oriole in its general behaviour. Males sit in a prominent position, high in a tree or near the edge of a tree, while singing. During the height of singing, the male may lean forward so as to be almost completely upside down. Singing males have also been noted to bounce rhythmically while singing (Sick 1993). Pairs perform antiphonal duets, the female answering the male's song with a slightly different version of the vocalisation (Belton 1985). The mating system is not known, but is most likely monogamous given this species' solitary nesting behaviour.

NESTING Nesting in Paraguay starts in late September. Nests are active in October in southern Brazil. In Argentina, nests with eggs have been observed between mid October–early December. This species is a solitary nester. It builds a hanging basket nest, 0.6–1 m in length; a very long nest given that this is a small cacique. Many, but not all, of the nests are woven with black fibres that are apparently the hyphae of a fungus of the genus *Marasmius* (Belton 1985). Nests are placed relatively low in trees, usually less than five meters off the ground. The clutch size is of three eggs. The eggs are white with chestnut, grey or brown spots concentrated on the wide end. It has been known to have been parasitised by Shiny Cowbird (102) in Argentina. However, the nests are probably too small to allow Giant Cowbird (98) to enter.

DISTRIBUTION AND STATUS Uncommon to common. It is found in two isolated populations. The western population ranges along the semi-humid lowland and montane forest zone on the east slope of the Andes and foothills from Santa Cruz, Bolivia, south to Tucumán, Argentina. The other population is found from Paraguay, excluding the far west; east through southern Brazil as far north as S Mato Grosso, São Paulo and W Rio de Janeiro; south to Uruguay and N Buenos Aires, Argentina; and found west to Santa Fé, E Chaco and E Formosa, keeping largely east of the dry chaco zone. It appears that these two populations are separated by the dry chaco zone, but it is not inconceivable that this species could be found in gallery woodlands along rivers crossing the chaco region. It has been known to range to 2000 m in Bolivia and Argentina.

MOVEMENTS Apparently sedentary.

MOULT Not well known. The complete definitive pre-basic moult occurs after breeding, roughly between February–March. The first pre-basic moult is partial, involving the body, wing-coverts and often the tertials and tail. Pre-alternate moults appear to be lacking.

MEASUREMENTS Males: (n=16) wing 102.5 (99.4–106.4); culmen 22.6 (20.0–23.8); tail 89.2 (85.0–94.3); tarsus 24.2 (23.1–25.6). Females: (n=10) wing 94.1 (88–105); tail 89.7 (82–96); culmen 21.1 (20–22); tarsus 24.1 (23–25).

NOTES This species and some of the other smaller caciques were initially considered to comprise a distinct genus, *Archiplanus*. When classified in that genus Golden-winged Cacique was known as *Archiplanus albirostris*, a name not available (due to complexities of zoological nomenclature) for it when placed in the genus *Cacicus*. An unusual male specimen housed in the Smithsonian Institution from Puerto Pinasco, Paraguay, shows scattered yellow feathers on the crown, nape and above the eye. The distribution of yellow feathering was not symmetrical, being more prevalent on the right side of the bird.

REFERENCES Belton 1985 [voice, nesting], De La Peña 1987 [nesting], Hardy *et al.* 1998 [voice], Lowery and O'Neill 1965 [measurements], Ridgely and Tudor 1989 [voice, nesting, distribution], Straneck 1990f [voice], Wetmore 1926 [voice, nesting].

18 MOUNTAIN CACIQUE *Cacicus chrysonotus* **Plate 7**
(Lafresnaye and d'Orbigny) 1838

The only black and yellow cacique likely to be observed in the subtropical and temperate zone of the Andes, north of Bolivia.

IDENTIFICATION A medium-sized black cacique with a yellow rump patch. In the north of the range they also sport a yellow wing patch. This is the only black and yellow cacique found in its montane range; however, it may occasionally come into contact with Yellow-rumped Cacique (13) and Golden-winged Cacique (17). Yellow-rumped Cacique is slightly larger and always shows yellow at the base of the tail. In addition, it has a yellow crissum unlike Mountain Cacique. The bill of Mountain Cacique shows a grey base, and a straight (or nearly straight) culmen, unlike the ivory-coloured bill with an obviously curved culmen of Yellow-rumped Cacique. The two species also differ vocally. Golden-winged Cacique is much smaller and slimmer than Mountain Cacique. The yellow on the rump of Golden-winged Cacique is restricted to a thinner band not extending to the lower back as on Mountain Cacique. The eyes of Golden-winged Cacique are most often yellowish or orange, but may be whitish-blue. Note that the area of likely contact between these two species is in Bolivia, where Mountain Cacique lacks yellow on the coverts, easily separating it from Golden-winged Cacique.

VOICE A noisy species, which utters largely unpleasant sounding notes.

Song: The song (*chrysonotus* in Peru) is a continuous series of hollow metallic whistles, some of which are repeated several times, before the next call syllable type is given. Meanwhile, another bird answers with a hawk-like, descending whistle *wheeeaaa*. This appears to be a duet of sorts. A full song may go on for several minutes. The varied song has also been rendered as *arr tjie tjie tiuee arr arr tiue kik...kik tjiue* (Fjeldsa and Krabbe 1990). A quickly delivered *waak-wee-wheehnk-wheehnk* is also given.

Calls: The common call is a frequently repeated, whining *wheehnk* or *whaak*, of variable pitch and frequency depending on the level of excitement of the bird. This call can sound jay-like. Hilty and Brown (1986) aptly described this call as gull-like. Sometimes utters a sharp, descending whistle *tsweeeeee*, or a modulated whistle that first descends in pitch and then rises, *tweeaaaweee*. Ridgely and Tudor (1989) state that the vocalisations of the northern *leucoramphus* group and the southern *chrysonotus* group are similar. However, recordings of the primary calls (the *wheehk* or *whak* call) appear to show a difference. The call note of southern birds sounds more high-pitched, metallic and faster; while northern birds sound lower-pitched, harsher and more powerful. More research needs to be done to elucidate similarities between the vocalisations of northern and southern forms.

DESCRIPTION A medium-sized, slim cacique with a rather thin bill, with a straight culmen and gonys. It is long-tailed, but not particularly long-winged. The tail is only slightly graduated, looking fan-shaped in the field. Wing Formula: P9 < P8 < P7 < P6 > P5; P9 ≈ P3; P7–P5 weakly emarginate. The form *leucoramphus* is described.

Adult Male (Definitive Basic): The bill is pale horn with a bluish base and a yellower tip. Eyes pale blue. The body is largely black with a dull gloss. The lower back, rump and some wing-coverts are lemon-yellow. Most of the lesser coverts are black, except for the lowermost inner ones. The median coverts are yellow except for the outer ones (nearest the wing edge), which are black. The inner greater coverts are also yellow. The result is a wing patch that does not reach the wing edge and is 'L' shaped. The underwings are entirely black. The tail is completely black, as are the uppertail-coverts. The legs and feet are blackish. There is a concealed white collar around the neck, upper back and breast formed by white bases to those feathers; if ruffled, this collar may be visible in the field.

Adult Female (Definitive Basic): Similar to the male, but slightly smaller and the plumage is a duller greyish-black.

Juvenile (Juvenal): Duller coloured than the adults, like the adult female but has a dark bill with a light tip. The eyes are dark. The body plumage is dark brown rather than black. The looser texture of the juvenal feathering is noticeable, particularly on the flanks.

GEOGRAPHIC VARIATION Three subspecies fall into two groups: *leucoramphus* [Northern Mountain Cacique] and *chrysonotus* [Southern Mountain Cacique]. These two forms were listed as separate species (Blake 1968), but more recently the two have been considered conspecific (Ridgely and Tudor 1989) mainly due to the presence of intermediate-looking specimens. Specimens of *peruvianus* from Auquimarca, Peru, show black fringes to the yellow-wing coverts (Bond 1953). However, the vocalisations do not appear to be similar, as has been stated (Ridgely and Tudor 1989). The two are certainly good phylogenetic species, and perhaps even good biological species. More work needs to be done before a decision as to the systematic placement can be made.

Northern Mountain Cacique is composed of two races.

C. c. leucoramphus occurs from NW Venezuela, south along the E Colombian Andes to E Ecuador. It is described above.

C. c. peruvianus inhabits the east slope of the Andes from Amazonas south to Junín. It is similar to *leucoramphus*, but has a heavier bill, with a more noticeably arched culmen. The bluish base to the bill is less extensive. In addition, the concealed white collar is thinner in this form than in *leucoramphus*.

Southern Mountain Cacique, ***C. c. chrysonotus***, is found south of Junín, Peru, to Cochabamba, Bolivia. It lacks yellow on the wing-coverts, but some individuals may show one or two yellow covert feathers or small yellow tips to several coverts. This is particularly true in the northern part of its range. However, some Bolivian *chrysonotus* specimens, collected far from any potential intergradation zones, showed yellow fringes to some of the wing-coverts (Bond 1953).

HABITAT Humid montane forest and cloud forest. Mountain Cacique is found both in subtropical and temperate zone forests (Remsen and Traylor 1989, Rasmussen *et al.* 1994). It is fond of bamboo (*Chusquea*) thickets. Mountain Cacique does not leave the forest and venture into cleared areas as does the sympatric Subtropical Cacique (16), but needs contiguous forest. However, it will forage near edge or in more open areas such as in the vicinity of narrow roads.

BEHAVIOUR Forages in small parties or pairs, often in mixed flocks with other medium-sized birds such as Turquoise Jay (*Cyanocorax turcosa*), Green Jay (*Cyanocorax yncas*), Hooded Mountain-Tanager (*Buthraupis montana*) and Grass-green Tanager (*Chlorornis riefferii*). These associations have been commented on by several authors and seem worthy of closer study. This species also flocks with confamilials such as Subtropical Caciques and Russet-backed Oropendolas (5). Flocks may contain up to 15 individuals, particularly if foraging with full-grown juveniles. Mountain Caciques are acrobatic and enjoyable to watch, often clinging from vines and probing into dry rolled leaves or into epiphytic bromeliads. Most of their foraging occurs in the upper or mid-levels of the forest, but they will come down to feed in bamboo thickets, probing for food at the leaf bases. When alarmed, this species cocks its tail while calling, as well jumping and rotating its body 180º, thus facing the opposite direction, nervously repeating this move several times (P. Burke pers. obs.). The courtship display is unknown.

NESTING In Colombia, birds in breeding condition have been collected between February–July (Hilty and Brown 1986). Fledglings have been recorded in January and March in NW Ecuador, and in October in NE Ecuador; in December in Puno, Peru, and in October in Cochabamba, Bolivia (Fjeldsa and Krabbe 1990). An active colony near Baeza, Napo, Ecuador, with six nests was observed in mid-February (P. Burke pers. obs.). The adults were observed entering the nests without food, and the nests were fully constructed by that time. Mountain Caciques have also been described as solitary nesters (Hilty and Brown 1986, Fjeldsa and Krabbe 1990). The nests are typical of caciques, a long, hanging basket (over 50 cm long) with the entrance at the top. They are made of coarse material, rather than fine strips. In the Baeza colony, four of the six nests were clustered at the tip of a branch of a tree, while the other two nests were isolated but within the same tree. The tree was situated next to a more open area with a dense stand of bamboo which lacked any overhanging of trees. The nests were situated in the open, overhanging the lower canopy of the down-sloping terrain.

DISTRIBUTION AND STATUS Usually uncommon, more common in the south. It is restricted to the Andes, largely at elevations of 1800–3300 m. Has a disjunct population in the east Andes of Colombia from Cundinamarca north to SW Venezuela (Tachira). Otherwise, found along the west and central Andes of Colombia from Antioquia south along the east slope (and the extreme northwest slope) in Ecuador, Peru and Bolivia reaching as far as Cochabamba and Santa Cruz. There appear to be breaks in the range, the main one caused by the Río Marañón valley (N Peru).

MOVEMENTS Apparently sedentary, but may undergo minor elevational movements as do many other montane birds.

MOULT The definitive pre-basic moult is complete. Pre-alternate moults appear to be lacking. Birds in Santander and Antioquia, Colombia, (*leucoramphus*) are in moult during July–September. However, moult timing varies due to differences in breeding seasons between different populations. In other parts of Santander, Colombia, moulting birds are known from February, while in Tolima, Colombia, December is the moult season, and in Cauca, Colombia, October is when the moult takes place. Specimens of *peruvianus* from Peru are in moult during May. Other specimens from June and July of *peruvianus* are moderately worn and not in moult. Breeding takes place in October to February for these forms, thus the definitive pre-basic moult likely occurs March–May. The southern form *chrysonotus* from Puno, Peru, moults during May. The extent of the first pre-basic moult in this species is not clear, but it certainly replaces the body feathers, and some or all of the primaries and secondaries are replaced.

MEASUREMENTS ***C. c. leucoramphus***. Male: (n=10) wing 156.5 (147–170); tail 144.2 (138–152); culmen 31.7 (30–34); tarsus 33.5 (31–35). Female: (n=10) wing 129.7 (121–136); tail 121.4 (115–128); culmen 27.1 (26–28); tarsus 29.5 (29–31).

C. c. peruvianus. Male: (n=7) wing 144.1 (127–156); tail 132.6 (117–142); culmen 31 (27–33); tarsus 33 (29–36). Female: (n=4) wing 126 (122–130); tail 128 (115–135); culmen 28.8 (27–30); tarsus 31.3 (30–33).

C. c. chrysonotus. Male: (n=7) wing 149.6 (145–155); tail 146.0 (142–152); culmen 32.9 (31–34); tarsus 35.0 (33–37). Female: (n=8) wing 128.6 (120–133); tail 128.7 (122–140); (n=7) culmen 28.7 (27–32); tarsus 31 (26–33).

NOTES A curious specimen of a cacique previously named *Cassiculus melanurus* Cassin was analysed and found to pertain to a *leucoramphus* Mountain Cacique which had yellow feathers glued on to the rump (Meyer de Schauensee 1945).

REFERENCES Fjeldsa and Krabbe 1990 [voice, nesting], Hardy *et al.* 1998 [voice], Hilty and Brown 1986 [behaviour, nesting, distribution], Meyer de Schauensee 1945 [notes, *C. melanurus*], Ridgely and Tudor 1989 [habitat, distribution], Zimmer 1924 [geographic variation, measurements].

19 SELVA CACIQUE *Cacicus koepckeae* — Plate 23
Lowery and O'Neill 1965

This is probably the least known blackbird, with no confirmed observation since 1965.

IDENTIFICATION This is a small, long-tailed cacique. It is black with a yellow rump. This species is outwardly similar to Golden-winged Cacique (17) but is larger and has no yellow on the wings. In shape and structure it more closely resembles Ecuadorian Cacique (20). The colour of the rump is more golden-yellow than on Golden-winged Cacique and the extent of the yellow is greater. The tail is more graduated on Selva Cacique. The bill is broader than in Golden-winged Cacique, particularly the base of the culmen which is expanded and rounded. The eye colour of these two species differs, being pale blue in Selva Cacique and orange-yellow in Golden-winged Cacique. Selva Cacique is smaller than Mountain Cacique (18) and is glossier black. The widespread Yellow-rumped Cacique (13) is the only one of the yellow and black caciques that presumably comes into contact with this species. Yellow-rumped Cacique has the yellow of the rump extending onto the tail base as well as onto the undertail-coverts, unlike in Selva Cacique. The culmen of Yellow-rumped Cacique is much more curved than on Selva Cacique. Finally, the similarly-shaped Ecuadorian Cacique may pose an unappreciated identification problem. A juvenile Ecuadorian Cacique specimen examined from Ecuador shows a brownish-black plumage and yellow tips to many of the rump feathers. It is unknown if all juvenile Ecuadorian Caciques show yellow on the rump. In any case, these yellow-rumped Ecuadorian Caciques could easily be mistaken for Selva Caciques. However, the juvenile Ecuadorian Cacique only shows yellow tips to the rump feathers, not a solid yellow rump. It is expected that juvenile Ecuadorian Cacique lacks the blue eye colour of Selva Cacique, this being an adult character.

VOICE Unknown, there are no descriptions or recordings of vocalisations available.

DESCRIPTION A medium-sized cacique with a long tail. It has a straight culmen and gonys, the culmen having a rounded base.

Adult Male (Definitive Basic): The bill is blue-grey with a paler tip. The straight culmen is broad and flat. The crown feathers are long, creating a crest of moderate length. The eyes are bluish-white. Selva Cacique is entirely glossy black except for the rump. The rump is golden-yellow. The black tail is slightly graduated. The legs are blackish. One of the known specimens has a single greater covert with a obscure yellow tip.

Adult Female and **Juvenile** These plumages are unknown.

GEOGRAPHIC VARIATION Monotypic.

HABITAT Humid, low-lying forests, perhaps restricted to river and lake-edge thickets (O'Neill pers. comm.). It has been observed flushing from the ground, but it is unlikely that this species forages on the ground.

BEHAVIOUR During the one sighting when the type specimen was collected, a party of six individuals were viewed bathing and drinking. This is the largest number ever observed. It is believed that this species may be most closely allied to Ecuadorian Cacique.

NESTING Unknown.

DISTRIBUTION AND STATUS This species is known from only two specimens. No aspects are known of its biology and it should be considered extremely endangered. The only locality it is know from is Balta, Loreto, Peru. This site is in the lowlands at 300 m. There are unconfirmed sightings from Manu NP, Madre de Dios, Peru.

MOVEMENTS Unknown.

MOULT Unknown.

MEASUREMENTS Males: (n=2) wing 109.3 (109.0–109.5); culmen 26.4 (26.3–26.4); tail 102.3 (100.8–103.8);tarsus 26.3 (26.1–26.4).

NOTES This species has been typically considered the closest relative of Golden-winged Cacique, based on Lowery and O'Neill (1965). However, it has recently been suggested that based on morphological and biogeographical considerations it is most closely allied to Ecuadorian Cacique (T. Parker in Cardiff and Remsen 1994), a notion seconded by O'Neill (pers. comm.) who considers his and Lowery's earlier ideas to be in error. O'Neill (pers. comm.) suggests that Selva and Ecuadorian Caciques possibly comprise a superspecies, or are even subspecies of one species, perhaps historically divided by the Amazon River. Two specimens of Ecuadorian Cacique, one from Ecuador and one from Peru which I examined, show yellow-tipped rump feathers which support this hypothesis (see Identification). A concerted effort should be made to find this species in the wild and study its ecology and conservation needs. This cacique is named in honour of Maria Koepcke, pioneer of Peruvian ornithology.

REFERENCES Cardiff and Remsen 1994 [notes], Collar *et al.* 1992 [status, habitat], Lowery and O'Neill 1965 [identification, description, distribution, measurements], Ridgely and Tudor 1989 [behaviour, distribution].

20 ECUADORIAN CACIQUE *Cacicus sclateri* (Dubois) 1887 — Plate 8

An all-black cacique with a very restricted range. Little is known about this species.

IDENTIFICATION A slim, black cacique with bright blue eyes, and a long-tail. This small, slender cacique is only likely to be confused with Solitary Cacique (21) which is also entirely black. However these two species differ in their proportions, habitat, behaviour and eye colour. Adult Ecuadorian Caciques have pale blue eyes whereas adult Solitary Caciques have dark eyes. Young Ecuadorian Caciques probably have dark eyes and appear similar to Solitary Caciques. However, Ecuadorian Caciques are smaller and very much slimmer, note the weaker legs and feet of Ecuadorian Cacique. Also, Ecuadorian Cacique is not a skulker or a bird that is particularly associated with water like Solitary Cacique. The bill of Ecuadorian Cacique is noticeably thinner than that of Solitary Cacique. Three other black icterids that are sympatric with Ecuadorian Cacique are Shiny Cowbird (102), Velvet-fronted Grackle (81) and Pale-eyed Blackbird (56). These species lack the blue eyes of the cacique and have shorter, stubbier bills, shorter tails and are less slim in appearance. Pale-eyed Blackbird has yellowish eyes.

VOICE Song: The song is short and sharp, lasting for just over one second. It begins with two low notes, and ends with a final, drawn-out, loud hollow whistle, *wha-kuu-CHEEAAA*. A ringing *pee-churr, pee-churr, pee-churr, chur-chur-chur* was described as the song by Servat and Pearson (1991), but this may actually be either a different song type or another category of vocalisation altogether.

Calls: A doubled *kip-pheew*, the first note on an even pitch while the second descends in pitch; perhaps this is the *pee-churr* described above by Servat and Pearson (1991). This call is given in situations of alarm. The first note of the pair, the *kip*, is given singly and repetitively at times. Also gives a series of mellow *tweeew* notes, sometimes doubled. These notes are pleasant and musical, unlike the harsh calls of several other caciques. However, a loud *tchak-tchak* is also given.

DESCRIPTION A small, slim, black cacique with a sharply pointed bill, having a straight culmen and gonys. The bill is rather short and stout for a cacique. The tail is relatively long and rounded. The primaries do not show noticeable tapering and the wings are somewhat rounded; the outermost primary ≈P9) is shorter than the next series of which P8, P7, P6 and P5 are similar in length. Wing Formula: $P9 < P8 < P7 \approx P6 > P5$; $P9 \approx P2$; P8–P6 emarginate.

Adult Male (Definitive Basic): Bill greenish-yellow or pale blue-grey with a paler tip. Eyes sky-blue. The plumage is jet black, showing a slight blue gloss, but without any red or yellow markings. The primaries may appear slightly browner than the rest of the plumage. Legs and feet black.

Adult Female (Definitive Basic): As in the male, but slightly smaller.

Juvenile (Juvenal): A juvenile in the collection of the American Museum of Natural History resembles the adults but the plumage is duller and browner, rather than black. Interestingly this bird shows yellow tips to many of the rump feathers. This appears to be the only juvenile known of this species. It would be interesting to determine whether all juveniles show these yellow-tipped rump feathers.

GEOGRAPHIC VARIATION Monotypic.

HABITAT Lowland tropical forest or forest edge, and will feed in plantations near forest. Tends to forage in the upper or middle strata of forest. Appears to have a preference for floodplain forest, and seasonally flooded forest (várzea).

BEHAVIOUR Forages singly or in pairs, usually high in the canopy but will come down to mid or low levels, particularly in disturbed areas or edge. It is not skulking like Solitary Cacique, which is much more of an understory species. It will come to flowers, presumably feeding on nectar and insects. Stomach contents have also included fruits, so like other caciques, this species appears to be somewhat of a generalist when foraging. The display is unknown. Joins mixed-species flocks of tanagers and flocks with Orange-backed Troupials (39).

NESTING Details unknown. Based on the timing of the definitive pre-basic moult and the presence of a juvenile specimen from April it appears that nesting has finished by February.

DISTRIBUTION AND STATUS Rare and little known, but reported to be locally common on the Samiria River, Peru. Ranges from 200–550m in elevation. Has been found in E Ecuador in the upper Rio Napo (Jatun Sacha, Misahuallí and Avila) in Napo Province, down to La Selva lodge, and in the Sucumbíos province at Rio Aguarico (Santa Cecilia) as well as in nearby N Peru in Amazonas (Huampami) and N Loreto. Probably occurs throughout areas between these sites. The most reliable and easily accessible site to see this species appears to be near Tena in Ecuador, particularly on the road to the Jatun Sacha reserve and perhaps near Misahuallí. It is unrecorded in Colombia, but it should be looked for in the south.

MOVEMENTS Unknown, but probably sedentary.

MOULT Largely unknown. Adult specimens in moult are found between February–May. These birds were undergoing the complete definitive pre-basic moult. There is no evidence of pre-alternate moults. The extent of the first pre-basic moult is unknown.

MEASUREMENTS Males: (n=10) 105.0 (93–111); tail 103.3 (95–111); (n=9) culmen 25.3 (23.9–27); (n=10) tarsus 25.6 (23–28). Female: (n=2) wing 95, 101; tail 92, 96; culmen 22, 23; tarsus 23, 24.

NOTES This rare species is little known and is perhaps best considered threatened. However, its status is still 'indeterminate'. This cacique is probably most closely related to the rare Selva Cacique (19) as was suggested by T. Parker in Cardiff and Remsen (1994), a notion seconded by O'Neill (pers. comm.). John O'Neill (pers. comm.) suggests that Selva and Ecuadorian Caciques possibly comprise a superspecies, or are even subspecies of one species, perhaps historically divided by the Amazon river. Two specimens of Ecuadorian Cacique, one from Ecuador and one from Peru, which I have examined, show yellow-tipped rump feathers which support this hypothesis. The juvenile is mentioned above, and a specimen in the Museum of Vertebrate Zoology from Rio Cenepa, Huampami, Amazonas, Peru, collected on Oct. 7 1972, shows one rump feather with yellow edges.

REFERENCES Dubois 1887 [description, measurements], Hardy *et al.* 1998 [voice], Moore 1996 [voice], Ridgely and Tudor 1989 [distribution, behaviour], Servat and Pearson 1991 [voice, status, behaviour, habitat].

21 SOLITARY CACIQUE *Cacicus solitarius* (Vieillot) 1816 — Plate 8

A skulking black cacique, often associated with water.

IDENTIFICATION Solitary Cacique is an entirely black cacique of the South American lowlands. It has a pale, whitish bill and a dark eye. There is no appreciable iridescence to the plumage. Its range overlaps with Ecuadorian Cacique (20) which is also black, with several other caciques which have red or yellow in the plumage, and with Giant Cowbird (98). Solitary Cacique is perhaps most similar to Yellow-billed Cacique (23), yet the ranges of these species do not overlap. Solitary Cacique has dark eyes, a duller-coloured bill and a shaggy crest which Yellow-billed Cacique lacks. Note also that Solitary Cacique is larger and slightly longer-winged. Note the differences in bill shapes, Yellow-billed Cacique having a chisel-shaped bill and the nostrils showing a flap of skin partly covering the opening (operculate nostril) while Solitary Cacique has an open nostril and a longer and more pointed bill. A sympatric and potential confusion species is the rare and local Ecuadorian Cacique. However, Ecuadorian Cacique is not a skulker and has vivid blue eyes as an adult. Immature Ecuadorian Caciques are dark-eyed and could be extremely confusing, but should be accompanying their blue-eyed parents at this stage. Note that Ecuadorian Cacique is slimmer and smaller than Solitary Cacique and in addition has longer wings which show up as a longer primary projection beyond the tertials on the folded wing. The bill of Solitary Cacique is relatively thick at the base, while that of Ecuadorian Cacique is noticeably slimmer, not so deep at the base. Giant Cowbird is black and large, like Solitary Cacique but as an adult the cowbird is much more glossy, has a dark bill and yellow or red eyes. However, juvenile Giant Cowbirds are less glossy and show pale bills and dark eyes. During this stage of its life, a juvenile Giant Cowbird is likely to be accompanied by its hosts.

Shape differences between the two species should be used to confirm the identification. Giant Cowbird has a different bill to the cacique, in particular an appreciably decurved culmen while the bill of Solitary Cacique has a straight culmen. The peculiar, small-headed look of Giant Cowbird is already noticeable on the juveniles; the cacique has a normally proportioned head. Other differences in proportions include the relatively longer primaries of Giant Cowbird (showing a long extension past the tertials) and the shorter and square-ended or notched tail. Solitary Cacique has a rounded tail tip and a longer tail. Red-rumped Cacique (14) and the rare Selva Cacique (19) may perch such that the colourful rump and lower back is hidden, making these species appear entirely black. However, Red-rumped and Selva Caciques have pale blue eyes. If there is any uncertainty to the identity of the bird, eventually a red or yellow rump will be visible on either of these species when it shifts position or flies away. Similarly, all largely black oropendolas can be identified by yellow in the tail, a feature that is visible both from above or below. Note also that Solitary Cacique is much more skulking and unobtrusive than any other *Cacicus* cacique. Finally, its varied repertoire of calls, particularly the cat-like 'mewing' aid in confirming an identification (see Voice).

VOICE Gives a variety of sounds, some most peculiar in quality. In the chaco of Paraguay, Wetmore (1926) mentioned that the locals called this cacique the 'que ve', in imitation of its common call; the Toba natives called it 'kom kom'.

Song: The song repertoire is extensive and birds will switch between different versions of the song quite often. The typical pattern is for it to begin with some very faint low notes followed by a loud series of staccato whistles or liquid squeals. Some examples are: *whup whup whup-TEEEEEWEEEEEu-uuu* the first notes almost inaudible and the terminal one loud and descending, *whoo-TEWEEE KAAAAAKUuuu, whup whup-TEW TEW TEW TEW TEW TEW TEW ti ti ti ti ti ti* with the final series faster and softer than the middle series of notes. Another song type begins with a low growl, followed by a doubled hollow vocalisation, *grrrr-KWHEAA KWHEAA*. It is known to be a mimic.

Calls: Typical is a doubled *quek-quek* or *quay-quay*. Another call is a loud mewing *wheeeeah*, as well as a harsh scold reminiscent of calls made by *Cyanocorax* jays, low growls, and odd frequency modulated whistles.

DESCRIPTION This is a broad-tailed, rounded-winged cacique. It has a short, shaggy crest that is often visible. Wing Formula: P9 < P8 < P7 ≈ P6 ≈ P5 > P4; P9 ≈ P1; P8–P5 emarginate. The eight primary is longer than the third, unlike in Yellow-billed Cacique.

Adult Male (Definitive Basic): The bill colour is this species' most obvious mark. It is a pale greenish-white, becoming slightly greyer at the base. The eyes are brown. The plumage is entirely black, with little gloss. The legs and feet are blackish. The tail is long and somewhat rounded, often the tips of the tail feathers are well worn.

Adult Female (Definitive Basic): Like male but slightly smaller.

Immature (First Basic): Similar to adults, but the black body contrasts with the browner primaries and secondaries. This contrast is most noticeable between the black wing-coverts and the browner greater primary coverts, primaries and secondaries.

Juvenile (Juvenal): Like adults, but with a greyer colour to the feathers and a duller bill. Gyldenstolpe (1945a) describes the bill as blackish except for a dull whitish tip, but is not clear if this is based on live birds or specimens, but we suspect the latter.

GEOGRAPHIC VARIATION Considered monotypic. However, Gyldenstolpe (1951) suggested resurrecting Spix's *Cacicus solitarius nigerrimus*. He noted that individuals from both sides of the Amazon from Amazonas province, Brazil, as well as from the Rio Beni of Bolivia were decidedly more black while those from Paraguay, the type locality for the species, were blackish-brown. More recent authors have not followed this treatment.

HABITAT Solitary Cacique frequents dense vegetation, both near the ground or at mid-levels. It inhabits heavy brush in disturbed areas in moist forest, and second growth. Also occurs in cochas (Amazonian oxbow lakes), várzea and gallery forest, as well as *Heliconia* thickets and tall grass by water. In Amazonia, this species is found on river islands of various ages, usually in second growth (Rosenberg 1990b). It is often associated with water and likes to clamber around in vine-festooned trees. The nests are commonly placed directly over water. In the south of its range it is invariably associated with gallery forests. However Hayes (1995) notes that in Paraguay this species is found in subhumid and scrub forests, presumably in thickets near water.

BEHAVIOUR Solitary Cacique is usually observed singly or in pairs. It does not flock with other members of its species, unless feeding fledglings. This is a skulker, often feeding in dense tangles and near the ground, but not infrequently moving up into the mid-story. In particular it likes to climb on lianas. Feeds by prying into rolled, dry leaves, bark and plant stalks. Often hangs upside down as it picks at leaves or bromeliads. This cacique is largely insectivorous, not being regularly observed nectaring at flowers or feeding on fruit. It is regularly a component of mixed-species flocks, particularly associating with *Cyanocorax* jays. It tends to sing from a partially hidden branch, but does not perform a noticeable display.

NESTING True to its name, this cacique does not nest in colonies. The nest is a long hanging basket about half a meter in length, with an entrance near the top. Nests are seldom placed higher than five meters off the ground. Placement is typically in the forked tip of a branch of a low tree or shrub, including low palms, and commonly overhanging water (Naumburg 1930). In Argentina, egg dates range from early October–mid November. In Mato Grosso, Brazil, a nest with eggs was collected in early January (Naumburg 1930). Clutch size is 2 eggs, white with small brownish spots and lines, the markings being concentrated at the broader end. It is not known to be a host of either Shiny (102) or Giant (98) Cowbirds.

DISTRIBUTION AND STATUS Common to uncommon, usually the latter perhaps due to its skulking nature. This species is only found in the lowland tropical zone (subtropical farther south) usually to 500 m, but it has been recorded to 800 m in Peru (Ridgely and Tudor 1989). Solitary Cacique is found along a thin band in the humid tropical zone at the east base of the Andes from SW Venezuela (Tachira, Barinas and Apure) south to Meta, Colombia, where it spreads out into the Amazonian lowlands; east along the Amazon river (including the north shore) to the Atlantic Ocean. It is not known to range much further north than the Amazon river in the Amazonian lowlands, but there is now a report from Maracá ecological station in Roraima, Brazil (Ridgely and Tudor 1989), which suggests that the species may be found in low densities throughout the northern Amazon basin. In Brazil, south of the Amazon river, Solitary Cacique may be found throughout the Amazon basin, Pantanal and in the moister parts of the dry caatinga region south to Alagoas, north-central Minas Gerais, and S Mato Grosso (absent from the Atlantic lowlands of SE Brazil). While it has yet to be recorded there (Belton 1985), Solitary Cacique may range farther south along the gallery forests of the Rio Uruguay in Rio Grande do Sul since it is known from the Argentinian side of that river. From Colombia it ranges south throughout the lowlands east of the Andes in Ecuador, Peru (not in the upper Marañón valley, however) and Bolivia where it is absent only in the departments of Chuquisaca, Oruro and Potosi (Remsen and Traylor 1989) to N Argentina. It ranges throughout Paraguay, in the chaco as well as the moister zones east of the Río Paraguay (Hayes 1995). In Argentina, the species may be found in the chaco region, west to Salta, south to Córdoba and east to the Río Uruguay, extending south through Entre Ríos to northernmost Buenos Aires province; it reaches its eastern range limit along the Uruguay river as far north as N Misiones, it is present on the east bank of the river in Uruguay.

MOVEMENTS Sedentary, no appreciable seasonal movements.

MOULT Little known. Adults have only one moult per year, the complete definitive pre-basic moult. The timing of this moult varies geographically, but occurs after breeding. The first pre-basic moult includes the body and wing-coverts; the juvenile tail, primaries, secondaries and primary coverts are retained. There is no evidence of a pre-alternate moult.

MEASUREMENTS Male: (n=10) wing 124.4 (110–134); tail 123.7 (111–135); culmen 31.9 (28–35); tarsus 32.9 (32–35). Female: (n=7) wing 116 (110–128); tail 112.3 (101–130); culmen 31.0 (27–35); tarsus 32.3 (30–35).

NOTES Previously classified in the genus *Amblycercus* due to its short wings and completely black plumage. However, Solitary Cacique and Yellow-billed Cacique do not appear to be closely related based on morphological, behavioural and genetic data. Solitary Cacique has also been put in *Archiplanus*, a genus that encompassed some of the smaller black or black and yellow caciques.

REFERENCES De la Peña 1987 [nesting], Gyldenstolpe 1951 [geographic variation], Naumburg 1930 [nesting], Rosenberg 1990 [habitat], Sick 1993 [voice], Wetmore 1926 [habitat, behaviour].

22 YELLOW-WINGED CACIQUE *Cacicus melanicterus* (Bonaparte) 1825 — Plate 6

This striking crested black and yellow cacique is practically endemic to the Mexican west coast. Perhaps a more appropriate English name for this species is Mexican Cacique, particularly since it is no more yellow-winged than several other congeners.

IDENTIFICATION Yellow-winged Cacique is a large black cacique with contrasting yellow patches on the wings, rump, crissum and tail. It has a whitish bill and a noticeable black crest. This is the only cacique in most of its range and no other sympatric bird shares its precise pattern of black and yellow. However, Yellow-winged Cacique could be confused with several black and yellow orioles. Note that no oriole has a pale yellowish bill, a crest or is nearly as large as the cacique. In addition, few orioles have such extensive amounts of black in the plumage and no sympatric oriole has a black belly and lower breast as in Yellow-winged Cacique. The large patches of yellow on the plumage distinguish the cacique from other large black birds such as cowbirds and grackles. Montezuma (7) and Chestnut-headed (6) Oropendolas, which are restricted to the Atlantic slope in Mexico, share the large size as well as a black and yellow tail. However, the oropendolas lack any yellow on the wings, rump or crissum. Yellow-winged Cacique is most like several South American Cacique species, with which it is allopatric. In particular, Yellow-rumped Cacique (13) resembles Yellow-winged Cacique in the position of yellow colouration in its plumage.

However, Yellow-rumped Caciques do not show entirely yellow outer rectrices. Furthermore, Yellow-winged Cacique is somewhat divergent from other black and yellow caciques in its slimmer body shape and its noticeably long and shallow-based bill. The long crest is not found in other caciques; this feature is more typical of several oropendola species.

VOICE This is a loud, conspicuous species.

Song: Has been described as a short rattle followed by a series of quiet notes ending with mechanical sounds, the last one being upwardly inflected (Howell and Webb 1995). Other versions include a series of soft, liquid sounds *whump-whump* followed by the sounds of a rusty barn door swinging open, or a *cluuk-trrrrrrr-KEEEEHHH*.

Calls: The common call is a liquid *huik* or *whik?* which strongly recalls the call of Brown-crested Flycatcher (*Myiarchus tyrannulus*) (S.N.G. Howell pers. comm.). There are various calls that have been described including mellow whistles, harsh crows, rattles and sweet bell-like notes. Some of the rattles have a mechanical quality, like that of a stick being broken or cloth being torn. A *caaeck* call is reminiscent of a similar call given by Yellow-billed Cacique (23), or even like that of a Clark's Nutcracker (*Nucifraga columbiana*).

DESCRIPTION This medium-sized cacique is the most conspicuously crested of the group, having a long and broad crest. On some adult males the crest can be longer than the bill. This cacique has long wings and a moderately graduated tail. The bill is long and pointed, and not very deep at the base. The culmen and gonys are nearly straight, adding to the spear-like shape of the bill. Wing Formula: $P9 < P8 < P7 \approx P6 > P5$; $P9 \approx P4$; P8–P6 emarginate.

Adult Male (Definitive Basic): The bill is a pale greenish-white. The eyes are brown. The body is mostly glossy black with yellow on the lower back, rump, uppertail-coverts, undertail-coverts, and most of the wing-coverts and the tail. The yellow wing patch includes most of the median coverts, but not the lesser coverts or the outermost greater coverts. Therefore the wing patch does not reach to the leading edge of the wing when extended, nor to the alula when folded. The tail is largely yellow, except for the black central pair of feathers, as well as the dusky outer edge of the outermost tail feather and the dusky tips of the other feathers. The legs are blackish.

Adult Female (Definitive Basic): Similar to the adult male, but noticeably smaller. In addition, the black is replaced by a dark olive-grey and the forehead and malar stripe may be streaked with yellow. The yellow tail feathers usually have olive fringes to them, as does the rump.

Immature Male (First Basic): Intermediate in appearance between the adult female and adult male. Similar to the male but slightly smaller, browner in colour and with a noticeably short crest. The primaries are not as long and pointed as on the adult male. Young males may look patchy (streaked or flecked) with contrasting black feathers on their bodies. Some immature males show a yellow forehead as in adult females.

Juvenile (Juvenal): Similar to adult female, but even duller and browner.

GEOGRAPHIC VARIATION Monotypic.

HABITAT This cacique is to be found from thorn scrub along the coast to moister tropical deciduous forest farther inland. In the mountains of Guerrero, Mexico, it only reaches to the semideciduous tropical forest zone, staying below 1000 m (Navarro 1992). It prefers forest edge and open areas, including agricultural plantations such as coconut or mango and other settled areas with larger trees. Also uses mangroves, particularly to roost (S.N.G. Howell pers. comm.).

BEHAVIOUR A noisy and noticeable species, usually foraging agilely in the mid- or upper levels. It is most frequently observed travelling in small flocks or pairs. However, it is known to roost in large congregations. The breeding system tends to be polygynous. The display behaviour is not well documented. One display involves a striking fanning of the crest (S.N.G. Howell pers. comm.).

NESTING Breeding begins early May–early June. Nesting is typically triggered by the onset of the rainy season (Rowley 1984, Schaldach 1963). The breeding period is finished by August. Yellow-winged Cacique may nest in small colonies or singly, unlike most caciques. Colonies have from three to 10 nests (sometimes more), smaller than those of several of the other colonial caciques. The nest is a hanging basket woven from plant fibres and vines, and often including green leaves in the construction. The pendant nests are usually less than 80 cm in length. They are placed at or near the top of tall trees, or even suspended from electrical or telephone wires. Nests are constructed 10 m–20 m above the ground. The clutch size varies from 2–4 eggs, which are pale blue with dark spots and blotches clustered around the broader. Yellow-winged Cacique is commonly parasitised by Bronzed Cowbird (99) (Dickerman 1960, Rowley 1984).

DISTRIBUTION AND STATUS Common to fairly common to 1500 m. Almost entirely a Mexican species. It ranges along the Pacific coastal lowlands from S Sonora, south along Sinaloa, Nayarit, W Jalisco, Colima and Michoacán, and east along the Rio Balsas valley to SW Mexico, continuing south along coastal Guerrero, Oaxaca and Chiapas (inland along the lowlands of the Isthmus of Tehuantepec) to easternmost coastal Guatemala. There is now a disjunct population near La Avellena in Guatemala. The population in Sonora may now be extirpated (Howell and Webb 1995).

MOVEMENTS Appears to be sedentary.

MOULT Pre-basic moult in adults (definitive) is complete and occurs after the breeding period. The moult period, based on specimens from various parts of Mexico, ranges from late July–mid October, peaking in September–October. The tail moult, which commences as the primaries are halfway through their moult, tends to be rather quick. Many rectrices are shed at the same time, beginning with the innermost (R1), such that a moulting individual may have as few as two or four full-grown old rectrices during the tail moult. The first pre-basic moult takes place during September. It is unclear whether this moult includes the tail and wing feathers. There may be a limited first pre-alternate moult in immature males which needs to be confirmed. Pre-alternate moults in adults appear to be lacking.

MEASUREMENTS Males are larger than females. Males: (n=11) wing 155.2 (147–165); tail 130.4 (121–138); culmen 40.4 (39–43.2); tarsus 35.2 (33.5–37). Females: (n=10) wing 130.6 (121.9–146.1); tail 113.7 (106.2–126.5); culmen 36.0 (33–40.6); tarsus 31.5 (30–34.8).

NOTES Formerly classified in its own genus *Cassiculus*. All of the other black and yellow caciques occur in South America and Panama. It would be interesting to determine the closest relative to Yellow-winged Cacique.

REFERENCES Davis 1972 [voice], Dickerman 1960 [nesting, brood parasitism], Hardy *et al.* 1998 [voice], Howell and Webb 1995 [voice, description, behaviour, nesting, distribution, moult], Rowley 1984 [nesting], Schaldach 1963 [habitat, nesting].

23 YELLOW-BILLED CACIQUE *Amblycercus holosericeus* (Deppe) 1830 — Plate 8

A skulking blackbird of dense undergrowth. In highlands it is a typical inhabitant of *Chusquea* bamboo thickets.

IDENTIFICATION A black cacique with a yellowish chisel-like bill and striking yellow eyes. Solitary Cacique (21) is quite similar but differs from Yellow-billed in its larger size, shaggy crest, duller, greener bill and its dark eyes. Also, Solitary Cacique is a bird of the Amazonian lowlands and does not come into contact with the South American population of Yellow-billed Cacique, which is restricted to the highlands. Similarly, Ecuadorian Cacique (20) is also all black, but it is a bird of the eastern lowlands and foothills. Ecuadorian Cacique's blue eyes separate it from Yellow-billed Cacique, as does its blue-grey or greenish bill. Although Mountain Cacique (18) does occur with Yellow-billed Cacique, and often feeds in *Chusquea* bamboo thickets, it has a bold yellow rump, and northern forms have yellow wing patches. In addition, the eyes of Mountain Cacique are blue and the bill is dark-based. Mountain Caciques are typically birds of the canopy, not of the understory. There are several black icterids that are superficially similar, such as Scrub Blackbird (86), Melodious Blackbird (85) Shiny (102) and Bronzed Cowbirds (99). These species, however, are found in open areas and all lack the pale eyes and pale bill of Yellow-billed Cacique. Habitat differences alone are enough to separate these species from Yellow-billed Cacique.

VOICE Like many other skulkers, Yellow-billed Cacique is often heard before it is seen.

Song. The race *holosericeus* sings a duet with males giving a doubled *whew-whew, whew-whew* or *chewee-chewo, chewee-chewo* which is answered by the female's single *wheee* or *wheee churr*. Females will answer the male's part of the song even while incubating. Although the songs are variable, this species' loud piercing whistles are distinctive. It is unknown if South American populations differ from Central American ones in their vocalisations.

Calls. Include a harsh *queeyoo* and a sweet *wreeeeoo*, along with a variety of whistles (males) and *churrs* (females). Pairs keep in contact by the male uttering a long ascending whistle that is immediately answered by the *churr* of the female. The male's whistle may last 1 second in duration, while the *churr* of the female is longer, often 3 seconds long. Skutch (1954) initially thought that the whistle was the sound of a single bird inhaling before giving the *churr* until he realised that each call was given by one of the members of a pair. Scolding *holosericeus* gives a nasal *ramp-ramp-ramp*. A loud *waak* is uttered by *australis*.

DESCRIPTION A black cacique with short and obviously rounded wings as well as a broad, graduated tail. At rest, the primaries may be almost completely covered by the tertials. The bill has a distinctly flattened culmen that is parallel-sided and thus chisel-tipped. The nostril in this species is partially covered by a flap (operculate), such that it looks slit-like in profile. True caciques lack this flap. No crest is present. Wing Formula: P9 $<$ P8 $<$ P7 $\approx$ P6 $\approx$ P5 $>$ P4; P8–P5 weakly emarginate. The nominate form is described.

Adult Male (Definitive Basic): The pale greenish-yellow bill is prominent, as is the pale yellow iris. The bill base may be slightly greyer, particularly on the lower mandible. The rest of the bird is dull black. The legs and feet are black.

Adult Female (Definitive Basic): Similar to male but smaller and slightly more slaty, not as pure black.

Juvenile (Juvenal): Similar to adults, but the black is duller, appearing greyish-black, and often looking patchily-coloured due to darker feather fringes. The plumage is looser, particularly on the undertail-coverts. Also, the eyes are dull brown or grey-brown. Recently-fledged juveniles have shorter bills than adults.

GEOGRAPHIC VARIATION Three subspecies that fall into two groups: *holosericeus* [Yellow-billed Cacique] and *australis* [Chapman's Cacique]. The nominate race has populations in both the lowlands and the highlands; the latter, in montane Costa Rica, may be bamboo specialists. Chapman's Cacique is entirely a bamboo specialist. In general, this species varies according to Bergmann's Rule; populations at higher latitudes, on either side of the equator, are larger than those closer to the equator (Kratter 1993). The *holocericeus* group includes two subspecies:

A. h. holosericeus is found from Mexico to northwest Colombia. It is described above.

A. h. flavirostris lives in the Pacific lowlands from Colombia to N Peru. It is smaller than *holosericeus*, and with a bright yellow bill.

The southern 'Chapman's Cacique' includes one subspecies, ***A. h. australis***. It occurs from the Andes and coastal Cordillera of Venezuela south along the Andes to W Bolivia, largely in highland (1500–3500 m) *Chusquea* bamboo thickets, at least in the southern portion of its distribution. The population from the Santa Marta highlands appears to belong to this form (Kratter 1993). Similar to *holosericeus*, but the bill is more slender and straw-yellow with a grey base. The wing is shorter and the tail longer than *holosericeus*. It is the smallest subspecies.

HABITAT Lives in dense thickets. In lowlands these may be in disturbed areas or abandoned fields, along rivers in canebrakes, as well as the dense understory of forests, particularly near the edge. It may also be found in grasslands if the grasses are adequately tall and dense. Dense *Chusquea* bamboo thickets appear to be the sole habitat for the highland forms. The highland bamboo specialist population, of Costa Rica, is most similar morphologically to its neighbouring non-bamboo lowland neighbours than to the highland South American bamboo specialist (*australis*), suggesting that specialisation on bamboo arose independently at least twice in this species (Kratter 1993). The *australis* subspecies is found between 1500–3500 m; similarly, highland populations in Costa Rica are found to 3000 m; lowland populations, however, are rarely found above 1000 m.

BEHAVIOUR Typically, this species is found roaming the understory in pairs, singly, or in family groups. They are skulking, and often difficult to observe. This species is inquisitive and can be brought into view by squeaking. Sometimes they may be found following mixed-species flocks in the forest. They poke and probe into rolled leaves, bark, bamboo stems and young shoots, and may even hammer woodpecker-like to get at insect prey. Often they may be seen hanging upside down, or in other odd positions to gain access to difficult to reach morsels. Yellow-billed Cacique has been observed following Army Ants on at least one occasion (Ridgely in Wetmore *et al.* 1984). During the breeding season, they are territorial and breed monogamously. The most common display is an antiphonal duet performed by the pair. The male whistles and the female answers with a *churr*, often while she is sitting on the nest (Skutch 1954). Like many antbirds and antpittas, Yellow-billed Cacique does not fly across open areas, but keeps to dense vegetation. Road building and habitat fragmentation are possibly negatively affecting the dispersal of these birds. On the other hand, road construction and cutting of forests creates the thickets that this species requires as breeding habitat, so perhaps it benefits from these activities.

NESTING The breeding season of South American *australis* is protracted, with fledglings observed between November and April, at various sites (Fjeldsa and Krabbe 1990). In Costa Rica (*holosericeus*), nesting takes place February–June (Stiles and Skutch 1989) while in the Panama Canal Zone, nesting has been recorded late April–mid September (Wetmore *et al.* 1984). Nests are not pendant baskets like in the true caciques, but rather a cup made of leaves and vines, placed less than two meters off the ground. The nest is usually fastened to several stalks of bamboo, cane or other stalks and can be quite a bulky structure. The cup-nesting habit of this species was one of the first clues that led ornithologists to believe that this species was not a true cacique. This has now been reaffirmed by genetic data. Nests are notoriously difficult to find as they are placed in very dense vegetation, usually less than one meter above off the ground (Skutch 1954). Clutch size 1–2 eggs which are pale blue with dark spots clustered around the broader end. All of the incubation is performed by the female. Shiny Cowbird (102) has parasitised the nests of this species, but this is not common and it has not been observed to successfully fledge from Yellow-billed Cacique nests. Similarly, Bronzed Cowbird (99) is known to parasitise nests of this cacique, but has not been observed to be successful (Friedmann and Kiff 1985).

DISTRIBUTION AND STATUS Uncommon to locally common. The *australis* subspecies is found between 1500–3500 m; populations in Central America are found to 3000 m; lowland populations, however, are rarely found above 1000 m. Its range is somewhat discontinuous, fragmented into three major populations, see Geographic Variation. Found from S Tamaulipas, Mexico, south along the Atlantic lowlands, crossing to the Pacific lowlands at the Isthmus of Tehuantepec, south through most of Central America to the Chocó region, west of the Andes in Chocó north to Bolivar, in Colombia. It is absent from the highlands from Mexico south to Nicaragua, but it does inhabit the highlands of Costa Rica; in Panama it is once again absent from highlands. Inhabits the Pacific lowlands of South America from S Chocó in Colombia south to N Peru in Tumbes department. The last population inhabits the temperate and subtropical zone in the Andes and is found in several isolated populations from NW Venezuela, including the coastal highlands of Aragua and the Federal District, south and west to Colombia in the east Andes, including the Santa Marta mountains; the range is more continuous along the east side of the Andes from Ecuador to Cochabamba, Bolivia.

MOVEMENTS Sedentary.

MOULT The complete definitive pre-basic moult occurs after the breeding season. Pre-alternate moults appear to be lacking, but there is little information available. Birds in Panama (nominate form) undergo their pre-basic moult between late November–mid January. The primaries and secondaries are the first feather groups to be moulted, the body moult occurs throughout the time the remiges are being moulted. Tail moult begins after the primaries are nearly all replaced. The tail and the face are the last feather areas to be moulted. The extent of the first pre-basic moult is not known.

MEASUREMENTS Males are only marginally larger than females. All measurements from Kratter (1993, unpub. data).

A. h. holosericeus. Males: (n=39) wing 99.2 (85.8–113.4); (n=38) tail 104.7 (91.3–118.7); (n=39) culmen 30.1 (27.4–32.6); (n=36) tarsus 32.3 (27.7–36.6). Females: (n=36) wing 92.4 (83.1–102.4); (n=34) tail 98.5 (90.5–11.6); (n=35) culmen 27.8 (24.5–31.3); (n=36) tarsus 29.9 (27.4–34.4).

A. h. flavirostris. Males: (n=2) wing 95.1 (92.1–98.0); tail 100.2 (98.7–100.7); culmen 28.5 (27.9–29.0); tarsus 28.9 (28.0–29.7). Females: (n=2) wing 88.1 (87.0–89.1); tail 93.4 (86.5–100.2); culmen 25.2 (25–25.4); tarsus 28.5 (28.2–28.7).

A. h. australis. Males: (n=22) wing 95.1 (89.6–101.3); tail 106.4 (91.5–116.9); culmen 28.4 (26.7–31.1); tarsus 29.8 (27.9–31.6). Females: (n=27) wing 87.7 (80.8–95.3); (n=26) tail 102.4 (88.0–114.1); (n=27) culmen 26.7 (24.4–28.1); (n=26) tarsus 28.9 (25.8–31.3).

NOTES This species is often classified in the true caciques (*Cacicus*) or *Cacicus solitarius* is included in the genus *Amblycercus*. Recent DNA sequence data, however, suggests the Yellow-billed Cacique belongs within its own separate group, distantly related to the *Cacicus* caciques and oropendolas (Lanyon *in litt.*). The nest of Yellow-billed Cacique is unlike that of any of the true caciques for which nests are known. Another recent study (Freeman and Zink 1995) of restriction sites of mitochondrial DNA of the blackbirds suggests that Yellow-billed Cacique is distantly related to the other caciques, and that the South American form ('Chapman's Cacique') is different from the nominate form, perhaps at the species level. More work is needed to accurately determine the species limits and generic placement of Yellow-billed Cacique. Called 'Pico de Plata' in Costa Rica, which translates to Silverbill (Skutch 1996).

REFERENCES Chapman 1915 [geographic variation],1919 [geographic variation] 1926 [geographic variation], Freeman and Zink 1995 [systematics], Hilty and Brown 1986 [voice, nesting], Kratter 1993 [geographic variation, habitat, measurements], Ridgely and Tudor 1989 [voice, identification], Skutch 1954 [voice, nesting, habitat, behaviour], Stiles and Skutch 1989 [voice, nesting].

24 MORICHE ORIOLE *Icterus chrysocephalus* (Linnaeus) 1766 — Plate 9

An appropriately named species as this oriole almost always breeds in moriche palm (*Mauritia flexuosa*) groves. Closely related to Epaulet Oriole, the two have been considered part of the same species in the past.

IDENTIFICATION A slim, long-tailed black oriole with striking yellow patches on the rump, crown, shoulders and thighs. Unlikely to be confused for any species other than Epaulet Oriole (25). The combination of a largely black plumage with a yellow crown, shoulders, rump and thighs is diagnostic. Epaulet Oriole may show yellow on the thighs and shoulder, but never on the crown or rump. The black and yellow Yellow-rumped Cacique (13) is larger, lacks yellow on the crown, has blue eyes and yellow on the underparts, unlike Moriche Oriole. Other South American orioles have more extensive yellow or orange on the body, particularly on the underparts. Moriche Orioles are easily differentiated from these species by their solidly (except the thighs) black underparts.

VOICE Song: The song is a long series of repeated whistles which are squeaky for an oriole and include short trills. In Guyana, the song is variable, suggesting that of a wren. The phrases include a clear slurred *heea-heea-heea, hee-chee-chwit-wit, heea-hwihoo-heheea,* and *hwit-hu-wit-heeuuweet-hwihuu-weet* (Snyder 1966). Individual syllables of the song are spaced widely enough apart that they sound discrete, not running together into a warble. Many of the whistles are sweet, and pleasant. Pairs of this oriole frequently duet.

Calls: Gives a chatter, *ch-ch-ch-ch* which is throatier than that of Yellow Oriole (27). Also utters a *chwut-chwut-chwut.*

DESCRIPTION A slim, long-billed and long-tailed oriole. Its slim bill is noticeably decurved at the tip. The tail is rather narrow and obviously rounded. Wing Formula: P9 < P8 ≈ P7 ≈ P6 > P5; P9 ≈ P5; P8–P5 emarginate.

Adult Male (Definitive Basic): The bill is black, tending not to show a grey base to the lower mandible. The eyes are brown. Moriche Orioles are largely black with bold patches of yellow. The black plumage does not show a pronounced gloss. The crown and nape are yellow to orange-yellow; the rump is yellow but the uppertail-coverts are black. The underparts are almost entirely black, except for the yellow thighs and scattered yellow feathers around the vent. The wings show yellow lesser and median coverts as well as yellow lesser underwing-coverts. The tail is entirely black. Legs and feet grey.

Adult Female (Definitive Basic): Indistinguishable from adult male.

Immature (First Basic): Resembles adult but with contrasting brownish-black wings. Retains the juvenal primaries, secondaries and greater coverts which appear browner and duller than the black body plumage and tertials. The immatures may show the lower wingbar of the juvenal, but this is often lost to wear. The yellow epaulet is bright, as in adults. However, as in juveniles, the rump feathers have dark bases so the rump patch does not appear solidly yellow, particularly in worn individuals.

Juvenile (Juvenal): Similar to adult, but obviously duller. The entire plumage is dark brown rather than black. As in adults, the cap, rump, epaulets and thighs are yellow but these are pale yellow on juvenile. The pale yellow rump feathers show dark bases, thus the rump colour appears patchy. The greater coverts are brown with yellow tips, forming a thin lower wingbar which is not observed in adults. The vent and undertail-coverts are yellowish.

GEOGRAPHIC VARIATION Monotypic. Birds from Ecuador may be slightly thinner-billed than those from the Guianas (Chapman 1926).

HABITAT Inhabits swamps and open areas near rivers or lakes in Amazonian forest, often foraging high in the canopy. In Surinam, it avoids the swamps near the coast and is found in dry sandy savanna where moriche palms grow. Moriche Oriole will persist in heavily disturbed forest sites which mimic savanna forest in that the trees are scattered. This species is greatly dependent on the presence of moriche palms wherever it is found.

BEHAVIOUR Moriche Oriole tends to be conspicuous, remaining high up in the trees or near their periphery, and always being visible. In lowland tropical forest it is tougher to see due to its canopy-dwelling preference. Generally found singly or in pairs high up in the canopy at forest borders or swamps; it does not flock with other Moriche Orioles. It sometimes joins canopy mixed-species foraging flocks. Moriche Oriole forages by gleaning insects from the foliage, particularly from the fronds of *Mauritia* palms. It is also fond of nectaring at large forest flowers. At the nest, visits during the early breeding stage are made by the two pair members arriving together. The male sings upon arrival at the nest.

NESTING Nests are known from February–June in Guyana; February, March, August and September in Surinam (Haverschmidt 1968); and February in NE Meta, Colombia; late July–early February in Manaus, Brazil (Oniki and Willis 1983). Commonly places its hanging basket nest directly beneath a palm frond, almost invariably *Mauritia* palms. The attachments for the nest are woven through the palm frond itself, usually along the mid-line of the leaf. Nests are woven from fibers garnered from green leaves of the *Mauritia* palm. One nest was described as 10 cm long and 7 cm at its widest point (Oniki and Willis 1983). The clutch tends to be of 2 eggs. In Surinam, the eggs are bluish-white with black spots (Haverschmidt 1968); in Brazil they are white with brown spots concentrated around the broader end (Oniki and Willis 1983). It is a known host of the Shiny Cowbird (102)(Friedmann and Kiff 1985).

DISTRIBUTION AND STATUS Locally common or uncommon. Found in northern South America, north of the Amazon, usually below 500 m. Resident east of the Andes in the lowlands of W and S Venezuela, SE Colombia, E Ecuador and NE Peru, reaching its southern range boundary on the north shore of the Amazon. Ranges east through the Guianas to E French Guiana and adjacent N Para, N Amazonas and Roraima, Brazil. Also present on Trinidad.

MOVEMENTS This species is known from Trinidad, but historically it appears not to have been present there. It is not unlikely that Trinidadian populations may have either recently become established or originated from escaped cagebirds.

MOULT Adults moult once a year, through the complete definitive pre-basic moult. This moult takes place after breeding has finished, therefore the timing varies geographically depending on the breeding season. In Putumayo, Colombia, the definitive pre-basic moult occurs in May, while in Guyana, moult occurs in April. The juvenal plumage is lost through the partial first pre-basic moult which involves the body, tertial and lesser covert feathers while retaining the juvenal tail, greater coverts, primaries and secondaries. Sometimes some or all of the wing feathers may be replaced. Pre-alternate moults are lacking.

MEASUREMENTS Males: (n=11) wing 106.8 (100–111); tail 107.9 (100–118); (n=10) culmen 22.6 (20–25); (n=11) tarsus 23.1 (22–25). Females: (n=10) wing 94.2 (90–100); tail 98.5 (85–112); culmen 21.8 (21–23); tarsus 23.2 (22–25).

NOTES Specimens that appear to be hybrids between this species and Epaulet Oriole do exist. These are known from Surinam, but are not very common, although the extent of hybridization has not been studied. Moriche Oriole is sometimes lumped with Epaulet Oriole and indeed it appears to be closely related to that species. However there are areas where both species are found sympatrically, as in Manaus, Brazil (Ridgely and Tudor 1989), where hybridisation is either lacking or very rare (M. Cohn-Haft pers. comm.) suggesting that treating these as two species is a more appropriate decision. Some Moriche Orioles show scattered yellow feathers that may show up anywhere on the body. The significance of these rogue yellow feathers is not known.

See Notes for Epaulet Oriole.

REFERENCES Belcher and Smooker 1937 [movements], Chapman 1926 [geographic variation], Haverschmidt 1951 [habitat, nesting], Oniki and Willis 1983 [nesting], Snyder 1966 [voice].

25 EPAULET ORIOLE *Icterus cayanensis* (Linnaeus) 1766 **Plate 9**

This is the least colourful of the orioles. It is an inhabitant of a large part of South America, wherever open woodland or savanna habitat is to be found.

IDENTIFICATION Epaulet Oriole is black with a yellow or chestnut epaulet. Throughout its range this oriole colour combination alone is diagnostic. Epaulet Orioles are sympatric with two other orioles, Moriche (24) and Yellow (27) Orioles as well as the troupials (37, 38, 39). Yellow Oriole and the troupials may be separated immediately from Epaulet Oriole by their much more extensive yellow or orange body plumage. The entirely black underparts of Epaulet Oriole are lacking on those orioles. Moriche Oriole is similar to Epaulet Oriole in its largely black plumage and yellow epaulets; however, Epaulet Orioles lack the yellow nape and rump patches of Moriche Oriole. Note that specimens intermediate between these two species exist from Surinam and Guyana. Most forms of Epaulet Oriole have black thighs unlike Moriche Oriole, but note that one (*tibialis*) does not.

Epaulet Orioles are more likely to be confused for any one of the largely black icterids of South America. In particular, male Yellow-winged Blackbird (58) resembles yellow-shouldered forms of Epaulet Oriole; however, the blackbird is sympatric with the chestnut-shouldered forms of the oriole. Nevertheless, the blackbird can be differentiated by its stockier shape, and shorter tail and bill, in addition to differences in behaviour, voice and habitat. Yellow-winged Blackbird has black underwings, while yellow-shouldered forms of Epaulet Oriole usually show some yellow on the wing linings. The chestnut-shouldered forms may be confused with entirely black icterids due to the dark colour of the epaulets, which look dark from a distance. Epaulet Oriole, however, lacks any obvious iridescence, is slim and shows a long tail and bill unlike most all-black icterids. Epaulet Oriole is arboreal and will not be found in marsh vegetation or foraging on the ground. The forest-based Velvet-fronted Grackle (81) may be the most problematic identification. Velvet-fronted Grackle will show obvious iridescence if seen well, and of course it lacks the chestnut epaulets of the oriole. Velvet-fronted Grackle has a stubby-looking bill, giving it quite a different look to the slender-billed Epaulet Oriole.

VOICE Song: The song is variable, sometimes sweet and musical and other times less pleasant, more squeaky and shrill in nature. Common to all songs is that the notes are delivered deliberately and clearly. Some songs are composed of single notes with obvious silences between notes, while others are composed of phrases (a series of notes) with silences between phrases. 'Phrase songs' are repetitive but pleasant with little change in the phrases that are used. Phrases may include sweet as well as harsh whistles. These songs are most similar to Moriche Oriole songs, but tend to be harsher in tone. In Argentina, songs may be unmusical and mechanical, composed of *chak, chak-chak* notes interspersed with chatters and an odd hollow nasal whistle *whzzheeeo* (Straneck 1990e). Brazilian songs are described as resonant and melodious, or staccato or flowing, with motifs such as *dlew-eet-trrreh* (Sick 1993). The song appears to be continuous, not having a predefined beginning or ending.

Calls: A sharp *kit, chik* or *spik* note is given quite commonly. This species has an ample repertoire of calls, including a clear *oint* or *oint-oint*, a descending squeal, a nasal *whhaaa*, a sweet high frequency whistle *twee-swee* and a low frequency whistle *wheee* (Remsen 1986). Sick (1993) notes a *kwou* note as part of this species' repertoire. Wetmore (1926) states that the male gives a *spick spick* while the female utters a mewing note or a harsh rattle. It needs to be determined if these different call types are actually sex specific. Both sexes of Epaulet Oriole are skilled mimics, at least in Argentina, southern Brazil and Bolivia. It tends to mimic alarm notes as well as calls of hawks Accipitridae and caracaras *Milvago*, but not as loudly as the original calls. These calls are not incorporated into the song, but tend to be given individually, or mixed with *kit* calls. Playbacks of recordings of these mimicked calls elicits a mobbing response by the orioles, but no response from the species mimicked (Remsen 1986, Fraga 1987, Sick 1993).

DESCRIPTION A slim oriole with a long, thin, rounded tail. The bill is thin, its length and degree of curvature is geographically variable. Wing formula: P9 < P8 < ≈ P7 < ≈ P6 > P5; P9 ≈ P2; P8–P5 emarginate. The form *pyrrhopterus* is described.

Adult Male (Definitive Basic): The bill is finely pointed, but shows no curve, and the culmen and gonys are perfectly straight. The bill is largely black with a slightly greyer base to the lower mandible. The eyes are dark brown, or dark reddish-brown. The body plumage is entirely black with a dull gloss. The lesser coverts are chestnut; this chestnut epaulet is difficult to see from a distance. The wing-linings are blackish, but may be tipped with yellow in some individuals. The tail is black and unpatterned. The slim legs and feet are bluish-grey, sometimes blackish.

Adult Female (Definitive Basic): Similar to male but slightly smaller.

Immature (First Basic): Similar to adult, but blackish tertials and wing-coverts contrast with the retained, browner juvenile primaries and secondaries. There may be similar contrasts with blacker and fresher central rectrices, contrasting with the rest of the tail.

Juvenile (Juvenal):Resembles adult in pattern, but duller in colour. The yellow shoulders (thighs and wing-linings depending on the form) are pale and washed-out. The body plumage is dull brownish-grey rather than black. The black greater coverts show grey tips. Juveniles of the chesnut-shouldered forms (*periporphyrus* and *pyrrhopterus*) are similar in that their epaulets are paler in colour than those of the adults and the body plumage is browner. The juveniles also show thin russet tips to the greater coverts which form an inconspicuous and usually broken, lower wingbar.

GEOGRAPHIC VARIATION Five races currently recognized.

I. c. cayanensis (Cayenne Oriole). Has yellow shoulders but black-tips to the yellow-wing linings. This race is found in most of Amazonia south to E Peru and N Bolivia (Beni, E La Paz), and east to northeastern Brazil (Pará) and the Guianas. Its thighs are black. However, rarely some *cayanensis* show yellow tips to the thighs. These appear to be variants as specimens showing this characteristic are known from areas distant from *tibialis* Epaulet or Moriche Orioles. In Guyana and Surinam, this form hybridizes with Moriche Oriole. This subspecies is cited to occur as far south as N Cochabamba and S Beni in Bolivia, approximately in the region of Rurrenabaque (including Susi and Chatarona) and upstream along the Río Beni as far as Chiñiri. Birds from Chiñiri show intermediate characters between this and *peryporphyrus*, having a darker epaulet colour than expected and dark underwings (Bond and Meyer de Schaunsee 1941). However both in Rurrenabaque and the Pilon Lajas reserve, both near Chiñiri, *cayanensis* is the most common subspecies encountered but sometimes it is seen in the company of birds fitting the description of *periporphyrus* with no evidence of intergradation (A. B. Hennessey pers. comm.). The small number of known intergrades and the presence of both races in the same area suggest that intergradation is rare and that a hybrid swarm does not exist; however, more work is required to clarify this situation. Both forms have been observed together during October and November, suggesting that the dark-shouldered forms are not migrants from farther south or east.

I. c. tibialis (Yellow-thighed Oriole). Ranges in eastern Brazil, from Maranhão, Piauí and Ceará south throughout Pernambuco and Bahia to Espírito Santo and Rio de Janeiro. It resembles *cayanensis* in having a yellow epaulet, but is smaller and has a shorter and thicker bill. The wing-linings and thighs are yellow, although this is variable and some have black-tipped thigh feathers (Hellmayr 1929). Birds from the north have more yellow on the thighs, while in the south (NW Bahia) some approach *valenciobuenoi* in having dusky wing-coverts.

I. c. valenciobuenoi (Valencio Bueno's Oriole). Found in S Goiás, W Minas Gerais, W São Paulo, and SE Mato Grosso in Brazil. Like *periporphyrus* but the epaulet is a paler ochre-orange. Also, the wing-linings are dusky but tipped with yellow. The thighs are black. This form is intermediate between the yellow-shouldered forms to the north (*tibialis* and *cayanensis*) and the tawny/chestnut-shouldered races to the south (*periporphyrus* and *pyrrhopterus*). In fact it has been suggested that it is best to drop the name *valenciobuenoi,* as formal taxonomic status should not be given to hybrid populations (Sick 1993); more work needs to be done on this matter.

I. c. periporphyrus (Tawny-shouldered Oriole). Found from NE Bolivia (E La Paz, S Beni) east through N Cochabamba and N Santa Cruz, Bolivia to WC Mato Grosso, Brazil. This race is most similar to *pyrrhopterus,* but the epaulet averages paler. Its underwings and thighs are black. In general, the coverts are cinnamon or tawny rather than chestnut. In addition, *periporphyrus* has a smaller and more slender bill than *pyrrhopterus.*

I. c. pyrrhopterus (Chestnut-shouldered Oriole). This form is described above. This subspecies is found from C Bolivia (S Cochabamba and S Santa Cruz) and SW Brazil (SW Mato Grosso) east through Paraguay to SE Brazil (Paraná) and south through S Bolivia (Chuquisaca and Tarija) and Uruguay to La Rioja, Córdoba, Santa Fé and N Buenos Aires in Argentina. This southernmost race is the darkest, having black-wing linings and a dark chestnut epaulet. Under typical field conditions, this form appears entirely black. Note that juveniles of this race and *periporphyrus* show paler epaulets than the adults, such that a juvenile *pyrrhopterus* appears more like an adult *periporphyrus.* To make things more confusing, Oberholser (1902) described two additional forms of this oriole: *compsus* of Mato Grosso, Brazil, was said to have ferruginous rather than chestnut epaulets while *argoptilus* of Buenos Aires, Argentina, was stated to be larger than *pyrrhopterus.* These two forms are not recognized, but the latter form suggests that there may be a cline in size in the species, with southern birds being larger than those in the north, which is worthy of further study.

HABITAT A species with a varied habitat choice. It is found in open forests, savannas, palm savanna, deciduous woodlands, forest edge, gallery forest, urban gardens, forest clearings with a few standing trees, cerrado woodlands and chaco woodlands. In Surinam, woodlands with scattered trees in sandy ground (Haverschmidt 1968). The general tendency is a preference for edge and open forests, and woodlands. However, in Bolivia, this species also inhabits denser forest Andean foothills of the upper tropical zone (A. B. Hennessey pers. comm.). In Bolivia, nominate *cayannensis* is found in moister forests than *periporphyrus,* but both may be found together at times (A. B. Hennessey pers. comm.).

BEHAVIOUR Typically, this oriole is quiet and is observed alone or in pairs, although small flocks have been reported. It is regularly observed joining mixed canopy flocks in the Andean foothills of Bolivia (Parker 1989, A. B. Hennessey pers. comm.), a behaviour not known from other parts of its range. Typically only one pair is present in a mixed foraging flock. Forages largely in the upper and mid-levels. Usually an insect eater but it will sip nectar from flowers. Dubbs (1992) has observed Epaulet Oriole slitting the long corollas of *Tabebuia spp.* in order to gain access to their nectar. It also feeds on fruit. Wetmore (1926) noticed a preference for foraging along vines or creepers. This is an active oriole, and commonly jerks its tail nervously as it forages. They are adept at hanging and will acrobatically reach for hidden food morsels. During the breeding season, the male will perch high in a tree and sing, sometimes joined by the female. No common or obvious displays have been published.

NESTING Breeds during October–December in southern Brazil, Paraguay, Uruguay and Argentina. The nest is a suspended basket as is typical of the genus. However, its nest is not as deep as that of other orioles, often being more of a suspended cup or hammock, rather than a deeper basket. Nests are often hung from large leaves, such as those of the Banana plant (*Musa acuminata*) or a palm. In Argentina, the nests are more often secured to a woody branch as palms are not as common. Palm fibers tend to be used in nest construction , but they are loosely woven such that the eggs may be seen if observed from directly below the nest. Both adults help to build the nest. The clutch size is 2–3 eggs (De la Peña 1987, Naumburg 1930). The eggs are white with lavender, chestnut-brown, grey-brown or dark spots and specks (Naumburg 1930, De la Peña 1987). Naumburg (1930) noted that the eggs were different from those of other orioles in lacking irregular lines and scrawls. In Alagoas, Brazil, Shiny Cowbird (102) has been confirmed to parasitise this oriole (Studer and Vielliard 1988), but successful fledging has not been seen. Brood parasitism by Shiny Cowbird is also known from Argentina (De la Peña 1987).

DISTRIBUTION AND STATUS Uncommon or locally common. To 1080 m in Bolivia (Remsen *et al.* 1987). Found in the Guianas but absent from the coastal plain, from W Guyana east throughout Surinam and French Guiana. In Brazil, found north of the Amazon river in E Amazonas, Pará and Amapá; south of the Amazon it is found in all of the Brazilian states from Amazonas in the west to Rio Grande do Sul in the south and Rio Grande do Norte, Paraíba, Pernambuco and Alagoas in the east. In Peru, found only in the lowlands east of the Andes from S Loreto south through Ucayali and Madre de Dios. The Bolivian range includes the lowlands east of the Andes in the departments of Pando, Beni, La Paz, Cochabamba, Santa Cruz, Cuquisaca, Tarija and Oruro. Occurs throughout Paraguay, both east and west of the Rio Paraguay. In Uruguay, Epaulet Oriole is found throughout the entire country. The Argentine range includes the lowlands east of the Andes in Jujuy, Salta, Formosa, Chaco, Corrientes, Misiones, Tucumán, Catamarca, Santiago del Estero, Santa Fé, Entre Ríos, south to E La Rioja, N Córdoba and N Buenos Aires as far south as the Bahía Samborombón.

MOVEMENTS Resident, no evidence of migratory movements even in temperate Argentina.

MOULT Not well known. Adults moult once a year through their complete definitive pre-basic moult which takes place after the breeding season. In Beni, Bolivia, *peryporphyrus* individuals are undergoing their moult in February while *cayanensis* birds from Peru are in moult during September. The juvenal plumage is replaced through the first pre-basic moult which replaces the body plumage, wing-coverts and some tertials but retains the juvenal wings and tail. In some cases the central tail feathers, or all of the tail, may be replaced. There is no evidence that pre-alternate moults occur in this oriole.

MEASUREMENTS ***I. c. cayanensis***. Males: (n=8) wing 101.8 (94–108); tail 100.5 (92–107); culmen 21.6 (19–23); tarsus 22.9 (22–25). Females: (n=8) wing 96.6 (93–102); tail 97.0 (86–103); culmen 20.8 (19–22); tarsus 23.0 (21–24).

I. c. tibialis. Unsexed (n=4) wing 92 (87–95); tail 99.5 (93–108); culmen 17.8 (17–19); (n=2) tarsus 20, 23. Male: (n=1) wing 92; tail 100; culmen 18; tarsus 22. Females: (n=7) wing 90.7 (85–97); tail 95.7 (90–108); culmen 19.3 (18–20); tarsus 23.0 (21–24).

I. c. valenciobuenoi. Males: (n=?) wing 94–100; tail 100–107; culmen 17.5–20. Females: (n=?) wing 88–91; tail 95–102.
From Naumburg (1930) — Males: (n=5) wing 92.1 (90–93.5); tail 94.8 (93–96); culmen 17.3 (17–17.5). Females (n=2) wing 85,87; tail 85.5, 92; culmen 16.5, 18.

I. c. periporphyrus. Males: (n=4) wing 92 (86–98); tail 96.3 (96–100); culmen 19.3 (18–20); tarsus 24.8 (24–26). From Gyldenstolpe (1945b) — Males: (n=6) wing (93–102); tail (93–99); culmen (18–19); tarsus (22–24). Females: (n=3) wing 86.7 (80–95); tail 91.7 (88–95); culmen 19.3 (19–20); tarsus 23.0 (21–25). From Gyldenstolpe (1945b) — Females: (n=3) wing (92–99); tail (87–90); culmen (18–18); tarsus (21–23).

I. c. pyrrhopterus. Males: (n=11) wing 90.2 (84–95); tail 96.7 (88–105); (n=10) culmen 19.8 (18–22); (n=11) tarsus 25.5 (21–25). Females: (n=10) wing 86.7 (82–90); tail 93.8 (88–102); culmen 17.9 (17–19); tarsus 23.1 (22–25). From Naumburg (1930) — Males (n=8) wing 86.5 (85–91.5); tail 91.1 (89.6–97); culmen 15.9 (15–16.5). Females (n=9) wing 84.5 (82.5–85); tail 87.9 (85.5–90); culmen 16.3 (15–16.5).

NOTES Both yellow-shouldered and chestnut-shouldered forms of Epaulet Oriole are sometimes observed together in parts of Bolivia, although they have not been found breeding sympatrically (A. B.Hennessey pers. comm.). The low number of presumed intergrades and the presence of presumed pure types in sympatry are intriguing, and suggest that there is little gene flow between the two forms. If this is confirmed to be the case, this implies that two species are involved. What complicates things is that the dynamics of intergradation between *pyrrhopterus* and *valenciobuenoi* may be different in Brazil, with the latter form perhaps best considered a hybrid swarm between yellow- and tawny-shouldered forms. Epaulet and Moriche Orioles are sometimes considered members of a single, highly variable species (i.e. Blake 1968) since there is confirmed hybridisation between the two species in Guyana and Surinam. (see Geographic Variation).

A recent comparison of mitochondrial DNA sequence in this complex suggests that the present species limits currently recognized are not correct as the cayanensis race of Epaulet Oriole may be more closely related to Moriche Oriole (Omland and Lanyon pers. comm.). These data suggest that the complex should be lumped as one variable species or split so that the dark-shouldered forms are in one species and the yellow-shouldered Epaulet Oriole are in another species which includes Moriche Oriole. Before this determination can be made, the remaining subspecies of Epaulet Oriole need to be analysed.

REFERENCES Bond and Meyer de Schauensee 1941 [geographic variation, intragradation], Fraga 1987 [voice, mimicry], Gyldenstolpe 1945b [measurements], Hellmayr 1929 [geographic variation], Naumburg 1930 [measurements], Remsen *et al.* 1987 [distribution], Straneck 1990 [voice].

26 YELLOW-BACKED ORIOLE *Icterus chrysater* (Lesson) 1844 — Plate 14

One of the few orioles with a yellow back, this species ranges higher up in elevation than most other orioles.

IDENTIFICATION A yellow oriole with a black throat bib, wings and tail. Adult Yellow-backed Oriole is most similar in colour and pattern to Yellow Oriole (27) with which it overlaps in N South America. Both of these orioles have fully yellow backs and black wings; however, Yellow-backed Oriole lacks the white wingbars of Yellow Oriole and the auriculars are more extensively black on the former species. The broadly sympatric Yellow-tailed Oriole (30) has a black back as an adult and yellow on the tail in adult plumage, as well as a yellow longitudinal stripe on the closed wings. Immature Yellow-tailed Orioles have largely green backs, and are thus more similar to Yellow-backed Oriole. However, these immatures have a faint but obvious yellow wing stripe, and yellow on the olive tail, as well as often a few darker feathers on the back. Similarly, immature Black-cowled Orioles (45) from Mexico and Central America are superficially similar to immature Yellow-backed Orioles; however, the former has black on the forehead and often the crown as well as entirely black auriculars. Black-cowled Orioles have thin, curved bills unlike the straight, triangular bills of Yellow-backed Oriole. Finally, Audubon's Oriole (49) is similar in overall pattern and structure but it is allopatric. Adult Audubon's Orioles have entirely black hoods, and eastern populations have extensive white on the wing. However, immature Audubon's Orioles are more similar, yet these have black auriculars and black spotting on the crown.

VOICE Sings throughout the year and both sexes are known to sing (Howell 1972).

Song: The song is a sweet series of clear whistles. One version has been described as a series of whistles that ascend and then descend the scale. Other songs consist of 4–6 whistles that sound somewhat off-key. In Panama, they may sound metallic, rather choppy and mechanical. There appears to be a change in the song as one proceeds southward. Mexican songs are the mellowest and sweetest, while in Panama and Nicaragua the songs are quicker, not sweet and sometimes warbled.

Calls: The main call is a musical *chert*. A *whink-whink-whink* is heard when alarmed. A whistling chatter is heard that sounds like *kzwee-kzwee-kzwee-kzwee* or *nyeh-nyeh-nyeh-nyeh*. Gives a wren-like chatter *dzat* or *dzzrt* which varies in speed (B. McKee pers. comm.).

DESCRIPTION A medium-sized oriole with a bill showing only a slight decurved culmen; it appears straight in most circumstances. The tail is slightly graduated.

Adult Male (Definitive Basic): Bill black with a blue-grey basal third of the lower mandible. Eyes dark brown. The forehead, face, chin, throat and breast are black, forming a black bib and mask. The crown, nape, posterior auriculars, sides of neck, back rump and uppertail coverts are orange-yellow. The crown is often speckled with black and the extent of the black forehead is variable. The scapulars and wings are solidly black without orange-yellow markings. The wing-linings are yellow. The underparts are orange-yellow other than the black bib, from the breast sides down to the undertail-coverts. The tail is rounded and entirely black. The legs and feet are blue-grey. Some males have black tips to a few of the back feathers, the significance of which is not understood.

Adult Female (Definitive Basic): Similar to male, but duller, with a greenish wash to the upperparts particularly the back and rump. Females of South American *giraudii* are bright and indistinguishable from the male.

Immature (First Basic): Somewhat variable in appearance. Like corresponding adults, but usually duller with a noticeable olive or greyish wash to the upperparts. The wing and tail feathers, as well as the greater coverts and primary coverts, are retained from the juvenile plumage and look brownish rather than black. The rate of replacement of these feathers is variable and some individuals may show a few contrasting black wing or tail feathers.

Juvenile (Juvenal): Like a duller version of the female, but lacks black on the face and throat. The head, nape and back are yellow with an olive wash. The crown is washed russet. The face is dull yellow and lacks black. The entire underparts are yellow, showing a russet wash on the breast, and becoming paler yellow on the vent. The thighs are pale yellow. The wings are brown with pale yellow edges to the tertials, primaries and secondaries. The coverts are brown with yellow-olive tips, forming inconspicuous wingbars. The wing-linings are pale yellow. The tail is blackish-brown with olive outer tail feathers.

GEOGRAPHIC VARIATION Four forms are recognized. In the older literature this species is referred to as *Icterus giraudii*, but Griscom (1932) clarified that the name *chrysater* had priority as it was published earlier and because both names referred to the same species.

I. c. chrysater ranges from Chiapas, Mexico east to C Honduras and NW Nicaragua. Females are duller than males and readily distinguishable from them, unlike South American *giraudii*, which is described above. The disjunct population in the Mosquitia of Nicaragua is not separable from *chrysater* of the mountains of W Nicaragua and El Salvador (Howell 1972). The formerly recognized subspecies *gualalensis* from Gualala, Guatemala, was described as showing more extensive black on the crown. However, this subspecies is almost certainly an odd specimen of *chrysater*. In addition, Howell (1972) describes one bright male in Nicaragua as showing a 'distinct wash or orange tingeing the back, sides of neck, and the mid-ventral region of the breast and abdomen,' confirming that there is a certain amount of variation in the colour of this species. This may account for the form *hondae*, mentioned below.

I. c. mayensis occurs only on the northern section of the Yucatán peninsula, Mexico. This is the smallest race of Yellow-backed Oriole. It is similar to *chrysater* and *giraudii* in colour. Specimens from Belize are closer to *chrysater*, but are intermediate between this subspecies and *mayensis*. Note that birds from the Yucatán, Belize and E Guatemala tend to have more extensive black on the face and head than the other forms (Griscom 1932).

I. c. hondae appears to be restricted to the upper Magdalena valley, Colombia. Previously, the race *hondae* was considered to extend from W Colombia to Panama; however, a recent analysis has concluded that these birds are not separable from *giraudii* from the rest of Colombia and Venezuela. It may be best to lump all of *hondae* with *giraudii*, but the type specimens (2) have been noted to be clearly different from that form. These birds are deeper orange than specimens of *giraudii* and also show a slightly finer, more slender bill and less black on the forehead. Miller (1947) noted that *giraudii* from the upper Magdalena valley are quite variable in colour saturation, and that furthermore the smaller size of the two *hondae* specimens can be explained by the fact that both specimens are young males. His conclusion is that *hondae* is a variant of *giraudii*. The type locality of *hondae* is within the tropical zone.

I. c. giraudii is found in Panama, W and NW Colombia and N Venezuela. It resembles *chrysater* but is smaller and may show more extensive black on the face. Females lack the greenish wash to the crown, nape and back of the nominate females and look indistinguishable from adult males. This form includes the dubious form *melanopterus* from Venezuela which was described as having a deeper bill (see Griscom 1932).

HABITAT A bird of forests, open woodlands and shrubby slopes, particularly in humid areas and in highlands, including cloud forests. It ranges to nearly 2700 m in Colombia, which is higher than other orioles. It is the only oriole that is found to the subtropical zone and is often common there. It is also known from the tropical zone in South America. In Mexico and Central America, it ranges from 500–2500 m, usually in the arid upper tropical zone, not often to the subtropical zone. Here it prefers dry pine-oak associations in particular, as well as scrubby woodland and second growth, and the edge of clearings. In Nicaragua, it is found either in pine savanna, and sometimes in oaks in the Mosquitia or in pine-oak in the central highlands.

BEHAVIOUR It is typically seen foraging in pairs or family groups, sometimes joining up in small flocks of up to a dozen individuals. Flocking is most common during the non-breeding season (Howell 1972). It commonly associates with *Cissilopha* jays and other oriole species in mixed-species flocks (Rand and Traylor 1954). Yellow-backed Oriole is largely an insectivore and is seldom observed nectaring at flowers. However, in places it can be a pest on bananas (*Musa acuminata*), spoiling many fruits for the farmer. In the pine savannas of coastal Nicaragua, this oriole forages mainly in the pines, even probing and flaking the bark, and also probes in epiphytes such as bromeliads or mistletoe (Howell 1972). Pairs remain together throughout the year and can be observed singing at all seasons. Both males and females sing.

NESTING In Central America, it breeds from February–May, later than most other icterids except Yellow-tailed Oriole. In Colombia, it nests from January–October. The nest is a shallow, bulky hanging basket, ranging up to 13 cm in length. However, it is typically a shorter nest than that of Yellow-tailed Oriole. Nests are often suspended below the cover of a palm frond. The clutch size is of 2–3 eggs which are white with purplish-brown blotching concentrated around the broader end of the egg. There are no recorded cases of cowbird parasitism.

DISTRIBUTION AND STATUS Uncommon to locally common, found to 2900 m, and ranges down to the lowlands. This species is found in two widely disjunct populations. It ranges north as far as the Yucatán peninsula and Chiapas, Mexico; east through SC Guatemala and northern highlands of El Salvador, SC Honduras and NW Nicaragua. A disjunct population exists in the Mosquitia region of E Nicaragua (Howell 1972). This species is apparently absent from S Nicaragua and all of Costa Rica! Also found from W Panama (Veragua) east into E Colombia south to Nariño and north to Bolívar, absent from the dry NW but present in the Santa Marta Mountains; ranges east of the Andes south to Meta and north to Santander and Boyaca; east into the Venezuelan Andes (Tachira to Lara) and coastal mountains to Miranda. It is unrecorded from Ecuador but may be present in the extreme north.

MOVEMENTS May be partially migratory, as it is most common in El Salvador during the winter months, the dry season of November to March. The details of these movements are unknown.

MOULT As in most other adult blackbirds there is one complete (definitive pre-basic) moult that occurs after the breeding season (roughly during September–October in Mexico and Central America). Pre-alternate moults are missing. Juvenal plumage is replaced by the first basic plumage through a variable first pre-basic moult. The body feathers are all replaced and a variable number of coverts, tertials, wing and tail feathers may also be changed. On most individuals, wing and tail feathers are retained from the juvenal plumage, with their replacement being variable. The timing of this moult is similar to that of adults; however, the first evidence of head moult is seen quite soon after fledging.

MEASUREMENTS ***I. c. chrysater***. Males: (n=18) wing 102.1 (91.4–110.3); tail 99.3 (95.8–108.7); culmen 24.4 (22.9–26.9); tarsus 26.9 (25.9–28.2). Females: (n=13) wing 99.2 (86.4–106); tail 99.7 (82.6–105); (n=3) culmen 22.6 (22.1–23.4); tarsus 24.9 (23.6–26.2).

I. c. mayensis. Males: (n=8) wing (95–99); tail (96–105); culmen (22.5–24.2).

I. c. hondae. Males: (n=2) wing 90, 92; tail 93, 91; bill length from the nostril 18, 17.

I. c. giraudii. Males: (n=10) wing 92.3 (85.0–97.0); tail 94.1 (87.5–103.5); culmen 22.3 (20.6–23.7); tarsus 24.5 (23.2–25.6). Females: (n=10) wing 87.8 (82.0–91.7); tail 90.7 (83.6–99.2); culmen 22.1 (20.5–23.7); tarsus 24.5 (22.3–26.2).

NOTES A specimen of a supposed hybrid between this species and the Yellow-tailed Oriole exists from Belen (245 m), Antioquia Department, Colombia (Olson 1983). In the older literature, Yellow-backed Oriole is referred to as *Icterus giraudii*, rather than the currently accepted *Icterus chrysater*, see Geographic Variation. In addition, an old common name for this species was Underwood's Oriole. A recent study comparing the mitochondrial DNA sequences of the orioles suggests that Yellow-backed Oriole is most closely related to Audubon's and Scott's Orioles (K. Omland, S. M. Lanyon pers. comm.).

REFERENCES B. Behrstock personal recordings [song], Cassin 1847 [geographic variation], Howell and Webb 1995 [identification, distribution], Miller 1947 [geographic variation, measurements], Olson 1981 [geographic variation], 1983 [hybrid], Rand and Traylor 1954 [behaviour, movements], Ridgway 1902 [measurements], Smithe 1966 [nesting], Thurber *et al.* 1987 [distribution], van Rossem 1938 [geographic variation, measurements], Wetmore *et al.* 1984 [measurements].

27 YELLOW ORIOLE *Icterus nigrogularis* (Hahn) 1819 — Plate 14

Most orioles are yellow, but this one is more yellow than the others due to the fact that black is restricted in extent.

IDENTIFICATION This oriole has the black restricted to the wings, tail and throat bib. It never shows a black back, which separates it from adult male Baltimore (40) and Orchard (43) Orioles as well as adult Epaulet (25), Moriche (24), Yellow-tailed (30) and Orange-crowned (31) Orioles as well as from Troupial (37); all of which are sympatric with Yellow Oriole. The most similar sympatric oriole to Yellow Oriole is Yellow-backed Oriole. Yellow Oriole however, has bold white edges to the tertials, a white lower wingbar and white edging at the base of the primaries which forms a white 'handkerchief'. Yellow-backed Orioles have entirely black wings, which at the most show some paler edging towards the primary tips. Note also that Yellow Oriole has yellow epaulets which Yellow-backed Oriole lacks. In addition, the facial pattern of these two species is different. Yellow Orioles have reduced black on the face, showing a black mask between the eyes and bill, as well as a thin black stripe down the center of the throat; the cheek, between these two areas of black is yellow. In contrast, Yellow-backed Oriole shows a black ocular area that is contiguous with the wide black throat bib; thus, the anterior auriculars of this species are mostly black. There are further differences in size and structure. Yellow Orioles are noticeably smaller than Yellow-backed Orioles. In addition, Yellow Oriole has a short and deep bill, unlike the longer and more pointed bill of Yellow-backed Oriole. Furthermore, Yellow Orioles are a lowland species while in South America Yellow-backed Oriole is a bird of the mountains, reaching the subtropical zone.

Juvenile and immature Yellow Orioles may also cause an identification problem, being largely yellow with darker wings and tail. Once the juvenile plumage is lost, immatures gain the black throat and lores typical of the adults. Immatures (first basic) are similar enough to adults for the above field marks to make identification possible. The briefly held juvenal plumage may confuse observers, however. Juveniles have bold, pale wingbars, the upper one usually more yellow than the lower one. Otherwise the body is entirely pale yellow below and green above. Juvenile Yellow-backed, Yellow-tailed and Orange-crowned Orioles may be confused with this plumage. Note the thick bill of Yellow Oriole, which helps to distinguish it from other orioles. In addition, the presence of only even one black feather on the back eliminates Yellow Oriole from consideration. Yellow-tailed Oriole has a different wing pattern, with yellow edging to the tertials and inner greater coverts, while Orange-crowned Oriole has an orange cast about the head, as well as a very dull upper wingbar. Yellow-backed Oriole is once again the most similar, but note the bill proportions, the lack of obvious wingbars, and the more olive back of juveniles of that species.

VOICE Song: The song is melodious and pleasant, but soft in volume. It has been described as being structured in pairs or trios of flute-like notes, which are simpler in pattern to that of Moriche Oriole. Long phrases combine flute like notes with throatier ones, *chwit-chwoot-cht-cht-cht-chwit-chu* (Snyder 1966). Voous (1983) describes the sweet, flute-like song as: *tjee-tjü tjee-tjü tjee-tjee.* The inclusion of buzzes, *chuk, tik* notes, and other unmusical sounds into the whistled song is quite distinctive. Another variation is *whoot-wheeo-whoot-whoot,* the second note higher and slightly rising. This species is reported to be a persistent mimic. It is not common to hear this species sing in the Netherlands Antilles, with singing only taking place in the morning. The song is weak and is often sung as a 'whisper song' (Voous 1983).

Calls: The most common call is a hesitant chatter, *chuck-uch-ch-ch* (Snyder 1966). This call is described as an unpleasant alarm call *chet-chet-chet* or *retch-retch-retch* Voous (1983). Also gives a cat-like whine which may be added to the chatter/alarm call.

DESCRIPTION A chunky oriole with a stout bill. The culmen and gonys are straight, and the bill is quite deep at its base. Wing Formula: P9 < P8 ≈ P7 > P6; P9 ≈ P4; P5–P8 emarginate.

Adult Male (Definitive Basic): The black bill shows a spot of grey restricted to the basal quarter of the lower mandible. Its eyes are dark brown. The black on the lores and throat of this oriole are not connected, unlike in related species. The lores and area directly surrounding the eyes are black, and there is a small spot of black at the base of the lower mandible. The malar area is yellow, extending to the bill base. The black throat bib extends from the chin to the upper breast; however, it is narrow and does not extend to the malar area. The crown and face (including almost all of the auriculars) are yellow, continuing to the mantle, lower back, rump and uppertail-coverts. The posterior scapulars are black with a yellow fringe; these are often hidden by the other scapulars, but when visible may look like a black band separating the mantle from the lower back. The underparts, other than the throat bib, are entirely yellow to the undertail-coverts with the yellow palest on the belly and vent. The wings are black with restricted yellow and white markings. The lesser and median coverts are yellow, while the black coverts show narrow white tips. This may appear as a thin white wingbar in the field, but it is often absent due to wear. The black tertials and inner secondaries show crisp, narrow, white fringes when fresh. The extreme bases of the outer primaries show white edges; when folded, these appear as a small white spot that contrasts with the black primary coverts. The primaries are also thinly-edged in white towards their tips. The wing-linings are yellow. The tail is black, with a restricted white tip to the outer two pairs of feathers (R5 and R6). The legs and feet are grey.

Adult Female (Definitive Basic): Similar to male in pattern. However, females tend to show a greenish wash to the back, which on close inspection contrasts somewhat with the brighter yellow head and rump.

Immature (First Basic): Similar to adults but duller. The black throat is narrowly tipped with yellow, while the lores are duller black. The crown and back are dull greenish, rather than bright yellow. The wings are grey-brown, rather than black, and the yellowish median coverts are black-based. The tail is blackish-olive.

Juvenile (Juvenal): Quite different from adults. The crown, neck and mantle are olive, becoming olive-yellow on the lower back, rump and uppertail-coverts. The face is yellow, lacking the black lores. The throat is pale yellow, becoming olive-yellow on the breast, continuing as pale yellow to the undertail-coverts. There may be an indistinct black wash on the throat. The wings are greyish-brown, rather than black. The coverts are grey-brown; the medians are tipped yellow while the greaters are tipped white. In the field, this appears as a dull yellow upper wingbar and a whitish lower wingbar. The tertials are edged white but show slightly more yellow tips. The primaries and secondaries are fringed greyish-white, forming a pale wing panel. However, the bases of the secondaries are dark, so that the pale panel does not extend to the greater coverts. The wing-linings are pale yellow. The tail is olive with greenish-yellow edges to the outer rectrices.

GEOGRAPHIC VARIATION Four races recognized.

I. n. nigrogularis is the most widespread race, being found from Colombia east through N Venezuela, the Guianas and N Brazil. It is described above.

I. n. curasoensis is found only on the Dutch Antilles: Aruba, Curaçao and Bonaire. This form resembles *nigrogularis*, but it has a longer, thinner bill. In addition, the white fringes on the wings are broader creating a larger wing patch and the yellow of the body is paler in colour. The juvenile is similar to the nominate, but shows a greyish wash on the crown and back.

I. n. helioeides is restricted to Margarita Island, Venezuela. Resembles *nigrogularis*, but is larger and has a larger, stouter bill with a straight culmen; the legs are weaker than in *nigrogularis*. The race *helioeides* is more orange-yellow than *nigrogularis* and has more extensive black on the throat. Its bill is stouter but shorter than that of *curasoensis*, and *helioeides* is larger, brighter and has more black on the throat.

I. n. trinitatis ranges from Trinidad and Monos Island to the extreme tip of the Paria peninsula of Venezuela. This race is most similar to *helioides*, but has shorter wings. The primaries are not edged white as in that form, and also its bill averages smaller. However, it has a larger bill than *nigrogularis* and longer wings than that form; *nigrogularis* also shows white edges to the primaries.

HABITAT Generally a bird of open woodlands, not closed forest. It occurs in a variety of habitats, as long as these are open in nature or have a forest edge component. In Venezuela, this species is commonly found in deciduous woodlands and scrub, as well as in urban areas. Yellow Orioles are not uncommon in towns and gardens. In the Netherlands Antilles, Yellow Oriole is most common in arid thorn and cactus scrub as well as in open fields with scattered thorny trees, in addition to mangroves (Voous 1983). Here it also visits gardens and fruit plantations, but not commonly. In Surinam it is found in swamps and wet gallery forests, particularly those near the coast. It is also partial to coastal mangroves (Haverschmidt 1951, 1968). In Colombia, it is found primarily in the arid coastal zone along the Caribbean shore. In Trinidad, populations are fond of forest openings as well as parks, gardens and mangroves.

BEHAVIOUR Usually observed singly or in pairs, but not encountered in flocks. Quiet during the non-breeding season but may be more vocal while breeding. Will often sing from a perch high in the trees. Forages in canopy of trees. Feeds mainly on large insects such as grasshoppers, cicadas and beetles. Visits flowers to both feed on nectar and to catch insects attracted to the flowers. Also feeds on fruits, but does not become a pest in fruit orchards. Visits bird feeders, but never becomes very tame (Voous 1983).

NESTING In Surinam, the breeding season extends from February–September, and June and July are the peak months for nesting (Haverschmidt 1951, 1968). In Trinidad, nesting may occur at any time of the year, but peaks between April–August. On the Netherlands Antilles, it has been observed nesting during January, May, November and December and it is not unlikely that this species breeds throughout the year (Voous 1983). In the Santa Marta region of Colombia, it is nests in April–May (Todd and Carriker 1922). It is a solitary nester. Observations of several old nests built only meters away from each other have suggested that Yellow Oriole is sometimes colonial; however, this does not appear to be the case. Pairs may nest in the same nest in subsequent seasons, but more commonly an old nest is torn up and the materials are then used to build a new nest (Voous 1983). Some old nests are retained in the nest tree and these may be used for sleeping and/or courtship (Voous 1983). These 'retained' nests explain the apparent colonies mentioned above. Pairs use a traditional nesting tree which they come back to season after season.

The nest is a hanging basket approximately 40 cm long, which is long for the size of the bird. Nests are built near the tip of an outer branch, and may be placed quite high in the tree. It is common for Yellow Orioles to build the nest over water. Sometimes nests are placed beneath a palm frond as in some other tropical orioles, but it is also common to see them suspended from other types of trees. The nest is constructed by both sexes, largely from grasses or palm fibers but they often incorporate anthropogenic materials such as paper, cloth, plastics and the like (Voous 1983, Jaramillo pers. obs.). The clutch size tends to be 3, sometimes only 2 eggs, and one clutch of 5 has been described (Todd and Carriker 1922). The eggs are pale green or greyish with dark spotting, blotching and scrawls, particularly around the broader end (Belcher and Smooker 1937). Eggs from Surinam are described as creamy-white with a few black spots (Haverschmidt 1968). In Trinidad, this species is a known host of Shiny Cowbird (102).

DISTRIBUTION AND STATUS Common, restricted to the lowlands below 500 m. Distributed throughout northernmost South America and some adjoining Caribbean islands. In Colombia, from the lower Sinú river in Córdoba, east along the Caribbean lowlands, north of the west and central Andes and west of the east Andes, to Guajira and south in the Magdalena valley south to at least S Santander. East of the east Andes in the Llanos from the Venezuelan border south to S Meta and E Vichada (Hilty and Brown 1986). In Venezuela, it is distributed widely north of the Orinoco River, except in the Andes from Tachira to Lara. South of the Orinoco it is found in Amazonas and Bolívar, but it is absent from the Tepui highlands in the south of the country. Yellow Oriole is found on offshore islands of Venezuela, such as Sucre, Patos and Margarita. In addition, it is a resident of Trinidad, but is absent from Tobago. This oriole can also be found in Aruba (very scarce), Curaçao (common) and Bonaire (uncommon), Netherlands Antilles (Voous 1983). On the mainland, the distribution continues eastward through Guyana, Surinam and French Guiana, with this species perhaps being absent from the southernmost sections of those countries but it is present in northernmost Roraima in Brazil. Also occurs in coastal Brazil, north of the delta of the Amazon river in Amapá.

MOVEMENTS No evidence of migration or seasonal movements, appears to be sedentary.

MOULT Little known. There appears to be only one moult per year in adults, the complete definitive pre-basic. Its timing is dependent on the time of nesting as it occurs subsequent to breeding. In Surinam, moulting specimens exist from October. In the Llanos of Venezuela, this moult is finishing by late November. Specimens of *trinitatis* are relatively fresh in February but worn by July, suggesting that the definitive pre-basic moult occurs shortly after July (August–September?). It appears that during this moult, the tail is moulted quickly and finishes growing before the outer primaries have been moulted. The extent of the first pre-basic moult includes the body, wing-coverts, tertials and sometimes the tail. At least in the Llanos of Venezuela, its timing is similar to that of the definitive pre-basic.

MEASUREMENTS ***I. n. nigrogularis***. Males: (n=10) wing 92.3 (88–96); tail 88.7 (87–92); culmen 20.7 (20–22); tarsus 26.6 (25–28). Females:

(n=10) wing 85.7 (82.0–89); tail 81.6 (76.5–89); culmen 21.2 (19–24.1); tarsus 26.4 (24.9–27.7).

I. n. curasoensis. Males: (n=7) wing 92.9 (90.2–96); tail 88.2 (84.6–93); culmen 26.4 (26–27.9); tarsus 26.5 (25.1–28). Females: (n=2) wing 87, 97; tail 88, 91; culmen 25, 26; tarsus 25, 28.

I. n. helioides. Males: (n=5) wing 98.6 (91–102); tail 95.8 (88–103); culmen 22.8 (22–24); tarsus 26.4 (24–29). Females: (n=4) wing 92.7 (88–94); tail 92.2 (83–99); culmen 23 (22–24); tarsus 24.8 (24–25).

I. n. trinitatis. Males: (n=10) wing (96–99.5). Males: (n=3) wing 95 (95–95); tail 92.7 (91–94); culmen 23 (23–23); tarsus 26.3 (24–28). Females: (n=2) wing 89, 92; tail 91, 90; culmen 22, 22; tarsus 26, 26.

NOTES The intensity of yellow, orange and red colours in orioles varies to a great extent. This is most likely due to differences in nutrition. One specimen from Venezuela showed a gorgeous persimmon-orange body colour, similar to the head colour of the 'Scarlet-headed' subspecies group of Streak-backed Oriole (35). A recent study comparing the mitochondrial DNA sequences of the orioles suggests that Yellow Oriole belongs to a group which includes Jamaican Oriole, Orange Oriole and Altamira Oriole (K. Omland, S. M. Lanyon pers. comm.).

REFERENCES Belcher and Smooker 1937 [nesting], Chapman 1917 [habitat], Clark 1902 [geographic variation, measurements], Hartert 1914 [geographic variation, measurements], Haverschmidt 1951 [habitat, nesting], Kroodsma and Baylis 1982 [voice, mimicry], Ridgway 1884 [geographic variation, measurements], Ridgway 1902 [measurements], Voous 1983 [voice, behaviour, distribution, nesting].

28 JAMAICAN ORIOLE *Icterus leucopteryx* (Wagler) 1827 — Plate 15

A distinctive olive-yellow oriole restricted to the Caribbean. Its distribution in the Caribbean is unusual as it was formerly found in three widely separated islands. Known as Auntie Katie in Jamaica.

IDENTIFICATION Jamaican Oriole is an olive-yellow oriole with a black mask and throat patch, and entirely white median and lesser coverts. It is sympatric only with the migrant Baltimore (40) and Orchard (43) Orioles. However, this distinctive oriole does not resemble either. The olive-yellow body colour is not found in any other oriole species. In addition, Jamaican Oriole has an olive-yellow back, not black like male Baltimore and Orchard Orioles. Females of those two species lack the black back, but neither have the bold white wing panel created by the white greater coverts of Jamaican Oriole. A large white wing panel like this is seen only in Bullock's (41), and Black-backed (42) Orioles, and Troupial (37), none of which is found sympatrically with Jamaican Oriole, albeit Bullock's could occur as a vagrant. Bullock's Oriole is unlike Jamaican Oriole in its black back, orange underparts and orange supercilium.

Juveniles may pose an identification problem, as they lack a completely white wing panel and the black throat and lores of the adults. In this respect they resemble female or immature Orchard and Baltimore Orioles. However, Jamaican Oriole is a rather thick-billed oriole, the bill being noticeably deep at the base, differing dramatically from the small-billed Orchard and the medium-billed Baltimore Orioles. While the white wing panel is lacking on juvenile Jamaican Oriole, there are two bold white wingbars, often with enough white edging on the greater coverts to approximate a wing panel. The wingbars of Orchard and Baltimore Orioles are not as obvious. In addition, Baltimore Oriole shows an orange wash to the plumage, particularly on the throat which Jamaican Oriole lacks. Orchard Orioles are smaller and noticeably slimmer than Jamaican Orioles.

The San Andres Island and the now extinct Cayman Islands subspecies of Jamaican Oriole show a more yellow, less olive, plumage than the Jamaican subspecies. Once again, the only possible confusion species in those areas are the two migrant North American orioles and differences listed above still apply.

VOICE Song: The true song is a series of quick whistles, sounding warbled as the individual notes are given in rapid succession. Overall it is a pleasant song, but not clear and sweet as that of many other members of the genus. The full song lasts just over 2 seconds, but it is repeated many times. The song is heard between October–June, but singing is most prevalent between February–March (Jeffrey-Smith 1972). Both sexes sing (Scott 1893).

Calls: A melodious 'Auntie Katie' or 'Auntie Bessie', which gives the species its local name in Jamaica. The call may also be interpreted as 'you cheat, you cheat' or 'Tom Payne, Tom Payne' or 'cheat-you'. Also gives a single whistle (Stewart 1984, Downer and Sutton 1990). A short chatter is sometimes heard.

DESCRIPTION A medium-sized oriole with a long, stout bill. The culmen is nearly straight, slightly decurved at the tip. The tail is rather square-tipped. Wing formula: P9 < P8 < P7 > P6; P9 ≈ P4; P8–P4 emarginate.

Adult Male (Definitive Basic): The black bill shows a grey basal half to the lower mandible. Eyes dark brown. Black on lores and around eyes, continuing on the anterior auriculars, chin, throat and upper breast forming a black mask and bib. The rest of the face, crown, neck, mantle, rump and uppertail-coverts are olive-yellow. Similarly, the underparts are entirely olive-yellow except for the black bib. The intensity of the olive-yellow varies, being most yellow on the belly and most olive on the head. On some individuals, there is a slight chestnut wash on the face and around the edge of the black bib. The wings are black with a large white patch created by the white lesser, median and greater coverts. The tertials and innermost secondaries are black with white outer edges, while the rest of the secondaries and the primaries are entirely black. The wing-linings are yellow. The tail is entirely black. The legs and feet are dark grey.

Adult Female (Definitive Basic): Similar to male, but on average the olive-yellow of the body is slightly duller.

Immature (First Basic): Similar to juvenile, but with a fully black throat, bib and lores. The underparts are brighter, and there is more extensive olive on the crown. Sometimes there is some dull blackish mottling on the crown. The upper- and underparts, while being brighter than those of the juvenile, are duller than the adult. On the wings, the greater coverts are widely-edged white, forming a white wing panel, rather than discrete wingbars. However, the outermost greater coverts show dark bases which break up the wing panel somewhat. The tertials are blacker than on the juvenile, and contrast with the browner, retained, juvenile primaries and secondaries. The tail may be olive as on the juvenile, and sometimes black flecking may be present on the rectrices. A variable number of primaries and secondaries may be black and adult-like.

Juvenile (Juvenal): The upperparts are olive from the crown to the uppertail-coverts; however, the mantle is slightly more grey. The face and underparts are yellow-olive, most yellow on the belly and more olive on the breast. The undertail-coverts are yellow. The wings are blackish with olive-grey lesser coverts and scapulars. The median coverts are widely-tipped white, while the greater coverts have bold white tips as well as thin white edges; the overall appearance is of two well-formed white wingbars. The tertials are also widely-edged white, while the secondaries and primaries show dull grey edges. The wing-linings are yellow. The tail is entirely olive.

GEOGRAPHIC VARIATION Three subspecies recognized

I. l. leucopteryx is found in Jamaica, and is described above.

I. l. bairdi was restricted to Grand Cayman Island, and appears to be extinct. It was similar to the nominate in pattern, but not colour. Olive-yellow was lacking in this form, being replaced by a vivid orange-yellow.

I. l. lawrencii is restricted to St Andrew's (San Andrés) Island, Colombia. It is similar to the nominate form but less olive on the upper- and underparts.

HABITAT As is typical of island species this species is not fussy in its habitat choice. It may be found in all available forest types, forest edge and wooded cultivated areas, as well as gardens. It ranges from lowland forests to montane, subtropical forests. It is rare in mangroves. The extinct subspecies *bairdi* inhabited gardens and woodlands (Bradley 1985).

BEHAVIOUR A confiding and noticeable species, especially due to its fondness for gardens. It is observed singly, in pairs or in family groups. Birds tend to perch and sing high in the trees, usually from a tall branch or treetop. Feeds on a variety of foods, including insects, fruit and necar from flowers. Fruits make up a significant component of the diet but do not outnumber insects; this species will also eat bananas (*Musa acuminata*), oranges (*Citrus sinensis*), pawpaw (*Asimina triloba*), pimento (*Pimenta officinalis*) and custard apples (*Ammona*) in particular (Stewart 1984). Jamaican Orioles may forage by probing bromeliads, flaking bark from trees, or by opening seed pods in order to extract insects (Downer and Sutton 1990). One study discovered that nearly two-thirds of the prey obtained by this species were insects garnered by probing and flaking bark from trees, with a small number of insects obtained from bromeliads; the remaining one-third of their food was composed of fruit and from the probing of flowers (Lack 1976). The use of this bark-flaking foraging technique is particular to this species amongst the orioles, perhaps accounting for Jamaican Oriole's rather thick bill. This last study did not include the use of introduced plants by this Oriole. The inclusion of *Erythrina* flowers and bananas amongst plant species used would have doubled the amount of flowers and fruit use by this species. In addition, Jamaican Oriole has been observed feeding from a hummingbird feeder. Jamaican Orioles forage mainly between three and 16 meters from the ground, occasionally lower. Typically, when gleaning from leaves, they forage on the outer part of the crown of the tree (Cruz 1978).

NESTING Individuals of the San Andrés subspecies were in breeding condition in April (Russell *et al.* 1979). Details of the nests and eggs of *bairdi* and *lawrencii* have not been published. The nesting season in Jamaica stretches from March–August. The nest is a hanging basket woven from dry *Tillandsia* (Spanish Moss), grass stalks, horse hair or other thin fibers particularly the strong strands garnered from Palmetto plants (*Sabal* sp.) as well as the fibrous strands from the bark of the Dodder or Love Bush. Nests sometimes include plastic threads or other artificial materials. The nest is suspended from two points, often at the fork of a branch or from two parallel branches (Stewart 1984, Downer and Sutton 1990). While the nest may sometimes be suspended from a palm frond, Jamaican Oriole does not weave the attachments of the nest through the palm leaf, as is the case with many other tropical orioles (Bond 1988). Both sexes aid in the building of the nest. The clutch size is commonly of 4 eggs, but ranges from 3–5, which are white with dark brown spots and scratches (Stewart 1984).

DISTRIBUTION AND STATUS Common in Jamaica, found from sea-level to the highest mountains but more common in the lowlands. It ranges throughout Jamaica, apparently present in all areas of the country. It was common in San Andrés Island, Colombia in 1972 (Russell *et al.* 1979), and it is likely that its status has not changed there. San Andrés lies in the Caribbean east of Nicaragua. Sadly it was formerly present on Grand Cayman Island but now appears to be extinct. This subspecies on Grand Cayman was apparently largely restricted to the north coast of the island (Johnston 1975). The last records were in George Town between 1965 and 1967 (Bradley 1985).

MOVEMENTS This insular species is strictly sedentary.

MOULT Adults undergo one moult per year, the complete definitive pre-basic moult which occurs subsequent to breeding. Moulting individuals may be observed between August–October. Juveniles lose their plumage through the variable partial first pre-basic moult. The first pre-basic includes the body plumage as well as the wing-coverts and usually the tertials. The primaries and secondaries are retained from the juvenal, but a small number of individuals moult some secondaries and primaries. Most do moult at least the central rectrices. During this moult, remiges and rectrices may not be changed symmetrically. Juvenal plumage appears to be short-lived, as black lore and throat feathers are obtained early in the bird's life. A specimen of the extinct subspecies *bairdi* was finishing its definitive pre-basic moult during early March.

MEASUREMENTS *I. l. leucopteryx*. Male: (n=10) wing 103.8 (98–108); tail 88.5 (81.3–95); culmen 23.2 (22–24.4); tarsus 24.3 (23–25.1). Females: (n=2) wing 101.6; (n=3) tail 85.4 (83.6–89); culmen 23.7 (22.9–25.1); tarsus 23.8 (23–24.9). Both sexes: (n=9) wing 99.1 (92–105); tail 85.2 (82–90); (n=8) culmen 22.4 (19–25); (n=9) tarsus 23.6 (22–26).

***I. l. bairdi*.** Male: (n=2) wing 95.3, 103; tail 76.2, 89; culmen 21.6, 24; tarsus 21.6, 24.

***I. l. lawrencii*.** Male: (n=4) wing 105.3 (97–111); tail 94.0 (91–97); (n=3) culmen 25.7 (25–26); (n=4) tarsus 24.3 (23–25). Female: (n=8) wing 98.5 (92–103); tail 86.6 (80–92); (n=7) culmen 25.3 (25–26); (n=8) tarsus 24.1 (23–25). Sex undetermined: (n=3) wing 102.1 (101.3–103); tail 86.0 (83.1–89); culmen 24.5 (23–25.4); tarsus 24.9 (24.6–25).

NOTES Beecher (1950) and Lack (1976) stated that Jamaican Oriole was most closely related to Streak-backed Oriole (35). The reasons for this supposed relationship were not given and we personally don't agree that this is the case. In several ways, Jamaican Oriole is most similar to Yellow Oriole (27). In any event, it does appear that the more slender and thin-billed orioles, the *Bananivorus* group suggested by Beecher 1950, which includes the Antillean forms of the Black-cowled Oriole complex and the three Lesser Antillean Orioles (St Lucia, Martinique and Montserrat Orioles) are more closely related to each other than is any to Jamaican Oriole. This has been recently confirmed through comparisons of the mitochondrial DNA sequences of these species. This research suggests that Jamaican Oriole belongs to a group which includes Yellow, Orange and Altamira Orioles (K. Omland, S. M. Lanyon pers. comm.).

REFERENCES Bond 1988 [nesting], Bradley 1985 [habitat, distribution], Downer and Sutton 1990 [voice, nesting, behaviour, habitat], Hardy *et al.* 1998 [voice], Lack 1976 [behaviour], Ridgway 1902 [measurements], Stewart 1984 [voice, behaviour, nesting].

29 ORANGE ORIOLE *Icterus auratus* — Plate 11

Bonaparte 1851

This oriole is no more orange than many other species; perhaps it should be renamed Yucatán Oriole as it is endemic to that peninsula.

IDENTIFICATION A small, slim oriole, similar to Hooded Oriole (36) in shape and structure. The male Orange Oriole is largely orange, with an orange back. The black wings show a white shoulder patch, white lower wingbars, a white patch at the base of the primaries and white edging on the tertials and secondaries (not reaching the base of the secondaries). The bill is straight, with a straight culmen. This feature alone is enough to distinguish it from Hooded Oriole, which has a noticeably curved culmen. Figure 29.1 illustrates the bill shapes of several similar oriole species, including Orange Oriole. In addition, male Orange Orioles have orange backs while Hooded Orioles are black-backed. All of the oriole species sympatric with Orange Oriole, except Yellow-backed Oriole (26), have black backs. The larger Yellow-backed Oriole is yellow, not orange, lacks white on the wings and has more extensive black on the face.

Immatures or females of Hooded and Orchard (43) Orioles are similar to females and immatures of Orange Oriole. Bill shapes are useful in making an identification: Orange Oriole has a stout bill with a straight culmen; Hooded Oriole has a long, thin bill with a curved culmen; Orchard Oriole has a thin bill with a straight culmen slightly curved at the extreme tip (Fig 29.1). Orange Oriole's wing pattern is distinctive, with a white patch at the base of the primaries ('handkerchief') and two wingbars being noticeable. Orchard and Hooded Orioles sport two wingbars, but lack the white patch at the base of the primaries. Females and immature male Orange Orioles always show black lores and a black bib; the juvenal plumage is the only one that lacks black on the face. In contrast, female Orchard and Hooded Orioles lack black on the face. Female and immature male Orange Orioles have black-tipped outer scapulars, which may be apparent given a good view.

VOICE Song: A series of widely spaced sweet whistles. Each note is given at an interval of 1–2 seconds. The individual whistles are short (less than a quarter of a second in duration) and they either rise or drop in pitch. Usually a rising whistle is followed by a falling whistle and vice-versa. The song is easily imitated by a human whistle.

Calls: Includes a harsh chatter, *chi-chi-chi-chi-chi.* The individual notes of this chatter are given slowly enough to be able to count each. Also gives a nasal *nyehk,* a *choo* and a nasal, drawn out *wheet,* also a whistled *chuchuchuchu* or *cheecheechee* (Howell and Webb 1995).

DESCRIPTION A small, slim oriole with a thin bill and longish tail. Its tail is slightly graduated. The bill has a straight culmen and gonys, and looks obviously pointed. Wing formula: P9 < P8 < P7 < P6 > P5; P9 ≈ P3.

Adult Male (Definitive Basic): The black bill shows a grey basal quarter to the lower mandible. The eyes are dark. The black lores, ocular area and throat bib, which extends to the upper breast, contrast with the orange head. The black face patch has an indentation of orange, reaching forward towards the bill base, just below the eye. The head and body are a brilliant orange, including the back, and both upper- and undertail-coverts. The orange is brightest

Figure 29.1 Head and bill shapes of several northern hemisphere orioles.

on the crown, face and breast. On the underparts, the orange is paler and more yellow on the belly. It also becomes paler on the lower back. The scapulars are black with orange tips or edges. The wings are black with orange epaulets, composed of the orange lesser and marginal coverts. However, the median coverts are white with black bases, which are often hidden in normal resting posture. This creates a wide, white upper wingbar. The greater coverts are black with restricted white tips which form a thin lower wingbar that wears away rapidly. A pale wing panel is formed by the white edges to the black tertials and secondaries. Note that the bases of the secondaries are entirely black, so the pale panel does not reach to the greater coverts. In addition, the primaries are edged white at their bases, creating a pale patch ('handkerchief') that contrasts with the entirely black primary coverts. A less noticeable amount of pale edging occurs towards the tip of the primaries. On the underwings, the linings are yellow-orange. The tail is black, tipped with grey. The legs and feet are grey. The tone of body colour varies depending on the individual; some are more scarlet, others more yellow.

Adult Female (Definitive Basic): Similar to male, but the body colour is duller, more yellow. The back and crown usually shows an olive wash. In addition, the wings and tail have a duller ground colour, blackish-brown rather than black. The white on the median coverts is reduced due to a greater amount of black at their bases. The pattern of white on the primaries, secondaries and tertials is as on the adult male. The tail is black.

Immature Male(First Basic): Similar to adult female, but more brightly coloured. However, the back is washed olive. The wings are brownish, showing a reduced but recognizable pattern of pale markings like the adults. The epaulet (lesser coverts) is brown, not yellow. The brown median coverts show small white tips, rather than being extensively white. The lesser coverts are brown and tipped orange-yellow. The tail is olive-brown with more yellow outer rectrices. This plumage is somewhat variable, some individuals may show more white on the coverts or less olive on the tail, for example.

Immature Female (First Basic): Similar to adult female and duller than an immature male. The tail is olive, unlike the adult female and the white tipping on the coverts is less extensive.

Juvenile (Juvenal): Similar to immature plumages, but duller yellow. Also, the black lores and throat are lacking. The juvenile wing shows pale yellow tips to the greater and median coverts, forming two dull yellow wingbars. The bill has a pink base to the lower mandible.

GEOGRAPHIC VARIATION Monotypic.

HABITAT Inhabits open woodlands and more open areas with scattered trees, ranging from arid to semi-arid in nature (Howell and Webb 1995). Sometimes occurs in flooded scrub woodland (Howell *et al.* 1992). Also noted to live in edge situations and second growth (Davis 1972) and abandoned farmland (Peterson and Chalif 1973). Nests over water, often in limestone sinkholes known as 'cenotes' (Howell *et al.* 1992).

BEHAVIOUR A typical oriole, being seen singly or in pairs. It may flock with other oriole species (Howell and Webb 1995). Behaves much like other orioles, except that it tends to be colonial (see below). It is not known if monogamy or polygyny is the operating mating system within these colonies. Very little else is published on this species.

NESTING Nests colonially and solitarily. This is the only oriole for which colonial nesting is common. Colonies have been found with 26–35 nests, spread out over a small area but in several trees. In one colony five nests were observed in one tree. Colonies are placed in low trees and shrubs in flooded scrub, with placement ranging from one to nine meters above the ground. Within colonies, the birds are not necessarily synchronized in their nesting stage. However, it is not clear if nests are clumped within colonies or if parts of the colony are synchronous. Single nests have been found, usually over water, in bushes or trees placed between one and 10 m above the ground. The nest itself is typical for an oriole, being a hanging basket woven from fine fibers. The nests are similar to, but longer than those of the sympatric Hooded Oriole. Unlike the nests of Hooded Oriole, those of Orange Oriole are conspicuous, not being hung under palm fronds as is typical of the former species (Howell *et al.* 1992).

DISTRIBUTION AND STATUS Endemic to the Yucatán peninsula, where it is fairly common. In the Mexican part of the peninsula it is found in the state of Yucatán, N and C Campeche and all of Quintana Roo except for the extreme SW. It has been recorded in extreme northern Belize (Ambergris Cay). It is not known if the species breeds in Belize; all reports are from the winter (Howell and Webb 1995).

MOVEMENTS Sedentary, no evidence of migration or local movements.

MOULT Adults undergo one moult per year, the complete definitive pre-basic, which occurs subsequent to the breeding season. Juvenal plumage is retained for a short period only; soon after fledging, black throat and lore feathers of the First Basic plumage appear. The first pre-basic moult replaces the body plumage as well as the wing-coverts and the tertials; on some (mainly males?) the rectrices are also replaced. It appears that all pre-alternate moults are lacking, yet this needs confirmation.

MEASUREMENTS Males: (n=10) wing 91.4 (88–94.7); tail 89.8 (84.6–96); culmen 20.5 (19–22.1); tarsus 24.3 (23.9–25.1). Females: (n=7) wing 91.4 (87–96); tail 89.4 (82–98); culmen 20.3 (19–21); tarsus 24.4 (22–25).

NOTES A recent study comparing the mitochondrial DNA sequences of these orioles suggests that Orange Oriole belongs to a group which includes Jamaican, Yellow and Altamira Orioles (K. Omland, S. M. Lanyon pers. comm.). Any resemblance to Hooded Oriole is thus due to convergent evolution.

REFERENCES Delaney 1992 [voice], Howell and Webb 1995 [habitat, distribution], Howell *et al.* 1992 [habitat, nesting], Ridgway 1902 [measurements].

30 YELLOW-TAILED ORIOLE *Icterus mesomelas* (Wagler) 1829 — Plate 13

This lowland oriole of Central and South America is well named as it is one of the only orioles with substantial amounts of yellow colouration on the tail.

IDENTIFICATION Adult Yellow-tailed Oriole is easily identified by the extensive yellow on the tail and by the longitudinal yellow stripe on the wing-coverts. The black back further separates it from many other orioles. The only other orioles with yellow or orange on the tail are Scott's (51), Baltimore (40), Bullock's (41) and Black-backed (42), yet the latter three have orange, not yellow on the tail. These three species are also largely orange, have black on the crown, and have a noticeable white wing panel on the greater coverts or obvious white wingbars. Scott's Oriole has a black hood and back as well as two bold white wingbars.

Yellow-tailed Oriole is broadly sympatric with Yellow-backed Oriole (26). Yellow-backed Oriole has a yellow back, an entirely black tail and no yellow on the wings, unlike Yellow-tailed Oriole. The immatures are more similar than the adults as young Yellow-tailed Orioles have greenish backs, but they still show the distinctive wing pattern that should identify them. Also note that even young Yellow-tailed Orioles may show darker speckling on the back.

A somewhat tricky identification problem involves White-edged Oriole (32) of coastal Ecuador and northern Peru. This problem is treated in more detail under the account for that species. The following points help in determining an identification. Yellow-tailed Orioles have noticeable yellow on the tail in all ages, whereas White-edged Orioles have black tails (brownish in juveniles and young) with white tips and largely white outer rectrices. Yellow-tailed Orioles have yellow inner greater coverts which link the yellow shoulders to the white tertial patch; the inner greater coverts are black on White-edged Oriole — this feature is not useful in juvenal plumage. The tertial patch of Yellow-tailed Oriole is made up of white edges to the tertials (in the South American subspecies) and is narrow, unlike the bold, wedge-shaped wing patch of White-edged Oriole which is created by wide white edges to the tertials and inner secondaries. This feature is useful at all ages. Finally, the yellow shoulders are entirely yellow on Yellow-tailed Orioles, being noticeably reduced in juveniles, while on immature and some adult White-edged Orioles the lower edge or innermost feathers are white.

Orange-crowned Oriole (31) is similar in overall pattern to Yellow-tailed Oriole. However, the two species differ quite dramatically in their structure. Orange-crowned Orioles are small, long-tailed and have long, thin, sharply pointed bills. Yellow-tailed Oriole is noticeably larger and bulkier, with a shorter tail and a deeper bill. Other than the yellow shoulders, Orange-crowned Orioles lack any other markings on the wings, unlike Yellow-tailed Orioles. The tail of the Orange-crowned Oriole is entirely black. As its name suggests, Orange-crowned Oriole has a contrasting orange nape and crown unlike Yellow-tailed Oriole. Juveniles and immatures of these two species may be separated by structural and tail pattern differences.

Several other orioles which have not been mentioned are sympatric with Yellow-tailed Oriole, but all of these lack yellow on the tail as adults. None of them show the unique wing pattern of Yellow-tailed Oriole. Several of these orioles (Yellow (27), Orange (29), and Audubon's (49)), have yellow backs, while others (Streak-backed (35), Hooded (36), Altamira (34) and Spot-breasted (33)) have entirely different wing patterns with much white. Finally some orioles (Black-cowled (45) and Orchard (43)) have entirely black heads. The immatures and juveniles of these species may be confused with those of Yellow-tailed Oriole particularly since this species lacks the black back as a juvenile. However, concentrating on wing, tail, and structural differences one may be able to identify these species. The much larger Yellow-rumped Cacique (13) may be confused with Yellow-tailed Oriole, but note that the cacique does not have yellow underparts and has pale blue eyes and an ivory-coloured bill.

VOICE Skutch (1954) considers this species to be one of the finest singers in Central America!

Song: The song consists of beautiful phrases of sweet whistles, each phrase repeated several times before moving on to a different phrase. The overall effect is similar to that of a *Thryothorus* wren. The repetitive nature of Yellow-tailed Oriole songs is characteristic. Trills are sometimes included in the song, but never any harsh-sounding notes. Female songs are similar to those of the male, but are shorter. The phrases used by males tend to be made up of 5–6 notes, while those of females range from 2–4 notes per phrase. Females may answer the song of the male in an antiphonal duet, or by overlapping the male song with her own, but not in any clear pattern. Songs from Ecuador are much harsher than those of Central America, often lacking the sweet nature described above. They are also choppy and repetitive.

Calls: Several calls are heard, including a *chick* and a *weechaw*. Sometimes a *chup-chup-cheet* or *chup-cheet* is given (Eisenmann 1957).

DESCRIPTION A medium-sized black and yellow oriole with a moderately stout bill that is not sharply pointed. The culmen is only slightly decurved and is straight in some specimens. The tail is long and moderately graduated. Wing Formula: P9 < P8 < P7 ≈ P6 > P5; P9 ≈ P1; P5–P8 emarginate. This species has a curiously bristly texture to the forehead feathers which is not seen in most orioles.

Adult Male (Definitive Basic): Bill black with the basal third of the lower mandible blue-grey. The eye is dark brown. There is a black mask created by the black lores, anterior auriculars and the area around eyes; it extends down to the chin, throat and upper breast as a black bib. The posterior edge of the mask descends vertically from the back of the eye, towards the bill base. The forehead and crown is orange-yellow with a light olive wash to the nape. There are a few black filoplumes on the nape which are not visible in the field, but which are quite obvious in the hand. The posterior auriculars and sides of the neck are orange-yellow down to the black back. The back and scapulars are black. The rump and lower back are yellow to the uppertail-coverts. The underparts (other than the black bib) are yellow to the undertail-coverts. The wings are black with yellow and white markings. The lesser and median coverts are yellow, while the greater coverts are black except for the innermost ones which are yellow on their outer half. This continues the yellow shoulder back as a longitudinal stripe on the closed wing. The black tertials and inner secondaries are edged with yellow on the outer half, continuing the wing stripe even further. The remaining secondaries are entirely black. The black primaries are thinly edged with white only on their outer half, creating a pale panel towards the primary tips. The bend of the wing and the wing-linings are yellow. The tail is black and yellow. The inner four rectrices (R1–R2) are entirely black; the next two (R3) are largely black with a yellow patch near the tip; the following pair (R4) are yellow with black on the basal third; finally the outer two pairs (R5–R6) are entirely yellow except for the extreme base which is hidden by the tail-coverts. It has blue-grey legs and feet.

Adult Female (Definitive Basic): Similar to male and not sexable in the field based on appearance.

Immature (First Basic): Similar to adult in pattern but overall duller yellow. The back is olive sometimes with some black mixed in. The base colour of the wings is brownish rather than black, but the pattern of pale edges and yellow markings is as in the adults. The tail is olive with yellow outer tail feathers. Before the breeding season some individuals obtain new blackish central tail feathers and the worn back shows more black.

Juvenile (Juvenal): Like immature but lacks the black mask and bib, and is even duller yellow. The lower mandible has a pinkish rather than blue-grey base. Head yellow with an olive wash on the crown. The back contrasts with the yellow head, being an olive-brown. The lower back and rump are yellow with an olive wash. The entire underparts are yellow being darker and showing a definite olive wash on the breast and becoming paler on the vent. The wings are brownish with yellow-tipped lesser, median and greater coverts. This creates two thin yellowish wingbars. The tertials are also edged yellow. The primaries are brown with pale grey edges. The wing-linings are yellow. The tail is olive, with the outer four pairs of rectrices (R3–R6) widely-edged yellow and the inner two pairs (R1–R2) solid olive.

GEOGRAPHIC VARIATION Four subspecies are recognized.

I. m. mesomelas ranges from Mexico south to Honduras. It is described above.

I. m. salvinii is found along the Caribbean lowlands of Nicaragua, Costa Rica and extreme NW Panama. It lacks the yellowish secondary edgings of *mesomelas* and *taczanowskii*. It is similar to *carrikeri*, but has a larger bill and more extensive black on the throat. On average, *salvinii* is more orange than *carrikeri*, but there is great overlap.

I. m. carrikeri is found from Panama (excluding the extreme NW) into Colombia and NW Venezuela. It lacks the yellowish secondary edgings of *mesomelas* and *taczanowskii*. It is similar to *salvinii*, but has a shorter bill, and less black on the throat. Its colour averages a paler orange, but there is overlap.

I. m. taczanowskii ranges along the Pacific lowlands of Ecuador to NW Peru to Piura and in the upper Marañón valley. It has a conspicuous white wing panel created by the white edges to the tertials, often yellowish at the base of the tertial edges. *Mesomelas* has these edges consistently yellow. The yellow of the tail averages more extensive than on *mesomelas*.

HABITAT Inhabits woodlands, clearings and thickets of the lowland Tropical zone, particularly where dense vegetation is present. Yellow-tailed Oriole does not shy away from agricultural land as long as there is some shrubbery or trees, and especially if these are in low-lying areas; banana (*Musa acuminata*) plantations are a favourite. It is almost always found near water, and the shores of lagoons, rivers and swamps (including mangroves) are preferred habitats. Bamboo thickets are also used (Eisenmann 1957). In Colombia, it is typically found in swamps but may venture out to shade trees in towns (Darlington 1931). Yellow-tailed Oriole avoids contiguous forest, but uses clearings within lowland forest areas. As a generalization, of all the Central American oriole species, Yellow-tailed Oriole is the one which is most attached to areas with dense understories, particularly in swamps or swamp forests.

BEHAVIOUR During the breeding season the male spends most of his time singing high up in trees; at times the female may answer his song in a duet. The female is known to sing back to the male while she is on the nest, and is the primary sex involved in nesting and taking care of the young. This oriole is commonly found nectaring at flowers, but is largely an insectivore. In flight, it appears lumbering and slow, often flashing its tail open while aloft. The wings often make a noticeable swishing sound, which may have some display function. Forages in pairs or small flocks, sometimes with other orioles. Yellow-tailed Orioles are somewhat skulking, keeping low to dense vegetation, much more so than in most sympatric orioles.

NESTING In Panama, it commences breeding in April, which is later than most other local blackbirds but at a similar time to Yellow-backed Oriole. In Oaxaca, Mexico, nests have been found in mid June. Nests have been noted in June in the Magdalena valley, Colombia, but adult males in breeding condition have also been collected during late November suggesting that the nesting period here is quite protracted (Boggs 1961). In other parts of Colombia such as the Santa Marta region, nests are known between September–November (Darlington 1931) and birds in breeding condition between January–April in S Córdoba and N. Antioquia (Carriker, in Hilty and Brown 1986). The nest is a woven hanging basket, roughly 12 cm long in some cases. However, Darlington (1931) notes that the nests he found were rather shallow in depth for an oriole. It is placed at a low or medium height often only a few meters from the ground. In Campeche and Tabasco, Mexico it is common to see nests of this oriole hanging from power or telephone lines (J. Arvin pers. comm.). One nest in Colombia was attached to the loose fibers at the end of an ornamental palm; these fibers were also used in the construction of the nest (Darlington 1931). It appears that this oriole does not ever attach its nest to the underside of a palm leaf as is typical of many tropical orioles. The clutch size is 2–3 eggs, which are whitish-blue with brown blotches at the broader end and with paler spots throughout. The young hatch after being incubated for 14 days; a further 13 days are needed for the young to fledge from the nest.

DISTRIBUTION AND STATUS Fairly common to common but rare in N Peru. It is a bird of the lowlands to 500 m in Central America and a similar elevation in South America. Found along the Caribbean lowlands from S Veracruz and N Chiapas, E through Yucatán, N Guatemala, Belize, N Honduras, south along E Nicaragua, E Costa Rica and Panama, also found along the Pacific slope of Panama from the Canal area eastward; in South America along the Pacific lowlands from Colombia south through W Ecuador to Lambayeque in Peru, as well as in the upper Marañón Valley in Cajamarca La Libertad, Peru; north in Colombia in the lowlands north to the Santa Marta mountains and up the Magdalena valley; also E Guajira and Norte de Santander east into NW Venezuela (around Lake Maracaibo in Zulia, Mérida, Tachira).

MOVEMENTS A non-migratory species, it appears to be sedentary.

MOULT There is an annual complete definitive pre-basic moult which adults undergo after nesting. In Central America, this occurs approximately during September. Specimens of *taczanowskii* from Peru are in moult during early March, probably due to difference in the timing of breeding. Pre-alternate moults appear to be lacking in adults. The juvenal plumage is lost through a first pre-basic moult that replaces the body feathers and wing-coverts and perhaps a few tertials. It appears that a variable limited first pre-alternate moult exists in some individuals which replaces some of the tail feathers and coverts.

MEASUREMENTS ***I. m. mesomelas***. Males: (n=10) wing 90.4 (86.9–94.5); tail 104.6 (99.3–113.3); culmen 22.1 (20.6–22.9); tarsus 29 (27.9–30.5). Females: (n=10) wing 86.0 (76–93); tail 103.9 (94.7–113); culmen 21.2 (20–22); tarsus 28.7 (26–31.0).

I. m. salvinii. Males: (n=19) wing 98.8 (87.6–104.5); tail 110.1 (100.3–117.4); culmen 25.5 (24.3–27.3); tarsus 30.9 (29.9–32.8). Females: (n=10) wing 93.3 (86.6–98.3); tail 107.8 (98.6–116.0); culmen 24.2 (22.9–26.0); tarsus 29.6 (26.1–30.5).

I. m. carrikeri. Males: (n=10) wing 95.1 (89.5–98.5); tail 108.0 (98.7–116.4); culmen 23.7 (21.3–25.4); tarsus 30.3 (25.3–32.9). Females: (n=10) wing 89.0 (86.4–92.0); tail 101.0 (97.3–104.2); culmen 23.0 (21.6–24.4); tarsus 28.7 (23.5–30.7).

I. m. taczanowskii. Males: (n=1) wing 92; tail 110; culmen 22; tarsus 30. Females: (n=4) wing 87.8 (87–89); tail 103.8 (100–110); culmen 21.3 (21–22); tarsus 28.3 (28–29).

NOTES A specimen of a hybrid between this species and Yellow-backed Oriole exists from Belen (245 m) Antioquia Department, Colombia (Olson 1983). A recent study comparing the mitochondrial DNA sequences of these orioles suggests that Yellow-tailed Oriole belongs to a group which includes the troupials, Spot-breasted Oriole and White-edged Oriole (K. Omland, S. M. Lanyon pers. comm.).

REFERENCES Binford 1989 [habitat, nesting], Hardy 1983 [voice], Hardy *et al.* 1998 [voice], Howell and Webb 1995 [habitat, moult], Olson 1981 [geographic variation], 1983 [hybrids], Ridgway 1902 [measurements], Schulenberg and Parker 1981 [distribution, status], Wetmore *et al.* 1984 [voice, habitat, behaviour, nesting, measurements].

31 ORANGE-CROWNED ORIOLE *Icterus auricapillus* Cassin 1848 — Plate 13

A small, attractive, brightly-patterned Oriole of northwest South America and easternmost Panama.

IDENTIFICATION A black-backed and black-bibbed oriole with a bright orange crown and nape. Overall, this small oriole is quite distinctive. It is small, slim and long-tailed, rather like a Hooded Oriole (36) in shape. The key field mark to look for is the contrasting orange crown which is brighter in colour than the yellow underparts and rump. Yellow-tailed Oriole (30) is the most similar sympatric oriole in general pattern, but it is larger and has a deeper bill. In addition, it has obvious yellow on the tail and a longitudinal yellow stripe on the folded wing. Orange-crowned Oriole also has a noticeably broad and extensive black bib on the throat and upper breast. Moriche Oriole (24) also has a contrasting yellow crown patch, but note that it is almost entirely black, with black underparts and rump; it is also only likely to be seen sympatrically with Orange-crowned Oriole near the mouth of the Orinoco River in Venezuela.

VOICE Song: The song is a single or short series of whistles which is repeated over and over again, often for 20 minutes without stopping. Has been described as *werr cheet-your-kurr* (Hilty and Brown 1986).

Calls: A buzzy *wheea* as well as a sharp *ze'e't* (Hilty and Brown 1986). Also commonly utters a short chatter. Recordings from Venezuela include a call which sounds very much like the imitation of the call of Roadside Hawk (*Buteo magnirostris*) (Hardy *et al.* 1998). This may parallel the behaviour of Epaulet Oriole (25) which frequently mimics calls of raptors, including Roadside Hawk (Fraga 1987). It is not clear if this hawk call is the same as the *wheea* described by Hilty and Brown (1986).

DESCRIPTION A small black and orange oriole with a slender and sharply pointed bill. The culmen is curved at the extreme tip. The tail is deeply graduated, the difference between the outer and innermost tail feathers is approximately 25 mm. Wing Formula: P9 < P8 > P7 > P6; P4 < P9 < P5; P6–P8 emarginate.

Adult (Definitive Basic): The bill is black without a grey patch on the lower mandible, and the eyes are dark brown. The forehead, lores and anterior auriculars are black as are the chin, throat and upper breast, creating a black bib. The bib is both wider and extends farther down the breast than in most other orioles. The crown, nape and sides of the neck are reddish-orange, noticeably brighter than any other part of the bird. The back and scapulars are black. The rump and lower back are orange, but paler than the crown, and the uppertail coverts are black. The underparts are pale orange or orange-yellow from the end of the black throat patch to the undertail-coverts. The tips of the longest undertail-coverts are black. The wings are black, with a yellow patch made up of the yellow lesser and median coverts. The greater coverts, tertials, secondaries and primaries are jet black without white edges. On the underwing, the wing-linings are yellow. The tail is entirely black. The legs and feet are blackish. Both sexes are similar, only differing in their size, but this is not apparent in the field.

Immature (First Basic): Intermediate between adult and juvenile. The body plumage is like a dull version of the adults with orange on the crown, usually washed olive, in addition to a black face and throat bib. The underparts are a duller yellow than on adults. Like adults, the immatures have a black back, but the feathers may be tipped green. The brownish primaries and secondaries are retained from the juvenal plumage and contrast with the black greater coverts and tertials. In some cases the central rectrices are replaced and these fresh, black feathers contrast with the duller olive juvenal tail feathers. It appears that the first pre-basic moult is protracted so birds with a patchy appearance may be observed for several months each year. In addition, this moult may be variable such that some individuals retain a sizable portion of the juvenal body feathering.

Juvenile (Juvenal): The basal half of the lower mandible is pink-orange, not grey. The crown and nape are olive-yellow, while the back is olive-buff. The rump is olive with a yellow wash. Brightest is the face, which is orange-yellow with an olive wash. The face lacks a dark eyeline, thus the dark eye stands out on the plain orange-yellow face. The underparts are pale yellow, with a green (or dusky) wash across the breast. The underparts are palest, nearly white, on the belly. The wings are brownish, with yellowish wing-linings. The median and greater coverts have yellow tips, creating two thin, dull yellow wingbars. In addition, the tertials are also edged pale yellow. The tail is brownish with olive fringes, the outermost feathers are mostly olive, the inner ones browner.

GEOGRAPHIC VARIATION Monotypic.

HABITAT A bird of humid forest, forest edge, open woodlands and gallery forest, but may also be found in drier woodlands. In N Colombia where it is found with Baltimore (40), Yellow-tailed and Yellow (27) Orioles, it tends to share the forest and forest edge habitat with Baltimore Oriole. It is much less common in the dry, open woodlands where Yellow Oriole is found. Orange-crowned Oriole can be particularly common along rivers and streams (Wetmore *et al.* 1984). Cleared areas are used as long as there is ample shrubbery present. It has an affinity towards palms.

BEHAVIOUR This oriole tends to travel in pairs, feeding high in the canopy. It appears to have a preference for foraging in flowering trees. During the breeding season, the male spends most of the morning sitting still and singing endlessly. Little information is available regarding nesting and behaviour.

NESTING Nests April–June in Santa Marta, Colombia; during January–September NW Colombia; and in June in NW Santander (Boggs 1961), Colombia. The nest is a short, hanging basket typical of the genus. It is secured to a branch or a palm frond. Todd and Carriker (1922) mention that nests are secured to the ends of a divided palm leaf. The nests are woven of palm fibers. The eggs are said to be indistinguishable from those of Yellow Oriole, being white with dark scrawls and spots largely around the broader end of the egg (Todd and Carriker 1922). This oriole is a known host of Shiny Cowbird (102) (Boggs 1961).

DISTRIBUTION AND STATUS Uncommon to locally common. A bird of the lowlands (tropical zone) to 800 m. Found in Panama from E Panama province east to the Darién. There is also a record of a flock from the Canal Zone (Ridgely and Gwynne 1989). In Colombia from the lowlands of Chocó, north of the C and W Andes to Guajira, but absent from the Guajira peninsula; and up the Magdalena valley to the department of Tolima. In addition, the Orange-crowned Oriole is found

along a thin strip along the immediate base of the eastern Andes in Colombia, south to Meta. In Venezuela throughout north of the Orinoco river east to Sucre and Monagas but absent from Delta Amacuro. Only found south of the Orinoco river in Bolivar, as far south as the lower Caura river valley.
MOVEMENTS Appears to be sedentary, there is no evidence that it is migratory or that it takes part in seasonal movements.
MOULT Adult Orange-crowned Orioles have a complete moult (definitive pre-basic) after nesting. Birds from Santander, Colombia, were in the process of this moult in late September. The first pre-basic moult is partial, and includes the body and covert feathers but birds retain the juvenal primaries, secondaries and rectrices. Sometimes the central pair of rectrices are moulted. In the Magdalena valley, Colombia, this moult occurs during March. In contrast, along the foothills of the East Andes in Colombia, the first pre-basic takes place in May and in the lowlands of the NW of the country (Bolívar) it takes place in July. This moult is protracted; therefore birds with patches of black and orange coming in over the olive and dull yellow of the juvenal plumage are observed for several months of the year. Pre-alternate moults appear to be lacking, but they may occur to some extent on immature birds (first pre-alternate). This needs further study.
MEASUREMENTS Males: (n=10) wing 91.0 (87.2–95.1); tail 83.3 (74.2–91.0); culmen 20.1 (19.2–21.0); tarsus 21.4 (19.6–23.3). Females: (n=10) wing 86.8 (84.4–89.0); tail 83.2 (80.4–85.7); culmen 19.0 (18.2–19.8); tarsus 21.6 (20.1–22.8).
NOTES Structurally resembles Hooded Oriole; perhaps the two species are closely related.
REFERENCES Hilty and Brown 1986 [voice, nesting], Wetmore *et al.* 1984 [voice, description, measurements].

32 WHITE-EDGED ORIOLE *Icterus graceannae* Plate 13
Cassin 1867

A restricted range oriole, found only in the dry Tumbesian region of coastal Ecuador and northern Peru.

IDENTIFICATION A black-backed oriole with a black throat bib and yellow-orange body plumage. It shows a yellow shoulder and a bold white wedge-shaped patch on the tertials. This oriole is sympatric with the very similar Yellow-tailed Oriole (30). Structurally, White-edged Oriole is slightly smaller and has a proportionately deeper bill than Yellow-tailed Oriole. The key differences are in the tail and wing patterns. White-edged Orioles have the tail tipped with white, and the outermost rectrices show a substantial amount of white, which is obvious when the tail is viewed from below. This is quite unlike the tail of Yellow-tailed Oriole which has the outer three to four pairs of rectrices entirely (or nearly so) yellow, making the tail appear largely yellow from below. If there is any yellow on the tail, one can safely rule out White-edged Oriole. The differences in the wing pattern between these two species are slightly more complex. White-edged Oriole has yellow shoulders (lesser and median coverts) and a bold, wedge-shaped white patch on the tertials and inner secondaries. The greater coverts are usually black, separating the white secondary patch from the yellow shoulders. The yellow epaulets sometimes show a white lower edge due to white on the median coverts, this is particularly common on the innermost median coverts. Three aspects of this wing pattern are not found on Yellow-tailed Oriole: the black greater coverts separating the yellow shoulders from white secondary patch; the obviously wedge-shaped white secondary patch; and the presence of any white on the shoulder patch. Yellow-tailed Oriole has yellow shoulders with the yellow extending onto the innermost greater coverts, becoming white on the tertial edges and forming a longitudinal white patch here. The white patch on the tertials is not wedge-shaped and often shows yellow at its base, whereas the patch of White-edged Oriole is always entirely white. Note also that the white wing patch of White-edged Oriole is created by white edging on the tertials and secondaries, whereas on Yellow-tailed Oriole this patch is almost always just due to white edging on the tertials only. Additionally, there is a small difference in the configuration of yellow and black on the face of these two species. On White-edged Oriole if one draws a line downward directly below the middle of the eye there is a noticeable intrusion of yellow in towards the bill base from this imaginary line, this is due to the narrower amount of black on the mask. If one goes through the same exercise on Yellow-tailed Oriole, this imaginary line basically defines the division between the yellow on the face and black mask; the yellow does not significantly intrude in towards the bill base on this species. In other words, White-edged Oriole has more extensive yellow on the auriculars than Yellow-tailed Oriole. Finally, in the hand the feathers of Yellow-tailed Oriole, particularly those of the head, have a peculiar soft and fluffy texture; in addition, there are noticeable black filoplumes present on the nape which are absent on White-edged Oriole. Vocal differences also aid in separating these two species in the field.

Separating the juveniles and immatures of White-edged and Yellow-tailed Orioles is also difficult. The slight differences in structure noted above always apply. In addition, the tail and wing pattern differences are present and are still the main characters that allow for field identification, but these are not as bold or as obvious as on adults. Furthermore, the juvenile tail is not black, but is brown or olive-brown in both of these species. Therefore, even though the outer tail feathers of Yellow-tailed Oriole are more yellow than the innermost, the contrast is not as noticeable as on the adult tail. Tails of juvenile White-edged Oriole may show some white tipping, but often lack this altogether. Nevertheless, identification is possible since juvenile White-edged Oriole should never show noticeable amounts of yellow on the tail. While the white tertial/secondary wing patch is present on juvenile White-edged Orioles, some of the markings typical of adult Yellow-tailed Orioles are missing on juveniles of that species. Most noticeably, the innermost greater coverts are black in juvenile Yellow-tailed Orioles, and they are therefore similar to White-edged Orioles in this respect. Furthermore, the white edges of the tertials may be absent on juvenile Yellow-tailed Orioles; however, this simplifies the identification since juvenile and immature White-edged Orioles always show a noticeable amount of white on the tertials and inner secondaries.

VOICE Song: A series of repeated, whistled phrases, given in quick succession with a very short interval between phrases. Reminiscent of some tropical wrens (*Thryothorus*), and Yellow-tailed Oriole. Sounds lively and energetic. In each song bout the phrase may be repeated five times or so. Each phrase is made up of a series of clear whistles, for example *wheeet-wheeo-wit-wit*.

Calls: One call is a warbled, 'phreeee', also a doubled *phreet-phreet*. Other vocalisations include largely short clear whistles, and conversational notes.

DESCRIPTION A medium-sized oriole with a stout bill, which is thick at the base, and has a straight gonys and culmen. The tail is moderately graduated. Wing Formula: P9 < P8 < P7 > P6; P9 ≈ P4; P8–P5.

Adult Male (Definitive Basic): The bill is black with a blue-grey patch on the basal quarter of the lower mandible. The eyes are dark brown. It has a black mask and bib composed of the black lores, area around the eye, anterior auriculars, chin, throat and upper breast. The entire crown from the forehead to the nape and base of the neck is orange-yellow, with this colour continuing to the side of the face and neck sides. Other than the black bib, the underparts are entirely yellow to the undertail-coverts; this species has a brighter orange tone on the breast which is not present on the belly and vent. The back and scapulars (mantle) are black which contrasts with the orange-yellow lower back, rump and uppertail-coverts. The base colour of the wings is black with white and yellow patterning. The lesser and median coverts are pale yellow, the median coverts being noticeably paler than the lesser coverts. In fact, the median coverts may appear white in the field. The greater coverts are black, sometimes with the innermost noticeably white on its outer half. The tertials and inner secondaries have white edges at their base, which in conjunction with the white on the inner greater coverts form a wedge-shaped white patch on the folded wing. The rest of the secondaries and the primaries are black, the latter thinly edged white near their tips, creating a pale panel at the folded wing tip. The wing-linings and the bend of the wing are yellow. The tail is black with a white edge on the outer rectrix (R6) and a white tail spot on the adjacent feather (R5). There is some variation in the tail pattern, with many birds showing white tips to the outer three pairs of feathers (R4–R6).

Adult Female (Definitive Basic): Similar to male. It is not field separable from the male by appearance even thought it is marginally smaller.

Immature (First Basic): Similar to adults in pattern and colour except the yellow colouration is duller. The back is blackish and is noticeably tipped with yellow or green when fresh. The wings and tail are retained from the juvenal plumage and are therefore brownish-black, but the tertials and coverts are usually new and show a pattern similar to that of adults except the white of the median coverts is more extensive and noticeable in the field. The white wedge on the secondaries and tertials is present, but is smaller and not as obvious as that of the adults.

Juvenile (Juvenal): Like adults but dull yellow with a green tone, never orange-yellow. The bib is dull black and less extensive than on the adults. It is unclear if young juveniles lack black on the bib altogether, as is typical of the genus. In any case, rather young birds already have a black bib. The bill has a horn or pinkish basal half to the lower mandible. The back is blackish-olive with olive tips. The white wedge on the folded wings is present, but is not as obvious as on adults. The base colour of the wings is dull black. The lesser and median coverts are blackish with yellow tips instead of being entirely yellow. The tail is dull black with the outer rectrix greyish-brown.

GEOGRAPHIC VARIATION Monotypic.

HABITAT Lives in xeric deciduous woodlands and second growth scrub. Also found in desert scrub and riparian thickets (Parker *et al.* 1982, Ridgely and Tudor 1989). This oriole shows a preference for the transition zone between dry and moister forest in E Ecuador. It is also quite tolerant of forest alterations, mainly any changes towards tree crops (Dusty Becker pers. comm.).

BEHAVIOUR Found singly or in pairs. Males sing from the top of a tree, while sitting in a vertical pose with the tail pointing straight down. This oriole becomes much less visible during the breeding season (Marchant 1958). Very little has been published regarding this species.

NESTING Nests during the rainy season, February–March. As rainfall varies annually, particularly during El Niño years, the timing and length of the breeding season also varies (Marchant 1959). The nest is a hanging structure typical of orioles. However, this species' nest is rather shallow, appearing more like a suspended cup than a deep basket. Its is constructed of plant stems and fibers which are woven together for strength. The nest is suspended from the rim from twigs of a tree (Marchant 1960). One nest was suspended from an Arrayan tree, 1.8 m from the ground (Marchant 1960). This oriole is a known host of Shiny Cowbird (102) both in SW Ecuador and Peru (Marchant 1960, Friedmann and Kiff 1985). The clutch size is unknown. However, at least two oriole young were observed in one nest with three Shiny Cowbird chicks (Marchant 1960), thus at least 2 eggs are laid. Both adults feed and care for the young. The length of the incubation and nestling periods is not known. Both adults help in feeding the young (Marchant 1960).

DISTRIBUTION AND STATUS Uncommon to locally common to 300 m. Restricted to the Tumbesian region, the arid coastal part of S Ecuador and N Peru. In Ecuador, found on the coastal slope from C Manabi, south including Guayaquil Bay, to Tumbes. It is found further inland in Loja (Ridgely and Tudor 1989). In Peru, from Piura south to La Libertad, once again restricted to the coastal lowlands. It is not as common as Yellow-tailed Oriole in Ecuador, but outnumbers it in N Peru (Schulenberg and Parker 1981).

MOVEMENTS It is not known if this species makes local movements during drought or heavy rainfall years. Presumed to be non-migratory based on the available evidence.

MOULT Adults have an annual complete (definitive pre-basic) moult after nesting. Birds from Piura, Peru, are finishing or have finished the moult by early August. In Ecuador, specimens in moult exist from February as well as from August. We assume that pre-alternate moults are lacking in adults, and perhaps immatures. The extent and timing of the first pre-alternate moult is not well known. It appears that the juvenal body plumage is lost entirely and that the tail, tertials and coverts are replaced as well. The primaries and secondaries are retained. In La Libertad, Peru, birds undergo the first pre-basic moult during May.

MEASUREMENTS Males: (n=10) wing 89.9 (88–91); tail 94.3 (87–98); culmen 21.6 (21–22); tarsus 27.0 (23–30). Females: (n=11) wing 85.3 (78–90); tail 91.8 (84–96); culmen 20.5 (20–21); tarsus 25.7 (25–27).

NOTES A recent study comparing the mitochondrial DNA sequences of orioles suggests that White-edged Oriole belongs to a group which includes the troupials, Spot-breasted Oriole, and Yellow-tailed Oriole (K. Omland, S. M. Lanyon pers. comm.). Note that this group includes all of the orioles which have expanded white on the tertials.

REFERENCES Marchant 1958, 1959, 1960 [nesting], Hardy *et al.* 1998 [voice], Ridgely and Tudor 1989 [distribution, status], Schulenberg and Parker 1981 [status].

33 SPOT-BREASTED ORIOLE *Icterus pectoralis* (Wagler) 1829 — Plate 12

This large oriole of Central America is the only one with obvious spots on the breast.

IDENTIFICATION Adult Spot-breasted Oriole is largely orange with a black back, white markings on the wings, and with a small black mask and bib, thus resembling Altamira (34), Hooded (36) and southern races of Streak-backed Oriole (35) in general pattern. The key field mark is the noticeable spotting on the breast sides of this species, but beware that spots may be absent in some immatures and may blend in to a solid mass in birds from SW Mexico (*carolynae*). The three previously mentioned orioles have a white lower wingbar that Spot-breasted lacks. Male Hooded Oriole also has a white median covert patch (shoulder). Hooded Orioles are further differentiated by their obviously thin and decurved bill, small size and graduated tails. Altamira Orioles are larger and bulkier and have very deep, massive bills for an oriole. Spot-breasted Orioles have broad white edges to the tertials, but not on the secondaries; this creates a white wedge-shaped patch on the folded wing. The other three oriole species have narrowly fringed tertials and secondaries, which do not form a wedge-shaped patch. Immature Altamira and Streak-backed Orioles are most similar to immature Spot-breasted Oriole. The distinctive wing pattern is present on young Spot-breasted Oriole, separating it from both of these species. The large bill of Altamira Oriole will identify that species. The slightly curved culmen of the Spot-breasted Oriole is a good feature in separating it from Streak-backed Oriole which has a straight culmen (Figure 29.1). However, some immatures of these three species, particularly when worn, cannot be reliably identified.

VOICE This species sings throughout the year, at least in Florida.

Song: A warbled set of whistles, at times rather repetitive. The song is lengthy and quite beautiful in quality with some whistles delivered slowly and clearly. There are two song styles that birds use. A motif of five to ten notes which is regularly repeated is a 'repetitive type' song. The other style is for notes to be used repetitively in songs, but their order is never quite the same each time. These songs are more variable in nature. Sometimes a certain note is doubled, which is characteristic of this oriole and tends to 'stand out' in the song. The song has been described as: *whi whew hi hew hew* and *chee-hee-oo hee-hee chee-chee-chee-chee-chee-chee...* (Howell and Webb 1995).

Females sing as well, but their songs are not as varied and complex as those of males.

Calls: The basic call is a nasal *nyeh*. Also a sharp *whip*. In addition, it gives a chattering *ptcheck* (Stevenson and Anderson 1994).

DESCRIPTION A large oriole with a moderately deep bill. The bill is sharply pointed and has a slightly curved culmen and gonys. The tail is only moderately graduated and appears rounded. The form *guttulatus* is described. Wing Formula: P9 < P8 < P7 > P6 > P5; P9 ≈ P2; P6–P8 emarginate.

Adult Male (Definitive Basic): Black bill with grey basal third to the lower mandible. The eyes are dark brown. It has a black mask and bib, and the lores, anterior auriculars, chin, throat and upper breast are also black. The rest of the head is orange. The crown, nape and sides of face are orange with the orange extending forward towards the base of the bill immediately below the eye, making an indentation on the black face. The orange colour continues to the base of the neck, where it is abruptly replaced by black. The back and scapulars (mantle) are black. The lower back, rump and uppertail-coverts are orange; here the orange has a noticeable yellow tone and is not as vivid as that of the crown. The underparts are orange becoming orange-yellow on the belly and undertail-coverts. The breast sides are adorned with obvious round black spots, adjacent to the black bib. Sometimes the spots continue below the bib, onto the center of the breast. In late summer many of the spots have worn off. Note that in the race *carolynae*, the spots are large and join together, linking the black of the bib with the black scapulars (see Geographic Variation). The wings are black with white and orange markings. The lesser and median coverts are orange-yellow, more yellow-orange than the orange crown. The greater coverts are entirely black. The black tertials and inner secondaries have wide white edges that form a wedge-shaped lengthwise white patch on the folded wing. The black primaries have obvious white edges at the bases, creating a small rectangular white patch on the folded wings that contrasts with the entirely black primary coverts. The primaries are slightly tapered, beyond the taper they are thinly edged white; this pale, distal wing panel is often not obvious in the field. The bend of the wing and the wing-linings are orange-yellow. The tail is black with a greyish tip to the outer two pairs of rectrices. The legs and feet are blue-grey.

Adult Female (Definitive Basic): Usually indistinguishable from adult male; however, they average duller in colour and may be separable when a direct comparison is available. The underside of the tail may sometimes have either an olive or brownish tone.

Immature (First Basic): Similar to adult in overall pattern, but the orange colour is duller and less intense. The facial pattern is as on adults. The black breast spots may be present, but are sometimes missing, making identification challenging. The back colour is variable, the blackish back and scapular feathers are widely fringed with olive-yellow. Some individuals, particularly early in the fall and winter will appear entirely olive-backed; other individuals show a blackish back with olive markings. As the tips wear off, the back becomes blacker. The base colour of the wings is blackish, not jet black. The coverts and tertials are newer and contrastingly blacker. The pattern of edging on the wings is as on adults. The tail is olive, unlike the adults, with greenish-yellow edges to the feathers which are most noticeable at the bases of the outer feathers. A limited pre-alternate moult of variable extent has been described. This increases the amount of black on the back of some birds, before May. As the juvenal plumage is lost, the crown tends to come in before the face. Often the black breast spots come in before the black throat has moulted in. However, these may wear off or be inconspicuous later on.

Juvenile (Juvenal): Like immature, but lacks black on the face and bib and the back is olive-brown without black bases. Breast spotting is always absent. The bill has a pink-coloured base to the lower mandible instead of blue-grey. Shows a dull yellow head, neck, throat and underparts. The back is olive-brown and contrasts with the yellowish head, rump and lower back. The yellow underparts are palest on the vent and show an olive wash on the breast. No black on the lores or throat. The wings are brownish. The lesser coverts have yellow tips as do the median and greater coverts; the latter form two thin wingbars. The outer primaries show some white at their bases, forming a white 'handkerchief', while the white edging on the folded secondaries and tertials forms a white wedge, not as extensive and bright as on the adult, but similar in pattern. The tail is brown with olive outer edges to all of the rectrices.

GEOGRAPHIC VARIATION Classified into four subspecies differing in colour intensity and spotting pattern on the breast.

I. p. carolynae is the northernmost race, found on the Pacific slope of the Andes in Guerrero and adjacent Oaxaca. It resembles *pectoralis*, and is a paler orange-yellow than *guttulatus*, but the spots on the breast are large, often coalescing such that the black mantle is connected to the black throat by a black band on the breast sides.

I. p. pectoralis is found from E Oaxaca to C Chiapas in Mexico. It resembles *guttulatus*, but paler on the crown and underparts.

I. p. guttulatus ranges along the Pacific lowlands from Chiapas, Mexico to NW Nicaragua. It resembles *espinachi* but is more orange on the head, breast and rump. It is described above.

I. p. espinachi is found in NW Costa Rica and likely adjoining Nicaragua. It is similar to *pectoralis*, but is more brightly coloured. It is yellower, less orange, than *guttulatus*.

HABITAT This species frequents open woodlands, preferring arid, shrubby places dominated by *Mimosa*. It also enters agricultural areas, commonly shade-grown coffee. Inhabits rural areas, particularly where large shade trees exist. It shares this habitat with Streak-backed and Altamira Orioles, but is less common than either. In Florida, this is a species of urban areas particularly those with an abundance of planted flowering trees and bushes. It is strictly a bird of the lowlands in Mexico, but is found to the upper tropical zone farther south.

BEHAVIOUR Forages in pairs or small parties, regularly with other orioles. They may also be seen travelling in small flocks during the non-breeding season. Probes dead leaves for insects and gleans from green leaves; also visits nectaring flowers. In Florida, Spot-breasted Oriole feeds largely on fruits, nectar and insects attracted to nectaring flowers (Stevenson and Anderson 1994). It has been observed to clip Hibiscus (*Hibiscus* sp.) flowers in order to reach the nectar (Terres 1980). Both sexes sing. When singing, the male perches high on a prominent bush or even a bare power line (Florida), from which he delivers his song. This is not a shy oriole, readily adjusting to suburban life both in the US and in its home range. The nest is built primarily by the female, but the male helps in the care and raising of the young. In much of Mexico and N Central America, this species is found sympatrically with the similar Streak-backed and Altamira Orioles; almost invariably Spot-breasted is the least common of this trio.

NESTING Nests from May–July in Oaxaca, Mexico; April–May in El Salvador, from May–July in Costa Rica, and June–July in the dry interior valleys of Guatemala. In Florida this species breeds-from April–late August and is seemingly double-brooded here (Stevenson and Anderson 1994). The nest is a hanging basket of woven grass and strips of vegetation, usually 50 cm in length. Its nest is not as long as that of Altamira Oriole but is longer than that of Streak-backed Oriole. However, the construction is much sturdier than that of Streak-backed Oriole, since it is woven tighter and of firmer material. The nests are commonly placed at the tip of a branch, secured to a fork. The thorny *Mimosa* is a preferred nest tree. The clutch size is usually of 3 eggs. The eggs are bluish-white patterned with dark spots and scrawls, most densely around the broader end. In Guatemala, Spot-breasted Orioles have been known to raise two broods, the male taking care of the first one as the female begins incubating the second (Skutch 1954). This oriole is a known host of Bronzed Cowbird (99) (Friedmann and Kiff 1985).

DISTRIBUTION AND STATUS Uncommon to fairly common in the tropical zone, found below 500 m in Mexico and Costa Rica, but up to 1500 m in El Salvador and Honduras. Ranges along the Pacific slope of Mexico as far north as Colima (recently confirmed from this state, Howell and Webb 1995) south through Guerrero, Oaxaca, Chiapas and into Guatemala. Further south it also is found in the interior, not only along the coast from E Guatemala, El Salvador and W Honduras (including the Atlantic slope) continuing along the Pacific slope of Nicaragua and NW Costa Rica south to Puntarenas. Introduced to the southern Atlantic coast of Florida and Cocos Island, Costa Rica. In Florida the oriole is present from Brevard Co. in the north to Dade Co. in the south.

MOVEMENTS Apparently sedentary. An individual banded and re-trapped in Homestead, Florida, achieved an age of 11 years and seven months, the longevity record for the species. Of the five or so recoveries of this species from southern Florida, all were banded nearby.

MOULT As in other orioles there is a complete moult in adults after nesting (definitive pre-basic) and this plumage is retained for the rest of the year as pre-alternate moults are lacking. This moult begins in mid August in El Salvador and Guatemala. The moult period continues into November (Oaxaca, Mexico and Honduras specimens). In Mexico, moult occurs in September. The introduced population in Florida undergoes the annual moult August–September. The first pre-basic moult of the juveniles occurs rapidly after they fledge, replacing the body plumage, tertials and wing-coverts but retaining the juvenal tail and flight feathers. There appears to be a variable but limited first pre-alternate moult in immatures which replaces some of the back feathers and sometimes the coverts and the tertials. Adult (definitive) plumage is achieved after the Second pre-basic moult.

MEASUREMENTS ***I. p. carolynae***. Males: (n=24) wing 105.6 (96–111); (n=21) tail 104.6 (100–111); (n=7) culmen 21.7 (20–24); tarsus 27.7 (26–29). Females: (n=8) wing 97.8 (96–100); (n=7) tail 97.3 (95–106); (n=2) culmen 22, 22; tarsus 27, 28.

I. p. pectoralis. Males: (n=13) wing 107.1 (90–114); (n=12) tail 105.5 (103–109); (n=6) culmen 21.5 (20–23); tarsus 28.0 (27–29). Females: (n=6) wing 100.8 (97–106); (n=4) tail 101.0 (98–104); (n=2) culmen 21, 22; tarsus 28, 30.

I. p. guttulatus. Males: (n=49) wing 106.5 (90–114); tail 103.6 (95–112); (n=8) culmen 21.5 (20–22); tarsus 28.6 (27–30). Females: (n=47) wing 101.8 (92–113); (n=41) tail 99.1 (85–112); (n=3) culmen 21.7 (21–22); tarsus 27 (25–28).

I. p. espinachi. Males: (n=12) wing 102.1 (95.8–106); tail 96.6 (89.4–102); (n=5) culmen 21.5 (19–23.4); (n=7) tarsus 26.9 (26–27.9). Females: (n=8) wing 94.2 (88–100); (n=7) tail 90.6 (87–93); (n=1) culmen 21; (n=2) tarsus 26, 27.9.

NOTES It has been introduced to both southern Florida and Cocos Island, Costa Rica and has established itself in both places. In Florida, this species was reported nesting as early as 1949 in Miami. The original introduction appears to have been from escaped cagebirds. By the late 1950s it had expanded its range to occupy an area roughly 40 km south and north of Miami. It now ranges north past Palm Beach. The population has fluctuated somewhat and dipped during the 1980s, but appears to be making a comeback. It is unknown which subspecies the Florida population originates from, but it appears to be one of the southern ones, either *espinachi* or *guttulatus* judging from their bright orange colouration.

A recent study comparing the mitochondrial DNA sequences of these orioles suggests that Spot-breasted Oriole belongs to a group which includes the troupials, White-edged Oriole, and Yellow-tailed Oriole (K. Omland, S. M. Lanyon pers. comm.). Note that this group includes all of the orioles which have expanded white on the tertials.

REFERENCES BBL Data [movements], Binford 1989 [nesting], Dickerman 1981 [geographic variation, measurements], Dickey and van Rossem 1938 [habitat, nesting, moult], Howell and Webb 1995 [voice, habitat, distribution], Lever 1987 [notes, introduction], Ridgway 1902 [measurements], Rowley 1984 [nesting], Skutch 1954 [habitat, nesting], Stiles and Skutch 1989 [voice, habitat, behaviour, nesting].

34 ALTAMIRA ORIOLE *Icterus gularis* (Wagler) 1829 — Plate 12

A large, bulky orange and black oriole with a thick, powerful bill. The long, beautifully woven nests are placed in obvious spots and are the first clue that this species is present in that area.

IDENTIFICATION Altamira Oriole is most similar to the *alticola* form of Streak-backed Oriole (35) and to Spot-breasted Oriole (33). These two species and the much smaller male Hooded Oriole (36) share with it a black back, and a largely orange head with black restricted to a face mask and throat patch. Apart from being smaller and longer-tailed, Hooded Oriole has a very thin and decurved bill, quite unlike the large bill of Altamira Oriole (Figure 29.1). In addition, Hooded Orioles have white lesser coverts, not orange-yellow. Spot-breasted Oriole has obvious spots on the breast, but note that these may be largely lacking on some immature birds. Also, this species lacks white on the greater coverts and on the secondaries, thus most of the white on the wings is restricted to the edges of the tertials. The most difficult identification challenge is with *alticola* Streak-backed Oriole, which seldom shows orange streaking on the back; however, the presence of streaks is a good characteristic for this species. Note also that Altamira Orioles with streaked backs have been observed in south Texas (G. Lasley pers. comm. and photos). It is not clear if these birds are all adults, or some are immatures. The bill size and colour are perhaps the best identification criteria. Altamira Oriole has a thicker bill (Figure 29.1) with less grey at the base of the lower mandible. Also, note that Streak-backed Oriole shows more white edging on the primaries than does Altamira Oriole.

Juveniles and immatures of these three larger species are quite confusing and many may not be safely identified in the field. Note the bill shape and size which can be the best field mark for Altamira Oriole. Vocalizations will also aid in the separation of these species. The distinctive wing pattern of Spot-breasted Oriole may be visible from an early age as will some breast spots.

VOICE Song: A series of loud musical whistles, sometimes with harsher notes interspersed between the whistles. The song has been likened to that of sounds made by an inexperienced human whistler. In the evening, a slightly different song is given; it is more metallic and the whistles are more widely spaced. As is typical of many tropical orioles, the song of Altamira Oriole tends to be composed of motifs that are repeated over and over. Typically, after a couple of repetitions Altamira Oriole stops before beginning another song bout, making its songs less repetitive than those of its near relatives. In Texas, the song is usually a stereotyped piping series of whistles, *DEE, DEE, dee-dee-DEE-dee* with the final phrase after the two first *DEE* calls sometimes repeated twice. Interestingly, in Tamaulipas, Mexico, birds of the same subspecies have a different and more variable song (J. Arvin pers. comm.)

Calls: The contact call is a nasal *ike*, or *yehnk*. A rasping alarm call is heard, but this species does not chatter. A double note, is used by adults when entering and leaving the nest. Other calls include a *chiu* a sharp *peen* and a longer *chu wee chu*.

DESCRIPTION A large oriole with a very thick bill, the culmen is straight showing no hint of a curve. It is not particularly long-winged, but has a relatively long tail. Wing Formula: P9 < P8 ≈ P7 ≈ P6 > P5; P9 ≈ P4; P5–P8 emarginate. The race *tamaulipensis* is described.

Adult Male (Definitive Basic): Thick bill mainly black, but shows a small gray patch at the base of the lower mandible. The eyes are dark brown. The head is orange-yellow with a black mask created by the black lores, supralores and area around the base of the bill. A thin black throat patch extends along the mid-line from the chin to the upper breast. Otherwise the neck, nape and underparts to the undertail-coverts are orange-yellow. The mantle and lower back are black, ending sharply at the base of the nape where it interfaces with the orange of the head. The rump and uppertail-coverts are orange-yellow. The wings are largely black with obvious white and orange markings. The marginal, lesser and some of the median coverts are orange, often more yellow than the rest of the body, creating an obvious yellow carpal patch. The greater coverts are widely tipped white creating a single, bold white wingbar. The bold tips become white edges on the innermost coverts, making the inner section of the wingbar appear less obvious. The bases of the primaries are white, showing as a small white square on the closed wing. The secondaries and tertials are boldly and crisply edged with white. The tail appears entirely black in the field, the extreme base which is hidden by the coverts shows some yellow as well as white shaft streaks. Legs and feet bluish-grey.

Adult Female (Definitive Basic): Exactly as male, perhaps having a marginally smaller throat patch. Females are slightly smaller than males.

Immature (First Basic): Like a dull version of adults and lacking the black back. The overall orange-yellow colour is duller than in adults. The back is olive with an orange tone. The wings and tail are dull blackish, with a reduced amount of yellow on the coverts. In addition, the greater coverts are edged olive and finely-tipped with white, appearing as a thin white wingbar. The white edging of the secondaries and tertials is not as bold as on the adults. The white square at the base of the folded primaries ('handkerchief') is very much reduced, but usually visible. The tail often has olive outer feathers.

The **Second Basic** plumage is like that of adults, but often some of the blacks are veiled or toned with olive.

Juvenile (Juvenal): Head, neck, rump and underparts yellow, lacking a black throat. The back and scapulars are olive. Wings and tail dull blackish, with greater coverts tipped with yellowish-white. The secondaries and tertials thinly fringed with white.

GEOGRAPHIC VARIATION Separated into six subspecies.

I. g. tamaulipensis ranges from S Texas south to Puebla and Campeche in Mexico. It is described above.

I. g. yucatanensis is found in the Yucatán peninsula, N Belize and Cozumel Island, Mexico. Similar to *tamaulipensis* but even smaller and much more intensely orange.

I. g. flavescens occurs along the coast of Guerrero, Mexico. This form is similar in size to *gularis*, but it is the palest and most yellow race. As well, it has slightly more extensive white edgings on the greater coverts and secondaries.

I. g. gularis ranges from Pacific coastal Mexico from Oaxaca to the interior of Guatemala and El Salvador. This form is paler, more yellowish and larger than *tamaulipensis* or *yucatanensis*.

I. g. troglodytes occurs from E Chiapas, Mexico to coastal Guatemala. Like *gularis* in colour, but much smaller. It is more orange-yellow than *tamaulipensis* and has a thinner bill.

I. g. gigas ranges from the interior of Guatemala (Río Negro valley) to Honduras and Nicaragua. This is the largest of the subspecies; in colour it is a paler yellow than *gularis*. A large but bright orange subspecies from C Guatemala (Griscom 1930) was formerly separated as *I. g. xerophilus*, probably a variant of *gigas* rather than a separate population.

HABITAT Lives in open arid woodlands, for example tropical deciduous forest and arid tropical scrub, as well as riparian zones. It may also be found in wetter (swamp) forest in parts of its range. It appears to have a fondness for trees of the *Mimosa* genus, and the clearing of forests and spread of *Mimosa* has likely helped in the spread of this species. It avoids contiguous tropical lowland forest, but may be found in openings within the forest and savannas.

BEHAVIOUR Usually observed in pairs, and may form feeding flocks during the winter. As the breeding season approaches, the pairs stake out a good nesting area and begin building the nest. There is no active defense of the territory, so conspecifics are not driven out from the nesting area. However, nests are solitary and coloniality is not known in this species. The mean distance between nests was 250 m in one Texas study (Pleasants 1981). The mating system is monogamous, with pairs staying together throughout the year. The nests are built only by the females, but feeding of the young is performed by both sexes. This oriole is primarily an insect feeder, but it will also eat fruit and nectar at flowers. The large bill of this species suggests that at some point in the year hard food (seeds, fruit pits?) may be important food sources. Brush (1998) notes that Altamira Orioles sometimes crunches branches with its bill before excavating into them, most likely this behaviour is associated with their large bill size and well-developed palatal ridge.

NESTING In Texas and N Mexico, the breeding season commences in late April and continues into July or August; breeds May–July in Oaxaca, Mexico. In Texas, the nest may be built as early as March (Brush 1998). One brood is raised per season. However, two may be fledged in the south of the range. A second nest is built for the second brood. The nest is a long, hanging basket up to 65 cm long, much longer than the nests of other local orioles. The nest is built by the female without help from the male. It may take 18–26 days to build a nest. Nests are often attached to the tip of a flexible branch or sometimes on power lines. Species of *Mimosa* are favoured nesting sites. Nest height averages 9.3 m, but they may be much lower or higher than this. Curiously, nests tend to be placed in extremely obvious situations, this species makes no attempt to hide its nest as do some of the smaller orioles. Trees chosen for a nest site are typically emergents from the adjoining canopy. The average clutch size varies according to the latitude, with birds in the south having smaller clutches: 4.9, Tamaulipas; 4.1, Veracruz; 3.7 Oaxaca. The eggs are pale bluish, with darker blackish or purplish spots and blotches. The incubation period lasts approximately 14 days. Bronzed Cowbirds (99) parasitise this species. However, up until recently there were no records of successful fledging of a cowbird from a nest of Altamira Oriole. Now two records of successful Bronzed Cowbird parasitism are known from Texas (Brush 1998). In Texas, Altamira Oriole tends to nest near Great Kiskadee (*Pitangus sulphuratus*) or Couch's Kingbird (*Tyrannus couchii*), presumably to gain some protection from predators as the kiskadees and kingbirds are aggressive in defending their nesting area (Pleasants 1981, Brush 1998).

DISTRIBUTION AND STATUS Common. Found from southernmost Texas (Rio Grande valley) south along the Caribbean slope of Mexico, including the Yucatán peninsula, to N Belize. Invades the Pacific slope along the Isthmus of Tehuantepec. Present along the Pacific coastal slope from S Mexico (C Guerrero, S Oaxaca and S Chiapas) east into S Guatemala, El Salvador, W

Honduras and W Nicaragua, nearly to the Costa Rican border. It has been recorded in the Petén of Guatemala, and the mountains of W Guatemale (Howell and Webb 1995).

MOVEMENTS Sedentary, no evidence of migratory movements. The range of this species has increased historically. Altamira Oriole was first collected in Texas in 1939. The population increased in southern Texas, with the first nest being located in 1951. Now it has increased enough to be considered the most common oriole along the Rio Grande Valley during the breeding season. There is anecdotal evidence that as this species' population increased, the numbers of the sympatric Hooded (36), Baltimore (40) and Audubon's (49) Orioles decreased.

MOULT The definitive pre-basic moult is complete. However, the first pre-basic moult is incomplete. It does not include the wings but none, some or all tail feathers may be moulted. This moult occurs in August and September in Texas. Prealternate moults are lacking except in some second calendar year birds (first pre-alternate), particularly young males which may moult some of the head and back feathers. The definitive basic plumage is not attained until the second pre-basic moult.

MEASUREMENTS ***I. g. gularis***. Males: (n=10) wing 125.0 (116.1–142.2); tail 107.4 (101.1–117.3); culmen 27.9 (26.2–30.2); tarsus 31.2 (29.7–32.5). Females: (n=10) wing 115.2 (108.0–126.2); tail 104.9 (91.7–116); culmen 26.6 (24.6–28); tarsus 30.7 (28.5–32).

I. g. tamaulipensis. Males: (n=10) wing 114.6 (108.2–117.9); tail 104.8 (97.3–111.0); culmen 25.5 (24–26.9); tarsus 30.5 (29.5–32). Females (n=10) wing 107.9 (103–114); tail 101.7 (95.0–105); culmen 24.4 (23.6–26); tarsus 29.4 (27–30.0).

I. g. yucatanensis. Males: (n=12) wing 113.0 (106.9–117.3); tail 103.1 (97.8–106.7); culmen 24.1 (22.9–25.4); tarsus 28.7 (25.9–30.0). Females: (n=6) wing 103.6 (99.3–107); tail 99.8 (96.5–103.1); culmen 23.2 (22.9–24); tarsus 28.4 (26.4–30.2).

I. g. flavescens. Males: (n=3) wing 122.1 (120–124.2); (n=4) tail 115 (106–122); (n=2) culmen 27, 27; tarsus 29, 30. Females: (n=3) wing 112.7 (109–116.2); (n=4) tail 108 (100–116); (n=2) culmen 26, 26; tarsus 28, 31.

I. g. troglodytes. Males: (n=31) wing 118.2 (110–121); (n=2) tail 107, 112; culmen 24, 26; tarsus 28, 28. Females: (n=2) wing 106, 109; tail 96, 100; culmen 24, 25; tarsus 28, 30.

I. g. gigas. Males: (n=28) wing 132.1 (122–138); (n=11) tail 117.4 (112–123); (n=10) culmen 26.9 (25–29); tarsus 32.0 (29–35). Females: (n=10) wing 119.0 (110–126); tail 105.4 (98–113); (n=6) culmen 26.0 (25–26); tarsus 31.5 (30–34).

NOTES Formerly known as Lichtenstein's Oriole and Black-throated Oriole. Altamira Oriole appears to have increased in abundance in its Texas range during the 1970s, then decreased in the 1980s and may be increasing again in the 1990s (Brush 1998). Originally this oriole was classified in its own genus *Andriopsar*, but was eventually merged into *Icterus*. However, *Andriopsar* was resurrected in 1919 based mainly on Altamira Oriole having a raised keel on the palate which is not observed in any other oriole, at least to that stage of development. This keel is of variable development on many blackbirds, being most well-formed on the grackles and close relatives. It appears that the keel is used as an aid in obtaining food (Wetmore 1919). For details of a mixed pairing between this species and Audubon's Oriole, see Notes section of that species. In Bentsen, State Park, Texas an unusually plumaged Altamira Oriole was observed showing a black face and extensive black on the breast and neck, and a paler orange colour than others. It was suggested that this bird could either be an unusually plumaged Altamira Oriole or a hybrid with an Audubon's Oriole (S. Bentsen pers comm.), photos suggest a hybrid origin.

A recent study comparing the mitochondrial DNA sequences of these orioles suggests that Altamira Oriole belongs to a group which includes Jamaican, Yellow and Orange Orioles (K. Omland, S. M. Lanyon pers. comm.). Any resemblance that Altamira Oriole has to Spot-breasted and Streak-backed Orioles is thus due to convergent evolution.

REFERENCES Berlepsch 1888 [geographic variation], Binford 1989 [habitat, nesting], Brush 1998 [behaviour, nesting, notes], Griscom 1930 [geographic variation, measurements], Hardy *et al.* 1998 [voice], Howell and Webb 1995 [identification, voice, distribution], Land 1962 [measurements], Phillips 1966 [geographic variation, measurements], Pleasants 1981 [voice, nesting, behaviour], 1993 [voice, habitat, movements], Pyle *et al.* 1987 [moult], Ridgway 1902 [measurements], Skutch 1954 [voice], Sutton and Pettingill 1943 [nesting].

35 STREAK-BACKED ORIOLE *Icterus pustulatus* (Wagler) 1829 — Plate 12

An extremely variable oriole in appearance, only the northern populations are always streak-backed.

IDENTIFICATION An orange to scarlet-orange oriole with a black throat patch and noticeable white wing-edging. It has a straight culmen and gonys. Northern populations of this species are easily identified by the obvious dark streaks on the yellow or orange back; however, individuals from the south are largely dark-backed, leading to confusion. In the US, this species is a rare visitor but has recently bred in SE Arizona. Here it may be confused for the more abundant Hooded Oriole (36) which shares with it a similar face pattern. Hooded Oriole has a black back, which is veiled with yellow-olive or pale orange in the winter, making the back appear superficially 'streaked'. Hooded Oriole has black centers to the back feathers and full pale fringes; as these wear, more of the black is exposed. The black never appears truly streaked (lengthwise) when observed in detail, rather the black centers appear as dark bars (crosswise) on the back. The pattern of dark colouration on the back of Streak-backed Oriole is always lengthwise, 'true' streaks in other words. Note also that the ground colour of the back of Streak-backed Oriole is the same as that of the nape; in Hooded Oriole the back is darker and contrasts with the nape (Kaufmann 1983). Furthermore, Hooded Oriole has a long, thin and curved bill (most easily noticed by concentrating on the culmen) unlike the thicker and straight bill of Streak-backed Oriole (Figure 29.1). Streak-backed Oriole has the black on the face (anterior auriculars) narrowed towards the bill, the black mask is more extensive on Hooded Oriole. Hooded Orioles are slimmer birds, with longer, graduated tails, and are also of a duller body colour than the nearly crimson *microstictus* Streak-backed Oriole.

A potential identification problem that has not been addressed elsewhere is the presence of some 'streak-backed' Altamira Orioles (34). These have been observed from time to time in southern Texas (G. Lasley pers. comm., photos), and are likely to occur throughout the range of that species. It is not known if this condition is age-dependent, consistent throughout a bird's life, or just a rare anomaly. Nevertheless these 'streak-backed' Altamira Orioles always show more black than orange-yellow colouration on the back, more like a southern subspecies of Streak-backed Oriole. The large, thick-based bill of Altamira Oriole may be used to identify these birds. Note that the greater thickness of Altamira Oriole's bill is largely due to the disproportionately thick base of the lower mandible which gives their bills a characteristic shape not seen on Streak-backed Oriole (Figure 29.1). In addition, Altamira Oriole has a much more restricted patch of blue-grey at the base of the lower mandible. The pattern of white on the wings also differs between these two species, with Altamira Oriole lacking a white upper wingbar as well as having largely black primaries which contrast with the white-edged secondaries; Streak-backed Orioles have a yellow upper wingbar and white-edged primaries.

The southern, solidly black-backed forms, of Streak-backed Oriole can be easily confused with the sympatric Altamira and Spot-breasted Orioles (33). Spot-breasted Oriole has black spots on the breast sides, a slightly curved culmen, and a wedge-shaped white wing patch on the folded tertials; the secondaries are not edged in white like on Streak-backed Oriole. Specimens of southern (*pustuloides, alticola*) Streak-backed Orioles exist, with three or four spots on each side of the breast. This situation is rare, and these birds may be identified by their straight bills (look at the culmen) and by the lack of a wedge-shaped wing patch. Immature Spot-breasted Oriole shows a dull version of the adult's wing pattern. Altamira Oriole may look extremely similar. While both species vary in colour, in the south they track each other's variation (co-vary), likely due to the effects of a common environment. However, Altamira Oriole always has a deep, strong-looking bill unlike the more standard-sized bill of Streak-backed Oriole (see above). In addition, dark-backed Streak-backed Orioles usually show some orange edging on the back feathers in fresh plumage. Immatures of these two species may be impossible to identify at times, but the bill shape differences should identify most individuals. Vocal differences between these species are also helpful in identification (see Voice).

Females and immatures of the northern subspecies of Streak-backed Oriole are likely to be confused with females or immatures of Bullock's (41) and Black-backed (42) Orioles. Most adult female *microstitus* and *pustulatus* Streak-backed Orioles show at least some dark streaks on the back; these are always absent in Bullock's and Black-backed Orioles. Female Streak-backed Orioles also typically show a black mask and throat patch; a black throat may be present but it is never connected to a mask on Black-backed Oriole, and Bullock's Oriole shows this pattern only as an immature male. Note that female Streak-backed Oriole shows a greater extent of white markings on the wings and a more solidly yellow belly than either Bullock's or Black-backed Oriole. Juvenile Streak-backed Orioles may be almost indistinguishable from some Bullock's and Black-backed Orioles but once the adult bill colours come in (pink lower mandible turns blue-grey) the species may be separated by bill colour. Streak-backed Oriole has an entirely black upper mandible (maxilla) and a black tip to the lower mandible; Black-backed and Bullock's Orioles both show an entirely blue-grey lower mandible and a blue-grey cutting-edge to the upper mandible. Juvenile Streak-backed Orioles show two thin and separate wingbars, the other two species often show a substantial amount of pale feather edging on the greater coverts, suggestive of the pale wing panel of adults.

VOICE Both sexes sing.

Song: The melodious, whistled song is similar to that of a Bullock's Oriole, but with a stop and start pattern reminiscent of a Vireo (*Vireo* sp.). Has been described as *chew-chewy, chip-cheer, chew.. chirp...weet-chewy*. Another example is, *chree chree chree chree chree-chu chee-chi* or *wheet'tchi-wee-chi-wee* (Howell and Webb 1995). The songs of Streak-backed Oriole are not as repetitive as those of some of the other tropical orioles. Males do not sing often, but usually when they do sing it is at the nest site. Females sing as well, but these songs are even more infrequent and less complex than the songs of the male. Slud (1964) noted that in NW Costa Rica, Streak-backed Oriole (*sclateri*) is a weak singer, giving 'individual oriole-like musical chirps' rather than a full melodious song.

Calls: The common call is a low *wrank*. Also performs a sweet *chuwit* similar to that of a House Finch (*Carpodacus mexicanus*). Also a short *whip*. The chatter is similar to that of a Baltimore (40) or Bullock's Oriole, but is faster. The chatters vary in length from approximately one to two seconds in length.

DESCRIPTION A medium-sized to large oriole. The bill is moderately deep and shows a straight culmen and a slightly curved gonys. The tail is weakly graduated, the outer feather being about 10 mm shorter than the innermost. Shows a significant amount of geographic variation, the form *microstictus* is described. Wing Formula: P9 < P8 < P7 > P6; P9 ≈ P3; P8–P6 emarginate. The form *microstictus* is described.

Adult Male Non-breeding (Definitive Basic, fresh): The bill is black, with the basal one-third to one-half of the lower mandible blue-grey. The eyes are dark. It shows black lores, anterior auriculars, chin, throat and upper breast creating a black mask and bib. The forehead and crown, and nape is fiery reddish-orange, veiled with olive. The auriculars, sides of the neck and breast, bordering the black bib, are reddish-orange. The base colour of the back and scapulars is orange-olive with distinct teardrop-shaped black central streaks on each feather. Overall, there is more orange than black on the back. The rump is orange and the lower back orange with a slight olive tone. The underparts, other than the black bib, are entirely orange becoming more yellow-orange on the belly and vent. The wings are jet black, patterned with white and orange. The lesser coverts are orange, while the median coverts are black with large white tips; these white tips become progressively smaller inwards. The overall effect is for this wide white upper wingbar to become thinner inward (toward the scapulars). The greater coverts are black with a whitish-grey outer edge and tip, creating a pale wing panel. The tertials and secondaries are widely edged whitish-grey. The primaries are also edged whitish, most noticeably at the bases where on the folded wing a rectangular white patch contrasts with the black primary coverts. Thus, the folded wings are boldly patterned. The bend of the wing and the wing-linings are orange. The tail is black, with a white tip and white outer edge to the outer rectrix (R6), the other rectrices are tipped greyish-white. The legs and feet are blue-grey.

Adult Male Breeding (Definitive Basic, worn): Similar to non-breeding male but the olive tipping to the nape and crown disappears and the pale fringes on the wings are narrower.

Adult Female (Definitive Basic): Soft parts as in male, but the bill base tends to show a more extensive amount of blue-grey at the base, commonly one-half of the lower mandible. The overall pattern is similar to the male, but is much more dull-coloured. The black mask and bib are as in the male, and is equally extensive. The forehead is orange, becoming olive on the crown and nape. The face is orange with a slight olive tone. The back and scapulars are olive-grey, thinly streaked blackish. The lower back is olive-grey like the mantle, but unstreaked, while the rump is a brighter olive-yellow. The underparts (other than the black bib) are yellow with an orange wash to the breast, and the flanks are washed with grey, and some individuals have whitish bellies. The wings are blackish-brown rather than black, and the pale edging is similar to that of the male in pattern. The lesser coverts are blackish-brown with a thin orange fringe and the median coverts are blackish-brown with a white edge which is wider on the outer vane. The greater coverts are blackish-brown, edged and tipped whitish-grey. The tertials, secondaries and primaries are widely-edged whitish, showing a bolder rectangular white patch at the base of the primaries. The wing-linings are yellow. The tail is olive-grey with yellow-olive outermost rectrices and bases to the others except for the central pair. The appearance of adult females is somewhat variable, particularly further south (*pustulatus*). Some are noticeably brighter orange than others, with more obvious back spots and a greater extent of wing-edging. These birds may also show blackish central two to four rectrices.

Immature Male (First Basic): Similar to adult female but the back spots are larger, darker, and more crisp and distinct. The underparts are brighter yellow, with less of an olive tinge. The tail is olive as in the female and the wings are similar to the female's. Some immature males are indistinguishable visually from adult females, particularly in S Mexico (*pustulatus*) where females are brighter than typical *microstictus*.

Immature Female (First Basic): Variable in appearance, but similar to a dull adult female. Often lacks a black bib or has a very indistinct one. The wings may appear less uniform, showing contrast between newer and darker coverts, and perhaps tertials, than the duller primaries and secondaries. The back streaks are dull or lacking.

Juvenile (Juvenal): Similar to a dull female, but always lacks the black bib and lores. The cutting edge of the bill and the mandible base is horn-coloured or orange-pink rather than blue-grey. The crown and nape is olive with a slight yellow tint. The back is much greyer than on a dull adult female with very indistinct back streaks, or none at all. The underparts are yellowish with a dull, dusky wash. The wings are blackish-grey and patterned like adults, but not as widely. The lesser coverts are greyish-black with olive tips.

GEOGRAPHIC VARIATION Classified into six subspecies which fall into three groups, the northern *pustulatus* group (Scarlet-headed Oriole), the southern *sclateri* group (Streak-backed Oriole) and the insular *graysonii* (Tres Marias Oriole). Perhaps more than one species should be recognized. Figure 35.1 illustrates the general back patterns of several subspecies.

SCARLET-HEADED ORIOLE. The northernmost group, with very small back streaks, noticeable sexual dichromatism, and a reddish tone to the colour of the head in males:

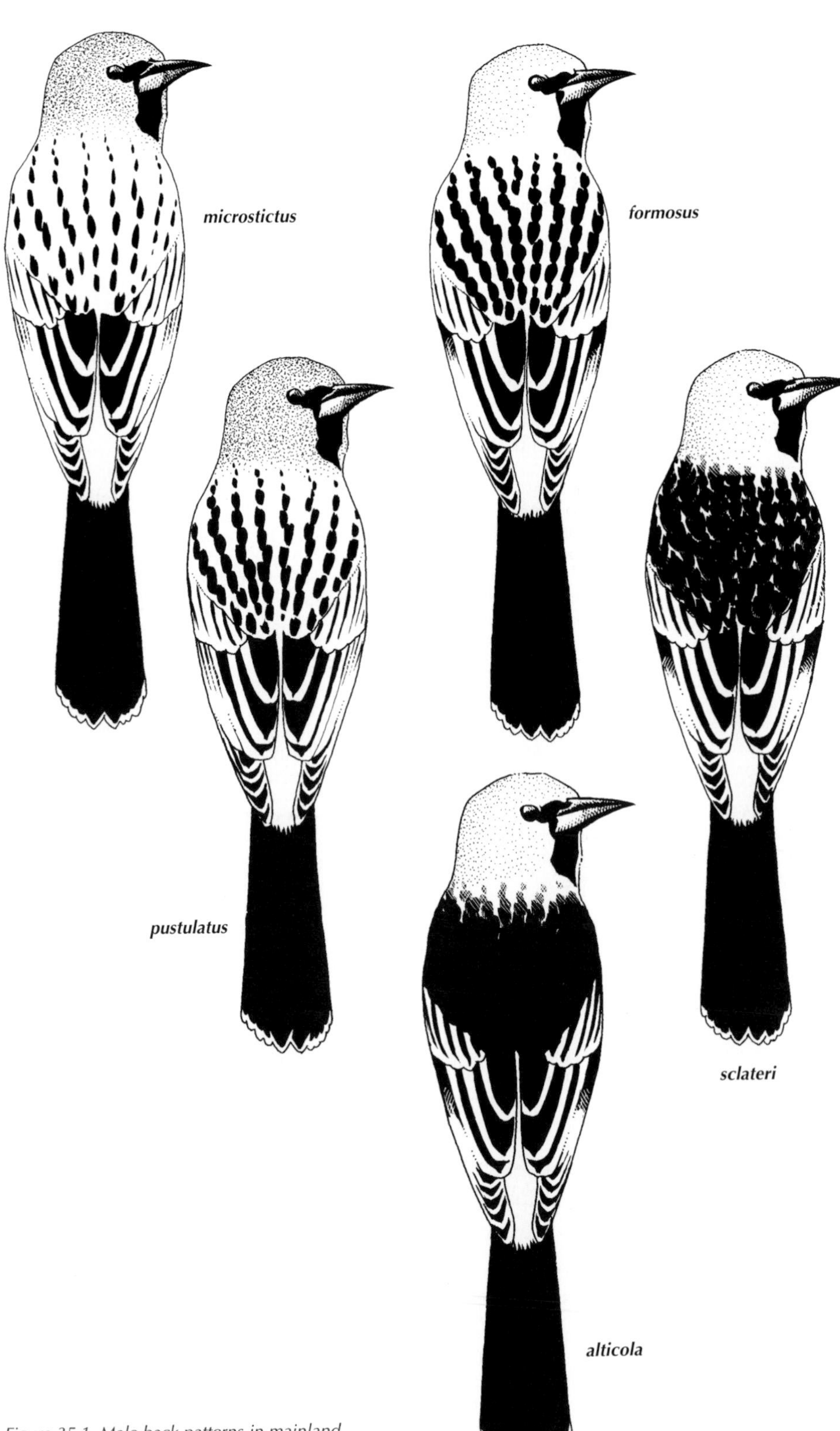

Figure 35.1 Male back patterns in mainland subspecies of Streak-backed Oriole.

I. p. pustulatus occurs in the tropical lowlands of Mexico from Colima in the west south to N Oaxaca and east from Guanajuato to Morelos, Puebla and W Veracruz. This form is slightly larger than *microstictus* and has heavier streaks on the back. The area of black on the back approaches 50%. The head does not tend to obtain the bright reddish tone of *microstictus*. The females are bright and may approach the appearance of the male. Note that a small proportion of *pustulatus* from Colima show characteristics of *microstictus*, suggesting introgression in this area (Schaldach 1963).

I. p. microstictus inhabits W Mexico from Sonora and Chihuahua south to Jalisco. This is the form that occurs as a rarity in the southwest US. It is described above. This form is characterized by the similarity to *pustulatus*, but has much smaller back streaks. It is also marginally smaller and usually has an obvious reddish tone to the head. The females are always noticeably duller than the males.

TRES MARIAS ORIOLE. An insular form that is large, yellowish and tends to lack back streaking:

I. p. graysonii is restricted to the Tres Marías Islands off the coast of Nayarit, Mexico. This race is larger and paler than *pustulatus* tending to be yellow to yellow-orange rather than orange. It usually lacks streaking on the back, or may have a few very narrow streaks on the scapulars. Females are similar to the males, but tend to have olive rather than black tails.

STREAK-BACKED ORIOLE. The southernmost group characterized by its large size, lack of a reddish-orange head, reduced pale edging on the wings, sexually monochromatism, and black colouration dominating the mantle. Note that Monroe (1968) found that Honduran highland and lowland populations did not vary consistently in size. While size varied on the whole, he did not find it to have a geographic pattern. He suggested that all Central American Streak-backed Orioles (*sclateri* group) be lumped into one form, *I.p.sclateri* until a proper review can be done. Our studies of museum skins suggest the same action should be taken particularly since the strength of oriole colouration appears to be highly sensitive to environmental factors. Below are the currently recognized forms:

I. p. formosus occurs in southernmost Mexico (S Oaxaca and Chiapas) to NW Guatemala. It is small and shows more yellow on the back than *sclateri*, but the colour of the yellow-orange body is similar to *sclateri*. This is the most similar form to *pustulatus* and appears to form a link between the *pustulatus* group and the *sclateri* group. Some additional differences from the *pustulatus* group are that it has white median coverts, unlike *pustulatus*, and more black on the back. The greater coverts are different in pattern, showing a wide white tip on the outer vane only, creating a wide white lower wingbar rather than a pale panel. The primaries are pale-edged only at their tips and base, the central part is entirely black. The females are like the males, but may show some olive on the outer tail feathers. The juveniles are similar to the *pustulatus* group, but with more noticeable black back streaks. The wings are similar in pattern, but tend to show two pale wingbars rather than a pale greater covert panel.

I. p. sclateri is the southernmost form, found along the Pacific slope of El Salvador, Honduras and Nicaragua and entering extreme NW Costa Rica. This form is larger than *pustulatus* and has much broader back streaks, with more black on the back than yellow-orange. The median coverts have less black towards the base. The head is yellow or orange-yellow, seldom the bright orange of *pustulatus*. It is similar to *formosus*, but averages more black on the back.

I. p. alticola is found in arid zones of Guatemala and the Atlantic coastal slope of Honduras. It is similar to *sclateri*, but the back is more extensively black, often solidly black; in addition, it tends to be brighter orange than *sclateri*. This form is also larger than both *sclateri* and *pustuloides*. Adult females are like adult males, but may show some olive on the outer tail feathers. The race *connectens* from interior El Salvador has been described as intermediate in size between *alticola* and *sclateri*, and is best treated as a population linking both of these forms but not as a subspecies. Van Rossem (1927) found that highland Streak-backed Orioles in El Salvador were not more extensively black on the back than those in the lowlands, and that there were no size differences between the two groups.

I. p. pustuloides is restricted to the area near Volcán San Miguel, El Salvador. It is smaller than *sclateri* and similar to it in pattern but has a much brighter, reddish-orange plumage much like the northern *microstictus* in its flame-coloured orange body.

I. p. maximus is restricted to the arid section of the Rio Negro valley in north-central Guatemala. It is similar to *alticola*, but is yellow not orange. The forms *pustuloides* and *maximus* are extremely range restricted and known from small series of specimens; further study may suggest that these are best lumped into one of the other races.

HABITAT Inhabits thorn forests, dry scrub and other dry deciduous woodlands. Also found in riparian corridors in desert, at least in the north of its distribution. It shows a preference for open woodlands, especially those with thorny bushes such as *Mimosa*. It also lives in second growth and near settled and agricultural areas, and in arid, pine-oak forests in the highlands of Mexico.

BEHAVIOUR Forages in pairs or family groups, sometimes gathering into larger groups. Mainly insectivorous, feeding on insects gleaned from foliage or dead, rolled leaves. Feeds on fruit and seeds to some extent. Flowering trees are frequented, both to nectar and to catch insects visiting the flowers. When visiting flowering *Pseudobombax ellipticum* trees in Morelos, Mexico, this oriole was observed to be submissive to Black-vented Oriole (44) in direct confrontations. However, Streak-backed Oriole would win out during confrontations with wintering Bullock's Orioles at the same tree (Eguiarte *et al.* 1987). Pairs tend to be monogamous, but polygyny has been recorded. During the breeding season most vocalisation activity occurs by the nest. Males rarely sing their full song during the nesting period (Corman and Monson 1995); perhaps, like Bullock's Orioles, singing frequency greatly decreases after pairing occurs.

NESTING In the north, the breeding season commences in early June and continues into August. Most nests contain eggs during May in the lowlands of Oaxaca, Mexico; higher up, birds nest in mid July (Rowley 1984). In El Salvador, *sclateri* nests late April–late June, while in the highlands, *alticola* nests about a month later. Apparently the female builds the nest without the aid of the male, and it takes just under a month to finish construction. The nest is a large hanging basket, up to 70 cm in length, which is secured to the tips of branches. Soft grasses are used to weave the nest and its construction does not look as sturdy as the nests of Altamira or Spot-breasted Orioles. The nesting site is often at a moderate height, and in the shade of the tree's canopy. It likes to secure its nest to a tree protected by thorns such as *Mimosa* or Bull's Horn Acacia (*Acacia*) which has the addition of an ant's nest (*Pseudomyrmex* sp.) for nest defense. Nests have been recorded hanging from power lines, in the same manner as nests of Altamira Oriole. The clutch size varies from 3–4 eggs and the eggs are pale blue with dark spotting that is evenly distributed. Incubation lasts 12–14 days and the young take approximately two weeks to fledge (Corman and Monson 1995). Both sexes share the task of feeding the young. This species is known to be parasitised by Bronzed Cowbird (99), but it has not been observed to successfully fledge young of that species (Friedmann and Kiff 1985).

DISTRIBUTION AND STATUS Fairly common to common. The lowland populations (*pustulatus* group, *formosus* and *sclateri*) are found below 500 m, occasionally higher; other populations inhabit elevations as high as 2000 m. Ranges from N Sonora and W Chihuahua (and recently SE Arizona, USA) in Mexico along the Pacific slope to W Guatemala; resident on the Tres Marías Islands, Nayarit; from E Guatemala along the Pacific Lowlands and interior highlands of El Salvador, west-central Honduras (including part of the Atlantic Slope) and continuing south along the Pacific lowlands of Nicaragua and extreme NW Costa Rica (head of the Golfo de Nicoya).

MOVEMENTS Partially migratory, particularly northern populations. It is also noted to be more common in El Salvador during the summer with small numbers in winter (Dickey and van Rossem 1938), although more recently Thurber *et al.* (1987) noted that the species is common throughout the year in El Salvador. Rand and Traylor (1954) state that *sclateri* and *alticola* are present in the breeding season (the wet season, May to September) but their numbers decrease markedly during the dry season. They concluded that the species is not migratory in that country. The dynamics of seasonal movements are still not well understood however. It is a rare fall and winter vagrant to California and Arizona and has recently bred in southern Arizona. In the Pacific states, the most surprising record was of a male that stayed from September 28 to October 1, 1993, at the Malheur NWR in SE Oregon. This is the northernmost record and the only US record outside of California and Arizona. However, there is one sighting from Nevada during May 1997, which is being reviewed by the state committee (J. Eidel pers. comm.). In California, there have been a total of five accepted records, two in fall, two in winter and one in spring. All of the records have been in the southern half of the state (San Diego, San Bernardino and Inyo Cos.). A new, recent record during the winter of 1996–97 in Orange Co., is the northernmost coastal record and has not been formally reviewed. Arizona has had nearly 20 records now, and recently this series has started to nest. The first nesting occurred in 1993 at Dudleyville, Arizona. One male was attending two separate nesting females that year. Two pairs nested there in 1994 and a pair nested at the same location in 1995. A third pair was observed nest-building along the Santa Cruz river, Pima Co., Arizona, in 1994.

MOULT Adults have one moult complete (definitive pre-basic) moult after the breeding season, and are in fresh plumage by October. The moult occurs mainly during September. Juveniles moult out of their plumage through a first pre-basic moult, which may continue into October. During this moult, the body plumage is replaced as are the wing-coverts and often the tertials. The juvenal wing and most of the tail is retained, with sometimes a few of the tail feathers changed. In some cases, some of the outer primaries may be replaced during the first pre-basic (Pyle 1997b). There is no evidence that pre-alternate moults exist.

MEASUREMENTS ***I. p. pustulatus***. Males: (n=10) wing 100.8 (98.6–106.2); tail 90.9 (87.1–97.3); culmen 21.3 (19.8–22.9); tarsus 24.9 (23.6–25.4). Females: (n=16) wing 90.5 (85–96.3); tail 88.0 (80.8–95); culmen 20.7 (18.8–22.9); tarsus 23.9 (22–25.4).

I. p. microstictus. Males: (n=11) wing 96.8 (88–105); tail 91.8 (81–104); culmen 21.1 (20–23); tarsus 24.2 (23–26). Females: (n=7) wing 90.6 (84–94); tail 87.6 (76–94); (n=6) culmen 21.6 (21–22); (n=7) tarsus 20.8 (23–25).

I. p. graysonii. Males: (n=16) wing 103.1 (99–107); tail 92.8 (90.2–98); (n=15) culmen 24.6 (24–26.7); (n=16) tarsus 26.2 (25–27). Females: (n=15) wing 94.3 (88–99.1); tail 88.4 (83.3–95); culmen 24.1 (22.9–26); tarsus 25.9 (25–27).

I. p. formosus. Males: (n=24) wing 105.7 (101–112); (n=9) tail 95.4 (88–100); culmen 22 (21–23); (n=8) tarsus 24.0 (23–25). Females: (n=6) wing 99.2 (96–103); tail 94.0 (90–97); culmen 21.8 (19–23); tarsus 24.4 (23–25).

I. p. sclateri. Males: (n=14) wing 107.2 (100–115.1); tail 93.6 (85–104.1); culmen 21.6 (19.8–23.1); tarsus 24.4 (22–26.4). Females: (n=6) wing 98.5 (94–102); tail 89.8 (81.8–100); culmen 21.3 (20.6–23); tarsus 24.8 (22.9–27).

I. p. alticola. Males: (n=12) wing 109.0 (98–115.5); tail 93.2 (85.5–112); culmen 20.2 (19–22); (n=1) tarsus 29. Females: (n=1) wing 100; tail 90; culmen 20; tarsus 25.

I. p. maximus. Males: (n=17) wing (111–118).

I. p. pustuloides. Size similar to *sclateri*, wing averaging 105mm.

NOTES The geographic variation in this species is substantial and perhaps the subspecies groups deserve species status. More work is needed to address this question. Two independent molecular studies suggest that the closest relative of the Streak-backed Oriole is the Bullock's Oriole (Freeman and Zink 1995, K. Omland and S. M. Lanyon pers. comm.). The striking similarity of *alticola* Streak-backed Oriole to Altamira Oriole may be more than a coincidence, perhaps there is some adaptive value to this convergence. The two species do not appear to be closely related (K. Omland and S. M. Lanyon pers. comm.).

REFERENCES Corman and Monson 1995 [nesting, movements], Dickey and van Rossem 1938 [identification, habitat, nesting, movements, measurements], Grant 1965 [geographic variation], Griscom 1934 [geographic variation], Hardy *et al.* 1998 [voice], Herlyn *et al.* 1994 [movements], Howell and Webb 1995 [voice], Kaufman 1983 [identification], Miller and Griscom [geographic variation, measurements], Ridgway 1902 [measurements], Rowley 1984 [nesting], Schaldach 1963 [habitat, geographic variation], Small 1994 [movements], Stiles and Skutch 1989 [voice, behaviour, nesting], Thurber *et al.* 1987 [movements], van Rossem 1927 [geographic variation].

36 HOODED ORIOLE *Icterus cucullatus* **Plate 11**
Swainson 1827

This is a slender, long-tailed oriole with a long and finely-curved bill which occurs in southwest US and Mexico. Look for it near palms.

IDENTIFICATION Males are orange with a black back and throat bib. They sport a bold white shoulder patch and white wingbar. The bill is long and noticeably decurved (Figure 29.1). Adult males are boldly marked but the pattern is not unlike that of several other species of sympatric orioles such as Streak-backed (35), Spot-breasted (33), Orange (29) and Altamira (34). The females and immature males are not as colourful, and this causes identification problems, particularly with female or immature Orchard/Fuerte's (43), Streak-backed, Orange and Bullock's (41) Oriole.

Winter male Hooded Orioles can be confusingly similar to Streak-backed Orioles. The culmen of Hooded Oriole is always curved while that of Streak-backed Oriole is nearly straight. Streak-backed Orioles have deeper bills than Hooded Orioles. Note the differences in the black mask and bib: on Hooded Oriole the black comes almost straight down from the eye, while on Streak-backed Oriole the black curves in toward the bill making an obvious sinuous back edge to the black patch. The orange colour on Streak-backed Oriole is most intense on the face, right at the edge of the black face patch, unlike the more evenly coloured Hooded Oriole. The buff or tan colour of the back contrasts with the orange or yellowish head of Hooded Oriole, but on Streak-backed Oriole there is no apparent contrast between the nape and back colour. The details of the back pattern also differ, with Streak-backed Oriole showing noticeably longitudinal streaks on the back, while as Hooded Oriole wears and begins to show black on the back, these markings are more like transverse spots or bars, not streaks. The similarly patterned adult Altamira Oriole is easily identified by its much larger size and by its thick, powerful-looking bill. The culmen of Altamira Oriole is straight, not curved. In addition, Altamira Orioles have orange lesser and median coverts, whereas the shoulder is white on the Hooded Oriole. The bases of the primaries on Altamira Oriole are widely-edged white for a short distance only, creating a white patch or 'handkerchief' immediately behind the greater primary coverts. On Hooded Oriole, this patch is absent, as this species shows narrow white edges to the primaries that extend from the base to the tip.

In the Yucatán, the small Orange Oriole may be confused with Hooded Oriole. Male Orange Orioles have orange backs and less black on the auriculars, and the culmen is straight not curved. The wing pattern is different with Orange Oriole showing an obvious rectangular white patch at the base of the primaries, the 'handkerchief'. Immatures are trickier to identify, but note the bill shape differences. Female and immature Orange Orioles have brighter yellow heads and backs as well as less black on the face than immature male Hooded Orioles.

Female Bullock's Oriole is similar to female Hooded Oriole, but the two species differ in shape with Hooded showing a thin curved bill (Figure 29.1) and a long and graduated tail. Bullock's Orioles have straight bills and a shorter, more squared-off tail shape. Additionally, Hooded Oriole is entirely yellow below, lacking the greyish-white belly that contrasts with a yellow breast on Bullock's Oriole. Females and fall immatures Hooded Orioles can look very similar to female or immature Orchard Orioles. Orchard Orioles are somewhat smaller than Hooded Orioles, it is not surprising that they are often initially mistaken for a warbler! Hooded Oriole, while small, should not be mistaken for a bird as small and compact as a warbler, largely due to its more elongated shape. Hooded Oriole's tail is proportionately longer and is noticeably graduated, whereas Orchard Oriole's tail is squarish and short for an oriole. Usually, the longer and more curved bill of Hooded Oriole will be noticeable; however, observers inexperienced with either species may find this feature difficult to judge and young Hooded Orioles may show a bill that is not fully grown. In general, Orchard Oriole is greener above and yellower below, while Hooded Oriole shows grey tones to the back and dull yellow underparts, and some overlap in these characters may occur. Immature males of these two species, showing black bibs, are also similar. Note that the black bib of a young male Orchard Oriole is not as extensive, nor extends as far down on the breast as that of Hooded Oriole. In addition, look for single chestnut feathers scattered on the breast and belly of Orchard Orioles. The calls of adults are diagnostic in these two species. Orchard Orioles give a sharp *chuck* note while Hooded Oriole has a whistled *wheet* that is similar to that of a House Finch (*Carpodacus mexicanus*). However, juvenile Hooded Orioles give a soft *tup*, or a louder *chuck* note but these are either not as loud as the Orchard Oriole's call or are lower pitched. The *chuck* note of Hooded Oriole is also slightly shorter in duration (C. Benesh pers. comm.).

VOICE Song: The song tends to be quick and abrupt. The individual notes are given rapidly and most lack the sweet whistled nature of many oriole notes; in contrast they sound springy, nasal and whiny. One song type recalls that of White-eyed Vireo (*Vireo griseus*) in the springy and explosive quality. Some songs last just over 1 second, others much longer. There is a substantial amount of variation in Hooded Oriole song. In addition, each male sings several different song types. Unlike some of the more tropical orioles, Hooded Oriole does not sound repetitive. It is unclear if Hooded Orioles also sing long songs of repeated motifs. This species does sometimes include mimicry within songs; in Arizona they have been noted to include Gila Woodpecker (*Melanerpes uropygialis*) and Ash-throated Flycatcher (*Myiarchus cinerascens*) calls in their vocalisations (S. Goldwasser pers. comm.). In northern California, Hooded Orioles rarely sing; however, the reason for this is not known.

Calls: Commonly gives a whistled *wheet* or *sweet*, similar to the flight call of a House Finch. A chatter is given as an alarm vocalization, which is somewhat weaker and shorter than the corresponding Baltimore (40) or Bullock's Orioles' call. In addition, the individual notes of the chatter are given more quickly creating a shriller sound. Some chatters only include two or three notes, making them sound much less like a 'true' chatter. Also gives a soft *chut* and a harsher *chuck* or *chet*, these notes are most common in juveniles but may be given by adults. The *chuck* notes resemble those of Red-winged (61) or Brewer's (95) Blackbirds. It is a widely-held belief that the *chuck* note of Orchard Oriole is diagnostic when trying to separate it from Hooded Oriole. However, this is not the case. However, it is clear that Orchard Oriole does lack the *wheet* note of Hooded Oriole.

DESCRIPTION A small oriole with a slim build and long tail. The tail is significantly graduated, the outer rectrix may be as much as 20 mm shorter than the innermost. The bill is slim, noticeably long and decurved, quite obviously so on the culmen. Wing Formula: P9 < P8 < P7 ≈ P6 > P5; P9 ≈ P4; P8–P5 emarginate. The race *cucullatus* is described.

Breeding Adult Male (Definitive Basic, worn): The bill is black, with the basal half or one-third of the lower mandible blue-grey. The eyes are dark. Most obvious is the black mask and bib created by the black lores, supralores, anterior half of the auriculars, chin, throat and upper breast. The mask reaches as far as the back edge of the eye and extends down to the upper breast largely in a straight, vertical line, usually bulging out behind the line of the eye below the auriculars; however, the shape of the black patch may change depending on posture. The rest of the head, nape, neck, lower breast and the underparts to the undertail-coverts are orange, most intensely so on the head. The back and scapulars are black but the rump and uppertail-coverts are orange. The wings are black patterned with white. The white is most obvious on the median coverts which are completely white, making a bold white upper wingbar. The greater coverts are thinly edged white and more broadly tipped white, creating a second white wingbar. The tertials, secondaries and primaries are edged in white, more broadly so on the tertials. The wing-linings are orange-yellow. The tail is entirely black, showing thin white tips when fresh. The legs and feet are blue-grey.

Non-breeding Adult Male (Definitive Basic, fresh): Like breeding male, but shows olive fringes on the back, sometimes wide enough to make the back appear entirely tan or tan with black spots or streaks. As well as all the white markings, the wings are broader and more noticeable, and the orange of the body is duller, particularly on the crown and nape where it is tipped with olive.

Adult Female (Definitive Basic): Head olive-yellow, greyer on the crown and nape. The mantle is olive-grey, contrasting with the greener upper back, head, lower back and rump. The rump is olive-green. Underparts yellowish, paler on the belly and greyer on the flanks. The brightness of the underparts varies depending on the individual. The underparts are brightest yellow on the breast and a few black feathers may be present on the throat. The wings are greyish-black, patterned with white as in the male. The lesser coverts are greyish-green, while the median coverts are white-tipped, forming a wingbar. The greater coverts are finely edged and more broadly tipped with white, creating a second wingbar and sometimes a paler wing panel depending on the extent of the pale edging. The tertials, secondaries and primaries are edged in white. The marginal coverts are yellow, while the wing-linings are grey. The tail is olive, edged with lime-green, thus appearing olive with greenish sides. The soft parts are coloured as in the male, but as much as the basal half of the lower mandible may be grey.

Immature Male (First Basic): Like the female, but with blackish lores and bib. The throat bib is not as extensive as that of the adult male, showing the yellow intruding forward, below the eye, toward the base of the bill. The extent of the black bib and lores varies, sometimes showing some green feathers on the chin, or black feathers on the anterior auriculars. Many young males obtain some bright orange on the breast sides. The upper wingbar is more obvious on young males than on females. The white edges of the primaries are more noticeable on young males than females. Wear and feather contrast as in immature female.

Immature Female (First Basic): Like adult female, but with darker, fresher coverts contrasting with the more worn flight feathers. They are quite worn during their second summer.

Juvenile (Juvenal): Like adult female, but with buffy wingbars and duller plumage. The overall colour is olive-brown above, and pale olive-yellow below. The underparts are much duller than on adult females, and they show a noticeably whitish belly unlike the females. In addition, there is less contrast between the grey flanks and the rest of the underparts. The bill tends to show a flesh base to the lower mandible, not grey. The bill of fledged juveniles may not be full grown, appearing uncharacteristic for a Hooded Oriole.

GEOGRAPHIC VARIATION Five subspecies are recognized varying largely in bill length and colour saturation in the adult male plumage. See Moult for Geographic Variation in that parameter.

I. c. cucullatus is found from the middle of the Rio Grande valley of Texas south through Mexico to Oaxaca and Veracruz, not including central and northern Tamaulipas. This is a very orange form, in fact the most orange in the US. Females are noticeably more orange on the underparts than female *sennetti, nelsoni* and *trochilioides*. It is described above.

I. c. sennetti occurs from the lower Rio Grande valley of Texas south along the coastal plain into C Tamaulipas. Similar to *cucullatus*, but males are noticeably paler, more yellow than that form. The females are also paler.

I. c. igneus occurs in southern Mexico from E Tabasco and the Yucatán peninsula south to N Chiapas and into Belize. It also occurs on islands off the coast, such as Cozumel and Mujeres. This is the brightest, most strongly orange form. The female is like *cucullatus* but more strongly orange-yellow on the underparts. It has also been reported that females of this race may show a greyish throat patch as opposed to the males black throat patch (Sutton 1948). Birds from the Cozumel Island population tend to be smaller, paler and have longer bills. Males from Mujeres have more black on the face, extending over the bill, and the greater coverts are not edged in white, just tipped with white. This race includes the previously recognized form *yucatanensis* from the Yucatán peninsula.

I. c. nelsoni is the western form, found from C California south to N Baja California, Mexico and east through C and SE Arizona, to SW New Mexico and south to S Sonora, Mexico. This race is noticeably more yellow than *cucullatus*, and even yellower than *sennetti*, looking largely yellowish-orange. The female is correspondingly duller than both of those forms. The race *nelsoni* has a longer and more slender bill than either *cucullatus* or *sennetti*, with longer wings and a shorter tail. The name *californicus* has also been given to this form, or part of it. The name *californicus* has precedence as it was described earlier; however, the type specimen is not racially identifiable and the collection locality is suspect, thus, the name *nelsoni* is now used.

I. c. trochiloides is found in Baja California Sur, Mexico. This race is yellowish-orange like *nelsoni*, perhaps a little yellower still; however, it has an even longer, more curved and thinner bill.

HABITAT A bird of dry open woodlands and riparian corridors in deserts and other dry areas. This species does not need extensive tree cover for it to be present. In California, its range has spread due to the planting of *Washingtonia* fan palms, which were native to the Colorado river in California; it is now a common member of suburban bird communities in parts of the state. The association with fan palms occurs in other parts of its breeding range, but perhaps not as strongly as in California. Winters in a variety of habitats, including urbanized areas, thorn forests, oak woodlands and arid pine-oak associations.

BEHAVIOUR Most often seen singly or in pairs foraging at low to medium heights along riparian vegetation. The long bill of this species is surely an adaptation to feeding on nectar and individuals are often observed perching on flowers (or hummingbird feeders) and draining them of their sweet resources. However, this species often robs the flower of its nectar by puncturing it at its base, and bypassing the pollination mechanisms of the flower. While wintering in Mexico, Hooded Orioles have been observed defending flowering trees from tanagers and other bird species. Furthermore, foraging on nectar alone of flowers of *Erythrina breviflora* trees in Mexico was calculated to be sufficient to provide the daily nutritional requirements for the orioles (Cruden and Hermann-Parker 1977). Little is known about the courtship of Hooded Oriole, but males will perform bow displays opening the wings and pointing the bill downwards in front of the female; this is likely a pre-copulatory display. Other displays include the male singing and bill tilting in unison with the female as he hops up and down on a branch; the significance of this display is not clear. Early in the breeding season, ritual chases are common, with the female flying off with one to several males chasing her. This appears to take place during the pair formation stage, before nests are constructed.

NESTING In the lower Rio Grande valley, males arrive in the last week of March to set up breeding territories. The breeding season extends from early April–July in Texas, mid May–August in Arizona, early April–mid August in California and early May–mid August in Baja California. Early accounts from the lower Rio Grande valley mention that most nests were constructed of Spanish Moss (*Tillandsia*), woven from the fibers and secured in a hanging mass of the plant. However, more recent reports tend to be of nests found secured to the undersides of palms or banana leaves. Locations and construction materials of nests appear to vary from locality to locality. In Arizona, nests are often woven of grasses and fastened to a tall tree, while in California nests are woven of palm fibers and usually stitched to the underside of a large palm frond. California Hooded Orioles are nearly obligate on *Washingtonia* fan palms or similar palms, including Banana plants (*Musa acuminata*), and their spread northward has been attributed to the planting of these trees north of their Colorado river valley native range. The nests are built under a live or dead fan palm frond and constructed of fibers garnered from green leaves of the palm. Non-palm Hooded Oriole nests in California have been found fastened to Pepper Tree (*Schinus molle*), oaks (*Quercus* sp.), Umbrella Trees, *Eucalyptus*, Sycamore (*Platanus* sp.) and man-made structures (Bailey 1910, Grinnell 1944, B. Bousman pers. comm.). The nests are a pendulous basket, much deeper than nests of Orchard Oriole, but shallower than the nest of the slightly larger Baltimore Oriole. The clutch size is commonly 4 eggs, but varies from 3–5. The eggs are whitish ranging to buffy-white or pale blue with dark blotches and spots, concentrated on the broader end of the egg. The incubation period lasts 12–14 days. The nestlings need approximately 14 days to fledge from the nest, and are fed by both parents. It appears that at least in Califonia, Hooded Oriole may have multiple broods within a season. Hooded Oriole is parasitised by both Brown-headed Cowbird (101) and Bronzed Cowbird (99); it has successfully fledged young of both species (Friedmann and Kiff 1985).

DISTRIBUTION AND STATUS Common, a bird of the lowlands found to 1500 m.

Breeding: Ranges from Northern California, as far north as Humboldt Co. and the northern extreme of the Sacramento valley, south to Baja California Sur, Mexico. Also found throughout S California to extreme S Nevada and SW Utah, though S Arizona, SW New Mexico and S Texas; south in Mexico along both slopes but absent from the Mexican plateau as a breeder; to N Sinaloa on the Pacific coast and south along the Atlantic coast to coastal Yucatán and N Belize. A resident on Cozumel Island.

Non Breeding: Along the Pacific coast this species winters regularly only in Baja California Sur, Mexico. In the US only regularly found in SE Arizona; in Mexico winters along both coastal slopes, in the west from Sinaloa to E Oaxaca, while in the Atlantic Slope found from S Tamaulipas to N Belize.

MOVEMENTS A short distance migrant. Banding records show 47 recoveries. An individual banded in New Mexico in summer was recovered in Mexico a few months later. An old report stated that a Hooded Oriole banded near Los Angeles, California, on Jan. 22, 1939, was recovered dead in Kansas on August 10, 1939 (Lincoln 1940). However, this is thought to be an error, but it is not clear how it occurred. The foot of the oriole is in the University of Kansas Museum and it appears not to be that of a Hooded Oriole (M. Thompson pers. comm.) The 45 other recoveries were all of birds banded during the breeding season in California and recovered in subsequent breeding seasons in California, the oldest of these a bird that lived to be at least six years old. Apparently specimens of *nelsoni* are rare in Baja California, implying that they migrate southeast to mainland Mexico for the winter.

This species has shown to be generally prone to vagrancy. The northeasternmost vagrant involved a male caught and banded at Long Point, north shore of Lake Erie in Ontario, Canada, on May 19, 1992 (Boardman 1992). It showed characteristics of the subspecies *cucullatus*. As stated above it has been reported from Kansas during August, but this may be in error. There is one record from Louisiana and two old specimen records (19th Century) from Cuba amazingly; one occurred in April. These Cuban records have been questioned and may pertain to escaped cagebirds. In the Pacific states, Hooded Orioles regularly stray north of their breeding grounds. There is one Washington record from late April 1992 from the coast. Hooded Oriole is now basically annual in Oregon, particularly in spring — there are many records. There are now four British Columbia reports. An adult male was observed in Sidney on May 6–7, 1996, and one was at the Esquimalt lagoon on July 19, 1997, both on Vancouver Island. There was a report of a male from the Vancouver area on September 19, 1997 and a male has been observed wintering at a feeder between November 19, 1997 at least to Christmas at Terrace in the interior (D. Cannings pers. comm.). In Mexico, there is an April 9, 1994 record from Isla Guadalupe, Baja California, and a February sighting from Isla Socorro (Howell and Webb 1995).

Once the commonest oriole in the lower Rio Grande valley, it has declined recently. This contrasts with the situation in California where Hooded Oriole has been slowly spreading northward as a breeder, likely as a result of the planting of *Washingtonia* palms. This spread appears to be ongoing, but proceeding rather slowly. This species has hybridised with Orchard Oriole in Mexico and with Bullock's Oriole in captivity.

MOULT First pre-basic moult is incomplete, changing the inner 3–5 secondaries and outer 4–7 primaries and a variable number of rectrices; this moult may begin while on the breeding grounds during July–September, but most moulting occurs in the non-breeding range (Pyle 1997b). Note that the first pre-basic of *nelsoni* averages more extensive than that of *cucullatus* and *sennetti* (Pyle 1997a, 1997b). The retained wings and tail become increasingly more worn, which helps in ageing birds during their first summer. In the late summer, both hatch year (hatched that summer) and second year (hatched the previous summer) birds will be present, both potentially in first basic plumage; however, second year birds will be much more worn. definitive pre-basic moults are complete and occur August–October, while in the winter quarters. Pre-alternate moults are lacking except in some young males which may moult some of the body feathers.

MEASUREMENTS ***I. c. cucullatus***. Males: (n=17) wing 84.6 (83–88); tail 92.9 (89.7–102); (n=13) culmen 19.5 (18–20.8); tarsus 22.6 (21.8–24). Females: (n=10) wing 80.1 (77–84); tail 87.8 (80–91); culmen 18.1 (16–19); tarsus 22.4 (21–23.1).

I. c. nelsoni. Males: (n=10) wing 88.4 (86.4–90.4); tail 89.9 (81.8–96); culmen 21.6 (20.8–22.1); tarsus 22.4 (21.6–23.4). Females: (n=10) wing 82.5 (79–88); tail 84.2 (78–90); culmen 20.2 (19–20.8); tarsus 21.9 (21.6–22.4).

I. c. trochiloides. Males: (n=16) wing 87.1 (84–90); tail 93.4 (85–101); culmen 21.8 (20–23); tarsus 22.3 (21–24). Females: (n=10) wing 81.2 (79–83); tail 85.1 (80–91); culmen 20.6 (19–21); tarsus 22.1 (21–24).

I. c. sennetti. Males: (n=13) wing 84.0 (80.5–87.0); tail 94.1 (87.9–99.1); culmen 19.7 (19.0–20.6); tarsus 22.3 (21.8–23.1). Females (n=10): wing 78.0 (76–81.3); tail 86.5 (80–93); culmen 18.7 (17–20); tarsus 21.6 (20–22.6).

I. c. igneus. Males: (n=12) wing 86.4 (81–90); tail 92.8 (89–96); culmen 19.8 (17–21.6); tarsus 22.4 (20.8–23.4). Females: (n=10) wing 77.7 (74–82); tail 82.4 (75–94); culmen 19.3 (17–20.3); tarsus 22.1 (20–24).

NOTES Hooded Oriole and Orange Oriole are outwardly quite similar; however, based on DNA sequence data the two are not closely related (K. Omland, S. M. Lanyon pers. comm.). Furthermore, this research has found that Hooded Oriole is perhaps most closely related to the mainland Black-cowled, Orchard and Black-cowled Orioles, and their relatives.

REFERENCES BBL Data [movements], Bailey 1910 [nesting], Barbour 1943[movements], Bent 1958 [behaviour, nesting], Boardman 1992 [movements], Cruden and Hermann-Parker 1977 [behaviour], Ewan 1944 [nesting], Grinell 1927 [geographic variation, movements, measurements], Grinnell 1944 [nesting], Howell and Webb 1995 [movements], Huey 1944 [nesting], Kaufman 1983 [identification], Pyle 1994 [movements], Ridgway 1885 [geographic variation], Ridgway 1902 [measurements], Sutton 1948 [geographic variation, measurements], Zimmer 1985.

37 TROUPIAL *Icterus icterus* (Linnaeus) 1766 — Plate 18

A large, well-known orange oriole of northern South America and parts of the Caribbean where it has been introduced. It is a common cagebird. Troupial is the national bird of Venezuela.

IDENTIFICATION The three troupials share the characteristics of being largely orange and black with a ragged lower edge to the black bib, due to pointed throat feathers, with yellow eyes surrounded by naked, blue skin. Troupial is characterized by a large white wing patch made up of a white shoulder and white secondaries and tertials. This is the only troupial species with white shoulders. Troupial is allopatric with the other troupials, but the black back and entire black hood easily differentiates it from Orange-backed Troupial (39). Campo Troupial (38) is extremely similar, but is most geographically distant from Troupial. Campo Troupial has orange on the greater coverts rather than white and has a very reduced eye patch. Most of the orioles sympatric with Troupial are yellow, and have yellow crowns as well as dark eyes. Most similar is adult male Baltimore Oriole (40) which has a full black hood, a black back, and orange underparts and rump. However, Baltimore Oriole has dark eyes, a straight lower edge to the black hood, the black colouration continuous with the crown, much less extensive white on the wings and orange on the tail.

VOICE Song: Sings a loud repetitive song, *cheer, taaw-cheer...*, or *cheer-toe, cheer-toe...* and other phrases with a similar structure. The individual notes are hoarse whistles, with distinct spaces between notes. Usually, Troupial songs are repetitions of two notes, one lower pitched and other higher pitched. However, Skutch (1969) describes the song as *come right heere, come right heere* or *come heere come heere* or *here come here come*. Voous (1983) describes the song in the Netherlands Antilles as being composed of 2–3 melodious syllables, *tee-oo, tee-oo* or *troo-pee-oo, troo-pee-oo*. Both sexes sing, but the male is the primary singer. The female may answer his song at times. Sings even during the middle of the day.

Calls: Gives mellow whistles as well as nasal sounds.

DESCRIPTION A large oriole, with a long tail which is slightly graduated. The bill is long and chisel-like, with a straight culmen and gonys. The wings are rounded. Wing Formula: P9 < P8 > P7 > P6; P9 ≈ P5; P6–P8 emarginate.

Adult (Definitive Basic): Bill black, with grey basal third to the lower mandible. The head is entirely black, including the neck, nape and upper breast. The eye is yellow or orange, with a large blue patch of naked skin surrounding it. The breast feathers are sharply-pointed, the 'hood' ends in a ragged line on the breast, without a demarcation between the black head and the rest of the underparts. The black head and back are separated by an orange band across the lower neck and upper back. The back and scapulars are black. The lower back, rump and uppertail-coverts are orange-yellow. The breast is black and the belly and flanks are orange, down to the undertail-coverts. The wings are black with a longitudinal, white wing patch and orange shoulders. This patch is created by orange lesser coverts, white median coverts and white outer greater coverts showing white bases, and progressively more white on the outer half of the inner greater coverts, the innermost showing a completely white outer half. The tertials and secondaries are black with a white outer edge, continuing the white patch far back on the wing. The primaries are black with a crisp white edge, creating a pale panel along the apical end of the folded wing. The wing-linings are yellow-orange. The tail is black with small white tips to the outer two pairs, and some individuals show a white patch on the inner half of the base of the outermost tail feather.

Immature (First Basic): Similar to adult, but shows a variable number of retained primaries and secondaries. Usually the juvenal secondaries are retained as are the inner primaries, while the outer three or more primaries are blacker and newer. The contrast between the retained and new feathers is most obvious between the black tertials and the brownish secondaries and may be visible in the field given a good view.

Juvenile (Juvenal): Similar to adult in general pattern but the orange colour is duller (more yellowish) and the black colours are tinged with brown. The bare eye patch is present, but is a grey-blue rather than bright blue. Juvenile *metae* lack a wing patch on the coverts, having the white patch restricted to the secondaries. The troupials are the only orioles that have a juvenile plumage so like the adult plumage, amongst other characteristics, suggesting that they are not closely related to the other orioles.

GEOGRAPHIC VARIATION Three subspecies are recognised.

I. i. icterus inhabits the Llanos of extreme NE Colombia (Vichada) and Venezuela from the SW (Amazonas) north to the coast from Carabobo east to Sucre. It is described above.

I. i. ridgwayi is found from the extreme NE of Colombia (Guajira and César) east to NW Venezuela as far east as Falcón and Lara. It is also resident on the islands of Aruba and Curaçao (Netherlands Antilles), as well as Isla Margarita (Venezuela). It has been introduced to Puerto Rico, St Thomas and Bonaire. This form is similar to *icterus* in colour pattern, but is larger, has a proportionately longer and stouter bill, and has stronger legs. The birds from Aruba are darker orange in colour than those from Curaçao (Voous 1983).

I. i. metae is found in the state of Apure in SW Venezuela, along the Meta river and in extreme NW Colombia in the province of Arauca. It differs from the other two subspecies in having orange extending onto the nape; however, the extent of this orange colour is variable with some individuals showing as much black on the nape as typical *icterus/ridgwayi*. The black on the back is less extensive on *metae*, because the orange of the rump extends further forward and the orange of the neck extends further down the back. The wing patch is also smaller, and the greater coverts are entirely black, dividing the wing patch into two sections: a white patch on the lesser and median coverts and a second on the secondaries and tertials.

HABITAT Lives in open savanna, llanos or dry scrub. It also frequents arid woodlands and gallery forest. Generally its preferred habitats are dry, and it avoids areas of heavy rainfall. Not usually found in mangroves. Troupials visit fruit plantations and can become a pest; they also inhabit gardens particularly those where fruit and flowers are available. In the Netherlands Antilles, it avoids the dry parts of Curaçao, but frequents very dry habitats on Aruba provided that larger cacti are present (Voous 1983).

BEHAVIOUR Forages as pairs or in family groups. Troupials are largely insectivorous, at least during the breeding season, but also feed on fruit. When the fruit of giant cacti is in season, they may feed almost entirely on this fruit. Most of their foraging is in trees or low bushes, but they will come to the ground to retrieve fallen fruit. In Puerto Rico, Troupials forage mainly in the herb layer (low down) and almost strictly on fruit (Post 1981a). Voous (1983) also notes that in the Netherlands Antilles, Troupial is largely frugivorous feeding on the following: mango (*Mangifera indica*), kenepa, sapodilla (*Achras zapota*), soursop (*Ammona muricata*), papaya (*Carica papaya*), dates (*Phonix dactylifera*) and *Malpighia* cherries. However, he also notes that they forage on insects as well as being adept nest robbers, eating bird eggs and nestlings. Visits bird feeders in gardens. The mating system is monogamous and it appears that pairs may remain together throughout the year. Apart from stealing Plain Thornbird (*Phacellodomus rufifrons*) nests to breed in, they also use them as sleeping chambers at night. Each member of the pair uses a separate nest to sleep in (Thomas 1983). Favourite sleeping chambers may eventually become the nest where the eggs are laid. Once the young fledge they will also sleep in a thornbird nest. The male does most of the singing, with the female joining in every so often. Singing primarily serves a territorial function and is most intense during the early morning. During singing, the hackles of the throat are prominently fluffed out, giving them a most distinctive appearance. Little is known about any other displays that these birds give.

NESTING The nesting season in South America seems to be March–September. In Venezuela, the main nesting season is May–June. In the Netherlands Antilles, Troupial has been recorded breeding in January, February, August and September but it is thought to breed throughout the year

(Voous 1983). Like all the troupials, this species pirates the nests of other species. It only uses covered nests, which are commonly the stick nests of Plain Thornbird or the nests of Great Kiskadee (*Pitangus sulphuratus*) when thornbird nests are not available. In older accounts, it is mentioned to also build a hanging nest of its own, but nest building has not been observed, and these were previous to the discovery that troupials are nest pirates. In all likelihood these basket nests probably belonged to orioles, caciques or oropendolas. Troupials may alter nests by removing sticks and enlarging the entrance, and they do tend to provide their own lining. Usually troupials take the lowest of a series of thornbird nests (Thomas 1983). Thornbirds may defend their nests against troupials by scolding; however, they may be violently attacked by the troupials (Thomas 1983). The clutch size is usually 3 eggs which are white, or pale pink, with dark spots and blotches concentrated on the broader end of the egg. The female incubates for 15 or 16 days. Nestlings are often heavily parasitised by bot flies (*Philornis* sp.) as is common of many tropical birds. Nestlings need 21–23 days to fledge from the nest, at which time they travel with the adults as a family party. Both of the parents share the task of feeding the young, both in and out of the nest. Troupial is a known host for Shiny Cowbird (102) in Puerto Rico.

DISTRIBUTION AND STATUS Uncommon to common, a bird of the lowlands it is seldom observed as high as 500 m. Ranges from NE Colombia (César and Guajira as well as Vichada) east through most of Venezuela, except the far SW, to the mouth of the Orinoco. Also found on the islands of Aruba, and Curaçao (Netherland Antilles) as well as Isla Margarita (Venezuela). It has been introduced to Puerto Rico and St Thomas, US Virgin Islands. Also introduced to Bonaire, Netherlands Antilles, in 1973; breeding was confirmed in 1974 and it has since become common (Voous 1983). The older literature mentions that Troupial was at one time introduced to Jamaica; if this is correct they had disappeared by the late 1890s (Scott 1893). Bond (1928a) noted that previous researchers had noted that Troupial had been introduced to Dominica, but he could not find it and it no longer occurs there.

MOVEMENTS Sedentary, no evidence of seasonal movements. Records of birds found out of range are more likely to be escaped cagebirds; these are often yellowish or pale orange. One was reported from St John, Virgin Islands, which may have been a vagrant from St Thomas, or perhaps an escapee (Robertson 1962). A report from Jamaica (Jeffrey-Smith 1972) is almost surely of an escapee.

MOULT Adults undergo one annual complete moult, the definitive pre-basic and it is timed to occur just after breeding. In the Llanos of Venezuela, *I.i.icterus* undergoes its definitive pre-basic moult during November. Pre-alternate moults are lacking. The juvenal plumage is replaced through the incomplete first pre-basic moult. All of the body plumage is replaced as well as the coverts and tertials. In addition, a variable number of primaries, usually the outer ones, and some of the secondaries are changed. In N Colombia (Magdalena), the first pre-basic moult occurs shortly before March. One specimen from El Sobrero, Venezuela, in the Smithsonian Museum appears to show a centripetal (towards center) tail moult.

MEASUREMENTS *I. i. icterus* Males: (n=7) wing 107.1 (106–109); tail 95 (91–100); culmen 29.4 (26–31); (n=2) tarsus 30,32. Females: (n=9) wing 103.2 (95–117); tail 90.9 (88–98); culmen 27.5 (25–31); (n=4) 29.8 (28–31).

I. i. ridgwayi. Males: (n=12) wing 119.4 (106–127); tail 109.0 (101–115); (n=10) culmen 32.9 (30–35); (n=12) tarsus 32.8 (30–35). Females: (n=9) wing 110.2 (105–114); tail 99.3 (95–103); culmen 30.4 (29–33); tarsus 31.9 (29–34).

I. i. metae. Males: (n=7) wing 105.8 (103–110); tail 91.2 (80–94); culmen 31.2 (30–33.5). Females: (n=6) wing 97.8 (93–101); tail 88.7 (84–97); culmen 29.2 (29–30).

NOTES The three troupials are often considered part of one variable species, albeit with some reservations (Ridgely and Tudor 1989). However, other authors conclude that the three are best separated as different species (Hilty and Brown 1986, Voous 1983, Sibley and Monroe 1990). *Metae* approaches Orange-backed Troupial in many respects, but is clearly within the Troupial (in the strict sense) species. In addition, *metae* is not found within at least one thousand km of Orange-backed Troupial; thus its intermediate nature in morphology is most intriguing. Campo Troupial is most similar to Troupial, but is the more distant taxa geographically than the more divergent Orange-backed Oriole. The differences in structure, plumage, ecology and reported dissimilarities in song suggest that these should be treated as three different species. The three are nonetheless closely related and may in fact be distantly related to the rest of the orioles; this is implied by their nesting behaviour, morphology and juvenile plumage as well as divergence of their mitochondrial DNA. However, a newer study of DNA sequence data shows that the troupials do belong with the rest of the orioles (K. Omland, S. M. Lanyon pers. comm.). This study suggests that there are at least two distinct species within the troupial complex, but not all forms were sequenced. The entire Icterid family obtains its name from the genus and species name of this oriole.

REFERENCES Freeman and Zink 1995 [systematics], Hardy *et al.* 1998 [voice], Hartert 1902 [geographic variation], Hilty and Brown 1986 [voice, habitat, nesting], Phelps and Aveledo 1966 [geographic distribution, measurements], Post *et al.* 1990 [brood parasitism], Skutch 1969 [voice, description, nesting, behaviour, notes], Thomas 1983 [nesting].

38 CAMPO TROUPIAL *Icterus jamacaii* (Gmelin) 1788 — Plate 18

A black-backed troupial of the xeric woodlands of eastern Brazil.

IDENTIFICATION The three troupials share the characteristics of being largely orange and black with a ragged lower edge to the black bib, and yellow eyes surrounded by naked, blue skin. Campo Troupial is very similar to, but is most geographically distant, from Troupial (37). Range alone is a criterion for identifying the species of troupial. Campo Troupial has orange rather than white on the greater coverts, therefore the white wing patch is restricted to the secondaries on this species. Campo Troupials have a reduced eye patch which may be lacking on Troupial altogether. Troupial, on the other hand, has a white wing patch extending from the greater coverts to the secondaries; it also has a large bare eye patch. Furthermore, the bill of Campo Troupial is finer and straighter than that of Troupial. Campo Troupial perhaps comes into contact with Orange-backed Troupial (39) in central Brazil, but this has not yet been documented. Orange-backed Troupial is the most divergent of the group, showing an orange back (or nearly so) as well as an orange crown and nape. The bare eye patch averages larger than on Campo Troupial. The white on the wings is restricted to the secondaries on Orange-backed Troupial, which is similar to that of Campo Troupial. However, the white wing patch of Campo Troupial is significantly larger than that of Orange-backed Troupial. In addition, Campo Troupial is larger and longer-billed than Orange-backed Troupial. No other largely orange orioles are found within Campo Troupial's range. Epaulet Oriole (25) is mainly black with yellow or chestnut shoulders.

VOICE Song: The song is sweet and composed of simple, whistled phrases. It is said to be similar in general structure to that of other troupials, but is significantly longer in duration than the song of Orange-backed Troupial. It may closely resemble the song of the sympatric Long-billed Wren (*Thryothorus longirostris*) (Sick 1993). Individuals kept in cages will apparently learn different tunes from their owners; one especially gifted singer was taught the Brazilian national anthem!

Calls: Utters a *crik* note. Will also utter harsh notes and imitate noises.

DESCRIPTION A large oriole with a moderately thick bill and a moderately graduated, rounded, tail. The bill is long (equal to the head length) and pointed, with a straight culmen and gonys. Wing formula: P9 < P8 < P7 > P6 ≈ 5; P9 ≈ P3.

Adult (Definitive Basic): The black bill displays a grey basal one-third to one-half of the lower mandible. The eyes are orange-yellow, and the blue-grey, naked, skin patch around the eyes typical of troupials is almost lacking in this species. It has a black hood and breast, the hood extends to the nape and as far as the pectoral area. The feathers of the throat and breast are somewhat pointed, but not strongly so. This creates a ragged lower edge to the hood. On the upperparts, the lower neck, upper back, lower back, rump and uppertail-coverts are orange in contrast to the black mantle and scapulars. The wings are black with orange lesser and median coverts, forming an orange epaulet. Then outer greater coverts are entirely black. There is a wedge of white on the folded secondaries, comprised of white secondary edges that narrow towards the tip of the feathers. A few of the inner greater coverts sometimes show white edges, and extend the white wedge forward on the wing. However, the orange epaulet and white on the wing are separated by black. The wing-linings are yellow-orange. The underparts are orange, from the lower breast to the undertail-coverts, palest on the vent and brightest on the breast. Its tail is entirely black and unpatterned. The legs and feet are dark grey.

Immature (First Basic): Like adult, but with contrasting brownish secondaries and primaries. The pattern of white and orange on the wing is as on the adult. However, the median coverts are often paler orange than the lesser coverts. The black greater coverts contrast with the brownish primaries and secondaries. The tail may retain brown (juvenal) rectrices which contrast with the blackish, newer feathers.

Juvenile (Juvenal): Similar to adult in pattern, but black is replaced by a duller brownish-black. Similarly, the orange of the underparts is a duller yellow-orange. It appears that juveniles have dark eyes.

GEOGRAPHIC VARIATION Monotypic.

HABITAT A species of xeric habitats. It lives in dry savanna and deciduous woodlands as well as other semi-open areas of the caatinga zone of Brazil. It is fond of shrubland with tall cacti. In addition, it inhabits agricultural clearings and it may be found along forest edge.

BEHAVIOUR Not a shy bird, often seen sitting high in a prominent situation. It is most often observed singly or in pairs, but after breeding, family groups are the norm. Both sexes sing, and there is a great deal of singing activity during the breeding season. Singing is often accompanied by a strange body posture than can only be interpreted as a display. The tail is raised and the neck is extended as the bill is pointed downward, prominently displaying pointed neck feathers. In flight this is a noisy species, making an audible swish similar to that of the larger caciques and oropendolas.

NESTING Like the other troupials this species is also a nest pirate. It does not make a nest of its own, but uses an old nest made by another species, or takes over the nest from its rightful owners. It prefers the domed, stick nests of various species of Furnariidae (Ovenbirds), particularly Firewood-Gatherer (*Anumbius annumbi*) and

Rufous Cacholote (*Pseudoseisura cristata*); it may also use the nests of Great Kiskadee (*Pitangus sulphuratus*) and Rufous Hornero (*Furnarius rufus*) (Pinto 1967). It breeds during February. This troupial is known to have been parasitised by Shiny Cowbird (102) (Friedmann and Kiff 1985).

DISTRIBUTION AND STATUS. Endemic to Brazil. A species of lowlands, found below 500 m. It ranges from Maranhão, Piauí and Ceará in the north, south to 18°S in Bahia and Minas Gerais. Also inhabits Rio Grande do Norte, Paraíba, Pernambuco, Alagoas, Sergipe and N Espírito Santo. Found in E Goiás but in the W of that state it is replaced by Orange-backed Troupial (Hellmayr 1929).

MOVEMENTS Sedentary, no evidence of migration or seasonal movements.

MOULT The only adult moult of the year is the complete definitive pre-basic moult, which occurs April–early May. There is no evidence for the existence of pre-alternate moults in Campo Troupial. The juvenal plumage is lost through the partial (to incomplete?) first pre-basic moult. The body plumage, as well as the tertials and wing-coverts are replaced along with a variable number of tail feathers. Most or all of the primaries and secondaries are retained. The first pre-basic moult occurs during April–May, the same period as the adult moult.

MEASUREMENTS Both sexes: (n=6) wing 106.5 (98–113); tail 108.3 (95–116); culmen 25.8 (24–28); tarsus 27.7 (23–31). Males: (n=2) wing 105, 107; tail 105, 112; culmen 27, 31; tarsus 29, 31. Female: (n=1) wing 107; tail 104; culmen 27; tarsus 29.

NOTES Introduced to the Brazilian island of Itamaracá in Pernambuco, roughly around 1928

REFERENCES Frisch and Frisch 1964 [voice], Pinto 1967 [nesting], Sick 1993 [voice, nesting, notes].

39 ORANGE-BACKED TROUPIAL *Icterus croconotus* Plate 18
(Wagler) 1829

An uncommon, largely orange, forest oriole of Amazonia and Mato Grosso.

IDENTIFICATION A largely orange oriole with yellow eyes, a black throat bib, orange epaulets and a white wing patch. Orange-backed Troupial is the only largely orange oriole found in its range, the other troupial species are allopatric. The other troupial species have black on the back and crown as well as wing patterns with extensive white. The naked, blue eye patch is larger in this species than in Campo Troupial (38) but smaller than in Troupial (37), and the throat feathers are less pointed than in both of these species. Yellow Oriole (27) may overlap with Orange-backed Troupial in Guyana, the former species is largely yellow, not orange like Troupial, and it has dark, not yellow eyes. In addition, Orange-backed Troupial has the black on the face extending to the forehead and the entirety of the auriculars, unlike the oriole which has a yellow forehead and yellow auriculars. Yellow Oriole also has a white lower wingbar and white edges to the tertials, these areas are entirely black on Orange-backed Troupial. The much larger Oriole Blackbird (53) has a naked blackish eye patch, but has dark eyes, is largely yellow and has a black back.

VOICE Song: A monotonously repeated series of whistles. Commonly it is broken up into series of 3 notes *tiew tiew tee?, tiew tiew tee?...*, the first two notes descending and the third slightly higher and doubtful, can be said to say *Its Tues-DAY Its Tues-DAY....* This 3-note song appears to be the primary song, and almost identical versions are heard from Ecuador to Peru, suggesting that there is little geographic variation in this one song type. Also given as two notes *tiew tweee?, tiew tweee?, tiew tweee?...*, the second note higher than the first. Also described from Guyana as a slow, rich, whistled *who-hu-chu-who-hu-chee-oo-oo*, the third note higher (Snyder 1966). Pairs frequently duet. Both individuals perform the same song and it is frequently impossible to tell whether or not a certain note is made by two, rather than one bird, due to the close synchronization within the pair.

Calls: Unknown.

DESCRIPTION A medium-sized to large oriole with a stout bill; the culmen is evenly curved while the gonys is straight. The wings are rounded. The tail is graduated, showing a difference of nearly 30 mm between the outer and innermost tail feathers. Tends to perch rather horizontally. Its flight is sluggish, and the bird appears somewhat humpbacked due to the drooping tail and head. Wing Formula: P9 < P8 < P7 > P6 > P5; P9 ≈ P3; P8–P5 weakly emarginate. The form *croconotus* is described.

Adult (Definitive Basic): Female is similar to male in plumage, but is slightly smaller. Bill black, with the basal third of the lower mandible grey. The eyes are yellow or orange and there is a small area of blue skin below and behind the eye. The forehead, lores, auriculars, throat and upper breast are black. The throat feathers are only slightly pointed; and the lower edge of the black hood is ragged, rather than a neat straight line. The crown, nape and neck sides are a vivid orange. The lower back and rump are a paler orange yellow, but the rump may be as bright as the mantle. The underparts posterior to the black upper breast are orange, deepest in colour on the breast, becoming paler towards the vent. The scapulars are black, contrasting with the orange back. The wings are largely black, but have patches of white and orange. The lesser and median coverts are yellow-orange, while the greater coverts are black. The squarish wing patch is formed by wide, white edges to the tertials, except the innermost, and the inner secondaries. The rest of the secondaries and the primaries are black and lack paler edges. The wing-linings are yellow. The tail is entirely black and noticeably graduated.

Immature (First Basic): Very similar to adult, but retains the brownish primaries, secondaries and tail from the juvenal plumage. Although sometimes the outer primaries are moulted during the first pre-basic moult, such that they are blackish and new, contrasting with browner and more worn inner primaries and secondaries. The tertials and coverts are black new and contrasting with the rest of the wing in most cases.

Juvenile (Juvenal): Similar to the adults but duller. The orange of the body is not as vivid and all black colouration is replaced by brownish-black. Orange on the epaulet is lacking on the juvenile. It appears that the eye colour is dark on juveniles.

GEOGRAPHIC VARIATION Two subspecies are recognized.

I. i. croconotus is found in the north of the distribution, including all of the Amazonian range south to SE Peru. It is described above.

I. i. strictifrons ranges from Bolivia, S Mato Grosso in Brazil, and N Argentina. It is similar to *croconotus*, but has more extensive white on the wings and has less extensive black on the forehead. A form from the Paraguayan chaco (*paraguayae*) was previously recognized. These specimens resemble *strictifrons* but sometimes have a thin black band on the lower back and less orange on the lesser coverts; apparently they represent extreme variants of *strictifrons*. The amount of black on the back is quite variable and many *strictifrons* lack this altogether. However, even black-backed *strictifrons* lack black on the crown.

HABITAT Inhabits the edge of lowland tropical forest, particularly along rivers or the shores of cochas (Amazonian lakes). It needs either edge or more open areas that are naturally afforded by rivers, and the like, but as forests are degraded Orange-backed Troupial is also being found along roads and man-made openings. However, they are not found in areas where forest is absent. In the extreme south of the range *strictifrons* is found in drier and more open habitats. It may be found in the taller and moister chaco woodlands, as well as in cerrado woodlands in the Mato Grosso region, and gallery forests and forest edge.

BEHAVIOUR Orange-backed Troupial is most regularly observed singly or in pairs. Orange-backed Troupials forage relatively high in the canopy, often near openings or edges of cochas (lakes) and rivers. Unless they are being vocal, they are inconspicuous and may be missed by the casual observer. They eat a great deal of fruit and will nectar at large flowers. While searching for insect prey, dead leaves are searched more frequently than green leaves. Orange-backed Troupials have been observed plucking *Tabebuia* sp. flowers and slitting them to obtain nectar (Dubs 1992). Pairs may counter-sing (duet) from a high perch. While singing, the crown feathers may be raised, making the head appear rounded.

NESTING Orange-backed Troupial does not build its own nest as far as is known. Instead, it uses an old nest of another species and refurbishes it for its own purposes. The most commonly reported nest used is that of Yellow-rumped Cacique (13). More than one pair of Orange-backed Troupials may nest within one Yellow-rumped Cacique colony; however, each pair tends to choose a nest in a separate cluster of cacique nests. The nesting season appears to commence between September and November in E Ecuador and N Bolivia, at a time when the caciques have finished nesting. In eastern Peru, Orange-backed Troupials nest at the same time as the caciques (July–February). In this case, the troupials actively pirate the nests of the caciques by harassing nesting females. The pair of troupials will attack females at the nest, and will peck and kill nestlings and remove eggs. They typically oust several females and then choose a nest within this, now empty, section of the cacique colony. The nest is afforded some protection from nest predators by being surrounded by empty nests, predators often give up after searching only a few nests in the colony (Pearson 1974, Robinson 1985). The southern form *strictifrons* nests in mid November in Brazil (Naumburg 1930). Orange-backed Troupial is a known host for Shiny Cowbird (102), but it has not been observed to fledge cowbird young (Friedmann and Kiff 1985).

DISTRIBUTION AND STATUS Uncommon to locally common, lowlands to 550 m. Found from SW Colombia (Putumayo, Amazonas) and E Ecuador (Napo, Pastaza) east largely south of the Amazon River with a tongue of distribution extending north through E Roraima, W Pará and E Amazonas to S Guyana; south along the lowlands E of the Andes of E Peru, N and E Bolivia to W Paraguay to NE Argentina (E Salta, Formosa and N Chaco); north along W Mato Grosso, S Goiás, and Amazonas E to the Rio Tapajós in Brazil.

MOVEMENTS Sedentary, no evidence of migrations or seasonal movements.

MOULT Adults undergo one complete moult a year, the definitive pre-basic moult which occurs subsequent to the breeding season. Specimens from Leticia, Colombia, and Beni, Bolivia, suggest that the main moulting period is in May. Other individuals from Beni, Bolivia, were moulting in March. In the Mato Grosso of Brazil, moult occurs during June. Timing may vary depending on geography, however. There is no evidence to suggest that adults have a pre-alternate moult. The juvenal plumage is lost through the first pre-basic moult (similar in timing to the definitive pre-basic?) where the body feathers, coverts and tertials are replaced. The retention of the primaries and tail feathers is variable. On some first basic individuals, the primaries, secondaries, tertials and rectrices are retained from the juvenal. In contrast, others may moult some of the tail and outer primaries such that there are older, worn wing and tail feathers that contrast with newer and blacker feathers. If any flight feathers are moulted it tends to be the outer primaries and some of the inner tail feathers. The possession of brown and worn primaries as well as fresher black feathers is diagnostic for the first basic plumage.

MEASUREMENTS ***I. c. croconotus***. Males: (n=7) wing 104.9 (98–111); tail 103.7 (95–111); culmen 25.4 (23.6–27); tarsus 29.2 (27–32). Females: (n=7) wing 94.4 (90–97); tail 97 (92–101); culmen 23.3 (22–24); tarsus 26.5 (24.8–27).

I. c. strictifrons. Males: (n=10) wing 102.9 (97–110); tail 103.0 (92–115); culmen 24.8 (23–26.5); tarsus 28.7 (25–33). Females: (n=9) wing 97.6 (88.5–103); tail 97.0 (85.5–109); culmen 23.9 (22–26); tarsus 26.5 (26–28).

NOTES The three forms of Troupial are commonly considered one variable species; however, plumage and vocal differences between the three forms are significant. We have chosen to regard the three as different species (see Troupial).

REFERENCES Brodkorb 1937 [measurements, *strictifrons*], Hardy *et al.* 1998 [voice], Moore 1996 [voice], Pearson 1974 [nesting], Robinson 1985 [behaviour, nesting], Todd 1924b [geographic variation].

40 BALTIMORE ORIOLE *Icterus galbula* (Linnaeus) 1758 — Plate 10

The commonest oriole of eastern North America, and the only orange oriole in most of its breeding range.

IDENTIFICATION The adult male is not typically an identification problem, but young and females are trickier. Adult Baltimore Oriole is the only oriole with a fully black hood and back that has orange on the tail. It is similar in overall pattern to male Orchard Oriole (43) which overlaps broadly with it in distribution; however, Orchard Oriole is not orange but chestnut. In addition, the Orchard Oriole is noticeably smaller and shorter-tailed. Adult female and 1st-summer male Baltimore Orioles are extremely variable but show various amounts of orange on the underparts, and black on the throat, hood or upperparts, as well as two bold white wingbars. The presence of orange is usually enough to differentiate these birds from female or immature Orchard Orioles.

Female, both adult and immature, Bullock's Orioles (41) are similar to immature female Baltimore Orioles, as the latter are greyish above and yellowish on the throat and breast. Baltimore Orioles tend to have more extensive yellow on the underparts and this has an orange wash to it. The yellow blends into the greyer belly gradually. On the Bullock's Oriole this change is more abrupt – some late fall immature Baltimore Orioles may approach the Bullock's pattern however (P. Lehman pers. comm.). Also note that female and immature Bullock's Orioles have more extensive yellow on the auriculars and supercilium, which contrasts with a darker eyeline. These areas are a darker olive-grey, contrasting with the yellow throat on Baltimore Oriole. On the upperparts, Baltimore Oriole usually shows darker centers to the mantle feathers, whereas those of Bullock's Oriole are nearly or entirely unicolored. In addition, the rump of Baltimore Oriole has a yellowish wash, while Bullock's Oriole shows a greyish rump; this feature needs more study and may be diagnostic for Bullock's Oriole. The wingbars on Baltimore Oriole are separate, while they are often connected by pale greater covert edges in Bullock's Oriole, forming a pale wing panel. A new and exciting field mark has been described concerning the edging of the median coverts (upper wingbar). It appears that Bullock's Orioles tend to have a pointed central dark region on these coverts while Baltimore Orioles tend to have a square-ended central dark region to the median coverts. In effect what one sees is a straight upper border to the upper wingbar on Baltimore Orioles but a jagged upper border in Bullock's Orioles (Lee and Birch In Press). This mark has not been widely field tested yet, and how age and the different extent of the first prebasic moults of the two species affects its usefulness is unclear. It appears that the undertail-coverts of young Baltimore Orioles are always yellow, whereas they are often grey, sometimes yellow, on Bullock's Orioles. A minority of very dull, fall female (most likely first basic) Baltimore Orioles are extremely similar to female Bullock's Orioles with restricted yellow on the underparts. With these individuals, concentrate on the face pattern, the colour of yellow on the breast, throat and face, the back pattern and the rump and undertail colours.

VOICE **Song**: A pleasant series of sweet, flute-like whistles. The song is short and abrupt (rarely with as many as 10 notes) and most males sing a personalised version, with few sharing an identical song. However, the frequency range of different male's songs is similar, averaging 2.1 kHz. A study in Michigan discovered that songs were composed of between 4–20 individual notes, with an average of 8.5, and that individual males sang several different forms of their song. The average song lasts 1.4 seconds, but they range from less than 1 second to nearly 3 seconds in length (Beletsky 1982b). When especially excited, males will sing in flight. The frequency of song delivery decreases after the male finds a mate. Songs given late in the season, after most other males have stopped singing, are invariably given by unmated, usually immature individuals. Females rarely sing; their song is simpler and not as melodic as the male's but is similar in duration (Beletsky 1982a). 'Female' song is sometimes given immediately after 'male' song, in a duet-like arrangement, which suggests that at least one function of female song is in pair bond maintenance (Beletsky 1982a).

Calls: A common call is a chatter, which is given by both sexes. Chattering appears to have an aggressive message and is given during conflicts, chases and when humans intrude into oriole territories. However, chatters are also heard from males approaching females and when bringing food to the nest. Different males give chatters of slightly different lengths. Female chatters are similar to those of males. Also frequently utters slurred whistles, roughly in the same frequency range as the song. These whistles and single note calls are rather variable.

DESCRIPTION A chunky oriole, with a straight and sharply pointed bill. Both the culmen and gonys are straight, showing no curve whatsoever (Figure 29.1). Wing Formula: P9 ≈ P8 ≈ P7 > P6; P9 ≈ P8; P8–P6 emarginate.

Adult Male Breeding (Definitive Basic, worn): Bill black with the lower mandible, and the cutting edge of the upper mandible, entirely blue-grey. Eyes dark brown. The head, down to the upper breast is black, as are the back and scapulars, all without any iridescence. The bases of the head and neck feathers are orange, and may be visible while the bird preens or if ruffled by the wind. The lower back and rump to the tail-coverts is orange. Note that the exact tone of orange varies from the most vivid orange to a yellowish-orange; these differences between individuals are probably related to diet. The black of the head cuts off sharply at the breast, usually lower down along the mid-line than at the sides of the breast. The rest of the underparts to the undertail-coverts are a vivid orange, brightest on the chest. The wings are black with a bold orange to orange-yellow shoulder patch created by the marginal, lesser and median coverts of that colour. The greater coverts are broadly-tipped white on the outer half, making a bold white lower wingbar. The primaries and secondaries are crisply-edged white while the tertial edges are wider and more noticeable, but only on the outer half. The tail is black and yellow, and the central two feathers are completely black (sometimes with a small yellow tip) while the outer ones are black at the base with the terminal half yellow on the outermost pair, becoming progressively less yellow inwards. Legs and feet blue-grey.

Adult Male Non-breeding (Definitive Basic, fresh): As the breeding male, but the black of the head and back is usually thinly tipped dull orange. In addition, the orange of the rump and underparts may be washed with olive. The white fringes of the wings are broader. Breeding plumage is obtained by the wear and loss of these dissimilarly coloured tips.

Adult Female Breeding (Definitive Basic, worn): Soft part colours as in male, perhaps showing more blue-grey on the bill. The plumage of females is extremely variable (Figure 40.1), with some bright individuals appearing superficially like adult males. Typical examples are not as bright as this though. Females caught and banded, and re-trapped a year later, are brighter in the 2nd year, therefore plumage brightness is (at least partially) age-related. Head olive-yellow to olive-orange, with greyish or dusky markings on the auriculars and crown. The back is olive with blackish bases to the feathers, the extent and darkness of the black on the back is variable. Rump yellow with an olive wash. The wings are blackish, but considerably greyer than those of adult males. The lesser coverts show a variable amount of orange or yellow. The median and greater coverts are pale-tipped, creating two obvious white wingbars. The tertials and flight feathers are fringed with white, particularly boldly on the tertials. The underparts are orange-yellow, palest on the vent and belly, and brightest on the chest. There is a variable amount of black mottling on the throat and neck. Tail olive-grey, the bases of the outer rectrices having a noticeable yellow or orange tone. During the breeding season, the wings and tail of females become increasingly worn as a function of going in and out of the nest.

Adult Female Non-breeding (Definitive Basic, fresh): As breeding female, but the upperparts are often veiled with grey.

Fall Immature (First Basic): Both sexes are similar in this plumage and are only separable through measurements while in North America. As adult female, but lacking all black on the upperparts and head. The dullest variants are likely all immature females. Upperparts from crown to back and scapulars largely olive-grey, without dusky feather centers, rump yellowish. The wings are more olive than those of the adult female; similarly the tail is entirely olive-grey without orange tones. Underparts dull yellowish, palest on the belly and vent. Upon arrival in the wintering area young males will begin to obtain black throat feathers.

Breeding Immature Male (First Alternate): Intermediate between fall immature and adult male. Shows a variable amount of black on the head, back and scapulars. Some bright orange colouration may be present on the rump and underparts. This plumage overlaps in appearance with that of the adult female, and thus, young males may not be reliably separated from females other than by measurements. On average, young males are brighter than females, but they have duller wings, due to greater wear. Exceptionally bright individuals are aged by their worn, brownish flight feathers, which are blackish on adult males.

Breeding Immature Female (First Alternate): As fall immature with a limited amount of black markings on the head and back. Shows more worn wings than the adult female.

Juvenile (Juvenal): Bill dark with a pinkish base to the lower mandible, eyes dark. Head and upperparts, to the tail-coverts, olive-brown with a yellow or orange tinge, particularly on the rump. Wings greyish-olive with the median and greater coverts edged with pale buff, creating two inconspicuous wingbars. The underparts pale olive with a yellowish tinge. Legs and feet greyish.

GEOGRAPHIC VARIATION Monotypic

Figure 40.1 Variation in plumage pattern of adult female Baltimore Oriole.

HABITAT Breeds in open woodlands and forest edge as well as isolated clumps of trees in open areas. A great deal of foraging is done in fields, meadows or other open areas and these are important to the orioles. Baltimore Oriole has adapted well to urbanization and is common in suburban areas with large trees. In the west of their distribution, they are almost exclusively a bird of riparian stands of Cottonwood (*Populus deltoides*). In winter, they can be found in a variety of habitats from dry deciduous forest to the canopy of wet rainforest. They are also fond of plantations, as well as agricultural areas where large shade trees are left standing. In rainforests they seldom come down to the dark interior, but remain in the canopy or along roads and more open places. They do come down to feed in lower shrubbery in open areas or at the forest edge. In winter, they roost in stands of tall grass or cane. Costa Rican wintering Baltimore Orioles are more abundant and occupy a more diverse number of habitats than resident orioles (Timken 1970).

BEHAVIOUR Upon arrival on the breeding grounds, males begin to sing continuously, with inter-song intervals as short as four seconds, until they obtain a mate. Singing decreases substantially once pairing occurs and by mid to late May, most birds singing continuously are unpaired first summer males. The mating system is monogamous, but males are opportunistically polygynous (Flood 1985). Apparently all females, including year olds breed, but few first summer males are successful in acquiring a mate. Those that do breed, tend to nest significantly later (five days) than adult males.

Courtship occurs for a short period in May. The male typically sits extremely close to the female and faces her such that his orange breast is obvious, while fanning the tail and holding the wings partially open. He then bows forward, first displaying the black of the upper breast and head, and eventually flashing the orange rump to the female. He may utter several monotonous sweet whistles as he does this. Females respond to courtship by either ignoring the male or by making a food begging-like trill accompanied by wing quivering and crouching. This is the pre-copulatory display of the female and it is followed by copulation under normal circumstances. Copulation events may be started by female display, not just by male courtship. Often a male will perform a post-copulatory flight song, which is a rapid and long song given as the male flies with exaggerated butterfly-like wingbeats (Edinger 1988). Copulation tends to occur while the female is finishing the outer shell of the nest or adding the lining. Observed copulations are most common between 6 am and 7:30 am (Edinger 1988).

Extra-pair copulations, those not involving the

male with which a female is paired are not unusual. Intruding males are most often chased out by the territorial male. However, sometimes intruding males arrive when the territorial male is not present. These intruder males may perform courtship to the female, which is often, although not always, ignored, or they may force the female to the ground and attempt to mate with her. Females may also be apprehended by males while the former are collecting nesting material away from the territorial male. Most intruding males are neighbours from nearby territories and most extra-pair copulation attempts occur during the most fertile period of the female. Paired territorial males deal with the threat of extra-pair copulations in several ways: the male may chase off any intruding male from the territory on sight, he may also follow his female around during her most fertile period, or he will also drive the female away from the territories of other males (Edinger 1988).

Immediately after nesting, Baltimore Orioles undergo their complete moult. This occurs in July, at which time the orioles become incredibly secretive, appearing to vanish until August when they become obvious once again. In fact, males in August may sometimes burst into song. As quickly as they 'appear' they will leave the area and begin their migration. Baltimore Orioles winter alone, or in small groups of up to a dozen individuals. They frequently follow mixed-species flocks. Previous reports that this species holds winter territories appear to be erroneous; however, they have been observed defending nectaring sites from other orioles and hummingbirds. At night they roost in larger numbers, often with other orioles or blackbirds.

For the most part, Baltimore Orioles forage by gleaning insects from leaves of trees. They have a fondness for caterpillars, and will eat many of the types that are hairy and most birds avoid. They also forage in grass and cultivated plants, picking insects from them. Berries and nectar are also included in the diet, subject to availability. In winter, they are often seen foraging high in trees, either on fruit or probing flowers, presumably for nectar. In Costa Rica, a detailed analysis of the diet of Baltimore Oriole determined that while fruits and nectar were commonly taken as food, butterfly larvae and other insects accounted for the majority of food taken in winter. Baltimore Orioles preferred foraging on trees with brightly coloured flowers or very dense foliage (Timken 1970). Unusual food choices include a male Baltimore Oriole that was seen to kill a Ruby-throated Hummingbird (*Archilochus colubris*), presumably to eat it (Wright 1962).

NESTING Breeding season May–late June. First year males seldom gets the opportunity to nest, but when they do it is often later in the season than most older males (Labedz 1984). The nest is a pendant basket roughly 15–20 cm long, with the entrance at the top. Nests are hung on trees, with the nest height dependent on the average height of the trees; however, they are often 10 m or more off the ground, occasionally much lower. There tends to be local preferences in the nest tree used. In the 1930s, elms (*Ulmus* sp.) were preferred in one site in Iowa (Nauman 1930), while in riparian zones in the Great Plains, cottonwoods (*Populus* sp.) are preferred. If at all possible, the nests are suspended from the tip of a thin branch, often one that droops, as far away from the trunk as possible. Nests of Baltimore Oriole have been found to vary geographically in their insulative properties. Nests of Canadian Baltimore Orioles are not as well insulated as those from the Great Plains, the hybrid zone, or nests of Bullock's Orioles (Schaefer 1980). Nests in the west of the species' range appear to have higher insulative values due to the higher temperatures there. The females are responsible for the major part of nest construction, with the male helping to some extent. A nest may be weaved in as little as two days (Nauman 1930) but it more commonly takes between eight and 11 days (Edinger 1988). Breeding concentrations can be quite dense; however, nests are distributed randomly in the habitat, showing no clumping tendency.

Egg-laying begins in mid to late May. Clutch sizes are commonly 4–5, and the eggs are greyish-white blotched and scrawled with brown and black, particularly around the broader end. Nests where the male was a first year male had significantly lower clutch sizes and fledging success in one Kansas study (Labedz 1984). Baltimore Oriole is parasitised by Brown-headed Cowbird (101); however, the oriole is able to recognize these eggs and it removes them. Unlike other cowbird egg-rejecting species, Baltimore and Bullock's Orioles spike the egg with their bill and carry it out, rather than taking it out held between the mandibles (Rothstein 1977c). Smith (1972) describes hearing a plopping sound below a nest of a breeding Baltimore Oriole in Nebraska, and noticing that the noise was made by a Brown-headed Cowbird egg that the female oriole had removed from her nest. After removing the egg, the oriole calmly preened while perching beside the nest.

The female abandons the male along with the brood several days before the young become independent, in order to begin her moult. This takes place in late June. Females will wander widely after breeding; however, males tend to stay put near their territory until they depart on their migration. Immature males (born the previous season) leave the breeding area early, while they are still in moult.

DISTRIBUTION AND STATUS A common species throughout the breeding range. Common in winter in Mexico and Central America, much less common in South America. It is a rare migrant in the Caribbean, see Movements.

Breeding: From NW British Columbia (Peace District) where it has been present only since the 1960s; NC and W Alberta; east through C Saskatchewan, C Manitoba, S Ontario, S Quebec, S New Brunswick, Prince Edward Island, to C Nova Scotia, south to N Texas, S Louisiana, S Mississippi, C Alabama, NW Georgia, W South Carolina, W North Carolina, W Virginia, Maryland and N Delaware. Breeds west to NE Montana, SW North Dakota, W South Dakota, W Nebraska, W Kansas, W Oklahoma and E Texas, where it hybridises with Bullock's Oriole along a narrow contact zone.

Non-Breeding: The northern limit of the wintering range extends from the Mexican states of S Nayarit, W Jalisco, C Michoacán, N Guerrero, C Puebla, E Hildago to E San Luis Potosí and S Tamaulipas. South of here this species winters throughout Mexico (except Yucatán) and Central America south into NW South America, W of the Andes: NW Colombia south to the upper Cauca valley (Valle), W Meta, W Casarne and W Arauca into NW Venezuela and the coast as far east as the Orinoco river delta and the forests east of El Palmar (Jaramillo, D. Beadle unpub. data). A few records exist from Trinidad and Tobago. Regular in winter along the coast of the SE US (North Carolina to Florida). It is rare but regular in winter in California. Rarely winters in the Greater Antilles and the Bahamas. There are a small number of winter records from the US Virgin Islands (Norton 1979).

MOVEMENTS Baltimore Oriole is a long distance migrant. Migration out of Central America begins in March, and most individuals have left by mid April. It is most abundant during the last half of March in Guatemala (Griscom 1932). Spring arrivals in the southern US show up in April, reaching the Canadian border in late April or more commonly in early May. Adult males precede the females by approximately one week. By the end of May, they have arrived in their northernmost outposts.

Baltimore Orioles are an obvious species in the spring and early summer, but as breeding winds down they become rather furtive. Young males begin leaving the breeding grounds as early as July and migration peaks in August. A small sample of tower-killed migrants suggests that adult males migrate in a narrower window than females and young males (Sealy 1986). Most Baltimore Orioles have left the breeding grounds by September. They arrive as early as mid September in Guatemala, and migration peaks in October (Griscom 1932). A small number linger on in the US and Canada, and they have attempted to winter as far north as Ontario. Migrants move south through Mexico, primarily along the Caribbean Slope to their wintering grounds. In Guatemala, they migrate largely along the central highlands and Caribbean Slope, being much rarer on the Pacific slope (Griscom 1932). A smaller number migrate through the Caribbean, primarily in the spring. It is a common migrant in the Bahamas and has been recorded from Bermuda. In Cuba, this species is not known as a fall migrant, but there is a spring movement. This oriole is a known transient through Grand Cayman, Cayman Brac and Little Cayman islands (Bradley 1994). Note that on the Netherlands Antilles (more South American than Caribbean really) there is a regular autumn passage on Aruba and Bonaire but it has been recorded during the spring only on Bonaire (Voous 1983). It is exceedingly rare east of Jamaica. Two records exist for Hispaniola, one male on Dec. 28, 1962 from Marfranc, Haiti, and a female W of Barahona, Dominican Republic, on Jan. 21, 1972 (Gochfeld *et al.* 1973). There is one record from Martinique, on Nov. 19, 1969 (Pichot 1976). In Barbados, it's a very rare migrant with three records, but it may be overlooked (M. Gawn pers. comm.). Arrival in Central America begins in September and peaks in October, much later than Orchard Orioles. At the southern end of their wintering distribution, in the Santa Marta region, Colombia, they are present as early as Oct. 13 and as late as March 10 (Darlington 1931).

The banding data records a total of 733 recoveries. In this sample, the oldest individual was a 12 year old, three birds were nine, three were eight and six were seven years of age. A captive male Baltimore Oriole lived to a few weeks short of 25 years (B. K. McKay pers. comm.). There are a total of 10 recoveries from Mexico and Central America. All were banded as migrants or breeders in the north and recovered during the winter in the south except for one July recovery in Guatemala (possibly an error?) and one May recovery in El Salvador. There is a total of five winter recoveries from Guatemala, these birds were originally banded in Ontario, Manitoba, Connecticut, Missouri and Wisconsin. Two recoveries from Honduras were banded in Michigan and Minnesota; one from Oaxaca, Mexico, and one from Veracruz, Mexico, were both banded in Massachusetts; finally, a bird recovered in El Salvador was originally banded in Michigan. Of nine recoveries of birds banded in Honduras, one each was recovered in summer in Ontario, Michigan and Wisconsin, and the remainder were winter recoveries in Honduras. A Mexican-banded bird was later found in Pennsylvania (BBL Data).

A small population of Baltimore Orioles winters in the southeast of the US; three banded recoveries of this wintering population include two caught wintering in Florida and two wintering in Washington D.C. (Lawrence and Brackill 1957) in subsequent years, and one banded in winter in South Carolina that was re-trapped the following spring in Quebec; obviously, this wintering population is surviving winters. A detailed study of a wintering population in North Carolina found that many banded individuals returned in subsequent years; one was trapped in six out of seven years (Erickson 1969). Note also that a small number winter in the Bahamas. Wintering in the southeast US is a relatively recent phenomenon (the last 40 years or so) and appears to be associated with an increase in bird feeders in urban areas. Most of these wintering birds are young males or females, with a minority of adult males, although males (of all ages) are equally abundant as females. A most interesting recovery involves a bird banded on Oct. 13, 1963 in Rhode Island that was recovered in a building a month later in Newfoundland! This is evidence that fall vagrants in Newfoundland are birds flying up from the south or southwest.

As a vagrant it has been observed several times in the Western Palearctic: records exist for Greenland, Iceland (3+), Netherlands (Oct. 1987), and Norway (Sept. 28, 1986). In Britain there are 16 records (including Scotland but not Ireland), with the majority falling between September–December, many in October. There is one spring record from May 6, 1970. Another individual, a 1st-winter female, wintered from Jan. 2–Apr. 23, 1989.

Baltimore Oriole was first found breeding in NE British Columbia in 1960, and has subsequently been observed to be locally common in the area. However, it was not noted to be present there in the late 1930's, suggesting that this population is of recent origin. It is a regular vagrant to the Pacific coast of North America. In S British Columbia, records include birds from Vancouver Island and the Okanagan valley on June 5, 1955 (Munro 1955). A Baltimore Oriole collected in Chilliwack, BC, on June 11, 1927 was actually a hybrid Baltimore × Bullock's Oriole (Brooks 1942). In Washington, there are two November records, one from Seattle and the other from Cle Elum in the east slope of the Cascades. In Oregon, there are over 30 records; it is annual mainly as a spring vagrant in the east of the state, and there are also summer records from Malheur NWR. It is much rarer on the Oregon coast, and one has wintered on the far southern coast (A. Contreras pers. comm.). In California, it is a regular fall rarity along the immediate coast, and occasionally winters in stands of Eucalyptus. In spring it is very rare, with most records coming from the interior desert oases, rather than from the coast. In Arizona, it is a very rare vagrant with over 12 records during spring and fall (including adult males). There is one specimen (Sept. 4, 1954) and one male paired with a Bullock's Oriole near the Colorado River, 16 km N of Blythe, California in 1977 (Monson and Phillips 1981). Most records are from the autumn, and there are no winter occurrences. There is one record from Utah, which was collected in Milford on June 27, 1964 (Hayward *et al.* 1976).

There are no records from Nevada (Alcorn 1988). It is a fall vagrant in Baja California, There are also records to the north of the range in N Manitoba, N Ontario (pair at Matchewan, June 1994) and N Quebec. It occurs as a fall vagrant in Newfoundland. There are several fall records for Sable Island, Nova Scotia. There is at least one record from Bermuda and one south of the wintering range in Pichincha, Ecuador.

MOULT Immediately after parental duties terminate, adults undergo their pre-basic moult, which is complete. Definitive pre-alternate moults are missing, or extremely limited, thus this is the only moult of the year, and accordingly, the slight change from non-breeding plumage to breeding plumage is due to wear. The moult takes place late June–early August. Females leave the brood before they are independent and are able to begin the moult several days before adult males, which stay with the brood. First-summer males (2nd calendar year immatures) moult at the same time as females, a few days earlier than adult males. The complete moulting process is rather quick, taking approximately five to six weeks. Moults occur on the breeding grounds, except for in 1st-summer males which may begin moving south as the moult is ongoing (Sealy 1979, Kelley 1960). Young males have a variable first pre-alternate moult which usually begins surprisingly early, upon arrival in the wintering area. This moult intensifies prior to the northward migration and may continue in 1st-summer birds on the breeding grounds. The minimal extent of this moult is for some of the throat feathers to be replaced by black feathers. Many birds moult much of the underparts and sometimes the inner secondaries. The varied appearance of 1st-summer males occurs due to the variability of this moult.

MEASUREMENTS Males: (n=12) wing 97.0 (91.4–102.1); tail 75.9 (70.6–80.0); culmen 18.3 (17.5–19.8); tarsus 23.9 (22.9–25.4). Females: (n=8) wing 88.9 (85.1–91.9); tail 68.6 (66.0–71.9); culmen 17.5 (16–18); tarsus 23.1 (22.4–24.1).

NOTES Recent molecular work has found that Baltimore and Bullock's Orioles are not each other's closest relatives. One analysis suggests that Bullock's Oriole is more closely related to Streak-backed Oriole (35), while Baltimore and Altamira Orioles (34) are more distantly related to the Bullock's/Streak-backed pair. However, the four species make up a clade (group with one common ancestor) (Freeman and Zink 1995). A more robust analysis, based on mitochondrial DNA sequence data, suggests that the Bullock's Oriole is more closely related to the Streak-backed Oriole, but that the Baltimore Oriole is perhaps most closely related to the Black-backed Oriole, a finding that needs to be studied more closely (K. Omland and S. M. Lanyon, pers. comm.).

The breeding bird survey (BBS) data shows that the population of Baltimore Oriole is stable. There are local decreases, some of them significant, but there is an equal number of areas that are witnessing increases in populations.

REFERENCES Barbour 1943 [movements], BBL data [movements, banding], BBS data 1995 [population, notes], Beletsky 1982a and 1982b [voice, behaviour], Crawford 1973 [movements, wintering], Edinger 1988 [behaviour], Erickson 1969 [wintering, North Carolina], Flood 1984 [nesting, behaviour] 1985 [behaviour], Freeman and Zink 1995 [notes], Kelley 1960 [migration], Labedz 1984 [nesting], Nauman 1930 [nesting], Norton 1979 [distribution], Ridgely and Tudor 1989 [movements, distribution], Rohwer and Manning 1990 [moult,], Rothstein 1977 [nesting], Schaefer 1980 [nesting, nest insulation], Schemske 1975 [behaviour], Sealy 1979 [movements, moult], 1985 [movements], 1986 [movements], Small 1994 [vagrants, movements], Timken 1970 [behaviour, food], Weber 1976 [movements, range in British Columbia].

40X HYBRID BALTIMORE × BULLOCK'S ORIOLES
Icterus galbula × bullockii

These intriguing birds are only likely to be observed where the ranges of the two parental species meet in the Great Plains of North America.

DESCRIPTION In all respects, the hybrids between these two species are intermediate between them. The farther away from the middle of the hybrid zone, the greater the similarity of the intermediates to the closest parental species. The change from intermediate to pure Bullock's (west from the hybrid zone) occurs more quickly than to pure Baltimore (east of the hybrid zone). Figure 40.2 illustrates three hybrid male individuals. A truly intermediate bird will be described.

Adult Male (Definitive Basic): Crown, nape, eyeline and center of throat black. The supercilium is mixed black and orange. The forehead is mixed black and yellow. The side of the neck is black and yellow, with the yellow noticeably extending up toward the auriculars, which are mixed black and yellow. The orange or yellow on the throat reaches anteriorly between the eye and the bill. The wings show two white wingbars with extensive white edging on the greater coverts, tending to unite these into a white patch, however some black is noticeable within this patch. The tail is variable, often showing completely yellow outer rectrices, or perhaps largely yellow with a dusky mark halfway along the feather length.

Adult Female (Definitive Basic): These tend to have yellow underparts with a noticeable, patch of greyish on the mid-belly. The undertail-coverts are less yellow than on Baltimore Oriole and more yellow than on Bullock's Oriole. The rump is yellowish-olive variably washed with grey. The head has a variable amount of black feathers on the crown and ear-coverts, but usually some on the throat. The back is mixed olive with grey, the

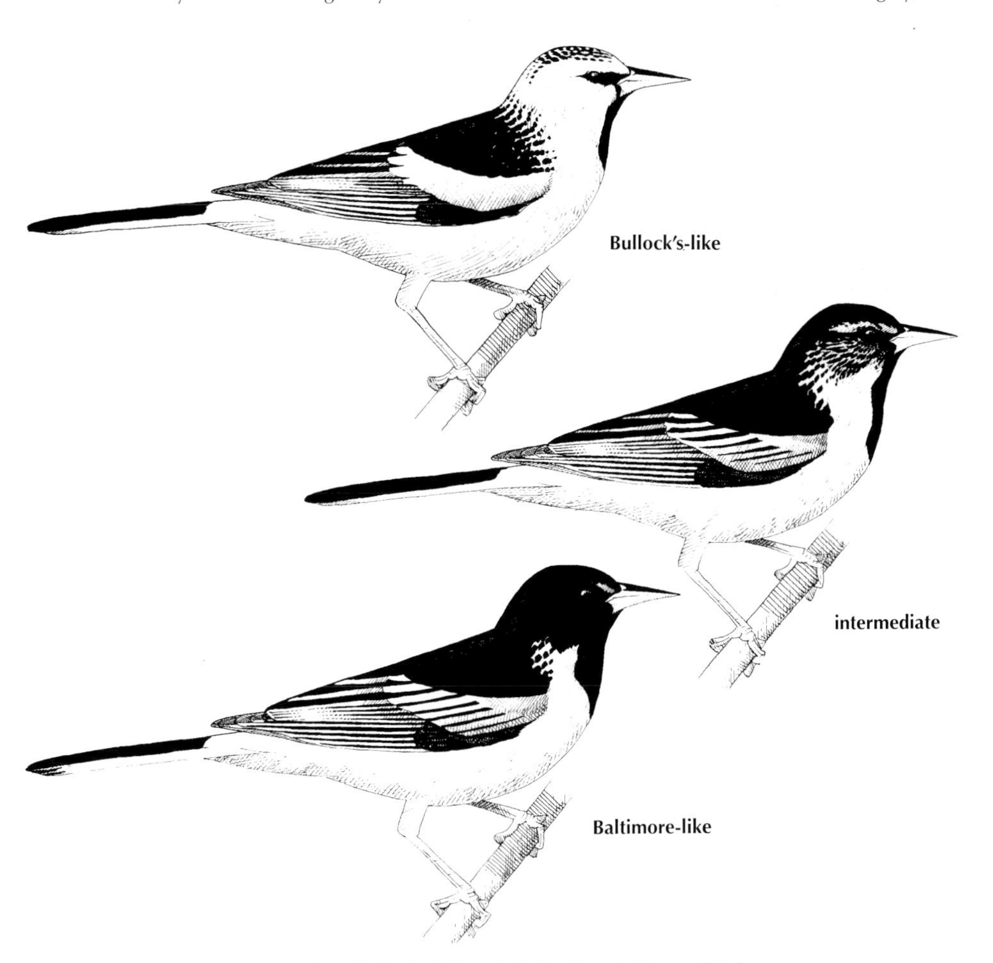

Figure 40.2 Representative examples of hybrid adult male Bullock's × Baltimore Orioles.

feathers not showing noticeable darker centers.

VOICE Song: The songs of hybrids tend to sound intermediate between those of the parental species (J. Rising pers. comm., Jaramillo pers. obs.).

STATUS AND DISTRIBUTION Hybrids are common along a narrow (160–240 km wide) strip along the interface between the breeding ranges of the two parent species.

Breeding: SE Alberta and SW Saskatchewan southeast diagonally from NC to SE Montana, SW North Dakota, W South Dakota, W Nebraska, E Colorado, W Kansas, W Oklahoma and the E edge of the Texas Panhandle.

Wintering: Specimens exist from Guatemala and Nicaragua. The winter range appears to be intermediate between that of the two parental species.

MOVEMENTS A record of a vagrant exists from Long Point, Ontario, May 2, 1992. In addition, the first published record of a Baltimore Oriole in British Columbia was actually an adult male hybrid as it was stated to show a faint yellow supercilium (Brooks 1942). There is a record of a hybrid individual banded on Southeast Farallon Island, California, on Nov. 15, 1980 (Evens and LeValley 1981).

NOTES Outside of the hybrid zone, the parental species do not show any sign of introgression. However, Bullock's-type characters persist farther east of the hybrid zone than do Baltimore-type characters west of the hybrid zone (Misra and Short 1974). If anything, there is evidence for an eastward invasion of Bullock's genes, but not the opposite. In the northern Great Plains, both orioles are present and only a few are intermediate in appearance, implying that the two species segregate there. The hybrid zone correlates quite well onto a gradient of rainfall, decreasing to the west. Bullock's Orioles live in drier habitats than Baltimore Orioles and are physiologically able to survive better in hotter temperatures. The hybrid zone is stable and did not change appreciably between the 1960s and late 1970s. However, in the N Great Plains there is evidence that Bullock's Orioles were more common a century ago; this was also a relatively dry period in that area. See the Notes section of Baltimore Oriole for information regarding the relationships between the two parent species.

The differences in moult timing are marked between these two species and are thought to be genetically determined. A hybrid female caught in early November in Chihuahua, Mexico, was undergoing a second wing moult. Perhaps this is not uncommon in hybrids, where some may moult early (like Baltimore) and then moult later on in the season (like Bullock's) (Rohwer and Johnson 1992).

REFERENCES Rising 1970 [voice, description, distribution] 1973 [notes], 1983 [notes], Rohwer and Johnson 1992 [moult], Sutton 1942 [distribution].

41 BULLOCK'S ORIOLE *Icterus bullockii* (Swainson) 1827 — Plate 10

The only oriole found in the west of North America, north of the arid southwestern states.

IDENTIFICATION Adult male Bullock's Oriole is distinctively patterned with a black back and crown, and an orange face with a black eyeline. Most similar is the closely-related Black-backed Oriole (42) of Mexico. The breeding ranges of these two species are largely allopatric, but they may be found together during the non-breeding season. Adult male Bullock's Oriole lacks the black auriculars, rump and flanks of Black-backed Oriole. Females and immatures are extremely similar, but Black-backed Oriole has a darker face contrasting with a yellow throat and faint yellow spectacles. Similarly, immature female Baltimore Orioles (40) have darker faces that contrast with the yellow throat, unlike the extensive yellow cheeks and supercilium, and contrasting darker eyeline, of female and immature Bullock's Oriole. On Bullock's Oriole, the dark crown contrasts with the pale supercilium. In addition, Baltimore Orioles tend to have more extensive yellow on the underparts and this has an orange wash to it. Bullock's Orioles have unicolored grey backs, whereas Baltimore usually has darker centers to the back feathers, appearing obscurely streaked from a distance. Grey undertail-coverts are a good feature (perhaps diagnostic?) for Bullock's; however, many Bullock's have yellow undertail-coverts as is always the case with Baltimore Oriole. The wingbars on Baltimore Oriole are separate, while on Bullock's Oriole they are usually connected by white greater covert edges, forming a pale wing panel. A new and exciting field mark has been described concerning the edging of the median coverts (upper wingbar). It appears that Bullock's Orioles tend to have a pointed central dark region on these coverts while Baltimore Orioles tend to have a square-ended central dark region to the median coverts. In effect what one sees is a straight upper border to the upper wingbar on Baltimore Orioles but a jagged upper border in Bullock's Orioles (Lee and Birch In Press). This mark has not been widely field tested yet, and how age and the different extent of the first prebasic moults of the two species affects its usefulness is unclear. The use of vocalizations as identification criteria needs to be studied. Female songs are similar to those of males, and are therefore readily identifiable, but chatters and whistled calls are much more similar. The Bullock's Oriole chatter appears to have the notes more evenly spaced than does Baltimore Oriole (J. Dunn pers. comm.).

Females of other western North American Orioles such as Hooded (36) and Scott's (51) have entirely yellowish underparts, unlike the whitish, or pale grey, belly of Bullock's. Note that a few Hooded Orioles may show pale (off-white) bellies. Shape is a good feature for separating Hooded Oriole. It is noticeably slimmer, with a thin and curved bill (Figure 29.1) as well as a long and graduated tail, unlike Bullock's Oriole. However, juvenile Hooded Orioles may show shorter bills, not having fully grown these yet. These birds have obvious pink or orange colouration at the base of the lower mandible. Scott's Orioles are darker olive on the upperparts and have two bold and clearly demarcated wingbars. Vocalizations are useful in identification. Hooded Oriole utters a sharp *wheep*, not the *pheew* of Bullock's Oriole; in addition, their chatters are shorter in duration than Bullock's. Scott's Orioles give a *chuck* note, unlike any heard from Bullock's Oriole.

VOICE Song: The song is musical, lively and short, ending in a sweet whistled note. Two descriptions are: *kip, kit-tick, kit-tick, whew, wheet* and *cut cut cudut whee up chooup.* In comparison to Baltimore Oriole, the songs are shorter, not as melodic and quite a lot less variable. Females sing, particularly early in the season, before and during nest building (Miller 1931). See Behaviour. The female song is similar to that of the male, but with harsher terminal notes that do not include as high a frequency range as the male song (Miller 1931).

Calls: A short rattle *chu-r-r-r-r* is commonly given by both sexes. A single sweet note is also heard, *kleek*, or *pheew*. Also gives a soft *chuk*.

DESCRIPTION A medium-sized oriole with long wings. The bill shows a straight gonys and culmen. It has a squarish tail, only slightly graduated. Wing Formula: P9 < P8 ≈ P7 > P6; P9 ≈ P6; P8–P6 emarginate.

Adult Male Breeding (Definitive Basic, worn): Bill black with blue-grey lower mandible. The eyes are dark. The crown, nape and back are black, and the forehead is usually black but it may be yellowish on some individuals. The face is orange-yellow with contrasting black lores and a black eyeline that connects with the black of the nape; this divides the orange supercilium from the rest of the orange face. There is a black, central throat patch down to the upper breast, which is narrow and positioned along the mid-line of the throat. The rest of the underparts are yellowish-orange, the exact colour variable depending on the individual; however, the orange is strongest on the face and breast, and much more yellow on the vent. The rump and uppertail-coverts are orange-yellow. The wings are blackish, heavily marked with white. The median and greater coverts are entirely white, creating a bold white wing patch; some of the innermost greater coverts may show black centers. The tertials, secondaries and primaries are boldly edged with white. The tail is largely black, with yellow at the base of the five outer rectrices. The legs and feet are blackish-grey.

Adult Male Non-breeding (Definitive Basic, fresh): Similar to breeding male, but the black on the back may be obscured by thin grey tips and the underparts may be tipped whitish. The white edging on the wings is wider and more noticeable.

Adult Female Breeding (Definitive Basic, worn): Females are variable in their appearance, but not nearly to the same extent as female Baltimore Orioles. The soft part colours are as on the adult male, perhaps showing more extensive blue-grey on the bill. The crown is greyish-olive, becoming grey on the back and scapulars, and continuing to the rump. The back feathers have darker centers making the back appear obscurely streaked at times. The face is yellowish with an olive tone, and a darker grey eyeline mimics the facial pattern of the male. In addition, a variable amount of grey or black feathers are present on the mid-line of the throat. The yellow of the face is most orange-toned at the anterior part of the auriculars and on the supercilium before the eye. The yellow continues to the upper breast, changing to whitish-grey on the lower breast, belly, flanks and thighs. The undertail-coverts are white or washed yellow as are the uppertail-coverts. The wings are blackish with grey edging. The median coverts are broadly tipped white, creating a bold upper wingbar. The greater coverts are edged and tipped with whitish-grey, visible as a less noticeable lower wingbar. The secondaries and tertials are edged with whitish-grey. The tail is olive with yellow on the basal portion of the outer feathers.

Adult Female Non-breeding (Definitive Basic, fresh): Like breeding female, but the edging on the wings is more noticeable.

Immature Male (First Basic): Apart from being slightly larger than the female, the immature male is similar. However, 1st-fall and 1st-summer males have black lores and a black throat stripe. Few young males undergo a first pre-alternate moult, therefore 1st-summer birds look similar to 1st-fall birds.

Immature Female (First Basic): Largely indistinguishable from the adult females, except the wings are more worn.

Juvenile (Juvenal): Like adult female, but lacks black on throat. In addition, the wing markings are buffier and the base colour of the wings is olive-brown rather than blackish. Juvenile females are slightly paler than on the males; however, this would be impossible to use in the field to sex individuals.

GEOGRAPHIC VARIATION Two subspecies are recognized. The nominate form is widespread north of Mexico and is described above.

I. b. parvus is a southwestern and Mexican race. This subspecies breeds in southern California, southern Nevada and central-western Arizona, south to N Baja California and NW Sonora, in Mexico. Its winter range is thought to be the southern part of its breeding range, and perhaps further south. It is smaller than the nominate subspecies. The separation of this subspecies obscures more complex patterns of size variation. Californian specimens tend to be smaller in culmen length, wing length and tarsus length. However, birds from the north have larger culmen measurements than all birds from the south, with Mexican birds having smaller culmen measurements than Californian birds. In terms of wing length, Californian examples have shorter wings than all others, including Mexican specimens. With regard to tarsus length, Californian birds have shorter tarsi than northern ones, but Mexican and southern Great Basin examples are intermediate. The separation of the race *parvus,* therefore, is not very useful (Rising 1970). In addition, Oberholser (1974) named an additional race of the Bullock's Oriole, *I. b.eleutherus* from W Oklahoma, N Texas and N Tamaulipas, Mexico. It was described as being more orange, less yellow than nominate *bullockii.* In fact, its colouration is well within the variation found in *bullockii,* and may be due to gene flow from Baltimore Orioles as the two species hybridize nearby, and therefore this subspecies should not be considered valid (Browning 1978).

HABITAT Bullock's Oriole breeds almost strictly in deciduous forest or forest patches. It is a bird of open woodlands and woodland edge. In many places, it is primarily found along riparian corridors through desert, grasslands or other unforested areas. In riparian areas, cottonwoods (*Populus* sp.) are preferentially used as nest trees and foraging sites. Otherwise other large, tall trees such as sycamores (*Platanus* sp.) are used. Open areas are important for foraging, including the edge of grassland and non-woody herbaceous vegetation. In the breeding season, this oriole is not adverse to using urban areas, but in towns it tends to be found in parks or riversides rather than in backyards, unless these are well treed. During the non-breeding season, on the other hand, Bullock's Orioles readily forage in urbanized areas, primarily on planted flowering bushes, as well as on hummingbird feeders. In California, introduced eucalyptus (*Eucalyptus* sp.), provides nectaring sites and sometimes nest sites. In addition, it winters in a variety of habitats ranging from thorn forest and pine-oak woodlands to moist tropical forests.

BEHAVIOUR Bullock's Oriole usually feeds in trees, gleaning insects from the foliage. However, it also frequently drops to the ground in forest openings or open grasslands to find insect prey. Throughout the year, this species will take nectar, and will even visit hummingbird feeders. Its use of nectar as a food source is more prevalent than in Baltimore Oriole. Of interest is an observation of a Bullock's Oriole feeding on a *Selasphorus* hummingbird on South Farallon Island, California (Ashman 1977). However, this predation event likely was caused by the limited foraging opportunities on the island. Bullock's Orioles have been observed to defend flowering trees from tanagers and other bird species of similar size. The orioles tend to be quite efficient at displacing these other species. Furthermore, the nectaring alone of flowers of *Erythrina breviflora* trees in Mexico was calculated to be sufficient to provide the daily nutritional requirements for the orioles (Cruden and Hermann-Parker 1977). However, in Morelos, Mexico, Bullock's Oriole was at the bottom of a dominance hierarchy involving two other oriole species at flowering *Pseudobombax ellipticum* trees. Streak-backed Oriole was dominant over Bullock's Oriole and Black-vented Oriole was dominant over both (Eguiarte *et al.* 1987).

After arrival in the breeding area, males begin singing persistently, awaiting the arrival of the females one to two weeks later. Both sexes will share duties in territorial defense; however, the female is less prone to chase off other males in the territory, being more likely to chase off females. The female song, which is given early in the season, appears to be territorial and aimed at other females. At this time, the females may sing more often than males (Miller 1931). However, females soon turn their attentions to nest building and spend less time and effort in dealing with intruders. Males do not aid in the building of the nest or in the incubation of the eggs. However, they do help in feeding the nestlings and fledglings.

During the ash fall caused by the May 1980 eruption of Mount St Helens in Washington State, banded Bullock's Orioles at a Yakima county site abandoned their breeding areas and disappeared. However, once the ash stopped falling, new arrivals (unbanded birds) recolonized and stayed to breed. It is unknown if these were new migrants or floaters that were in the area at the time (Butcher 1981).

NESTING Most commonly a solitary nester, but the spacing of nests varies depending on the ecological characteristic of the habitat (Pleasants 1979). Both the abundance of nest sites in riparian areas and the abundance of food contribute to producing a variety of nesting situations from territorial and solitary nesting to non-territorial and semi-colonial nesting (Pleasants 1979, Rising 1970). Breeding occurs between early May and late June, but depends on the latitude. For example, nest building may occur as early as late April in the San Francisco Bay area, but seldom occurs before mid May in the Okanagan valley, British Columbia (Cannings *et al.* 1987).

The nests are a hanging basket averaging 15 cm in length, with the opening at the top. They are most similar to the nests of Baltimore Oriole, but may average shallower and less pendant. In addition, nests of Baltimore Orioles are not as well insulated as those of Bullock's Orioles, a species that lives in hotter areas (Schaefer 1980). As a rule, Bullock's Orioles build a new nest each year but there are observations of adults refurbishing an old nest (Cannings *et al.* 1987). Usually nests are placed from 2–5 m, and occasionally higher. Bullock's Oriole nests are often clumped in their distribution, unlike those of Baltimore Oriole. There is also anecdotal evidence that Bullock's Orioles nest near Western Kingbirds (*Tyrannus verticalis*), presumably to benefit from that species' aggressive nest defense. This oriole is single-brooded, but a replacement clutch may be laid if the first is destroyed. Egg dates range from late April–mid July: April 22–June 11 in California; April 30–June 25 in Texas; May 27–June 17 in Utah; May 16–July 12 in the Okanagan, British Columbia (Bent 1958, Cannings *et al.* 1987). The clutch size is 4–5 eggs which are greyish-white, occasionally bluish, with dark spots, blotches and scrawls. Incubation is performed by the female and takes roughly 14 days. The young leave the nest in just under two weeks. One brood is raised per season. Brown-headed Cowbird (101) parasitises this oriole, but the orioles are adept at spotting cowbird eggs. Unlike other cowbird egg-rejecter species,

Bullock's Oriole spikes the egg with its bill and carries it out, rather than taking it out held between the mandibles (Rothstein 1977c). The eggs of Bronzed Cowbird (99) have also been found to be rejected by this oriole (Friedmann and Kiff 1985).

DISTRIBUTION AND STATUS Common throughout its range.

Breeding: Ranges from S British Columbia, S Alberta and SW Saskatchewan in Canada, south through NE Montana, SW North Dakota, W South Dakota, W Nebraska, W Kansas, W Oklahoma into central Texas, and into Mexico as far south as N Cohahuila, west through N Durango, Sonora, into N Baja California, the western limits of the range are the coastal ranges, but it breeds casually on the coast side of these ranges.

Wintering: Mainly in Mexico (S Sinaloa to Puebla) south to W Guatemala where it is rare (Griscom 1932). The species has been recorded at least twice on the Pacific slope of Costa Rica and there are two winter records for El Salvador (Thurber *et al.* 1987), but these may be best thought of as vagrants that far south.

MOVEMENTS A short to medium distance migrant. In spring, it arrives as early as mid March in the south (California), not reaching British Columbia until early May. During some years arrival in the north may be delayed by as much as two weeks (Cannings *et al.* 1987). In the Great Plains, arrival begins in early May but many birds do not arrive until late May. Female arrival is one to two weeks after that of the males. Departure occurs early, sometimes in July but more often in August, a couple of weeks earlier than in Baltimore Oriole. Adult birds are the first to leave. By September adults have left the northern part of the range. There is good evidence that after breeding, Bullock's Orioles migrate to the southwest deserts to moult during that region's rainy season. Arrival in Central America averages September–October. A minority try to winter well to the north, with records even in British Columbia (Cannings *et al.* 1987), but many of these are not observed past December, presumably because they die.

A total of 210 banding recoveries exist. Of these birds the oldest was a bird aged six years and one month old, and there were six five year olds in the sample. All were banded on the breeding grounds in summer and were subsequently recovered on the breeding grounds near the place they were banded either later that year, or in later years. Three recoveries were made in winter: a bird banded in California was recovered in Baja California, Mexico, and a bird banded in Texas was recovered in Michoacán, Mexico. The remaining bird was banded within the hybrid zone (North Dakota) and was recovered in Costa Rica, south of where Bullock's Orioles typically winter; this was likely a hybrid.

This species is a rare, but regular vagrant on the Atlantic coast, particularly in fall and winter. Bullock's Orioles have also occurred in most of the states east of their normal range, again largely in the fall and winter. However, determining the correct status of this species is a difficult task. Due to the many years during which this species was lumped as Northern Oriole with Baltimore Oriole, record-keeping on stray Bullock's to the east was slack. Many rare bird records committees would not vote on these birds as they were regarded merely as a subspecies. In addition, the separation of pale immature Baltimore Orioles from Bullock's Orioles has been an under-appreciated problem, and many Bullock's Oriole records from the east may in fact refer to pale Baltimore Orioles. The authors recommend that record-keepers in the east should re-evaluate all vagrant Bullock's Oriole records using the most strict criteria, in order to confirm the identification and presence of this species in the east.

The majority of records in the east come from the Atlantic provinces and states. There are no reports from Newfoundland or Prince Edward Island. There are several reports (mainly October–November but also May) of Bullock's Orioles from New Brunswick, but currently, the New Brunswick Bird Records Committee is grappling with the status of this species in the province, as even photos do not always conclusively secure a correct identification. In Nova Scotia, there are several records from the fall, including photographed birds. In Maine, there are several reports, including one specimen of an immature male from Sorrento. In Massachusetts, there are at least 10 records of males or specimens, from fall, winter and spring (Veit and Petersen 1993). Bullock's Oriole has been adequately documented in Rhode Island several times by sight records, late September–early February, but there is still no specimen (D. Ferren pers. comm.). There are no Rhode Island records of adult males, and some previously accepted records of immatures have turned out to be pale Baltimore Orioles. In Connecticut, there are a few records, and at least one individual has overwintered. In New York state, there are several reports, of which four involve specimens or birds which were banded for confirmation of the identification. These include an immature male present from Dec. 12, 1963–March 4, 1964 which was banded and photographed; an adult male present from Jan.–April 2, 1966 which was photographed, and a female collected on November 30, 1969. This last bird may have been a hybrid with Baltimore Oriole. All three records are from Long Island (Bull 1974). A specimen from Ondondaga Co. (upstate New York) from May 17, 1875, is now missing. Of the remaining reports, most are of immature birds in fall/winter, but adult males have been reported from upstate New York in spring. There are nearly 30 reports of Bullock's Orioles from New Jersey. Four of these involved males (of unknown age), but with inadequate details to establish the identification. Due to the uncertainty of identification of these records, all have been recently rejected by the New Jersey Bird Records Committee and the species has been taken off the state list (L. Larson, P. Lehman pers. comm.). In Delaware, Bullock's Oriole has been reported at least twice during winter. There are several Maryland reports, but only one accepted record, from Sept. 25, 1994 (P.

Davis pers. comm.). There are no accepted records from Virginia. In North Carolina, it has only recently been accepted as occurring in the state, even though it has been reported a number of times. Most reports, including published reports, lack any substantive details; records are distributed mainly throughout the fall and winter season. There are no specimens from North Carolina (H. Le Grand pers. comm.). It has occurred in South Carolina, mainly in winter. Adult male records exist for the winter as well as for the spring. In Georgia, there are two specimens of immature males, one collected on Feb. 5, 1947 and the other on Nov. 22, 1948 (Stoddard 1951). Presumably, Bullock's Oriole will turn out to be a regular fall and winter vagrant to Georgia, as fieldwork increases. In Florida it is regular in the fall and winter, sometimes with multiple birds present at one site, usually a feeder.

There are at least six specimens of Bullock's Oriole taken in Florida 1956–1967; three are from the Miami area, from Oct. 12–Jan. 11, and three from the panhandle, Oct. 1–Feb. 4 (fide P.W. Smith). East of the regular range of Bullock's Oriole, but away from the Atlantic coast, it has been reported from Saskatchewan, and may have bred in Yorkton and Carlyle in the SE; Manitoba, where there is a report of a pair in Brandon which appears to have bred in 1940, the details for the male are convincing but it is not impossible that the female was a Baltimore Oriole; Minnesota (one record); Wisconsin (more than one report); Michigan (one record); Ontario (many reports, but only four accepted records, all of females, two in fall, one in winter, and a June record from Thunder Bay in the north); Quebec has six reports, including a spring bird from Rimouski; Iowa (more than one record); Illinois (two reports including one of two spring males and the other of a spring female); Ohio (at least two records); Pennsylvania (one accepted record, fide N. Pulcinella); Missouri (one record, Robbins and Easterla 1992); Arkansas (at least five records exist, all in winter at feeders, and at least one of these birds was an adult male fide M. Parker); Tennessee (first recorded during the winter of 1983, two other records since then, both during the winter, have included one male. There are no accepted records from Indiana and West Virginia. In Kentucky, it had not been recorded into the late 1960s (Mengel 1965), but there may be more recent records. One suspected female Bullock's Oriole wintering in Kentucky was collected and proved to be a pale Baltimore Oriole (Mengel 1965). In the Gulf Coast states, Bullock's Oriole is quite regular, particularly right along the coast. In Louisiana, it is rare but regular in the fall and winter, primarily along the coast. There are specimens (Stoddard 1951) and many records of adult males. In addition, there have been wintering birds which appear to have returned to the same wintering site on several consecutive years. Their status varies, and during the 1990s it appeared that Bullock's Oriole was much rarer than in the late 1970s and early 1980s (N. Newfield pers. comm.). However, in the winter of 1997/1998 many reports of Bullock's Orioles wintering in Louisiana were received; the increase in numbers may perhaps be related to the recent split of this species. In Alabama, it is a rare visitor along the coast in fall and winter and very rare inland (Imhof 1976). The status in Mississippi is likely very similar to that of Louisiana and Alabama, but due to the small number of observers there, few are reported. A female was reported from Vieques, Puerto Rico, on Dec. 22, 1993. This is the only record for the West Indies. It has been recorded at least twice on the Pacific slope of Costa Rica and there are two winter records for El Salvador (Thurber *et al.* 1987).

MOULT Adults have only one moult per year, the complete definitive pre-basic moult (September–early October). Definitive pre-alternate moults are lacking, all seasonal changes in plumage occur through wear. Unlike Baltimore Oriole, Bullock's Oriole does not moult on the breeding grounds. Bullock's Orioles apparently migrate to the southwestern states (Arizona and New Mexico) and NW Mexico, during the fall rains, and moult there before proceeding to their wintering grounds (Rohwer and Manning 1990). The first pre-basic moults also occur south of the breeding range; these replace the body feathers, but retain juvenal wing and tail feathers. A few 1st-year males have a limited pre-alternate moult of the throat feathers.

MEASUREMENTS ***I. b. bullockii***. Males: (n=12) wing 99.8 (97–102.4); tail 78.7 (75.7–81.8); culmen 18.5 (16.5–20.6); tarsus 24.9 (24.1–25.4). Females: (n=12) wing 93.7 (89.4–98.3); tail 74.4 (69.3–79.2); culmen 18.3 (17.0–19.8); tarsus 24.6 (23.4–25.4).

I. b. parvus. Males: (n=42) wing 96.3 (94–99); tail 75.5 (72–79); culmen 19.5 (17.3–20.4). Females: (n=15) wing 90.6 (85–93); tail 72.6 (68–76); culmen 18.3 (17.6–19.0).

NOTES Bullock's Oriole hybridises to a limited extent with Black-backed Oriole of the Mexican plateau, along the Río Sestín in northern Durango. The series from the Río Sestín are obviously intermediate; however, they are closer to Bullock's than Black-backed Oriole (J. Rising pers. comm.). This hybrid zone has not been studied, but it does not appear to be extensive. Given the fact that this species also hybridises with Baltimore Oriole, which is not its closest relative, and that these two species are thought to be different, we feel it is appropriate to also consider Black-backed Oriole as a different species. Bullock's Oriole has hybridised with Hooded Oriole in captivity. See Notes under Baltimore Oriole for genetic relationships with that species.

REFERENCES BBL data [movements, banding], Bent 1958 [voice, behaviour, nesting], Butcher 1981 [behaviour], Cruden and Hermann-Parker 1977 [behaviour], Miller 1906 [hybridisation, notes], Rising 1970 [geographic variation], Rohwer and Johnson 1992 [moult], Rohwer and Manning 1990 [moult, movements], Rothstein 1977 [nesting], Schaefer 1980 [nesting, nest insulation], Schaldach 1963 [habitat], Thurber *et al.* 1987 [distribution], van Rossem 1945 [geographic variation, moult].

42 BLACK-BACKED ORIOLE *Icterus abeillei* (Lesson) 1839 — Plate 10

Another Mexican endemic, this species was formerly lumped as a subspecies of Northern Oriole which also included Bullock's (41) and Baltimore (40) Orioles. It is also known as Abeille's Oriole.

IDENTIFICATION A largely black oriole with orange underparts, a narrow black throat bib, orange spectacles, bold white-wing markings and an orange base to the tail. Black-backed Oriole is most similar to Bullock's Oriole in all plumages. Adult male Black-backed Oriole is readily identified by its black rump, flanks and ear-coverts. Females and immatures of these two species pose a formidable identification challenge and many females may not be reliably identified. Structurally, the two species are almost identical. Female Black-backed Orioles may show a ghost pattern of the male's face pattern, with darker ear-coverts that contrast with the yellow malar area. Female Black-backed Orioles show darker lores than Bullock's Orioles. Unlike Bullock's Oriole, female Black-backed Oriole lacks the yellow wash to the face which contrasts with a darker eyeline and crown. In addition, the yellow of the underparts is more extensive on female Black-backed Orioles, usually extending farther down on the breast, and often showing a more orange colour than that of Bullock's Oriole. The yellow of the throat and breast blends in more gradually with the greyish or pale yellow belly, unlike on Bullock's Oriole which shows a more discrete change from the yellow throat to the grey belly. In addition, along the midline of the throat, Black-backed Orioles either show a pale, whitish stripe, or black throat feathering, unlike most Bullock's Oriole females. Some female Black-backed Orioles also show obvious dark centers to the back feathers. The crown of Black-backed Oriole tends to be a darker olive, rather than a greenish-grey. Mimicking the pattern of the male, the females tend to show darker olive-yellow or olive-grey, rather than grey, flanks which contrast with the paler grey belly.

The juvenal plumage of this species and Bullock's Orioles are quite similar. Black-backed Oriole averages darker, particularly on the underparts and flanks. The flanks are brownish-grey on Black-backed but silvery-gray on Bullock's, which lack contrast with the belly. Black-backed has bright yellow underparts with a slight green wash; however, Bullock's are pale yellow without green tones. The crown of Black-backed is olive and darker than the back, whereas in Bullock's the back and crown are concolorous. The face pattern is different in these two species. Bullock's Orioles have yellowish auriculars (ghost pattern of the adult male) matching the throat colour, and contrasting with a darker eyeline, while on Black-backed Oriole the cheeks are olive with a yellow tone like the crown, and contrasting with the throat colour.

The face pattern of immature and female Black-backed Oriole is like that of immature and female Baltimore Oriole in that the former species has dark cheeks that contrast with a yellowish throat. However, note that Baltimore Oriole lacks the clearly set-off yellow supralores (supercilium anterior to the eye) of Black-backed Oriole. The underpart orange-yellow of Baltimore Oriole is usually more extensive, most commonly extending from the chin to the undertail-coverts, but there is some overlap. However, Baltimore Oriole lacks the darker flanks of Black-backed Oriole. Baltimore Orioles typically show dark feather centers to the back, separating them from Black-backed Orioles, but some (probably only juvenile females) have nearly unicolored back feathers.

VOICE No known recordings.

Song: Superficially similar to that of Bullock's Oriole (S.N.G. Howell pers. comm.).

Calls: Includes a Bullock's Oriole-like chatter.

DESCRIPTION A medium-sized oriole with the bill showing no curvature, the culmen and gonys are straight. Wing formula: P9 $<$ P8 $\approx$ P7 $\approx$ P6 $>$ P5, P9 $\approx$ P5; P8–P5 emarginate. Like Baltimore and Bullock's orioles, this species shows a long primary extension. The tail is square-tipped.

Adult Male (Definitive Basic): The bill is black with the cutting edge of the upper and the entire lower mandible pale grey. The eyes are dark brown. The head is black, except for an orange supraloral line that connects the bill and the eye. In addition, there is an orange lower eye crescent which may not be obvious in the field. This creates a spectacled appearance. The black of the head continues to the rest of the upperparts; the crown, nape, back, rump and uppertail-coverts are black. The central throat patch (bib) is black. There is a broad orange malar area between the black throat and auriculars. Most of the underparts are orange, most intense on the breast and more yellow on the belly and crissum. The black flanks and sides of the neck contrast with the orange underparts. The wings are black with white markings. The lesser coverts are black, while the median and greater coverts are entirely white. The tertials, secondaries and primaries are black with white outer edges, forming an obvious wing panel. The underwings show yellow wing-linings. The tail is black and yellow. The central four rectrices (R1 & R2) are black, the next (R3) is black, with its basal outer half yellow, and the outer three rectrices (R4, R5 and R6) are yellow with a black tip. The legs and feet are blue-grey.

Adult Female (Definitive Basic): The soft parts are as in adult male. The face is mostly yellow, with a yellow forehead but an olive-grey crown. The back, scapulars and rump are olive-grey, while the uppertail-coverts are a brighter olive-yellow. Often the back feathers show darker centers, giving an obscurely streaked appearance. The underparts are orange-yellow down to the lower breast, although the center of the throat is whitish. The flanks, belly and vent are pale grey, with a pale yellow crissum. Often the flanks are olive-yellow rather than grey. The wings are brownish with whitish markings. The median coverts are white-tipped while the greaters are edged white, forming a whitish wing panel. The tertials are white-edged, as are the secondaries and primaries. The underwing shows yellow wing-linings. The tail is olive, more yellow on the outer rectrices.

Immature Male (First Basic): Similar to female but the back feathers have dark centers, making the mantle appear mottled. In addition, there is a variable black bib on the throat that includes the lores. On adult females, this black bib is usually lacking and does not extend to the lores. The immature male shows a more extensive amount of olive-yellow on the head than does the female. Finally, young males have entirely white median coverts, unlike females.

Juvenile (Juvenal): Similar to female, but duller and with a loose plumage texture. The underparts are washed with buff and the yellow of the throat and breast is less extensive. The flanks are buff. The greater coverts are not as broadly edged white as on the adult female.

GEOGRAPHIC VARIATION Monotypic, formerly considered a subspecies of Bullock's Oriole, or part of Northern Oriole complex which included both Baltimore and Bullock's Orioles.

HABITAT Commonly breeds in riparian zones, thus open forest with abundant edge habitat. It may also be found breeding in orchards and gardens. Winters in a variety of habitats ranging from montane forests in the Boreal Oyamel Fir (*Abies religiosa*) zone, or in pine-oak forests in the highlands, as well as in riparian sites, urban areas and other semi-open sites with trees.

BEHAVIOUR Typically observed in pairs or small flocks. Its general behaviour patterns resemble those of Bullock's Oriole. Black-backed Oriole forages mainly for insect prey, but it also comes to nectar at flowers. Paynter (1952) suggested that based on behaviour he witnessed at a flowering tree in November, that pairs remain together throughout the year, not merely during the breeding season. However, his observations can also be explained by general flocking behaviour. Retention of pairs in winter is very unlikely in a migratory species such as this one. Details on the mating and social system of this species are not known.

Curiously, Black-backed Orioles are one of two species that regularly feed on the toxic Monarch butterfly (*Danaus plexippus*), at the butterflies' winter congregations in Mexico. The other predator species is Black-headed Grosbeak (*Pheucticus melanocephalus*). Most individuals of this butterfly species are distasteful or toxic to birds. The cardenolide poison is stored mainly in the cuticle of the butterfly. Black-backed Orioles feed on the butterfly by splitting them open and stripping out the less poisonous soft 'insides' of the insect, largely the abdomen and thoracic muscles. Nevertheless, the orioles commonly vomit after feeding on the butterfly. Orioles and Black-headed Grosbeaks account for more than 60% of the mortality at these winter butterfly roosts. The orioles eat more Monarch Butterflies on cold days. In addition, the orioles feed on Monarch Butterflies in a four to seven day cycle, depending on the year. Thus, foraging begins at high levels before dropping, and then peaking before the cycle begins again. The cause of this cycle is not known, but it is thought that it may be due to physiological limits on how much cardenolide toxin can be ingested in a given period of time (Arellano *et al.* 1993).

NESTING No details published. Presumably similar to Bullock's Oriole.

DISTRIBUTION AND STATUS Endemic to Mexico, found from 1000–3200 m, but breeds above 1500 m (Howell and Webb 1995). **Breeding**: An inhabitant of the central Mexican plateau. Found as far north as Durango, Zacatecas, S Nuevo León and ranges south to Michoacán and Veracruz. **Non-Breeding**: From the central volcanic belt (Michoacán east through Mexico, Puebla to west-central Veracruz) south to Oaxaca (Howell and Webb 1995).

MOVEMENTS A short distance migrant with the majority moving to the southern end of the breeding range or slightly farther south. No information is available on its migratory routes or the timing of migration, but it is present in the breeding areas April–August and in the wintering areas August–April (Howell and Webb 1995).

MOULT Largely unknown, probably similar to Bullock's Orioles with only one complete moult a year. Adult specimens are in complete moult as early as late June. Fall and winter birds are in fresh plumage. Juveniles are assumed to have an incomplete first pre-basic moult, where body feathers and perhaps a few coverts are changed. The extent or presence of a first pre-alternate moult is unknown, but definitive pre-alternate Moults appear to be lacking.

MEASUREMENTS Males: (n=10) wing 104.4 (100–107); tail 85.3 (80.0–92); culmen 18.1 (17.3–19); tarsus 23.7 (22–25.4). Females: (n=6) wing 93.9 (93–98); tail 82.4 (76–97); culmen 18.3 (17–20); tarsus 23.7 (20–25.1).

NOTES A series of specimens confirms that this species hybridizes to a limited extent with Bullock's Oriole in N Durango, Mexico. The extent and dynamics of this hybrid zone has not been studied. Bullock's Oriole maintains its integrity even though it hybridises throughout an extensive area with the distantly-related Baltimore Oriole. A recent comparison of the mitochondrial DNA sequence of orioles suggests, quite surprisingly, that Black-backed Oriole is more closely related to Baltimore Oriole (K. Omland, S. M. Lanyon). This relationship needs to be studied more closely. This species is also known as Abeille's Oriole.

REFERENCES Arellano *et al.* 1993 [behaviour, feeding on Monarch Butterflies], Miller 1906 [hybridisation, notes], Rising 1973 [distribution, notes], Rohwer and Manning 1990 [identification, notes]

43 ORCHARD ORIOLE *Icterus spurius* (Linnaeus) 1766 — Plate 9

A small, migrant oriole breeding in the east of North America, the male is the only largely chestnut icterid.

IDENTIFICATION The male has a black hood, back and tail, and chestnut underparts, rump and epaulet. The wings are edged with white. Adult males of this species should not pose an identification problem, their chestnut colouration is unlike any other oriole. Females and fall immatures can look very similar to female or immature Hooded Orioles (36). Usually, the longer and more curved bill of Hooded Oriole is noticeable (Figure 29.1); however, observers inexperienced with either species may find this feature difficult to judge and young Hooded Orioles may show a bill that is not fully grown. In general, Orchard Oriole is greener above and yellower below while Hooded shows gray tones to the back and dull yellow underparts, although some overlap in these characters may occur. Orchard Orioles are quite a bit smaller, shorter-tailed, and can often be initially mistaken for a warbler! Hooded Oriole, while small, should not be mistaken for a bird as small as a warbler. In addition, Hooded Oriole has a proportionately longer tail that is noticeably graduated, which Orchard Oriole's tail is squarish and short for an oriole. Immature males of these two species, showing black bibs are also similar. Notice that the black bib on a young male Orchard Oriole is not as extensive, nor does it extend as far down on the breast as that of Hooded Oriole. Also, look for single chestnut feathers scattered on the breast and belly of Orchard Orioles; these may appear quite early on in the fall. The calls are diagnostic for these two species. Orchard Orioles give a sharp *chuck* note while Hooded Oriole has a whistled *wheet*. However, at times juvenile, and rarely adult, Hooded Orioles may give a soft *tup* note, or a louder *chuck* note, but these are either not as loud as Orchard Oriole's call or are lower pitched and slightly shorter in duration (C. Benesh pers. comm.).

In eastern North America, Orchard Oriole is one of two regularly occurring orioles. Baltimore Oriole (40) is larger, and more orange in all plumages. These two species are not easily mistaken. They may nest side by side in some places. At a distance, the jerky flight of Orchard Oriole will give away its presence. In flight this species often flicks its tail, unlike the more steady flight of Baltimore Oriole.

As mentioned above, female Orchard Orioles are often mistaken for warblers, at least initially. The small size, white wingbars and general colouration may also suggest a female Western Tanager (*Piranga ludoviciana*). This is particularly problematic in California where Orchard Oriole is a rare but regular fall vagrant and the tanager is a common migrant. However, the tanager has a shorter and stouter bill, usually with an orange base, not the grey or flesh-coloured base of the oriole. The tanager is a brighter yellow below and is greener above as well. The calls are a further aid to identification.

VOICE Song: The male's song is a lively warble of varied notes, with no two males singing exactly alike. There appear to be two types of songs, long ones and short ones. The long songs are reminiscent of an American Robin (*Turdus migratorius*), or perhaps a *Pheucticus* Grosbeak. These songs last 3–5 seconds and are more clearly whistled than short songs. Typically, there is a downslurred note at the end of the songs, which is characteristic of this species. The short songs are springier and faster, perhaps more like that of a House Finch (*Carpodacus mexicanus*), or a *Passerina* Bunting. Short songs are not so clearly whistled, and include sharper notes. They average approximately 2 seconds in duration. Males may sing while in flight. Female Orchard Orioles are also known to sing (Beletsky 1982a). Statistical analyses of adult versus immature male songs finds them to be significantly different in several respects (Clawson 1980), but they may not sound noticeably different to the human ear. Females, however, can identify if a male is mature by the song that male sings (Clemson 1980).

Calls: The common call is a sharp *chuck*. This *chuck* note is widely believed to be a clear way to separate fall Orchard from Hooded Oriole; however, note that juvenile (and some adult) Hooded Orioles also utter a similar note. As in other orioles, there is a chatter given as an alarm call. In addition, a single whistle is used to communicate between parents and young.

DESCRIPTION A small oriole with a short squared, off-tail and a small, slender bill. The culmen is only slightly curved (Figure 29.1). Wing Formula: P9 < P8 > P7 > P6; P9 ≈ P6; P8–P6 emarginate. The nominate form is described here.

Adult Male Breeding (Definitive Basic, worn): The bill is black with the basal half of the lower mandible blue-grey, sometimes extending to cover almost all of the lower mandible except for the very tip. The eyes are dark brown. The head, neck, upper breast, nape, back and scapulars are black. The rump and uppertail-coverts are rich chestnut. The underparts from the breast down to the undertail-coverts are a rich chestnut, the same colour as the rump. The wings are largely black, except for the lesser and median coverts which are chestnut like the underparts, creating a chestnut shoulder (epaulet). The greater coverts are tipped and edged with white, forming a broad white wingbar, on some worn individuals only the tips of the feathers are white. The primaries, secondaries and tertials are edged on their outer vanes with white as well. These white edges form a pale panel on the closed wings, most prominent on the secondaries. Note that the uppermost tertial is edged chestnut. Often the chestnut edges are also present on the other two tertials but are restricted to the bases, the terminal part of the feather being edged white. On the underwing, the wing-linings are chestnut. Tail entirely black, narrowly tipped with brownish white. The legs and feet are blue-grey.

Adult Male Non-Breeding (Definitive Basic, fresh): Like breeding male, but the underparts are tipped yellowish. In addition, the back and scapulars and sometimes the head and neck are tipped olive to pale chestnut, obscuring the black back and crown. The throat seldom shows pale tips. The edges of the tertials and coverts are wider, showing a buffy tint. The tail feathers often show noticeable grey tips. This plumage is usually not seen north of Mexico.

First Fall Immature Male (First Basic): Similar to adult female but underparts a brighter yellow. In addition, the throat, and/or the lores may show some black feathering.

First Summer Immature male (First Alternate): Like female, but shows black lores, and black feathers around the base of the lower mandible and throat, forming a black bib. Usually these individuals show some chestnut on the lower edge of the black throat. A smaller number may show scattered black feathers on the back. This is a variable plumage, ranging from female-like birds with the black restricted to the lores and patchily on the throat, to birds showing a full black bib, with extensive chestnut on the underparts and black feathers on the upperparts. These advanced individuals may show blackish outer rectrices.

Adult Female (Definitive Basic): Yellowish-olive above, more yellow on the rump and uppertail-coverts. The back is slightly grayer, usually with duskier centers to the feathers. Entirely yellow on the underparts, more olive on the flanks. Wings dark grey, with the greater and median coverts noticeably tipped with white, forming two white wingbars; the greater coverts are also thinly edged with white. The primaries, secondaries and tertials and edged with olive. Tail olive. Soft part colours as in the adult male.

Immature Female (First Basic-First Alternate): Immature females are similar to adult females but are browner above and brighter yellow below. Plumage changes from first basic to first alternate plumage are slight and may not be noticeable in the field. First alternate (1st-summer) females are browner on the underparts than first basic females, but have more evenly coloured and browner upperparts than adult females.

Juvenile (Juvenal): Like adult female, but duller and with buffy or yellowish wingbars. Bill with a pink or pale orange base, particularly on the lower mandible and looking more greyish-pink as the birds age. The crown is olive with a yellow wash, becoming olive-brown on the nape and back. The face and underparts are lemon-yellow, brightest on the chest and becoming buffier on the flanks and vent. The wings are brownish with yellow-tipped lesser coverts. The median coverts are boldly tipped buff, forming a buff upper wingbar. The greater coverts are tipped and edged buff creating a buff lower wingbar and pale panel. The wing-linings are off-white. The tertials are also edged buff. The tail is olive with yellow feather edges.

GEOGRAPHIC VARIATION Three subspecies are recognized, including *fuertesi*, which is treated in a separate account below. The two remaining subspecies, which comprise the *spurius* subspecies group are treated here.

I. s. spurius is the form that breeds in the US and Canada, south to E Chihuahua, N Coahuila, N Nuevo León and N Tamaulipas in Mexico. It is described above. There is a gradual cline in size in this form, with the southwestern populations being the smallest (Graber and Graber 1954); these were given the name *I.c.affinis* in the past. There is great overlap in measurements and a lack of any other characteristic to separate this population from *spurius*.

I. s. phillipsi is found in the central plateau of Mexico, from C Durango south through W Zacatecas, Aguas Calientes, to SE Jalisco, W Guanajuato, N Michoacán, N Mexico and W Hildago. It resembles *spurius*, but is larger, particularly when compared to the southernmost populations of *spurius* which are the smallest of that subspecies. The juveniles are paler below and grayer above than *spurius*. The females are indistinguishable on plumage (Dickerman and Warner 1962). Note that Monroe (1968) found *phillipsi* to be indistinguishable from *spurius*. It may be prudent to lump *phillipsi* with *spurius*.

HABITAT In summer, it prefers open woodlands, savanna woodlands and open second growth, such as riparian areas, pecan plantations, orchards, tree nurseries, golf courses and edge habitat near agricultural lands. Overall, there is a preference to nesting near water, either rivers or lake shores. While there is a preference for open forests or woodlands, the nest trees chosen tend to be dense or closely spaced for cover. Orchard Oriole is one of the few eastern Texas species which is more common in the narrowest riparian strips left after clear-cutting of the adjoining forest, whereas most bird species are found in their lowest abundance in these narrow forest strips (Dickson *et al.* 1995). On the Gulf coast, they have been known to set up colonies in *Phragmites* marshes. During migration they may be found almost anywhere. However, Orchard Orioles do appear to avoid forests with closed canopies and coniferous forests. Second growth and scrub vegetation is favoured during migration. It is not uncommon to observe migrant Orchard Orioles foraging on open lawns during the spring. Winters in a variety of situations from dry, deciduous woodlands to moist or wet forests; however, it is almost restricted to lowlands while in Central America. Orchard Orioles prefer light woodlands, dry forest, thickets, scrub forest, suburban areas and moist tropical forest edge. It is fond of plantations (i.e. banana) and may cause some economic damage to these crops. In winter it roosts in aggregations, often in tall grass or marshes but sometimes in small, dense trees (Bent 1958, Dickey and van Rossem 1938).

BEHAVIOUR Southbound migrants in Central America often sing, up until October. They begin singing again in earnest in March, usually before they leave their winter quarters. At this time, the young males begin training their voices, often singing more than their adult counterparts (Skutch 1971). Wintering Orchard Orioles often flock, and appear to me to be more social than other orioles. They also roost in groups, often numbering in the hundreds. Roost sites tend to be used winter after winter. In winter, Orchard Orioles forage to a great extent on nectaring flowers. The flower of *Erythrina fusca* is the same colour as the oriole and is apparently only opened properly, such that the pollen clings to the chest of the bird, by the Orchard Oriole. It has been suggested that Orchard Oriole and *Erythrina fusca* have coevolved into a mutualistic relationship (Wetmore *et al.* 1984).

Orchard Oriole is weakly territorial, only defending the area right around the nest site. This allows for several oriole pairs to nest in the same tree in some high density nesting areas. These aggregations have been likened to true colonies (see Nesting). The nearest neighbour distance is significantly less in Orchard Oriole than in Baltimore Oriole (Scharf and Kren 1996). In marginal habitat and particularly in the northeast of the continents, Orchard Orioles nest solitarily. During nesting, males may be aggressive toward each other. Curiously, immature males elicit an equally aggressive response as fully adult males, contrary to what is generally found in other species that delay plumage maturation (Enstrom 1992a). Orchard Orioles may nest in the same tree as Bullock's or Baltimore Orioles, particularly in the Great Plains, and they do not appear to show any aggression toward each other. However, the *chuk* and chatter vocalisations of Orchard Oriole are answered by the other two species (Scharf and Kren 1996).

The mating system appears to be monogamous in this oriole. Males sing primarily to attract mates and songs do not appear to perform a territory defense function (Scharf and Kren 1996). Females have been described as giving three types of displays. 'The Bow' is a lowering of the head given in the direction of the male, while 'the Seesaw' is an alternate lowering and raising of the head and tail, and finally 'Begging' is a stereotyped passerine display where the female rapidly flutters the wings and gives a high trill (Enstrom 1993). Males may 'Bow' and 'Seesaw' to females, but they also give a flight display where movements similar to those given during 'the Seesaw' are performed during a slow, butterfly-like flight (Enstrom 1993). Females display mainly to adult males, less so to immature males (Enstrom 1993). Pair formation occurs shortly after females arrive on the breeding grounds. Both adult and immature males are present on the breeding grounds simultaneously and are in direct competition with each other for females. After the young fledge and become more independent, both adult and 1st-summer males leave the breeding area and begin their migration. Females and the year's fledglings remain on the breeding ground longer, typically forming into larger foraging flocks (Scharf and Kren 1996).

NESTING In the north, the nesting season is mid May–July; this is earlier further south and may be as early as mid April. In the far north of their breeding range, in Manitoba, males arrive in late May and females in early June, with nesting beginning upon the arrival of the females (Sealy 1980a). In this same Manitoba population, banded nesting birds returned on consecutive years. In addition, a significant proportion of the males breeding in this new northern breeding outpost were males still in their first basic (immature) plumage (Sealy 1980a).

Nesting is commonly solitary, particularly in the east and north of the distribution. Where population densities are high, more than one nest may be found in a tree. Up to 20 nests per tree have been found in Arkansas (Thomas 1946) but typically multiple nest numbers on a single tree are much lower (Scharf and Kren 1996). These situations do appear to be colonial in nature, but they have not been studied in any detail. Nests tend to be low or at mid-levels, seldom to 10 m off the ground. The nest is a deep, hanging cup, basket-like but not a long hanging basket as in other orioles. It is usually woven with grasses and is attached to a fork of a branch or to hanging Spanish moss (*Dendropogon usneoides*) (Bent 1958). Nests are built in six days (Sealy 1980a) primarily by the females (Scharf and Kren 1996). Schaefer (1976) discovered that western nests have significantly wider diameters presumably to allow for greater heat dissipation. Some observers have noticed an association with Eastern Kingbird (*Tyrannus tyrannus*), probably to gain some nest protection, as the kingbird is aggressive in its defense of the nest site from predators. The clutch size is variable; commonly 4 eggs are laid but it can range from 3–7 eggs. The eggs are bluish-white with brown, or purplish spots, scrawls and blotches, concentrated on the broader end of the egg. Incubation lasts 12 days, sometimes as much as 14 days and it is carried out solely by the female (Scharf and Kren 1996). The young fledge 11–14 days after hatching (Sealy 1980a). Both adults share the responsibility of feeding of the young. Only one brood is raised in a season.

Orchard Oriole is a common host for both Brown-headed (101) and Bronzed (99) Cowbirds, and it has been observed to fledge young of both cowbird species (Friedmann and Kiff 1985). Parasitism rates as high as 28% have been reported but they are typically much lower than this (Scharf and Kren 1996). The overall effect is the reduction of fledging success for the oriole. Recently, Bronzed Cowbird has started to nest in parts of coastal Louisiana. It appears that cowbirds are parasitising Orchard Oriole quite heavily and the abundance of nesting orioles has decreased (D. Purrington pers. comm.).

DISTRIBUTION AND STATUS Uncommon to common, locally distributed in the north of the range.

Breeding Range: North to SE Saskatchewan, SW Manitoba, S and W Minnesota, S Wisconsin, north-central Michigan (absent from the Upper Peninsula), S Ontario (south of the Canadian Shield), C New York (south of the Adirondack mountains), W and S Vermont, SE New Hampshire, S and E Massachusetts, and extreme S Maine. South along the Atlantic coast to N Florida (Citrus, Sumter, Lake and Orange Co.). Breeds south to the Gulf coast of the Florida panhandle, Alabama, Mississippi, Louisiana, Texas to northernmost Tamaulipas, Mexico, but is absent from the central Rio Grande valley in Texas. Continues south into the central plateau of Mexico, eastwards to E Coahuila, E Zacatcas, W San Luis Potosí, to W Querétaro; south to S Guanajuato, and N Michoacán; west to C Jalisco, E Nayarit, W Durango, and NC Chihuahua (Howell and Webb 1995). Breeds irregularly in E Sonora. The western border of the breeding distribution ranges to E Wyoming (first confirmed nest in 1986), E Montana, E Colorado, W Oklahoma (the panhandle), NW Texas (along the W border of the panhandle), and up the Pecos river valley of SE New Mexico to Roswell. See also Fuerte's Oriole (43X).

Non-Breeding Range: From Mexico to northern South America, much rarer in the latter. Winters along both coasts of Mexico from C Sinaloa on the Pacific and S Tamaulipas on the Gulf coast, restricted to the coastal slopes as far south of the Isthmus of Tehuantepec where they begin to be found in the interior except in the highlands of C Chiapas. Also present on the Yucatán peninsula, including Belize. In Guatemala, it winters in the lowlands of the north and the Pacific coast, but is absent from the highlands (Griscom 1932). Winters throughout El Salvador, Honduras (rarer or absent from highlands in the south?), Nicaragua, Costa Rica to 1500 m, where it is more common on the Pacific slope (Stiles and Skutch 1989), and the lowlands of both slopes in Panama (Ridgely and Gwynne 1989). Ventures south into NW South America in Colombia and Venezuela. Occurs in the Colombian lowlands (to 500 m) west of the Andes south to Valle (Hilty and Brown 1986), in the lowlands around the Santa Marta area and up the major river valleys (Cauca and Magdalena), east of the Andes in the lowlands south to W Meta (Hilty and Brown 1986). Also found in extreme W Venezuela (Zulia, Maracay and Aragua)(Meyer de Schauensee and Phelps 1978). There are several confirmed winter records from Florida, but these are extremely rare (Stevenson and Anderson 1994), and a few winter records also exist from Louisiana (Lowery 1974).

MOVEMENTS A long distance migrant. This species vacates the wintering grounds in Central America by early April, with most of the migration occurring in March. Orchard Orioles begin arriving in the north in late April–early May, but are present in the southern US from late March. At the northernmost breeding outposts, arrival may not occur until mid to late May. Many birds fly directly north over the Gulf of Mexico, making landfall on the coasts of the Gulf coast states. A smaller number of migrants go through Florida, primarily in the spring, and these tend to be found along the western coast or the Keys. During cold fronts, numbers of spring migrants have been grounded on the Florida Keys and in the Dry Tortuga Islands (Stevenson and Anderson 1994). In general, Orchard Orioles are more common along the Mississippi flyway than along the Atlantic coast during spring. In Cuba, they are a spring migrant and are rare. Males, both adults and immatures, arrive in the breeding grounds first, with females returning later. After nesting, Orchard Orioles, particularly males, leave the breeding grounds early and many are gone by mid July. Overall, it is rare to see an Orchard Oriole in northeastern North America past August. This is not the case in California, where fall vagrants are most regularly found in September. Fall migration appears to take place farther west than in the spring. This species is rarer in Florida and Cuba in fall than in the spring. Perhaps the fall migration takes place largely over land, or perhaps on a more westerly route over water. In the Bahamas, Orchard Oriole is considered a vagrant as only two records exist, one from Cay Sal and the other from Nassau. In the wintering grounds, fall arrival is early, beginning in early or mid July in parts of Mexico. It is one of the first migrants to arrive in Central America on its southbound migration.

Orchard Oriole is a regular vagrant on the Pacific coast, mainly in California. Washington state has had one record, Samish Dec. 15, 1991, which is the northernmost record on the Pacific coast. They are more regular in Oregon, which has four records, one from each season, the first during September 1981 (Gilligan and Irons 1987). The summer record, June 4–7, 1991 in Fields, Harney Co., is the only one from the interior of Oregon. In California, they are extremely rare in spring (March–July), and rare but regular in fall (August–October). An average of 10–12 individuals are found each fall along the coast. A small number winter annually in the south of the state (Small 1994). Many of the Rocky Mountain states have no records, and this species should be looked for there. However, there is one record (June 27, 1964) from Utah (Worthen 1973) and three records from Nevada, two in summer and one in fall (Alcorn 1988). It is a rare vagrant in Arizona where over 25 records exist, mostly from May and June, with one winter record (Monson and Phillips 1981).

In the Northeast it is a regular vagrant in Nova Scotia, almost strictly as a spring overshoot. The first report was of a female collected at Three Fathom harbour on 6 Sept. 1890 (Piers 1894). Saunders (1902) collected a young male on Sable Island on 15 May 1901. This species was not reported again until 1958. Since then, it has been reported quite regularly. Most birds have been recorded in spring, for which there are some 40 reports of 52 birds. The earliest was on 15 Apr. 1971 in Yarmouth Co., and the latest, from Sable Island was on 22 June, 1975. A male in Shelbourne Co. on 28 June 1966, was suspected of breeding, as was a pair that spent two days in an orchard near Wallace, Cumberland Co., in June 1982. Fall vagrants are not as regular, with 14 reports of 18 birds between 20 Aug.–3 Oct. (Tufts 1986). It is less than annual in New Brunswick, but there are many records, also mainly for the spring. In Newfoundland it has occurred twice, in Oct. 1991 and in November 1991. Quebec has had over 30 observations, mostly in May but none have included breeding birds yet. There are two northern Ontario records from Thunder Bay, one in late May, and the other in late September. It is casual in spring in N Michigan and N Minnesota.

The longevity record is held by a banded Orchard Oriole which was retrapped at the age of nine years and seven months old (Klimkiewicz and Futcher 1989). One bird banded in Honduras in December was recovered in April of the following year in Alabama, while another banded in Belize on March 20, 1961, was killed by a cat in Alabama on May 26, 1961 (Imhof 1976). Of 74 recoveries in the Bird Banding Lab. database, almost all were of birds subsequently recovered during the breeding season in the same state where they had been originally banded. One individual banded in April 1964 in Louisiana was recovered in March 1969 in Honduras (BBL Data).

Orchard Orioles were not historically known to breed in Saskatchewan, having colonized that province from the late 1970s. Colonisation began in the southeast of the province and advanced both north and west. A similar increase in population and northward extension of the breeding distribution has occurred in Manitoba (Sealy 1980a). Recently, numbers have decreased, but the reason for this is unclear (Bjorklund 1990). It has also increased southward in Florida as a breeder and westward in Colorado (Scharf and Kren 1996).

MOULT definitive pre-basic moult is complete and occurs on the wintering grounds (October–January). This contrasts greatly with the moult of Baltimore and Bullock's (41) Orioles, also neotropical migrants, which moult either on the breeding grounds or in an intermediate locality. Fall-arriving Orchard Orioles in Central and South America are often noticeably worn. Most neotropical migrant passerines undergo their complete moult on the breeding grounds, thus the moult pattern of Orchard Oriole is unusual. Definitive pre-alternate moults are limited or sometimes absent, involving some of the body feathers and greater coverts, and this type of moult occurs on the wintering grounds (January–March). This moult is more extensive in males than females, but the latter increase the rate of moult later on in the winter (Enstrom 1992b). The extent of the first pre-alternate moult is variable in young males which accounts for the variable appearance of first pre-alternate (1st-summer) male plumage. Its timing is similar to that of adults. The first pre-basic moult is incomplete, usually involving the replacement of the inner 3–6 secondaries and the outer 5–7 primaries, and a variable number of tail feathers (June–August). The first pre-basic moult is sometimes complete, but perhaps in males only (Pyle 1997). This moult occurs largely on the breeding grounds.

MEASUREMENTS ***I. c. spurius***. Males: (n=51) wing 77.4 (73–82); (n=47) tail 69.9 (65–75); (n=30) culmen 16.3 (15.0–17.5); tarsus 21.6 (20.6–22.9). Females: (n=27) wing 73.8 (69–81); (n=26) tail 67.2 (63–74); (n=9) culmen 15.7 (15.2–17.3); tarsus 21.6 (20.6–23.1).

I. c. phillipsi. Males: (n=38) wing 79.3 (76–81); (n=36) tail 73.4 (70–79); (n=7) culmen 15.5 (?–?); tarsus 21.6 (?–?). Females: (n=11) wing 74.0 (71–81); (n=10) tail 68.6 (65–73); (n=3) culmen 15.7 (?–?); tarsus 21.6 (?–?).

NOTES This oriole has experienced a 1.8% annual decrease in numbers based on the breeding bird survey data between 1966 and 1996 (Sauer *et al.* 1997). This decrease is significant, and the causes are unclear. However, certain regions such as Canadian survey routes have shown increases in Orchard Oriole populations.

The origin of the specific name *spurius* is interesting. Originally, the female Baltimore Oriole was thought to be the male of this species, resulting in the name *spurius* or 'bastard' Baltimore Oriole when the situation was clarified.

A recent study of mitochondrial DNA sequence in orioles has illuminated that Orchard Oriole belongs in the group that includes Black-cowled Oriole, Lesser Antillean oriole species, Black-vented Oriole, Epaulet/Moriche oriole complex, and Hooded Oriole (K. Omland, S. M. Lanyon pers. comm.). Any resemblance in the plumage pattern between Orchard Oriole and Baltimore Oriole is merely due to convergence.

REFERENCES Beletsky 1982a [voice], Bjorklund 1990 [movements], Dickerman and Warner 1962 [geographical variation, measurements], Howell and Webb 1995 [distribution], Klimkiewicz and Futcher 1989 [age], Lehman 1988 [identification], Pyle *et al.* 1987 [moult], Ridgway 1902 [measurements], Scharf and Kren 1996 [behaviour, distribution, nesting, moult], Sealy 1980a [nesting, distribution, movements], Wetmore *et al.* 1984 [behaviour], Wormington and Lamond 1987 [vagrants, movements].

43X 'FUERTE'S ORIOLE' *Icterus spurius fuertesi* Chapman 1911 — Plate 9

Also known as the Ochre Oriole, this pale form of Orchard Oriole (43) was once given species status.

IDENTIFICATION The females of this and Orchard Oriole are essentially unidentifiable. However, there is a tendency for females of 'Fuerte's Oriole' to be paler below and slightly smaller. Males resemble Orchard Oriole, but the chestnut is replaced by pale ochre. For identification from other species see the section under Orchard Oriole.

VOICE Song: Similar to that of Orchard Oriole,

but softer and not as bold and varied. It is known to sing while in flight.

DESCRIPTION Structurally similar to the *spurius* group of Orchard Oriole. Wing Formula: P9 < P8 < P7 > P6; P9 ≈ P5; P8–P6 emarginate.

Adult Male (Definitive Basic): Similar to adult male Orchard Oriole in pattern, wing edgings and tail colour, only differing by having the chestnut parts replaced by ochre. Non-breeding males in fresh plumage are tipped with buff on the upperparts similar to winter Orchard Orioles.

Adult Female (Definitive Basic): Like female Orchard Oriole, perhaps averaging slightly paler and smaller.

Immature Male (First Basic/Alternate): Like corresponding ages of Orchard Oriole and likely not separable with any certainty except if ochre body feathers are present.

Juvenile (Juvenal): Similar to juvenile Orchard Oriole, but noticeably paler, and differs in the following ways: the head and upperparts are greyish-buff, becoming pale buff on the rump, the tail is also greyish-buff, the wingbars are narrow and whitish, and the underparts are pale yellow, buffier on breast.

GEOGRAPHIC VARIATION Monotypic.

HABITAT Typically breeds in dense shrubbery of coastal sand dunes. This form prefers tangles of Hibiscus (*Hibiscus tiliaceus)* and Mangle Negro *(Conocarpus erectus)* which build up on the inland side of the dunes. Further inland they are also found in edge habitat near agriculture, in urban areas and particularly in scrubby bull's horn acacia thickets (*Acacia* sp.). It is likely that the spread to this habitat type has been recent as forests have been cleared, and that originally this was a bird of the coastal dunes.

BEHAVIOUR Males sing from exposed perches to attract females. However, no territories are maintained and little aggression is seen between males during the breeding season. Males may even sing from the same tree.

NESTING The breeding period stretches from May–August. Nests are shallow, hanging basket-like structures as in Orchard Oriole, generally placed less than 6 m off the ground. In coastal sites, the nests tend to be constructed with emergent grasses such as *Spartina* and *Eleocharis*, and unlike the nests of Orchard Orioles, these are not lined with softer material. The clutch size tends to be two eggs, which are white with dark speckling and spotting on the larger end. They are single-brooded. Bronzed Cowbirds (99) are known to parasitise this form (Friedmann and Kiff 1985); however, it is not known if they successfully fledge from their nests. Both sexes feed the young.

DISTRIBUTION AND STATUS Uncommon to common.

Breeds along the E coast of Mexico from S Tamaulipas to S Veracruz.

Winters along the Pacific slope of Mexico from Guerrero to Chiapas.

MOVEMENTS A short distance migrant. It appears to arrive in the breeding grounds in May and stays into August, with most birds having departed by the end of this month. It has been reported from southern Texas.

MOULT Unknown, presumably similar to those of Orchard Oriole. The complete pre-basic moult occurs in August and September.

MEASUREMENTS Males: (n=11) wing 74.5 (68.5–78); tail 69.4 (67–73); (n=7) culmen 15.6 (13–17); (n=4) tarsus 21 (19–25). Females: (n=4) wing 73.5 (71–76); tail 66.9 (65–70); (n=3) culmen 16.5 (14.5–18); (n=2) tarsus 22, 22.

NOTES Also known as Ochre Oriole. This species was originally described as a different species from Orchard Oriole but was lumped with it in more recent lists. Both forms are extremely similar, except for the plumage of the adult male. However, there is some variation in the colouration of male Fuerte's Orioles and the northernmost (closest to Orchard in range) are darker than the rest. There are no known intermediates, but this tendency for northern birds to be darker, along with the extreme similarity of these two forms in most respects, suggests that they are best treated as conspecific.

REFERENCES Chapman 1911 [description, measurements], Graber and Graber 1954 [voice, behaviour, nesting, distribution, measurements], Howell and Webb 1995 [distribution, voice].

44 BLACK-VENTED ORIOLE *Icterus wagleri* Plate 16
Sclater 1857

This Mexican and Central American oriole is the only one in its range with a black vent and black uppertail-coverts.

IDENTIFICATION A black and yellow oriole with a black hood, back, wings, vent and tail. This species is most similar to *prosthemelas* Black-cowled Oriole (45) in colour pattern, however the two species are allopatric. Adult Black-vented Oriole is differentiated from male Black-cowled Oriole by its black uppertail-coverts and black vent. In addition, it is a larger bird with a thicker and straighter bill. Black-vented Oriole has less extensive black on the chest than Black-cowled Oriole, and shows no sexual dichromatism (the sexes look alike). Immatures of the two species are similar, but note that Black-vented Oriole does not have entirely black auriculars and shows more black on the breast. Scott's Oriole (51) shows bold white wingbars in all plumages, with the males resembling Black-vented Oriole in pattern but note that they have yellow at the base of the tail. Note that very worn (midsummer) Scott's Oriole may appear to show little or no white on the wings and could be confused with Black-vented Oriole (C. Sexton pers. comm.). The presence of a yellow vent and yellow tail base will identify these birds as worn Scott's Orioles. Male Bar-winged Oriole (50) is also similar in pattern, but has a bold, white, lower wingbar and yellow vent and uppertail-coverts. Females and immatures of this species are more olive above than immature Black-vented Oriole, and have a thinner curved bill and show a pale lower wingbar. Their black throats and masks have a crisp border, unlike the more ragged pattern shown by immature Black-vented Orioles.

VOICE The primary song of this species is not well documented, and has not been widely recorded.

Song: It has been described as a gurgling warble with a mix of nasal and squeaky notes (Howell and Webb 1995). Marshall (1957) describes a whisper song resembling that of Hooded Oriole (36).

Calls: A nasal *nyeh* is the most characteristic call, also a *coo-nyah-ra*. The *nyeh* notes are often given in succession, particularly when alarmed. Also utters a mechanical chatter.

DESCRIPTION A slim oriole with a rounded (graduated) tail and a long and shallow, sharply-tapered bill. The culmen and gonys are slightly decurved at the tip. The difference in length between the outer (R6) and inner (R1) tail feathers may be as much as 30 mm. Wing Formula: P9 < P8 ≈ P7 ≈ P6 > P5; P9 ≈ P5; P8–P5 moderately emarginate.

Adult Male (Definitive Basic): The bill is black with a blue-grey base to the lower mandible. The legs are greyish and the eyes are brown. The head, back and breast are black as are the wings and tail. The wings are black except for the orange-yellow lesser and median coverts, which form a noticeable shoulder patch or epaulet. In flight, the wing-linings are orange-yellow like the epaulets. The belly and lower breast are also orange-yellow, most orange immediately below the black of the upper breast. Sometimes, even in the nominate race, there can be a trace of chestnut where the orange-yellow and the black of the breast meet. The rump and lower back are orange-yellow like the belly. The orange-yellow of the rump and belly contrasts with the black vent, undertail-coverts and uppertail-coverts. The tail is entirely black. Some individuals show some yellow along the rachis of the outer one or two rectrices (R5 and R6). It is unknown if this is age-related.

Adult Female (Definitive Basic): Similar to male. However, on average the females are slightly duller. More specifically, they average more pure yellow rather than orange-yellow but there is a good deal of variation.

Immature (First Basic): Soft part colours are as in adults. The head is olive-grey on the crown, with a wash of yellow on the forehead becoming yellow to orange-yellow on the face. The back and nape are olive-grey like the crown, perhaps slightly less olive than the crown. The back often has dark shaft streaks on each feather. The rump and lower back are greyish-yellow, noticeably more colourful than the upper back. The lores are black. The underparts are bright yellow from the chin to the belly, while the lower belly and undertail-coverts are greyish-yellow. Immatures lack the black vent typical of adults. The throat shows a variable amount of black, sometimes with only a few contrasting black feathers while on other individuals both the chin and malar patch are black, as well as patches of black throughout the throat. The difference in the extent of the throat patch appears to be due to individual variation; the black does not appear to increase during the first spring. The base colour of the wings is grey-brown. The lesser coverts are tipped with yellow and the median coverts are tipped with pale grey, forming a greyish upper wingbar. The greater coverts are edged grey-buff, creating a paler wing panel. The primaries and secondaries are thinly fringed grey-buff. The wing-linings are pale yellow. The brown tail has the outer three (R4 to R6) rectrices olive-yellow at their bases but tipped with brown, while R2–R3 show an olive-yellow outer edge towards their bases. As the immature moults into adult plumage, the extent of the black on the throat and back increases and the wing and tail are moulted into the adult black colour. Individuals appear a patchy mix of black and brown during this stage.

Juvenile (Juvenal): Similar to the immature, but showing several differences. The bill is pale pink towards the base. The black on the lores and throat is entirely lacking, but note that it begins to 'moult in' quite early in life. The crown and upper back are more olive-buff at this age, as opposed to the olive-grey of the immature (first basic). The underparts are duller yellow than an immature. The rump is olive-buff, like the crown, on juveniles. The wings of the juvenile show almost no yellow tipping on the lesser coverts, and in addition the median and greater coverts are buff-tipped, showing up in the field as two buffy wingbars.

GEOGRAPHIC VARIATION Two subspecies are recognized.

The form ***I. w. castaneopectus*** occurs in the north of the range, in the NW of Mexico in S Sonora, N Sinaloa and Chihuahua.

The nominate form, ***I. w. wagleri***, is found south of *castaneopectus*, covering the rest of the distribution. *Castaneopectus* is larger than *wagleri*. It also has a more extensive band of chestnut where the black of the breast meets the orange-yellow underparts. The change in form between these two races is gradual and clinal, and in central Mexico most specimens are intermediate (Griscom 1932). In addition, female *castaneopectus* sometimes show only a limited amount of chestnut on the chest. Summer individuals of both forms are slightly duller in colour than fresh birds and this may account for some of the variability that has been noted (Griscom 1932).

HABITAT Occurs in open woodlands and scrub forest in the south and in dry scrub and in arroyos of semi-desert and desert areas in the north. It does not shy away from agricultural areas and may be commonly found in coffee plantations. Open areas with isolated trees are also used. However, its preference is for shrubby sites, particularly dry subtropical scrub, often deciduous woodlands. In addition, it ranges into the oak and juniper scrub, dry oak forests, pine thickets and arid pine-oak associations. Also inhabits dry desert canyons with scattered shrubs, visiting adjoining riparian vegetation (Marshall 1957).

BEHAVIOUR Forages singly, in pairs, and sometimes in flocks, but is not generally a conspicuous species. Pairs are maintained during the entire year. Black-vented Oriole may join mixed flocks with other orioles and feeds at flowering trees. In a study of pollinators visiting flowering *Pseudobombax ellipticum* trees in Morelos, Mexico, the behaviour of Black-vented Orioles varied depending on the density of flowers. When few flowers were available a pair of Black-vented Orioles defended the entire tree against other birds, particularly orioles. As the flower density increased, the Black-vented Orioles dropped their defense of the tree and allowed access to the flowers to other orioles species, Streak-backed (35) and Bullock's (41) in this case. However, during conflicts, Black-vented Oriole was dominant over the other two oriole species (Eguiarte *et al.* 1987). In riparian areas this species has been observed foraging high in the crown or sycamores (*Platanus* sp.). Here it feeds on insects, including caterpillars (Marshall 1957). During the early part of the breeding season this species becomes more prominent as males begin to sing.

NESTING In Oaxaca, Mexico, it has been found in breeding condition June–July (Binford 1989). Makes a hanging nest typical for an oriole, except it tends to be rather shallow and hammock-like. Nests are positioned at a low or medium height, often under a banana leaf or palm frond with the supports of the 'hammock' woven into the leaf. A bird collecting fibers from a Yucca (*Yucca* sp.) plant (Marshall 1957) was likely using these as nesting material. Three eggs are laid (Salvin 1859). This oriole is a known host of the Bronzed Cowbird (99) (Friedmann and Kiff 1985).

DISTRIBUTION AND STATUS Common or fairly common. Ranges from nearly sea level to 2500 m in Mexico (Howell and Webb 1995); 1750–1900m in Honduras and 900–2200m in Guatemala (Thurber *et al.* 1987). Found from S Sonora and W Chihuahua, in Mexico, south along the Pacific slope to E Oaxaca; inland in the S Mexican plateau from S Zacatecas to W San Luis Potosí, and S Nuevo León; south along W Veracruz and Puebla to Oaxaca. Along the highlands from W Chiapas through C Guatemala, N El Salvador, S Honduras and NW Nicaragua. It is not recorded as nesting in El Salvador, and most records are from the fall and winter suggesting that it is only a non-breeding visitor to that country. In the non-breeding season it is found in the coastal lowlands from S Sonora to C Guerrero, Mexico, areas where it does not breed. The irregular wanderings of this species are still not fully understood (Howell and Webb 1995).

MOVEMENTS Wanders to an unknown extent and may be partially migratory. Some movements may be altitudinal as lowland sightings are irregular and appear to be seasonal. In El Salvador, it is present mainly late November–early February, the dry season, although breeding is unrecorded (Dickey and van Rossem 1938, Rand and Traylor 1954). Schaldach (1963) detected this oriole during most of the year in Colima, Mexico, but failed to find it in February; he suggested that they may not winter in that state at least not in the numbers present during the breeding season. It is a vagrant to Texas, US. An individual sighted for the first time in Sept. 27, 1968 at Rio Grande village returned for two summers (April–October) in 1969 and 1970. Another was found in Kingsville on June 17, 1989 and remained until early October. There is a sight record of an individual that remained for a week at Cave Creek canyon, Chiricahua mountains, Arizona during mid July (Monson and Phillips 1981).

MOULT Adults undergo a complete (pre-basic) moult after the breeding season, August–October. Pre-alternate moults appear to be lacking in adults. The first pre-basic moult results in the change from juvenal to the immature (first basic) plumage. This moult replaces the body plumage, wing-coverts and perhaps some tertials, but not the wings and tail. The first pre-basic moult usually occurs in September; it happens early in life beginning soon after fledging. Immatures (first basic) have been detected to be in moult November–December (Dickey and van Rossem 1938), it is unclear if this means that the first pre-basic moult is long and ongoing or some head and body feathers are moulted in a limited first pre-alternate moult. However, my study of specimens has not detected a general increase in black face and throat feathering as the winter progresses, which is what one would expect if a limited first pre-alternate moult is present. There is a great deal of variation in the extent of black on the throat of first basic birds, but it appears to be due to individual variation in the extent of the first pre-basic moult.

MEASUREMENTS *I. w. wagleri*. Males: (n=19) wing 104.9 (97.8–113.8); culmen 24.6 (23.4–25.4); tail 108.2 (101.1–122.2); tarsus 24.9 (24.1–25.9). Females: (n=13) wing 96.5 (93–100.1); tail 99.0 (79.5–107); culmen 22.2 (20–24.4); tarsus 24.8 (23.4–26).

***I. w. castaneopectus*.** Males: (n=8) wing 103.8 (97–110); tail 106.7 (99–116); (n=7) culmen 23.3 (20–25); (n=8) tarsus 25.3 (24–27). Females: (n=5) wing 99 (95–107); tail 100.4 (83–111); culmen 22.8 (22–24); tarsus 25.4 (25–27).

NOTES While Black-vented Oriole is outwardly similar to Black-cowled Oriole complex it does not appear to be its closest relative, based on mitochondrial DNA sequence data (K. Omland, S. M. Lanyon pers. comm.). However these two species belong to the same group (clade). See also NOTES for Orchard Oriole.

REFERENCES Alvarez del Toro 1971 [voice, habitat, nesting], Binford 1989 [nesting, habitat], Brewster 1888 [geographic variation, measurements], Dickey and van Rossem 1938 [geographic variation], Eguiarte *et al.* 1987 [behaviour],Hardy *et al.* 1998 [voice], Howell and Webb 1995 [description, voice, nesting, movements], Marshall 1957 [habitat, nesting, behaviour], Rand and Traylor 1954 [movements], Ridgeway 1902 [measurements], Schaldach 1963 [movements].

45 BLACK-COWLED ORIOLE *Icterus dominicensis* (Linnaeus) 1766 — Plate 16

A slender, black and yellow Caribbean and Central American oriole lacking white markings on the wings.

IDENTIFICATION A geographically variable, black and yellow oriole. All male forms have a black hood, back, wings and tail, the amount of yellow on the underparts varies. On the mainland, the adult male is most similar to Black-vented Oriole (44) in its general plumage pattern and its unmarked wings, other than the yellow shoulder. The mainland Black-cowled Oriole has a much thinner, slightly decurved bill and lacks the black vent or uppertail-coverts of Black-vented Oriole. However, these two species are allopatric and therefore should not be found together. Confusion is most likely with the females or immatures of the two sympatric yellow and black orioles, Yellow-tailed (30) and Yellow-backed (26). Note Yellow-tailed Oriole's yellow inner greater coverts as well as the pale fringes on the tertials. It lacks the black auriculars of female Black-cowled Oriole and has a deeper, stouter bill at all ages. Similarly, Yellow-backed Oriole also has a stouter bill and lacks black auriculars. Any black on the forehead, above the eyes, is diagnostic for Black-cowled Oriole excluding these other two species. Finally, the allopatric Bar-winged Oriole females and immatures are quite similar to the respective plumages of Black-cowled Oriole; however, they show a pale, whitish or white wingbar on the greater coverts that Black-cowled lacks.

In the Greater Antilles, Black-cowled Oriole only overlaps in distribution with the rare migrant and winter visitant Baltimore Oriole(40), the rare migrant (largely in Cuba) Orchard Oriole(43) as well as the introduced Troupial (37) in Puerto Rico, and perhaps Bullock's Oriole (41) as a vagrant; all are quite different in appearance. However, Yellow-shouldered Blackbird (65) is superficially similar in having a largely black body and yellow shoulders. Unlike the oriole, the blackbird lacks yellow on the underparts or rump, and has a completely dark bill, not blue-grey at the base.

VOICE Song: Songs of the insular forms are all significantly different (P. W. Smith pers. comm.) but detailed information is lacking. In the Dominican Republic, the nominate form sings a clear and pleasant whistled song (Wetmore and Swales 1931); however, Stockton de Dod (1978) stated that this form lacked a melodious song as is heard on other islands in the Caribbean and described the song as weak. The song of *northropi* is made up of eight or nine sweet whistles. The race *portoricensis* gives a song composed of high whistles, with songs lasting just over two seconds in length. The songs are similar to those of *melanopsis*, but differ in their higher frequency (shrill) notes and slower tempo. Apparently it rarely sings during the middle of the day, but prefers the early morning or evening. The form *melanopsis*, has a clear whistled song, lasting approximately 2–3 seconds and comprised of under 10 notes. In Cuba (*melanopsis*), this species is known as a frequent cagebird due to its pleasant song (Gundlach 1876). Songs are separated by a period that is slightly longer than the song itself, as they are in *portoricensis*. Central American *prosthemelas* is said to give a scratchy warble or a sweet, soft warble that is not audible at a great distance.

Calls: A hard *keek* or *check*. The nominate form in the Dominican Republic utters a *chur-r-churr-r* (Wetmore and Swales 1931, Stockton de Dod 1978). *Prosthemelas* delivers a scolding *chuh-chuh* or *cheh-cheh-cheh-chek* (Howell and Webb 1995) as well as sharper monosyllabic notes. *Northropi* emits a plaintive double note. The race *portoricensis* gives a single, musical *chup*. Cuban *melanopsis* gives a sharp *chip* and a nasal *wheenk*.

DESCRIPTION A slender-billed Oriole, with an obviously decurved culmen. The form *prosthemelas* will be described. Wing formula: $P9 < P8 < P7 > P6 > P5$; $P9 \approx P4$.

Adult Male (Definitive Basic): Bill black with bluish-grey basal half to the lower mandible. The eyes are dark brown. Head and body black with yellow underparts from the pectoral line back to the undertail-coverts. On the upperparts the lower back, rump and uppertail-coverts are yellow. The longest undertail- and uppertail-coverts are tipped black. The yellow of the breast has a chestnut tone at the interface with the black of the upper breast, the width and depth of colour of this band is variable. The wings are black with yellow bend of wing and epaulet (lesser and median coverts). The wing-linings are yellow. The long, graduated tail is black and unpatterned. The legs and feet are blue-grey.

Adult Female (Definitive Basic): The females are quite variable in appearance, with some resembling the male and others being much duller. There is a tendency for southern birds have more male-like plumages than populations from the north, with intermediate appearances in between. A dull female is described here. Upperparts olive, with a yellowish wash on the nape and rump, sometimes with some black flecking on the scapulars. The face, including the lores, forehead, auriculars and throat are black. The rest of the underparts are yellow. The wings and tail are black as in the male, and the epaulets are yellow or black with yellow tips. Usually, the base colour of the wings is a duller blackish-brown rather than the jet black of the male. Soft part colours as in the male. Females that are more male-like show a black back, sometimes tipped with olive. In bright females, the nape is the last spot on the upperparts to retain any olive. These females have fully yellow epaulets as well as blacker wings.

Immature Male (First Basic): Like adult female, but with brownish rather than black wings, which contrast with blacker greater coverts. Usually the olive juvenal tail is retained, but commonly, at least the central rectrices are replaced with black ones. The black on the head is usually more extensive than on adult females, particularly on older immatures.

Immature Female (First Basic): Similar to adult female, but with browner primaries and secondaries which contrast with blacker greater coverts. The tail is olive.

Juvenile (Juvenal): Like immature but even duller, with black restricted to the lores and upper throat. It is not clear if young juveniles lack black on the face altogether. The brownish wings have yellowish tips to the coverts and olive edges on the tertials. The tail is olive.

GEOGRAPHIC VARIATION There are six subspecies recognized. The amount of variation shown in this species, along with biogeographic and ecological considerations suggest that it should be divided into several species; one on the mainland and each one of the island forms separated out. The genetic differences between these forms are at a magnitude similar to that of other taxa which are considered different species (K. Omland pers. comm.). The differences in the extent of black and yellow on adults is illustrated on Figure 45.1.

BLACK-COWLED ORIOLE: This form includes two subspecies.

I. d. prosthemelas is the Central American and Mexican form found north of *praecox* and is described above. Some specimens from Honduras and Nicaragua may show more extensive black on the breast, and sometimes on the flanks. Wing Formula: P9 < P8 < P7 < P6 > P5; P9 ≈ P3; P8–P6 emarginate.

I. d. praecox is found on the Caribbean slopes of Costa Rica and Panama. It does not differ from *prosthemelas* in the adult plumage apparently. However, the juveniles have much more extensive black on the throat, extending to the lower breast. The back is black rather than olive-green. Wing Formula: P9 < P8 < P7 > P6 > P5; P9 ≈ P2; P8–P5 emarginate.

BAHAMA (NORTHROP'S) ORIOLE:

I. d. northropi inhabits Andros and Abaco islands, Bahamas. It is nearly, if not completely, extirpated on Abaco, but is uncommon on Andros Island. Unlike all of the other Caribbean forms, it has extensive amounts of yellow on the body, like the Central American *prosthemelas*. In addition, it has a stouter bill than *dominicensis*. It is black, with a brown or greenish wash to the back, and the underparts from the mid-breast to the vent, thighs, rump, lower back and epaulets (lesser and median coverts) are yellow. The wing-linings are yellow. The greater coverts and primaries are thinly fringed with white. The outer tail feathers have small white tips. The bill is black with a blue base to the lower mandible. The legs are grey-blue and the eyes dark. The female is similar to the male, but the black is duller, not as dark, and females have dusky rather than black tips to the longest undertail-coverts. Immatures of this taxon are different from the other forms as well. The upperparts are olive-grey, more yellowish on the head. The underparts, lesser wing-coverts, rump, lower back and tail-coverts are greenish-yellow, being most vivid on the abdomen and rump. The median coverts are pale sulfur yellow and the greater coverts are brown with thin white fringes. The lores are black. Older immatures have a black throat patch which extends on to the auriculars, chin and throat. Eventually, black feathers appear on the head, breast and back. The wings and tail are brownish. The juveniles are olive above, with a yellowish rump and uppertail-coverts. The throat is yellowish, while the breast is dull olive and the flanks, belly and vent are dull yellow. The dusky wings have yellow tips to the median and greater coverts, creating two yellowish wingbars. The wing-linings are yellow. The tail is dusky with olive fringes. Wing Formula: P9 < P8 < P7 ≈ P6 > P5; P9 ≈ P4; P8–P6 emarginate. Todd and Worthington (1911) note that birds from Abaco were perhaps slightly more deeply coloured than those from Andros.

CUBAN ORIOLE:

I. d. melanopsis occurs in Cuba and the Isle of Pines. This subspecies is most similar to *portoricensis*, but is even less yellow below;, in fact it is the form with the least amount of yellow as an adult. On the underparts, yellow is confined to the thighs and tips of the undertail-coverts. Apparently, it shows the largest patch of blue-grey at the base of the lower mandible of all of the subspecies. The immature is olive above with a black chin, throat and lores, similar to *dominicensis* but unlike *portoricensis*. The rest of the underparts are olive-green, becoming more yellowish on the belly and vent. Juveniles are similar, but lack the black on the throat. The juvenal plumage is like that of *dominicensis* (see below) but the crown and back are olive, lacking a russet wash. Also, the breast is olive and does not show a russet colour. Wing Formula: P9 < P8 < P7 > P6; P9 ≈ P4; P8–P5 emarginate.

HISPANIOLAN ORIOLE:

I. d. dominicensis inhabits the island of Hispaniola which includes Haiti and the Dominican Republic, and the nearby islands of Gonave and Tortue. The adult female appears similar to the adult male. Of the three Greater Antillean forms of this oriole, this one shows most yellow, but still far less than *northropi* or the mainland Central American/Mexican forms. On the underparts, the thighs, crissum and hind flanks are yellow. The bill is longer and thinner than in Cuban Oriole. Females are similar to the males. Immatures have black lores and throat, which are not cleanly demarcated. The forehead is chestnut, becoming olive on the crown, nape and mantle. The rump is more yellow. On the underparts, the chest is chestnut, becoming olive-yellow on the belly. The blackish wings show olive-grey edging; the lesser and median coverts are greenish-yellow. The tail is olive, with the outer rectrices showing some yellow unlike the inner ones. There is a certain amount of variability in this plumage, largely depending on the extent of the first pre-basic moult. Other birds may have black feathers on the belly or back, as well as black tertials and a variable number of black remiges. Juvenile (Juvenal): The crown is russet-brown, becoming more olive on the back, and yellowish-green on the rump. Shows black lores. The throat is olive, but quickly changes to russet on the breast and becomes greenish-yellow on the lower breast, belly, vent and undertail-coverts. The wings are brown with the lesser and median coverts showing obvious yellow tips which form a solid yellow upper wingbar. The greater coverts are thinly edged but not tipped with yellow, so there is no lower wingbar. The primaries, secondaries and tertials are edged

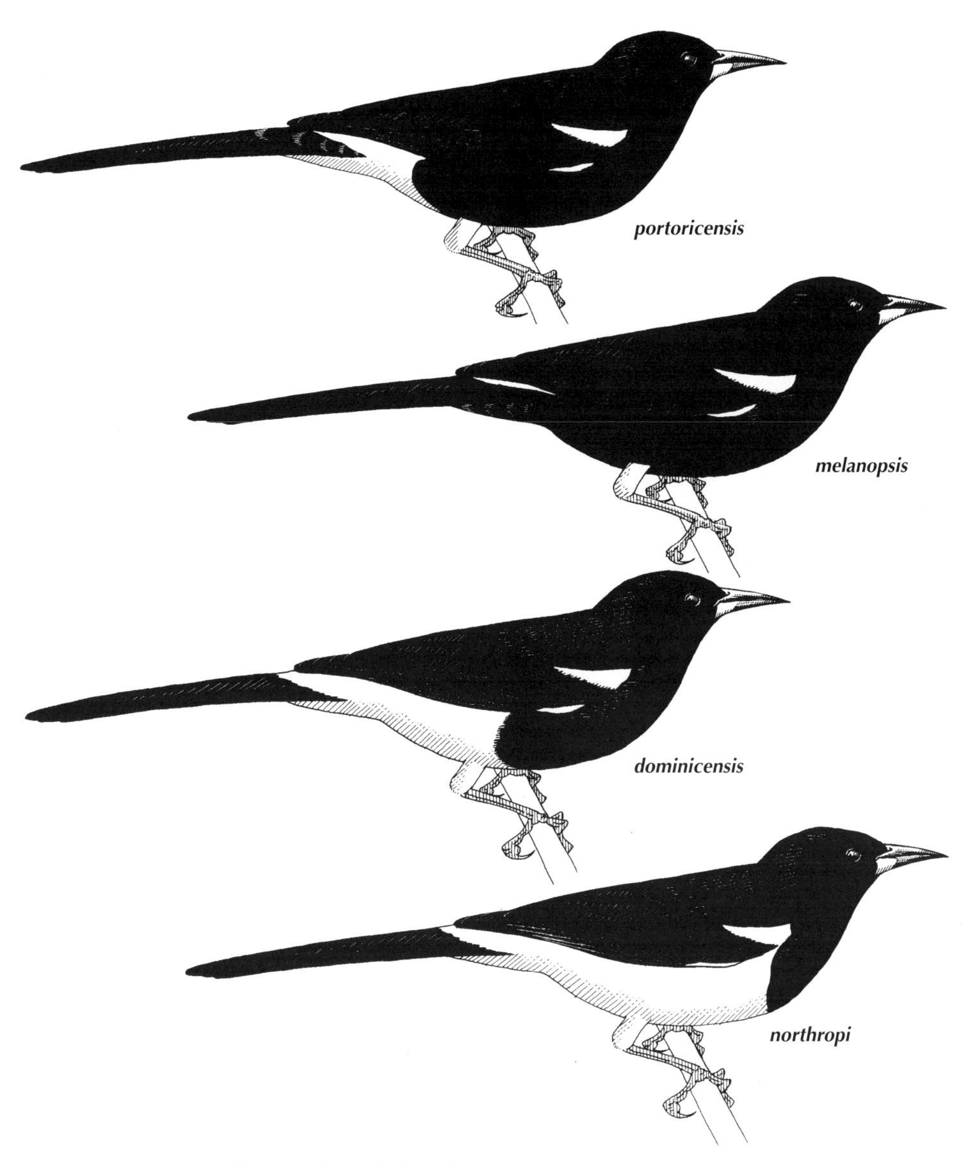

Figure 45.1 Races of Caribbean Black-cowled Orioles.

with greyish-yellow. The bend of the wing is yellow, as are the wing-linings. The tail is olive-brown. Wing formula: P9 < P8 >» P7 > P6; P9 ≈ P6; P8–P5 emarginate.

PUERTO RICAN ORIOLE

I. d. portoricensis is found on Puerto Rico. It is similar to *dominicensis* but the black is more extensive, and the bill is stouter and longer. Most of the uppertail-coverts, except for the yellow tips and the crissum, are black. On the rump and lower back, the yellow is less extensive on this form than on *dominicensis*. The yellow on the underparts is restricted to the crissum and thighs. Adult females are similar to the male. The immature is olive above, with a reddish tint which is brightest on the head, and the rump is yellow. The underparts are yellowish, with a rufous wash on the breast. The chin and lores are sometimes black, but often yellowish. Wing Formula: P9 < P8 < P7 ≈ P6 > P5; P3 < P9 < P4; P8–P5 emarginate.

HABITAT It inhabits open forests or edge habitats, and shows a preference for palms (Puerto Rico). Away from the coast (Puerto Rico) it often is found in citrus groves or coffee plantations. On Andros Island, Bahamas, the preference for palms is also marked. The planting of palms in residential neighbourhoods has allowed the oriole to colonize settled areas in the Bahamas. In Cuba, where it is common, this species is particularly fond of foraging on *Erythrina* trees planted as hedgerows. In Cuba, it is known to show a diversity in habitat choice, being found in areas ranging from gardens and plantations to dense native forest (Gundlach 1876). Occupies a variety of wooded habitats in Hispaniola, and does not avoid arid mesquite woodlands. However, it appears to be rare in pine (*Pinus occidentalis*) forests in the highlands of the Dominican Republic (Latta and Wunderle 1996). In Central America, its habitat preferences are similar, liking open woodlands or plantations (particularly banana, *Musa acuminata*) and showing an affinity for palms. In forested situations, it shows an association with rivers.

BEHAVIOUR Not shy, but it often forages in dense vegetation, making it difficult to observe. Commonly perches high in the crown of a tree, particularly palms. Pairs travel together gleaning from leaves and eating fruit, and visiting larger flowers. Cuban birds are known to drill holes at the base of *Erythrina* flowers, robbing them of nectar without helping to pollinate them (Barbour 1943). In Cuba, a preference for nectaring from *Erythrina, Hibiscus*, banana, *Agave* and *Citrus* has been noted while soft fruits such as banana and anon (*Anona squamosa*) are also noted as a food source (Gundlach 1876). Insects are also taken. In Haiti, large flocks of up to 50 have occurred (Wetmore and Swales 1931). They often hang upside down to reach food or insects on the undersides of leaves. In Puerto Rico, they sometimes forage by probing and gaping in epiphyte (*Tillandsia recurvata*) clumps (Post 1981a). After breeding they travel with the young in a family flock. In Cuba, birds in immature plumage are sometimes observed breeding with an adult-plumaged bird. The sex of the immature-plumaged breeding individuals is not known. However, in Hispaniola, a female in first basic plumage (immature) was collected and discovered to be in the egg laying stage (Wetmore and Lincoln 1933).

NESTING Nests are often on palms (Puerto Rico, Hispaniola), palms or banana plants (Cuba) and coconut palms (*Cocos nucifera*). In the Bahamas, the form *northropi* is very closely tied to the presence of coconut palms. The nominate form is known to nest April–July (Bond 1928b, Wetmore and Swales 1931), but may have a more extensive nesting period. In Puerto Rico, nesting may occur throughout the year, but the main nesting season is March–June. Bahamian *northropi* nests May–June. Cuban *melanopsis* begins nesting in February (Gundlach 1876) and continues into June (Todd 1916). At times, birds in immature (first basic) plumage have been observed nesting, at least in Cuba (Gundlach 1876) and the Bahamas (Baltz 1997). The nest is a hanging basket made of string-like fibrous material garnered largely from palm leaves. It is usually attached to the underside of a palm leaf, woven through holes made in the leaf itself, with the nest being suspended from two points. This species may also suspend the nest from the tip of a woody tree branch (Bond 1928b). Both sexes help to construct the nest (Gundlach 1876). Clutch size tends to be 3 eggs which are white, unmarked with a blue tint (Puerto Rico, Bahamas); however, Puerto Rican eggs have also been described as whitish with a green or chocolate wash, and blackish or brown spots! Eggs of *dominicensis* are white to pale blue with russet spots (Wetmore and Swales 1931). Eggs of *melanopsis* are greenish-white with lilac-grey or olive spots and scrawls, particularly around the broader end (Gundlach 1876). The Caribbean forms are parasitised by Shiny Cowbird (102), particularly in Puerto Rico, where most nests near the coast contain cowbird eggs; parasitism levels are lower in the interior of Puerto Rico. In one study in Puerto Rico, Black-cowled Oriole was found to be the 'highest quality' host for Shiny Cowbird, fledging and an average of 3.6 cowbirds per nest (Wiley 1985). On the Bahamas, the recent presence of Shiny Cowbirds does not bode well for the small populations of *northropi* Black-cowled Orioles, and cowbird trapping has been suggested as a conservation strategy (Baltz 1997). There are no known cases of parasitism by Bronzed Cowbird (99) on nests of *prosthemelas* (Friedmann and Kiff 1985).

Central American *prosthemelas* also hangs its basket nests from the underside of a palm leaf, stitching it to the leaf using small holes created by the bird. They are attached at two ends, similar to a hammock. In Central America, nesting takes place March–July. One pair was observed building a nest near a colony of Orange Oriole (29) during July in Quintana Roo (Howell *et al.* 1992). The clutch size is 3. The eggs of mainland forms appear to be unknown.

DISTRIBUTION AND STATUS Found in Mexico and Central America as well as the Bahamas and Greater Antilles. Uncommon to common in most of its range. Common and generally distributed in Hispaniola (*dominicensis*), but absent from the higher elevations. Uncommon to rare on Andros Island and possibly extirpated, last confirmed sighting in 1993 (Baltz 1997) on Abaco in the Bahamas (*northropi*). Fairly common in Mexico and Central America. Wanders to 1000 m in Mexico and Central America. The continental range extends along the Caribbean slope from S Veracruz and SE Puebla in Mexico south through N Oaxaca, Tabasco, the entire Yucatán peninsula (absent from the north coast?) and NE Chiapas. It is absent from Cozumel Island. The range extends south through N Guatemala (the Petén and tropical zone north of the highlands) throughout Belize, the lowlands of N Honduras extending furthest inland in the NW and NE where the mountains do not come so close to the sea, E Nicaragua, Caribbean lowlands of Costa Rica and W Panama (to Bocas del Toro). In the West Indies it is found on Andros, and formerly (perhaps still) Abaco and Little Abaco in the Bahamas, on Cuba and on the Isle of Pines, on Hispaniola including the nearby islands of Tortue, Gonave and Île-à-Vache, and on Puerto Rico.

MOVEMENTS Sedentary, with no evidence of local movements in the non-breeding season. No confirmed reports of vagrancy are known in this complex, particularly for the insular forms. There is one published report from Seal Island, Nova Scotia, Canada, on May 24, 1971 (Doane 1971). The identification of this bird was far from conclusive, and our review of the record leaves us with no option but to reject it as a valid record. The species has been recently left off the Birders Journal list of Canadian Birds for the same reasons (Anon 1996).

MOULT. In adults the only moult is the complete definitive pre-basic which occurs after the fledglings have become independent. The timing of this moult is dependent on the race and its breeding schedule. Specimens of *prosthemelas* from Mexico are in definitive pre-basic moult during October, while September specimens are very worn. Individuals of *portoricensis* are undergoing moult in July, while *melanopsis* moults mainly September–October. Pre-alternate moults are missing in adults. The first pre-basic moult is incomplete, with immature birds retaining the juvenile flight feathers. The general timing of the first pre-basic is similar to that of adults. However, *northropi* often retains a yellowish throat late into the winter and early in spring; perhaps the black throat is not obtained until the first pre-alternate moult, or the first pre-basic moult is delayed in this form. In *dominicensis* (perhaps other forms?), the first pre-basic moult is variable. The body plumage and wing-coverts are replaced and a variable number of secondaries and tertials are also changed; some individuals retain all of the juvenal remiges. It is more likely that juvenal remiges are replaced in this moult than juvenal rectrices. First basic flight feathers are black, contrasting with the brownish juvenile feathers. It appears that a limited and prolonged first pre-alternate moult occurs. This is implied by immature specimens (*prosthemelas*) from March which consistently show more black on the upper and underparts suggesting that a body moult occurs during the winter or early spring. The presence and extent of this moult needs to be clarified. Todd and Worthington (1911) state that both *northropi* and *prosthemelas* undergo a partial first pre-alternate moult and that it is more extensive in the latter.

MEASUREMENTS ***I. d. prosthemelas***. Males: (n=21) wing 86.6 (78.5–93.5); tail 93.2 (84.6–100.8); culmen 20.6 (19.1–21.3); tarsus 22.9 (22.1–23.6). Females: (n=10) wing 83.2 (80.3–85.1); tail 88.3 (84.1–95); culmen 20.6 (19.6–21.6); tarsus 23.2 (22.1–24).

I. d. praecox. Males: (n=10) wing 88.3 (85–91.0); tail 91.6 (85–97); culmen 20.7 (20–21); tarsus 22.4 (20.5–24). Females: (n=10) wing 82.3 (78–87.5); tail 86.0 (82.2–90.6); culmen 20.3 (19–21.9); tarsus 22.5 (21.1–24).

I. d. northropi. Males: (n=3) wing 96.7 (91–100); tail 91.5 (89.4–95); culmen 21.8 (21–23.4); tarsus 24.7 (23–26). Females: (n=3) wing 91.1 (90–92.2); tail 89.5 (87–91.4); (n=2) culmen 21, 22, (n=3) tarsus 23.8 (21–25.4).

I. d. melanopsis. Males: (n=10) wing 95.2 (94.2–96.0); tail 90.3 (84.3–94); culmen 21.1 (20.0–22.4); tarsus 23.7 (21–24.4). Females: (n=10) wing 92.1 (89.4–94.7); tail 88.1 (81.3–92.7); culmen 21.2 (20–22.1); tarsus 23.8 (22.9–25).

I. d. portoricensis. Males: (n=10) wing 95.2 (88.9–100); tail 90.2 (84.1–96); culmen 23.9 (20–26); tarsus 22.9 (20–24). Females: (n=10) wing 90.7 (88–94); tail 84.4 (79.8–89); culmen 23.4 (21–25.1); tarsus 22.9 (21–24).

I. d. dominicensis. Males: (n=26) wing 96.2 (88.9–101.0); tail 86.5 (82.0–94.2); culmen 21.8 (19.2–24.4); tarsus 23.6 (21.7–24.9). Females: (n=10) wing 90.2 (87.0–93); tail 83.8 (82.3–86); culmen 20.6 (19.9–22.1); tarsus 23.6 (22.6–24.5).

NOTES The mainland forms (*prosthemelas* and *praecox*) are often separated from the others as a different species. The island forms would then be referred to as *Icterus dominicensis*, Greater Antillean Oriole, and the mainland birds (*Icterus prosthemelas*) would retain the English name, Black-cowled Oriole. In the past, all of the island forms were also considered separate species, probably not an unreasonable organisation (see Geographic Variation). A recent study comparing the mitochrondrial DNA of the orioles suggests that more than one species is involved in what is now known as Black-cowled Oriole (K. Omland and S. M. Lanyon pers. comm.). In fact, *prosthemelas* Black-cowled Oriole may be more closely related to Orchard Oriole. In addition, *portoricensis* may be more closely related to the Lesser Antillean orioles (St Lucia, Montserrat and Martinique) than to the other members of the Black-cowled complex. Thus at least three species are suggested to exist. Cuban *melanopsis* and Bahamas *northropi* were found to be more closely related and perhaps they should be treated as subspecies (K. Omland and S. M. Lanyon pers. comm.). However, nominate *dominicensis* was not included in this study and until it is sampled it is premature to finalize a new systematic and taxonomic arrangement for the species in this complex.

The conservation status of *northropi* needs to be carefully assessed. Baltz (1997) gave a liberal estimate of 150–300 birds on north and south Andros. All available evidence suggests that the population from Abaco is now extinct. In Puerto Rico, Black-cowled Oriole has been one of the species which is most heavily parasitised by Shiny Cowbird. Given the small population in the Bahamas, and the recent arrival of Shiny Cowbird there, the parasite may have a significant effect on the remaining population of *northropi* Black-cowled Orioles. Baltz (1997) has suggested a concerted effort to trap cowbirds as a management technique.

REFERENCES Allen 1890 [voice, geographic variation], Anon 1996 [Nova Scotia sighting], Baltz 1977 [nesting, distribution, status, conservation, Bahamas], Biaggi 1983 [nesting, voice, status, habitat], Bryant 1866 [geographic variation], Doane 1971 [report from Nova Scotia], Hardy *et al.* 1998 [voice], Howell and Webb 1995 [description, voice], Garcia 1987 [nesting, behaviour, Cuba], Latta and Wunderle 1996 [habitat], Moore 1994 [voice], Perez-Rivera 1986 [nesting], Phillips and Dickerman 1965 [geographic variation], Raffaele 1989 [voice, nesting, habitat], Ridgway 1902 [measurements], Skutch 1954 [nesting, behaviour], Wetmore *et al.* 1984 [measurements, nesting], Wetmore and Swales 1931 [behaviour, nesting, voice, measurements].

46 MONTSERRAT ORIOLE *Icterus oberi* Plate 15
Lawrence 1880

This island oriole was declared the national bird of Montserrat in 1982. It is the island's only endemic species and under serious threat of extinction due to the damaging volcanic eruptions on the island.

IDENTIFICATION A slim oriole unlikely to be confused for anything else on its island home. The male is black with russet belly, thighs, vent and rump. The wings and tail are entirely black. Of the three endemic Lesser Antillean orioles, Montserrat Oriole is the only one which is obviously sexual dimorphic. The male is largely black, with ochre-yellow or tawny rump, underparts from the breast to the undertail-coverts, and wing-linings. Like its close relatives, this species lacks white wingbars or bold white edging to the tertials and secondaries. The female is olive above and yellow on the underparts. The face may show a slight chestnut wash. No other oriole occurs on Montserrat, but migrant Orchard (43) and Baltimore (40) Orioles could occur as vagrants. Males of both of these migrant species are easily separated from Montserrat Oriole; Baltimore by its more orange colour, extensive white edging on the wings, and orange on the tail; Orchard Oriole shows chestnut rather than orange or tawny on the body. Female Baltimore Orioles are obviously orange below, with streaking on the back unlike female Montserrat Oriole. Some female Baltimore Orioles are yellowish below, showing few orange tones, but these show a greyish or pale belly, they are not solidly yellow on the underparts as is Montserrat Oriole. Female Orchard Orioles are solidly yellow on the underparts, however. Note that Montserrat Oriole is larger and longer-tailed than Orchard Oriole. In addition, Montserrat Orioles show a chestnut wash on the ear-coverts. Female Orchard Oriole do not show any chestnut on the plumage; immature (first basic) males may show variable amounts of chestnut on the underparts. These young males also show a black throat patch, while female Montserrat Oriole do not have any black on the throat. For differences between the Montserrat and the closely-related Martinique (47) and St Lucia (48) Orioles, see the accounts for those species.

VOICE Song: Heard only in the breeding season, the song is a loud series of melodious whistles. It does not sing frequently, and even in the height of the breeding season it may be difficult to find one singing. The song is slow and methodical; in tempo it sounds like a slow Red-eyed Vireo (*Vireo olivaceous*). Single notes are given at the rate of approximately one every 2.5–3 seconds. The notes themselves are composed of a single syllable or two syllables, usually short, sharp whistles or lower pitched gurgled whistles. One predominant note is a sharply descending whistle *tseew*.

Call: The call sounds like a sharp *chic* (Siegel 1983), or a sharper *chuck*.

DESCRIPTION A slim oriole with a thin bill. The culmen is only slightly curved, while the gonys is straight. Wing Formula: $P9 < P8 < P7 > P6$; $P9 \approx P4$; P8–P6 emarginate.

Adult Male (Definitive Basic): Bill black with the basal half grey, the eyes are dark. The head, mantle, breast and wings are black. If ruffled by the wind, or while preening, the yellow bases of the neck feathers may be revealed. The black breast ends abruptly, turning tawny on the lower breast down to the vent and undertail-coverts. The underpart feathers are yellower towards their base. The belly is slightly more yellow than the rest of the underparts. The lower back, rump and uppertail-coverts are yellow with a tawny wash, contrasting with the black mantle. The wings are entirely black from above, although the wing-linings are yellow. The tail is entirely black and unmarked. The legs and feet are black.

Adult Female (Definitive Basic): Soft part colours as in the male. The crown is yellow-olive, becoming olive on the nape and mantle. The rump is olive with a yellow wash, contrasting slightly with the olive back. The face is yellowish, with dark lores and a chestnut wash to the auriculars. The underparts are entirely olive-yellow from the chin to the undertail-coverts. The brownish wings are patterned as follows: the lesser and median coverts are tipped olive-yellow, while the greater coverts are edged tawny; the tertials, secondaries and primaries are also edged tawny. The underwings show yellow linings. The tail is olive with a yellow wash.

Immature Male (First Basic): Similar to adult female, but the back is darker. In addition, a variable number of black throat feathers may be present.

Juvenile (Juvenal): Similar to adult female, but lacking black on the lores. The upperparts are olive, with a yellow wash to the crown and rump. The face is greenish-yellow, while the underparts are yellow with a tawny wash and more olive flanks. The brown wings have the coverts and tertials fringed buff, and the linings are yellow.

GEOGRAPHIC VARIATION Monotypic.

HABITAT Lives in moist montane forest, keeping high in the canopy. Appears to prefer the zone above 460 to 610 m, in cooler conditions where mountain-palms, filmy ferns and other abundance of epiphytic growth is found (Grisdale 1882).

BEHAVIOUR This species is wide-ranging, with a large foraging range. Observed either singly or pairs, unless with fledged young. Apparently a strict insectivore, it is not known to feed on nectar. It gleans the undersides of leaves to obtain insect prey. Early collectors of this species would attract it by pishing (Grisdale 1882).

NESTING Nests June–August. As is typical of this genus, the nest is a hanging basket-shaped nest, woven from vegetable strands. It is typically secured to one of the lower leaves of a forest palm tree or perhaps a banana tree. Clutch 2 spotted eggs. The eggs range from having a white background to a greenish-white background, and the spotting is chocolate or grey-brown. Only the female incubates the eggs, but both adults feed and care for the young.

DISTRIBUTION AND STATUS Found at 400–900 m. Endemic to the Lesser Antillean island of Montserrat, a British dependency. Rare inhabitant of the Centre and Soufrière Hills. May be found at Chance Peak (historically known from this site), Galway's Soufrière and Runaway Ghaut. The recent volcanism on Montserrat has seriously damaged most of the available habitat, and its current range is greatly restricted.

MOVEMENTS Sedentary, no known movements.

MOULT Adults have one complete moult a year, the definitive pre-basic. There is no evidence of a pre-alternate moult. Juveniles lose their plumage through the partial first pre-basic moult, involving only the body plumage. The timing of these moults is not known.

MEASUREMENTS Males: (n=13) wing 92.6 (88–98.5); (n=12) tail 99.7 (93.5–104); (n=11) culmen 24.2 (21.8–26); tarsus 25.9 (23.9–30). Females: (n=5) wing 84.6 (80–94); tail 87.9 (81–93); (n=4) culmen 22.9 (22–25); (n=5) tarsus 24.8 (23–27).

NOTES The Soufrière Hills volcano, at the south end of the island, became active in July 1995 after several centuries of calm. The many irruptions of ash and lava have eliminated most of the available oriole habitat on the Soufrière Hills. In addition, Hurricane Hugo devastated parts of the island in 1989. It is clear that the oriole's population has suffered immensely, and is on the brink of extinction. Currently, members of the Sustainable Ecosystems Institute, based in Oregon, are planning to capture the remaining orioles and maintain a breeding stock both in Jersey Zoo (UK) and in San Diego Zoo (US). This project is ongoing.

REFERENCES Bond 1939 [nesting], Grisdale 1882 [habitat, behaviour], Hardy *et al.* 1998 [voice], Lawrence 1880 [description], SEI unpublished [conservation], Siegel 1983 [voice, habitat, nesting, distribution, behaviour].

47 MARTINIQUE ORIOLE *Icterus bonana* — Plate 15
(Linnaeus) 1776

A rare oriole of the island of Martinique, a department of France; sadly its existence is seriously imperiled at this point in time.

IDENTIFICATION Martinique Oriole is a small, slim oriole with a distinctive plumage pattern. It is tawny-orange on the rump, epaulet and underparts, and has a chestnut head. The back, wings and tail are black. No other oriole has a dark chestnut head. No other oriole is expected to be found on Martinique, but migrant Orchard (43) and Baltimore (40) Orioles could occur. Male Baltimore Oriole is largely orange, with orange on the tail and a black head. Female Baltimore Oriole is also largely orange or orange-yellow on the underparts, with variable amounts of dark on the head, and dark feather centers on the back. This species should pose no confusion with Martinique Oriole. Male Orchard Oriole shares some chestnut plumage with Martinique Oriole, but note that chestnut is restricted to the underparts in the former. Male Orchard Orioles have black heads, not chestnut as in Martinique Oriole. Female and immature Orchard Orioles are largely yellow below, not tawny-orange as in Martinique Oriole. The other Lesser Antillean orioles, St Lucia (48) and Montserrat (46) are similar in size and shape, but not plumage, to Martinique Oriole. These three species are not migratory and do not overlap in range. However, the following differences should be noted. The closely-related Martinique and Montserrat Orioles should never overlap with St Lucia Oriole, but it is worth emphasising plumage differences. Both of these species have black heads. Adult St Lucia Orioles have most of the underparts black, with orange restricted to the belly and vent. Adult male Montserrat Orioles are similar to St Lucia Oriole, but the black is more restricted on the underparts, extending only to the breast. In addition, they are tawny-yellow rather than orange or orange-chestnut. Female Montserrat Orioles lack black, being olive above and yellow below.

VOICE Song: The song is apparently not harmonious, and its shrill nature may sound similar to that of a Carib Grackle (93)(Pichot 1976). In contrast, Babbs *et al.* (1987) describe the species as quiet and unobtrusive, giving a soft warbling song. Song was considered so infrequent as to make it unusable for census purposes.

Calls: A harsh *cheeo* or doubled as a *cheeo-cheo.*

DESCRIPTION A slim oriole with a thin bill. The culmen shows a slight curve while the gonys is straight. The long tail is slightly graduated. Wing formula: P9 < P8 > P7 > P6; P9 ≈ P5; P8–P7 weakly emarginate.

Adult Male (Definitive Basic): The bill is black with the basal half of the lower mandible pale grey. The eyes are dark brown. The head, including the chin and upper breast, is chestnut, becoming tawny on the lower breast down to the vent and crissum. Note that the longest undertail-coverts are tipped black. The thighs are yellow. The mantle is a contrasting black, while the lower back and rump is tawny and the uppertail-coverts are a darker chestnut. The longest uppertail-coverts are tipped black. The wings are blackish with tawny markings. The lesser and median coverts are tawny, creating a tawny epaulet; the rest of the wings are unmarked. The underwings show tawny-yellow wing-linings. The tail is entirely blackish and unmarked. The legs and feet are dark grey. The black on this species typically fades to a dark brown.

Adult Female (Definitive Basic): Similar to the male although marginally smaller and slightly duller in colour.

Immatures (First Basic): Unknown.

Juvenile (Juvenal): Unknown.

GEOGRAPHIC VARIATION Monotypic.

HABITAT Found in forests below 700 m. As is typical of many island species, Martinique Oriole is somewhat of a habitat generalist using mangroves, dry and wetter forests, forest edge, dense scrub, as well as urban areas and agricultural lands, particularly plantations. It is not found in cloud forest however. Dry forests and mangroves have been suggested to be the most important habitat for this species (Babbs *et al.* 1987). Given its wide habitat choice, it is not likely that forest clearance is the main cause of the species' population decline.

BEHAVIOUR This is not one of the more sociable orioles as it does not forage in flocks, but rather as family groups, singly or in pairs. Foraging for invertebrates and fruit takes place mainly in the forest canopy. Birds forage by looking beneath dead leaves, and probing into hollow stems, particularly on lianas (Babbs *et al.* 1987). While nesting, most of the foraging occurs very close to the nest site itself. The nestlings are fed insects almost exclusively. Territory defense is minimal, with adults only concerning themselves with the immediate location of the nest. However, they do defend the nest against Carib Grackles, a potential nest predator, and Shiny Cowbirds (102), a known brood parasite. Both sexes aid in nest defense. During the breeding season, Martinique Oriole sings, but not vigorously or often. It is not known if both sexes sing (Babbs *et al.* 1987).

NESTING It nests in all forest types (except cloud forest) from December, but more usually in February, and winding up in July, as by then most young have fledged. The nest is the typical hanging pouch of the genus, but is shallow. It is placed between 2–4 m off the ground, sometimes as high as 10 m (Babbs *et al.* 1987). Its placement is similar to that of many tropical orioles, preferring a large leaf such as Banana (*Musa acuminata*) or other large leaved-species, often palms, to hang the nest from. Babbs *et al.* (1987) report the use of Banana, Baliser *(Heliconia caribaea)*, Trumpet Wood (*Cecropia peltata*) and Raisinier Grand-feuilles (*Coccoloba grandifolia*) as nest trees. The nest is attached at two points from the underside of the leaf, with these attachments woven directly through the leaf (Pichon 1976). The construction incorporates strong fibers which are often taken from coconut palms (*Cocos nucifera*). Typically, 3 (but sometimes 2) eggs are laid; the eggs are white with a pale blue wash and brown spotting that is restricted to the wide end of the egg (Pichot 1976). The incubation period lasts 14–18 days (Babbs *et al.* 1987). Nestlings fledge at around 15 days after hatching (Babbs *et al.* 1987). This oriole is a known host for Shiny Cowbird. In fact, the population decline in this oriole has been attributed to parasitism by the cowbird. The cowbird is a recent invader on Martinique, having established itself as late as the 1940s. Today an average of 75% of Martinique Oriole nests are parasitised by eggs of Shiny Cowbird. The effects of this parasitism have not been studied in detail, but they probably account for most of the decline in this species. In addition, Carib Grackle has been insinuated to be causing heightened mortality at oriole nests due to its nest predatory behaviour. The grackle is a native of Martinique, but its population has increased immensely due to forest clearing and urbanization.

DISTRIBUTION AND STATUS. Only found on the island of Martinique, Lesser Antilles, a French dependency. Its population (and distribution?) has seriously decreased in recent years and it is now listed as a red data (endangered) species. It is given legal protection on Martinique. Previously, it was found throughout the island to 700 m. More recently, it has not been observed higher than 520 m (Babbs *et al.* 1987). Since the north of the island is more mountainous, it is thought that the stronghold for this species is in the south. In the 1970s this species was considered to be common both in the north and south of Martinique (Pichon 1976), and locals remember the species as much more common than it is today (Babbs *et al.* 1987). No population estimates exist. It has been observed recently in the extreme north, in the area north and northeast of Fort-de-France, along the bay south of the airport, at Morne Dore and Cite Canal, at L'anse a L'ane in the southwest, near Rivière Pilote in the south and on the Caravelle peninsula in the northeast (Babbs *et al.* 1987).

MOVEMENTS This insular species is not know to make any seasonal movements. Apparently, post-breeding dispersal of adults has not been observed in the species.

MOULT Adults change their plumage through the complete definitive pre-basic moult. As is typical, this occurs after the breeding season. In the case of this species, the definitive pre-basic moult occurs in October. The timing and extent of the first pre-basic is not known. However, the extent is likely similar to that of St Lucia and Montserrat Orioles. There is no evidence for the existence of pre-alternate moults.

MEASUREMENTS Males: (n=2) wing 83, 86.9; tail 83.8, 91; (n=1) culmen 21; (n=2) tarsus 22.9, 25. Females: (n=3) wing 77.7 (76–80); (n=2) tail 81, 84.6; (n=3) culmen 21.6 (21.0–22.1); (n=4) tarsus 22.5 (21.8–23.1).

NOTES This species is listed as Endangered by the IUCN and the Red Data Book. It was first suggested to be listed as vulnerable in 1987 (Babbs *et al.* 1987), a recommendation followed by the French Red Data Book (Thibault and Guyot 1988), now it has been upgraded once again to endangered (Collar *et al.* 1992). The primary reason for the decline in this species appears to be the invasion by the brood-parasitic Shiny Cowbird. Habitat loss and illegal hunting does not appear to have had a major effect on Martinique Oriole populations. It is recommended that a cowbird control program be established and tested to protect this oriole (Babbs *et al.* 1987). Population counts of the Martinique Oriole are not available.

REFERENCES Babbs *et al.* 1987 [voice, habitat, distribution, status, behaviour, nesting, conservation], Collar *et al.* 1992 [status, conservation, distribution, behaviour], Hardy *et al.* 1998 [voice], Lawrence 1878 [habitat], Post *et al.* 1990 [brood parasitism], Ridgway 1902 [measurements].

48 ST LUCIA ORIOLE *Icterus laudabilis* Sclater 1871 — Plate 15

This oriole is endemic to St Lucia and is not currently considered critically threatened like its close relatives.

IDENTIFICATION This slim, long-billed oriole is unlikely to be misidentified for anything else on its home island of St Lucia. It is largely black, with orange shoulders, rump, belly and vent; the female is similar to the male. There are no other resident orioles in St Lucia; however, migrant Baltimore (40) or Orchard (43) Orioles may occur as vagrants on the island. Note that both of those species have the black of the hood reaching only to the breast on the underparts. In addition, male Orchard Oriole is chestnut rather than orange. Male Baltimore Orioles also have orange on the tail, and both of these species have white wingbars. St Lucia Oriole has no wingbars and an entirely black tail. Immatures show chestnut on the plumage, like Orchard Oriole, but note that the chestnut extends to the neck and nape, and is not restricted to the underparts. The closely-related Martinique (47) and Montserrat (46) Orioles should never overlap with St Lucia Oriole but it is worth pointing out differences. Adult Martinique Orioles have a chestnut head and lack black on the throat or breast. Adult male Montserrat Orioles are similar, but the black is more restricted on the underparts, reaching only to the breast. In addition, they are tawny-yellow rather than orange or orange-chestnut. Female Montserrat Orioles lack black, being olive above and yellow below.

VOICE Song: The song is composed of a short (2 second) series of sweet, varied whistles which are repeated several times. It is not known if many different variations of the series are given, or if the same phrase is repeated again and again. Bond (1928a) states that the song of this oriole is similar to that of Orchard Oriole, but is weaker.

Calls: Gives a harsh *chwee*. Also a soft *chup*.

DESCRIPTION A slim oriole with a thin bill. The culmen is straight. Wing formula: P9 < P8 < P7 < P6 > P5 > P4; P9 ≈ P4; P8–P6 weakly emarginate.

Adult Male (Definitive Basic): The black bill shows a grey basal quarter to the lower mandible. The head, neck, mantle and underparts from the chin to the lower breast are black. An orange-yellow (or brighter crimson-orange) lower back and rump contrast with the black mantle. The black on the breast ends abruptly, posterior to this the underparts are orange to the undertail-coverts. Some individuals show a tawny wash to the orange immediately posterior to the black breast. The wings are black with an orange (sometimes tawny) epaulet created by the lesser and median coverts. Otherwise the wings are unmarked, although the wing-linings are orange-yellow. The tail is black and unpatterned. The legs and feet are blue-grey.

Adult Female (Definitive Basic): Similar to male, but the underpart colour is slightly paler.

Immature Male (First Basic): Like immature female, but with more extensive black on the face, crown, throat and mantle. Black extends to the breast, but note that these feathers are chestnut-tipped. The neck is chestnut, dividing the black-based nape feathers from the blackish mantle. The rump and belly are similar to the females; however, the flanks are orange-olive. There is a variable number of blackish wing (coverts and tertials) and tail feathers, unlike the brown of the adult female. The lesser coverts are tawny and the median coverts are tawny-yellow, while the greaters are black with olive-buff tips. The blackish tertials have olive edges. The tail is blackish with olive tips, particularly on the outer rectrices; sometimes the tail is largely made up of the retained juvenal (olive) rectrices with blackish central feathers.

Immature Female (First Basic): Soft part colours as in the adults. The forehead, lores and throat are black. The rest of the crown is chestnut with black spotting, continuing to the neck. The mantle is olive, while the rump and lower back are olive with a yellow wash. The breast is chestnut, becoming tawny-yellow on the belly and undertail-coverts. The flanks are yellow with a buff wash. The wings are brown with tawny tips to the lesser coverts and tawny-yellow tipping to the median coverts; the greater coverts are narrowly edged buff as are the tertials. The brownish primaries show olive edges. The wing-linings are yellow. The tail is brown with olive outer rectrices.

Juvenile (Juvenal): The bill shows a pinkish base to the lower mandible. Cinnamon crown, darker on the forehead. The cinnamon becomes a warm brown on the nape and back. The auriculars are cinnamon, but the black lores and anterior part of the auriculars are black. The throat is black, while the breast is cinnamon, becoming tawny posteriorly on the belly. The undertail-coverts are tawny-yellow. The wings are blackish with buff edges to the primaries and secondaries. The median and greater coverts are blackish-brown with cinnamon tips, creating two thin cinnamon wingbars. The lesser coverts are cinnamon, while the bend of the wing is yellowish. On the underwing, the linings are yellow. The tail is blackish with olive outer rectrices.

GEOGRAPHIC VARIATION Monotypic.

HABITAT It is found in all types of forest, but particularly in humid forests. While it is present in the dry coastal scrub forests of St Lucia, it appears to prefer the taller canopy afforded by the humid forests of the highlands. It was encountered at a rate of 1.2 birds per hour in humid forest but only 0.5 birds per hour in dry forest (Diamond 1973). It inhabits both primary and secondary forests as well as the periphery of plantations such as banana, citrus and coconut (Babbs *et al.* 1988). It is not known if the plantations are used for nesting (Babbs *et al.* 1988). In addition, this oriole is present in mangroves, particularly when adjoining coastal scrub.

BEHAVIOUR Most regularly found in pairs foraging up in trees. However, small groups are sometimes observed of up to 10 individuals (Semper 1872). A more recent study observed orioles almost always in ones or twos, with a maximum of four adults observed at one time (Babbs *et al.* 1988), so the days of larger flocks may be a past memory. St Lucia Oriole feeds on a variety of foods including fruit and insects (Babbs *et al.* 1988). It has been observed stripping bark while foraging, in the manner of Jamaican Oriole (28).

NESTING Makes a typical oriole nest, a hanging basket woven of plant fibers. This species sometimes builds its nest under the large fronds of banana plants, weaving the supports of the nests directly into the leaf. Also nests in coconut palms (Bond 1928a). Juveniles, some being fed by adults, have been observed between late June–August (Babbs *et al.* 1988) suggesting that nesting occurs April–early June. A total of 14 family groups included either one or two juveniles (Babbs *et al.* 1988), implying that the clutch size is of only 2 eggs. It is a known host for Shiny Cowbird (102). St Lucia Oriole has been seen feeding fledgling cowbirds in the absence of fledgling orioles (Post *et al.* 1990).

DISTRIBUTION AND STATUS Uncommon and likely declining, a density of 4.9 individuals per square kilometer has been reported (Post *et al.* 1990). Restricted to the main island of St Lucia, Lesser Antilles, where there is forest. Found at least to 600 m in elevation.

MOVEMENTS Appears to be sedentary.

MOULT Adults moult once a year, which occurs after breeding through the complete definitive pre-basic moult. The juvenal plumage is replaced through the first pre-basic moult which replaces the body feathers as well as the wing-coverts, alula, some tertials and often the central tail feathers. The primaries, secondaries and most of the tail feathers are retained from the juvenal plumage. It is unclear if juvenile females also moult their coverts and tail feathers during the first pre-basic moult. The presence or extent of first pre-alternate moults is not known. The period of moult for both adults and immatures appears to be in August and September.

MEASUREMENTS Both sexes:(n=6) length 197.9 (188–205.7); wing 96.5 (88.4–102.9); tail 93.5 (84.3–101.6); culmen 24.6 (23.6–26.2); tarsus 25.1 (24.4–25.9). Males: (n=8) wing 98.7 (90–105); tail 98.1 (90–105); (n=6) culmen 24.4 (23–26); (n=8) tarsus 25.3 (23–27); (n=3) (24–27). Females: (n=14) wing 94.0 (85–103); tail 90.1 (82–102); (n=9) culmen 22.6 (21–24); (n=14) tarsus 24.7 (24–26).

NOTES This species is not currently considered threatened and the population is secure; although the recent arrival of Shiny Cowbird to the island may put it at risk. However, the oriole is present in the humid forests of the highlands where the cowbirds are absent. In addition, the loss of forest habitat poses a conservation challenge to this oriole (Babbs *et al.* 1988).

REFERENCES Babbs *et al.* 1988 [habitat, behaviour, nesting, distribution, conservation], Bond 1928a [voice, nesting], Diamond 1973 [habitat, behaviour], Hardy *et al.* 1998 [voice], Post *et al.* 1990 [status, brood parasitism], Ridgway 1902 [measurements].

49 AUDUBON'S ORIOLE *Icterus graduacauda* — Plate 17

Lesson 1839

Largely a Mexican species, in the US this species can only be observed in southernmost Texas. It is the only black-hooded oriole.

IDENTIFICATION A large, stout, yellow oriole with a black hood; the white wing edging and brightness of the bird vary geographically. This monomorphic oriole is the only yellow oriole with a contrasting black head (hood); however, some immatures of other species can approximate this pattern. Of the sympatric yellow orioles, perhaps Scott's (51) and Black-cowled (45) are the most similar. Scott's Oriole differs in its black back, bold white wingbars and yellow-based tail. In addition, Scott's Orioles are found in arid zones. Black-cowled Oriole also has a black back, and it is also slimmer and smaller than Audubon's Oriole. A third similar species, Yellow-backed Oriole (26), is probably not sympatric. Yellow-backed Orioles have black on the head restricted to the face and throat. Immatures of Black-cowled, Black-vented (44) and Scott's Oriole may be confused with immature Audubon's Oriole due to their largely yellow plumage and restricted amount of black on the head, and largely olive-green back colouration. There are noticeable shape differences between these orioles, with Black-cowled and Black-vented being slimmer, longer-tailed and thinner-billed; Black-cowled Oriole has a noticeable curve to the culmen. The culmen is almost straight on Audubon's Oriole. By the time an appreciable amount of black is gained on the face of immature Black-vented Oriole, the back also shows noticeable amounts of black. Audubon's Orioles do not show black on the back, except in rare cases (see Description). Immature Scott's Orioles not only show dark or blackish streaking on the back, unlike Audubon's Oriole, but they also possess two white wingbars which further aid identification. Immature Audubon's Orioles show dull brownish-black wings, while immature Black-cowled and Black-vented Orioles tend to show obviously blackish wings.

VOICE Song: A rather long song that has been described by Flood (1990) as 'a long, slow-sounding, whistled song that seems to drift lazily from one note the next'. Both sexes sing, in fact their rates of song delivery are not significantly different, nor are their song types. The songs are composed of a sequence, up to 5 seconds long, of single whistled notes that are strung together in a set order. The order of notes does not appear to change, but individuals may alter their songs by truncating them before they have completed the full series. In comparison with songs of Baltimore Oriole, the song of Audubon's is longer, lower frequency, simpler and lacking frequency modulated notes (Flood 1990). The final point is perhaps the most important for field recognition of the song; all of the notes sound flat and in the same pitch, there is little range in the songs of Audubon's Oriole.

Calls: The alarm note is a nasal *nyyyee*, which is repeated at a rate proportional to the state of excitement. Also includes a melancholy whistle and gives a high frequency buzz. Calls of this oriole are similar to those of Altamira Oriole (34), but they are higher pitched, more nasal and more metallic than those of Altamira Oriole (J. Arvin pers. comm.).

DESCRIPTION A medium-sized oriole with straight, stout bill. The tail is long, narrow and obviously graduated; the difference between the longest and shortest tail feathers is approximately 30 mm. Wing Formula: P9 < P8 < P7 > P6 > P5; P9 ≈ P3; P8–P5 emarginate.

Adult Male (Definitive Basic): The bill is black with a grey basal third to the lower mandible. The eyes are dark brown. A largely yellow oriole with a black hood. The entire head, to the nape, and throat and upper breast are black. The edge of the hood is not clearly demarcated, but ragged. On some birds the lower border of the hood, on the breast, finishes in a series of dense streaks. The mantle, rump and uppertail-coverts are olive-yellow, brightest on the rump and lower neck. The underparts are lemon-yellow, as far as the undertail-coverts. The black wings contrast with the yellow body. The lesser and median coverts are yellow, forming a yellow epaulet. The greater coverts are black, tipped on their outer vane with white, creating a narrow, lower wingbar. The tertials are edged and tipped white, and there are narrow white edges to the secondaries and primaries; on fresh birds these white markings may show up as an obvious pale wing panel. The wing-linings are yellow. Audubon's Orioles have entirely black tails (Figure 49.1), and sometimes a small white tip is present on the lateral tail feathers. The legs and feet are grey. A minority of the specimens of southern *dickeyae* show a variable amount of black spotting or feather edging to the yellow back feathers. This does not appear to be related to age.

Adult Female (Definitive Basic): Similar to adult male in pattern, but duller overall in colour. The back typically shows an obvious green tone that is stronger than on the male. In addition, the wing colour is a slightly duller black, not quite the jet black of the male, although this varies depending on wear. Finally, the outer tail feather typically shows an olive tip or extensive amounts of olive on the female, and sometimes the other lateral rectrices are also olive-tipped (Figure 49.1).

Immature (First Basic): Appears intermediate between juvenile and adult female. Compared to the juvenile, the main difference is the immature's black hood, which is as extensive as on adults. During the first pre-basic moult, the black hood emerges, beginning gradually on the crown and

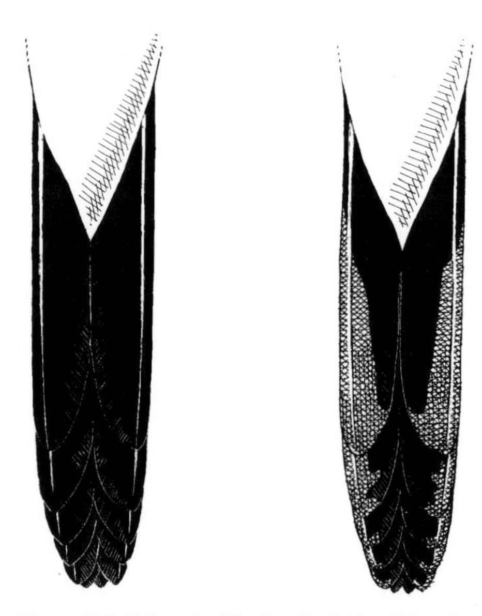

Figure 49.1 Tail underside of male (left) and female Audubon's Oriole.

breast, and proceeding towards the bill. The wings are dull brownish, as in the juvenile, but with contrasting blackish inner tertials and wing-coverts. The tail is retained from the juvenal plumage, but often the central tail feathers are replaced, and look contrastingly darker than the rest of the tail. Unlike in the adults, the back has a strong olive tone and the black hood may be tipped with olive, particularly on the lower edge.

Juvenile (Juvenal): The bill has an orange base to the lower mandible. Overall, juveniles are quite different from adults, mainly since black is lacking from their plumage. The upperparts are olive, greyer on the crown and nape, but yellowish on the rump. The face is greenish-grey. Underpart colour is yellow, with a grey wash on the throat and slightly more olive flanks. The wings are brownish-grey with yellowish edging on the tertials. The coverts are greyish, with the greater coverts showing pale greenish-yellow fringes. The primaries are edged pale grey. The tail is olive with slightly darker central rectrices.

GEOGRAPHIC VARIATION Four forms are currently recognized.

I. g. audubonii is found from S Texas, USA to Nuevo León and Tamaulipas in Mexico. It is similar to *graduacauda* but is larger and shows broad white edges to the inner secondaries, and often white tips to the greater coverts. It is described above. The forms *audubonii* and *graduacauda* apparently merge into each other and may best be lumped under one subspecies.

I. g. graduacauda is found in the southern section of the Mexican plateau, in the states of San Luis Potosí and Veracruz. It resembles *audubonii* but is smaller and lacks the white edges to the inner secondaries and the white tips to the coverts.

I. g. nayaritensis is found in Nayarit and NW Jalisco, Mexico. It is similar to *graduacauda* but the male is paler and a brighter yellow. The females are more extensively olive on the upperparts and a duller yellow on the underparts, therefore showing noticeable sexual dichromatism.

I. g. dickeyae ranges along the Sierra Madre in Guerrero and Oaxaca, Mexico. It is similar to *graduacauda* but it is larger and has a proportionately thicker bill. The yellow of the body is brighter, particularly on males. The borders of the black hood are neater, more clearly cut, than those of *graduacauda*. The yellow epaulets are restricted to the lesser coverts on this race. In addition, the wings lack the bold edging of *audubonii*. A few specimens show a variable amount of black tipping or spotting on the back. The consistent differences between this taxon and the others is intriguing; closer scrutiny of this complex may conclude that more than one species is involved.

HABITAT This oriole is a species of edge areas adjoining dense woodlands, specially riparian thickets. It prefers to keep within the dark shade of these dense sites but will venture into clearings. Inhabits evergreen or partially evergreen forests, and is also known from pine-oak woodlands, oak woodlands and tropical deciduous forest (Howell and Webb 1995, Schaldach 1963). In Guerrero, Mexico, this species inhabits semideciduous tropical forest as well as the lower edge of the cloud forest (Navarro 1992). At a site in Tamaulipas, this oriole was found in all forest types except those of the highest elevations; this included tropical, semi-evergreen, tropical deciduous, thorn woodlands, cloud forests, humid pine-oak and dry pine-oak habitats (Flood 1990).

BEHAVIOUR A shy and retiring oriole that tends to behave rather inconspicuously. Most often it is observed in pairs, birds foraging together in the depth of shade throughout the year. Mellow contact whistles aid individuals in maintaining contact with each other. This oriole joins mixed species flocks composed of large species such as other orioles, tanagers (*Thraupidae*) and jays (*Corvidae*). During the nesting period, males spend a large proportion of their time near the nest while the female is incubating. The pair keep in contact with each other by giving single notes or singing. Both sexes sing. Females may give single note calls from the nest and the male usually responds with the same type of vocalization. Some pairs are much less vocal than others (Flood 1990).

NESTING The breeding season extends from early April to mid-June, the rainy season. Two broods may be raised in a season. The nests are often placed at mid-height, 3–6 m off the ground. However, nest height varies depending on the habitat as Flood (1990) found that nests ranged from 12.9 to 33.0 m off the ground. Nests are cup-shaped, often semi-pensile, being attached at the rim of the nest to the branch rather than from below. Unlike many orioles which place the nest at the extremity of a branch, Audubon's Oriole places its nest closer to the trunk. The nests of this oriole are reported to be difficult to find (Flood 1990). These nests are woven from fine grasses or *Palmetto* fibers. The nest appears to be rather small relative to the size of the bird. The clutch size ranges from 3–5 eggs. Eggs are pale blue or greyish-white with variable dark markings. On some eggs there are a few dark scrawls and specks, evenly distributed on the surface. Other eggs are blotched heavily with brown, chestnut or purplish-brown; markings can be heavy enough to almost entirely obscure the ground colour. These spots tend to be clustered more heavily on the broader end of the egg. Bronzed Cowbird (99) is a common brood-parasite of this species and it has been observed to fledge from nests of this oriole (Friedmann and Kiff 1985). Brown-headed Cowbird (101) is also a known brood-parasite of this species, but it has not been observed to fledge from nests of Audubon's Oriole (Friedmann and Kiff 1985). Incubation is performed by the female, but both sexes will feed the young. There is no difference in the number of trips males and females make to the nest to feed the young; both take an equal part of the work. The nestlings fledge after 11 days in the nest.

DISTRIBUTION AND STATUS Uncommon in its US range and uncommon to common further south. Found to 2500 m, usually above 500 m and avoids sea-level (Howell and Webb 1995). Found only in the US and Mexico. Restricted to the lower Rio Grande valley of Texas, US. Ranges south from there through Tamaulipas, Nuevo León, E San Luis Potosí, NW Veracruz, E Hildago and N Querétaro. An allopatric population is found along the Pacific slope. This population rages from S Nayarit, Jalisco, Colima and Michoacán through south Guerrero and S. Oaxaca. Perhaps found as far east as W Chiapas (Howell and Webb 1995).

MOVEMENTS Resident and non-migratory, although limited northward wandering is known during the winter. Has wandered north to San Antonio, Texas.

MOULT Adults moult once each year. The complete definitive pre-basic moult occurs after breeding (July–September for *audubonii*); pre-alternate moults are lacking. Southern *dickeyae* shows similar timing, with moulting specimens found mid-July–early September; the same is true for the first pre-basic moult. The juvenal plumage is lost through the partial first pre-basic moult (July–September in *audubonii*). The body plumage, coverts, upper tertials and often the central tail feathers are replaced during this moult, the remiges and other rectrices are retained from the juvenal plumage.

MEASUREMENTS ***I. g. graduacauda***. Males: (n=10) wing 95.5 (90–98.3); tail 97.0 (90.2–102.4); culmen 22.5 (22.0–23.4); tarsus 26.0 (24.9–27.4). Females: (n=4) wing 91.4 (89.4–95.8); tail 95.2 (88.9–102); culmen 23.1 (22.4–23.6); tarsus 26.4 (25.4–27).

I. g. audubonii. Males: (n=14) wing 100.2 (95–106); tail 106.0 (99–116); culmen 25.2 (23–28.2); tarsus 26.6 (25–28). Females: (n=10) wing 98.4 (94–102); tail 104.6 (99.6–111); culmen 24.1 (21.8–26.4); tarsus 26.4 (24–28).

I .g. dickeyae. Males: (n=9) wing 99.3 (94–102); tail 102.3 (97–110); culmen 23.1 (22–24); tarsus 26.0 (25–27). Females: (n=10) wing 95.6 (92–100); tail 98.6 (88–105); culmen 22.5 (21–24); tarsus 25.5 (25–27).

NOTES Also known as Black-headed Oriole. The form *dickeyae* is significantly different in appearance from the *graduacauda* group. More study is needed to ascertain the relationships between these forms. A male Audubon's Oriole was observed paired with a female Altamira Oriole near Ricardo, Texas in 1988. The Altamira Oriole built two nests but no eggs were ever laid (P. C. Palmer unpublished data). Audubon's Oriole belongs in a group which includes Yellow-backed and Scott's Orioles (K. Omland and S. M. Lanyon pers. comm.).

REFERENCES Bent 1958 [behaviour, nesting, voice], Flood 1990 [voice, behaviour, nesting, habitat], Howell and Webb 1995 [distribution], Ridgway 1902 [measurements], Schaldach 1963 [habitat, distribution], van Rossem 1938 [geographic variation, measurements].

50 BAR-WINGED ORIOLE *Icterus maculialatus* — Plate 17

Cassin 1848

An oriole with a restricted range, found only from southern Mexico to El Salvador. It is aptly named, as the clear-cut, parallel-sided wingbar is an obvious characteristic of this species.

IDENTIFICATION Male Bar-winged Oriole is a black-hooded, black-backed oriole with yellow underparts, a distinctive bold white lower wingbar and yellow epaulet. The female and immature is olive above, yellow below and shows a contrasting black face and throat bib. Bar-winged Oriole is most likely to be confused with the sympatric Black-vented (44), Yellow-backed (26) and Audubon's (49) Orioles. However, the allopatric *prosthemelas* Black-cowled Oriole (45) and Scott's Oriole (51) are the most similar in plumage. Black-vented Oriole male lacks a white lower wingbar and possesses a black crissum, unlike Bar-winged Oriole. In addition, Bar-winged Oriole is yellow below, not orange-yellow with a russet wash on the breast. Older immature Black-vented Orioles show black throats, but unlike female/immature Bar-winged Orioles, the edge of the throat both on the face and breast is ragged rather than crisp.

Black-vented Oriole has a long, thin bill, with

the length accentuating the downcurve of the culmen; Bar-winged Orioles have shorter bills with a less obvious downcurve. Adult Yellow-backed and Audubon's Orioles have yellow backs, unlike the black-backed Bar-winged Oriole. Yellow-backed Oriole lacks white markings on the wing. Immature Audubon's Orioles are similarly coloured to immature and female Bar-winged Orioles. However, once Audubon's Orioles obtain a substantial amount of black on the throat and face they will also show many black feathers on the crown and nape but never on the back as is the case with immature male Bar-winged Orioles. Female Bar-winged Orioles lack black on the crown and nape at all times. Moulting male Bar-winged Orioles which show a substantial amount of black on the crown will also show some black back feathers. Audubon's Orioles have a straighter, thicker bill than Bar-winged Orioles.

Female *prosthemelas* Black-cowled Oriole may be confused with immature or female Bar-winged Oriole. Note the extent of black on the hood, and the colour of the wings and the tail. Female Black-cowled Oriole has extensive black on the forehead and crown, and additionally the black extends farther down on the breast. Bar-winged Orioles (females and immatures) have olive tails and brownish wings, quite unlike the black wings and tail of female Black-cowled Oriole. Immature and female Bar-winged Orioles show noticeable wingbars, unlike in Black-cowled Oriole. Note also that the nape, back and rump of Black-cowled Oriole averages a brighter yellow than the more olive upperparts of Bar-winged Oriole. Male Scott's Orioles, which are found north of Bar-winged Oriole's range, are superficially similar. However, Scott's Oriole has white median coverts and yellow on the tail. It is also a larger and bulkier species. Immature Scott's Orioles may show a black throat and mask like a female/immature Bar-winged Oriole, but the black on the throat has a ragged border on Scott's Oriole, not a neat, crisp border, and the upperparts of Scott's Oriole are always noticeably streaked.

VOICE Song: The song is a slow warbled series of sweet whistles. As in Audubon's Oriole, the songs of this species may be quite prolonged. However, Bar-winged Oriole's song has a greater frequency range, thus it does not sound so lazy. Songs may switch tempo suddenly, largely through an increase in the inter-note period of silence.

Calls: Gives a dry chatter, *grrrr* or *ahrrr* (Howell and Webb 1995).

DESCRIPTION A medium-sized oriole, with a slightly downcurved thin bill. The bill ends in a sharp tip. Wing Formula: P9 < P8 < P7 ≈ P6 ≈ P5 > P4; P9 ≈ P3; P8–P5 emarginate.

Adult Male (Definitive Basic): In all adults the bill is black with a blue-grey basal third to the lower mandible. The legs and feet are blue-grey, the eyes dark brown. The head and upper chest are black, as is the mantle. The rest of the underparts to the undertail-coverts are deep yellow, becoming more orange-yellow on the breast. The black hood is sharply defined from the yellow underparts on the upperbreast. The rump and lower back are yellow. The lesser and median coverts are orange-yellow, contrasting with the largely black wings. There is an obvious white lower wingbar due to the white tips to the greater coverts. These are tipped largely on the outer vane, thus the wingbar appears to be composed of individual white spots rather than being continuous. The black tertials are conspicuously fringed white. In addition, the terminal half of the outer vanes of the primaries are edged white, creating a thin, white wing panel running lengthwise along the closed wing. The wing-linings are yellow. The tail is black and unpatterned; however, when fresh there are tiny white tips to the outer two pairs of rectrices (R5 and R6).

Adult Female (Definitive Basic): The upperparts are olive-green, with a yellow wash on the rump as well as the nape. The forehead, auriculars, throat and upper breast are black. The rest of the underparts down to the undertail-coverts are yellow, brightest on the breast. The wings are brown with olive-green epaulets (lesser and median coverts). The greater coverts have noticeably pale tips, which may show up as a pale lower wingbar. The wing-linings are yellow. The tail is entirely olive.

Immature Male (First Basic): Similar to adult female, but may show isolated black feathers on the head and back. Often the forehead is olive, rather than black as in the female. The lesser coverts are olive, while the median coverts are blackish with bold yellowish tips. Both of these sets of coverts are newer and contrast with the more worn greater coverts. The tail is olive as in the female.

Immature Female (First Basic): Like adult female, but the black on the face is more restricted, particularly on the rear auriculars. The black often extends only to the lores and throat. The lower edges of the black throat feathers are usually tipped with olive. Also, the flanks are washed with olive.

Juvenile (Juvenal): Like immature female, but lacks black on the face altogether. It also has browner wings and lacks any trace of the pale lower wingbar.

GEOGRAPHIC VARIATION None described, monotypic.

HABITAT In El Salvador, it is largely restricted to oak woodlands (Thurber *et al.* 1987). Howell and Webb (1995) state that apart from inhabiting arid and semi-arid oak scrub, in Mexico it also occurs in semi-deciduous woodlands. In addition, open areas with scattered trees (particularly oak trees) are also used.

BEHAVIOUR Little known. Often found in pairs and also observed in small family parties, once the juveniles have fledged. Sometimes occurs in flocks as large as five to 20 birds, mainly comprising immatures and females after the breeding season has ended (H. Gomez de Silva Garza pers. comm.). It is occasionally associated with other icterids such as Yellow-backed, Black-vented and Streak-backed Orioles at flowering trees (S.N.G. Howell pers. comm.). It is shy and keeps to the canopy, often hiding in the darkest foliage (Rand and Traylor 1954). The breeding system and display patterns have not been described.

NESTING The nest and eggs of this species appear to be undescribed. In El Salvador, birds in physiological breeding condition have been collected during May and June. Adults have been recorded accompanying juveniles in November (Thurber *et al.* 1987).

DISTRIBUTION AND STATUS Rare to uncommon, and scarcer in El Salvador than in southern Mexico. Found in highlands from 500–1800 m (Howell and Webb 1995). Distributed along the central highlands, from easternmost Oaxaca through Chiapas, Mexico, eastwards through the highlands of Guatemala and N El Salvador (Howell and Webb 1995). This species has not yet been recorded from Honduras, even though it has been sighted near the border in El Salvador (Thurber *et al.* 1987). It ought to occur in the extreme south of Honduras; survey work is required to confirm this. Note that this species was not recorded from Mexico until as recently as the 1950s (Paynter 1954).

MOVEMENTS Appears to undergo seasonal movements which as yet are not well understood. Howell and Webb (1995) note its presence in El Sumidero, Chiapas, April–October but its absence here December–March.

MOULT Adults undergo a complete moult (definitive pre-basic). Specimens from the presumed breeding period are in worn plumage, therefore it appears that the definitive pre-basic moult occurs after the breeding season, as would be expected. Juveniles lose their plumage through the first pre-basic moult which includes the body feathers and the lesser and median coverts, and sometimes the outer primaries. It is unknown if there are pre-alternate moults in any age/sex combination.

MEASUREMENTS Males (n=4) Wing 101.4 (98.5–103); tail 99.4 (93.5–105); culmen 21.2 (21–21.8); tarsus 24.5 (23–26).
Females (n=1) Wing 91; tail 94; culmen 20; tarsus 23.

NOTES Griscom (1932) considered Bar-winged Oriole to be most closely related to Black-vented and Black-cowled Orioles. He noted that the latter two species were entirely allopatric, replacing each other in the different faunal/climatic areas where they occur. Therefore, he found it unusual that Bar-winged Oriole should be found sympatrically with the closely related (in his view) Black-vented Oriole. It is not unusual for closely related species to share the same habitat as long as there is a partitioning of the niche such that both taxa may survive. Another possibility is that the closest relative is the allopatric Scott's Oriole, this was eventually Griscom's (1932) conclusion. The relationship of Bar-winged Oriole is not clear, based on DNA data, but it seems not to be related to Scott's Oriole and to be more closely related to the group that includes the Black-vented and Black-cowled Orioles (K. Omland and S. M. Lanyon pers. comm.).

REFERENCES Griscom 1932 [notes], Howell and Webb 1995 [distribution, nesting, movements], Thurber *et al.* 1987 [nesting, distribution]

51 SCOTT'S ORIOLE *Icterus parisorum* — Plate 17
Bonaparte 1838

A black and yellow oriole of the deserts of Mexico and the US.

IDENTIFICATION A large oriole with a long, stout bill (Figure 29.1). The male has a black hood, breast and back; the rest of the body is lemon-yellow. The black wings show two bold white wingbars and the black tail has a large amount of yellow at the base of the outer rectrices. The yellow at the base of the tail separates Scott's Oriole from most other orioles. Black-vented Oriole (44) is superficially similar in colouration, but lacks yellow on the tail, and lacks wingbars. Black-vented Orioles have black vents and uppertail-coverts, both of which are yellow on Scott's Oriole. Other sympatric or near-sympatric oriole species, such as Bullock's (41), Baltimore (40), Altamira (34), Streak-backed (35), Hooded (36) and Orchard (43) are not yellow, but orange, orange-yellow or chestnut; all except Bullock's and Baltimore Orioles have entirely black tails. Yellow-tailed Oriole (30), which should not be found with the Scott's Oriole, also shows yellow on the tail, but this is not restricted to the base as on Scott's Oriole. Furthermore, Yellow-tailed Oriole lacks obvious white wingbars and has a yellow crown and nape. Scott's Orioles are birds of arid and desert areas, and are not found in moist forests as are some of the above-mentioned orioles.

Female and immature Scott's Orioles are rather variable in appearance. The variation is caused by differing amounts of black on the throat and face on individual females, and the intensity of dark back streaks also varies. Immature (first basic) males, on the other hand, always show black throats and auriculars. Scott's Orioles are yellow below, but with an appreciable olive tone, particularly on the flanks. The back and crown are grey to olive-grey with large blackish feather centers, which creates a streaked appearance. Most females and immatures of sympatric orioles have solidly-coloured backs. One exception is Baltimore Oriole, which often shows dark centers to the back feathers. However, Baltimore Oriole has an orange-yellow throat, and often shows a pale vent and belly. In addition, it is smaller and less bulky than Scott's Oriole. The call is a useful identification aid; Scott's Orioles give a loud *chuck* note. Note that very worn individuals (mid summer) of this species may appear to show little or no white on the wings and could be confused with Black-vented Oriole (C. Sexton pers. comm.). The presence of a yellow vent and tail base will identify these birds as worn Scott's Orioles.

VOICE **Song**: A persistent singer early in the breeding season, even during the heat of midday. Females are known to sing as well; their songs are softer and weaker than those of the males but are similar in pattern. The male's song has a particular fluty, warbled quality; this, coupled with the notes it performs gives the song a peculiar resemblance to that of Western Meadowlark (72). The songs are short, usually just under two seconds in length, but are repeated every several seconds.

Calls: Gives a harsh *chuck* as the common contact note. Also performs a harsh scolding *cheh-cheh*, with a nasal quality (Howell and Webb 1995). In addition utters a quiet *huit* (Howell and Webb 1995).

DESCRIPTION A stocky oriole with a rather straight culmen, only slightly downcurved at the extreme tip (Figure 29.1). The gonys is straight. The tail is only slightly graduated. Wing formula, P9 < P8 > P7 > P6; P9 ≈ P7.

Adult Male (Definitive Basic): Bill black with a grey-blue basal third to the lower mandible. Eye dark. The head, neck, mantle, throat and breast are black. The black extends rather low on the breast. The rump and lower back are lemon-yellow. The underparts, other than the breast, are also lemon-yellow. The wings are black with pale yellow lesser and median coverts, forming a yellow epaulet. The bases of the black greater coverts are white, but largely hidden by the yellow epaulet; however, at times they may show as a white lower border to the yellow shoulder. The greater coverts are also tipped with white, forming a noticeable white lower wingbar. The tertials and flight feathers are finely fringed with white, which wears relatively rapidly. The wing-linings are yellow. The tail is black with yellow bases to the outer feathers. The visible part of the central two to four rectrices (R1–R2) are black, but in fact have restricted yellow bases that are fully hidden by the tail-coverts. The outer eight to ten rectrices (R2–R6) are yellow with a black terminal portion roughly one third of the length of the tail. The legs are greyish-blue. Fresh males show a wider white wingbar, more obvious white on the tertials and sometimes some white fringes to the lowermost back feathers.

Adult Female (Definitive Basic): Soft part colours as in male, except the blue-grey patch at the base of the lower mandible averages larger. The head is olive-brown, particularly on the crown and sides of the face. The centers of the crown and nape feathers are darker, creating a dull, streaked appearance. The mantle also appears subtly to noticeably streaked due to the dark brown centers to olive feathers. The overall appearance is of an olive-grey back with dark streaks. The lower back and rump are yellow-olive, contrasting with the mantle. The underparts are greenish-yellow from the throat to the undertail-coverts, and the throat may be slightly greyer than the breast. There is a variable amount of black spotting on the throat of females, which at times is restricted to a few dusky feathers, while some females will show an entirely black throat patch. The wings are brownish-black, with olive-green edged lesser coverts. The median and greater coverts are tipped with white, forming two obvious wingbars. The upper wingbar is wider than the lower one. The primaries, secondaries and tertials are thinly edged with white. The wing-linings are yellow. The olive tail shows greenish-yellow bases to the four outer feathers (R3–R6).

Immature male (First Basic): Similar to adult male, but yellow parts are much duller, washed with olive. The crown and back are dark brown with olive edges, thus these individuals look largely olive-backed when fresh, but reveal the dark underlying colour as they wear. The tail may be olive with yellowish bases to the outer feathers or may show some dusky tips to the rectrices.

Immature male (First Alternate): Similar to first basic male but with more black on the head and back. On some, the back feathers may be a mix of olive feathers with dark centers, creating a streaked look, along with some entirely black feathers. The head is much more solidly black than on a first basic male, and this increases as the spring progresses. There is a certain amount of variability in immature male plumages, but they should be readily identifiable from both adult males and females. The lesser coverts are yellow at this age, but often show noticeable dark centers. In their first spring, Scott's Orioles show rather worn, brownish flight feathers.

Immature Female (First Basic): Most similar to adult females, but do not show any black on the throat. In addition, the brownish wings contrast with darker, fresher, primary coverts. This contrast may be visible in the field. Young females have comparatively more worn flight feathers than adult females.

Juvenile (Juvenal): The bill shows a pink base to the lower mandible and along the cutting edge of the upper mandible. Patterned largely like the female, but the upperparts appear more uniform, not streaked. The crown and back are olive, with the back showing an obvious grey wash. The face is olive, while the throat is yellowish, becoming yellow-olive on the breast. The rest of the underparts are yellow, thus the yellow underparts appear divided by a poorly demarcated olive breast band. This contrast between the breast and the belly is not obvious on adult females. The flanks are slightly more olive than the center of the belly. The wings are brownish with obvious white wingbars and noticeable wide white tertial fringes. The primaries and secondaries are crisply fringed with white. The wing-linings are yellow as in the older individuals. The olive tail has yellow-olive bases to the outer rectrices.

GEOGRAPHIC VARIATION No subspecies or geographic variation described.

HABITAT An inhabitant of desert areas, particularly along the interface between low desert and mountain areas. It is found in the pinyon pine-juniper belt at mid elevations, on dry slopes in the mountains and higher elevation arid plains and intermontane valleys. It tends to avoid the true low elevation Sonoran desert zone, preferring slightly higher elevations and increased humidity. In most of its range it has a great affinity for taller species of *Yucca,* specially as a nest tree. The *Washingtonia* fan palms, along riparian zones and oases, are also preferred nesting trees. In the north of its range it breeds in Utah junipers (*Juniperus osteosperma*) interspersed with Big Sagebrush (*Artemisia tridentata*). In S Arizona and Sonora, Mexico, it has a preference for south-facing slopes which is exactly where Agave is most common (Marshall 1957). The territories are very large, often encompassing the south-facing slope as well as adjoining pine-oak or pine forest. The orioles will forage and use the pine forests, but the primary habitat appears to be the more open south-facing slopes (Marshall 1957).

BEHAVIOUR Migrants arrive in the southwest US in early April or late March. Males arrive on the nesting grounds roughly one week earlier than females. First-year males appear to arrive with the females. At this time, males sing vociferously and can be quite conspicuous. The territories are large, so adjoining pairs are quite distant from each other. The species is monogamous; polygyny has not been reported. Scott's Orioles are largely insectivores but will take fruit, often Prickly Pear Cactus (*Opuntia* sp.) fruit when available. They are also known to forage at flowers, presumably taking nectar. Yuccas (*Yucca* sp.) and Agave (*Agave* sp.) are important, providing nectar as well as insects for the orioles. Occasionally, Scott's Oriole has been observed feeding on the toxic Monarch Butterfly (*Danaus plexippus*), at the butterflies' winter congregations in Mexico. However, they do not do this regularly (Arellano *et al.* 1993).

NESTING Breeding season April–late July, with egg dates as early as late April and as late as late June. The nest site varies depending on the geographic locality. Similarly the height of the nest depends on the available plant species; nests may be only 1–2 m from the ground or up to 20 m or more in taller trees. Where they are present, plants of the genus *Yucca,* including the Joshua Tree (*Yucca brevifolia*) are preferred nesting sites. Often these nests may be placed rather low (1.5 m) off the ground. Typically, the nest is woven into the dead leaves that hang down along the trunk of the yucca. It is constructed of dry fibers of the yucca mixed with dry grass; the texture and colour is such that the nests are quite well camouflaged by the dry leaves of the plant. The nests are a shallow (10–15 cm) hanging-basket, as is typical of the genus. However, in many cases the nest is sewn along the length of one of the dry Yucca leaves such that one edge of the basket is largely composed of the Yucca leaf itself. Other nest trees include Candlewood (*Fouquiera columnaris*), sycamores (*Platanus* sp.), oaks (*Quercus* sp.), junipers (*Juniperus* sp.) and pines (*Pinus* sp.) among others. The clutch size is typically 3 eggs, varying from 2–4. The eggs are pale blue with dark (black, brown, chestnut or grey) streaks and blotches about the broader end of the egg. The incubation is performed by the female and lasts approximately 14 days. The nestling period also lasts approximately 14 days. Both the male and female assist in foraging for and feeding the young.

Two broods may be raised in some areas. A new nest is used for any second brood. These second nests are typically in another nest tree. Bronzed Cowbird (99) and Brown-headed Cowbird (102) are known to parasitise Scott's Oriole, but neither has been observed fledging from a nest of this species (Friedmann and Kiff 1985).

DISTRIBUTION AND STATUS Uncommon to common, decreasing in abundance in the far south and far north of the breeding range. Found from sea-level to 3000 m.

Breeding: In the US, the northernmost breeding sites appear to be part of an isolated population in southwest Wyoming (Sweetwater Co.) northwest Colorado (Moffat Co.) and NE Utah (Dagget Co.). Otherwise this species occurs from C and S Nevada and C and SW Utah south through SE California (east of the Sierra Nevada and peninsular ranges. White-Inyo mountains south into the Mojave desert zone and Colorado desert), Arizona, S and C New Mexico (absent from the north) and W Texas, east to the Chisos mountains. In Mexico, it breeds in Baja California Norte and Sur. It also breeds in E Sonora, Chihuahua, C and W Coahuila south through Durango, Zacatecas, Aguas Calientes, SW Nuevo León, San Luis Potosí, C and E Jalisco, Guanajuato, Queretaro, Hildago, N and C Michoacán, Mexico, Distrito Federal, Morelos, Tlaxcala, Puebla and N Oaxaca.

Non-Breeding: Ranges south to Baja California Sur along the Pacific coast. Also occurs in the interior of Mexico from C Sonora, south to N Sinaloa, and extending to coastal Sonora. Also from S Durango, Zacatecas, SW Nuevo León, San Luis Potosí, E Jalisco, Aguas Calientes, Guanajuato, Querétaro, Hildago, N Michoacán, Mexico, Distrito Federal, Morelos, Tlaxcala, W Puebla, NW and C Oaxaca, C and E Guerrero. A small number winter in extreme southern Arizona, along the western edge of the Colorado desert in California and rarely in coastal southern California.

MOVEMENTS A medium distance migrant, most US breeders leave the country and spend the winter in Mexico. Spring arrival occurs late March–early April in Arizona and California. Further north, in Utah and Nevada, arrival does not occur until early May. Some birds remain in the wintering grounds of Mexico until late April. Note that Scott's Orioles migrate north, at least partially, along the Pacific coast of Baja California but appear to move inland along the mountains further north as they are rare along the coasts of southern California. The rare spring records from coastal S California are largely from San Diego Co.

Birds start to leave the breeding grounds in mid July–August, arriving in the southern wintering grounds in late September. By mid September, most

Scott's Orioles have left the US, but a small number do winter. This species shows up with some regularity as a fall migrant along the southern California coast from Santa Barbara Co. to San Diego Co. They are extremely rare on the Channel Islands.

A bird banded and re-trapped in Nogales, Arizona, was six years and five months old, the longevity record for the species (Klimkiewicz and Futcher 1987).

Scott's Oriole is prone to vagrancy, both to the north and northeast as an overshoot. In Washington, it has occurred once, a male was present from Feb. 11 to Mar. 1980 at Chehalis, Lew Co. (Roberson 1980). In Oregon, a female was present between June 4–7, 1991 at Fields, Harney Co. In California it is a vagrant in the northern part of the state, west of the Sierras. There are approximately 12 records for Northern California, from spring and fall as well as winter. In addition, it has occurred in summer in Alturas, Modoc Co., on July 13, 1988 (fide John Sterling). Away from the south and desert interior, records exist from Modoc, El Dorado, Marin, San Francisco, Santa Clara, Monterey, Fresno and Tuolumne Cos. In Southern California, most coastal records, where this species is a regular rarity, come from fall and winter. There are sporadic reports from Idaho with several detected W of Stone, May 25 1991. It appears that this species may be of regular occurrence there, rather than a vagrant. In Nevada, Scott's Oriole strays to the north of the state, and has been reported as a spring overshoot in Reno. In Utah, it has strayed to the Salt Lake City region. In Colorado there is a small breeding population in the W part of the state, but it is a rare stray to the Denver area. In Kansas, a female was collected in Morton Co., on 16 April, 1977 (Thompson and Ely 1992). Farther north in Nebraska there are two sight records of males: one from May 20–June 24 1975 in Hall Co., this bird appeared to have been a first basic male (Stoppkotte 1975); and in the spring of 1978, in McPherson Co. There is one record from Minnesota, this bird was present in Duluth from May 23, 1974 and remained into June. It was caught, banded and released. In Wisconsin, there is a record of a wintering individual from January–February 1996 at a feeder. In Ontario, there is a record of a male on Nov. 9, 1975 at Silver Islet, Thunder Bay District, and this was the first Canadian record (Denis 1976). In Texas, it is regular in the southwest, but has occurred as far N as Buffalo Lake, S of Amarillo. An individual well east of the species' range was at Brownsville, on March 1–2, 1995 (B. McKinney pers. comm.). It is a stray to NC Texas. In Louisiana, it has occurred on at least 10 occasions, all records from fall, winter or spring. One occurred in the north of the state, while all other records have been from the south.

MOULT Juveniles undergo a partial first pre-basic moult (July–August) which replaces the body feathers, wing-coverts and some or all of the tail feathers. In some birds, the fifth and sixth secondary may also be replaced as well as a variable number of outer primaries (Pyle 1997a, 1997b). A variable and partial first pre-alternate moult occurs in young (first basic) males February–April; this moult replaces some of the head and back feathers. The extent of the female first pre-alternate moult is not well known, and it may be lacking, although some young females do appear to acquire a limited number of black feathers on the throat in their first spring. Adults undergo a complete pre-basic moult after nesting, July–August. The extent of the pre-alternate moult in adults is not well understood.

MEASUREMENTS Males: (n=16) wing 104.4 (98.6–106.7); tail 88.4 (79.2–91.9); culmen 22.9 (20.8–24.6); tarsus 23.9 (22.9–25.4). Females: (n=63) wing 97.8 (94.5–102.1); (n=10) tail 84.3 (81.3–88.4); culmen 21.3 (20.3–22.9); tarsus 24.1 (23.4–24.9).

NOTES Scott's Oriole appears to be most closely related to Audubon's and Yellow-backed Orioles, based on mitochondrial DNA sequence data (K. Omland and S. M. Lanyon pers. comm.)

REFERENCES Arellano *et al.* 1993 [behaviour, Monarch Butterflies], Bent 1958 [behaviour, nesting, voice, movements}, Findholt and Fitton 1983 [distribution, habitat], Klimkiewicz and Futcher 1987 [age], Pyle *et al.* 1987 [moult], Small 1994 [distribution].

52 JAMAICAN BLACKBIRD *Nesopsar nigerrimus* (Osburn) 1859 — Plate 33

A uniformly black icterid of the forests of Jamaica, this blackbird is an epiphyte specialist.

IDENTIFICATION This black icterid is unlikely to be misidentified in its native Jamaica due to a combination of plumage and behavioural characteristics. It has been described as behaving like an all-black oriole (*Icterus*). It is entirely black with a dull gloss, and has dark eyes. It is a bird of montane forests being commonly found feeding near the canopy. Greater Antillean Grackle (92) is also black but has a yellowish eye and a much longer, wedge-shaped tail. The grackle is not a bird of deep forest. Similarly, Shiny Cowbird (102) is not found in the forest. It is black, like the Jamaican Blackbird and has a dark eye; however, the cowbird has a shorter bill and a longer tail, and is more iridescent than the blackbird.

VOICE Song: The songs are made up of three or four buzzy notes, similar to that of *Agelaius* blackbirds, lacking musicality or sweetness. It has been described as *zwheezoo-whezoo whee* (Downer and Sutton 1990). There are three different song types, differing in the exact notes uttered, but similar to each other in general structure. Both males and females sing, often with one sex responding to the song of the other pair member. The song is often given in a song flight over the forest, primarily by males.

Calls: A *dzik* or *check* call is given regularly by both sexes, but primarily by males conducting the 'patrolling display'. At such times, the male will give several *dzik* calls in repetition. A quiet, trilled *zeenk* is given near the nest, commonly when making nest visits. In alarm, a series of *chet* notes, a chatter really, are given or a thin, high *seee seee.* Mainly from Wiley and Cruz (1980).

DESCRIPTION A small blackbird with a proportionately long bill. The bill is conical, but not very deep, with the culmen and gonys straight. In general, the bill is *Agelaius*-like; however, the culmen is noticeably flattened. The wings are short and rounded while the tail is also short and slightly rounded. Wing formula: P9 < P8 > P7 > P6; P9 ≈ P5; P8–P6 emarginate. The legs of Jamaican Blackbird are proportionately short.

Adult (Definitive Basic): The bill is black and the eyes are dark. The entire plumage is black with a slight blue gloss. The gloss is most noticeable on the head, upperparts and coverts, less so on the breast and is entirely lacking on the belly. The primaries appear slightly browner than the rest of the plumage. The legs and feet are blackish.

Immature (First Basic): Not known if it is different from the adult.

Juvenile (Juvenal): Similar to the adult, but with a looser texture to the plumage. In addition, the black plumage lacks the gloss and appears more brownish.

GEOGRAPHIC VARIATION Monotypic.

HABITAT Restricted to montane forests with high rainfall, where there is an abundance of epiphytic vegetation. This species needs forests which have not been greatly disturbed and it avoids openings. These blackbirds live in the lower montane rainforests and mist forest, but avoid the elfin or ridge forest of the higher elevations and the more exposed sites. Mist forest found in the flatter areas that are sheltered from wind is preferred by Jamaican Blackbirds; this forest type was profusely covered by epiphytes (Wiley and Cruz 1980). They avoid forests on exposed slopes where the common epiphyte is *Usnea* lichen, preferring bromeliads and *Phyllogonium* moss (Lack 1976).

BEHAVIOUR Jamaican Blackbirds spend most of their day foraging near the forest canopy. They search epiphytic vegetation, particularly bromeliads as well as ferns, lichens, *Phyllogonium* moss and tree ferns. Any dry leaves hanging from the base of the epiphyte are searched most vigorously; however, they are not meticulous in their searches and move from one plant to the next rather quickly, covering a lot of ground as they forage. They are quite acrobatic when feeding, often hanging sideways from dead leaves or twigs to get at a specific morsel of food, and will land on vertical trunks and crawl up them while foraging. They rarely strip bark as does the sympatric Jamaican Oriole (28). As the birds search for food they vocalise frequently, particularly when flying between trees.

Jamaican Blackbird is monogamous, with each pair maintaining an exclusive territory. The members of the pair are rarely found together, unless nesting, and tend to feed solitarily, and may subdivide the territory between themselves. During incubation, the male tends to perch near the nest, vigilant for predators or any other danger. He will chase other birds away from the immediate nest site at this time. Once the young have hatched, both the male and the female forage to bring them food, with the male accounting for roughly half of the food brought to the nest. After fledging, the young travel with the parents for an extended amount of time, at least two months, after fledging.

Both sexes sing and perform song flights. The song is not usually accompanied by a 'song spread' display. The singing bird will commonly tilt its bill upward and move its head sideways as it sings. Rarely, the wings are raised and slowly moved as the bird sings, while the tail is also fanned. The most common display is the song flight which is usually performed by the male, but may also be performed by the female. Jamaican Blackbird regularly flies above the forest canopy while giving one of the typical songs. As it sings, the bird slowly descends from its maximum altitude and flaps its wings slowly and deliberately in a 'butterfly flight' display; after the song is completed the bird closes its wings and dives back to the forest below. Flight displays are exceedingly rare in forest-dwelling birds. The song flight is conducted throughout the day, but at irregular intervals and rarely more than once per hour. Another unusual display conducted by the male is the 'patrolling flight'. At dawn, Jamaican Blackbirds do not perform a dawn chorus. Instead the male flies back and forth across the centre of the territory in a rapid, noisy whirring flight and uttering a *dzik* call.

NESTING The nesting season starts in mid May and continues into July by which time the young have fledged. The nest is a bulky cup placed approximately 8m off the ground in the lower canopy. Nest sites were in trees that slanted so that the trunk was almost horizontal. Nests are placed against the trunk, and are sometimes supported by branches. They are constructed from rootlets and epiphytic orchids, resembling nests of *Agelaius* blackbirds. The clutch size is of 2 eggs, and incubation lasts roughly 14 days. Both parents help to feed the young.

DISTRIBUTION AND STATUS Uncommon to locally common, but only where mature rainforest exists. It is endemic to Jamaica, and usually found above 500 m. It is almost certain that Jamaican Blackbird was formerly widespread throughout the highlands of Jamaica. Currently this interesting blackbird has a much more spotty distribution in some of the larger remaining forest patches. It may be found in Newcastle, Worthy Park, Kew Park, Cockpit Country, Hardwar Gap and John Crow mountains (Downer and Sutton 1990). Jeffrey-Smith (1972) lists Port Royal and Blue mountains as areas where this species is present.

MOVEMENTS Apparently sedentary, maintaining a territory year-round. However, it may disperse to lower elevations in winter, having been observed as low as 210 m (Lack 1976).

MOULT Largely unknown, but assumed to be similar to many other icterids with a complete moult (definitive pre-basic) that occurs after breeding. The timing of this moult is unclear, but it starts as early as September and lasts as late as early January. The extent of the first pre-basic moult is not known.

MEASUREMENTS Males are about 5% larger than females, a small size difference for an icterid. Males: (n=6) wing 98.4 (92–102); (n=5) tail 75.5 (73.2–80); (n=6) culmen 26.9 (22–30.5); tarsus 23.5 (22.0–25). Females: (n=7) wing 96.5 (94.5–101); (n=6) tail 71.0 (63.5–73); culmen 25.5 (25.1–26.7); (n=7) tarsus 23.5 (23–25).

NOTES A recent molecular taxonomic study suggests that Jamaican Blackbird is not closely related to the Caribbean and North American *Agelaius* group (Lanyon 1994). This is contrary to what was proposed earlier by Bond (Lack 1976). In fact, based on mitochondrial DNA sequence data, Jamaican Blackbird appears to be the sister group to the 'typical' blackbirds (grackles, cowbirds, *Agelaius, Curaeus, Amblyramphus,* Marshbirds, *Oreopsar, Dives* and Euphagus) (Lanyon 1994).

REFERENCES Downer and Sutton 1990 [voice, distribution], Hardy *et al.* 1998 [voice], Lack 1976 [habitat, behaviour], Lanyon 1994 [systematics, notes], Ridgway 1902 [measurements], Wiley and Cruz 1980 [voice, habitat, behaviour, nesting].

53 ORIOLE BLACKBIRD *Gymnomystax mexicanus* (Linnaeus) 1766 — Plate 18

An unmistakable, large black and yellow icterid of northern South America.

IDENTIFICATION The combination of large size and striking black and yellow plumages makes this a difficult species to misidentify. Notice that the entire head is yellow and that the tail is entirely black. At close range, the naked blackish skin around the eye and along the malar area is diagnostic. Several orioles may be confused with this species, but all of these are very much smaller and have black on the throat, and yellow or orange backs, as well as white markings on the wing. Yellow-rumped Caciques (13) are also large, but they are largely black, with black underparts and yellow at the base of the tail; in addition, they show a pale blue eye. Male Yellow-hooded Blackbird (59) has a yellow head, but no yellow on the coverts or underparts, and it is much smaller. The allopatric and local Saffron-cowled Blackbird (55) is similar in pattern but is not found within the range of this species. Note that Saffron-cowled Blackbird is smaller, slimmer and has a yellow rump.

VOICE Most vocalisations are scratchy and unmusical.

Song: Has been described as *chaa chaa chrick chaa*, the *chrick* note rising in pitch and the 'chaa' notes sounding quite nasal. In some respects, the nasal notes are muffled, and in that way are similar to the songs of Common Grackle (91). Another interpretation of the song is *ting-ting-wreg-wreg-gri-gri*. Both sexes sing the same song, but male songs tend to be longer.

Calls: The common call is a long drawn-out screech, resembling the sound of a rusty gate. Other calls include a *chrick chaa* when taking off from a perch; the flight call has also been described as *wreg-kreg*. Adults give a high, metallic *cleek*. Females at the nest utter a *cluck* or *tuc-titit*. The alarm calls of the parents when nestlings are present is a sharp series of *chip* notes.

DESCRIPTION This is a large, robust blackbird with a stout bill and strong legs. The bill is roughly as long as the length of the head. The culmen is entirely straight as is the gonys. The long tail is square-tipped. Wing formula: P9< P8 ≈ P7 ≈ P6 > p5; p9 ≈ P4; P8–P6 emarginate. The sexes are alike and are treated together.

Adult (Definitive Basic): Bill deep and stout, entirely black. The eyes are brown. The face has a bare patch around the eye, extending forwards to the base of the bill, with a bare skin extension pointing back along the malar area. Otherwise the entire head to the upper back is yellow-orange, as are the underparts. The back and wings are flat black, except for the lesser coverts and the bend of wing which are yellow. The tail is entirely black as is the rump and uppertail-coverts. The legs and feet are blackish grey.

Immature (First Basic): Most are indistinguishable from adults. However, some immatures retain some black on the back of the crown. The bare facial skin is black.

Juvenile (Juvenal): Resemble adults but the bare face patch is pinkish-grey. Unlike the adults, juveniles show a clean-cut black cap. In addition, the back has an obvious brown cast. The secondaries and primaries have thin buff or yellowish edges.

GEOGRAPHIC VARIATION Monotypic.

HABITAT Frequents open areas with scattered trees and palms, often moist marshy localities. It is fond of cattle-grazed grasslands and agricultural zones, as well as native grassland and savannas, particularly if moist. Also frequents the edge of forest and gallery woodlands. Oriole Blackbird avoids lowland tropical forest, keeping to the edge of large openings. Along the Amazon river, it is found in open wetlands and on river islands as well as cleared areas.

BEHAVIOUR Usually observed in pairs or small flocks, seldom more than a dozen per flock. Roosts in larger groups, sometimes in the hundreds. Most of their foraging occurs on the ground, slowly walking and picking, or probing into the ground. Some food, mainly fruit, is secured from trees. They frequently raid cornfields, making pests of themselves (Skutch 1967). Oriole Blackbird typically perches on trees or structures at low elevations, such as fence posts. When alarmed it may retreat to a dense tree, making it difficult to see (Wetmore 1939). The display that accompanies the song is not as spectacular as that of many other blackbirds. The male fluffs out the feathers of the back of the neck and mantle while fanning the tail and drooping the wings, elevating the bill to a vertical position while he sings (Skutch 1967). While the female is at the nest the male perches in a conspicuous position in a nearby tree. Males sing infrequently from the nest area. They do not act territorially, but if a pair of conspecifics approaches, the resident pair will fluff out their plumage and sing until the intruders leave, but overt aggression is not observed (Skutch 1967). The behaviour of this species suggests it is monogamous.

NESTING The breeding season may be tied to the wet period of the year, which is June in Venezuela. Nests between August–December in S Guyana, and in March in E Peru. Nesting is not colonial in this species. The nest is built primarily by the female, with some help from the male. It is placed at medium height (6–8m), usually in a tangle or a dense area of a tree or palm. The nest itself is a bulky cup made of straw and other coarse vegetation. The clutch size is 3 eggs, which are pale blue with blackish or purplish spots and scrawls, particularly around the broader end (Skutch 1967). Shiny Cowbird (102) has been known to parasitise, but not fledge from, nests of this species (Friedmann and Kiff 1985). The female incubates alone, and incubation takes approximately 18–20 days.

DISTRIBUTION AND STATUS Common, a bird of the lowlands, ranging to 950 m. It appears to live in two separate populations, one in the north of South America, the other along the Amazon river, but it may also occur in the intervening area. Found from the Llanos of NE Colombia, east of the Andes through most of N Venezuela in the Llanos, the Orinoco river and its delta as well as river courses in the scrub zones; it is absent south of the Orinoco. It is also found in northernmost, coastal Guyana where it is rare. Otherwise it is found along the Amazon river and some of the major tributaries from S Colombia (Leticia), E Ecuador (Napo river, as far up as the junction with the Coca river) and the Amazon headwaters in NE Peru (Loreto, San Martín and Ucayali) east to the Amazon delta in Brazil and north to E French Guiana. In the Amazon Basin, Oriole Blackbird appears to be largely restricted to the Amazon river itself as well as the lower reaches of its tributaries on the south shore; it is largely absent from the north shore and may be completely absent from blackwater areas of N Amazonia (Ridgely and Tudor 1989). However, it has recently been discovered well north of the Amazon in Roraima, Brazil (Ridgely and Tudor 1989).

MOVEMENTS Apparently sedentary, but moves short distances to reach communal roosting sites.

MOULT Largely unknown. Adults undergo one complete annual moult, the definitive pre-basic. Wetmore (1939) noticed that in northern Venezuela, Oriole Blackbirds were undergoing definitive pre-basic moult in mid November. Specimens from the Llanos of Venezuela confirm that November is the main month for moult in that area. In eastern Ecuador, moulting birds are known from mid December. Like grackles and some other icterids, the tail moult in this species is quick. In many cases, the tail feathers are cast simultaneously, rendering the bird tail-less for a short period of time. There is no evidence that pre-alternate moults occur in Oriole Blackbird. The juvenal plumage is lost through the first pre-basic moult which is incomplete (sometimes complete?). It includes all of the body feathers as well as the tail, and most or all of the flight feathers. The timing of the first pre-basic moult is similar to that of the definitive pre-basic, November in N Venezuela.

MEASUREMENTS Males: (n=10) wing 130.8 (120–135); tail 116.2 (104–125); culmen 30.2 (28–32); tarsus 34.2 (30–39). Females: (n=9) wing 126.2 (121–141); (n=8) tail 115 (110–130); (n=9) culmen 29.6 (26–31); tarsus 33.7 (30–36).

NOTES One specimen was examined which had a single yellow feather on the rump. These scattered yellow feathers are observed in several species of black and yellow icterids. Two separate molecular data sets classify this species as a close relative of Velvet-fronted Grackle (81) (Lanyon 1994, Freeman and Zink 1995), the former paper also noting that Colombian Mountain-Grackle (83) belongs in that group.

REFERENCES Goeldi 1897 [voice, habitat, nesting], Goodfellow 1901 [description, distribution], Ridgely and Tudor 1989 [habitat, distribution], Skutch 1967 [voice, behaviour, nesting], Wetmore 1939 [habitat, behaviour, moult].

54 YELLOW-HEADED BLACKBIRD

Xanthocephalus xanthocephalus **Plate 24**

(Bonaparte) 1826

An attractive black and yellow species typical of western North American marshes. Like Red-winged Blackbird, it makes its presence known and local people usually recognise it by name.

IDENTIFICATION Both the male and female are largely unmistakable in their range. No other North American (including Mexico) bird is black with a yellow head and white wing patch like male Yellow-headed Blackbird. There are some largely black and yellow birds of a similar size, such as orioles (*Icterus*) and Yellow Grosbeak (*Pheucticus chrysopeplus*), but all of these species are found in different habitats and have the yellow extending to the belly and vent. Beware of leucistic blackbirds which may show off-white heads and may superficially resemble Yellow-headed Blackbird. These birds are unlikely to show the dark lores (mask) of male Yellow-headed Blackbird. The brown female is easily identified by the yellow patch on the breast and the yellow supercilium. No other sympatric bird resembles this pattern. Female Red-winged Blackbirds (61) are obviously streaked both above and below, and lack yellow on the throat. Female 'Bicolored Blackbirds' (61X) are more solidly coloured, but still show some streaking towards the throat, and tend to lack a pale supercilium in worn plumage. If the throat of female Red-winged Blackbird is any colour other than white, it will be pink, peach or buff, never yellow. Flocks of female Yellow-headed Blackbirds look very much like other blackbirds in flight, or at a distance. Listen for Yellow-headed Blackbirds gruff *chak* call. Yellow-headed Blackbirds appear long-winged in flight, are noticeably bulky and have only a slightly undulating flight. Red-winged Blackbirds undulate more noticeably in flight than Yellow-headed Blackbirds. On the breeding grounds, birders may encounter the unfamiliar looking juvenile Yellow-headed Blackbird. Unlike adults, juveniles have two wide cinnamon wingbars, and a cinnamon head with a darker crown and ear-coverts. No other sympatric species closely resemble juvenile Yellow-headed Blackbird but those unfamiliar with this plumage type may not know what species it belongs to at all! The bold cinnamon, or creamy, wingbars are a good field mark for this plumage. Yellow-hooded Blackbird (59) is not found near the range of Yellow-headed Blackbird but it is quite similar in overall colour pattern. Note that Yellow-hooded Blackbird is very much smaller, more so than a Red-winged Blackbird in fact. Yellow-hooded Blackbirds lack a white wing patch and the black face mask is less extensive.

VOICE Song: The male's song has been likened to the sound of the opening of a rusty gate. Under no circumstances could one call its song beautiful, but it is distinctive. Songs are usually accompanied by the 'song spread display'. There are two song types. The most common is the 'buzzing song' which lasts for several seconds, beginning with a few clucking notes and followed by a long vibrating, nasal buzz *kuk, kuk, kuuk..whhh-haaaaaaaaa*. The second song type is the 'accenting song', which lasts for approximately two seconds, and is made up of several rich notes sometimes followed by a trill or buzz *kuuk-ku, WHAAA-Kaaaa*. The female song is a raspy, nasal chatter that may be given with a 'song spread display' *chee..chee..chee..chee*

Calls: The common call is a single throaty *chek* or *chuck*, accompanied by a tail flick. The *chuck* of Yellow-headed Blackbird is deep and hollow-sounding, unlike the sharper *chuck* of Red-winged Blackbird. During alarm, or the presence of a predator, males give a harsh rattle, while females perform a screaming chatter. During 'sexual chases', males give a deep, hollow growl. While mating and performing the pre-copulatory display, both sexes utter a high *ti-ti-ti-ti-ti*, like many other blackbirds. Begging juveniles are loud and in mid-summer their notes are a noticeable element of the soundscape in Yellow-headed Blackbird marshes. The begging notes are a shrill *prrerererereeeeee*.

DESCRIPTION A stocky and large blackbird with a thick conical bill. It has long, pointed wings. The culmen and gonys are straight. Wing Formula: P9 > P8 > P7 > P6; P9 > P8; P8–P6 emarginate.

Adult Male Breeding (Definitive Basic, worn): The bill is black and the eyes dark brown. The head down to the nape and breast is yellow with a saffron wash, particularly prominent on the nape. A black mask comprising of blue lores, the area around the eyes, a thin line above the bill and the chin contrasts with the yellow head. The yellow of the breast tapers to a point along the mid-line of the body, and often shows black spotting towards its posterior end. The rest of the body is jet black without a noticeable gloss. The feathers around the cloaca are also yellow, but these are usually hidden while the bird perched. The wings are black with an obvious white wing patch on the greater primary coverts and outermost greater coverts. This wing patch appears as a white rectangle at the leading edge of the wing while in flight. The tail is entirely black. The legs and feet are black. Birds in winter and summer are not appreciably different.

Adult Male Non-Breeding (Definitive Basic, fresh): Resembles breeding male but the crown and nape feathers are tipped yellowish-brown.

Adult Female (Definitive Basic): The bill is black and eyes dark. Largely a brown bird with yellow on the breast. The forehead and crown are brown, extending to the nape and neck. The auriculars are a similar brown colour with the lores more greyish. Shows a thin, pale yellow supercilium and a pale yellow submoustachial which often curls up around the auriculars so that it almost touches the supercilium. The throat is white, as is a patch on the sides of the neck; a dark malar stripe is usually visible. The breast is entirely yellow except for brown spurs extending from the breast sides towards the mid-line. The lower breast is whitish with dark streaking, quickly becoming solid brown on the belly, vent and undertail-coverts. The entire upperparts to the uppertail- coverts are brown. The wings are also brown, sometimes showing slightly paler fringes to the coverts and flight feathers; the edge of the alula is often white. The tail is brown. The legs and feet are black.

Immature Male (First Basic): Similar to adult female but noticeably larger. Has darker lores, a brighter yellow breast, and white fringes to the primary coverts and outer greater coverts. In summer, appears blackish rather than brown, and the yellow of the breast is more vivid.

Immature Female (First Basic): Similar to adult female but the yellow is more buffy. The primaries are duller and more brown. First-year females during the breeding season have lighter coloured breasts, throats and faces.

Juvenile (Juvenal): Similar to immature plumages, but paler in colour and with cinnamon or yellow-buff extending to the forehead, crown and nape as well as on the breast. The throat is usually whitish. The brown back is tipped with cinnamon. The dull, brownish wings have cinnamon edges to the tertials and whitish tips to the coverts, which form two dull wingbars. The primary coverts are tipped white and the bend of the wing is white. The legs are dull brown rather than black.

GEOGRAPHIC VARIATION Monotypic. There is a cline in body size, with birds from the south east being the smallest and populations from the northwest the largest.

HABITAT Nests strictly in marshes or other wetlands (meadows, aspen thickets and lake edges). In marshes it prefers using the deeper parts, in tules (*Scirpus* sp.), *Phragmites* or cattails (*Typha* sp.). This species nests in the interior of marshes, well away from the edges, particularly where these are forested. It forages in wetlands, but often flies to nearby grasslands, agricultural fields or other open habitats. Roosts in marshes. In winter it forages largely in open agricultural fields, pastures or grasslands. Favours marshes for roosting, but may use other dense vegetation.

BEHAVIOUR Adult males arrive on the breeding grounds up to two weeks before the females. They defend a territory from other males through a variety of displays and vocalisations. Territories range in size from 116–4072 m^2, depending upon the habitat quality and the locality. Yellow-headed Blackbirds arrive on the nesting marshes after Red-winged Blackbirds, and often evict them, taking over their territories upon arrival. Males are highly polygynous, regularly breeding with up to six females. Many males do not breed; this is invariably the case with immature males. Females are not monogamous, however, and will solicit matings from males other than their primary mate.

There are various male displays. The 'song spread display' is highly stereotyped and resembles that of other blackbirds. However, Yellow-headed Blackbirds perform two different 'song spreads', each accompanied by one song type. The 'symmetrical song spread' is accompanied by the 'accenting song' (see Voice). In this display, the white wing patches are flashed by partially spreading the wing, the bill is pointed upwards as the male sings, and the tail is lowered and spread. During intense versions the wings are held in a 'V' above the body. The 'asymmetrical song spread' is accompanied by the 'buzzing song' (see Voice). In this 'song spread', the bill is pointed upwards but the head is twisted to the left while the wings and tail are partially spread. Females sing and perform a 'song spread' that is most similar to that of the male's 'asymmetrical song spread'; female 'song spreads' are most commonly aimed at other females and communicate aggression. Males also execute two types of song flight. In one, the 'flight stall', the body plumage is fluffed out as the male flies in a slow 'butterfly-like' manner while holding the fanned tail downwards. The 'bill-up flight' differs in being undulating, quicker, with the bird pointing the bill upwards while keeping the body feathers flat to the body and the tail not spread. This latter display is an aggressive one used against invading males. Female flight displays are aimed at other females and also tend to be aggressive in nature. These are performed with the body plumage sleeked, the legs dangling, the bill pointing upwards and are accompanied by shallow flapping of rapidly vibrating wings.

While establishing the pair bond, the pair may partake in a 'sexual chase'. These are often initiated by the female which swiftly flies close to the male; he then takes off and accelerates while the female twists and turns, finally he may catch up, grabbing at her rump, at which point the female drops into cover and the chase ends. Prior to copulation, the male gives a 'crouch display' where he fluffs up the body plumage, fans and lowers the tail, and crouches with the bill pointed downwards. The female may then give her pre-copulatory display where she assumes a crouched position with the tail raised while at the same time tilting the bill upwards. The male mounts and mating occurs.

Yellow-headed Blackbirds actively exclude Marsh Wrens (*Cistothorus palustris*) from the vicinity of their nests. This may be due to competition for damselflies (Zygoptera) as food, or more likely because the wren is known to puncture the eggs of blackbirds and other species which it nests close to (Verner 1975).

NESTING The breeding season lasts from May (sometimes late April)–July, and there is only one brood per season but females will re-nest if the first clutch is destroyed. Second-year females arrive on the nesting grounds, and begin nesting later than adult females (Crawford 1977). The nest site is chosen by the female and is located within the territory of the male. It is built by the female from strips of marsh vegetation, and is a bulky cup attached to stalks of emergent vegetation. The nest is built over water, usually within 50 cm of the surface. The clutch size is commonly 3–4 eggs, sometimes 5. The eggs are pale grey or pale green, speckled and spotted with darker marks, particularly around the wide end of the egg. Incubation begins after the second egg is laid, which creates a noticeable size hierarchy in the young, as two hatch earlier than the rest. All of the incubation is performed by the female, and likewise the young are fed largely by the female. The role of the male in feeding the young depends on several factors, such as how much of his time is spent trying to mate with other females and whether this is the first (primary) or a later nest. Males are more likely to help out with primary nests. Incubation lasts 12–13 days and the young take nine to 12 days to fledge from the nest. Parasitism by Brown-headed Cowbird (101) is very rare, and has not been observed to be successful. Yellow-headed Blackbirds are not aggressive towards cowbirds and readily accept their eggs (based on experiments) but up to one third of Red-winged Blackbird eggs added to a Yellow-headed Blackbird nest were rejected! However, these Red-winged Blackbird eggs were put into nests where Brown-headed Cowbirds were actively parasitising the Yellow-headed Blackbirds; thus it is reasonable to assume that in this situation the blackbirds have been selected to recognise foreign (almost certainly cowbird eggs in the wild) eggs and reject them (Dufty 1994). Yearling (1st-year) females nest later than adult females, and lay a smaller clutch of smaller eggs. The smaller nestlings have lower survival rates, therefore young females fledge fewer young than older females (Crawford 1977).

DISTRIBUTION AND STATUS Common to uncommon, less so in the eastern and western extremes of its breeding range.

Breeding: Nests mainly in the Great Plains, Prairies and Mountain states and provinces, with isolated colonies in the far west and the east. The main breeding range lies east of the Cascades from C British Columbia to NW California. A small population breeds near Vancouver, British Columbia, and in coastal Washington; there are also isolated colonies along the Columbia river valley and in W Oregon. There are isolated populations in California in the Central Valley as well as along the southern coast between Los Angeles and San Diego, and along the Colorado river valley to NW Baja California Norte, Mexico. The breeding distribution ranges north to C British Columbia (N to Prince George, Hazelton, also including the Peace River District where it is of recent origin, M. Phinney pers. comm.), east through N Alberta, NC Saskatchewan, SC Manitoba, extreme W Ontario (Rainy river), N Minnesota, N Wisconsin, and W Upper Michigan. The distribution lies W of Lake Michigan except for isolated populations in SW Ontario, Lower Michigan and N Ohio. Note that this easternmost spur of the breeding range may be a recent development, as this species was not known to breed in Ontario before 1965 (Sawyer and Dyer 1968). The southern edge of the distribution lies from NW Indiana, west through N Illinois, S Iowa, NW Missouri, NE diagonally to SW Kansas, the Oklahoma panhandle, the W Texas panhandle, SC New Mexico and extending to WC Arizona but not NW New Mexico or SW Colorado. There are isolated colonies in SC Arizona. The southern fringe of the range continues from SW Utah, N and NW Arizona, through most of Nevada except for the SW with a tongue extending into the Mono Lake region of California.

Non-breeding: Winters almost completely south of the breeding range, mainly in Mexico and the southwestern states of the US The main wintering distribution ranges north to C Arizona, C New Mexico and W Texas to the Rio Grande valley; it extends south through E Sonora as far south as coastal Jalisco, Mexico and reaches its southern extreme in Mexico, from C Jalisco, W Michoacán, NE Guerrero, Morelos, W Puebla, Hildago, E San Luis Potosí to S Tamaulipas. Small numbers winter in Central America regularly as far south as Costa Rica and accidentally in Panama (see Movements). The winter distribution of Yellow-headed Blackbirds south of Mexico is poorly understood. It is not known if the species regularly winters in, or only erratically visits sites south of Mexico. Isolated individuals occur in eastern North America either as migrants in large blackbird flocks or wintering at feeders.

MOVEMENTS A medium to long distance migrant. Failed breeders may begin returning south as early as July, but most do not leave until late August–September. Most have left the breeding grounds by late September–October. Arrival in Mexico ranges from August–November and birds remain as late as May, although most depart before then. Adult males arrive on territory before females or immature males, late April in Wisconsin, March–April in Washington, early May in North Dakota (Crawford 1978). Adult females arrive one to two weeks after the adult males. In one North Dakota study, adult males arrived before adult females and 2nd-year immature females arrived last of all. There were few 2nd-year males observed at this site, and their arrival time was similar to that of adult females (Crawford 1978). This species is a diurnal migrant, with males usually travelling separately from females. Females migrate further than males and make up most of the population in S Mexico. Males mostly winter in SW. US.

Yellow-headed Blackbird often flocks and roosts with other blackbirds. The oldest known Yellow-headed Blackbird reached the age of 11 years and 8 months (Klimkiewicz 1997). A male banded in North Dakota was re-trapped in early May in Apatzingan, Mexico, at the age of 10 years and eleven months (Klimkiewicz and Futcher 1987).

This species is a regular autumn migrant in the east of North America either singly or in small flocks of two or three birds, and many of these remain to winter. Yellow-headed Blackbird appears to have been recorded in all of the eastern states and Canadian provinces, north to Quebec, New Brunswick, Nova Scotia and Newfoundland. Most records are from coastal locations, in fall, along the Atlantic and Gulf coast states, as far south as Key

West, Florida. In spring, they may appear almost anywhere in the east, not necessarily along the coasts. Inland from the coastal states, Yellow-headed Blackbird is perhaps more common during the spring than in the fall. North of the regular distribution there are few records. In Alaska, there are several records and only two specimens (Gibson and Kessel 1997); one of these was a female collected on July 3, 1975, at Barrow, Alaska. The second was a female found on Herschel Island, Yukon Territories, from August 12–14, 1996. In N Quebec, it is very rare, and it has occurred at least once at Radisson, in the James Bay lowlands.

This species has a tendency for vagrancy. Caribbean records include two from Cuba, including a report of two individuals from Guantanamo, as well as an individual observed being sold in the Havana market (Ramsden 1912); three from the Bahamas, the first on October 16, 1965, the second on October 20, 1966, both on Grand Bahama (Kale *et al.* 1969), and the third from San Salvador, Nov 1976; one from Grand Cayman Island (Bradley 1994), and amazingly one from Barbados. There are three fall records from Bermuda. A fall vagrancy pattern is evident for the Caribbean sample. Yellow-headed Blackbird has also overshot its regular wintering grounds winding up in NW Costa Rica and in Panama. There are November and January records (including a flock of nine) from Panama (Wetmore *et al.* 1984). In the Western Palearctic, there is a July record from Iceland, several in Great Britain, an early fall record in France, an October record from Denmark, a May record from Norway and a May record from Sweden. Not all European records are regarded as being of wild origin. In Mexico stray records exist from S Baja California Sur, S Oaxaca, C Veracruz and Yucatán.

MOULT The complete definitive pre-basic moult of adults occurs July–August, after breeding is finished. This moult occurs close to the breeding grounds. There is a limited pre-alternate moult (January–April), but this appears to be restricted to immature (first basic) males, and it occurs in the non-breeding grounds. The juvenal plumage is replaced through the first pre-basic moult (July–August) which replaces most feathers except for some of the underwing-coverts and sometimes the tertials.

MEASUREMENTS Males are substantially larger than females and may weigh nearly twice as much. Note that even as adults males become slightly larger as they age (Searcy 1979). Males: (n=11) wing 141.2 (135.1–145.5); tail 102.6 (93–108.5); culmen 22.9 (21.1–25.1); tarsus 35.8 (33.3–37.1). Females: (n=10) wing 113.4 (108–117.9); tail 82.7 (78.7–90); culmen 20.4 (19–22); tarsus 30.6 (29.7–32).

NOTES Data from the Christmas Bird Count (CBC) 1959–1988 does not show a significant change in the population wintering in North America north of Mexico. The overall trend in numbers was negative (decreasing population) but this was not statistically significant. The wintering population in Arizona did show a more striking decline than other wintering populations. However, little can be inferred from these data in terms of the total population since Yellow-headed Blackbird winters largely south of the US; also note that birds wintering north of Mexico are more likely to be adult males. Breeding birds surveys (BBS) 1966–1994 show the North American population to be stable or slightly increasing (1.9% per year), but this increase is not highly significant statistically. In general, there appears to have been a small increase 1966–1979, but this increase did not continue 1980–1994. The Yellow-headed Blackbird is most closely related to the meadowlarks and Bobolink, based on DNA sequence data (K. Omland, S. Lanyon pers. comm.)

REFERENCES Barbour (1943) [movements], BBS Data [population notes], Bond 1988 [movements], CBC Data [population notes], Crawford 1977 [nesting], 1978 [movements], Kale *et al.* 1969 [movements, vagrancy], Klimkiewicz and Futcher 1987 [age], Lewington *et al.* 1992 [vagrants, movements], Miller 1978 [Bahama records], Orians and Christman 1968 [voice, behaviour], Pyle *et al.* 1987 [moult], Ridgway 1902 [measurements], Twedt and Crawford 1995 [voice, habitat, nesting, behaviour, distribution, movements].

55 SAFFRON-COWLED BLACKBIRD *Xanthopsar flavus* (Gmelin) 1788 — Plate 29

A gorgeous and rare species of the moist grasslands and marshes of southern South America. The populations of this species have plummeted in the last century.

IDENTIFICATION Male Saffron-cowled Blackbird is a brightly-coloured and distinctive bird. It shows a yellow head, underparts, epaulet and rump. The nape, tail, back and wings are black. A small black mask is present. However, it is possible to confuse this bird with either of the marshbirds (73, 74). Saffron-cowled Blackbird is noticeably smaller and slimmer than either of the marshbirds and shows a thin, spike-like bill. The yellow head and black lores are diagnostic; both of the marshbirds have dark heads and breasts. In addition, Brown-and-Yellow Marshbird (74) lacks the yellow rump of Saffron-cowled Blackbird. More easily confused is the female (and immatures) of Saffron-cowled Blackbird which are brownish-grey above, yellow below, with a dark eyeline and crown. Once again, the marshbirds are a confusion problem, but the structural differences listed above still apply. Immature marshbirds may show yellow throats, but these also have noticeable streaking on the breast and entirely dark faces. Female Saffron-cowled Blackbird has a yellow face and supercilium, and an entirely yellow breast without streaks. Much closer in shape and structure is female Unicolored Blackbird (57), even though it is not as slim and has a deeper-based bill than Saffron-cowled Blackbird. The female of the sympatric form of Unicolored Blackbird is yellowish below, but shows faint streaking on the breast. In addition, it has entirely dark cheeks, not just a dark eyeline, and noticeable chestnut edging to the wing feathers. Female Saffron-cowled Blackbird shows only a dark eyeline, no breast streaks and buffy or greyish edges to the wing feathers. In addition, Saffron-cowled Blackbird has a whitish crissum, while female Unicolored Blackbird has a dusky vent and does not show the yellowish rump of Saffron-cowled Blackbird. The presence of yellow on the lesser coverts of Saffron-cowled Blackbird further differentiates it from female Unicolored Blackbird.

VOICE Song: Performs a rather simple, shrill and high-pitched song made up of squeaky syllables and a final longer buzz. For example *twee-chwee-CHEEOO* the final note downslurred and drawn out, or a *ch-weeko-CHWEEee*. The terminal buzz is not unlike that given by some of the South American *Agelaius* species such as Chestnut-capped (60) and Yellow-hooded (59) Blackbirds; it also resembles some of the final notes of South American meadowlark (*Sturnella*) songs. Between most songs, birds utter a constant stream of *chuck* and *chick* notes. The females sing a similar, albeit softer, song than the males.

Calls: A doubled *chewp-chap* or *dwat-dwat*. The common contact call is a sharp *chup*, sometimes several *chup* notes are run together into a chatter.

DESCRIPTION A small and slim blackbird with a sharply-pointed, conical bill and a long tail. The bill shows an entirely straight culmen and gonys. Saffron-cowled Blackbird has long wings and a square-tipped tail. Wing formula: P9 < P8 > P7 > P6; P9 ≈ P8; P8–P6 emarginate. Sometimes lumped into *Agelaius*, but many authors separate it from that genus on the basis of its thinner bill and more pointed wings (see Notes).

Adult Male (Definitive Basic): The bill is jet black and the eyes are dark brown. Its head is orange-yellow, veiled with grey on the nape, with contrasting black lores. The orange-yellow of the face continues to the breast sides and underparts, except for the belly which is whitish and the thighs which are black. The colour of the underparts varies, being most intense orange on the breast and dullest yellow on the lower breast. The undertail-coverts are yellow. On the upperparts, the lower nape, hind neck, mantle and lower back are jet black, although the rump is yellow and the uppertail-coverts are black with grey tips. The wings are black with yellow epaulets (lesser, median and marginal coverts) and yellow wing-linings. When fresh, the primaries show thin yellowish edges. The tail is black and unmarked. The legs and feet are black. Some specimens are typical except that the crown is tipped with tawny and there are chestnut tips on the back, and noticeable chestnut edges on the tertials. It appears that these may be subadult males (first basic, or alternate?).

Adult Female (Definitive Basic): The soft part colours are as in the adult male. The face is yellow with black lores, a dark eyeline and a noticeable yellow supercilium. The crown is olive-brown with a yellow wash, but the nape is noticeably greyer. The back, however, is a similar olive-brown but shows blackish feather centers, making it look obscurely streaked. A greyish-yellow rump is obvious in flight. The uppertail-coverts are brown. The scapulars are blackish with brown tips. The wings are blackish with yellow lesser coverts and blackish median coverts with buffy tips. The rest of the flight feathers are blackish with buff edges. The greater coverts are edged and tipped buff. The wing-linings are pale yellow. The underparts are yellow, saffron on the breast and more whitish-yellow on the crissum. However, the thighs are brown. The tail is blackish.

Immature Male (First Basic): Similar to adult female, but the face is entirely yellow, lacking the dark eyeline. In addition, the back has a variable number of black feathers. The primaries, secondaries and greater converts are retained from the juvenal plumage.

Juvenile (Juvenal): Similar to adult female. Juveniles are more extensively brown on the upperparts, showing obvious olive edges to the feathers, appearing more streaked than adult females. The wings are more noticeably buff-fringed. Also, the dark eyeline is more contrasting on the juvenile's face. The thighs are olive, not blackish as in the adult female.

GEOGRAPHIC VARIATION Monotypic

HABITAT A species of open regions, particularly marshes and grasslands. Also found in shrubby areas, pastures and agricultural fields. Apparently, the presence of *Eryngium*, a plant of the carrot family that is morphologically similar to a bromeliad due to its fleshy, thorny leaves, is important to Saffron-cowled Blackbird for foraging and nesting. Foraging occurs in a variety of habitats, as does nesting, but as in the marshbirds, nesting usually occurs in moister sites. Roosts in dense habitats, such as tall stands of grass.

BEHAVIOUR A sociable species, it was historically observed in flocks, sometimes of hundreds of birds. Flocking behaviour is not restricted to the non-breeding season, while nesting pairs often feed with the resident flock. It will often flock with other icterines, particularly the marshbirds, meadowlarks, Chopi Blackbirds (79) or even Baywings (96), and forage or roost with them. Particularly odd is the reported association between Black-and-white Monjita (*Xolmis dominicana*) and Saffron-cowled Blackbird, particularly in Rio Grande do Sul. Groups of blackbirds follow the monjitas around (Belton 1985), presumably this behaviour is of mutual benefit to both parties. It has been suggested that the monjita gains increased foraging success by catching the insects roused by the blackbirds while the blackbirds decrease the probability of predation due to the monjita's watchfulness and alert behaviour. This association is not obligatory, and both species are regularly observed by themselves, particularly in Argentina. What is known is that blackbirds seek out monjitas, but the reverse is not the case. All foraging by blackbirds is carried out on the ground or by pecking at clusters of low-growing plants, such as *Eryngium* or bunch grasses. During the breeding season, males perch conspicuously to sing and preen. Territorial defense is minimal, other males and pairs are allowed to approach closely; males usually do not drive these intruders away until they are less than five meters away from him or the nest. Yellow-rumped Marshbirds (73) are also not tolerated near the nest. The female may also aid in nest defense by scolding intruders.

NESTING Breeds during the austral spring, with egg dates known mainly from October and peaking in November, but also known from as late as December. A colonial nester, but may nest singly particularly where populations have declined extensively. The colonies form in areas of dense bushes, *Eryngium* or other low-growing and thick vegetation. Nests are cup-shaped and are placed near the ground, usually less than one meter off the ground. They are constructed of coarse, dry grass and have a deep cup. The entire construction of the nest is performed by the female, usually while the male perches nearby. The eggs are pale and densely spotted brick-red; clutch size is typically 4 but may be as high as 5 (Pereyra 1933). It is a known host for Shiny Cowbird (102), but no records of fledging exist (Friedmann and Kiff 1985).

DISTRIBUTION AND STATUS Rare and declining. Reaches the northern extent of its range in Villa Hayes, Paraguay, being found mainly east of the Río Paraguay in the extreme southwest of the country. Recently, flocks observed in the grasslands of San Rafael NP in SE Paraguay (Madroño *et al.* 1997) suggests that this is the stronghold of the species in that country. In Argentina, it has been found west of the Río Uruguay in S Misiones, Corrientes, extreme eastern Chaco and Formosa, and south into Entre Ríos and Buenos Aires as far south as the Sierra de Ventanas. In Uruguay, it has been recorded in the southern perimeter of the country along the Río Uruguay, near the coast and in the extreme east by the Brazilian border. In Brazil, it is known only from Rio Grande do Sul, near the coast in the south, and along the highlands of the northeast, and Santa Catarina, mainly in the lowlands along the Atlantic coast. Largely absent from the highlands of southern Brazil, E Paraguay and Misiones, Argentina. This species has suffered a serious population decline, being rare in places where it was said to be abundant or common in the late 1800s. In recent years, declines have still been detected, for example in southern Brazil where populations in the late 1980s appeared to be smaller than in the 1970s. In addition, its range has contracted noticeably, particularly in Buenos Aires province, Argentina. Recent records are all from the extreme northeast of the province, and it appears to be nearing extirpation there. The bulk of the population appears to be found in parts of Corrientes and Entre Ríos in Argentina, western Uruguay and southern Brazil.

MOVEMENTS Not confirmed to be migratory, but apparently makes irregular movements or irruptions as it is often absent from some areas of suitable habitat only to return years later. The extent and cause of these movements is not known.

MOULT Adults experience the complete definitive pre-basic moult after the breeding season. Birds of the year undergo the first pre-basic moult which is partial, only involving the body plumage and retaining the juvenal wings and tail. No evidence of pre-alternate moults, but there is a possibility that these exist in young males.

MEASUREMENTS Males: (n=9) wing 105.9 (100–111); (n=8) tail 76.9 (72–81); culmen 23.4 (21–25); (n=9) tarsus 26.7 (24–28). Females: (n=8) wing 102.6 (97–109); tail 73.8 (67–78); culmen 22.3 (20–24); tarsus 26.8 (25–28).

NOTES Listed as vulnerable in the red data book (Collar *et al.* 1992). The reasons for the population decline and range contraction are not understood. Generally it is thought that habitat alterations are to blame, particularly cattle ranching, pesticide use and perhaps burning of grasslands or their conversion to other habitats (i.e. pine plantations). This odd species superficially resembles *Agelaius* in structure, but aspects of its behaviour, habitat and to some extent appearance, suggest an affinity with the marshbirds (*Pseudoleistes*). A study of mitochondrial DNA sequence patterns suggests that the relationship with the marshbirds is indeed real, and that they may in turn be related to Yellow-hooded (59) and Chestnut-capped (60) Blackbirds (Lanyon 1994).

REFERENCES Belton 1985 [nesting, behaviour], Collar *et al.* 1992 [conservation, status, distribution, ecology], Hardy *et al.* 1998 [voice], Sick 1993 [voice].

56 PALE-EYED BLACKBIRD *Agelaius xanthophthalmus* Short 1969 — Plate 19

A rare and little-known marsh icterid of Peru and Ecuador. Both the male and female are entirely black with pale eyes.

IDENTIFICATION Adults of both sexes are entirely black with yellowish eyes. No all-black icterid within this species' range has pale eyes. The most likely icterid confusion species is dark-eyed Solitary Cacique (21), which may also be found skulking along the edge of a cocha. There are puzzling reports of pale-eyed Solitary Caciques which would be difficult to identify. The cacique is much larger, with a pale greenish bill and a somewhat crested, or shaggy-headed look. The bill of the cacique is proportionately larger than on Pale-eyed Blackbird, being noticeably deeper and wider. The cacique has the habit of climbing around in lianas and foraging in trees, something that the blackbird does not do. Shiny Cowbird (102) and Velvet-fronted Grackle (81) could also be confused for this species, although both have dark eyes. The male cowbird is much glossier, shorter-billed, and unlikely to be found in wetlands except when roosting. The grackle inhabits marshy habitats, but seldom forages right amongst the emergent vegetation. In addition, the cowbirds and grackles tend to travel in flocks, unlike Pale-eyed Blackbird. At close range, the 'plush' crown feathers may be visible on the grackle. Velvet-fronted Grackle has a stubby-looking bill and a long tail.

VOICE Song. A loud, piercing *tew-tew-tew-tew* resembling that of Black-capped Donacobius (*Donacobius atricapillus*), but not nearly as variable. The song has also been likened to that of Northern Cardinal (*Cardinalis cardinalis*) (Short 1969).

Call A *chek* reminiscent of the call of the Red-winged Blackbird (61).

DESCRIPTION A small blackbird with short, rounded wings. The 9th primary is shorter than P3 with P4–P8 roughly equal in length. The bill is sharply tapered to a fine point with entirely straight gonys and culmen. The bill is roughly as long as the head and shows a somewhat flattened culmen. The eyes are straw-yellow.

Adult Male (Definitive Basic): The bill is glossy black and the eyes are yellow. The body is entirely black with a slight greenish-blue gloss that is not noticeable under most field conditions. The tail is long and somewhat graduated, appearing rounded at the tip, and often the tail tip is noticeably frayed, further accentuating this rounded tail appearance. The individual tail feathers are somewhat pointed. The sturdy legs and feet are black.

Adult Female (Definitive Basic): Like the male, but slightly less glossy and smaller; however, the bill is somewhat deeper. Soft part colours as in the male.

Juvenile (Juvenal): Quite unlike the adults and show some similarities to Unicolored Blackbirds (57). The upperparts are brown with thin buff tips to some of the tertials and back feathers. The underparts are brown with golden-buff streaks from the throat down to the breast. There is no visible gloss; the wings and tail are brown and wear quickly. The extent of golden-buff on the underparts may vary as some individuals have been observed which show yellowish breasts, streaked with brown.

GEOGRAPHIC VARIATION Monotypic

HABITAT Inhabits wetlands bordering Amazonian oxbow lakes or cochas, particularly those with thickets of floating grasses or sedges. Its preferred marshes appear to be those with tall, emergent grasses and shrubbery.

BEHAVIOUR Display of male is unspectacular. He spreads his tail and sings without noticeably spreading wings. The female responds by sitting in an alert position with the head slightly raised. Most observations are of pairs or single individuals skulking in aquatic vegetation; this species is more skulking than other marsh blackbirds. Stomach contents from specimens have revealed insects to be main food item. Usually observed in pairs, not flocks (Ridgely and Tudor 1989).

NESTING The co-types collected in August for the original description were both in pre-breeding condition and probably at pair formation stage. The nest and eggs are unknown.

DISTRIBUTION AND STATUS Rare and patchily distributed, this may be an effect of patchy observer coverage in this part of South America; it is our opinion that the patchy nature of this species' habitat does give rise to a real patchiness in distribution. It is still known only from a handful of sites. Found up to 650 m (Ridgely and Tudor 1989). Pale-eyed Blackbird has been recorded from Limoncocha and La Selva, both in Napo province, Ecuador, from Rioja, San Martin dept., Peru, from Manu, Tambopata reserve and Hacienda Amazonia, Madre de Dios dept., Peru and from Tingo Maria, Huánuco dept., Peru.

MOVEMENTS Unknown, but probably sedentary, moving to other sites depending on water levels.

MOULT Almost unknown. A Dec. 8 specimen from Madre de Dios, Peru was completing its definitive pre-basic moult, which appeared to be complete. Tail moult appears to be centripetal.

MEASUREMENTS Male: (n=1) wing 96.9; culmen 23.2; tail 89.0; tarsus 28.4. Female: (n=2) wing 83.3, 90; culmen 22.7, 25; tail 79.9, 83; tarsus 28.2, 28.

NOTES The original description gives the English name as Pale-eyed Marsh Blackbird (Short 1969). Recent mitochondrial DNA sequence comparisons strongly suggests that the closest relative to this rare species is the much more widespread Unicolored Blackbird. Yellow-winged Blackbird (58) appears to be the sister species to the Unicolored/Pale-eyed blackbird pair (Lanyon 1994).

REFERENCES Hardy *et al.* 1998 [voice], Parker 1982 [voice, description, habitat, distribution]; Short 1969 [description, measurements].

57 UNICOLORED BLACKBIRD *Agelaius cyanopus* Vieillot 1819 — Plate 19

A nondescript, black marsh icterid that is often overlooked. It is one of the few species where the female is more distinctively plumaged than the male.

IDENTIFICATION Males are easily confused with other entirely black icterids such as Shiny Cowbird (102) and Chopi Blackbird (79). The habitat preference and behaviour of this blackbird are important clues in establishing identification. Unicolored Blackbird is almost always found in wetlands, seldom venturing even to nearby fields. They perch on marsh vegetation and drop down to feed at the edge of the water or mud. Male Yellow-winged Blackbirds (58) can conceal their yellow epaulets and look deceivingly similar, if in doubt, wait for them to fly. Unicolored Blackbird lacks the bold gloss of Shiny Cowbird and is slimmer, longer-tailed and noticeably longer-billed. Chopi Blackbird is a stockier bird that lacks an obvious gloss. The bill is much more stout than that of Unicolored Blackbird and shows obvious grooves on the lower mandible at close range. Chopi Blackbird is noisy, often giving its distinctive *Shop-wee* call. However, the identity of male Unicolored Blackbird is usually revealed by the presence of the distinctive females of the species. The chestnut edges of the upperpart feathers and the yellow on the underparts is diagnostic, however, note that there is geographical variation in the female plumage.

VOICE Song: A whistled *tew tew tew tew* as well as a long gurgling noise. This song has been likened to that of Black-capped Donacobius (*Donacobius atricapillus*) and is similar to that of Pale-eyed Blackbird (56). Note that in S Brazil, Unicolored Blackbird is referred to as 'Carretão' which is an ox-cart with squeaky wheels (Belton 1985); this is undoubtedly a reference to the quality of the species' vocalisations.

Calls: A *chek* and a *jep* note resembling that of House Sparrow (*Passer domesticus*).

DESCRIPTION A small, slender, somewhat long-tailed blackbird. The bill is slightly shorter than the length of the head. The culmen and gonys are straight. The bill is sharp and finely pointed. The culmen of this species is distinctly flattened, more so than in other *Agelaius* blackbirds. The nominate form is described here. P9 < P8 < P7 ≈ P6 > P5; P9 ≈ P3; P8–P5 emarginate.

Adult Male (Definitive Basic): The black bill is sharply pointed with no visible curve to the culmen or gonys. The eye is dark brown. The plumage is entirely black, with a slight blue gloss which is not visible other than in very close views. Freshly moulted birds do not have paler fringes to the feathers. Legs and feet black. Has an undulating flight

Adult Female (Definitive Basic): The bill is black with a blue-grey base, particularly on the lower mandible. The crown and upperparts, to the rump, are brown with obvious darker streaking. On the face, a yellowish supercilium is present as well as dusky auriculars, creating a dark ear patch. The eyes are dark brown. The brown wings are boldly fringed with rufous, particularly on the greater coverts, tertials and secondaries. The wing-linings are yellow. The tail is blackish, with thin pale fringes on each feather. The underparts are pale yellow, brightest on the throat, with noticeable dusky streaks confined to the flanks and breast sides. Thin streaks may be present on the breast, which are most obvious on worn individuals. The legs and feet are blackish.

Immature Male Fall (First Basic): Like female, but noticeably larger given a direct comparison. Compared to the adult female, the crown and nape are paler, more yellowish, allowing the dark streaks to stand out more noticeably. The breast is not streaked as in the female. Older immature males will begin to show contrasting black feathers on the underparts or face.

Immature Male Spring (First Alternate): Like 1st-fall male, but has black feathers on the throat.

Immature Female (First Basic): Like adult female, but more heavily streaked on the upper and underparts, particularly in the center of the breast.

Juvenile (Juvenal): Like immature female, but the colour of the underparts is much more buffy-yellow. The underpart streaking is most extensive in juveniles.

GEOGRAPHIC VARIATION Four subspecies are recognized which differ largely in the plumage of the females. Males differ only in size. The two Brazilian subspecies are disjunct populations from the main range of the species. The nominate form inhabits E Bolivia, S Brazil and Paraguay, south to Buenos Aires in Argentina. It is described above.

A. c. xenicus, the most disjunct form, is found in NE Brazil, within the states of Pará and Maranhão. The male is indistinguishable from *cyanopus* males, except that it has a smaller, thicker bill. The females are distinct, however. The head, breast and upperparts are black with the rump black with an olive wash. The back feathers are fringed narrowly with a paler dark brown. The rufous wing-edgings of other forms are brown in this race. The flanks down to the undertail-coverts are blackish with an olive tint, and the belly is dark olive, sometimes showing indistinct paler streaks closer to the lower breast. The young male is similar, but the back tends to be fringed with a more chestnut colour and the lower breast may have some chestnut mixed in with the olive.

A. c. atroolivaceus occurs in the coastal region of E Brazil, in the state of Rio de Janeiro. Males are like *cyanopus*, but with a longer culmen. Females and immatures are much darker than *cyanopus*. Adult females show olive-green underparts. The upperpart base colour is a darker brown, that shows less contrast with the darker streaking than on *cyanopus*. The rufous of the wings in this form is not as obvious as in the nominate. The underwing-coverts are dark olive or blackish, like *beniensis* and unlike *cyanopus*. A flock of 20 birds seen in Belo Horizonte, Minas Gerais, is geographically intermediate between *atroolivaceus* and *cyanopus*, their subspecific identity is unknown and it is not known if this is a record of a vagrant flock or a new population.

A. c. beniensis is found near the Rio Beni, in N Bolivia. This is the largest race. Females and immatures are darker on the underparts than *cyanopus*, but not as dark as *atroolivaceus*. The flanks and undertail-coverts are duskier than in *cyanopus*. Adult females are very dark on the upperparts, some even darker than *atroolivaceus*; however, immatures are approximately as dark as *cyanopus* on the upperparts. The underwing-coverts are blackish or dark olive, as in *atroolivaceus*.

HABITAT A bird of wetlands, such as marshes or reed-fringed ponds. Primarily inhabits marshes with tall emergent vegetation and standing water, including rushes (*Juncus* sp.), sedges including tules (*Scirpus* sp.), but unclear if it uses cattails (*Typha* sp.). It seldom leaves wetland habitats to forage, but when it does it will forage in agricultural areas including plowed fields.

BEHAVIOUR Forages amongst aquatic vegetation, by gleaning or gaping into the plant stalks as well as by walking on floating aquatic vegetation. It is found in small groups, never in large flocks. Canevari *et al.* (1991b) state that Unicolored Blackbirds are found singly, in pairs or in small flocks, and do not appear to be as gregarious as other South American *Agelaius* species. Performs a 'song spread (ruff-out)' display while it sings. The body plumage is ruffled and the tail is slightly spread, but the wings are held closed. Also sings in flight (Canevari *et al.* 1991b).

NESTING Breeds in loose colonies. Nesting in the south (S Brazil and Argentina) peaks in November, but may occur October–January. The cup-like nest is placed either in a low bush or on emergent marsh vegetation, usually less than 2m off the ground. Structurally the nest resembles that of Red-winged Blackbird (61) (Naumburg 1930). There does appear to be a preference for nesting in bushes, rather than in herbaceous vegetation emerging from the water. The nest is secured in a crotch or upright fork in the branches. The clutch size is typically 3 eggs which are greenish-blue with grey speckles scattered throughout, and darker spots and scrawls near the larger end of the egg (de la Peña 1987). It is known to be parasitised by Shiny Cowbird (102) (de la Peña 1987, Friedmann and Kiff 1985).

DISTRIBUTION AND STATUS Locally common. There are three disjunct populations. The northernmost inhabits NE Brazil in the region of the mouth of the Amazon river, in the states of Amapá, Pará and Maranhão. The major portion of the distribution ranges from N Bolivia in the lowlands, east through Paraguay (present throughout except in the Alto Paraná region in the extreme east of the country) (Hayes 1995) and SW Brazil (C Mato Grosso and W Goiás), south through W São Paulo, Paraná and Rio Grande do Sul (few records, only known from the São Donato marsh in the extreme west of that state) (Belton 1985), to NE Argentina (south to N Buenos Aires province), west to W Santa Fé, NE Santiago del Estero and E Salta. The final population lives in coastal SE Brazil from Rio Janeiro to Espírito Santo. It has been recently recorded from Belo Horizonte and Minas Gerais, Brazil. There are no records for Uruguay, but it almost surely must occur there.

MOVEMENTS Sedentary.

MOULT The definitive pre-basic moult is complete. There is no evidence that a pre-alternate moult occurs in adults, changes in appearance are a result of feather wear. Nominate *cyanopus* undergoes the definitive pre-basic moult after breeding, in January–March approximately. The obvious wear on specimens of *xenicus* from mid June suggests that the definitive pre-basic moult occurs in July approximately. Birds of the form *beniensis* are finishing moult in late September. The juvenal plumage is lost through the first pre-basic moult, but it is not known if this moult involves the flight feathers. A limited first pre-alternate moult seems to occur; some young males of the nominate race obtain some blackish feathers on the face in their first spring. This first pre-basic moult needs more study, particularly with respect to its extent, and whether or not it occurs in females and other subspecies.

MEASUREMENTS Males are larger than females.

A. c. cyanopus. Males: (n=20) wing 92.8 (89–98); (n=9) tail 83.7 (77–93); (n=20) culmen 24.4 (23.5–25.5); (n=9) tarsus 25.9 (23–29). Females: (n=17) wing 85.0 (80–90.5); (n=6) tail 73.7 (70–79); (n=16) culmen 22.5 (21.5–23.5); (n=6) tarsus 23.2 (20–25).

A. c. atroolivaceus. Males: (n=4) wing 94.1 (92–95); (n=3) culmen 25.5 (25–26). Females: (n=4) wing 86.25 (84.5–88); culmen 23.2 (22–24).

A. c. beniensis. Males: (n=6) wing 99.7 (97–102); (n=3) tail 91.7 (88–97); (n=6) culmen 24.5 (23–26); (n=3) tarsus 26 (24–28). Females: (n=8) wing 90.4 (88–93); (n=3) 82.3 (80–85); (n=8) culmen 23.4 (22–25); (n=3) tarsus 26.3 (26–27).

A. c. xenicus. Males: (n=2) wing 97.5 (96–99); (n=3) culmen 23.3 (23–23.5). Females: (n=2) wing 88.0 (87–89); culmen 22.5 (22–23).

NOTES Recent molecular work suggesting that this species is most closely related to the rare Pale-eyed Blackbird is also supported by their similar songs. See Notes under Pale-eyed Blackbird. It is likely that more than one species may be involved in what is now known as the Unicolored Blackbird.

REFERENCES Babarskas and López Lanús 1993 [nesting], De la Peña 1987 [nesting], Parkes 1966 [description, geographic variation, measurements, moult], Pearman 1994 [geographic variation, distribution].

58 YELLOW-WINGED BLACKBIRD *Agelaius thilius* (Molina) 1782 — Plate 20

The common marsh icterid of temperate South America. The male is distinctive with its black body and bright yellow epaulets. Females are brown and boldly streaked, showing less obvious yellow epaulets.

IDENTIFICATION The male's combination of a completely black body with yellow epaulets is diagnostic within its range. Adult males often hide the yellow epaulets, looking entirely black, and may be confused with males of Unicolored Blackbird (57), Chopi Blackbird (79) or Shiny Cowbird (102). Shiny Cowbirds are glossier, and Chopi Blackbird is larger and bulkier. Male Unicolored Blackbird is extremely similar, but eventually Yellow-winged Blackbirds will call or show their yellow epaulets, revealing their identity. The vocalisations of Yellow-winged and Unicolored blackbirds differ substantially, see Voice. Females and immatures are streaked and sparrow-like, with a noticeably paler supercilium (except birds from the altiplano of Bolivia and Peru). They are more streaked, especially below, than any other blackbird within their range. Female Yellow-winged Blackbirds may be confused with female Unicolored Blackbirds, but the latter have distinctly yellow underparts and rufous edges to the tertials and wing-coverts, as well as rufous fringes on the back. Female and immature Chestnut-capped Blackbirds (60) and Shiny Cowbirds are less streaked. Females and some immature males show yellow epaulets as in adult males. To the observer familiar with female Red-winged Blackbird (61), female Yellow-winged Blackbird appears like a petite version of that species. There are many streaked passerines in South America that could be confused with females, but note Yellow-winged Blackbird's habitat and sharply-pointed, typically icterid, bill.

VOICE Song: The male's song includes a harsh, descending *treeelayyyy* or *chree-layy*, the second part nasal in quality and descending in frequency. The song sounds like the Spanish pronunciation of 'Chile', and some accounts maintain that the country was named Chile on hearing the abundant calls of Yellow-winged Blackbird! During the full song, the *treeelayyyy* sound is preceded by a series of sharp *chip* calls and a frequency-modulated whistle. The end product is a *chip chip chip weeuu...treeelayyyy*. In many other cases, the *trelayyyy* call is given by itself. Other calls include a sharp *chika-wee* and a nasal buzz, similar to the *treeelayyy* call but not descending and slightly less nasal. Often many males will sing simultaneously at the colony, creating a raucous series of whistles, chucks, *treeelayyy* and chip notes. Songs of the highland form *alticola* are similar in that the same song types are given, but these differ in quality from the southern populations, most noticeably they are drier in quality and less nasal.

Calls: The call is *jak* or *check* and a *chik* is also given. The flight call is a sharp *chet*. A harsh *d-rrrrr* alarm call has been described.

DESCRIPTION A small blackbird with a sharply-pointed bill of moderate length, roughly as long as the head. The culmen and gonys are straight. Wing Formula: P9 < P8 < P7 > P6 > P5; P9 ≈ P5; P8–P5 emarginate.

Adult Male Breeding (Definitive Basic, worn): The bill is black and the eyes dark brown. The plumage is largely black with no gloss. The wing-linings, lesser, and marginal coverts as well as the edge of the alula are lemon-yellow. The undertail-coverts and vent usually show some pale greyish tips. The legs and feet are black.

Adult Male Non-Breeding (Definitive Basic, fresh): Like breeding adult, however, the plumage is veiled with paler coloured tips. In general, the upperpart tips tend toward chestnut-buff, particularly on the back, while the underpart tips tend to be whitish-buff. The crown shows an olive-brown cast and a noticeable buff supercilium. The wings are also pale fringed; the coverts are fringed with buff, forming two indistinct wingbars, similarly, the tertials are boldly buff-fringed.

Adult Female (Definitive Basic): Streaked throughout, most noticeably on the underparts. The head and upperparts are brown with darker streaking, the crown being slightly darker than the nape and back. The auriculars are the same colour and as dark as the crown. The pale buff supercilium is long and bold, and widens behind the eye; a short, paler, median crown stripe is often present. The brown upperparts are more chestnut on the lower scapulars and lower back, while the rump is a colder olive-brown. The wings are dark brown, usually with some yellow on the marginal coverts and wing-linings. The yellow wing-linings are often more extensive than on the male, but the yellow is not as vibrant. The coverts, tertials and secondaries are noticeably edged paler: olive-yellow on the lesser coverts, buff on the median coverts, outer greater coverts, primaries and secondaries, chestnut on the inner greater coverts and tertials. The underparts are buff with blackish-brown streaks from the throat to the belly. The undertail-coverts are pale-fringed and do not appear streaked. Darker malar stripes are present. The tail is blackish with thin chestnut edges to the rectrices. Soft parts as in adult male. Fresh females (austral autumn) are more heavily striped, but become darker and more uniform with wear.

Immature Male (First Basic): Appears intermediate between female and non-breeding adult male. Boldly streaked and shows a pale supercilium like the female, but has a full yellow epaulet. The base colour of the feathers, particularly on the underparts and face, is blackish unlike the brown of the female. The wingbars are obvious, white on the median coverts and white, becoming chestnut, toward the body (medially) on the greater coverts. The tertials are boldly fringed with chestnut and the primaries show thin yellowish outer edges. **Juvenile (Juvenal)**: Like female, but buffier throughout and the plumage has a loose texture.

GEOGRAPHIC VARIATION Three subspecies are recognized.

A. t. thilius ranges from the Chilean provinces of Atacama to Llanquihue.

A. t. petersi is found from N Argentina, S Paraguay and S Brazil south, including Uruguay, to Chubut and Santa Cruz, Argentina, and also infiltrating into Chile in Aisén and Magallanes.

A. t. alticola is a disjunct population from the altiplano of Peru and Bolivia.

Subspecies differ mainly in size, *A. t. petersi* is the smallest and *A. t. alticola* the largest. However, there is a cline in the size of *petersi*, with the southwesternmost (Chubut) birds being the largest. Female and immature *A. t. alticola* are visibly different from the other forms. The supercilium is reduced to a thin greyish line at most, and is often lacking. Overall, *alticola* is darker, the underparts show a darker background colour, making the streaks less noticeable than on *petersi* or *thilius*. The chestnut of the back is more extensive. Female *alticola* also tend to show more yellow on the epaulets than the other forms, approaching males in this respect.

HABITAT Breeds in wetlands with some standing water. Nests in taller reedbeds. In Argentina, nests largely on cattail (*Typha* sp.) and pampas grass (*Cortaderia selloana*), *Cyperus* sp. or tall grasses, rather than the more abundant tule (*Scirpus* sp.), but it will forage in the latter (Orians 1980). Highland populations use mainly *Scirpus* marshes, often on the shores of lakes or ponds. In Chile, where *Scirpus* marshes predominate, this is the most common habitat used. Yellow-winged Blackbirds will forage on muddy shores as well as agricultural land near wetlands, the latter particularly outside the breeding season.

BEHAVIOUR A gregarious species, usually found in groups even while nesting, and in the non-breeding season flocking is the rule. Flocks during the breeding season are composed of a high proportion of 1st-year males. Winter congregations may be quite large. During most of the year it is typical for roosting congregations to form. In Buenos Aires province, this species is largely monogamous. Pairs form before nest site selection takes place. Males may act aggressively towards each other while competing for females, but territories do not appear to be defended. During breeding, males spend a large proportion of their time perched up on tall vegetation and singing. Often feeds at the edge of standing water, the birds either walking on floating vegetation or on dead, emergent stalks. However, much food is extracted by gaping into stems of the marsh vegetation (Orians 1980). The 'ruff-out (song spread') display is similar to that of Red-winged Blackbird (61). The male sings with the body plumage ruffled, the wings partially spread showing the yellow epaulets, and the tail slightly spread and held pointing downwards. This display is given in varying degrees of intensity, in low intensity displays, the plumage is held sleeked and the wings are held closed against the body.

NESTING Groups break up into pairs September–early October and the female begins nest building. During the breeding period, males are not territorial and breeding can be considered colonial. Nests are placed mainly in clumps of pampas grass or cattails in the pampas of Argentina and Paraguay. The typical clutch is 3 eggs, but clutch size ranges from 2–4. Eggs are white or pinkish with black spots and scrawls, clustered around the wide end. Shiny Cowbird (102) is known to parasitise Yellow-winged Blackbird; however, there is evidence that Yellow-winged Blackbird only accepts unblemished cowbird eggs, as spotted ones are rarely found in their nests. Incubation lasts roughly 13 days. Both sexes feed the young, but only females incubate.

DISTRIBUTION AND STATUS Locally common. An isolated population exists in the highlands of southern Peru (Cuzco and Puno) and W Bolivia (La Paz, Cochabamba and Oruro). Otherwise found in N Argentina from Jujuy and Salta east through Chaco and Corrientes, southernmost Paraguay (Misiones, Canendiyu with a northern record from Alto Parana), S Misiones (Argentina), S Brazil (coastal Rio Grande do Sul and Santa Catarina) and Uruguay; south to Chubut and probably Santa Cruz in Argentina. In Chile, it is found in the lowlands from Atacama south to Llanquihue, with an isolated population in Aisén, Magallanes and nearby Santa Cruz, Argentina. Found in the altiplano of Peru and Bolivia to 4300 m, otherwise seldom seen above 1500 m. Note that there are no breeding records for Paraguay which appears to be strictly a wintering locality.

MOVEMENTS Partially migrant, Paraguayan records are only of birds in the non-breeding season (Ridgely and Tudor 1989, Hayes 1995). It has been recorded in Paraguay late July–early September (Hayes 1995). The details of its migration are unknown, however most populations are seemingly sedentary.

MOULT The complete definitive pre-basic moult occurs after breeding, in February–April (*thilius* and *petersi*). Specimens of *alticola* are moderately worn in June–August, suggesting a similar timing. Pre-alternate moults are lacking, the breeding plumage is obtained through wear. The first pre-basic moult appears to be almost complete, with some individuals retaining the juvenile tail. In addition, some brownish juvenal underwing-coverts are sometimes retained.

MEASUREMENTS Males are larger than females.

A. t. thilius. Males: (n=13) wing 94.0 (91–96); tail 75.9 (73–80); culmen 23.0 (22–24); (n=6) tarsus 26.2 (25–27). Females: (n=10) wing 86.2 (82–90); tail 67.8 (63–72); culmen 19.9 (19–21); (n=6) tarsus 24.5 (23–25).

A. t. petersi. Males: (n=64) wing 84.2 (81–90); (n=53) tail 69.1 (66–73); (n=10) culmen 21.4 (20–23); tarsus 24.4 (22–27). Females: (n=10) wing 79.4 (74–89); (n=40) tail 64.0 (59–73); (n=7) culmen 19.3 (18–20); (n=10) tarsus 25.0 (21–24).

A. t. alticola. Males: (n=3) wing 100.3 (96–107); tail 81 (79–83); culmen 23.7 (23–24); tarsus 25.7 (25–26). Females: (n=4) wing 90.5 (83–98); tail 75.7 (74–77); culmen 21.5 (20–23); tarsus 23.8 (22–25).

NOTES The appearance and general habits of this species and Red-winged Blackbird provide an interesting example of convergent evolution, as the two species are not closely related. A recent comparison of mitochondrial DNA sequences showed that Yellow-winged Blackbird is the sister species to the group which includes Unicolored and Pale-eyed (56) Blackbirds (Lanyon 1994). The other species which are presently classified in the genus *Agelaius* are distantly related.

REFERENCES Araya and Millie 1991 [geographic variation], Camperi 1988 [measurements], De la Peña 1987 [nesting], Fjeldsa and Krabbe 1990 [voice,geographic variation], Johnson and Goodall 1965 [measurements], Orians 1980 [behaviour, nesting], Ridgely and Tudor 1989 [voice], Sick 1993 [voice], Todd 1932 [geographic variation, measurements], Wetmore 1926 [description].

59 YELLOW-HOODED BLACKBIRD *Agelaius icterocephalus* (Linnaeus) 1766

Plate 19

A small marsh blackbird of northern South America. The colouration of this bird is reminiscent of the unrelated and much larger Yellow-headed Blackbird (54).

IDENTIFICATION The male is distinctive with its black body and contrasting yellow head. It should not be readily confused with any other species in its range. Oriole Blackbird (53) is also yellow and black, but it is much larger and has yellow underparts. Also note Oriole Blackbird's bare facial skin. The North American Yellow-headed Blackbird is entirely allopatric with this species, the overall similarity in plumage and the English name of these two species tends to cause confusion. Yellow-hooded Blackbird can be differentiated from Yellow-headed by its lack of a white wing patch and its smaller size. Female Yellow-hooded Blackbirds are largely brown but show an obvious yellow throat which is visible from a distance. No other marsh passerine of northern South America shows this pattern. Immature Pale-eyed Blackbird (56) may be yellowish below, but not as extensively as Yellow-hooded Blackbird. Note also that Pale-eyed is a rare bird of Amazonian lakes, not open marshes as is Yellow-hooded Blackbird. Females of the northernmost race (*xenicus*) of Unicolored Blackbird (57) are blackish with olive undersides, quite unlike female Yellow-hooded Blackbirds. Female Chestnut-capped Blackbirds (60) are similar in shape, size and upperpart colouration to female Yellow-hooded Blackbirds. However, female Chestnut-capped Blackbirds lack any obvious yellow on the throat or supercilium.

VOICE Song: The primary song is a wheezy, drawn-out note similar to that of several other South American marsh blackbirds. There is often a starting *cluck* or *chuck* before the wheezy note, and it is usually terminated by an accented *cluk* or *tik*. For example *took-TOOWEEEEEZ,tik* or *TWOWEEEEEZ-tik* or *cluck-CHEEWIIIIZ-tik*. Songs last approximately one second. The song may be followed by a more warbled *te-tiddle-de-de-do-dee*. Sweet whistled notes, *chucks* and other notes are liberally sprinkled in before song bouts.

Calls: The common call is a dry *chek* or *ship*. Also gives a descending whistle *tieeewww*.

DESCRIPTION A small blackbird with a sharply pointed bill, the culmen and gonys are nearly straight, the culmen showing a slight downcurve. The short tail is rounded. Wing Formula: p9 < p8 > p7 > p6 > p5; P9 ≈ P5; P8–P7 emarginate.

Adult Male Breeding (Definitive Basic, worn): The bill is glossy black. Eyes dark. The head, upperbreast and neck are a vivid lemon-yellow. The lores and subloral area (immediately below the lores) are black, forming a contrasting mask. The rest of the body is black, with a moderate blue-green gloss. The yellow hood is crisply demarcated from the black of the body. The legs and feet are black.

Adult Male Non-Breeding (Definitive Basic, fresh): Resembles breeding male, but has small yellow tips to the black back and breast feathers. The yellow head feathers are tipped olive, particularly on the crown and nape. There are buff tips on the tertials.

Adult Female (Definitive Basic): The bill is blackish, but shows a paler base to the lower mandible. The eyes are brown. The head is greenish with yellow flecks on the crown and an indistinct yellowish supercilium. The darker malar streaks are thin and obscure. The back is olive with darker feather centers, appearing obscurely streaked, while the rump is darker olive. The throat to the upperbreast is yellow with a green tone, showing some orange overtones on the breast. The yellow of the throat and breast ends rather abruptly, not blending in with the colour of the rest of the underparts. This yellow colour extends up the sides of the neck as a semicollar. The rest of the underparts are vivid olive-green with faint darker streaking, becoming olive-grey on the crissum. The thighs are olive. The wings are blackish-brown, the coverts and tertials are thinly edged with olive, and the underwings are dark. Tail blackish-brown. Legs and feet blackish.

Immature Male, Fall (First Basic): Similar to adult female, but shows a more extensive and brighter yellow throat, no malar streaks, and a more noticeable yellow supercilium and forehead. The upperpart colouration is as in the female, but a few black feathers may be present. The underparts are blacker, lacking the green or yellow wash seen on the female and may show some entirely black feathers. The belly and vent is dull blackish, while the thighs are olive. The flight feathers and tail are more blackish than on the female. Soft parts as in adults.

Immature Male, Summer (First Alternate): Young males have a variable appearance during their first breeding season. Some resemble the 1st-fall males described above, largely like a bright female, while others are much closer to adult plumage. Prior to and during the first breeding season some young males appear to undergo a slow moult, gaining some yellow on the head with blackish feathers replacing the brownish body colouration. The first set of blackish feathers are tipped olive and are not as dark as those of adult males. At this time, young males may fall anywhere between these two extremes, and plumage variability is further enhanced in regions where the breeding season is especially prolonged.

Juvenile (Juvenal): Resembles the female but duller. The crown is brownish-buff lacking yellow speckling. On the face, yellow is confined to the supercilium. The throat is yellow, but very much duller than that of the female. This washed-out yellow colour merges more gradually with the yellowish-buff underparts, not showing an abrupt colour change as in the adult female. The underparts have an obvious buff tone, with obscure grey streaking. The upperparts are a paler greyish-olive than those of the adult female. Juvenile males are larger than females, but are similarly coloured, possibly showing more grey and less yellow-buff on the underparts than on juvenile females.

GEOGRAPHIC VARIATION Two subspecies are recognized. The nominate form is described above.

The race ***A. i. bogotensis*** is found in the marshes of the Bogota plateau in the highlands of eastern Colombia. *Bogotensis* is larger than *icterocephalus,* and the males are indistinguishable other than on size. Females however, are quite different, *bogotensis* being noticeably darker than *icterocephalus* with less yellow on the head and more grey on the underparts. The upperparts are less noticeably streaked as the base colour is darker. The belly and vent of female *bogotensis* is blackish, not dark olive as in *icterocephalus.* The nominate form occurs throughout the rest of the range. Individuals from north of the Amazon are indistinguishable from birds found south of the river (Gyldenstolpe 1951).

HABITAT It nests in marshes and other wetlands including rice paddies. Emergent aquatic vegetation is needed for nesting, which may include sedges (*Eleocharis* sp.), cattails (*Typha* sp.), broad-leaved ginger (*Thalia geniculata*), wild rice (*Oryza perennis*), or various species of wetland shrubs. During the breeding season it may feed away from the marshes in nearby agricultural land, tall grasslands, or rice paddies. Females are more apt to be found foraging away from marshes during the breeding period. During the non-breeding season, this blackbird is often found foraging in pastures and agricultural areas as well as wetlands. Marshes are chosen as the common roosting site, throughout the year.

BEHAVIOUR Commonly observed in small flocks while foraging both in the breeding and non-breeding seasons. Once the nesting season begins, the adult males form colonies in marshes and begin constructing nests. The males attempt to attract females to these nests by singing and displaying at the nest site. Most males are polygynous. However, some males defend isolated territories away from the colony, in lower quality habitat. Once a polygynous male attracts a female to his nest, he will attend to her up until the second day she is incubating. At this point, the male will construct another nest and shift his attentions to finding another female. Males may mate with up to five females, and each one has to be provided with a nest built by the male. Obligate male nest construction is not known in other icterids, but note that in the closely related Chestnut-capped Blackbird the males sometimes build the nest. First-summer (immature) males do not breed and remain in wandering flocks at foraging sites. Males attending a newly built nest are forever on the lookout for patrolling females. Once a female approaches, the male will fly to her with a distinctive fluttering flight. Females that are receptive to a male's advances follow him to his nest. Once she has accepted the male, she will remain near the nest for most of a day often perching near the nest or preening in the vicinity. She will lay the first egg between two and five days later. While the male constructs the nest, the female is in charge of lining it, this occurs immediately previous to her laying. The more common display, the 'ruff-out' or 'song spread' is usually directed at other males and appears to be aggressive in nature and used to stake claim to the territory. It is stereotypic, and similar to that of the more well known northern *Agelaius.* The male perches high on a stalk, ruffles his plumage and spreads his tail as the song is given. During a high intensity display, the wings are raised and fluttered. Most data from Wiley and Wiley (1980).

NESTING In Trinidad, the season begins as the rains commence late May–early June and lasts until October or November, peaking July–August. The key nesting trigger is adequate stands and densities of aquatic emergent plants. The breeding season varies according to locality: in Venezuela August–November, in Bogota, Colombia it is January, in N Colombia May–July, in Cali, Colombia, it is February and in Surinam, March–September (Haverschmidt 1968, Hilty and Brown 1989). In the Llanos of Venezuela, rainfall is extremely seasonal and for a large part of the year the marshes are dry and the blackbirds are absent. Breeding occurs during a short window while the rains fall, October–November, at which time the marshes are flooded. Males construct the cup-shaped nest made of the abundant aquatic vegetation. It is anchored to several of the growing stalks of marsh vegetation, usually about half a meter from the water's surface. The clutch size is commonly 2–3 eggs which are pale blue with darker spots and scrawls. Incubation lasts 10–11 days. It begins after the first egg is laid, creating a size hierarchy in the nestlings, as the first egg hatches earlier than the others. In the Llanos, where the breeding season is extremely short, the females are more synchronized in their laying than in other places where the season is longer. Also, male territories are smaller and more densely distributed. The nestlings take 11 days to fledge. Males help to feed the young, particularly later on in the nestling period, but the females perform most of the work. Many nests are lost to predators, and a significant number of nestlings starve to death. This contrasts with marsh-nesting blackbirds from North America where nestling mortality is rare as food is abundant here, but it is much more difficult to procure in tropical marshes. Yellow-hooded Blackbirds commonly feed orthopterans (katydids and grasshoppers) to their brood.

This blackbird is a common host of Shiny Cowbird (102). The cowbirds parasitise the first nests of males, and non-colonial males, more often than subsequent nests. Females abandon a nest if three or more cowbird eggs are laid in her nest. In colonies where the rate of cowbird parasitism is high, the entire colony may abandon after the egg-laying period, due to these parasites. Yellow-hooded Blackbird does have some defenses against the cowbirds, however. Some females can apparently identify cowbird eggs and remove them from their nests within 24 hours of their appearance. Other blackbird females may cover cowbird eggs by re-lining their nests. Males are extremely aggressive towards female cowbirds and effectively keep them away from nests. When a male is off foraging, neighbouring males may intrude into his territory to chase off cowbirds. Female blackbirds do not chase cowbirds away. Few cowbirds successfully fledge from nests of this blackbird.

DISTRIBUTION AND STATUS Locally common. The highland subspecies is found to 2600 m, while the lowland race is seldom seen above 500 m. In Colombia, found in the NW lowlands, from N Magdalena south to N Chocó, but extending as far south as Valle along the Cauca valley and to Tolima along the Magdalena valley. An isolated population is found in the highlands near Bogotá. East of the Andes it inhabits the Llanos lowlands N of the Río Guaviare in Colombia and east through Venezuela, Trinidad, the Guianas to the mouth of the Amazon river in Brazil (to Belém). Ranges west along the Amazon river, including the lower reaches of major tributaries, to the Amazon headwaters (Río Ucayali) in NE Peru. It is not present in S Colombia (S of Río Guaviare), Ecuador, S Venezuela (S Amazonas and S Bolivar), N Amazonas and W Roraima, Brazil, and Tobago.

MOVEMENTS Largely sedentary, however it has been recorded from Bonaire and Curaçao (Voous 1985) suggesting that some dispersing individuals may stray away from the expected range. The three Netherlands Antilles records are as follows: a male on Aug. 15, 1972 at Malpais, Curaçao, a male on Jan. 18, 1977 at Tera Corá, Bonaire and a male on Mar. 9, 1979 at Kralendijk, Bonaire (Voous 1983). In the Llanos of Venezuela, where many marshes are seasonal, Yellow-hooded Blackbird is only present when these are flooded. During the dry season, the blackbirds retreat to permanent lakes, so here at least regular seasonal movements do occur.

MOULT Adults undergo a complete body moult after the young become independent. Changes from non-breeding to breeding plumage occur strictly as an effect of wear and fading in adults. Juveniles undergo a partial moult, the first pre-basic, in which the primaries, secondaries and rectrices are retained. Pre-alternate moults are lacking except in 1st-year males. These young males undergo a prolonged moult into a more adult-like plumage. The extent is variable depending on the individual. However, full adult (definitive) plumage is not achieved until after the second pre-basic moult. The timing of moults in different populations is dependent on the timing of the breeding season.

MEASUREMENTS ***A. i. icterocephalus***. Males: (n=18) wing 86.4 (80–93); tail 68.8 (60–76); culmen 19.2 (18–20); (n=10) tarsus 25.8 (25–27). Females: (n=10) wing 72.1 (68–77); tail 56.4 (51–62); culmen 17.0 (15–18); tarsus 23.2 (22–25).

A. i. bogotensis. Males: (n=10) wing 90.2 (84–92); tail 72.9 (66–80); (n=9) culmen 19.9 (19–21); (n=10) tarsus 26.0 (25–27). Females: (n=7) wing 79.3 (79–80); tail 67 (65–76); culmen 17.9 (17–18); tarsus 23.7 (22–24).

NOTES Apparently an introduced population was present south of Lima, Peru, during the 1960s and 1970s, but may have been extirpated subsequently. For systematic relationships see Notes under Chestnut-capped Blackbird.

REFERENCES Chapman 1914 [measurements], Cruz *et al.* 1990 [behaviour, nesting], Hilty and Brown 1986 [voice, nesting], Gyldenstolpe 1945a [measurements], 1951 [geographic variation, measurements], Ridgely and Tudor 1989 [distribution, notes], Voous 1985 [movements], Wiley and Wiley 1980 [description, habitat, behaviour, nesting, movements, moult].

60 CHESTNUT-CAPPED BLACKBIRD *Agelaius ruficapillus* Plate 20
Vieillot 1819

A locally common marsh blackbird of South America. The males are unmistakable with their black bodies, chestnut cap and throat.

IDENTIFICATION Male Chestnut-capped Blackbirds are black with neatly off-set chestnut on the crown and throat. The females are largely olive with indistinct streaking on both the upper- and underparts, and with a yellowish wash to the throat. Males may appear entirely black at a distance. The chestnut cap and throat are diagnostic. The bill of this species is shorter and more conical than the finer and longer bills of the sympatric Yellow-winged (58) and Unicolored (57) Blackbirds; however, this difference is difficult to see at a distance. Note that the vocalisations of Chestnut-capped Blackbird do show some similarities with those of Yellow-winged Blackbird, but Chestnut-capped Blackbird includes sweeter and more melodic sounds. Unicolored Blackbird sounds quite unlike the other two species (see Voice for that species).

Female and immature Chestnut-capped Blackbirds are dull and nondescript, and look remarkably like juvenile or female Shiny Cowbirds (102). Chestnut-capped Blackbird can be differentiated from the cowbird by its smaller size and its proportionately shorter tail. The throat of Chestnut-capped Blackbird is pale, often with a beige or yellowish wash, that contrasts with a darker and more olive belly and breast. Female cowbirds are more evenly brownish-grey below, with a slightly paler throat which is not yellowish. The crown of Chestnut-capped Blackbird is more noticeably streaked than the largely solid cap of Shiny Cowbird. Fresh-plumaged Chestnut-capped Blackbirds show an olive wash on the crown, back and underparts unlike in Shiny Cowbird. When worn, some of this olive colour may fade, but it is usually retained on the supercilium and crown. Vocalisations and habitat choice differ between the two species (see Voice and Habitat). Female Yellow-winged Blackbirds are distinctly streaked both above and below, and show a crisp pale supercilium, unlike in Chestnut-capped Blackbird. Female Unicolored Blackbirds are noticeably yellow below and show bright rufous edges to the wing feathers. In the Amazon delta region and environs, Chestnut-capped Blackbird comes into contact with Yellow-hooded Blackbird (59). Female Yellow-hooded Blackbird resembles female Chestnut-capped Blackbird in its brownish-grey plumage with obscure streaking; however, Yellow-hooded Blackbird has a bright yellow throat and a distinctly yellow supercilium.

VOICE **Song**: Similar to that of Yellow-winged Blackbird, in that it has a long, drawn-out terminal screech that descends in pitch. The screech is preceded by soft notes, usually two. Thus, a typical song may sound like *teep teep...tcheeeeerrrrrrr* or *si, si, si-grahh*. Songs are accompanied by the 'song spread display'. A typical song lasts approximately 1.5 seconds. Songs vary slightly in length, pitch and the exact nature of the preliminary notes. There is a shorter song, which is higher in pitch, *tic-tic-WHEEE* which may serve a different function to the primary song. In addition, Chestnut-capped Blackbird also performs a canary-like series of sweet whistles and trills. Some of these whistles are given in one pitch, are nearly one second in length and sound quite striking. In comparison to a colony of Yellow-winged Blackbirds, Chestnut-capped Blackbirds sound sweeter and more musical. Males of this species are known to sing while they parachute-flight down over the marsh.

Calls: A sharp *chee* and a high *tip*, as well as a *pewt*, *tsiew* and *chat*. Sometimes a slow chatter *che-che-che-che-che* is performed.

DESCRIPTION A small blackbird with a sharply pointed bill that has a straight culmen and gonys. The tail is square-tipped. Wing formula: P9 $<$ P8 $>$ P7 $>$ P6; P9 $\approx$ P6.

Adult Male Breeding (Definitive Basic, worn): The bill is black, and sharply pointed with a straight culmen and gonys. The eyes are dark brown. Most of the body is black with a faint bluish gloss, except for the cap and throat which are dark chestnut. The cap and throat patches are crisply delineated from the black plumage. On distant views or in bad lighting conditions, the chestnut head markings may not be obvious. The legs and feet are black.

Adult Male Non-Breeding (Definitive Basic, fresh): Similar to breeding male, but the back, breast and flank feathers are tipped olive.

Adult Female (Definitive Basic): The bill is blackish, slightly greyer at the base. The eyes are dark brown. The face and upperparts are olive-brown with slightly darker feather centers, thus looking obscurely streaked from above. The face lacks a distinctive pattern, no obvious supercilium is present and at most it may show a short, dark malar streak. The wings are brown with fine buffy-brown fringes to the coverts, tertials and flight feathers. The tail is blackish-brown. The underparts are warm buff on the throat, becoming greyer towards the back. The whole of the underparts is obscurely streaked. The legs and feet are black.

Immature Male (First Basic): Intermediate between adult male and female. When fresh, it appears more like the adult female, but is larger and with darker centers to the upperpart feathers. Immature male plumage is quite variable due to differences in moult, wear and fading between individuals. Some young males may look nearly identical to females, while others may look like adult males but with an olive crown once worn and faded. Most individuals have blackish body feathering with large olive tips, which wear away to reveal more of the black. The crown and throat are dull chestnut, veiled with olive tips. The immature wings are blackish-brown with buff edges to the primaries, secondaries and coverts. The soft part colours are as in the adult.

Immature Female (First Basic): Indistinguishable from adult female.

Juvenile (Juvenal): Similar to female but noticeably paler in colour. The face, breast and back are buffy-yellow. Juveniles are also more noticeably streaked than adult females, particularly on the head and back. The chest shows a tawny wash, while the belly and lower breast is greyer and not as dark as on adult female. The greater coverts, primaries and secondaries are widely edged with buff, forming a pale panel on the closed wing.

GEOGRAPHIC VARIATION Two subspecies recognized. The nominate subspecies occurs in the south part of the range, in Bolivia, S Brazil, Uruguay, Paraguay and Argentina. It is described above.

The race ***A. r. frontalis*** is found in French Guiana, and north and central Brazil. It is less strongly marked with brighter chestnut markings, especially on the crown. The differences between these two forms are minimal.

HABITAT Breeds in lowland marshes and reedbeds. Forages in a variety of habitats including marshes, roadside ditches, wet meadows, grassland and agricultural fields. Rice plantations are favoured (Ridgely and Tudor 1989). Of the three southern marsh *Agelaius*, this is the least attached to wetlands and may be commonly found foraging in fields well away from water, particularly during the non-breeding season. In the south, this species may nest side by side with Yellow-winged Blackbirds in the same types of marsh.

BEHAVIOUR Gregarious at all times. During the non-breeding season this species will flock in large numbers, raiding agricultural areas, in particular rice fields. Belton (1985) mentions that in Rio Grande do Sul, wintering flocks may number thousands of individuals. Chestnut-capped Blackbirds sometimes separate out into single sex flocks during the winter, or they may join up with other blackbirds such as Shiny Cowbirds (Belton 1985). Marshes are used for roosting throughout the year. Males perform a 'song spread (ruff-out)' display, singing from a stalk with the tail spread, the plumage fluffed out and with wings vibrated. They also perform a short flight song, flying a few meters into the air and parachuting down again to the aquatic vegetation, with the tail well spread. Males may flutter their wings in front of females as a pre-copulatory display. Males spend most of their time singing and displaying during the breeding period, and take on little of the responsibility of caring for the nest. The mating system is unknown. At the colony, males are extremely aggressive towards intruders of other species such as Great Kiskadees (*Pitangus sulphuratus*) or even Yellow-winged Blackbirds, and will chase after and attack such intruders.

NESTING Breeds colonially. Having been established, colonies may be abandoned for a few years, probably due to changes in water levels. Nesting in the south of the range occurs October–January, but peaks in November–December. Nesting at a colony is not particularly well synchronized. Males rarely help (Belton 1985) the females to construct the cup-shaped nest which is anchored to emergent vegetation or bushes, usually about one meter off the water's surface. Nests are often within one meter of each other. The clutch size is commonly 3 and the eggs are pale blue with dark spots clustered around the broader end. Shiny Cowbird is known to parasitise the nests of this blackbird. In Córdoba, Argentina, 22.5% of 213 nests were parasitised by the cowbird (Salvador 1983). Chestnut-capped Blackbird has successfully fledged young of Shiny Cowbird (Friedmann and Kiff 1985). Incubation takes approximately 10 days and commences before the clutch is finished. Thus, the first laid egg hatches before that of its siblings (asynchronous hatching). Fledging occurs 10–13 days after hatching. Almost all of the care of the young is performed by the female.

DISTRIBUTION AND STATUS Locally common to abundant. A species of lowlands, found below 500 m. Widespread in eastern South America. Its distribution reaches its northernmost outpost in E French Guiana. Most of the range lies within Brazil where north of the Amazon river it is restricted to a narrow fringe along the coast of Amapá. South of the Amazon river, it is found as far west as the Rio Xingu in Pará and in NW Mato Grosso, and it ranges east to the Atlantic coast and south to Brazil's western and southern border. In Bolivia, it is restricted to the lowlands of the east in Santa Cruz, Chuquisaca and Tarija. Recorded throughout Paraguay. In Argentina, it is found east of the Andes as far west as E Jujuy, E and S Salta, E Tucumán, E Catamarca and NE La Rioja, and as far south as C Córdoba, S Santa Fé and N Buenos Aires, and throughout all of the provinces to the east and north. Recorded from all parts of Uruguay.

MOVEMENTS Migratory, but the details are not well understood. In northern parts of its range, it is not considered migratory, but it does take part in erratic movements, depending on water levels or the rice cycle. Here, it is best considered nomadic as it will appear in areas to breed or just pass through, and it may disappear for months or years. However, in Buenos Aires province, Argentina, it is much rarer in winter than in summer suggesting a large scale movement north and out of the province (Narosky 1985, Narosky and Di Giacomo 1993).

MOULT Pre-basic moult occurs after breeding and is a complete moult. Pre-alternate moults are lacking in adults. Adult plumage changes between winter and summer are caused strictly by wear and fading. The first pre-basic moult is incomplete; it includes the body and wing feathers but retains scattered juvenal feathers, particularly on the underwing. Plumage changes in young males occur largely through wear and fading, but a limited pre-alternate moult may also occur. The pre-basic moults occur March–June.

MEASUREMENTS ***A. r. ruficapillus***. Male: (n=10) wing 94.7 (88–100); tail 72.9 (66–81); culmen 19.7 (16–20); (n=9) tarsus 26.5 (25–28).

Females: (n=3) wing 83 (82–84); tail 61.7 (58–65); culmen 16.3 (16–17); tarsus 23.7 (22–25).

A. r. frontalis. Male: (n=10) wing 93.8 (90–98); tail 71.2 (62–80); culmen 18.6 (18–21); tarsus 25.9 (24–27).

NOTES In captivity, it has hybridized with Baywing (96). A female Baywing mated with a male Chestnut-capped Blackbird, producing four clutches over a period of two years. In each case, the male Chestnut-capped Blackbird built the nest. The nests failed due to predation, but the final one was successful in fledging young. The hybrid young were curiously described to be dirty brown with white wings (Shore-Baily 1928). Astonishingly, this species has also been known to successfully father chicks with two domesticated female canaries (*Serinus canaria*) while in captivity. Only one of the seven chicks lived longer than eight months. This female chick resembles a female Chestnut-capped Blackbird. A recent comparison of the mitochondrial DNA sequence patterns of this species shows that it is most closely related to Yellow-hooded Blackbird (Lanyon 1994). The other South American *Agelaius* are more distantly related, while the Caribbean and North American *Agelaius* are not closely related whatsoever (Lanyon 1994). Interestingly, the Chestnut-capped and Yellow-hooded Blackbirds are the only two species currently classified in the genus *Agelaius* where the males help or entirely construct the nests.

REFERENCES Belton 1985 [habitat, behaviour, nesting], Klimaitis 1973 [behaviour, nesting], Ridgely and Tudor 1987 [geographic variation], Shore-Baily 1928 [notes, hybrid], Straneck 1990 [voice], Sick 1993 [voice, moult, description].

61 RED-WINGED BLACKBIRD *Agelaius phoeniceus* (Linnaeus) 1766 **Plate 22**

This is one of the most abundant birds of North America, as well as one of the best known. For many, the 'redwing' on a marsh is a sure sign of the arrival of spring.

IDENTIFICATION In North America, adult male Red-winged Blackbirds are basically unmistakable, outside of California. No other bird is black with red and yellowish shoulders. The closely related Tricolored Blackbird (62) poses a problem in California, Oregon, N Mexico and surrounding areas. Tricolored Blackbird differs in shape from 'Redwing', having a longer, thinner bill as well as a more squared-off tail. The main feature to look for is the white median coverts of Tricolored Blackbirds, rather than buff or yellow; note that fresh fall Tricolored Blackbirds may show a buff tone to the epaulet borders and that worn Red-winged Blackbirds may show white epaulet borders. However, fresh Tricolored Blackbirds have cold greyish-buff feather edges to the body plumage, unlike the cinnamon, warm buff, or chestnut colours of 'Redwings'. The black plumage lacks a noticeable gloss on Red-winged Blackbird, but has a faint bluish gloss on Tricolored, which can be obvious in good light. The Caribbean Tawny-shouldered Blackbird (63) is patterned like a male Red-winged Blackbird. Note that the former is smaller, thinner-billed and has a tawny epaulet with a buff border, not a red epaulet. Also see the Red-shouldered Blackbird (64).

A fresh fall male Red-winged Blackbird may be so widely edged with buff and rusty markings that it may be confused with Rusty Blackbird (94). Note that Rusty Blackbird has a yellowish eye; however, some early migrating 1st-fall Rusty Blackbirds may show a dark eye. Rusty Blackbird also lacks a red epaulet in flight, and the bill is thinner and longer-looking than that of Red-winged Blackbird. Often a perched Red-winged Blackbird covers the epaulets with his black scapulars, making the entire bird appear black. Such a bird could be confused with some of the other black icterids, but eventually the red will be exposed and its identity revealed. Note the different shape of the cowbirds and Melodious Blackbird (85) compared to Red-winged Blackbird.

Females are more problematic to identify as they are not so boldly coloured as males. Female Tricolored Blackbird is very dark, having a solidly dark belly and vent. The upperparts usually look unstreaked, particularly in summer, and the pale supercilium is often lacking. Female Red-winged Blackbirds are always obviously streaked and show a pale supercilium. The body plumage of female Tricolored Blackbirds is edged in grey or buffy-grey colours which are cold in tone, while female Red-winged Blackbird shows warmer buff, rusty, or chestnut tipping and edging. Females of the central California form of Red-winged Blackbird, 'Bicolored Blackbird' (61X), are much more difficult to separate from Tricolored Blackbird. This identification problem is treated under 'Bicolored Blackbird's' account. The separation of females of typical Red-winged from those of 'Bicolored' and Red-shouldered Blackbirds is treated in the sections pertaining to the latter two forms.

In flight, the obvious field marks of Red-winged Blackbird may not be apparent. During migration and winter, they are often seen flocking with other blackbirds, adding to the problem of identification in flight. Note that Common Grackle (91) is obviously long-tailed, even the females, and often shows a wedge shape to the tail. Grackles do not undulate, but fly level and straight. Red-winged Blackbirds undulate somewhat in flight. Brown-headed Cowbird (101) has wings that look more pointed and longer in proportion to the body than those of Red-winged Blackbird. Rusty and Brewer's Blackbirds (95) do not undulate in flight, and the latter has a long and thin-based tail. Tricolored Blackbirds and Red-winged Blackbirds cannot be reliably separated on the wing, even though the former has a slightly more pointed wing shape.

VOICE Both sexes sing and call, and some of these vocalizations are sex specific while others are not. **Song**: The characteristic male song of the redwing is often rendered as *oak-a-lee*, the terminal *lee* is a ringing, scratchy, buzzy trill. The speed of the terminal trill varies as do the opening notes, sometimes a short note is added after the trill. Each individual male sings several different variations of the song; the repertoire size for a male ranges from 2–8 variations, but is commonly 4–5. There is little geographic variation in song in the north of the range, but in the south, where more populations are sedentary, dialects are more pronounced. Female songs are composed of a series of *chit, teer, ti, hee* or *check* notes strung together in a scolding chatter. A typical female song is *chit chit chit chit chit chit chit cheer teer teer teerr.* Songs with a high component of *teer* notes are more aggressive. These component notes may grade into each other, some being intermediate between two of the five note types. Only the descending and buzzy *teer* note type is distinct and never intermediate (Armstrong 1992). At times females answer the male's song in what could be termed a duet. During a territory boundary dispute or when an intruding male arrives near a male's territory, the territorial male answers the songs of the intruder shortly after the intruder sings. This may be a way of communicating that the resident male is aware of the intrusion and ready to defend the territory (Smith and Norman 1979).

Calls: The common call, used by both sexes in a variety of contexts is a throaty *check.* Males give a whistled *cheer* or *peet* in alarm and in response to the presence of a predator; males also give a *seet*, a growl as well as a *chuck*, and *cut.* Females issue a short chatter and a sharp scream. During the pre-copulatory display and mating, both sexes utter a *ti-ti-ti* call. When leaving or entering his territory the male utters a complex series of calls in flight. The full vocalization may last up to 6 seconds, beginning with a low and clear note, and then repetitions of higher frequency notes.

DESCRIPTION A medium-sized blackbird with a sharply pointed conical bill that has a straight culmen and gonys. It has a moderate length tail that is square-tipped. Wing Formula: P9 < P8 ≈ P7 > P6; P9 < P6; P8–P5 emarginate.

Adult Male Breeding (Definitive Basic, worn): The bill is entirely black, and the eyes are dark. Other than the wing-coverts, the plumage is entirely black. The black is neither iridescent in the field nor in the hand. The lesser coverts are salmon-red to cherry-red, while the median coverts are yellowish-buff, forming a noticeable border to the red shoulders. In late summer, when worn, the epaulet border can fade to white. The legs and feet are blackish.

Adult Male Non-Breeding (Definitive Basic, fresh): Similar to breeding male, but the plumage is heavily tipped with buff or chestnut. The paler tips are wider on the upperparts than on the underparts, in fact often lacking below, and they wear off as the winter progresses. The epaulet is tipped with buff, obscuring the red lesser coverts, while the greater coverts, tertials, secondaries and primaries are fringed with pale chestnut.

Adult Female Breeding (Definitive Basic, worn): A noticeably streaked, brownish bird. The bill is blackish-grey with a paler grey base to the lower mandible. The face is boldly patterned with a dark crown, pale supercilium, dark cheeks, pale sub-moustachial and dark malar stripe. The crown is blackish-brown with paler brown streaks, the same pattern continuing to the nape. The pale buff supercilium is wider behind the eye and terminates at the posterior border of the auriculars. The auriculars are brown, sometimes showing a pale spot below the eye. The back and scapulars are heavily striped, with the feathers being dark brown with paler brown edges. The rump is similar in colour, but not so obviously streaked. The uppertail-coverts are brownish-black with thin paler fringes. The underparts are whitish with heavy blackish streaks. The chin and throat are usually unstreaked, and tend to be washed with pink or buff. There is a dark brown malar stripe bordered above by white or pinkish-white. The base colour of the underparts is white with dark central streaks, which become wider towards the belly and vent. The increase in width of the stripes causes the belly to look dark with white stripes, while the breast appears white with dark stripes. The undertail-coverts are dark brown with pale fringes. The wings are dark brown, with the extent of red on the epaulet variable depending on the individual. Some individuals show a few red-tipped feathers on the lesser coverts, while others have these mostly red. The median coverts are dark brown with white fringes. The greater coverts are dark brown with a buff edge and paler tip; these patterns may appear as two indistinct pale wingbars. The tertials, secondaries and primaries are dark brown with thin buff fringes. Similarly, the tail is dark brown with paler edges to the rectrices. The legs and feet are blackish.

Adult Female Non-Breeding (Definitive Basic, fresh): Similar to the breeding plumage, but is brighter above due to the wider and brighter edges to the back, scapulars and wings.

Immature Male (First Basic): Similar to adult male, but variable. Some young males are much more female-like, while others are almost the same as non-breeding adult males. In fresh fall and winter plumage, young males have much wider buff and chestnut fringes to the body plumage and wings than adult males. In summer, they still show some pale tipping and have duller epaulets. The epaulets tend to be more orange and usually show noticeable black bases or tips to some of the lesser and median coverts.

Immature Female (First Basic): Similar to adult female, but the chin and throat are always white, never pink or buff. Young females only rarely show any red on the epaulets.

Juvenile (Juvenal): Similar to immature female, but the loosely textured plumage has a yellowish base colour rather than white. Therefore the supercilium and upperparts are pale yellowish or buff. The upperparts and wings are widely edged buff, never chestnut or cinnamon. The bill often shows a horn or pinkish base to the lower mandible.

GEOGRAPHIC VARIATION The distinctive subspecies that lack a yellow epaulet border and in which females are dark, are treated below in their own account as 'Bicolored Blackbird'. At least 26 subspecies of Red-winged Blackbird are recognized. Up until 1997, Cuban Red-shouldered Blackbird was considered a subspecies of Red-winged Blackbird, but is now regarded as a separate species and does not appear to be closely related to Red-winged Blackbird.

For the most part, subspecific divisions are based on the colouration of adult females, as these show more geographic variation in plumage than males, in addition to differences in size. The large number of named subspecies makes understanding geographic variation in this species rather difficult. This is the result of attempting to fit a rather simplistic system (the trinomial or subspecies) on to a rather complex expression of geographic variation shown by this species. Red-winged Blackbird is grossly over split at the subspecies level. Based on published information (Power 1970a) we have lumped several subspecies. However, the remaining races should be further lumped together as they appear to be merely points on morphological clines. For example, Stevenson and Anderson (1994) suggest that the two peninsular Florida subspecies should be lumped with nominate *phoeniceus*. Furthermore, experiments have shown that morphological differences in this species have a significant environmental rather than genetic cause between populations. Eggs were transferred from northern to southern Florida, and vice versa, and from Colorado to Minnesota. In all cases, the nestling's morphology (size of body parts, rather than colour) shifted toward the direction of the foster population (James 1983).

A study of the variation of mitochondrial DNA from individuals of various populations of Red-winged Blackbird discovered surprisingly small genetic differences between populations. In addition, there did not appear to be any geographic pattern to the DNA variation. In conclusion, it was determined that morphological differentiation between populations has occurred without major genetic differentiation, gene flow appears to be high and unrestricted in this species (Ball *et al.* 1988). This study concluded that the non-migratory Mexican populations were slightly more differentiated than populations in the north; the study did not include California 'Bicolored Blackbird' within their samples. A more recent study of allozyme variation in this species came to similar conclusions. Allozymes are different forms of an enzyme that are coded by a unique version of the gene for that enzyme. This study found extremely little genetic difference between populations, and did not support the large number of subspecies that have been described. However, populations from the Bay area of California (*mailliardorum* 'Bicolored Blackbirds') and from the Salton Sea, California (*sonoriensis*), were quite different from those from the rest of North America (Gavin *et al.* 1991). These California populations likely experience reduced gene exchange with those from the rest of North America, but genetic differences do not necessarily map on to described subspecies, or even between a separation of 'Bicolored' and typical Red-winged types.

Forms found in North America, north of Mexico and including the Caribbean:

A. p. phoeniceus is described above. Under this form I have followed Power's (1970a) suggestion and lumped in the subspecies *arctolegus* and *fortis*. It is likely that other US forms should be lumped in here as well, for example *mearnsi* and *floridanus* (Stevenson and Anderson 1994). However, it should be pointed out that even after the publication of Power (1970a), new subspecies of Red-winged Blackbird were named. These subspecies are *stereus* from North Dakota and the east slope of the Rocky Mountains in Montana and Colorado, and *zasterus* from Idaho and western Montana (Oberholser 1974, Browning 1978). Nevertheless, we are more comfortable with Power's (1970a) analysis and do not think that the addition of these two new forms aids in understanding the geographic variation of this species. The distribution of this all-inclusive *phoeniceus* thus ranges from Alaska, southeast of the Rockies to central Texas, east to the Atlantic coast, but not including coastal areas of the Gulf coast and Florida, north to the treeline, including the Maritimes and eastern Canada. Geographic variation is clinal, and gradual between these three previously recognized forms. The general trends are that the longest, stoutest-billed forms are found in central Canada, with individuals becoming thinner-billed to the southeast and shorter-billed to the south, such that individuals from the western Great Plains have stout but short bills while birds from the southeast US have long, but thin bills. The wings and tails are longest on birds from the western Great Plains, becoming shorter to the east with the birds of the southeast Great Plains having the shortest tails and wings. Females are darkest in the southeast Great Plains, where rainfall is highest, and palest in the more arid areas of the western Great Plains. Northern populations of this race move south in the winter. The remaining North American subspecies were not analysed by Power (1970a), and are here left as valid races.

A. p. mearnsi is found in Florida, not including the very tip of the peninsula. It is smaller than *phoeniceus*, with a longer, more slender bill. The females are brownish, less blackish above, and are buffier, less whitish below. It has been suggested (Stevenson and Anderson 1994) that differences between *mearnsi*, *phoeniceus* and *floridanus* are so slight, and clinal, that the three forms should be lumped.

A. p. floridanus is restricted to the tip of the Florida peninsula and the Florida Keys. In measurements and proportions this race is like *mearnsi*, but females are paler, less brownish than that form. The supercilium and underparts are whitish and the upperparts are paler brown. Males are like *phoeniceus*, but they have a deeper buff fringe to the epaulets. Stevenson and Anderson (1994) found that birds from the Everglades were smaller than those from other parts of the state. They suggested lumping *phoeniceus*, *floridanus* and *mearnsi*, but wondered if with a larger sample size the smaller Everglades birds might belong in their own subspecies. Males are extremely similar to *bryanti* of the Bahamas, in their plumage and measurements.

A. p. bryanti is found only on the Bahamas. It is like *floridanus*, but the bill is slightly longer, however Todd and Worthington (1911) state that males of both forms cannot be safely differentiated. The females of this race are the whitest of all of the forms, lacking the pinkish throat and they have pure white underpart colouration. It is commonly found on New Providence, Andros, Grand Bahama, Eleuthera and Abaco. It has also previously been reported from the Bimini and Berry Islands, but its status there is not confirmed.

A. p. littoralis inhabits the coastal plain of the Gulf coast from the central Texas coast to the Florida panhandle. Compared to *phoeniceus*, the females are darker above and below. In addition, the wings and tail are shorter and the bill is more slender. It is darker than *mearnsi*, with a shorter, thicker bill. *Littoralis* is the darkest eastern race. Stevenson and Anderson (1994) suggested lumping all forms of Red-winged Blackbird found in Florida, except *littoralis* which they noticed was significantly darker than the rest.

A. p. nevadensis is the subspecies of the Great Basin, extending north to interior British Columbia, south to southern Nevada and east of the Sierra Nevada in California. In bill shape, *nevadensis* is stouter than *caurinus* or *sonoriensis*, but less stout than *neutralis*. The female is not as richly coloured as *caurinus* and is more darkly streaked on the underparts than *sonoriensis*. Females of this form and *neutralis* are not always distinguishable with regards to colour. At the north end of the Sacramento valley, *californicus* abruptly changes to *nevadensis*, there is likely little gene flow between these two forms. In winter, this form may be found near the coast from San Francisco to northernmost Baja California, often in flocks of 'Bicolored Blackbirds'.

A. p. caurinus occurs in the Pacific Northwest west of the Sierra Nevada and Cascades, from southwestern British Columbia to northern coastal California (Humboldt Co.). This is a beautiful, richly-coloured subspecies with the rusty fringes of the upperparts of the females being reminiscent of patterns seen on small *Calidris* sandpipers! The buff epaulet fringe of the male is also much more deeply coloured than in other forms. It has a longer, thinner bill than *nevadensis* and *sonoriensis*. This form is a short distance migrant and is regular in winter south to the San Francisco Bay area and Modesto in the San Joaquin valley, California.

A. p. aciculatus is found in Kern Co., California. It is like *neutralis*, but larger. It has a long, slender bill, more so than other US subspecies, and resembles Tricolored Blackbird in this respect. In colour, females are intermediate between *neutralis* and *nevadensis*. Some male *aciculatus* and *neutralis* show black tips to the yellowish median coverts, a trait characteristic of 'Bicolored Blackbirds'.

A. p. neutralis occurs in coastal southern California south to NW Baja California in Mexico. It resembles *sonoriensis*, but is smaller with a thicker bill, and the females are darker. Northern males may show an extensive amount of black on the median coverts, females become darker northward also. However, the changes in the males northward are gradual, while the female shows a more discrete change into *californicus*. In summary, this form does appear to show some introgression with California 'Bicolored Blackbird', but more so in the male than the female plumage.

A. p. sonoriensis is found from Arizona, SW Nevada and SE California south to N Sonora, Mexico. The females of this race are the palest of the western races. Their bills are more slender than those of *nevadensis*. Males in fresh plumage are widely and extensively fringed with pale brown, and these edges may be retained as late as early spring, even on old males. Males have immaculate buff median coverts, never showing black.

Coastal Caribbean forms of Mexico and Central America:

A. p. megapotamus is found from coastal and southern Texas, to N Veracruz in Mexico. The female is paler than *littoralis*. Compared to *phoeniceus*, it has a longer wing and shorter bill, and the female is paler. The breast and flanks are dull buff, not ochraceous as in *richmondi*; also this is a larger race than *richmondi*. The juveniles are richly coloured like *richmondi* but are more heavily streaked. In northern Veracruz, *megapotamus* slowly blends in with *richmondi*, birds becoming richer in colour and smaller in size as one travels south.

A. p. richmondi inhabits the coastal slope from N Veracruz, Mexico south to N Costa Rica where it is found in the interior. The females of this form are the reddest, richest ochre, particularly on the breast and flanks. The streaking is narrow and pale. There is a cline in body size, with the southern populations being smaller than northern ones.

A. p. pallidulus is found in the north part of the Yucatán peninsula. It is similar to *megapotamus*, but is warmer buff, less grey, as well as being more heavily streaked. It is paler than *richmondi*, and more heavily streaked on the underparts. Note that the geographic pattern of variation parallels that of Great-tailed Grackle (87) in that Yucatán birds are more similar to the southern Texas/NE Mexico form rather than the adjacent subspecies. The population on Cozumel Island is intermediate between *pallidulus* and *richmondi*. It is sometimes included in *richmondi*.

A. p. arthuralleni is restricted to the Petén region of Guatemala. The female is like *pallidulus*, but is darker and more richly coloured, with blacker, heavier streaking on the underparts. It is less reddish than *richmondi*, but darker above and more strongly streaked below.

Pacific lowland and interior forms from Mexico south:

In general these are large, heavier-billed birds than the populations from the Gulf coast and Caribbean. The juveniles of these various forms do not show much variation in colour as do the forms from eastern Mexico.

A. p. nelsoni is found in Morelos and Puebla, Mexico. It is like *nayaritensis*, but is larger with longer wings and longer tail. It is much larger than *richmondi*. The female is inseparable from female *nayaritensis* in colour, and is paler than female *sonoriensis* and less richly coloured than *richmondi*. It is apparently expanding into the range of *gubernator* 'Bicolored Blackbird' and hybridizing with it.

A. p. nayaritensis is found in the Pacific lowlands from Nayarit, Mexico to western El Salvador, including the highlands of Chiapas, Mexico and Guatemala. There are gaps in the distribution of this subspecies, particularly where the highlands meet the sea, as in Colima, Michoacán and Oaxaca, Mexico. Individuals from the highlands are larger than those from the coast, suggesting that perhaps it would be prudent to lump *nelsoni* and *nayaritensis*. The females are darker and more richly coloured than *sonoriensis*, but inseparable from *nelsoni* in colour. While smaller than *nelsoni*, *nayaritensis* is larger than eastern Mexican birds.

A. p. grinnelli ranges from E El Salvador to Guanacaste province in Costa Rica. The females are like *nayaritensis*, but with more reddish fringes on the back, the crown is paler and the bill more slender. In dorsal colouration, they are intermediate between *nayaritensis* and *richmondi*. However, they are paler on the underparts and with more narrow streaking than *nayaritensis*, and are much larger than *richmondi*. The southernmost birds of this form, from Costa Rica, are the smallest.

HABITAT Breeding: It is a typical bird of cattail (*Typha* sp.) marshes, tending to remain in the shallower parts of the marsh when found with Yellow-headed Blackbirds (54). Also uses tule (*Scirpus* sp.) wetlands, salt marshes and sedge marsh. An extremely adaptable bird, it is also found breeding in upland habitats especially wet shrubby fields, edge of second growth and even urban parks. Use of these upland habitats appears to be a recent phenomenon. One Michigan study found that the order of breeding habitat preference is: wetlands, hayfields, old fields and finally pastures. Adults firstly used wetlands and old fields, followed by hayfields later in the season and pastures even later than this (Albers 1978). In Costa Rica, it breeds in marshes that have large populations of herbivorous insects. The marshes chosen are variable in structure ranging from cattail and bunch grasses (*Coix* sp., *Paspalum* sp.) without shrubs, to those where *Mimosa* shrubs offered the only potential nesting sites. Adults forage both in the marshes, and in nearby pastures and agricultural fields like their northern relatives. During the moulting period, marshes are once again the favoured habitat. **Non-Breeding**: During winter, most foraging occurs in open fields and agricultural areas where Red-winged Blackbirds are considered a pest species. Marshes are not as important to these birds during this period, other than as roosting sites. Other roost sites include any thick vegetation, such as dense conifers or tangles and thickets. Winter flocks are particularly attracted to grain-growing areas or to those where livestock are abundant.

BEHAVIOUR Adult males arrive on the breeding grounds first and start defending a territory against other males. An average territory is 2000 m^2 in area, but this varies according to territory quality and habitat type. Upland territories are larger than those in marshes. Females appear to chose a male based on the quality of the male's territory, rather than due to attributes of the male. Male pairing success is related to the male's size, experience, physiological condition and ability to devote a large amount of time to territorial defense, all of which probably serve the male in obtaining the best territories (Searcy 1979b and 1979c). Males are highly polygynous, with some males breeding with as many as 15 females. The female which settles first in a male's territory tends to be dominant over any other females that may later join that male's harem (Roberts and Searcy 1988). DNA fingerprinting studies confirm that females do mate with males other than the territory holder (Gibbs *et al.* 1990), a behaviour that was entirely missed in studies observing colour banded individuals. These extra-pair copulations are understood to be solicited by the female by traveling to another male's territory, not by a male invading the territory of that female's mate. Many males do not get to mate, this is especially true of young males. Almost all females mate, including young ones. It is interesting to note that the most aggressive males are not necessarily the ones that obtain territories (Yasukawa 1979). Instead, successful males tended to spend more time on the territory and foraging on their territory than the unsuccessful males, and relied on less intense displays. It appears that a male's experience has more of an effect on successful territory ownership than aggression (Yasukawa 1979). Males which had their epaulets artificially darkened tended to lose portions or the entirety of their territories (Peek 1972). Immature (one year old) males are physiologically able to reproduce (producing sperm) during their first breeding season; however, few establish territories and fewer still are successful in breeding. Their peak sperm production is roughly three weeks later than that of adults, and their nesting efforts are similarly delayed. Young males vary considerably in the appearance of their plumage, some being more female-like and others closer to adult males, the degree of plumage brightness is not correlated with the development of the testes during the first breeding season (Wright and Wright 1944).

In Costa Rica, males will defend a territory for many months before breeding takes place. The average size of the territory is roughly two to four times that of a temperate 'redwing'. These are defended mainly in the early morning and late evening only, but full-time once nesting commences. The breeding season arrives quickly and somewhat unpredictably as soon as the first rains begin.

In summer, the most noticeable display is the 'song spread'. The male fluffs up his body plumage, raises the epaulets, and spreads his tail as he sings. Depending on context, the 'song spread' may be weak or intense. Intense displays are usually aimed at another male and involve spreading the wings, in an arc, with the epaulets directed towards the other individual. Weaker displays occur spontaneously during the course of territory maintenance by the male. Experimental tests have found that the role of the red epaulets is to threaten other males and keep them away from the territory, they do not function in attracting females. Larger epaulets are even more threatening to Red-winged Blackbirds than normal sized ones. A male will perform a flight display in which he flies in a distinctive fluttering style, at the same time spreading the tail and raising the epaulets. Song is given during the flight display.

Females also give a 'song spread', which is similar in form to that of males, and tends to be directed at other females. These are aggressive displays. Territoriality has also been thought to occur in females, each female defending a sub-territory within a male's territory. However, some studies have found otherwise. Although this is still unresolved, aggression towards other females is common particularly early in the season. When kept away from the nest full of young due to a disturbance, the female performs the 'wing flip' display where she holds one wing up towards the cause of the disturbance (usually the male).

Gregarious roosting occurs throughout the year. During the height of the breeding season, roosts are composed mainly of young and nonbreeding males. After the breeding season winds down, more birds join these roosts and eventually begin their moult. Red-winged Blackbirds migrate in small groups and form large or huge roosts in the winter quarters. Some roosts have been estimated to number millions of birds, these are usually mixed roosts that also contain Common Grackles (91), Brown-headed Cowbirds (101) and European Starlings (*Sturnus vulgaris*). The blackbirds may forage up to 80 km from the roost site, but the average is less than 20 km. The majority of blackbird control is conducted while in wintering roosts.

NESTING There is variation in the timing of the breeding season depending on the geographic locality, with variation greatest in the tropical forms. Costa Rican 'Redwings' from the Guanacaste province begin to breed when the main rainy season commences in May–June. Individuals inland in Alajuela province begin nesting in July. Some individuals may attempt a second nesting after the moult, approximately December–January. In El Salvador, *grinelli* nests late April–July with some birds apparently raising two broods in a season. Throughout most of North America, north of Mexico, the breeding season extends from April–early July.

The nest is a bulky cup made of dry emergent vegetation, typically cattail in the north. The nests are woven into stalks of cattail or tules, usually within 50 cm of the water's surface. Recently, Red-winged Blackbirds have begun nesting away from marshes in the eastern part of North America. These nest in edge habitats including brushy fields or meadows, and even urban areas. In the tropics, they often use bushes in the marsh to nest in, as well as the emergent vegetation. The clutch size averages 3.3 north of Mexico, 2.96 in Costa Rica and 3 in El Salvador; this clutch size is smaller than in the US and Canada which fits the pattern seen in many other birds. Tropical populations have smaller clutch sizes. The eggs are pale greenish-blue with dark spots, blotches and scrawls. The incubation is conducted by the female and lasts up to 13 days. The young need up to 12 days to fledge from the nest. Feeding of the nestlings and fledglings is done mainly by the female, but males often help, particularly with first nests and later on in the season. Yearling (1st-year) females nest later than adult females, and lay a smaller clutch of smaller eggs. The smaller nestlings have lower survival rates, therefore young females fledge fewer young than older females (Crawford 1977).

Brown-headed Cowbird commonly parasitises 'Redwings', but this varies geographically. The cowbird punctures an egg of the host, decreasing its nesting success in this way, but the young cowbird is raised alongside the foster siblings with little negative effect on them. Bronzed Cowbird (99) is also known to parasitise Red-winged Blackbird, albeit unsuccessfully (Friedmann and Kiff 1985).

DISTRIBUTION AND STATUS Common to locally abundant, it may be the most numerous songbird in North America. The central California and Mexican plateau populations are treated as the 'Bicolored Blackbird' below, but are included in this composite distribution.

Breeding: It is found nesting throughout the continent up to the treeline. Breeds in Alaska north to SE Alaska (Anchorage), and SC Alaska (Fairbanks); throughout British Columbia except in the highest elevations; SC Yukon; NW Northwest Territories (Great Bear Lake); NE Saskatchewan; NC Alberta; N Ontario (Moosonee); C Quebec (Lake Abitibi) along the north shore of the Gulf of St Lawrence to the extreme eastern end; also in SW Newfoundland. The southern edge of the breeding ground includes N Baja California Norte, Mexico; S Arizona; NE Sonora, Mexico; the Atlantic lowlands of Mexico to Belize including coastal but not interior Yucatán; the Sierra Madre Occidental and S Mexican plateau; the Pacific lowlands and the highlands of Mexico from Guerrero south to Guatemala; and isolated populations in W El Salvador, NW Honduras and NW Costa Rica. There is at least one record from Nicaragua, at Klupi on the Rio Wanks, but it has not been found in the eastern savannas of the country since then (Howell 1972). It is not clear if this blackbird is regularly observed in any part of Nicaragua.

Non-Breeding: Most Mexican and Central American populations are sedentary. Northern North American populations retreat southward, but the southernmost populations in the US are likely resident. The northern edge of the regular wintering range extends from coastal British Columbia to the S Alaska panhandle, through S British Columbia, C Wyoming, NW South Dakota, C North Dakota, C Minnesota, C Wisconsin, S Michigan, S Ontario, S Quebec (along the St Lawrence to Montreal), NE New York, S Vermont, S New Hampshire, coastal Maine and S Nova Scotia. Wintering 'Redwings' are found in S Baja California, Sonora, W Chihuahua, Sinaloa, W Durango and N Nayarit, Mexico; places where they are not known to breed.

MOVEMENTS A short to medium distance migrant north of Mexico. Migration is obvious as they are diurnal migrants, often seen passing over with other blackbirds and European Starlings. There is some sexual segregation in the winter due to differential migration. Adult females tend to migrate farther south than adult males. Immatures also migrate farther than adult males (James *et al.* 1984). Thus, at the northern part of the wintering distribution, males are more common than females.

While there is a substantial number of banding recoveries of this species, a continental overview of the migration is difficult to produce as there have been few birds banded or recovered from the Great Plains and the north. During the breeding season, birds spread out throughout the continent, but then migrate to wintering areas where they may concentrate into huge (numbering thousands to millions) flocks. Individuals tend to return to the same general area in the summer, year after year. They are not so site faithful to their wintering flock and may winter several hundred kilometers from where they wintered in a previous year, however, most do winter in the same general place. Fall migration begins after the complete moult (October) in most populations; however, northern populations move an average of 700 km south in August–September before the moult is finished. Winter roosts begin to fill up during November. Migration north commences February–March, except in the Great Plains, Rocky Mountains and northern Canada; these birds return a little later. Many of the birds from the Gulf coast appear to be year-round residents. Breeding populations do not winter together in the same area, but instead they disperse and winter over a broad area. General movements are north-south except in the Great Plains where there is a southeast-northwest tendency, and along the Atlantic coast where there is a southwest-northeast tendency, the latter two movements are likely due to physical barriers such as the Rocky Mountains, Appalachians and the Atlantic coast. Birds breeding in the northern Atlantic areas winter along the coast but further south, not crossing the Appalachians; however, birds breeding in the Great Lakes region will winter in the southern Atlantic region or the Gulf coast region, therefore crossing the Appalachians. Winter recoveries from Florida have come mainly from two distinct regions, the northeast from New Hampshire south to N Virginia, and the Midwest from Michigan, SW Ontario, Indiana and Ohio (Stevenson and Anderson 1994). In the west, there is largely a north-south migratory tendency but a good proportion of the birds move from the interior to the Pacific coast, adding an east–west component to their movements. Populations that breed in Mexico and Central America are little studied but appear to be largely resident. A banded 'redwing' was recovered 15 years and nine months later (Klimkewicz and Futcher 1989), and there is at least one recovery of a 14 year-old male.

Red-winged Blackbird has benefited from the activity of people, particularly with agricultural land providing so many new foraging opportunities. It appears to be expanding in Alaska somewhat, being rarer there 20 years ago than it is now. In Mexico, *nelsoni* appears to be a recent invader into the Mexican plateau. Given the huge population of this bird and that it is migratory it is surprising that it has not been found as a vagrant in Europe. There are reports from England and Italy, but these are thought to refer to escaped captive birds. There are records for both Greenland and Iceland, probably true vagrants. To the north, they have occurred out of range in Alaska north to Barrow, well beyond the Arctic Circle. It has occasionally been recorded during winter on Bermuda.

MOULT There is one complete moult (definitive pre-basic) in adults, which occurs in July-September. Pre-alternate moults are missing in adult males, thus all change in plumage from winter to breeding occurs through wear and fading. Adult females undergo an extremely variable pre-alternate moult that includes some of the body plumage, often some median and greater coverts, but no flight feathers. This moult occurs March–early April. The juvenal plumage is lost through the first pre-basic moult (July–October) which replaces almost all of the body and flight feathers, and most commonly the underwing-coverts and the tertials are retained from the juvenile plumage. First pre-basic moult begins 45–65 days after leaving the nest and is completed 60–70 days later. The pre-basic moults occur near the breeding grounds, usually within 200 km of the breeding site; however, northern populations begin migration before the moult is finished due to the quicker approach of winter here. There is a variable first pre-basic moult in immatures which occurs March–June. For the most part, this is restricted to the body plumage, but may include some of the tail feathers (in yearling males) or some of the wing-coverts (usually the greater coverts) in both sexes. The first pre-alternate moult is more extensive in yearling females than males. The pre-alternate moult is also more extensive in yearling females than in adult females, the latter sometimes only moult a few greater coverts and no body feathers (Miskimen 1980).

Detailed studies of the first pre-basic moult have revealed that individuals which begin moult earlier in the summer, moult more slowly than those that begin later, such that both finish their moults at roughly the same time. The end of moult is also synchronized between the sexes. Males are larger and have longer flight feathers, and therefore need more time to grow in their feathers. Apparently males moult at a faster rate than females, largely by moulting more feathers concurrently, and this is how they are able to keep up with the female's moult schedule. The timing and rate of moult differs geographically. One study determined that the regression of the testes, which signifies the termination of the breeding season for males, is closely correlated to the beginning of the definitive pre-basic moult (Wright and Wright 1944).

As is typical, Costa Rican populations moult after breeding, mainly between the second week of September–mid October in Guanacaste, and November–December in Alajuela. Pre-basic moults begin in September in El Salvador.

MEASUREMENTS Males are larger than females, sexual dimorphism in size varies geographically.

A. p. phoeniceus. Males: (n=50) wing 120.7 (113.5–128.5); tail 91.3 (86.0–98.0); culmen 23.7 (21.9–25.0); tarsus 29.0 (27.0–30.3). Females: (n=40) wing 98.9 (95.0–104.0); tail 72.7 (67.0–78.0); culmen 19.9 (18.1–21.4); tarsus 25.8 (24.5–27.7).

A. p. littoralis. Males: (n=13) wing 114.5 (110.0–118.0); tail 88.8 (85.0–93.0); culmen 25.2 (23.2–27.0); tarsus 29.2 (28.3–30.2). Females: (n=23) wing 92.6 (90.0–96.0); tail 70.5 (68.0–73.0); culmen 21.0 (19.4–23.0); tarsus 25.7 (25.0–26.9).

A. p. mearnsi. Males: (n=35) wing 113.1 (109.0–118.5); tail 87.4 (81.0–93.5); culmen 25.0 (23.8–27.2); tarsus 28.7 (27.2–30.0). Females: (n=20) wing 92.8 (89.0–96.5); 70.9 (66.5–74.5); culmen 21.0 (19.4–22.2); tarsus 25.2 (24.2–26.0).

A. p. floridanus. Males: (n=26) wing 113.2 (107.5–119.0); tail 86.2 (76.5–92.5); culmen 25.0 (23.8–27.2); tarsus 28.7 (27.2–30.0). Females: (n=18) wing 93.2 (89.0–96.5); tail 70.2 (65.5–74.5); culmen 20.3 (19.2–21.5); tarsus 24.8 (23.8–25.8).

A. p. bryanti. Males: (n=5) wing 114.3 (112.3–120.7); tail 86.1 (83.8–88.1); culmen 25.9 (25.7–26.2); tarsus 29.2 (27.9–30.2). Females: (n=5) wing 92.5 (89.9–95.8); tail 68.6 (64.8–71.6); culmen 19.3 (18.5–20.3); tarsus 24.9 (24.4–25.4).

A. p. caurinus. Males: (n=10) wing 123 (120.0–127.0); tail 91.4 (88.0–95.5); culmen 26.0 (24.6–27.5); tarsus 29.3 (28.5–30.0). Females: (n=7) wing 101.8 (97.5–104.0); tail 76.4 (73.5–78.0); culmen 21.6 (20.5–22.2); tarsus 26.2 (25.1–26.9).

A. p. nevadensis. Males: (n=15) wing 124.3 (121.0–128.0); tail 91.7 (88.0–96.5); culmen 25.1 (23.9–26.2); tarsus 29.5 (28.2–30.9). Females: (n=6) wing 102.3 (99.5–105.0); tail 74.5 (73.0–77.5); culmen 20.8 (20.1–21.4); tarsus 26.3 (25.4–27.4).

A. p. sonoriensis. Males: (n=45) wing 125.4 (120.0–130.5); tail 92.8 (86.3–97.5); culmen 24.8 (22.5–26.9); tarsus 29.5 (27.5–30.6). Females: (n=18) wing 104.5 (101.5–112.0); tail 76.7 (72.5–80.5); culmen 20.6 (19.3–21.4); tarsus 265.1 (24.8–27.3).

A. p. neutralis. Males: (n=13) wing 125.5 (121–130.5); tail 92.0 (86.5–98.5); culmen 23.2 (22.3–24.1); tarsus 30.1 (28.4–31.2). Females: (n=5) wing 103.7 (100.5–105.5); tail 74.3 (72.5–77.0); culmen 19.8 (19.3–20.5); tarsus 26.2 (25.5–26.6).

A. p. aciculatus. Males: (n=21) wing 126.2 (122.2–131.1); tail 92.4 (83.6–98.9); culmen 27.2 (25.1–31.4); tarsus 29.5 (28.8–30.4). Females: (n=11) wing 113.9 (111.4–115.5); tail 76.3 (72.8–78.3); culmen 22.9 (21.2–23.9); tarsus 25.7 (25.1–27.1).

A. p. megapotamus. Males: (n=7) wing 115.3 (113–120); tail 86.8 (83–92); culmen 22.6 (22–24); tarsus 29 (26–31.5). Females: (n=9) wing 93 (84–98); tail 68.5 (63–72.5); culmen 19.5 (18.6–22); tarsus 25.6 (24.5–26.5).

A. p. richmondi. Males: (n=54) wing 110.6 (104–118); (n=49) tail 85.6 (72.8–98); (n=69) culmen to nostril 17.1 (15.4–19.3); (n=9) tarsus 28.8 (26.4–31.1). Females: (n=52) wing 87.9 (81–94.5); (n=42) tail 67.8 (61–74); (n=62) culmen to nostril 14.3 (13.2–15.7); (n=8) tarsus 25.5 (24.5–26.9).

A. p. pallidulus. Males: (n=18) wing 114.0 (112–120); (n=17) tail 87.7 (83–93); (n=22) culmen from nostril 17.3 (16.1–18.6). Females: (n=12) wing 92.4 (89–94); tail 70.8 (67–76); (n=18) culmen from nostril 13.8 (12.9–15.3).

A. p. arthuralleni. Males: (n=13) wing 109.9 (105–114); (n=11) tail 87.7 (81–96); culmen from nostril 17.5 (16.9–18.3). Females: (n=19) wing 89.3 (87–93); tail 69.3 (66–73); (n=18) culmen from nostril 14.5 (12.0–15.5).

A. p. nelsoni. Males: (n=17) wing 139.7 (133–144); (n=23) tail 103.0 (98–109); (n=3) culmen 23.7 (23–24); tarsus 32.2 (32–32.5). Females: (n=30) wing 110.3 (106–117); tail 80.2 (75–86); (n=2) culmen 20 (20–20); tarsus 29.5 (29–30).

A. p. nayaritensis. Males: (n=72) wing 126.8 (122–139); (n=66) tail 97.7 (91–110); (n=77) culmen from nostril 17.2 (15.3–19.5). Females: (n=109) wing 102.2 (94–112); (n=105) tail 77.0 (68–87); (n=110) culmen from nostril 14.1 (12.1–16.0).

A. p. grinnelli. Males: (n=44) wing 121.5 (115–132); (n=27) tail 93.3 (88–100); (n=45) culmen from nostril 16.8 (15.2–18.5). Females: (n=31) wing 95.5 (90–102); (n=28) tail 73.1 (61–83); (n=38) culmen from nostril 13.7 (12.4–15.0).

NOTES Red-winged Blackbird is often cited as the most abundant bird in North America which is true amongst passerines. One estimate of the population of this species arrived at the number of 190 million, during the mid 1970s. Population densities are lower in the boreal-forest zone, and parts of the mountainous west, particularly where the forest is largely unbroken and at high elevations.

Albinos are quite frequent in this species, with leucism being the most common form. Adult males that are leucistic on the throat and neck show pinkish or reddish tones to these feathers. This implies the presence of red colours which are not expressed due to the overriding effect of the black plumage. Similarly, leucistic females show a red tint to the epaulettes; however, this is only partially masked on normally-coloured adult females (Nero 1954). A bird observed in Easton, Pennsylvania, during the spring of 1994 was reported as a possible Red-winged × Yellow-headed Blackbird hybrid. It showed both a red and a white epaulet, along with yellow blotches on the head. Leucism appears a better explanation for this odd-looking individual. A most odd aberration, reported by Nero (1957) was of a dead male Red-winged Blackbird showing vestigial claws on both wings!

The systematics are treated under Geographic Variation as well as in the accounts for 'Bicolored Blackbird'. Currently, the data does not support a split between 'Bicolored' and typical Red-winged Blackbirds, and the relationship between California and Mexican 'Bicolored' Blackbirds is not clear. However, it is worth noting that until recently Red-shouldered Blackbird was treated as a well-marked subspecies of Red-winged Blackbird. The available evidence suggests that it is not likely even the closest relative of Red-winged Blackbird.

REFERENCES Albers 1978 [habitat], Armstrong 1992 [voice], Ball *et al.* 1988 [geographic variation, genetics], Bent 1958 [behaviour, nesting, geographic distribution], Browning 1978 [geographic variation], Burtt and Giltz 1977 [movements], Crawford 1977 [nesting], Dickerman 1965a, 1974 [geographic variation, measurements], Dolbeer 1978 [movements] 1982 [movements], Howell and Van Rossem 1928 [geographic variation, measurements], Gavin *et al.* 1991 [geographic variation, genetics], Gibbs *et al.* 1990 [behaviour, cuckoldry], Greenwood *et al.* 1983 [moult, pre-alternate], James 1983 [geographic variation, environment and morphology], James *et al.* 1984 [migration], Klimkewicz and Futcher 1989 [age], Kroodsma and James 1994 [vocalizations], Linz 1986 [moult, first pre-basic], Orians 1973 [nesting, behaviour, moult], Meanley and Royall 1976 [populations, notes], Miskimen 1980 [moult], Monroe 1963 [measurements, geographic variation], Nero 1954 [notes, albinos] 1957 [notes, mutant], Oberholser 1919a [measurements, geographic variation], Orians and Christman 1968 [voice, behaviour], Pyle *et al.* 1987 [moult], Røskaft and Rohwer 1987 [behaviour], Van Rossem 1926 [geographical variation, measurements], Yakusawa and Searcy 1995 [voice, distribution, habitat, behaviour, nesting, notes].

61X 'BICOLORED BLACKBIRD'
Agelaius phoeniceus gubernator
(Wagler) 1832

Plate 23

A Red-winged blackbird lacking a border on the red epaulet, it has a disjunct range in California and central Mexico.

IDENTIFICATION Both sexes differ from typical Red-winged Blackbirds (61). Male 'Bicolored Blackbird' lacks a yellow or buff border to the epaulet. Females are similar to typical female Red-winged Blackbirds, but are much darker. Their bellies are entirely dark and unstreaked, the supercilium is faint or lacking and the upperparts may look only slightly streaked. In this respect, female 'Bicolored Blackbird' is much more similar to female Tricolored Blackbird (62). 'Bicolored Blackbird' has a deeper and shorter bill than Tricolored Blackbird, as well as a more rounded wing (Figure 61.1) and a more notched tail, all features which are difficult to assess in the field. Male Tricolored Blackbirds are glossier than 'Bicoloreds' and show an obvious white border to the epaulets; the colour of the epaulets themselves is a deeper red than the more orange-red of 'Bicolored Blackbird's' epaulet. Females are extremely difficult to differentiate, however. Female plumages are easiest to identify early in the fall, when fresh, but become difficult as they wear. Female 'Bicolored Blackbirds' have rusty edges to the upperparts and look warmly coloured, while female Tricoloreds are colder and edged with greyish-buff. The thin fringes on the uppertail-coverts, rump and uppertail-coverts are buff or brownish on female 'Bicolored Blackbird' and greyish or greyish-white on Tricolored Blackbird. In general, the white tips of the median coverts are more extensive on Tricolored Blackbird. The greater coverts are tipped white on Tricolored Blackbird and not noticeably edged, while on 'Bicolored Blackbird' the greater coverts are edged rusty or brown and tipped buff or white. Tricolored Blackbird thus appears to show a thin lower wingbar and a wide upper wingbar, with greyish-black inbetween, while on 'Bicolored Blackbird', the lower wingbar is buff or white and is separated from the upper wingbar by a brown or rusty panel. Some older female 'Bicolored Blackbirds' show a distinctive pinkish wash to the throat which is lacking on female Tricolored Blackbird. Many of these differences wear away by the spring, when the females can be extremely difficult, if not impossible, to identify. Female 'Bicolored Blackbirds' may retain some of the rusty edges to the upperparts into March. During the summer months, when these two species are most difficult to identify, look for the thicker bill and slightly more notch-tipped tail of the 'Bicolored Blackbird'.

VOICE Song: Males have a repertoire of four to five different songs. There is marked geographic variation in the songs of this form. The general sound and quality of the songs is similar to that of a typical Red-winged Blackbird but they are not as musical, lacking the ringing quality of typical 'Redwings'. In addition, the song pattern is different. Frequently the song is a *ka-leeeee-ooo* not the *oak-a-leee* typical of eastern birds. In Mexico, *gubernator* songs are simple and unmusical *galleeeaaa*. The songs of

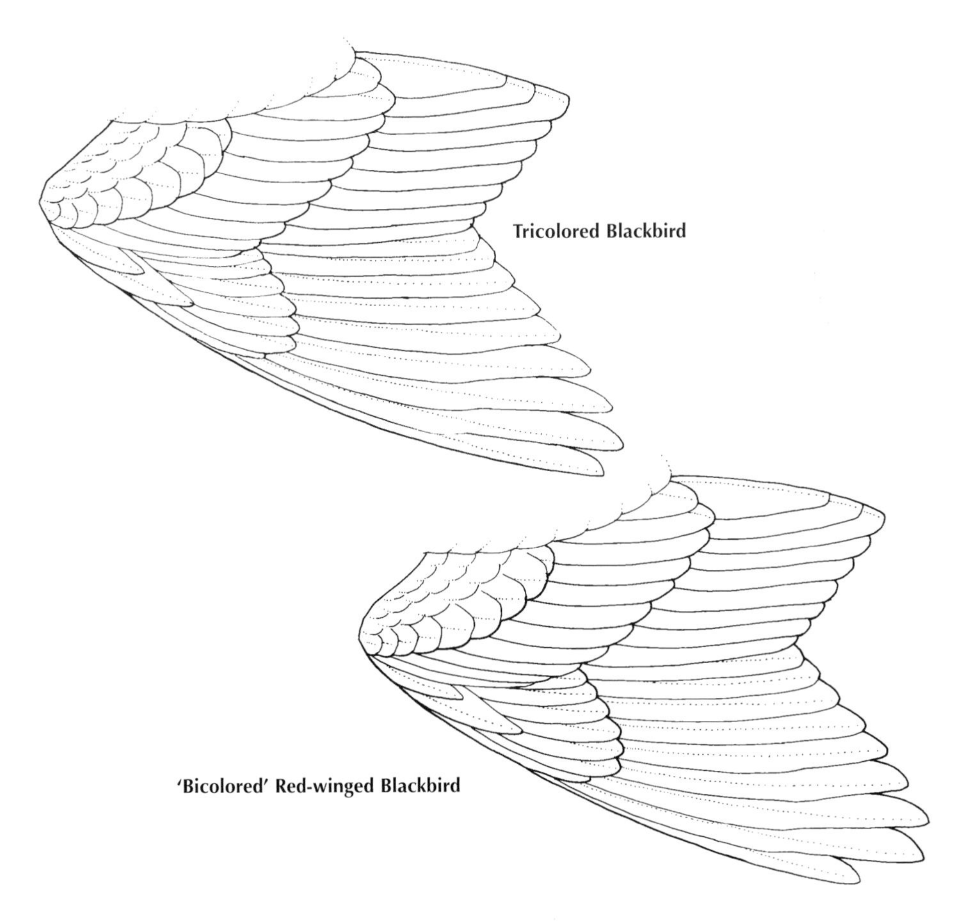

Figure 61.1 Wing shape differences between Tricolored and 'Bicolored' (Red-winged) Blackbirds.

maillardorum and *californicus* 'Bicolored' Blackbirds are much more similar to songs of typical Red-winged Blackbird than to those of *gubernator* 'Bicolored' Blackbird. Where *gubernator* hybridises with *nelsoni* Red-winged Blackbird, the songs of hybrids are intermediate between those of the parental types (Hardy and Dickerman 1965, Hardy 1983). Females of the California populations are seldom heard singing, unlike in typical Red-winged Blackbirds.

Calls: The common call of the California populations is a *chuck*, somewhat lazier and slower sounding than the similar call of typical Red-winged Blackbird. In California, the chuck note of Brewer's Blackbird (95) is more similar to that of typical Red-winged Blackbird than is the call of 'Bicolored Blackbird'. A *tee-wrrr* is heard in alarm, as is a sharp *tieewww*. Many of the other vocalisations are as in typical Red-winged Blackbirds.

DESCRIPTION A chunky blackbird with a relatively stout bill. The bill is pointed with a straight culmen and gonys, but is deeper and wider than on other red-winged blackbirds. Wing Formula: P9 < P8 ≈ P7 > P6; P9 < P6; P8–P6 emarginate (Figure 61.1). The California form *mailliardorum* is described.

Adult Male Breeding (Definitive Basic, worn): The bill is black and the eyes are dark brown. The body plumage is black with almost no gloss. The lesser coverts are orange-red, usually with a few orange-yellow feathers on the anterior lower edge of the epaulet. The rest of the wings and the tail are black.

Adult Male Non-breeding (Definitive Basic, fresh): Similar to breeding male, but has rusty edges to the tertials, some of the coverts and rusty-brown tips to the upperparts. These tips wear throughout the winter, revealing the breeding plumage.

Adult Female Breeding (Definitive Basic, worn): This plumage is obtained through wear of the brighter edges of the non-breeding plumage, so females are darker and more unicolored as the summer progresses. The bill is usually solidly blackish. The upperparts tend to be solidly blackish-brown, sometimes showing a paler supercilium, but this is often missing by the middle of the summer. The wide chestnut tips of the upperparts are largely worn away, but some may still remain into summer, particularly on the lower scapulars. The underparts are whitish (or pink) on the chin and throat, with blackish streaks; the rest of the underparts are solidly blackish-brown. The wings are blackish-brown, rarely showing thin white tips to the median coverts and sometimes showing some red on the epaulet. The tail is blackish and the legs and feet are black.

Adult Female Non-Breeding (Definitive Basic, fresh): The bill is black, with a horn-coloured base to the lower mandible. The eyes are dark brown. The crown is warm brown with an indistinct paler median stripe. The buff supercilium is indistinct and finely streaked darker. The auriculars and nape are warm brown. The back and scapulars are blackish with wide chestnut-brown fringes, making the mantle appear largely chestnut-brown with indistinct darker streaking. The rump is blackish with pale brown tips to the feathers. The chin and throat are whitish or sometimes pinkish on the chin. The white of the throat is streaked blackish. Below the breast, the underparts are largely solid blackish, with brown tips to the breast and greyish tips to the belly feathers. The undertail-coverts are edged pale brown, or buff. The wings are blackish-brown, with the lesser coverts tipped brown and not tending to show much red. The extent of red on the female coverts is variable. The median coverts are tipped buff or white, while the greater coverts are edged rusty and tipped whitish or buff. This appears as two buff wingbars, the lower one narrower, or two whitish wingbars with a rusty panel (the rusty edges) between them. The blackish tertials are fringed chestnut. The tail is blackish with thin brown edges. The legs and feet are black.

Immature Male (First Basic): As in adult male, but with wider brown edgings to the body feathers. Most obtain red epaulets in their first fall and may be almost indistinguishable from adult males during the first summer, except that the young birds have more worn (browner) primaries and retain some juvenile underwing-coverts or tertials. Other immature males have obviously paler epaulets than adults and may show some black intermixed there.

Immature Female (First Basic): Similar to adult female, but retains some juvenile underwing-coverts which contrast with the darker first basic feathers. Young females never show a pink wash to the chin.

Juvenile (Juvenal): Similar to adult female, but the plumage is more loosely textured and buffier overall, with the streaking on the underparts not as crisp. The bill tends to show a pinkish base.

GEOGRAPHIC VARIATION A total of three subspecies fall under this form. The range is disjunct with a population in Mexico and one in California. The relationship between the California and Mexico forms is not clear.

A. (p.) gubernator is resident in the Mexican central valley, found from Durango south to the trans-Mexican volcanic Bbelt (Jalisco and Puebla). It is the largest and thickest-billed of the three forms.

A. (p.) californicus is resident in the Central Valley of California. The females are slightly more streaked than *gubernator*. It is smaller than *gubernator* with a thinner bill.

A. (p.) mailliardorum is found in coastal California from Humboldt Co., south to Monterey Co. It is similar to *californicus* but has a smaller bill and the females are darker than that form and are longer-winged. It is described above.

HABITAT California birds choose habitats similar to typical 'Redwings', marshes with emergent vegetation in which to breed as well as roadside ditches and old field habitats away from water. Winters in flocks, feeding in agricultural fields and pastures. *Gubernator* breeds in meadow bogs of sedge (Juncaceae), not tall tules (*Scirpus* sp.) and cattails (*Typha* sp.).

BEHAVIOUR Similar to typical Red-winged Blackbird.

NESTING Similar to typical Red-winged Blackbird.

DISTRIBUTION AND STATUS Common. In Mexico, this is a form of the high Central Valley, from 1000 to 2500 m, while in California, it is a bird of the lowlands, seldom to 1000 m. Two disjunct populations exist, one in California and the other in Mexico. In California, it is found from S Humboldt and N Mendocino Cos. south to N Monterey Co., along the coast and inland in the coastal mountains to Lake and Napa Cos. and south to San Benito Co.; it is absent from the highest mountains. Also found in the Sacramento and San Joaquin valleys, including the Sacramento River delta. The Central Valley distribution ranges as far north as Glenn and Butte Cos., and as far south as S Kern and N Ventura Cos. The Mexican population inhabits the central Mexican plateau from S Durango south to Jalisco, Mexico, Tlaxcala and Puebla.

MOVEMENTS Little known, but there appears to be a migration towards the coast by some individuals of the California population. Otherwise appears to be largely sedentary, but joins up in large flocks in the winter, often away from the breeding grounds.

MOULT As in typical Red-winged Blackbirds.

MEASUREMENTS ***A. g. gubernator***. Males: (n=22) wing 137.7 (124.5–147); tail 93.9 (85.9–101.6); culmen 21.7 (20.3–24); tarsus 32.4 (30–34.3). Females: (n=10) wing 110.8 (104.1–119); tail 74.1 (64.8–82); culmen 18.6 (17.3–20); tarsus 28.1 (26.4–29.2).

A. g. californicus. Males: (n=29) wing 125.9 (121–131); tail 91.7 (84–97.5); culmen 22.7 (20–24.8); tarsus 29.5 (28.2–31.6). Females: (n=4) wing 103.8 (101.5); tail 74.9 (73.0–77.5); culmen 19.2 (18.9–19.7); tarsus 26.5 (26.0–26.7).

A. g. mailliardorum. Males: (n=15) wing 124.4 (120.0–131.5); tail 87.0 (82–92); culmen 21.9 (20.1–23.4); tarsus 29.2 (27.6–30.0). Females: (n=10) wing 105.7 (104.0–108.5); tail 73.4 (70.5–76.5); culmen 19.0 (18.0–20.1); tarsus 26.2 (25.4–27.4).

NOTES Mexican *gubernator* previously bred sympatrically with the typical Red-winged Blackbird in the Lerma marshes of Mexico, showing no hybridization; in addition, the two forms have different vocalizations and habitat choices. This implies that the two had differentiated to species status; however, at another site the two were seen to hybridize readily (Hardy and Dickerman 1965). The relationship between these two forms needs to be studied further in Mexico. In California, 'Bicolored Blackbird' tends to maintain its integrity, but apparently mixes with typical Red-winged Blackbirds in the northeast of the state and toward the south of the Central Valley. The hybrid zone between these two forms has not been studied. The genetic distance (difference) between coastal *mailliardorum* and Sacramento *californicus* was 10 times more than between populations in Florida and Oregon (Gavin *et al.* 1991). This suggests that there is a major barrier to gene exchange between *mailliardorum* and other Red-winged Blackbird populations. *Californicus,* on the other hand, experiences considerable gene exchange with typical Red-winged Blackbirds (Gavin *et al.* 1991). Further work is needed to resolve the species limits between typical Red-winged Blackbirds and 'Bicolored Blackbirds'. An open question is whether California 'Bicolored Blackbirds' and Mexican 'Bicolored Blackbirds' are actually most closely related to each other, or if they are most closely related to nearby typical Red-winged Blackbirds. If the two 'Bicolored Blackbirds' are indeed each other's closest relatives, this suggests that this form was once much more widespread and that it has been slowly engulfed by the typical Red-winged Blackbird.

REFERENCES Dickerman 1965a [geographic variation], Gavin *et al.* 1991 [notes, genetics], Hardy 1983 [voice, habitat], Hardy and Dickerman 1965 [hybridisation], Howell and Webb 1995 [distribution], Kroodsma and James 1994 [vocalizations], Mailliard 1910 [notes], Nelson 1897 [measurements, geographic variation], Van Rossem 1926 [geographic variation, measurements].

62 TRICOLORED BLACKBIRD *Agelaius tricolor* (Audubon) 1837 — Plate 23

Extremely similar to Red-winged Blackbird (61) in appearance, this near endemic to California is actually quite different from it in breeding behaviour.

IDENTIFICATION The identification of this species is complicated due to the overall similarity of this species to Red-winged Blackbird, in addition to the variability of that species. Tricolored Blackbird may be found, often in the same flock, with typical Red-winged Blackbirds and 'Bicolored' Red-winged Blackbirds (61X). Male Tricolored Blackbird is more similar to males of typical Red-winged Blackbird, while females are similar to female 'Bicolored Blackbirds'. In general, Tricolored Blackbirds are slimmer-billed than their sympatric 'redwing' forms and tend to show more square-tipped tails. In addition, the wing formula differs, with Tricolored Blackbirds having a slightly more pointed wing (Figure 61.1), but this is difficult to assess in the field.

The males are more glossy than male Red-winged/'Bicolored' Blackbirds, but this is only obvious in direct comparison. Throughout the major part of its range, Tricolored Blackbird is found with 'Bicolored' forms of Red-winged Blackbird, making the white fringe to the epaulet an obvious field mark. 'Bicolored Blackbirds' tend to hide their red epaulets when foraging, appearing entirely black, whereas Tricolored usually hides the red but not the white of the epaulet, thus appearing black with a white slash on the wing. Adult Tricolored Blackbirds tend to be more heavily tipped brown on the upperparts than adult 'Bicolored Blackbirds'; this is quite obvious early in the fall. Typical Red-winged Blackbirds with their buffy yellow median coverts may cause an identification problem early in fall when Tricolored Blackbird's median coverts are light buff. Note that on Tricolored Blackbird, the red of the epaulet is more intense, the median coverts are buff without a yellow tone and the upperpart fringing is grey-brown rather than rusty-brown.

Females are most likely to be confused with the extremely similar females of 'Bicolored Blackbird'. Females of typical Red-winged Blackbird are decidedly paler and more obviously streaked, showing an obvious pale supercilium and tending not to show the solidly dark belly of female Tricolored and 'Bicolored' Blackbirds. The plumages of these two forms parallel each other, being brightest in the early fall and darkest (least streaked) in the late summer. The identification is most straightforward early in the season, and may be impossible late in the summer other than by looking at the males that the females are traveling with. During the fall and winter, female Tricolored Blackbirds have grey rather than rusty fringes to the back, scapulars and tertials. A fresh 'Bicolored Blackbird' appears warmly coloured above, whereas a Tricolored always appears cold coloured above. The thin fringes on the uppertail-coverts, rump and uppertail-coverts are greyish or greyish-white on female Tricolored Blackbird, and buff, brownish, golden or chestnut on female 'Bicolored Blackbird'. In general, the white tips of the median coverts are more extensive on Tricolored Blackbird. The greater coverts are tipped white on Tricolored Blackbird and not noticeably edged with any colour, while on 'Bicolored Blackbird' the greater coverts are edged rusty or brown and tipped buff or white. Tricolored Blackbird thus shows a thin lower wingbar and a wide upper wingbar, with greyish-black in between, while on 'Bicolored Blackbird' the lower wingbar is buff or white and is separated from the upper wingbar by a brown or rusty panel. In addition, some older female 'Bicolored Blackbirds' show a pink wash to the throat, which is lacking on Tricolored Blackbirds. Many of these differences wear away by the spring, when the females can be extremely difficult, if not impossible to identify. Female Bicolored Blackbirds may retain some of the rusty edges on the upperparts into March. During the summer months, when these two species are most difficult to identify, look for the more slender bill and square-tipped tail of Tricolored Blackbird.

VOICE While Tricolored and Red-winged Blackbirds are quite similar in appearance, their vocalisations are very different.

Song: A nasal, drawn-out *guuuaaaak*. It can be described as a low-pitched growl that lasts 1–1.5 seconds. Once the eggs have been laid, singing dies down dramatically. In winter and outside the early part of the breeding season, singing is heard, but is never accompanied by a 'song spread'. The female song is a chatter, similar to that of female Red-winged Blackbirds. It is only given by females and is analogous to the male song. It sounds like a hollow chatter or rattle.

Calls: The common alarm and flight calls are a nasal *chwuk* and *churr*, which are given by both sexes. Many of the calls are similar to those of Red-winged Blackbird but tend to be lower in frequency. When taking flight, during the breeding season, males give a nasal *keet* or *cheep*. Short growls are used by males against intruders and during sexual chases. When leaving the breeding area, male Tricolored Blackbirds give a complex flight call that begins with a low and clear note which is followed by a series of higher notes and repetition of the introductory note; this series of calls lasts for several seconds. Other calls heard from Tricolored Blackbirds include *prit*, *chek* and *kwik*. During the pre-copulatory displays, both sexes utter a repeated *ti-ti-ti-ti-ti-ti*. No specific hawk alarm calls are given, unlike in Red-winged and Yellow-headed (54) Blackbirds.

DESCRIPTION A slim blackbird with a sharply pointed bill. On the wing P9 usually longer or equal to P6 (shorter or equal to P6 on Red-winged Blackbird). Wing Formula: P9 < P8 > P7 > P6; P9 > P6; P8–P6 emarginate (Figure 61.1). The tail is square-tipped.

Adult Male Breeding (Definitive Basic, worn): The bill is black and the eyes are brown. The body is entirely black with a dull blue gloss, except for the epaulets. The lesser coverts are dark carmine, with the median coverts a contrasting white, forming an obvious border to the dark red. The white median coverts are visible even when the red epaulets are hidden by the scapulars. The rest of the wings and the tail are black, with a dull blue gloss. The legs and feet are black.

Adult Male Non-Breeding (Definitive Basic, fresh): Similar to breeding male except the body plumage is tipped with pale greyish-brown. The pale tipping is densest on the nape and back, and is not as heavy on the underparts, the face (lores, anterior auriculars and throat) is black without paler tips. The undertail-coverts are fringed white, while the rump is thinly tipped with grey-brown. The median coverts may be white, but are often pale buff. The carmine epaulet is often tipped dusky, obscuring this patch of colour, particularly when the bird is perched. The greater coverts may be thinly-tipped white, creating an indistinct white lower wingbar. The tertials are fringed buff. All these pale tips and fringes are lost during the winter, through wear, revealing the shiny black breeding plumage.

Adult Female Breeding (Definitive Basic, worn): The bill is black and the eyes are dark. Overall this is a very dark plumage, appearing largely blackish except for the paler throat. The head is blackish-brown, showing a faint supercilium on some individuals, but this is often lacking. The upperparts are blackish, faintly streaked paler on the back. The underparts are whitish on the chin and throat, quickly turning solid blackish-brown below the breast. The whitish throat is streaked with dark brown. The wings are blackish, showing no pale edging except on the median coverts which usually have small white tips. The lesser coverts may be touched with red, but this is variable. The tail is blackish-brown and the legs and feet are black.

Adult Female Non-Breeding (Definitive Basic, fresh): Soft part colours as on breeding adult female. The head is brown with a paler supercilium. The brown crown is concolourous with the nape and upper back. The back is blackish, edged grey-brown. The rump and lower back are blackish with grey tips, while the uppertail-coverts are fringed grey. The underparts are palest on the chin and throat, which is almost white. The throat is streaked dark, while the breast is grey-brown with darker centers. The lower breast and belly is blackish with grey tips, the undertail-coverts are fringed grey. The wings are blackish, with grey paler fringes or tips to the feathers. The lesser coverts are blackish with grey edges, some are washed with red, but this is variable depending on the individual. The median coverts are blackish with wide white tips, creating a whitish upper wingbar. The greater coverts are a paler blackish-brown, thinly edged with olive and tipped white, forming a thin whitish lower wingbar. The tertials are blackish with a thin brown fringe, and the primaries and secondaries are blackish. The tail is blackish and the legs and feet are black. The dark summer plumage is obtained through wear of the paler tips of this plumage.

Immature Male (First Basic): Similar to adult male, but retains juvenile underwing-coverts and often some juvenile wing-coverts or tertials. The black plumage may be less glossy than an adult male, and the red epaulet may be paler in colour or mixed with black or brown feathers. When fresh, 1st-fall males typically show wider grey-brown tips to the upperparts than do adults. In addition, these young males show paler tips to the underparts and face. Adult males rarely show pale tips to the lores or anterior auriculars. Juveniles in their first pre-basic moult appear brown with blackish patches, these are most often seen in the early fall, but fall hatched birds may undergo this plumage change much later on in the winter.

Immature Female (First Basic): Similar to adult female, but may show contrasting worn underwing-coverts retained from the juvenile plumage.

Juvenile (Juvenal): Similar to adult female, but duller and buffier.

GEOGRAPHIC VARIATION Monotypic.

HABITAT Breeds in freshwater marshes with tall emergent vegetation such as cattails (*Typha* sp.) or tules (*Scirpus* sp.), as well as moist bramble (*Rubus* sp.) tangles, nettles (*Urtica* sp.), thistles (*Cirsium* and *Centaurea* spp.) or willow (*Salix* sp.) thickets (Neff 1937). A change in breeding habitat has occurred in the recent past. In the 1930s 93% of colonies were in marshes, while in the 1970s only 53% of colonies were using marshes, a greater portion of nesting colonies now use dense thickets of the invasive Himalayan Blackberry (*Rubus discolor*) (Neff 1937, Beedy and Hamilton 1997). Forages in agricultural areas such as pastures, alfalfa fields, rice fields and animal feedlots. In winter, prefers open agricultural areas, particularly where livestock is present and the grass is short. Winter concentrations have even been observed at large city dumps. At night, it shows a preference for roosting in marshes.

BEHAVIOUR Tricolored Blackbirds forage on the ground, mainly on insects. However, adult males feed on a greater proportion of seed food than females or nestlings (Skorupa *et al.* 1980). In addition, seeds and grain are consumed in higher proportions during the non-breeding season. Orians (1961a) suggested that Tricolored Blackbird was a 'grasshopper follower', exploiting outbreaks of these insects. In fact, it appears that Tricolored Blackbirds are opportunistic and will forage on any insect outbreak that may be available (Beedy and Hamilton 1997).

This species is gregarious throughout its life, including the breeding season. Males are not strictly territorial like 'Red-wings', but will chase other males from their immediate singing area. Females are not territorial whatsoever and adjacent nests may even touch each other. In winter, they usually flock in single species flocks, but other blackbirds will join them at times. It is less usual for single, or small numbers, of Tricolored Blackbirds to join up with other blackbird flocks.

Unlike other North American marsh-nesting blackbirds, males and females arrive at the breeding site together. Territorial defense, pairing and nest building occur almost synchronously over a period of only a few days. Males defend a home area but are not exclusively territorial; in any case, all territorial-type behaviour occurs over a period of only a week, after which time males are more likely to ignore each other. The mating system is polygynous, with males copulating with many females. Females solicit copulations by adopting the stereotypic posture, bill up, tail up, wings drooped and quivering while the bird crouches. The male responds by singing, drooping and quivering the wings, before he mounts. Both sexes give the *ti-ti-ti* call during this display.

The full 'song spread' is similar to that of Red-winged Blackbird; however, this display is common only for about a week early in the breeding season. Males ruffle the plumage and sing while exposing the epaulets, spreading the tail and opening the wings somewhat. The back feathers are raised conspicuously, unlike in Red-winged Blackbird. 'Song spreads' are not delivered from a conspicuous perch, but from low in the vegetation, such that only the nearest neighbours see the display. Females sing, but do not give 'song spreads'. There are no aerial displays in Tricolored Blackbird, and sexual chases are very rare and only observed when females are nest building. As in Red-winged Blackbirds, females perform the 'wing-flip' display in situations where they are kept from the nest during the nestling stage. Most often this is given in response to a male that is interfering in the female's actions. The female raises one wing, revealing the underside to the cause of the disturbance.

Males sing mainly during the early stages of breeding; once the clutch has been laid they cease to sing. A distinct lowering of the volume of sound from the colony indicates that clutches are complete (Beedy and Hamilton 1997). Once incubation is underway, males will often leave the colony and forage nearby in all-male flocks. Once the young hatch, some males may return to the nest and assist in feeding the nestlings.

NESTING The breeding season is typically early April–July. The start of breeding depends on the habitat, with birds in the foothills starting first, followed by valley-nesting birds and finally birds which nest in rice-growing areas. Sometimes fall breeding occurs October–November (Orians 1960).

This species is highly colonial. Nests are placed more closely together than other colonial North American blackbirds. Single bushes may have a dozen nests and there can be one nest to every square metre of cattail marsh. Colonies with an estimated 100,000 nests were located on a marsh only four hectares in extent (Neff 1937, Beedy and Hamilton 1997). In the past, some colonies were estimated to have 200,000 nests (Neff 1937). Larger colonies may return to the same general area year after year, but on the whole this species is fickle regarding colony location, and often colonies are not established in the same area the following year. Water levels and the availability of nearby foraging areas influence the decision to situate a colony in any particular place. Once a nesting site has been chosen, the birds are closely synchronized in their activities, from nest-building to egg-laying. In larger colonies different sections may not be synchronized with each other, but may be closely synchronized within the section. Colonies may be suddenly abandoned during the nesting period, usually because of large predation events due to inadequate feeding opportunities nearby. However, colony abandonment is probably overstated in the literature and it is likely quite rare (W.J. Hamilton pers. comm.).

Females build the nests without help from the males. The nests are usually woven onto stalks of an aquatic emergent, and they resemble those of Red-winged Blackbird. A clutch size of 4 eggs is typical, but clutches range from 5–6 eggs. The eggs are pale green with brownish-black spots and scrawls. Incubation lasts 11–14 days (Emlen 1941, Orians

1961a, Payne 1969a), and different researchers have noticed slightly different incubation period lengths. Incubation begins before the clutch is completed, creating an asynchronous hatching schedule and a size hierarchy in the nestlings (Beedy and Hamilton 1997). The young require 13 days, sometimes as little as 9 days, to fledge. The females primarily care for the young, but males will often help. During the incubation period, males leave the breeding area, but come back to help with feeding the young. Once the young fledge, they may congregate in a group (crèche) sometimes, between the nest sites and the foraging areas (Payne 1969a, Beedy and Hamilton 1997).

Fall breeding occurs after the rains. Fledglings may be out of the nest as early as late October–late November. These fall colonies are not as closely synchronized as those in the spring and summer. Clutch size is as in summer, but success rates are low. The adults breed in fresh 'winter' (definitive basic) plumage. It is unclear if fall breeding is historical or has been induced by human changes to the landscape such as rice production and irrigation (Orians 1960). Brown-headed Cowbird has been known to parasitise Tricolored Blackbird (Friedmann and Kiff 1985), but parasitism rates are low. Most cowbird eggs in Tricolored Blackbird nests do not hatch (Beedy and Hamilton 1997).

DISTRIBUTION AND STATUS Common to rare, but always local. Tricolored Blackbird is nearly a California endemic, breeding mainly in the Central Valley and other points west of the Cascades and Sierra Nevada. It is a bird of the lowlands, but it has bred to 1300 m in the Klamath area and along the W side of the Sierras. Breeds from Oregon (as far north as Wasco Co., and in Portland, both near the Washington border), also in Multomah, Umatilla, Wheeler and Jackson Cos. (Beedy and Hamilton 1997). The stronghold in Oregon is Klamath. The Oregon population may be expanding, there were historical reports from many places in N Oregon but Gabrielson and Jewett (1940) felt that these were unreliable. Thus, the specimens from the Klamath area (Neff 1933) are the only reliable historical records for that state. It appears that the northern Oregon populations are thus of recent in origin. This species should be searched for very carefully in southern Washington. In California, it is most abundant in the Sacramento and San Joaquin valleys with strongholds in Sacramento Co. and vicinity, Merced Co., NW Kern Co. and vicinity (Beedy and Hamilton 1997). In northern California, Tricolored Blackbirds regularly breed in the Klamath Area (Siskiyou Co.), with colonies occurring sporadically between this area and the northern portion of Sacramento valley. Coastal colonies are rare, but occur sporadically between Humboldt and Monterey Cos., in N California. It also breeds, in small numbers, in the San Francisco Bay area, particularly in the South Bay (Santa Clara and Alameda Cos., continuing south along the valley to San Benito Co.). Colonies regularly nest along the Salinas River valley from Monterey Co. south into San Luis Obispo Co. Tricolored Blackbird is absent from the California Transverse ranges, in fact these mountains appear to divide the species into two separate populations (Beedy and Hamilton 1997). In southern California, Tricolored Blackbirds are present largely south of the Transverse Ranges and west of the Peninsular Ranges in S Ventura, Los Angeles, Orange, SW San Bernardino, W Riverside and SW San Diego Cos. This population ranges as far south as NW Baja California Norte, Mexico. A colony has recently established itself in Douglas Co., in extreme W Nevada. Breeding east of the Sierra and Cascades mountains is not unknown, it occurs in Lassen, Shasta and the Klamath region of N California and Oregon.

MOVEMENTS The species is not typically considered a migrant as it can be found throughout the breeding range during the winter; however, Tricolored Blackbird is migratory. The pattern of movements may be confused by a general nomadic tendency in this species. In central California (Marin to San Mateo Cos.), there is a marked increase of Tricolored Blackbirds along the coast in the fall, with most of these birds wintering there. Thousands may be seen along the coast in winter, however extremely few nest near the coast in this part of the state. The birds arrive on the coast after completing their moult, in September, and remaining until late March. Banding returns show that birds banded as nestlings in Merced Co. disperse mainly to the north and northwest in fall and winter (Neff 1942). It appears that breeders from the San Joaquin Valley winter largely to the north and west of their breeding sites. The coastal (San Francisco to San Luis Obispo Cos.) wintering population and birds overwintering in the San Francisco Bay area likely originate from breeding populations in the San Joaquin Valley. Note also that during the breeding season, movements also exist. Once breeding finishes in the Jan Joaquin valley and Sacramento Co., some individuals move farther north in the Sacramento valley and establish breeding colonies there (Beedy and Hamilton 1997). No banded individual from the S California population has ever been recovered from N California and vice versa. It is likely that the S California and Baja California population is separate and distinct from the N California population (Beedy and Hamilton 1997).

The oldest individual was a banded male recovered at the age of 13 years and three months (Klimkiewicz and Futcher 1989).

This species is not prone to vagrancy. However, it has been observed on several occasions in Nevada, initially as a vagrant but since 1996 a small breeding colony has established. It has yet to be observed in Washington State, and should be looked for in the south part of the state particularly since there are now populations in Oregon very close to the border. The Oregon population has been present since the 1930s and recently appears to have been expanding.

MOULT Adults have a complete pre-basic moult after nesting, from June–September. The pre-alternate moult is lacking. Therefore plumage change from winter to breeding occurs through wear in adults. Juveniles undergo a variable first pre-basic moult, usually incomplete, but sometimes complete, the timing variable depending on the time of birth. Summer-born juveniles moult July–October, but fall-hatched birds moult December–February. It has not been confirmed if the first pre-alternate moult is lacking; if it does occur it is not obvious.

MEASUREMENTS Males, (n=9) wing 121.2 (117.6–123.7); tail 88.1 (84.3–95.3); culmen 23.4 (22.1–24.1); tarsus 29.7 (28.7–30.7). Females, (n=10) wing 106.7 (104.4–109.7); tail 75.4 (74.2–80.3); culmen 20.1 (19.8–21.1); tarsus 26.4 (25.4–26.9).

NOTES In 1992, the population of this species was estimated at 250,000 adults. Historically, single colonies reached that size, and it does appear that numbers of this species have declined greatly during the last century. A study in the late 1930s (Neff 1937) stated that fears that the species was rare and declining were not borne out by their data which estimated a total of 1,500,000 nests! Presently, the Christmas bird count (CBC) and breeding bird survey (BBS) data do not show any significant population changes. The population appears to be stable. However, the restricted range of this species coupled with its need to nest in marshes makes it vulnerable to land use changes, pollution and drought. In Califonia, it is listed as a species of management concern.

REFERENCES BBS data, Beedy and Hamilton 1997 [habitat, behaviour, nesting, movements], Bent 1958 [behaviour, nesting], CBC data, Klimkiewicz and Futcher 1989 [age], Lack and Emlen 1939 [behaviour, nesting], Neff 1937 [distribution, status] 1942 [movements], Orians 1960 [nesting, autumnal breeding], 1961 [nesting], Orians and Christman 1968 [voice, behaviour], Pyle *et al.* 1987 [moult], Small 1994 [notes].

63 TAWNY-SHOULDERED BLACKBIRD *Agelaius humeralis* (Vigors) 1827

Plate 21

A small blackbird with a shoulder patch, found in open woodlands and forest edge from Cuba and Haiti.

IDENTIFICATION A small blackbird with tawny epaulets and a yellowish epaulet border. Within its limited distribution, Tawny-shouldered Blackbird is unlikely to be confused for any other species. Only Red-shouldered Blackbird (64) is at all similar within Cuba. However, Tawny-shouldered Blackbird is noticeably smaller and has a shorter bill. The main difference is the colour of the epaulets. Tawny-shouldered Blackbirds have tawny lesser coverts and paler buffy median coverts. This is quite unlike the red lesser coverts and buffy or yellowish median coverts of Red-shouldered Blackbird. Note also that pairs of Tawny-shouldered Blackbirds are not obviously dimorphic, while female Red-shouldered Blackbird lacks the red epaulets of the male. Habitat choice also differs between the two species. Breeding Tawny-shouldered Blackbirds are not restricted to marshes, as are Red-shouldered Blackbirds, and are often found in lightly wooded or urban habitats.

Tawny-shouldered Blackbird has previously occurred on the Florida Keys. In the US, Red-winged Blackbird (61) is the most similar species to Tawny-shouldered Blackbird. Once again, the epaulet colour is the main difference. Tawny-shouldered Blackbirds have tawny lesser coverts and yellowish or buff median coverts while male Red-winged Blackbirds have red lesser coverts and yellow or buff median coverts. The difference in colour between the epaulet border and the epaulet is slight on Tawny-shouldered Blackbird, but striking on Red-winged Blackbird. However, bleached or leucistic Red-winged Blackbirds may appear to show tawny epaulets and the size and structure of such a bird should be examined to positively identify it. Tawny-shouldered Blackbird is decidedly smaller than Red-winged Blackbird. Male Tawny-shouldered Blackbird is slightly smaller than female Red-winged Blackbird; in a mixed blackbird flock, this size difference should be apparent. Lone birds may cause more of a problem. In addition, Red-winged Blackbird, particularly the Florida subspecies, has a relatively longer bill than Tawny-shouldered Blackbird. Also note that the faint iridescence of Tawny-shouldered Blackbird tends towards green, while it tends toward blue on Red-winged Blackbird. Finally, the vocalizations of Tawny-shouldered Blackbird are quite unlike those of Red-winged Blackbird (see Voice).

Tawny-shouldered Blackbird is very similar to Yellow-shouldered Blackbird (65), except it has tawny or dull chestnut wing-coverts rather than yellow. Apart from colour, Tawny-shouldered Blackbird differs from Yellow-shouldered Blackbird in its bill proportions, having a flatter culmen that widens toward the base. The bill of Yellow-shouldered Blackbird is noticeably longer than that of Tawny-shouldered Blackbird and comes to a sharp tip. In addition, Tawny-shouldered Blackbird is smaller.

VOICE Song: The primary song is a muffled, buzzy drawn-out note. It may be interpreted as *whaaaaaaaaaa* or *zwaaaaaaaa*, and lasts just over one second in length. Sometimes the song is preceded by a higher pitched buzz *preeee-whaaaaaaa*. Both sexes sing and they may sing back to back in a duet, similar to that of Red-shouldered Blackbird. At colonies, a full chorus may comprise a noisy cacophony of buzzes interspersed with metallic notes, chucks and a myriad of other noises.

Calls: Gives a *chuck* note much like that of many other blackbirds, similar notes include a doubled *chup-chup*, a nasal *whaap* and a *nhyaap*. Also gives a metallic *pleeet* and *tweeep*, the latter high and piercing.

DESCRIPTION A small blackbird with a sharply pointed, conical bill which is as long as the head. The culmen and gonys are both entirely straight, without a noticeable curve. The medium tail is short and neatly square-ended. Wing formula: P9 < P8 ≈ P7 ≈ P6 > P5.

Adult Male (Definitive Basic): The bill is glossy black, and the eyes are dark. The entire plumage is black with a dull blue gloss which is not generally noticeable in the field. The lesser and median coverts (epaulets) are pale tawny; a cream colored band borders the lower edge of the epaulet. The entire underwings are black. The legs and feet are black.

Adult Female (Definitive Basic): Similar to male, but smaller and with reduced tawny on the epaulets. Usually the median coverts are black with buff tips, or buff with a noticeable black center. The body plumage has a brown wash, unlike in the male.

Immature Male (First Basic): Similar to adult male, but lacks the cream lower border to the epaulet.

Immature Female (First Basic): Similar to immature male, but smaller and with smaller epaulets.

Juvenile (Juvenal): Similar to immatures, but the tawny epaulet feathers (mainly lesser coverts) are tipped black. The median coverts are often entirely black.

GEOGRAPHIC VARIATION Two subspecies are recognized.

A. h. humeralis occurs in Cuba and Haiti, and is described above. This form has strayed to Florida. However, Wetmore and Swales (1931) state that *humeralis* from Haiti is quite variable in its epaulet colour, unlike Cuban populations.

A. h. scopulus is found on Cayo Cantiles, south of the Zapata peninsula, Cuba. It is significantly smaller with relatively shorter wings. It has a longer, thinner and sharper bill than *humeralis*. Furthermore, the epaulet is smaller in this form. It is rare and considered an endangered population.

The Haitian bird was originally described as a new species, *Agelaius quisqueyensis*. However, the noted ornithologist James Bond determined that this 'species' was in fact identical to Tawny-shouldered Blackbird of Cuba (Wetmore in Bond 1928b).

HABITAT This is not a marsh blackbird, being rather arboreal on the whole. However, it is not found in dense forest, but occupies open areas and deforested zones. It also may be found in edge habitats, open woodlands and settled areas particularly farmland with scattered trees. In winter, flocks are found in more open habitats, often farm fields. Rice paddies are a favourite place for foraging. Cayo Cantiles is a rocky island with open forests and mangroves, providing habitat for this blackbird. Curiously, the habitat on this island is most similar to that of the Guanahacabibes peninsula and the southern Isle of Pines, two areas where Tawny-shouldered Blackbird is absent (Garrido 1970). In Hispaniola, it has been seen in low mesquite forest adjoining both dry pasture and a river (Wetmore and Lincoln 1933).

BEHAVIOUR Like many other icterids, Tawny-shouldered Blackbird often occurs in large flocks, particularly in the non-breeding season. It is known to frequent rice paddies but probably causes little economic impact. It also feeds on other grains and will nectar at flowers, a habit more commonly associated with orioles (*Icterus*) than *Agelaius* blackbirds. It forages both on the ground and arboreally. This species is not territorial during the breeding season, but only defends the area immediately around the nest site. Nests may be as close as eight meters apart (Whittingham *et al.* 1996). Flicks and jerks tail in the manner of an alarmed Red-winged Blackbird.

NESTING Nests in Cuba in April–May through to July. The timing appears to be similar in Haiti, where fledged young are observed in July (Wetmore and Swales 1931). Nests are placed in a dense tree or bush, as well as in palms. The nest is a cup made of dry grasses and Spanish Moss (*Tillandsia* sp.). Clutch size is typically four, but sometimes three eggs (Whittingham *et al.* 1996). The eggs are white with a greenish-grey or ash-blue tint, and sparsely spotted with large brown spots, particularly on the large end of the egg (Gundlach 1876, Wetmore and Swales 1931). A recent study in Cuba found that 11 nests were placed either high (c. 50 m) in Eucalyptus trees on branches overhanging water, or much lower down in the tops of outdoor lamps, where the bulbs were not functioning. Nest building was by both sexes, but with the majority of work performed by the female. Both the male and female provision the young with food, at roughly similar rates (Whittingham *et al.* 1996).

DISTRIBUTION AND STATUS In Cuba, it is found in all six provinces, but is absent from the Guanahacabibes Peninsula and the Isle of Pines (Isla de la Juventud). Curiously, this species is present on several offshore cays, such as the Archipelago de los Jardines de la Reina, and Cayo Cantiles in the Archipelago de los Canarreos (Garrido 1970). In Haiti, it is restricted to the Port-de-Paix area and the mouth of the Artibonite River in the north of the country (Wetmore and Swales 1931).

MOVEMENTS Not known to migrate, but it has been recorded as a vagrant to the US. Two adult Tawny-shouldered Blackbirds were collected at the Key West lighthouse on February 27, 1936. Both had been foraging for several days with a flock of Red-winged Blackbirds (61). This was the first record for North America (Demeritt 1936). There have been two subsequent sight reports, but these were not considered conclusive by Stevenson and Anderson (1994).

MOULT Adults moult once a year, the complete definitive pre-basic moult which occurs in September, after the breeding season. The juvenal plumage is changed through the first pre-basic moult which may be complete or incomplete. It includes the body, wings and tail. If any feathers are retained from juvenal plumage, they are likeliest on the underwing-coverts. The first pre-basic moult occurs between September and October.

MEASUREMENTS ***A. h. humeralis***. Adult Males: (n=10) wing 103.7 (± 0.6); tail 80.0 (± 0.6); culmen 17.9 (± 0.1); tarsus 23.36 (±0.2). Immature Males: (n= 2) wing 102, 103; tail 78, 79; culmen 17.7, 17.7; tarsus 22.2, 22.9. Females: (n=8) wing 95.8 (± 0.9); tail 75.5 (± 0.8); culmen 16.7 (± 0.2); tarsus 21.9 (± 0.5). (Whittingham *et al.* 1996). From Garrido (1970), Males: (n=30+), wing 103.7 (100–108), culmen 16.7 (14.5–16.5).

A. h. scopolus. Adult Males: (n=5) wing 98.0 (96.5–99.5); culmen 17 (16–18); (n=1) tarsus 24. Female specimens are not known (Garrido 1970).

NOTES Appears to be most closely related to Yellow-shouldered and Red-shouldered Blackbirds (Whittingham *et al.* 1996). A comparison of the mitochondrial DNA sequences of many icterids clearly showed that Yellow-shouldered and Tawny-shouldered Blackbirds are sister species and they, in turn, are the sister group to Red-winged and Tricolored (62) Blackbirds (Lanyon 1994). Red-shouldered Blackbird was not included within this study. The South American *Agelaius* are not closely related to the Caribbean/North American group (Lanyon 1994).

REFERENCES Barbour 1943 [voice, nesting], Barnes 1945 [description], Garrido 1970 [geographic variation], 1985 [geographic variation], Gundlach 1876 [habitat, behaviour, nesting], Wetmore and Lincoln 1933 [habitat, behaviour], Wetmore and Swales 1931 [distribution, nesting, measurements], Whittingham *et al.* 1996 [description, habitat, nesting, behaviour, measurements].

64 RED-SHOULDERED BLACKBIRD *Agelaius assimilis* Lembeye 1850 — Plate 21

A marsh blackbird endemic to Cuba, it was until recently treated as a subspecies of Red-winged Blackbird (61) but it does not appear to be closely related to that species.

IDENTIFICATION Male Red-shouldered Blackbird is black with a yellow-bordered, red epaulet. Females are blackish. Males of this form and Red-winged Blackbird are probably not safely distinguishable in the field, even given Red-shouldered Blackbird's smaller size and proportionately longer legs (Whittingham *et al.* 1992). However, Red-winged Blackbirds have not been recorded in Cuba, the closest they come is the Bahamas. Given their migratory tendency it is not impossible that a Red-winged Blackbird may one day occur in Cuba. An identification would only be possible (with any degree or confidence) through their vocalizations, which are quite different (see Voice). The streaked plumage of female Red-winged Blackbird is quite unlike any plumage of Red-shouldered Blackbird however. Female Red-shouldered Blackbirds are entirely black and are not streaked.

The only sympatric species that resembles Red-shouldered Blackbird is the closely related Tawny-shouldered Blackbird (63). These two species may be found in the same flock, but Tawny-shouldered tends to be found in open woodlands and woodland edge rather than marshes or fields. Tawny-shouldered Blackbirds are smaller than Red-shouldered Blackbirds and have a petite, spike-like bill. They sport tawny epaulets with a narrow paler border, and the epaulets never show any red. However, an immature or bleached Red-shouldered Blackbird could show paler (orange) epaulets approaching those of Tawny-shouldered Blackbird. Immature Red-shouldered Blackbirds should show some black on the epaulets. Otherwise note that the pale edge of Tawny-shouldered Blackbird's epaulets is much narrower than that of Red-shouldered Blackbird. Structural differences still apply and vocalizations are quite different between the two species (see Voice).

The all-black female Red-shouldered Blackbird may be confused for any other all-black Cuban icterid such as Cuban Blackbird (84), Greater Antillean Grackle (92) and Shiny Cowbird (102). These latter three species all show some iridescence when seen well, while in good light female Red-shouldered Blackbird lacks a gloss. Note also that Red-shouldered Blackbird has a short, thick bill, and a relatively short tail. Both the grackle and Cuban Blackbird are larger, longer-billed and longer-tailed. The grackle's tail, even on females, is noticeably graduated and wedge-shaped. Adult grackles have yellow eyes. Shiny Cowbird is more similar in shape and size to Red-shouldered Blackbird, and apart from its noticeable gloss it shows a short and more finch-like bill while Red-shouldered Blackbird has a longer, more spike-like bill which may show a flattened culmen if seen closely. The blackbird spends its time in marshes during the breeding season, while the cowbird is more a bird of open woods and pastures.

VOICE **Song**: As in Red-winged Blackbird, both females and males sing; however, in Red-shouldered Blackbirds the songs of the male and female are largely indistinguishable. The songs are similar in general pattern to those of eastern Red-winged Blackbirds but Red-shouldered Blackbird songs are shorter and of a higher frequency. Pairs of Red-shouldered Blackbird usually duet, that may be started by a member of either sex. The two songs are sung back to back, with some overlap.

Calls: A *chuck* as in the typical Red-winged Blackbird.

DESCRIPTION A small to medium-sized blackbird with a short square tail, but long legs. The bill is sharply pointed, having a straight culmen and gonys. The culmen is slightly flattened. The bill length is shorter than that of the head. Wing Formula: P9 < P8 ≈ P7 (> P6; P9 ≈ P5; P8–P5 emarginate.

Adult Male (Definitive Basic): The bill, legs and feet are black. The eyes are dark brown. The body plumage and wings, excluding the epaulets, are solidly black with a dull blue gloss. On the epaulets, the lesser coverts are red while the median coverts show buff to their bases. This creates a red epaulet with a buff border.

Adult Female (Definitive Basic): Similar to male, but duller and lacking the colourful epaulets. The body plumage is entirely blackish-brown as are the wings. The crown, back and breast feathers may show indistinct brown tips. The underparts are blackish-brown and unicolored, not showing any streaks. There is no trace of red on the coverts of the female. Soft part colours as in male.

Immature Male (First Basic): Similar to adult male, but the epaulet is smaller in extent and duller in colour. Epaulets of immature males tend to be more orange than red and usually show noticeable amounts of black within them. The yellow marginal band of the epaulets, on the median coverts, is lacking in immature males. In addition, immature male is a dull black colour.

Immature Female (First Basic): Similar to adult female, but shows olivaceous feathering on the chin. In addition, immature female has pale edges to the median and greater coverts as well as the undertail-coverts. Also differs from adult female in being browner.

Juvenile (Juvenal): Similar to immatures of respective sexes but appears crisper, fresher on the remiges and has looser textured body plumage. Juvenile female lacks red or orange-red on the epaulets, while juvenile males already show the distinctive epaulet colour.

GEOGRAPHIC VARIATION Monotypic, but two subspecies were previously recognized. The nominate was found in western Cuba. The other form, *subniger*, is found on the Isle of Pines (Isla de la Juventud) and was said to differ in having a rounder culmen, rather than the flat culmen of *assimilis*. Both sexes of *subniger* were a blackish brown rather than a glossy black as in *assimilis*. A newer analysis discovered that the two specimens on which *subniger* was based were in fact first basic (immature) birds, explaining their brown colouration. In addition, the population from the Isle of Pines did not differ significantly in measurements or culmen morphology. The species is now considered monotypic (Garrido 1970).

HABITAT Breeds in wetlands, with tall emergent vegetation, particularly cattails (*Typha dominguensis*) or *Phragmites* sp. as well as arrowhead *(Sagittaria lancifolia)* (Whittingham *et al.* 1996). At one site, the species was observed along the edge of a mangrove lagoon (Garrido 1970). During the non-breeding season, forages in marshes as well as in agricultural fields and pastures.

BEHAVIOUR During the non-breeding season, this is a sociable species, roosting in large flocks, often in the same site year after year. Sometimes flocks and feeds with other blackbirds, including migrant Bobolinks (103) (Gundlach 1876). In contrast, this species is territorial during the breeding period (Whittingham *et al.* 1996). However, immature males may remain in flocks during the breeding season. The mating system is apparently monogamous (Whittingham *et al.* 1996). Both sexes aid in territorial defense. Territories vary in area from an average of 700 m^2 in a densely vegetated area to 900 m^2 in a spot with more water and less emergent vegetation (Whittingham *et al.* 1996). While singing, the male tends to perch slightly higher than the female on the reeds. Unlike Red-winged Blackbirds, the male does not spread his red epaulets when singing. The wings are kept closed, but are drooped, and the epaulets are kept covered. While performing the 'song spread' the tail is spread and the feathers of the back are raised. Only in flight, or flight displays does the male show off his red shoulders. While singing or perching, the sexes are difficult to identify! (Whittingham *et al.* 1992). In winter, seeds and rice are eaten (Gundlach 1876).

NESTING Breeding may start in April, but is usually May–July. Nests in tall reedbeds, with nests being built roughly 20 cm above the water (Whittingham *et al.* 1996). The nest is woven into the reeds or placed in a low bush within the marsh. Nests are built by the female, with material that is secured from within the breeding territory (Whittingham *et al.* 1996). The clutch size is typically 2–3 eggs (Whittingham *et al.* 1996), but sometimes 4 eggs which are white with a blue tint, marked with brown or lilac spots (Barbour 1943). Only the female incubates. Nestlings appear to be prone to overheating and females are often observed standing over them in the nest, shading the nestlings from the sun (Whittingham *et al.* 1996). Both males and females forage for and feed the nestlings at similar rates.

DISTRIBUTION AND STATUS Uncommon to fairly common but local on mainland Cuba. Restricted to the W half of Cuba, from the Guanacahabibes peninsula (La Jaula) east to the Zapata swamp. It is particularly common in the Zapata swamp which is the stronghold for this species. This blackbird is also known from the Isle of Pines (Islá de la Juventud), where it is rare. Here it may be found largely in the Ciénaga de Lanier, particularly south of La Vega (Todd 1916, Garrido 1970).

MOVEMENTS Sedentary.

MOULT Adults have one complete moult per year, the definitive pre-basic moult. Juveniles undergo a first pre-basic moult which replaces the body plumage and wing-coverts, it is not clear if this moult also involves the wings and tail. Pre-alternate moults appear to be lacking.

MEASUREMENTS Males are larger than females.

A. a. assimilis. Adult Males: (n=23) wing 110.5 (± 0.3); tail 83.9 (± 0.5); culmen 23.84 (± 0.3); tarsus 29.8 (± 0.5). Immature Males: (n=10) wing 105.2 (± 0.9); tail 80.0 (± 1.0); culmen 22.7 (± 0.3); tarsus 28.6 (± 0.4). Females: (n=9) wing 96.4 (± 0.4); tail 73.3 (± 0.6); culmen 20.7 (± 0.3); tarsus 27.37 (± 0.5). Immature Females: (n=8) wing 94.9 (± 0.5); tail 72.3 (± 0.8); culmen 20.82 (± 0.2); tarsus 27.02 (± 0.7).

A. a. subniger. Males: (n=1) wing 105; tail 79; culmen 24; tarsus 27. Females: (n=1) wing 90; tail 72; culmen 19.5; tarsus 25.5. (see Geographic Variation regarding this form)

NOTES Red-shouldered Blackbird was until quite recently considered a subspecies of Red-winged Blackbird. It differs significantly from Red-winged Blackbird in behaviour, morphology, juvenile plumage, colouration and vocalizations amongst other things. It appears to be more closely related to Tawny-shouldered and Yellow-shouldered (65) Blackbirds, rather than Red-winged. It was first described as a separate species, but was mistakenly included as a subspecies of Red-winged Blackbird, due to a mix up of specimens from Cuba and those of 'Redwings' of comparable material from the US. Thus *assimilis* was initially classified as a subspecies of Red-winged Blackbird was based on examinations of specimens from the US, not Cuba! This error was eventually corrected, and it was not until 1947 that Bond again reclassified this species as a form of the wide-ranging Red-winged Blackbird (Garrido 1970).

For relationships within *Agelaius*, see Notes under Tawny-shouldered Blackbird.

REFERENCES Bangs 1913 [geographic variation, measurements], Barbour 1943 [nesting, geographic variation], Garcia 1987 [nesting, behaviour], Garrido 1970 [geographic variation, habitat], Gundlach 1876 [behaviour], Whittingham *et al.* 1992 [vocalization, behaviour, measurements] *et al.* 1996 [nesting, behaviour, systematics, measurements].

65 YELLOW-SHOULDERED BLACKBIRD *Agelaius xanthomus* (Sclater) 1862

Plate 21

A black bird with yellow shoulders restricted to Puerto Rico and Mona Island, it is now very rare and endangered.

IDENTIFICATION Adult Yellow-shouldered Blackbird is a black icterid with a yellow epaulet. In Puerto Rico, the only potential identification pitfall is to confuse the blackbird with Black-cowled Oriole (45) which is also black and yellow. Apart from yellow shoulders, Black-cowled Oriole has a yellow rump, undertail-coverts, underwings and thighs, as well as a thin and slender bill, unlike the blackbird. The blackbird has an entirely black bill, while the oriole shows a blue-grey base to the lower mandible. The longer-tailed, slimmer body proportions and larger size of the oriole should be enough to separate it from Yellow-shouldered Blackbird. Yellow-shouldered Blackbird is most similar to Tawny-shouldered Blackbird (63) of Cuba and Haiti. The two species are extremely unlikely to ever be found together; however, given that Tawny-shouldered Blackbird has strayed to Florida it could potentially reach Puerto Rico. Apart from colouration, Yellow-shouldered Blackbird differs from Tawny-shouldered Blackbird in its bill proportion, having a narrower and rounder culmen. The bill of Yellow-shouldered Blackbird is both longer and shows more of a curve to the culmen than that of Tawny-shouldered Blackbird. The names of these two species accurately describe them; Yellow-shouldered has yellow shoulders rather than tawny.

VOICE Song: The song is a nasal rasp, sometimes descending in frequency slightly at the beginning. A song lasts between one and one and a half seconds. It can be described as a *nhyaaaaaaaaa* or *ttnyyaaa*. The song is accompanied by the 'song-spread' display and often while giving the 'wing-raise' display (Post 1981a). Both sexes sing. Sometimes a bird will sing in response to the song of a nearby bird; this need not have to be a duet between members of the pair, and sometimes it involves several birds in what could be considered to be a group chorus. Sometimes an individual will give a series of calls while taking off, the 'flight series'. These are quite variable and include *cut-zee*, *queea*, *pee-puu* and *chwip* notes described below. The 'flight series' does not seem to be homologous to a true flight song (Post 1981a).

Calls: A *check* or more nasal *chwip* are the common calls, and tend to be accompanied by a 'tail-flip', a quick movement of the tail. These notes may be given quickly, almost like a chatter. A rasp, *vvvt*, similar to the song but lower in pitch and less resonating is given during aggressive encounters. A scolding, nasal *greeah*, which resembles the alarm note of Red-eyed Vireo (*Vireo olivaceous*) is performed near the nest if the bird is disturbed. An alarm vocalisation is *cut-zee*, the first part short and sharp, and the second longer and descending in pitch. It tends to be given while mobbing predators around the nest, but may also be heard away from the breeding areas. A descending *queeaa* is sometimes repeated in a series and often accompanies the 'wing-raise' display. Like *cut-zee*, it is given while mobbing predators, and when females are pursued by males in sexual chases. A sweet *pee-puu,* the first note rising, the second falling, appears to function as a contact call between members of a pair. When the young have fledged they communicate with the adults by giving a sharp *pink* similar to the flight call of Bobolink (103), the adults will also use this note to reply to the juveniles. When exited, birds at the colony may give rapid metallic squeals (Post 1981a).

DESCRIPTION A small, slim blackbird with a straight and pointed bill. Both the culmen and gonys are straight, and both taper to a fine point. Wing Formula: P9 < P8 < P7 > P6; P9 ≈ P5; P8–P6 emarginate.

Adult Male (Definitive Basic): The bill is black and the eyes are dark. The body is entirely black in this species, with slight blue iridescence, except for the epaulet. The epaulet is yellow and includes all of the lesser and median coverts, sometimes the median coverts are white-tipped.

Adult Female (Definitive Basic): Resembles male, but slightly smaller and tending to show less extensive yellow on the shoulders. There is a great deal of overlap in this character though. In fact, sexing adults in the field is not possible, and is only possible in the hand by measurements or physiological characters such as brood patches (females) or cloacal protuberances (males).

Immature (First Basic): Similar to the adult, but duller and brownish. The lesser coverts are yellow but show blackish-brown feather centers. The median coverts are blackish with yellow tips, not solid yellow.

Juvenile (Juvenal): Similar to adults, but the yellow of the shoulders is more restricted and of a paler colour. The body colour is less black, tending to dark brownish-grey. In addition, the yellow epaulet feathers are tipped dark greyish-brown. Juveniles usually show a restricted amount of pale yellow on the chin or malar area.

GEOGRAPHIC VARIATION Two subspecies are recognized. The nominate form (described above) is restricted to Puerto Rico.

A. x. monensis is the form found on Mona Island, west of Puerto Rico. It is characterized by having paler yellow shoulder patches than *xanthomus.* Also, the median coverts are even paler, sometimes approaching white. These forms are similar in measurements.

HABITAT Older accounts mention that Yellow-shouldered Blackbird nested in a variety of habitats from mangroves, offshore mangrove cays, open forest, palms in cleared areas, suburban areas and probably forest and forest edge. It appears to have always been most common along the coast. Nesting is now largely restricted to mangroves and salt flats near dry scrub forest on Puerto Rico. Birds still nest in isolated trees in pastures, provided that these are close to mangrove habitat (Post 1981a). Dry cactus scrub and subtropical dry forest is preferred habitat on Mona Island and most of the population nests in this habitat on seaside cliffs. On Mona Island, Yellow-shouldered Blackbird is absent from pine forests (Barnes 1946). During the non-breeding season, fields and other agricultural areas, particularly livestock corrals, are important for foraging.

BEHAVIOUR Yellow-shouldered Blackbirds forage in communal feeding sites, which are often quite a distance away from breeding areas (Post 1981a). During the winter, large flocks form at these sites. During the breeding season, flocks at the foraging sites are smaller. They also roost communally, and these roosts are used throughout the year (Post 1981a). On Puerto Rico, it usually forages in the upper or mid-strata of mangrove or other woodlands, gleaning and probing epiphytes, bark and leaves usually in the outer parts (periphery) of the tree. Much of the arboreal foraging involves 'gaping' into crevices or 'probing' into holes or other difficult to reach spots. They may cling vertically, woodpecker-like, on trunks while foraging, and often hang acrobatically in the manner of orioles. Most of the food taken comprises of arthropods, particularly during the nesting season. It also nectars from various flowers, particularly in January and February when *Aloe vulgaris* blooms (Post 1981a). Fruit is taken when available; the height of the fruiting season is in the fall during the rains. Recently independent juveniles forage mainly on fruits (Post 1981a). It will also descend to feed on the ground in open areas and will forage from feed trays. Most food taken while on the ground is vegetable (rice, grain or seeds). On Mona Island, it most commonly forages low down in vegetation. Fresh water is scarce on Mona Island and absent other than as rainfall on Monito Island; here, fruits appear to be used to obtain liquids. Yellow-shouldered Blackbird has been observed 'anting', rubbing ants on their bodies presumably to rid themselves of feather parasites (Post and Browne 1982).

The breeding system in this blackbird is monogamous, there is no evidence for polygyny (Post 1981a). Sex ratios are not significantly skewed. Pairs form as the spring rains arrive, and several weeks (six or so) before nesting begins. Winter flocks begin stopping off at breeding sites after leaving the communal roosting site and before arriving at foraging areas. It appears that pair formation may take place during these nest site visits, in late March. Males display at a specific area, usually a nesting site from the previous season, and persistently follow females. The male will sing from the nest site, and sometimes peck at the old nest while singing-it appears that the nest site is chosen by the male. During this period, males also vigorously defend females from other males, and much aggressive behaviour may occur. Breeding behaviours, including 'mate guarding', do not take place away from the colony. In communal foraging areas, males cease to defend females or to sing. It is common for pairs from the previous season to re-form the next year, and to remain together during re-nesting within the season. Once nesting takes place, the male is largely responsible for defense of the nest site, until the young hatch, at which point he will begin foraging for them. Defense against conspecifics is limited to the immediate area around the nest, roughly 3 m from the nest (Post 1981a). Colony members cooperate in nest defense and will jointly mob predators or human intruders.

The 'song spread' is the most noticeable display during the breeding season, and it occurs only at the nesting site. It may begin with the bill pointed upwards, but sometimes this part is omitted; once the bird starts to lower its bill, it begins to raise the wings. As the wings are raised higher, the tail is spread and the body plumage is fluffed; eventually the beak is lowered onto the breast and the belly plumage is fully ruffled. The wings are held and rotated forward so the epaulets are in full frontal view for 2–3 seconds, and then are lowered, the head is raised and the bill is pointed upwards (Post 1981a). 'Song spreads' are always accompanied by song. A similar display, which may be a less intense version of the 'song spread', is the 'wing raise'. Unlike the 'song spread', the bill is not held up before and after the display. The wings are raised, and during the most intense form are elevated almost as high as during the 'song spread' and rotated forward; at other times the wings may only be raised a small amount. The body plumage and tail are not so fully ruffled or spread. 'Wing raises' may occur in quick succession and tend to be accompanied by the song or by *queea*, *pee-puu* or *cut-zee* calls (Post 1981a). The 'bill-tilt', where a bird points the bill upwards is a general icterid aggressive display that Yellow-shouldered Blackbird also performs. In addition, Yellow-shouldered Blackbird may give a 'bill-down' display, which also appears to function as an aggressive signal (Post 1981a). A 'wing-flutter' is used by females begging for food from mates. It also may be used as a pre-copulatory display, if the tail is raised. Males may respond with a 'head-in' display, where they crouch and retract the head towards the body. The 'wing-trail' display is given by parents when there is a disturbance near a nest with young. The bird will walk slowly, with the wings lowered and trailing such that they touch the ground. Several different vocalisations may accompany this display. An odd moth-like flight, with slow and shallow wingbeats, is performed by females around the nest. Rather than being a sexual display, this appears to be a type of distraction display to lead predators away from the nest (Post 1981a). During an aggressive situation, a bird may give a 'head-forward' display, where the head is extended forward toward an opponent with the bill open. The 'head-in' display, described above, may also be performed during aggressive displays. Full fighting does occur, and birds will jab at each other with their claws and beat their wings on each other. When excited and about to fly, Yellow-shouldered Blackbird give a tell-tale 'sleeked' plumage posture. Finally, during pair formation, 'sexual chases' are sometimes observed. This is when several males chase a female, often to the ground (Post 1981a).

NESTING The nesting season is variable, depending on the rainy season. The formation of pair bonds is timed to coincide with the onset of the spring rains, which are not as heavy as the fall rains (Post 1981a). On Puerto Rico, nesting usually occurs April–September, while on Mona Island it may begin as early as February and continue as late as November. On average, the east Puerto Rico population nests earlier than the southwest population, usually by a month. Nestlings are in the nest during the dry summer period, but as they fledge and become independent they are able to take advantage of the fruiting and flowering period brought on by the fall rains (Post 1981a). Pairs may nest up to three times during one season, but once or twice is more common.

Nesting tends to be colonial, with several pairs nesting in trees in close proximity to each other. At times more than one pair may nest in the same tree (Post 1981a). Yellow-shouldered Blackbirds may nest in cavities or in open cup nests. Cavity nests appear to be preferred, but cavities are not widely available. Cavity nesting is most common in dead trees bordering salt pans. Two types of cavities are used, either holes in dead trees or holes at the top of a stump or a broken-off dead tree (Post 1981a). Open cup nests are most common in suburban habitats, live mangroves and scattered trees in pastures. In pastures, Black Olive (*Bucida buceras*) trees are used as a nesting substrate, particularly on the main branches or in a fork or crotch, nests ranging from 4.6–7.6 m in height (Post 1981a). In small cays (offshore islands in shallow water), the blackbirds nest only on islands which are free from introduced rats (*Rattus* sp.) and those that are covered mainly by Red Mangrove (*Rhizophora mangle*). In these situations, Yellow-shouldered Blackbirds place their open cup nests on the main branches or in a crotch of the tree, at an average of 2.14 m above the water (Post 1981a). In a suburban situation, Royal Palm (*Roystonea borinquena*) was used as a nest tree, with open cup nests being woven under the midrib of the palm fronds; these nests were placed highest from 10.7–19.8 m above the ground (Post 1981a).

The open cup nests are typical for a blackbird, being a woven structure of grass and stems, placed on a platform of leaves, grass or garbage such as paper, string or plastic bags. The lining is of a fine grass. Cavity nests tend to be built on top of old blackbird nests, and are also made and lined with grass. The cay nests were larger and bulkier, than mainland nests, particularly since sargassum was used to make the nests. In some cases, nest material appears to be in short supply and females may rob each other of material (Post 1981a). As a management technique to attempt to increase fledging rates, nestboxes have been put up for the blackbirds; currently at some sites most Yellow-shouldered Blackbirds use nestboxes. These nestboxes have been successful in increasing survivorship of nests as they are not predated as commonly as open nests. On Mona Island, most nests are placed on cliffs; however, earlier reports mention that tall cacti were the preferred nesting sites (Barnes 1946). Nests in offshore cays have higher success than mainland nests (Post

1981a). In addition, cavity nests are up to three times more successful than open cup nests in the same habitat (Post 1981a).

Clutch size 1–4, but averages 3.03 eggs per nest (Post 1981a). The eggs are light blue with purplish-brown spots, more densely concentrated at the broader end. Incubation begins after the second egg is laid and thus the hatching of the young is not synchronous. Incubation lasts 12–13 days and is performed strictly by the female (Post 1981a). During the incubation period, males leave the nesting area at night to roost at their communal roosts. Once the nestlings do not need to be brooded, the female will join the male at the roost (Post 1981a). Both parents feed the young at equal rates. Food is brought in the bill, but adults also regurgitate food from their crops. The latter behaviour has likely been selected for, due to the long distances between the nest and foraging sites. The young need an average of 14.6 days to fledge, ranging from 13–16 days (Post 1981a). After fledging, the young remain in the vicinity of the nest for several days, while being fed by the adults, and they may not leave the nesting area altogether until several weeks later (Post 1981a).

Shiny Cowbird (102) which has recently arrived in Puerto Rico takes a toll on this bird. The cowbird is probably largely responsible for the recent population declines of this blackbird, even though nest predation by rats and Pearly-eyed Thrashers (*Margarops fuscatus*) also has an effect. Overall habitat loss may have initially made this species vulnerable to cowbird parasitism. In parts of Puerto Rico, Shiny Cowbird uses the Yellow-shouldered Blackbird as its major host and 80% of blackbird nests are parasitised. Nests on offshore cays are not as highly parasitised as those on the mainland (Post 1981a). Parasitised nests fledge half as many young as do non-parasitised nests, the loss in productivity is caused by pecking and removal of eggs of the host by the cowbirds. Yellow-shouldered Blackbird nests may contain as many as six cowbird eggs (Post 1981a). The host young are larger than those of the cowbird and successfully compete with the parasite while in the nest. Cowbird control programs in some of the breeding areas have helped to increase the fledging rate of blackbirds. Removal of males or removal of all cowbirds are effective in increasing blackbird success. In experiments, parasitism rates decreased from 92% to as low as 30% through cowbird removal. Cowbird control has allowed nest success to increase from 35% to 71% (Wiley *et al.* 1991). Birds on Mona Island and on offshore cays experience low levels of parasitism, either due to the inaccessibility of cliff nests or due to their distance from the shore.

DISTRIBUTION AND STATUS Endemic to Puerto Rico. Once abundant, now rare and endangered. Historically it ranged throughout Puerto Rico and nearby Mona Island, including Monito Island. Now it is restricted to Mona Island, extreme SW Puerto Rico and near Ceiba in easternmost Puerto Rico. The Boquerón Commonwealth Forest is one of its strongholds on the mainland and is a good place to observe this species. See Notes for details on the disappearance of some populations.

MOVEMENTS Sedentary, the same communal roost sites are used throughout the year by the same birds. However, they range widely as they use separate areas for nesting, foraging and roosting; these may often be several km from each other. There are also reports of dispersal to Monito Island from Mona, roughly a 5 km journey.

MOULT Adults undergo a complete definitive pre-basic moult early August–late December; pre-alternate moults are missing (Post and Browne 1982). Moult is at its heaviest October–November (Post 1981a). It appears that some individuals begin their definitive pre-basic moult while still nesting (Post 1981a); this overlap between breeding and moult is seen in other tropical species but not in temperate ones. The moult period is timed with the arrival of the fall rains, which provide an increase in food supplies. The juvenal plumage is lost through the first pre-basic moult, which occurs in the autumn of the year of hatching, its timing being similar to that of the adults. Since some adults may begin moult before the young have fledged, it is likely that on average the first pre-basic moult occurs slightly later than the definitive pre-basic moult. The extent of the first pre-basic moult is incomplete, replacing the body, wings and tail but retaining scattered juvenal feathers, particularly on the wing-linings. A specimen of *monensis* was in first pre-basic moult during late August, suggesting that this race's moult timing is similar to that of birds from Puerto Rico.

MEASUREMENTS ***A. x. xanthomus***. Males: (n=10) wing 104.0 (91–108); tail 86.0 (78–93); culmen 20.5 (19–23); tarsus 24.1 (22–26). Females: (n=10) wing 95.6 (93–97.3); tail 78.5 (74.2–82); culmen 19.9 (19.0–20.3); tarsus 23.9 (23.0–25).

A. x. monensis. Males: (n=10) wing 103.6 (96–106.4); tail 81.1 (77.0–84.4); culmen 21.4 (20–22.8); tarsus 25.9 (23.7–28.3). Females: (n=6) wing 96.6 (95.5–99); tail 76.4 (73.4–85); culmen 20.0 (19.7–21); tarsus 24.2 (23–24.9).

NOTES This is now listed as an endangered species in the US. In 1976, the population was estimated to be 2400 birds, currently it is estimated at 1250 individuals, while as late as the 1940s this species was still widespread and common. The eastern population, that near Ceiba, is dangerously close to extirpation. In the 1970s an estimate of 200 birds was made from this area, while in 1986 a total of only 31 was found. The interior breeding site of San Germán which numbered a dozen pairs in 1975 appeared to be extirpated in 1982. On Mona Island, the population is stable with a population of up to 900 individuals. A conservation program instigated in 1980 has been relatively successful in decreasing the proportion of nests parasitised by Shiny Cowbird (8% now, 87% in the past), and increasing fledging rates (2.3 vs. 0.3 blackbirds per nest). Habitat destruction is still a major concern and may be thwarting an increase in the population of this blackbird. For relationships within the genus, see Notes under Tawny-shouldered Blackbird.

REFERENCES Barnes 1945 [description, geographic variation, measurements], 1946 [habitat, nesting], Collar *et al.* 1992 [behaviour, nesting, notes], Hernández-Prieto and González 1992 [behaviour, habitat, distribution], Post 1981a [voice, habitat, behaviour, nesting, movements, moult], Post and Browne 1982 [behaviour, moult], Post and Wiley 1976 [distribution, nesting, notes], 1977a [nesting], Raffaele 1989 [voice], Ridgway 1902 [measurements], Wiley *et al.* 1991 [nesting, conservation, parasitism], Wiley and Cruz 1991 [conservation, notes].

66 RED-BREASTED BLACKBIRD *Sturnella militaris* (Linnaeus) 1758 — Plate 25

The common red-breasted meadowlark of Central America and most of northern South America. Often a conspicuous species of cleared areas.

IDENTIFICATION The primary identification challenge is to separate this species from its close relative, White-browed Blackbird (67). The two species may be found together in NW Bolivia and SE Peru; it is unknown if they hybridize. Males are similar in all respects, except that Red-breasted Blackbird lacks the white supercilium of White-browed Blackbird; this supercilium is visible from a distance. In general, Red-breasted Blackbirds are slightly larger than White-browed Blackbirds. Females are much more difficult to identify, and most are not identifiable in the field. Red-breasted Blackbird has a longer bill than White-browed Blackbird (Figure 66.1) and is slightly smaller and shorter-winged, otherwise they are similar. It appears that female Red-breasted Blackbirds may average brighter red on the underparts, while White-browed averages more heavily streaked on the breast and vent. The reliability of these characteristics needs to be accurately assessed in the field. Juveniles of these two species are not safely separable. The songs and calls of both species appear to differ, but it is not known if these differences are obvious in the area of sympatry. Red-breasted Blackbird has not been discovered west of the Andes in Ecuador, so it is unlikely to be found within the range of Peruvian Meadowlarks (68) unless the latter species spreads northward to the Colombian Pacific coast. However, Peruvian Meadowlark is a noticeably larger species with more streaking on the upperparts and has white wing-linings. Also, Peruvian Meadowlark has a bill as long as the head, while Red-breasted Blackbird's bill is shorter than the head length and more conical. Finally, females and juveniles may be confused with non-breeding Bobolink (103). The pointed tail feathers of Bobolink are diagnostic, but are seldom seen in the field. Otherwise, Bobolink differs by lacking red or pink anywhere on the body, by having streaking on the underparts restricted to the flanks, by its bold white central crown stripe, its short, pinkish bill and by its *boink* flight call.

VOICE Song: The most noticeable part of the song is the drawn-out buzz that terminates it. The most variable part of the song, however, is the notes given previous to the buzz. This introductory sequence tends to be composed of three to six (probably more in some cases) notes which are high and metallic; the entire series may last 1–1.5 seconds, often longer than the terminal buzz. The terminal buzz, tends to be given on an even frequency, however a few songs may have an initial descending portion before the buzz levels out. The song may be described as *ti-ti-pee-pee-*

Red-breasted Blackbird

White-browed Blackbird

Figure 66.1 Heads of female Red-breasted (left) and White-browed Blackbirds.

KWWAAAAHHH, pip-ti-ti-tewdle-WHAAAAAA, pee-too-teee-twoo-CHAAAAA. Published descriptions include the following: in Brazil it is a drawn out *tsi-li-li-EE* (Sick 1993); in Valle, Colombia, a *chert-zeeeeee-e-e-e* (Hilty and Brown 1986). However, in Meta, Colombia, there are two song types. The first is similar to the ones already described, a *zit-zit—toweeezzzzz* while the other is a two part *keet-dear* (Hilty and Brown 1986). The song is often delivered on the wing as the bird parachutes down to the ground with epaulets flared. In Colombia, the song is not usually given while in flight (Hilty and Brown 1986).

Calls: These have been described as an *ee* which are the most commonly rendered call (Sick 1993). The alarm call is a *pist*. There are several other calls whose purpose is not fully known, these include the following: *chit-chit, cha*, and a *kwak* (Sick 1993). Also common is a chatter, quite similar to that of a female Shiny (102) or Brown-headed (101) Cowbird.

DESCRIPTION A small, stocky meadowlark with a short, almost conical bill and short tail. Wing Formula: P9 < P8 ≈ P7 ≈ P6 > P5; P9 ≈ P7 ≈ P6; P8–P5 emarginate.

Adult male breeding (Definitive Basic, worn): The bill is black with a grey base, largely on the lower mandible. Head blackish-brown, unstreaked. The eyes are blackish. The upperparts, including the back, scapulars and rump are blackish and unstreaked. The wings are similarly dark, without banding on the feathers, except for the red marginal coverts which appear as a red patch at the bend of the folded wing. The wing-linings are blackish. The underparts, including the throat, breast and upper belly are salmon-red. The red on the throat is 'pinched in' by the extent of dark on the sides of the face. The rest of the underparts, largely the undertail-coverts and the lower belly are blackish. The tail is blackish without any white or buff. The legs are blackish.

Adult Male Non-Breeding (Definitive Basic, fresh): This is the fresh plumage and shows more patterning than worn, breeding adult. It often shows a subdued buff supercilium, while the face remains largely black. The back and scapulars have olive-grey fringes. The tertials have olive-brown bars at the extreme tip and are fringed pale buff. The wing-coverts are boldly fringed buff. The secondaries and primaries are unbarred, but do show thin pale fringes. The rump feathers are tipped olive as are the undertail-coverts. The blackish tail feathers have thin brown tips.

Adult Female Breeding (Definitive Basic, worn): Much more patterned than male, overall rather like a non-breeding Bobolink. The crown is blackish-brown with faint brown streaks and a warm brown median stripe. Most obvious is the pale buff supercilium on the dark face; it is bordered below by a dark eyeline starting from the eye. The lores and cheeks are buffy-brown. The nape is blackish with warm brown edges, creating a brown-streaked appearance. The back and scapulars are similarly coloured, but the brown markings form a complete fringe, making the back more scaled than streaked-looking. The dark rump has brownish tips. As in the male, the wings have red marginal coverts. The overall ground colour of the wings is dark brown. The lesser and median coverts are tipped buff-brown. The greater coverts are similar, but with complete buff-brown fringes as well as faint darker barring. The tertials are blackish-brown with thin brownish fringes and indistinct darker barring. The primaries and secondaries are unbarred and with a pale buff leading edge. The underwings are dark as in the male. The underparts are totally unlike the male. The throat is buff, with a reddish wash. There are faint brown streaks on the upper breast. The lower breast and upper belly are reddish, but not as vivid as on the male; also, the colour is obscured by buffy tips. The flanks are buff with black streaks, the thighs brownish. The undertail-coverts are brown with thin black barring. The blackish tail has olive bars on all of the feathers.

Adult Female Non-Breeding (Definitive Basic, fresh): More boldly streaked and patterned than breeding female, as this is the fresh plumage. **Immature Male (First Basic)**: largely indistinguishable from the female on the upperparts, showing broad brown or buff feather edges and an obvious buff supercilium. There is a buff median crown stripe, bordered by cinnamon-brown lateral head stripes. On the underparts, the red is almost as extensive as on adult male, except the feathers are tipped buff, being darkest on the breast. The flanks and belly are widely edged brown, hiding the underlying black colour.

Juveniles (Juvenals): Similar to adult female, but lacks red on the underparts and the breast is streaked.

GEOGRAPHIC VARIATION Considered monotypic. Birds from Panama and Trinidad are smaller than individuals from elsewhere. It has also been noted that the population from lower Amazonia, east of the Rio Negro north of the Amazon, and east of the Rio Madeira south of it, are larger and may deserve recognition as a different subspecies, *S.m.erythrothorax* (Gyldenstolpe 1951). Rather than separating this form as a different subspecies it seems more prudent to recognize that there is noticeable size variation in this species and that it may be clinal.

HABITAT Like the other meadowlarks, this is a bird of open country. It inhabits moist grasslands, moist pastures and agricultural fields including rice paddies; however it will use much drier sites. It is probably most commonly found in cultivated areas, as these are some of the most abundant open areas in the neotropics. The presence of bushes does not seem to be a prerequisite for this bird, but it likes to use fence posts. Recently cleared tropical forests are readily accepted by this species if these areas are open enough, savannas and meadows are also used even when grazed by cattle. Scrub vegetation, given that it is not too dense is also frequented, particularly when moist or wet.

BEHAVIOUR Most commonly observed foraging on the ground or perching on posts or even telephone wires. They feed largely on insects and seed, but may take berries from low bushes (Skutch 1996). In the winter, flocks are commonly seen, which may number up to 100 birds (Wetmore *et al.* 1984). During the breeding season, Red-breasted Blackbird is often seen performing its flight song. The display may start from the ground or from a perch; the male ascends to approximately 10 m, then parachutes down on half-open wings, singing with the red chest puffed out. The red epaulets of the male are flared as he drops. Males also commonly sing from a fence post or small bush. The mating system appears to be polygynous.

NESTING Breeding occurs January–April in the lowlands of the Magdalena valley in Colombia, but is later (June) higher up in the valley. In the upper Cauca valley, Colombia, birds were in breeding condition during October (Miller 1960). In Trinidad, nests March–November but May is the peak of the season. In Surinam, the nesting season stretches from January–November (Haverschmidt 1968). In Costa Rica, Red-breasted Blackbird nests during May (Kiff 1975). Panamanian birds nest June–September (Wetmore *et al.* 1984). Nesting is apparently colonial in W Colombia. The nest has been described as an open basket, and is placed on the ground. The nests are typically covered by overhanging grass and there is an entrance tunnel created by the birds as they push their bodies through the dry grass. Clutch 2–4 eggs, average 3, are cream-coloured, with reddish-brown blotching, densest at the wide end of the egg. Eggs from Surinam are pale blue with dense reddish spots and blotches (Haverschmidt 1968). This species is frequently parasitised by Shiny Cowbird (102).

DISTRIBUTION AND STATUS Uncommon to common in most of its range, but can be local, particularly in areas where forest clearing is just beginning. Found to 1740 m in Colombia (Lehmann 1960). Reaches the northernmost extension of its range in the Golfo Dulce region (Punta Arenas province) of SW Costa Rica, regularly to Palmar, rarely to Volcán (Kiff 1975, Stiles and Skutch 1989); it is a recent arrival to the country and continues to spread northward. In Panama, it is a resident along the Pacific slope from Chiriqui east to Darién. Recently it has invaded the Caribbean slope in Colon (Ridgely and Gwynne 1989). The distribution in Colombia is fragmented due to the Andes. It is found west of the West Andes from the lower Atrato valley (Antioquia) south to Cauca, wrapping around north of the Andes to the base of the Santa Marta mountains (Magdalena) and south along the Cauca valley to Valle and the Magdalena valley to Huila (Hilty and Brown 1986). Present in E Guajira and east of the Andes from Boyaca south to Meta in Llanos habitat and forest openings, it is absent from eastern Colombia (Amazonas) where forest cover is still extensive but it will likely invade here as soon as larger clearings are created. In E Ecuador occurs in cleared areas of Napo province, where it is a recent arrival. Found throughout Venezuela, except extreme S Amazonas and the Tepui Highlands of S Bolívar. Breeds in Trinidad, but it is absent from Tobago. It is present throughout Guyana, Surinam and French Guiana. In Brazil, it is found north of the Amazon from Roraima and E Amazonas east to the delta, and south of the Amazon from coastal Maranhão west through S Para and S Amazonas to Acre. Continues west to NE Peru in Loreto and also found in extreme N Bolivia in Pando.

MOVEMENTS Apparently partially migratory, but there is little data. This is inferred from the fact that the species may disappear for the winter from some southern sites. However, in the far north of its range in Costa Rica, birds do move away from nesting areas during the non-breeding season (Kiff 1975); however, it is not known if these are true migrations or merely local movements. A noteworthy increase in the range of this species has occurred due to the large amounts of open habitat created from forest clearing. Range expansion has been noted in Colombia, Ecuador, Brazil and Bolivia. In Colombia, it was first noted in the upper Cauca valley (Cali) in the spring of 1956 (Lehmann 1957) and was reported to be breeding in the area in 1958 (Miller 1960). Noticed to be invading the southeast of Colombia late in the 1950s (Lehmann 1960).

MOULT The definitive pre-basic moult of adults is complete and is the only moult of the year. All plumage changes between the non-breeding and breeding season occur through wear. The pre-basic moult occurs after the breeding season. In E. Panama, the moult takes place during January–February. In Guiana, moult occurs in September. Pre-alternate moults are lacking. The first pre-basic moult is complete or nearly so. In Bolívar, Venezuela, the first pre-basic moult occurs in March.

MEASUREMENTS Males are noticeably larger than females. Males: (n=18) Wing 95.0 (88.9–99.1); tail 55.4 (48.8–67.8); culmen 20.3 (19.8–22.9); tarsus 30.5 (27.9–32.8). Females: (n=10) wing 85.0 (81–89.9); tail 56.4 (51.0–61.0); culmen 18.8 (17.8–20); (n=9) tarsus 28.2 (25–31).

NOTES Sometimes considered conspecific with White-browed Blackbird (i.e. Wetmore *et al.* 1984). The meadowlarks were previously classified into three genera (*Sturnella, Pezites* and *Leistes*). This species was then called *Leistes militaris* which was confusingly similar to Long-tailed Meadowlark's (70) name, *Pezites militaris*. Also known as Red-breasted Meadowlark, Military Blackbird and Northern Marsh Meadowlark.

REFERENCES Belcher and Smooker 1937 [nesting], Rodriguez 1982 [nesting], Gyledenstolpe 1951 [geographic variation, measurements], Hilty and Brown 1986 [voice, nesting], Parker and Remsen 1987 [distribution, movements], Ridgway 1902 [measurements], Sick 1993 [voice], Short 1975 [identification].

67 WHITE-BROWED BLACKBIRD *Sturnella superciliaris* (Bonaparte) 1851 — Plate 25

White-browed Blackbird is the common small meadowlark of southern South America. The beautiful, butterfly-like, aerial display is an exciting sight to behold in the Pampas.

IDENTIFICATION The main identification problem is separating this species from the similar Red-breasted Blackbird (66), see that species for discussion. In the southern part of the range, it comes into contact with two other meadowlarks, Long-tailed (70) and Pampas (69). Both of these species are larger and have noticeably longer bills than White-browed Blackbird. The Long-tailed differs further in having a noticeably long tail, white wing-linings in flight, and more brown edging to the upperparts, making it look much more streaked than White-browed Blackbird. Pampas Meadowlark is more similarly coloured to the White-browed Blackbird, but can be differentiated by the already mentioned size and shape differences. Female Pampas Meadowlark has a blackish belly which separates it from female White-browed Blackbird which has a brownish belly. Females and juveniles may be confused with non-breeding or female Bobolink (103). Bobolinks are buff or cream on the chest, never showing traces of pink or red. On Bobolinks, streaking is restricted to some bold streaks on the flanks, unlike in the blackbird. Finally, Bobolink has a crisp, noticeable, white, central crown stripe and diagnostic pointed tail feathers that the blackbird lacks.

VOICE Song: The song is given in two parts, one is a drawn-out buzz and the other is a variable series of stuttering single notes. The most noticeable part of the song is the buzz, which begins the song. The terminal series of notes tend to be sharp chucks. The buzz lasts for one second, but the terminal series of notes may last for over two seconds, often less. Typically, there is also a sharp *teee* note that finishes the song and precedes the terminal series; perhaps this should be considered a separate, intermediate element. Thus a song sounds like *TCHEEE-teee-chuk-chuck-chak-chak* or *TZZZZZZZZ-teee-chu-chu-chak-chak*. This song is given in flight as part of the song flight display. The buzz is given just as the bird reaches the top of its ascent in its flight, while the stuttering terminal notes are given as the bird drops back to earth. Songs of White-browned Blackbird differ from those of Red-breasted Blackbird in that they have a terminal series of single notes, not an introductory series of notes as in the latter species. Also, the buzz of White-browed Blackbird has the energy focused on a higher frequency, therefore it sounds higher pitched.

Calls: Include a harsh *pshee* as well as other squeaky notes. The typical call is a low *chuck*.

DESCRIPTION A chunky, small, short-billed meadowlark. The bill is short with the culmen showing a small amount of curvature. Wing Formula: P9 < ≈ P8 ≈ P7 > P6; P9 ≈ P8; P8–P6 emarginate.

Adult Male Breeding (Definitive Basic, worn): The bill is black with a greyish base, not as long as the head. The eyes are dark brown. Head black with a bold white supercilium starting at the eye and reaching to the nape, the supraloral area is black. Back, rump and upppertail-coverts black. The wings are blackish with red marginal coverts forming a red bend of the wing. The coverts and tertials may be thinly fringed with brown. The wing-linings are black. The chin, throat and breast is a beautiful crimson colour. The rest of the underparts are black, sometimes with some pale fringes on the undertail-coverts. The short tail is blackish. The legs and feet are blackish-grey.

Adult Male Non-Breeding (Definitive Basic, fresh): Resembles breeding male, but with more extensive pale feather edging particularly on the back, wings and flanks. At all times, the white supercilium is obvious.

Adult Female Breeding (Definitive Basic): A small streaky brown bird. The soft part colours are similar to those of the male, the bill perhaps more horn rather than grey. Crown, nape, back and rump brown with darker streaking, and a paler central crown stripe is often visible. There is a noticeable pale supercilium as well as a darker eyeline. The ear-coverts are brown and finely streaked, similar to the nape. The chin and throat are off-white, becoming buffier further down. A pale pink wash is present on the lower breast. The rest of the underparts are buffy, becoming brownish on the vent and they are streaked from the breast sides and flanks to the vent and crissum. The wings appear brownish at a distance, but on close views can be seen to be intricately patterned. The feathers are brown with paler buff-brown notching, and a fine buff-brown fringe to the feather is also present. The pale notches do not reach the mid-line (rachis) of the feather. The overall pattern is most pronounced on the tertials, secondaries and greater coverts, and it creates a weakly barred look. The underwings are buffy. The tail is brown with paler barring created by pale buffy-brown notches on the feathers.

Adult Female Non-Breeding (Definitive Basic, fresh): Similar to breeding female, but body plumage even more widely edged buff and the pink breast patch is always obscured by brownish feathers.

Juveniles (Juvenal): Similar to female, but lacks red on the breast. Slightly more streaked below than adult females.

GEOGRAPHIC VARIATION Monotypic.

HABITAT Lives in open, grassy areas, takes well to agricultural fields. It tends to prefer wetter fields to dry ones. Like Red-breasted Blackbird it does not require bushes in the area. It has an affinity for livestock, and will forage about them in flocks, reminiscent of European Starlings (*Sturnus vulgaris*). The available habitat for this species has grown immensely due to forest clearance in the northern part of its range.

BEHAVIOUR Flocks disband in the spring (August–September) and individuals will seek territories to defend. Males sing in parachute display flight down from a moderate height (20 m) over their territory. They may also sing from the ground. It is also typical to observe males surveying their territory from an open patch of ground or from a telephone pole or fence post. Foraging takes place away from the breeding territory. Females are much more difficult to observe and keep close to the ground. Both sexes, but particularly males, nervously flick their tails when alarmed. Chases are often observed in the breeding territories. In the south of their range, they maintain non-overlapping territories with Pampas Meadowlarks, defending their territories against this related species. Long-tailed Meadowlark, however, is tolerated and its territory may overlap that of White-browed Blackbird. In winter, the apparent abundance of this species changes radically as birds congregate into flocks concentrating them into smaller home areas. Winter flocks can be substantial in size.

NESTING Nests October–January. The nest is placed on the ground concealed by grass, including some draped over the nest itself such that the eggs are not visible from above. Clutch commonly 3–5. The eggs are greenish with soft ochre markings distributed on the surface of the egg. This species is commonly parasitised by Shiny Cowbird (102). Pereyra (1933) reports finding a nest of this species with one blackbird egg and 19 cowbird eggs and another with two blackbird eggs and 12 cowbird eggs! Details of the incubation and nestling period are not known.

DISTRIBUTION AND STATUS Common to uncommon in most of its range. Found to 4000 m in Bolivia (Gyldenstolpe 1945b). Found in two allopatric populations, one of which is restricted to the Atlantic lowlands of NE Brazil, from Ceará south to São Paulo. The main distribution includes the northernmost records (non-breeding) from SE Peru (Madre de Dios). Ranges throughout the E Lowlands (except Pando) and up to the subtropical zone in the mountains. It is listed as absent from the Altiplano (S La Paz, Oruro and Potosi) (Remsen and Traylor 1989), but has been recorded at 3000 m in Cochabamba and 4000 m in La Paz (Gyldenstolpe 1945b), perhaps these specimens refer to stray individuals or migrants. In W Brazil it ranges as far north as Rondonia, C Mato Grosso and S Goías, also in Rio Grande do Sul. It is distributed throughout Paraguay and Uruguay. In Argentina throughout the northern half of the country, as far west as Jujuy, SW Salta, Tucumán, E La Rioja, E San Juan, C Mendoza, SW La Pampa and NE Rio Negro which is also the southern limit of the distribution.

MOVEMENTS White-browed Blackbird is apparently migratory with birds recorded in SE Peru being migrants from further south, as breeding does not take place there. However, at least some of the northern populations in Bolivia appear to be resident. At breeding localities, numbers fluctuate from year to year suggesting that shifts within the breeding range do occur. In addition, it is largely a breeding visitor to Buenos Aires being rarer in the winter (Narosky 1985, Narosky and Di Giacomo 1993). There is one record of a stray, from Baños del Toro, Coquimbo, Chile (Johnson and Goodall 1965).

MOULT Adults have one complete moult a year, the definitive pre-basic which occurs after the breeding season. Breeding plumage is obtained through wear and fading as pre-alternate moults are lacking. In Argentina, this species moults February–March; it is unknown if most moult in the breeding or wintering grounds. The first pre-basic moult replaces the juvenal plumage, and it is also a complete or nearly complete (incomplete) moult. Its timing appears to be similar to that of adults.

MEASUREMENTS Males: (n=23) wing (95–107); tail (11.2–14.1); culmen (18.0–20.4); tarsus (29.2–32.4). Females: (n=4) wing 90.5 (89–91); tail 55.3 (54–57); culmen 20.0 (18–19); tarsus 28.5 (27–31).

NOTES Sometimes considered conspecific with Red-breasted Blackbird. Also known as White-browed Meadowlark and Southern Marsh Meadowlark.

REFERENCES Belton 1985 [voice, habitat, behaviour], De la Peña 1987 [nesting], Gochfeld 1979 [behaviour], Gyldenstolpe 1945b [distribution, measurements], Short 1968 [measurements], Sick 1993 [voice], Wetmore 1926 [voice].

68 PERUVIAN MEADOWLARK *Sturnella bellicosa* Plate 26
Filippi 1847

The only meadowlark along the arid Pacific coast of South America, its brilliant plumage adds a much needed splash of colour to some of the bleak landscapes it inhabits.

IDENTIFICATION This is the only meadowlark within its range and thus it does not pose any identification challenges. However, in the north of its range it approaches the range of Red-breasted Blackbird (66) and in the south it comes close to the range of Long-tailed Meadowlark (70). This is a smaller, shorter-tailed bird than Long-tailed Meadowlark, with a proportionately shorter bill but thicker legs. On the other hand, it is larger and longer-billed than Red-breasted Blackbird with a less solidly black body, including a pale supercilium which the blackbird lacks. The bill of Red-breasted Blackbird is noticeably shorter than the head, while it is equal to the head length in Peruvian Meadowlark. The overall plumage is more solidly black than on Long-tailed, and it has white rather than brownish thighs, in both sexes. Male Red-breasted Blackbirds have black thighs. The red of the underparts is more intense than on Long-tailed Meadowlark. In many respects, Peruvian Meadowlark is like Pampas Meadowlark but unlike both it and Red-breasted Blackbird, it shares the white wing-linings of Long-tailed Meadowlark. Differences in song between Peruvian and Long-tailed Meadowlarks are outlined below.

VOICE Songs are quite variable in this species, and each male may sing several song types.

Song: The song is similar to that of Long-tailed Meadowlark in its structure, being composed of a set of introductory notes and then a drawn-out, nasal buzz. In Peruvian Meadowlark, the initial notes tend to be short, sweet whistles, sometimes frequency modulated (slurred) and often resembling the song of North American races of Eastern Meadowlark (71). Usually 4–8 whistles are given, sometimes more. The terminal buzz spans a wide range of frequencies (2–10 kHz), with most of the energy concentrated in the 4-6 kHz range. The buzz is just under half a second in length and does not show a decline in frequency (down slur). The harsh song can be described as a shrill *tee-sweeaa-ko TRRRRRRR, sweeako-tit-tk-TRRRRRRRR* or *piu-piu-CHRRRRR* for example. The flight song is longer and more varied in pattern. Compared to Long-tailed Meadowlark, the primary song averages differences in several characteristics. The final buzz is shorter and higher in frequency (more shrill) in Peruvian Meadowlark than in Long-tailed Meadowlark. Usually, the buzz of Long-tailed Meadowlark is slurred, as it declines in frequency towards the end, which is not the case in Peruvian Meadowlark. Finally, the introductory notes tend to be sweeter and more pleasant in Peruvian Meadowlark, and probably average less in number.

Calls: A buzzing *tzp* or a *chup*.

DESCRIPTION A medium-sized meadowlark with a short tail, and thick, strong legs. Wing Formula: P9 < P8 ≈ P7 ≈ P6 > P5; P9 ≈ P5; P8–P6 emarginate.

Adult Male Breeding (Definitive Basic, worn): The bill is greyish with a black culmen and tip, roughly equal in length to the length of the head. The head is blackish with a bold supercilium, red in front of the eye and white posterior to it. The eye is dark brown. Some faint brownish streaking may be present on the crown. The nape and back are blackish with individual feathers narrowly edged brown, creating a brown streaked appearance continuing to the rump. The uppertail-coverts are greyish with darker barring. The scapulars are like the back, but often with paler grey-brown notches near the tip. The tertials show the same grey-brown notching as do the scapulars. The coverts are blackish with grey-brown notches near the tip and a pale grey fringe around each feather. This fringing and notching is most extensive on the greater coverts. The primaries and secondaries are blackish with a grey edge and notching on the leading edge. The wing-linings are white, and the marginal coverts are red, creating a red bend to the folded wing. The most noticeable characteristic is the vermilion red chin, throat and breast. The black of the cheeks continues down to the breast sides and lower breast, framing the red underparts. The flanks are broadly fringed whitish-grey, and the thighs are white. The belly and vent is black with thin whitish edges. The tail is mainly blackish, often with the central pair of feathers boldly notched with grey-brown. Most other tail feathers have the notching restricted to the outer edge, particularly near the tip. The legs and feet are blackish.

Adult Male Non-Breeding (Definitive Basic, fresh): Similar to breeding plumage, but much fresher-looking and more heavily patterned due to the greater amount of notching and fringing on the upperparts in particular.

Adult Female Breeding (Definitive Basic, worn): Noticeably smaller and duller than male. The upperparts, including the head are much more streaked due to wider brown fringes to the feathers. The supercilium is buff, rather than white, but is often tinged with red in front of the eye. On the underparts, the red is much more restricted. There are two dull red patches, one on the throat and the other on the lower breast. The upperbreast is brown with dark streaks, separating the two red areas of the underparts. Sometimes the red is extremely restricted, nearly lacking or just noticeable immediately below the breast band. The flanks, belly and vent are greyish-brown with bold dark streaks. Like the male, the underwings are white. The tail tends to be more heavily patterned, showing more notching than the male's.

Adult Female Non-Breeding (Definitive Basic, fresh): Similar to breeding plumage, but even more strongly patterned. Often the red on the underparts is obscured by buff tips.

Juvenile (Juvenal): Resembles adult female, but does not show any red on the underparts. Streaking is also more extensive below.

GEOGRAPHICAL VARATION Two subspecies are recognized. The nominate subspecies is found in the northern part of the range, from Ecuador south to central Peru. The southern ***S. b. albipes*** is found in SW Peru, north to Ica, and northern Chile. It is similar to the nominate, but smaller. In addition, *albipes* males have duller scarlet underparts and the females may lack pink on the underparts altogether, unlike *bellicosa* (Bond 1953).

HABITAT Peruvian Meadowlark is found in open habitats including cultivated areas, arid native grasslands, open woodlands, desert oases and riparian zones in the arid Pacific Coast. In most of its range it is restricted to river valleys where vegetation grows, as it is absent from the desert itself. In the north (N Peru and Ecuador) it is in a much less arid situation than in nothern Chile and southern Peru, and this species is not restricted to oases and river valleys. In general, the few areas in the coastal

Peruvian/Chilean desert which do have enough water available for vegetation to grow, have been severely altered in order to grow crops and fruit. Fortunately, these habitats are often suitable for Peruvian Meadowlark, and this species is now often associated with agriculture. Irrigation has increased the available habitat for this meadowlark. In urban situations, it will also frequent lawns and parkland. Breeding and wintering habitats are similar, but Ridgely and Tudor (1989) state that wintering flocks will venture into brushier habitat than would be used during the breeding season.

BEHAVIOUR This species is typically seen on the ground feeding in typical meadowlark fashion. It also perches on telephone poles, lines or fence posts. When singing, it typically perches high on a bush or fence post. This species is more prone to sing from a perch than from the ground. The flight display is similar to that of some of the other South American meadowlarks. The male flies low over the bush tops, before ascending to a peak and then singing as he parachutes steeply downwards to a perch. Aspects of its mating system and other behaviours are not known.

NESTING In Chile, nests contain large young by early November. In Ecuador, the season begins after the rains, from March–May. The nests are on the ground as is typical for meadowlarks; nests are carefully woven with grass and have a domed roof as is typical in the genus. Nests are often placed near a bush or grass clump. Clutch size 3–5, commonly 4 eggs. The eggs are cream or pinkish with darker brown or purplish smudges, sometimes with black squiggles, similar to those of Long-tailed Meadowlark, but smaller. Incubation lasts approximately 14 days and the nestlings need 12 days to fledge. Shiny Cowbird (102) is known to parasite the meadowlark's nests. Only one brood is raised per season.

DISTRIBUTION AND STATUS Common, a bird of the lowlands found to 2500 m. Distributed from Esmeraldas, Ecuador south along the arid coastal slope as well as intermontane valleys in Chimborazo, Azuay and Loja in Ecuador (Ridgely and Tudor 1989). Found in the Pacific slope of W Peru, from Piura south to Tacna, and also present in the upper Marañón valley. Occurs south to northernmost Chile, as far south as the province of Antofagasta. The southernmost locality where this species can be found regularly is Quillagua.

MOVEMENTS Not known to be migratory, but it does form larger flocks that may wander during the non breeding season. In SW Ecuador, it is present in small numbers throughout the year but it becomes abundance during the breeding season suggesting that some local movements may occur (Marchant 1958). Its presence in isolated oases in Chile suggests that some dispersal must take place.

MOULT Adults undergo one moult per year, the complete definitive pre-basic moult. Pre-alternate moults are lacking. All plumage differences between non-breeding and breeding plumage occur as an effect of wear and fading. No information is available on the extent of first pre-basic moult.

MEASUREMENTS ***S. b. bellicosa*.** Males: (n=20) Wing 113.7 (107–121); tail 71.8 (64.8–80); culmen 26.9 (24.8–28.7); tarsus 36.2 (32–38.5). Female: (n=5) wing 99.2 (94–105); tail 59.4 (57–70); culmen 25 (22–28); tarsus 33.4 (32–36).

NOTES Also known as Peruvian Red-breasted Meadowlark, White-thighed Meadowlark and Pacific Red-breasted Starling.

REFERENCES Bond 1953 [geographic variation], Fjeldsa and Krabbe 1990 [voice], Johnson and Goodall 1965 [nesting], Marchant 1960 [nesting], Short 1968 [identification, measurements].

69 PAMPAS MEADOWLARK *Sturnella defillippi* Plate 26

(Linnaeus) 1771

Previously a common bird of the grasslands of the Pampas, its numbers have seriously declined as these native grasslands have been altered by agriculture.

IDENTIFICATION This species overlaps with two other meadowlarks, Long-tailed Meadowlark (70) and White-browed Blackbird (67). Long-tailed Meadowlark often breeds in adjoining territories and can be an identification problem, they were previously thought to be conspecific! Pampas Meadowlarks are significantly smaller than Long-tailed Meadowlarks. As the name implies, Long-tailed Meadowlark has a long tail, which is noticeable in the field, giving them a different shape to Pampas Meadowlark. However, females can be closer in tail lengths. Other structural differences include Pampas Meadowlark's shorter and thinner bill, and its shorter, weaker legs than Long-tailed Meadowlarks. While this species is smaller than Long-tailed Meadowlark, its wings are nearly the same length, giving it a long-winged appearance. The wings are also more pointed and the primaries are more noticeably emarginate on Pampas Meadowlark. Pampas Meadowlark has black wing-linings which differentiates it from Long-tailed Meadowlark, but resembles the smaller White-browed Blackbird. The overall plumage pattern is much blacker than in Long-tailed Meadowlark, particularly on the flanks and upperparts. This difference is most obvious on worn birds; fresh individuals of both species have brown fringes to much of the plumage and can thus appear quite similar. On the underparts, the red is darker on Pampas Meadowlark than Long-tailed Meadowlark. In addition, the tail of this species is less barred than that of Long-tailed Meadowlark. Pampas Meadowlark has a tendency to show a collar of small streaks on the breast. Long-tailed Meadowlark and Red-breasted Blackbird does not show this tendency; however, these streaks may be lost through wear on Pampas Meadowlark. Females are duller than males but still differ in many of the same ways as the males do. Fresh-plumaged females would be the hardest to separate, the dark wing-linings are diagnostic for Pampas Meadowlark. Long-tailed Meadowlark appears to lack a flight song which is commonly given by Pampas Meadowlark. In addition, Pampas Meadowlark does not sing from high perches, as does Long-tailed, but instead is often partially hidden in the grass when singing. Vocally, Pampas and Long-tailed Meadowlarks are quite different and identification can be made on vocalisations alone (see Voice). White-browed Blackbird is even blacker than this species and is much smaller and shorter-billed. The different size and structure should be enough to distinguish these two species.

VOICE Song: The primary song lasts about two seconds and has two parts. There are several introductory notes followed by a whiny *zheet* note. The initial notes are choppy and short, unlike the slurred whistles of Long-tailed Meadowlark. This song may sound like *tika-tika-tika-zheet*. The secondary song is the flight song, which is often more commonly performed than the primary song. The flight song consists of three parts, beginning with series of short whistles, followed by a steady unmoderated buzz and terminated by a lower frequency series of short notes as a liquid trill. For example, *zwee-BZZZZ-trltrltrl*. The similar Long-tailed Meadowlark differs in its primary song mainly by its longer series of slurred introductory whistles and the much more grating and nasal downslurred buzz which terminates the song.

Calls: include a *peet* call given by both sexes, similar to that of Long-tailed Meadowlark. Pampas Meadowlark also utters a rasping *zheet* call. Females have been heard to give a chatter.

DESCRIPTION This medium-sized meadowlark is not particularly long-billed, and has long, pointed wings. In length, the bill is equal to the head length or slightly shorter. The primaries are pointed (emarginate) and the outer (P9) is as long or longer than the next three and tends to be longer than P5. The tail is short. Wing Formula: P9 < P8 > P7 > P6; P9 ≈ P6 or P7; P8–P5 emarginate.

Adult Male Breeding (Definitive Basic, worn): The bill is silvery-grey with a dark tip and culmen, the legs are dark brownish, and the eyes brown. Crown blackish with brown streaks. The nape is black with few pale brown streaks. The face shows a red supraloral area that continues as a bold white supercilium behind the eye, reaching the nape. The lores are black as are the ear-coverts. The black of the face continues down to the sides of the breast, pinching in the red of the underparts. The sides of the neck are blackish. The underparts are largely salmon-red from the chin to the upper belly. The flanks, thighs, belly and undertail-coverts are black. The back is blackish with restricted brown edges, thus not appearing boldly streaked. This pattern continues to the rump; however, the uppertail-coverts are barred. When worn, the rump and uppertail-coverts appear brown with darker bars. The scapulars are similar to the back but the brown continues all the way around as a fringe. The tertials are blackish-brown with indistinct darker bars that do not reach the feather edges; each feather is fringed buff. On the wings, the marginal coverts are salmon-red and visible on the folded bend of the wing. The coverts are black with thin white or grey fringes. The blackish-brown primaries have a crisp, thin white or grey fringe. The outermost primary (P9) has a bold and obvious white leading edge, more so than on the other primaries. The wing-linings are black. The tail is blackish with olive bars near the end of each feather.

Adult Male Non-Breeding (Definitive Basic, fresh): Differs from breeding male in being boldly streaked above due to wider brown feather edges on the upperparts. The crown, face, breast sides and flanks which are black on the breeding adult are streaked with brown in this plumage. The supercilium is buff. The red on the underparts is veiled by olive or buff feather tips and there may be some blackish spotting on the breast. The undertail-coverts are blackish with olive-grey fringes. The lesser coverts are grey-tipped, the median coverts buff-tipped and the greater coverts show buff edges and tips. There is a small white spot at the base of the mandible, which may also be visible on breeding birds.

Adult female (Definitive Basic): Similar to male, but smaller and duller. The total amount of red on the body is reduced, particularly on the face, throat and breast. In addition, the red colour is not as vivid as on the male. In general, the upperparts are more heavily tipped brown, thus not appearing as black as on the male. The black of the flanks and belly is obscured by brown fringes, thus appearing paler than on males. As in males, fresh birds (non-breeding plumage) are warmer in colour and more heavily streaked. However, since females are streaked at all times, the changes in appearance through the year are much less obvious.

Immature Male (First Basic): Similar to adult female, but larger, and show a greater amount of red on the breast.

Juvenile (Juvenal): Resembles female, but lacks red on the underparts.

GEOGRAPHICAL VARIATION Monotypic.

HABITAT Inhabits open grasslands, particularly the better-drained ones, but will use moister sites. Avoids bushy areas or forest edge. Unlike Long-tailed Meadowlark, Pampas Meadowlark will nest in open grasslands lacking perching sites such as bushes or fence posts. The grasslands (meadows) where this species breeds in S Buenos Aires are diverse in their species composition, having a mix of both short and medium length grasses and sedges, as well as a variety of short herbaceous plants. It appears that these fields are 'mature' in that they are not in the early stages of succession which heavy grazing produces, and likely need several years of no grazing or low grazing impact to achieve the stage that appears to suit the meadowlarks. The impact of grazing on this species has not been adequately addressed; it is not unlikely that a managed grazing regime would allow for the production of good meadowlark habitat as well as healthy forage for animals. However, much of the suitable land is quickly being converted to crops such as sunflower, where the meadowlarks are absent (Collar *et al.* 1992).

BEHAVIOUR This species sings from the ground or from low grass perches when singing the 'primary' song. At times it may perch on fence posts or similar height structures, but it does not sing from tall utility poles or wires. Hudson (1920) noted that it never perched on trees. The most notable display given by this species is the nuptial flight song and display. These are often performed directly above the female and can enable the observer to locate the female. During the display, a male flies off the ground at 45° to approximately 5–10 m in the air, sings at the peak of the ascent and then drops to the ground with the wings held stiffly over the body, giving the terminal whine as it does so. Flight singing by one male often stimulates other males to perform the display. It is also not uncommon to observe chases between males during the breeding season. During breeding, Pampas Meadowlark will act aggressively toward nearby White-browed Blackbirds with the two species defending non-overlapping territories (interspecific territoriality). Long-tailed Meadowlark is accepted to a much greater degree, and its territories may overlap those of Pampas Meadowlark. In contrast, this species is highly sociable during the non-breeding season. Non-breeding flocks were large when populations were greater, but sadly such flocks are becoming a rarer sight.

NESTING In the southern Pampas, nest building begins in mid November. The nest is a woven grass nest placed on the ground. Clutch typically 4 eggs. The eggs are white or pinkish-grey, with dark purplish-brown spots and lines clustered around the broader end. Eggs of this species are almost identical to those of Long-tailed Meadowlark, differing only in their smaller average size. Pampas Meadowlark is not used as a host by the parasitic Shiny Cowbird (102) while in the same areas, Long-tailed Meadowlark may suffer heavy parasitism.

DISTRIBUTION AND STATUS Now uncommon and declining, but previously common.

Breeding: Most recently breeding has only been detected in southern Buenos Aires province, particularly north of Bahia Blanca. It likely also breeds in La Pampa province to the west and in southernmost Córdoba. Formerly it was a common bird throughout Buenos Aires province, with nesting confirmed in northern Entre Ríos and southern Santa Fé provinces. It likely bred in western Uruguay formerly. There are also specimens taken to the west in San Luis during the breeding season.

Non-Breeding: The northernmost observations in winter are from Brazil, where it has been recorded in Paraná, Santa Catarina (as recently as 1991) and Rio Grande do Sul. In Uruguay, it has been recorded from various sites in the past, scattered throughout the country. Most of these historical records are from the fall or winter months (April–July); recently (July 1993) a flock of 100 was located in Tacuarembo department. Historically, in Argentina, wintering birds were recorded from as far south as southern Buenos Aires to Santa Fé and Entre Ríos. There are many mentions of wintering flocks near the coast of Buenos Aires province, the most recent winter record from the province is of one collected in June 1971 near General Lavalle, a coastal site. It is unknown if a record from Corrientes, Argentina, was from the nesting or winter season. Old records from Paraguay are erroneous.

MOVEMENTS Partially migratory. All of the Brazilian observations have been made during the non-breeding season suggesting that at least part of the population moves northward in the winter. Breeding has not been confirmed in Uruguay, and most records are from the non-breeding season, so these observations likely pertain to migrants or wintering birds. Hudson (1920) observed large flocks of Pampas Meadowlark walking northward during the autumn. He hypothesized that they were in active migration walking over the flat stretches of the pampas! Hudson noted how appropriate the name Military Starling was when one compared these massive movements to that of military troops on the move.

MOULT Adults go through one plumage change per year, the complete definitive pre-basic moult. It occurs after the breeding season, during March. However, it is not clear if the moult takes place before or after the northward migration. The pre-alternate moults are lacking. All plumage changes between breeding and non-breeding plumage occur as an effect of feather wear and fading. The first pre-basic moult appears to be incomplete, involving the body, wings and tail, retaining only scattered juvenal feathers. Its timing is not known.

MEASUREMENTS Males are larger than females. Males: (n=20) wing (118–127); culmen (24.1–28.7); tail (73.4–81.8); tarsus (28.0–33.6). Females, (n=5) wing 106.6 (102–110); tail 69 (63–71); culmen 23.4 (22–25); tarsus 29.2 (29–30).

NOTES Also known as Lesser Red-breasted Blackbird. Felix de Azara, the early naturalist who explored Paraguay and nearby Argentina called this meadowlark the 'degollado' (Hudson 1920), a Spanish word that translates to the beheaded, as the red breast of this bird suggested to him the blood of someone who had been beheaded.

REFERENCES De la Peña 1989 [nesting], Gochfeld 1979a and 1979b [habitat, behaviour, nesting], Gochfeld 1978 [behaviour], 1979b [behaviour], Hardy *et al.* 1998 [voice], Pearman 1994 [winter distribution], Short 1968 [voice, habitat, behaviour, measurements], Wetmore 1926 [behaviour].

70 LONG-TAILED MEADOWLARK *Sturnella loyca* (Bonaparte) 1851 — Plate 26

The common meadowlark of southern Argentina and Chile. It is similar to other red-breasted meadowlarks, but is set apart from them by its long tail. This is one of the best known birds of Chile, there are few rural people who will not recognize the 'Loica'. In the monotonous landscapes of Patagonia, Long-tailed Meadowlark is easily the most colourful bird around.

IDENTIFICATION Long-tailed Meadowlark overlaps with two other meadowlarks, the similar Pampas Meadowlark (69) and White-browed Blackbird (67). Male Long-tailed Meadowlarks are true to their names and have a much longer tail (in relative and absolute terms) than either of the two other southern meadowlarks. Pampas Meadowlark is slightly smaller and much more solidly black than Long-tailed Meadowlark, particularly obvious on the head and sides. White-browed Blackbird is even more extreme in this respect, being almost completely black except for the white supercilium and red breast, and it is much smaller. The bill of White-browed Blackbird is short, rather conical and obviously shorter than the length of the head. In flight, both Pampas Meadowlark and White-browed Blackbird have black wing-linings, rather than white as in Long-tailed Meadowlark. Long-tailed Meadowlark has salmon-red underparts unlike the other two species, which possess bold red chests. Behaviorally, the three species are quite different. Between the more similar Long-tailed and Pampas Meadowlarks there are a couple of key characteristics to look for. In the zone of contact between the two, a meadowlark singing from the ground is never a Long-tailed, while one singing from a tall utility pole or wire is always a Long-tailed.

VOICE Song: At least two song types are given. The first is a longer song, lasting over 2 seconds. This vocalisation begins with a series of several whistles followed by two longer, distinctive, modulated whistles, the first rising and the second falling (sometimes reversed), followed by a terminal whine (the terminal buzz) that falls at the end. For example, *tip-tip-tip-tip-tp-tp-tp-sweetweeo-TZZZRRRRRR*. The second song type is shorter. It ends with the same terminal buzz as in the first song but the introductory notes differ. These introductory notes span a higher frequency range and sound like long down-slurred whistles. An example is *twip-twip weeo-swee-TZZZRRRR* or *twip-twip twee-Tweeep-TZZZRRRR*. Apparently the subspecies *catamarcana* gives slightly different songs from southern birds, including a flight song which is not usual from nominate Long-tailed Meadowlarks. On the Falkland Islands, the song averages seven notes, falling then rising and ending in a low note (Woods 1988). The insular *falklandica* quite often gives a song flight while gliding over the territory, this song is longer and more elaborate than the regular songs; unlike Pampas Meadowlark, this meadowlark does not tend to parachute down while performing the song flight. Overall, songs of *falklandica* differ from the nominate in the greater complexity of the introductory notes, as well as in the more nasal quality of *falklandica's* vocalisations. At least in Tierra del Fuego and the Falklands, the females sing in response to the males; this song is squeakier than the male's (Woods 1988). Singing also occurs sporadically in the winter.

Calls: include a *peet*, or *pimp*, given by both sexes, which resembles a similar call of Pampas Meadowlark. A short *twick* has also been heard. In the Falkland Islands, an explosive *cheeoo* is the most common flight call, and a *chook* or *chink* is given while foraging (Woods 1988).

DESCRIPTION Differs from all of the other meadowlarks in having a noticeably long tail. The bill is long, about 1.25 times the length of the head. The wings are somewhat rounded, with the outer primary (p9) being roughly equal in length to p5, and to p6–p8. Wing Formula: P9 < P 8 < ≈ P 7 ≈ > P6 > P5; P9 ≈ P5; P8–P5 emarginate. When taking off it may jump into the air and fly directly upwards for a few meters before beginning level flight. The flight is purposeful and lacks strong undulation.

Adult Male Breeding (Definitive Basic, worn): The crown is blackish with brown streaks, and a pale brown median stripe. The nape is streaked black and pale brown, appearing paler than the lateral crown stripes. The face shows a salmon-red supraloral area that continues as a bold white supercilium behind the eye that reaches to the nape. The lores are black as are the ear-coverts, the latter with thin olive fringes. The white malar stripe is short and only reaches to below the eye. The black of the face continues down to the sides of the breast, pinching in the red of the underparts. The sides of the neck are blackish with pale streaks. The underparts are largely salmon-red from the chin to the upper belly. The flanks are black with pale fringes, as are the thighs and belly, but with broader fringes. The undertail-coverts are black with white fringes. The back is blackish with brown edges, slightly paler on each feather tip. This creates a brown streaked appearance on the blackish back. The scapulars are similar to the back but the brown continues all the way around as a fringe. The tertials are brownish-olive with indistinct darker bars that do not reach the feather edges, each feather is fringed buff. The rump is similar to the back but each feather has a brown tip, making it appear browner. The uppertail-coverts are olive-brown with distinct black bars. On the wings, the marginal coverts are salmon-red and visible on the folded bend of the wing. The lesser coverts are black with thin white fringes. The median coverts are black-centered with brown outer edges and an off-white inner edge. The greater coverts are blackish with a pale buff outer edge, creating a pale panel on the closed wing. The secondaries are brownish-olive with indistinct darker bars and a thin pale fringe. The blackish-brown primaries have a crisp thin white fringe; they are emarginate on the outer vane of p5–p8. The primary coverts are blackish-brown with a brown fringe contrasting with the much more bold blackish greater coverts. The wing-linings are white with grey undersides to the primaries and secondaries, but with slightly paler bases. The tail is blackish with olive bars near the end of each feather, with the central two feathers barred with olive to the base. The outer feather on each side has a crisp white fringe. The long bill is silvery-grey with a dark tip and culmen, the legs are brownish-grey, eyes brown.

Adult Male Non-Breeding (Definitive Basic, fresh): Similar to breeding male, but the red underparts are obscured with buff tips, and upperpart patterns appear crisper.

Adult Female (Definitive Basic): Similar to male, but smaller. The red on the supraloral area is much reduced and the red on the underparts is obscured by buff fringes. The face is paler with more brown streaks. The area above the malar and on the breast sides is blacker than the rest of the surrounding areas creating a blackish half collar, similar in position to the 'V' of the yellow-breasted Meadowlarks. The black of the flanks and belly is obscured by wider olive fringes, thus appearing paler than on males. Upperparts and wings as in male but the lesser coverts are not blackish. The tail is the same colour as the males, but is proportionately shorter.

Juvenile (Juvenal): Similar to female but the colour of the underparts is greatly reduced to a pink patch in the center of the breast. In addition, the scapulars and tertials show crisp pale fringes, giving a scaly appearance. Adult female has a diffuse pale fringe that is paler at the tip and buffier on the edge. Similarly, the alula and primary coverts of juveniles show a crisp white fringe, unlike the buff of adult females. Juveniles show a more densely streaked breast.

GEOGRAPHIC VARIATION Four subspecies are recognized.

S. l. loyca is the Chilean and Argentine Patagonian subspecies. Some migrate north to Córdoba during winter. This form is described above.

The breeding race along the Andean foothills of Argentina from Jujuy to Salta is ***S. l. catamarcana***.

The isolated sierras of Córdoba and San Luis are the breeding range for ***S. l. obscura***.

The resident subspecies on the Falkland Islands is ***S. l. falklandica***.

Catamarcana is paler and greyer than *loyca*. The belly is not as blackish as on *loyca*, but more of a dark chestnut with whitish fringes to each feather. This form is smaller than southern *loyca*, but more similar in size to northern specimens; however, *catamarcana* appears to have proportionately longer wings. It may differ vocally as well. The insular *falklandica* is larger and even longer-tailed than *loyca*; it also has a long bill with a tendency to have a wider bill tip. It also consistently shows a white outer edge and white tip to the inner vane of the outer rectrix.

HABITAT Found in well-drained open areas: grasslands, fields, pastures, Andean grasslands, matorral, Patagonian steppe and tussac grass meadows. In most of its range it prefers grasslands interspersed with small shrubs or trees that it uses as perches. Apparently, perches are a prerequisite for breeding. On the Falkland Islands, it uses a variety of habitats including white grass camp near shores, short turf, open heathland, and tussac grass paddocks with clearings of short turf, where densities are higher than on open heathland (Woods and Woods 1997). In the southern Pampas, where this species breeds, it has been proposed that Long-tailed Meadowlark is a recent invader as the presence of shrubs in these grasslands is a phenomenon associated with human disturbance of the habitat. Two higher elevation areas in the Pampas, the Sierra de la Ventana and Tandil, which are more Patagonian in character may have harbored Long-tailed Meadowlarks before European colonization. At some coastal sites it may be found in the tussock grasslands stabilizing sand dunes.

BEHAVIOUR This is a peculiar meadowlark. While it usually behaves like a typical meadowlark, it also often prefers to perch in trees or bushes in flocks, all the while calling and behaving more like typical blackbirds (*Agelaius, Curaeus,* black icterids in general). It is often seen in flocks in the winter (up to 60 individuals) but this is also not unusual during the breeding season. Most of the time Long-tailed Meadowlark will be observed foraging on the ground, picking at seeds and insects. In the breeding season, males sing from exposed perches (bushes, posts and fences) well above the grass, unlike Pampas Meadowlark. Males seldom sing from the ground, except in the Falkland Islands.

NESTING Breeding commences in September in central Chile and continues until at least January; October is a more typical breeding period in the far south. Timing is similar in Argentina. However, a May nesting is known from Atacama, Chile, from an area where nesting typically does not extend past January. On Falkland Islands, nests late August–November (Woods 1988). The nest is located on the ground and is domed, except in Falkland Islands (Woods pers. comm.), and usually has an entrance tunnel of up to one meter in length (Woods 1988), like those of its North American relatives. On Falkland Islands, nests may be up to one meter above the ground, built on a flat pedestal in tussac grass (Woods 1988). The nests are built from well-woven grass. Nest building and incubation is performed strictly by the female, although males will help in feeding the young. Clutch 3–5: eggs buffy, pale blue or greyish-pink with chestnut, purple or black spots and scrawls concentrated at the broader end of the egg. During a good season some females may be double-brooded. In southern Buenos Aires province, this is a common host for Shiny Cowbird (102), unlike Pampas Meadowlark. However, cowbirds are seldom observed to successfully fledge from Long-tailed Meadowlark nests. Breeding densities on Falkland Islands estimated at 0.57–0.95 males per ha (R. Woods Unpub. Data).

DISTRIBUTION AND STATUS Common throughout its range. Found to 2500 m. Endemic to Argentina and Chile. In Chile, it is found from Atacama south to the Beagle Channel Islands in Tierra del Fuego. In Argentina, restricted to the foothills and Andean intermontane valleys in the north from Jujuy south to Mendoza, including the isolated mountains in Córdoba (Sierras de Comechingones) and San Luis (Sierras de San Luis). Distribution spreads to the steppes from C La Pampa and S Buenos Aires south to Tierra del Fuego. Resident on both East and West Falkland Islands as well as smaller offshore islands; the only breeding icterid in the Falkland Islands. Not recorded from Bolivia, but it should be looked for in the extreme south.

MOVEMENTS Most populations appear to be resident, but some of the Patagonian birds move north into Córdoba and Buenos Aires. Southern Chilean birds apparently displace northward, augmenting the populations of the central regions. The southernmost populations in Tierra del Fuego and the Falklands are present year round. One vagrant record exists for South Georgia Island, from Prince Olav harbour, on April 9, 1987 (Curtis 1988).

MOULT Definitive pre-basic moult late February–mid April, the only adult moult of the year and is complete. The extent and timing of first pre-basic moult in juveniles is not clear. This moult involves the body plumage, but whether the wings and tail are also shed is not known.

MEASUREMENTS ***S. l. loyca***, both sexes mixed: Wing 117.3 (1.36); culmen 33.3 (0.75); tail 98.4 (2.14). Males (n=32): Wing 121.1 (115–129); tail 101.5 (85.8–106); culmen 33.9 (28.3–36.4); tarsus 36.1 (31.6–38). Females: (n=10) wing 115.2 (104–122); tail 89.8 (85–94); culmen 29.5 (25–33); tarsus 34.4 (33–35).

S. l. catamarcana. Males: (n=4) Wing (126–131); culmen (32.3–34.1); tail (93–100.7); tarsus (35.1–36.0). Females: (n=3) Wing 118 (116–120); culmen 31.5 (31–32); tail 95.7 (95–97); tarsus 36.3 (35–37.5).

S. l. falklandica. Males: (n=12) wing 131.8 (129–137); (n=6) tail 96.0 (89–105); (n=10) culmen 38.2 (35–42); tarsus 32.9 (30–35). Females: (n=14) wing 120.8 (114–128); (n=4) tail 87.0 (84–92); (n=13) culmen 35.7 (32–39); tarsus 31.2 (28–36).

NOTES Has also been called Greater Red-breasted Meadowlark, Military Meadowlark and Military Starling (Falkland Islands).

REFERENCES De la Peña 1987 [nesting], Gochfeld 1979 [habitat, behaviour, nesting], Hardy *et al.* 1998 [voice], Humphrey *et al.* 1970 [voice, behaviour, moult], Johnson and Goodall 1965 [movements, measurements], Millie 1938 [nesting], Short 1968 [voice, habitat, behaviour, measurements], Woods 1988 [voice], Woods unpublished data [*falklandica* measurements], Woods and Woods 1997 [habitat, nesting], Zotta 1937 [geographic variation, measurements].

71 EASTERN MEADOWLARK *Sturnella magna* (Linnaeus) 1758 — Plate 27

A common bird of open country in the east of North America and in the Neotropics.

IDENTIFICATION In North America, the identification of the meadowlarks is one of the most challenging problems, unless the distinctive vocalisations are heard. North American Eastern Meadowlarks have a slurred, whistled song that does not resemble the flute-like song of Western Meadowlark (72). The songs and calls are consistently of a higher frequency than those of Western Meadowlark. The common call of Eastern Meadowlark is a characteristic *dzert* quite unlike the *chup* of Western Meadowlark. Vocally, 'Lilian's Meadowlark' (71X), possibly a species distinct from Eastern Meadowlark, is similar to Eastern Meadowlark, but its song may sound intermediate between Eastern and Western Meadowlark. The song of 'Lilian's Meadowlark' is lower in frequency than that of Eastern Meadowlark. However, the calls are similar if not identical to those of typical Eastern Meadowlark.

Eastern Meadowlark is darker than both Western and 'Lilian's Meadowlark', and like 'Lilian's' it lacks yellow on the cheek (above the malar area) that is present on Western Meadowlark. This latter feature is unreliable during the non-breeding season as buff feather tips may obscure yellow on the cheek. Also note that male Eastern Meadowlark has more extensive yellow on the throat than does female; two individuals showing differences in the extent of yellow on the throat are then not necessarily of different species. The key feature is whether the yellow extends well above the malar area (Figure 71.1). A useful feature for separating Eastern from Western Meadowlark is the colour of the greater coverts. Eastern Meadowlark has a noticeable warm wash to the coverts, making them appear warm brown at a distance, while Western Meadowlark has a grey wash to the coverts making them look greyish-brown to pale grey at a distance. 'Lilian's Meadowlark' is intermediate in its greater covert colour or more similar to that of Western Meadowlark. The flanks and undertail-coverts of Eastern Meadowlark are rich buff, while on Western Meadowlark they

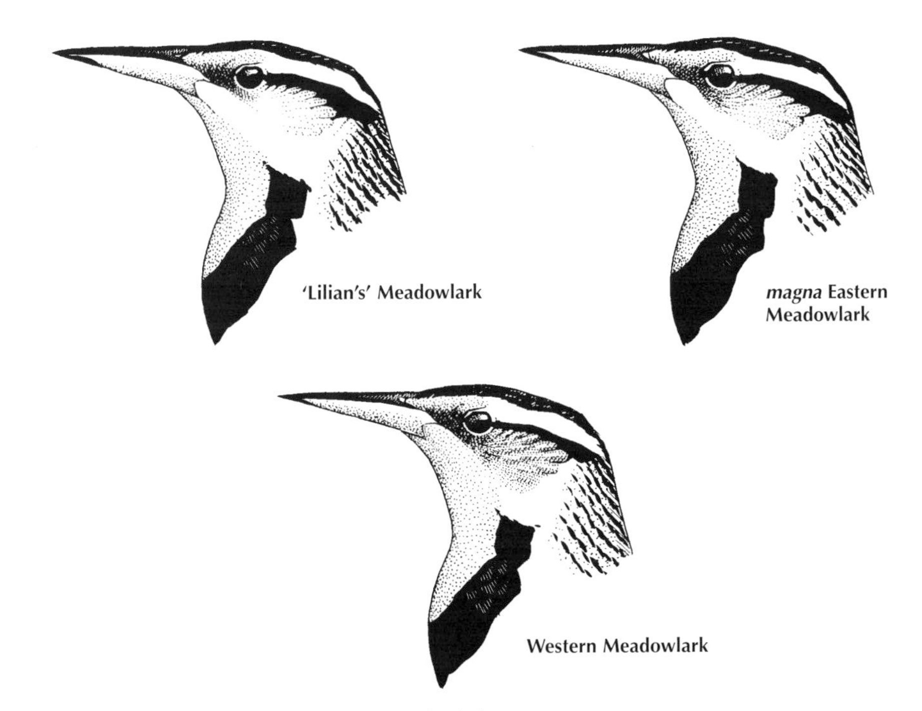

Figure 71.1 Face pattern in North American meadowlarks.

are greyer (J. Dunn pers. comm.). Eastern Meadowlark has more white on the tail than Western Meadowlark, but less than on 'Lilian's Meadowlark' (Figure 71.2). The outer two rectrices of Eastern Meadowlark have slightly darker shaft streaks, but much less noticeable than on Western Meadowlark. The extent of white on the third tail feather from the outside (R4) is also important; Eastern tends to have this feather mostly white, while on Western this largely dark and on 'Lilian's' is usually completely white. Eastern Meadowlark can be separated from most Western Meadowlarks by looking at the pattern of barring on the tail and tertials: the bars are separate from adjoining bars and parallel-sided on Western Meadowlark, but neighbouring bars join up along the center of the feather (shaft) on Eastern Meadowlark. Note however that some Western Meadowlarks, primarily those breeding in the Pacific Northwest, have confluent bars on the tail feathers and tertials. The pattern is intermediate in the 'Lilian's Meadowlark', but most show separate bars on the tail and tertials.

There are some slight shape differences between these three meadowlarks which may be visible in the field. Eastern Meadowlark is on average the largest of the three, but has the shortest wings (particularly the neotropical forms) and longest legs. Note also that Eastern Meadowlarks show streaking on the face which lessens the contrast between the dark eyeline and the paler cheeks, on 'Lilian's Meadowlark' the cheeks are much less streaked and therefore show a greater contrast with the dark eyeline. The facial pattern of Western Meadowlark is streaked, as in Eastern Meadowlark. Where Western and Eastern meadowlarks winter together, the latter favours taller grass than the former (J. Dunn pers. comm.). In both North and South America, Eastern Meadowlark may be confused with Dickcissel (*Spiza americana*) which shares its yellow underparts and black 'V' on the breast. Dickcissel should be easily separated from the meadowlark by its small size (like a House Sparrow *Passer domesticus*) and by its thick conical bill. Also note that Dickcissel has a dark tail and chestnut shoulders, among other differences. Other Neotropical meadowlarks are red-breasted and not easily confused with Eastern Meadowlark. Outside of North America, Eastern Meadowlark is the only yellow-breasted meadowlark and can be identified on distribution alone.

VOICE Song: The primary song is composed of a series of 3–5 (sometimes more) loud, descending, sliding whistles lasting approximately a second and a half. It is clearly different from the song of Western Meadowlark in its simplicity and whistled nature. Eastern Meadowlark's song never sounds flute-like or jumbled as does the song of the western species. The song differs according to location and there is individual variation and an ample number of different song types sung by any one male. Individual males may sing up to 100 different versions of their songs (d'Agincourt and Falls 1983, Lanyon 1995) which is an outstandingly large repertoire. Males repeat one song type many times, then switch to another. A male may require between one and two hours to go through its entire repertoire even during times of frequent singing (Lanyon 1995).

S. m. argutula often sings slightly different songs from those of its northern relatives, more musical and more warbled like Western Meadowlark. Cuban *hippocrepis* has a weak, wheezy song that has been likened to that of Dickcissel! Lowland Venezuelan birds begin with a series of slurred whistles but end with a jumbled warble, somewhat reminiscent of Western Meadowlark. Similarly, Colombian *meridionalis* is said to sing like a typical Eastern Meadowlark with notes suggestive of Western Meadowlark. Lanyon (1995) states that highland Venezuelan (Mérida) Eastern Meadowlarks *meridionalis* have longer songs which include 'steep slope' notes which are like those of Western Meadowlark; however, *meridionalis* from the Magdalena valley, Colombia, have more typi-

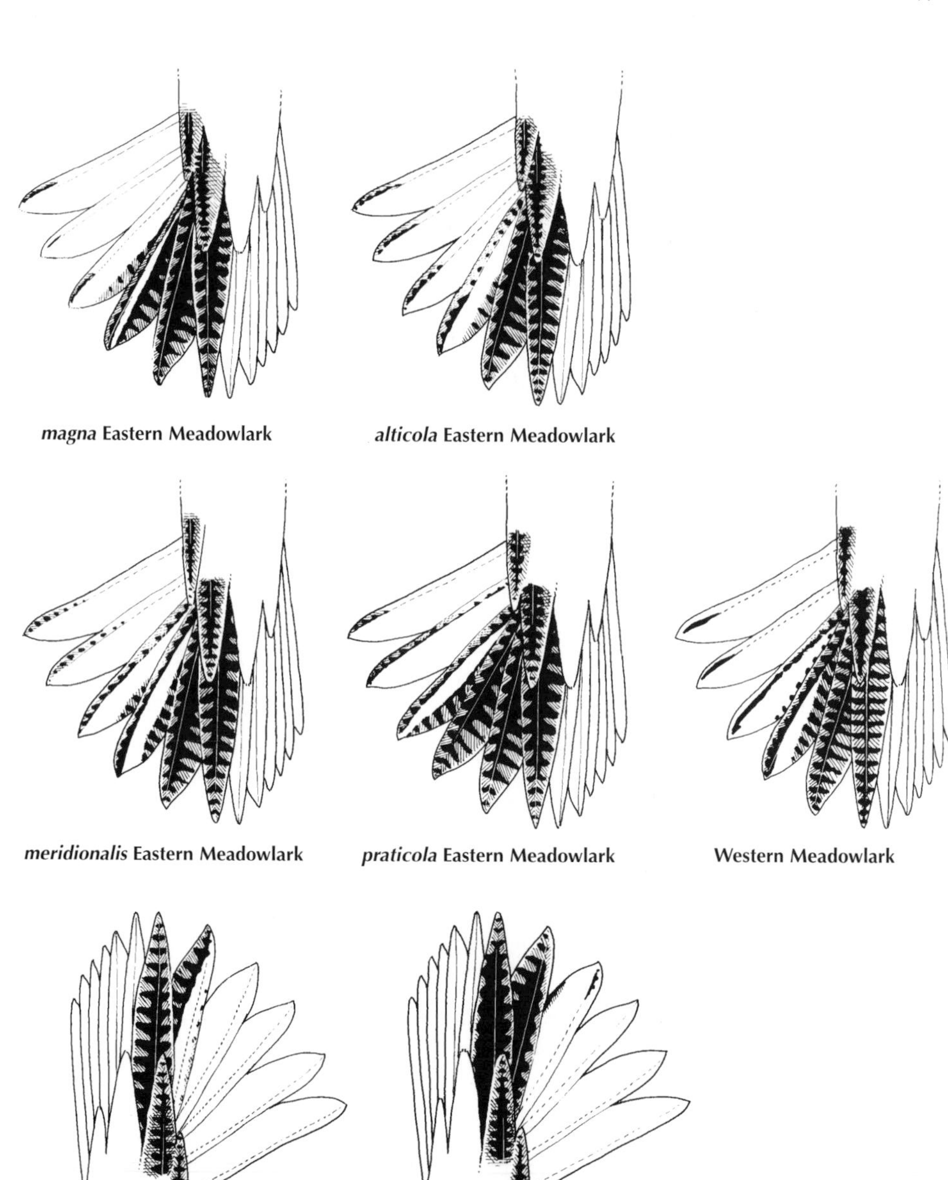

Figure 71.2 Tail patterns of yellow-breasted meadowlarks.

cal Eastern Meadowlark songs. A local Mexican name for Eastern Meadowlark is 'tortilla con chile', referring to the Spanish phrase it appears to say in its song. The calls of South American *S. m. praticola* are described as *chit, chit,* while the song is a short undulating whistle. Obviously there is great need to study these vocal differences and see if they hold any significance systematically.

In eastern Nicaragua, pairs of Eastern Meadowlarks perform a song duet. The female rattles or chatters immediately after, or overlapping the end of the male's whistled song (Howell 1972). Females of the Cuban race, *hippocrepis,* are also known to answer the songs of males either with a female song or with a rattle (Gundlach 1876). Female song has been noted from banded females in North America (nominate race) but it is exceedingly rare (Lanyon 1995). Female songs in the US are similar to male's but weaker (Lanyon 1995). Their presence is difficult to determine since sexing Eastern Meadowlarks cannot be achieved with reasonable accuracy in the field.

The flight song of Eastern Meadowlark is extremely rare, much more so than Western Meadowlark, but it is similar to that species, a jumble of notes sounding similar to Bobolink (103) song. The flight song is fast and jumbled beginning with a series of short whistles, it lasts 8–12 seconds (Lanyon 1995).

Primary songs are learnt, while calls are innately 'known' (genetic) in this species. Songs that sound intermediate between the two species, 'hybrid songs', are usually due to a meadowlark learning aspects of the wrong species' song rather than as a sign of hybrid origin for that individual. Young Eastern Meadowlarks learn their songs by listening to males during a critical period between four weeks of age and their first winter (Lanyon 1995). During this time, they can learn incorrect songs which they will retain for the rest of their lives. In one instance, in Ontario, an Eastern Meadowlark gave a typical song as well as that of a Northern Cardinal (*Cardinalis cardinalis*) (J. Bruce Falls in Lanyon 1995). In another situation, a captive Eastern Meadowlark learned a fair rendition of its species song by listening to an Eastern Meadowlark imitation given by a local European Starling (*Sturnus vulgaris*) (Nice 1952).

Calls: The common call note is a diagnostic *dzert,* which is given by both sexes. Another call given by both male and female is 'the whistle', rendered during times of excitement or alarm. 'The whistle' rises slightly in frequency. The 'chatter call' is also given during times of excitement, it is equivalent to 'the rattle and roll' of Western Meadowlark, but unlike in that species, both sexes of Eastern Meadowlark give the same call. The Eastern Meadowlark 'chatter' is higher in frequency than the 'rattle' of female Western Meadowlark. The *weet* call is shared by both sexes and cannot be differentiated from the corresponding call of Western Meadowlark. The *weet* call is the one commonly given by migrant and wintering Eastern Meadowlarks. Females give a *tee-tee-tee* call during the period before incubation. All of the above calls have a similar counterpart in the Western Meadowlark. Eastern Meadowlark also has a variety of uncommon calls with no counterpart in Western Meadowlark, such as the *zeree* and *dzert-tut-tut* calls. (mainly from Lanyon 1995).

DESCRIPTION A stocky meadowlark with a long bill and short tail; the legs are thick and strong. Wing Formula: $P9 < P8 \approx P7 \approx P6 > P5$; $P9 \approx P5$; P8–P6 emarginate.

Adult Male Breeding (Definitive Basic, worn): The long, pointed bill is roughly the same length as the head. The crown is blackish with a buff central stripe, the dark lateral crown stripes fringed with chestnut. It shows a long, pale supercilium, yellow before the eye, greyish-white behind the eye. The same greyish-white colour continues on the auriculars and to the malar area. The dark post-ocular stripe is coloured like the lateral crown stripes. The lores are greyish, and the eyering whitish. The upperparts appear striped and barred. The back feathers are blackish with a paler brown tip, each feather is bordered by rich buff on both sides. The buff edges give the dark back a buff-streaked look. This pattern continues on the rump, and includes the scapulars. Uppertail-coverts blackish with rich buff edges. The marginal coverts are yellow, showing as a yellow bend of the wing. Lesser and median coverts grey-brown with buffy fringes, the inner edge of which is scalloped in pattern. Greater coverts similar, but toward the tip the buffy notches become confluent creating a pattern of dark bars separated by buff. The tertials are again similar, mainly dark brown feathers with yellow-buff notches along the edge, very few of which reach the mid-vein, except for a few at the tip. The primaries are blackish, showing paler bars particularly on the outer vane of each feather; each primary shows a pale grey fringe. The secondaries are similar but with wider paler brown bars. The tail is mainly dark with white edges. The central two pairs of feathers are brownish-black with olive-brown notches along the edge, which do not reach the mid-vein. The outer three pairs are mainly white with a dark spot along the shaft, on the outer vane of each feather. The next feather in (rectrix 3) can be mainly dark, but usually shows a white inner vane or more extensive white coloration. There is some variation in the tail pattern. The underparts are mainly yellow from the chin down to the belly, interrupted by the obvious black 'V' on the breast. The yellow on the throat does not extend up above the malar area. Flanks buff down to the undertail, with some streaks on the hind flanks. Eye dark brown, bill bluish-grey with a dark culmen and tip, bluish-grey legs and feet.

Adult Male Non-Breeding (Definitive Basic, fresh): Similar to breeding male, but much fresher and crisper. Upperpart feathers completely fringed with buff, giving a scaly appearance to the back. The underpart colours are also masked with duller fringes. The black 'V' looks duller due to buff fringes as does the yellow which is fringed with duller yellow.

Adult Female Breeding (Definitive Basic,

worn): Similar to male but slightly duller in colour. The extent of yellow on the throat is reduced.

Adult Female Non-Breeding (Definitive Basic, fresh): Indistinguishable from non-breeding male.

Juvenile (Juvenal): Similar to non-breeding adult but pale fringes on the back, and the scapulars are paler and are complete, thus appearing more scaly. Head stripes are duller. The throat is pale and the breast 'V' is replaced by dark streaking in the same area. The yellow on the underparts is not as intense.

GEOGRAPHICAL VARIATION Eastern Meadowlark appears to be 'over split' at the subspecific level and may actually comprise more than one species, but more research needs to be conducted to clarify the relationships within the species (See 'Lilian's Meadowlark'). A total of 15 subspecies are currently recognized. It is interesting to note that there are gaps in the distribution of Eastern Meadowlarks, isolating some of the subspecies or groups of subspecies from their nearest neighbours. Further work is required to sort out the exact relationships between these different races. The subspecies *lilianae* ('Lilian's Meadowlark') is treated below in its own account. Tail patterns for several forms are illustrated in Figure 71.2.

S. m. magna is the race in the NE US and in Canada, it is described above.

Southern Meadowlark, ***S. m. argutula***, can be found south of the range of *magna*. It ranges from S Oklahoma east to Georgia and south to SE Texas and southernmost Florida. Compared to *magna*, it is smaller, darker above and with a more vivid yellow breast. The darkness of the upperparts results from a mixture of darker colours and narrower pale edging and notching.

S. m. hippocrepis is restricted to Cuba and the northern part of the Isle of Pines, and appears to be most closely related to *S. m. argutula*. It is smaller than *argutula* and is more streaked on the underparts. Its song is reported to be simpler than *magna*, but still made up of clear whistles.

S. m. hoopesi is found from SE Texas to N Tamaulipas, Mexico. It is much paler and greyer than *argutula*, with tail bars usually separate, not confluent, as in Western Meadowlark (72). Saunders (1934) noted that *hoopesi* and *argutula* intergrade in Texas.

In the highlands and on the Pacific slope of Mexico and Central America, there is a group of subspecies characterized by being large, paler in back colouration and showing more white on the tail than the Caribbean lowland forms. This group includes three subspecies.

S. m. auropectoralis is found along the western highlands of Mexico from Durango and Sinaloa south to Puebla, and Michoacán; found down to sea-level in Nayarit. It is similar to 'Lilian's Meadowlark' in the extent of white on the tail (completely white third rectrix, except in 1st-year birds), the dark colour of yellow of the underparts, and the short legs and tail, but it is much darker above. The yellow of the underparts is even darker than 'Lilian's', tending to orange-yellow. It has a shorter wing, and the bars on the tail and tertials are confluent rather than separate. *S. m. auropectoralis* has white rather than greyish cheeks. In non-breeding plumage, the yellow breast is heavily veiled with buff.

S. m. saundersi ranges in the Pacific lowlands of SE Oaxaca. It is most similar to *auropectoralis*, but paler, less rufous above, as well as shorter-winged and shorter-billed.

S. m. alticola inhabits the southern Mexican highlands of Chiapas south to the highlands of Guatemala, Honduras, Nicaragua and Costa Rica. *S. m. alticola* is similar to *auropectoralis* in the darkness of the upperparts, including the black-centered uppertail-coverts, but has shorter wings, longer legs and less white on the longer tail, yet more white than on lowland *mexicana*. *Alticola* looks greyish due to greyish-tipped upperpart feathers and its greyish flanks. Apparently, it sometimes shows some yellow extending onto the cheeks like a Western Meadowlark. The third retrix (R3) is black bordered, so is only partially white.

The next group occurs along the Caribbean lowlands of Mexico and Central America, except for *subulata* which is a Pacific lowland bird that has recently arrived on the Caribbean coast. Compared to the last group, these are smaller and darker, with less white on the tail. Four subspecies are involved.

S. m. mexicana lives in the Atlantic coastal lowlands of se. Mexico, Veracruz, NE Oaxaca, Tabasco and Chiapas; across the base of the Yucatán peninsula and the Petén of Guatemala, and into Belize. Within this group it has long legs, a long tail and short wings, yet the tail is shorter than on highland forms. It is small and richly coloured with rufescent on the upperparts and has a narrow chest 'V'. The tail has the three outer tail feathers (R6–R4), white with a small dusky streak, the fourth tail feather (R3) has more white than dusky on the inner web.

S. m. griscom is restricted to the N Yucatán peninsula, along the coast. It is most like *mexicana*, but larger and paler, and with unstreaked, black crown stripes.

S. m. inexpectata is found in the Caribbean lowlands of Honduras south to Nicaragua. It is the smallest Eastern Meadowlark. It is also much shorter-winged than the other Mexican races; otherwise it is like *mexicana*, but darker, with northernmost birds blending into *mexicana*. This form tends to have spots, rather than streaks on the breast sides. The tail is dark, the fourth rectrix from the outside (R3) usually has the white reduced to a central streak, the next feather out (R4) has a substantial amount of dark on the inner web often with white restricted to the outer vane.

S. m. subulata occurs on the Pacific slope of Panama, and recently to the Caribbean side. This is a small and dark meadowlark, similar to the more northern *inexpectata*. However, the tail pattern is more similar to the larger, paler *alticola*; thus it has much more white on the tail than *inexpectata*. In structure, this meadowlark differs from the others – it is long-billed yet short-tailed and short-legged. The bill is finely pointed. Morphologically, this form does not link the Central American races with the South American forms as its geography may imply.

There are four subspecies in South America, birds in the highlands tend to be larger than the ones from the lowlands; they vary in the amount of white shown on the tail.

S. m. meridionalis ranges in the Andes of Colombia from Santander and Cundinamarca to NW Venezuela, along the Andes up to Trujillo; an isolated population is found at the head of the Magdalena valley in Colombia. It is a bird of the temperate zone. This is a large, dark form, with a long bill and long tail. It is the longest billed Eastern Meadowlark. This form shows a noticeable chestnut wing panel, made up by the chestnut-edged secondaries and coverts. The dark barring on the rectrices and tertials is wider and darker than on *paralios*. The yellow of the throat invades the lower cheeks, as on Western Meadowlark. The fourth rectrix from the outside (R3) is typically dark on the inner web. At least three outer rectrices always show some white. The nape tends to look greyish and the 'V' on the underparts is broad.

S. m. paralios occurs in the coastal lowlands of N Colombia, west of the Eastern Andes, as well as the Santa Marta highlands and east to the savannas of N and C Venezuela south to N Amazonas. This is also a large race, but smaller, paler and browner than *meridionalis* and with a shorter bill than that form. The nape is buffy, not grey. The cheeks on this form are pale, contrasting well with the dark eyeline. The fourth rectrix from the outside (R3) tends to be almost completely white and the fifth (R2) almost always has some white.

S. m. praticola inhabits the lowlands (Llanos) of central Colombia east of the Andes through southern Venezuela, east to the Guianas at least to Surinam. This lowland form is small, smaller than *meridionalis* or *paralios*. It is paler than *meridionalis* and similar to *paralios*, but unlike *paralios* it has dark inner webs to the fourth rectrix from the outside (R3). The upperparts are more chestnut than *paralios* or *meridionalis*. This form shows some yellow extending to the cheeks as on Western Meadowlark. The birds from Mount Roraima are slightly larger and darker, and are sometimes given the name *S. m. monticola*.

S. m. quinta occurs in Surinam and NE Brazil, to the mouth of the Amazon. It is most like the distant *paralios* as it is greyer than the neighbouring *praticola*. However, the pale fringes on the coverts and secondaries are grayer than in either form.

The relationship between the Mexican forms and 'Lilian's Meadowlark' is not known. *Hoopesi* apparently interbreeds to some extent with 'Lilian's' as well as with *mexicana*, but the extent of hybridization has not been examined. Few descriptions of songs and calls exist for Neotropical forms, vocalization data will likely elucidate the relationships between these forms as would a detailed molecular analysis.

HABITAT Typically found in pastures, meadows and old field habitats as well as native grasslands (tall grass prairie). Eastern Meadowlark is not adverse to using agricultural areas as habitat, particularly alfalfa and hay fields as well as the shrubbier edge of wheat and corn fields (Lanyon 1995). Any open area with grass is used, including golf courses and airports (Lanyon 1995). Overall Eastern Meadowlark shows a preference for grasslands with a good growth of green grass as well as ample amounts of dead litter (Wiens and Rottenberry 1981). It prefers moister sites than Western Meadowlark. In the Great Plains, at the western boundary of its range, Eastern Meadowlark is largely confined to the moister floodplain grassland along major rivers. Upland (drier) sites are all taken by Western Meadowlarks. In the overlap zone in the northeast part of the Western Meadowlark's range, there appears to be little difference in the nesting habitat, partially because of the lack of dry prairie. Winters in similar habitats, but will take muddy fields and farmland where little vegetation is available. The northward extent of the wintering range is not delimited by habitat as much as by temperature. Eastern Meadowlarks winter south of areas where the mean minimum winter temperature is above –9ºC and most winter south of areas where this temperature is above –1ºC (Root 1988). Mexican birds live in mountain meadows and steppes as well as lowland areas. In eastern Nicaragua, Eastern Meadowlark inhabits pine savanna, being absent when the trees become too dense or in treeless grassland (Howell 1972). Cuban *hippocrepis* is found in savanna, dry grassland, cattle pastures and open agricultural zones (Gundlach 1876), and in grasslands at the edge of pine woods on the Isle of Pines (Todd 1916). South American birds inhabit open areas ranging from rangeland and savanna to Andean meadows.

BEHAVIOUR Eastern Meadowlarks feed on the ground and can be quite inconspicuous unless singing. They take to the wing in a fluttery flight on bowed wings, frequently flashing the white of the tail. Males arrive in the northern part of the range in late March and begin to proclaim their territory by singing from trees or fence posts. During territory settlement, males frequently fight over boundaries. Males also defend their territories against Western Meadowlark males (interspecific territoriality) in areas of sympatry, but they do not respond to songs of Western Meadowlarks where the two are allopatric (Falls and Szijj 1959). The size of territories does not differ between Eastern and Western Meadowlarks. In Ontario, male Western Meadowlarks respond to songs of either Eastern or Western Meadowlarks. However, only Eastern Meadowlarks with Western Meadowlark neighbours will respond to songs of the other species. This implies that meadowlarks learn to react to songs of the other species. At this particular site, where Western Meadowlarks are rare, only experienced Eastern Meadowlarks learn to react to the former's songs; this likely also applies to Western Meadowlarks. Females arrive later than males and are intensely courted. Courtship behaviours involve the male following the female through the grass and by 'bill tilting', 'tail flashing', 'wing fluttering' and fluffing out and displaying the yellow and black of the breast. The male may do a 'jump display', often flashing the white on the tail, during times of intense excitement. The mating system involves both polygynous males and monogamous males, roughly half of each in one study in Ontario. Breeding success of females mated with polygynous males appears to be higher than for those mated monogamously, the polygynous males may be older more experienced individuals. Females are not territorial and nest completely within the territory of their mate. Males of both species will apparently mate with females of either species indiscriminately. The females are choosy with regard to which male's territory they settle in, and therefore mating with the correct species is based entirely on the behaviour of the female. In winter, Eastern Meadowlarks flock in small (5–10) or very large numbers (300+).

The social structure in the non-migratory populations is less well known. Howell (1972) noted that birds in eastern Nicaragua sang throughout the year, even when not in physiological breeding condition. Flocks are observed in the non-breeding season, but pairing begins early in the year (February–April). A song duet, with females 'rattling' in response to the male's song was heard during March and April.

NESTING North American populations nest April–July. Highland Mexican birds nest late February–July. Cuban *hippocrepis* nests March–June (Todd 1916). In eastern Nicaragua, nesting begins after late April (Howell 1972). The nests are on the ground, and most are domed grass structures, often with a covered tunnel that leads to the entrance. Nest construction time ranges from three to 18 days, and the male appears to help the female. Clutch size 3–7 eggs, but 5 is standard, and 4–5 eggs in Cuban *hippocrepis* (Gundlach 1876). The clutch size of second nesting attempts tends to be smaller. Eggs are white with heavy brownish markings, which are more concentrated on the broader end. Only the female incubates the eggs, which hatch in 14 days. The young stay in the nest for 12 days, at which time they walk out of the nest. Fledglings fly very little during the first few days out of the nest, instead they keep out of sight walking in the grass.

DISTRIBUTION AND STATUS A common bird in eastern North America. Uncommon to fairly common in Central America and South America. Found to 1900 m in Colombia (Lehmann 1960).

Breeding: Found north to NW Ontario (Rainy River), NE Minnesota and Wisconsin east through N Michigan (Upper and Lower peninsulas) and S Ontario (to Sudbury and Timmins), S Quebec east to the north shore of the Gulf St Lawrence, Anticosti Island and the base of the Gaspé peninsula, continuing southeast in SE Maine, S New Brunswick and W Nova Scotia. The eastern and southern limits to the distribution in the US are set by the Atlantic coast and Gulf coast, the range includes all of the Florida peninsula. Breeds west to North Dakota where it has been noted on territory at one site each in Hope and Grant Co. in 1994, C Minnesota, W Iowa and along river valleys to SW South Dakota, W Nebraska, W Kansas, W Oklahoma (to Beaver Co. at the base of the panhandle), C Texas panhandle, Texas east of the Llano Estacado and Edwards plateau (see 'Lilian's' Meadowlark). The distribution continues south into Mexico, in the coastal lowlands of Tamaulipas, Veracruz, Tabasco to W Campeche. Also the highlands of the Sierra Madre Occidental and parts of the central plateau (absent from the N central plateau) as far north as Durango, C Zacatecas and N Guanajuato; west to W Nayarit, W Jalisco and E Colima; south to S Michoacán, N Guerrero, N Oaxaca and N Chiapas; east to Veracruz and E Chiapas and the highlands of SC Guatemala. Also resides in isolated population along the coastal lowlands of N Campeche and N Yucatán, in the Yucatán peninsula. Another isolated population occurs in the lowlands of Belize and extreme E Guatemala. Another population inhabits the highlands of N El Salvador and S Honduras to adjoining NE Nicaragua. It may occur in S Nicaragua. In Costa Rica, it is widely distributed, both in the interior and on both coastal slopes. In Panama, it is found along the Pacific slope from Chiriqui east to Panamá province, extending up to the highlands and recently invading the Caribbean slope along the Canal Zone in Colon province (Ridgely and Gwynne 1989). The Colombian range is disjunct with a population in the Santa Marta highlands, the East Andes from Norte de Santander south to the the head of the valley in Huila and extending to the E slope of the Central Andes in SW Huila. Also in the lowland Llanos of NE Colombia as far south as Meta and S Vichada and east to the Orinoco (Hilty and Brown 1986). In Venezuela, throughout the Llanos and lowlands north and south of the Orinoco river; also in the Tepui highlands (Gran Sabana), and adjoining Roraima, Brazil. Widely distributed in Guyana, Surinam and French Guiana, the range continues east along the coastal lowlands of Amapá in Brazil to the mouth of the Amazon river, including the south shore in E Pará. Also found in Cuba, the Isle of Pines and Cayo Coco, but absent from the Bahamas.

Non-Breeding: The populations in most of Mexico, Central America, South America and Cuba are sedentary. However, the northern population in the US and Canada does migrate south in winter. Winters mainly in the southern US. Vacates the following areas in winter: Minnesota, Wisconsin, Michigan, Ontario, Quebec, Nova Scotia, New Brunswick, Maine, Vermont, New Hampshire and the mountains in W New York, C Pennsylvania and Virginia. The northern edge of the winter distribution varies depending on the severity of the winter, in particular snowfall.

MOVEMENTS Tropical populations and southern US *S.m.argutula* are sedentary, but northern ones are migratory. *S.m.magna* arrive in the breeding grounds late February in their southern range and typically early March farther north. Spring arrival varies depending on whether or not snow remains on the ground. Fall migration begins early September–late October. It is mainly a diurnal migrant. Northern birds migrate an average of over 1000 km to their wintering areas (Lanyon 1995).

Birds from a given breeding population do not necessarily winter in the same area (Lanyon 1995). The oldest banded individual was initially caught as a juvenile in Pennsylvania and recovered aged 8 years 8 months while wintering in North Carolina (Klimkiewicz and Futcher 1987). The longest distance banding recoveries are as follows: two individuals banded in the breeding grounds in Minnesota, one of which was recovered wintering in Oklahoma while the other was recovered in Alabama; one banded in the breeding grounds in Wisconsin was later recovered on northward migration in Mississippi; another Wisconsin breeder was found wintering in Arkansas; an Ontario breeder was recovered wintering in South Carolina (BBL data).

This meadowlark is unrecorded west of its regular range. This may be partly due to a general tendency not to wander or become lost, but more likely the perceived rarity is due to the immense identification problem posed by attempting to separate non-vocal Eastern and Western meadowlarks. Some western records appear to be little more than breeding range extensions or overshoots. In Colorado, this species has been confirmed from the far NE part of the state. Singing birds have been reported as far west as Fort Collins (T. Leukering pers. comm.). In North Dakota, Eastern Meadowlarks have been noted on territory at two sites in 1994, in Hope and Grant County, they are rare, indeed accidental, this far west. In Manitoba, they are exceedingly rare but have attempted to oversummer. A record from June 25, 1993 was the first record in the province since 1988. Eastern Meadowlark has been suspected at least twice in California, once in May and once in Oct. as a vagrant to Southeast Farallon Island; both were not accepted by the state rare bird committee. By contrast, in the northeast, north of breeding areas, Eastern Meadowlark is a regular stray. Vagrant records exist for Newfoundland, the first record was on Jan 4, 1938 (Peters and Burleigh 1951). It has strayed in spring to Sable Island, Nova Scotia (Tufts 1986). Also recorded from N Ontario, NW Quebec and Prince Edward Island. It has occurred twice in Bermuda, both times in winter (Lanyon 1995). Note that there are fossil Eastern Meadowlarks from New Providence Island (Lanyon 1995), therefore this species existed there at some time in the past. There is one historical record from the British Isles, a specimen of the race *magna* was collected in March 1860 in Thrandeston, Suffolk (Sclater 1861a). Eastern Meadowlark, probably the race *paralios*, has been recorded as a straggler to Bonaire, Netherlands Antilles. The Bonaire individual was seen on Nov. 2, 1977 (Voous 1983).

It has been expanding in distribution in Costa Rica, Panama and Colombia due to the expansion of agriculture and creation of suitable habitat. For example, the mountains of Huila, Colombia, were quickly colonized after the forests there were cut down (Lehmann 1960)

MOULT In North America north of Mexico, juveniles have a complete first pre-basic moult in August–October while on the breeding grounds. After this moult, immatures are indistinguishable from adults. Adults undergo a complete definitive pre-basic moult on the breeding grounds, between August–October. The brighter breeding plumage is acquired through wear, thus fall birds are fresh and late summer birds are most worn. It is not known if the extent of these moults is similar in the neotropical races. Moult timing differs depending on the subspecies and local conditions which affect the timing of nesting. *Meridionalis* of Colombia is in pre-basic moult June–July, specimens from October–November are in fresh plumage. Specimens of *paralios* from Santa Marta, Colombia, are in fresh plumage July–August (Todd and Carriker 1922). Mexican *alticola* are undergoing the pre-basic moult during October and November. In contrast, Costa Rican *alticola* specimens show evidence of moult in June.

MEASUREMENTS Males are slightly larger than females.

S. m. magna. Males: (n=21) wing 122.4 (114.3–129); culmen 34.3 (30.7–36.8); tail 78.7 (67.6–86.4); tarsus 41.9 (38.9–46.2). Females: (n=8) wing 107.4 (104.6–113.8); culmen 30 (26.9–31.5); tail 67.6 (65.8–73.7); tarsus 37.3 (36.1–40.6).

S. m. argutula. Males: (n=16) wing 111.8 (104.6–118.9); culmen 32.8 (30.5–35.1); tail 72.9 (67.6–78.5); tarsus 41.4 (39.4–43.9). Females: (n=8) wing 99.1 (95.3–101.6); culmen 28.7 (27.9–30.2); tail 63 (60.2–67.1); tarsus 38.4 (36.3–40.6).

S. m. hippocrepis. Males: (n=9) wing 102.1 (96.5–107.2); culmen 32.3 (31.5–34); tail 65.8 (61.2–69.9); tarsus 38.6 (36.8–40.4). Females: (n=5) wing 93.7 (88.9–101.1); culmen 30.7 (29.2–32.5); tail 60.2 (53.1–67.8); tarsus 36.1 (35.1–39.4).

S. m. hoopesi. Males: (n=27) wing 117.1 (113.0–126.0); culmen 33.3 (30.2–35.6); tail 72.4 (63.5–79.2); tarsus 41.0 (37.8–44.4). Females: (n=9) wing 106.4 (100.3–109.7); culmen 31.7 (30.5–33.5); tail 68.3 (64.0–73.7); tarsus 37.8 (36.8–39.4).

S. m. auropectoralis. Males: (n=12) Wing 116.1; culmen 30.9; tail 67.5; tarsus 40.8. Females: (n=4) Wing 105.0; culmen 29.0; tail 62.0; tarsus 37.6.

S. m. alticola. Males: (n=45) Wing 110.5; culmen 30.5; tail 74.7; tarsus 37.6. Females: (n=21) Wing 98.2; culmen 28.5; tail 65.3; tarsus 38.7.

S. m. mexicana. Males: (n=10) Wing 104.5; culmen 29.6; tail 69.5; tarsus 41.1. Females: (n=3) Wing 97.1; culmen 29.5; tail 67.0; tarsus 39.6.

S. m. saundersi. Males: (n=14) Wing 114.3 (111–119); (n=13) tail 76.9 (74–82). Females: (n=11) Wing 102.0 (99–107); (n=10) tail 67.8 (63–72).

S. m. inexspectata. Males: (n=15) Wing 96.3 (94–99); (n=9) tail 63.9 (59–67). Females: (n=10) Wing 88.0 (86–89); (n=7) tail 57.9 (54–60).

S. m. griscomi. Males: (n=19) Wing 107.3 (103–112); (n=7) tail 73.1 (69–79). Females: (n=18) Wing 96.9 (94–101); (n=10) tail 64.5 (61–71).

S. m. subulata. Males: (n=13) wing 104.7 (100–110); (n=11) tail 69.8 (67–75); (n=13) tarsus 38.1 (35.1–40.2). Females: (n=16) wing 92.6 (91–107); (n=9) tail 62.2 (58–71); (n=16) tarsus 36.9 (34.8–39.1).

S. m. meridionalis. Males: (n=10) wing 112.2 (106–117); tail 72.2 (66–76); tarsus 40.0 (38.7–41.6). Females: (n=9) wing 102.8 (95–110); (n=6) 66.8 (64–73); (n=9) tarsus 38.0 (35.7–40.4).

S. m. paralios. Males: (n=18) wing 109.2 (102–115); (n=7) tail 72.6 (66–78); (n=4) culmen 33.6 (32.6–34.6); (n=18) tarsus 39.9 (38.0–41.9). Females: (n=8) wing 101.6 (91–109); (n=3) tail 64.3 (62–66); (n=1) culmen 32.0; (n=9) tarsus 36.5 (34.5–38.2).

S. m. praticola. Males: (n=13) wing 102.8 (98–107); tail 66.8 (64–70); tarsus 38.7 (34.9–43.0). Females: (n=6) wing 95.5 (92–101).

S. m. quinta. Males: (n=6) wing 104.3 (102–106); tail 68.3 (61–71); tarsus 37.7 (36.1–39.2). Females: (n=5) wing 94.8 (93–98); tail 61.8 (57–67); tarsus 36.4 (35.7–37.0).

NOTES It is not unlikely that neotropical forms of Eastern Meadowlark may deserve species status, but not enough is known at this time. For hybridization information, see Western Meadowlark.

REFERENCES Bangs 1900 [geographic variation, measurements], Bent 1958 [behaviour, migration], Chapman 1917 [voice], 1931 [geographical variation], Chubb 1921 [geographic variation], Dickerman 1989 [geographic variation], Dickerman and Phillips 1970 [geographic variation, measurements], Dunning 1993 [weight], Falls and Szijj [behaviour], Griscom 1934 [geographical variation, measurements], Gundlach 1876 [voice, habitat, nesting], Haverschmidt 1949 [distribution], Klimkiewicz and Futcher 1987 [age], Knapton 1988 [behaviour], Lanyon 1956 [behaviour], 1957 [vocalizations, behaviour], 1995 [voice, geographical variation, measurements], Pyle *et al.* 1987 [moult], Rohwer 1972a [distribution], 1972b [hybridization, identification], 1976 [identification], Saunders 1934 [geographic variation, measurements], Sclater 1861 [movements], Sick 1993 [vocalizations].

71X 'Lilian's Meadowlark' *Sturnella magna lilianae* (Oberholser) 1930 — Plate 27

A meadowlark of desert grasslands. Usually considered part of Eastern Meadowlark species, but recent research has suggested that this may not be the most appropriate classification.

IDENTIFICATION Note that 'Lilian's Meadowlark' is the smallest of the North American meadowlark trio but it has longer wings than Eastern Meadowlark (71). Several other shape differences exist, including shorter legs than Eastern Meadowlark and a shorter tail than the other two species. All of these shape differences are not obvious in the field and should be used only with the utmost care. 'Lilian's Meadowlark' has the most white on the tail, and the least amount of yellow on the throat. Compared to Western Meadowlark (72), 'Lilian's Meadowlark' is similar in that it is pale and has separate (not confluent) bars on the tail and tertials. However, 'Lilian's Meadowlark' has a white, not yellow malar area. The face pattern differs further in that the auriculars of 'Lilian's' are pale and unstreaked, and contrast strongly with the blackish eyeline (Figure 71.1) (Zimmer 1985). On Western Meadowlark, the auriculars are lightly streaked and do not contrast as strongly with the dark brown eyeline. There is some variation in the face pattern of Western Meadowlarks and it is not clear if this face pattern difference also applies to winter individuals. Furthermore, 'Lilian's Meadowlark' has much more extensive white on the tail, a feature that easily separates it from Western Meadowlark in flight when seen well (Figure 71.2). During the breeding season, 'Lilian's Meadowlark' is found in desert grasslands while Western Meadowlark is found in moister agricultural areas and riparian zones. The song and calls are the most reliable way to identify this form from Western Meadowlark. In Arizona, wintering Western Meadowlarks are more likely to be found in roadsides, towns and other disturbed sites while 'Lilian's Meadowlarks' tend to prefer native grasslands. Furthermore, Western Meadowlarks tend to be in larger flocks, while 'Lilian's Meadowlarks' may be alone or in small flocks (K. Kaufman pers. comm.).

VOICE Song: A series of clear whistles, similar to that of Eastern Meadowlark. However, some songs are more reminiscent of Western Meadowlark, being more varied and complex. 'Lilian's Meadowlark' has a smaller syllable repertoire than Eastern Meadowlark, but a larger one than Western Meadowlark (Barlow *et al.* 1994, Barlow pers. comm.). In addition, the song of this form averages lower in frequency than that of Eastern Meadowlark (Barlow *et al.* 1994). The song can be described as 'tortillas con chile'.

Calls: The calls are similar to those of Eastern Meadowlark (J. Barlow pers. comm.).

DESCRIPTION A stocky, short-tailed meadowlark with a long bill. The culmen is straight, with a slight curvature at the extreme tip. Wing Formula: P9 < P8 ≈ P7 > P6; P9 > P8; P8–P5 emarginate.

Adult Male Breeding (Definitive Basic, worn): The crown is blackish-brown with a cream median crown stripe. It shows a bold cream-coloured supercilium, which is yellow before the eye. The blackish-brown eyeline contrasts with both the pale supercilium and the creamy ear-coverts and neck sides. The nape is cream with brown streaks. The back is blackish with brown bars, and cream or buff streaks. This pattern is produced by blackish-brown feathers with a brown (paler than the base of the feather) terminal bar, and some lateral brown barring basal to that, in addition to bold cream or buff edges. The rump is like the back, but slightly paler. Each uppertail-covert is creamy-buff with an irregular blackish-brown central stripe. The underparts are largely yellow with an obvious black 'V' on the breast. The throat, breast and belly are yellow, broken by the black 'V' on the center of the breast. Note that the yellow of the throat does not invade the cheek, but stays below the malar area. The flanks are pale creamy-buff with brown terminal spots on each feather, creating a loosely streaked or spotted pattern. The crissum is similar to the flanks in colour, but instead of a terminal spot there is a brown central streak on each feather. The wings are as complexly patterned as the body plumage. The lesser coverts are creamy-grey with darker bases, while the median coverts are warmer creamy-buff with more obvious dark feather centers. The greater coverts are creamy-buff with dark bars across each feather, the bars are separate, not confluent along the rachis. The tertials are patterned as the greater coverts. The primaries and secondaries are creamy-brown, with darker bars particularly on the outer vane and the primaries show a thin pale fringe that shows up as a weak pale panel on the closed wing. There is an obvious yellow bend of the wing created by yellow marginal coverts. The underwings are pale, showing white wing-linings. The tail is largely white (Figure 71.2). The outer four rectrices (R3 to R6) are entirely white, sometimes there is a small dark streak along the rachis of R6, the outermost rectrix. The central tail feather, R1, is the only one which is entirely dark. It is creamy-brown with separate darker brown bars. The second pair of rectrices from the inside, R2, shows the inner vane patterned like R1 and the outer vane usually white; however, there is a great deal of variation in the extent of white on this feather. In some cases, only the outer three pairs of feathers (R4 to R6) are entirely white, R3 being largely white but with a dark central streak and R1 and R2 entirely dark and barred. The legs are greyish-pink.

Adult Male Non-Breeding (Definitive Basic, fresh): This is the fresh plumage which wears into the breeding dress. It is similar to the breeding plumage, but the black 'V' of the breast is obscured by buff feather tips. Similarly the yellow of the underparts is veiled under creamy or buffy feather tips. Due to the smaller and more concentrated feathers on the throat, the feather tips obscure the yellow throat entirely, making it appear creamy or buffy. The face (auriculars mainly) is dark cream, darker than on the breeding bird and contrasting less with the dark eye-line. The back and scapular pattern is crisp and scaly, due to complete cream-coloured feather fringes. The overall upperpart colouration appears warmer, less cold and greyish, than in the breeding season.

Adult Female Breeding (Definitive Basic, worn): Similar to male but slightly duller in colour. The underparts are not as bright yellow.

Adult Female Non-Breeding (Definitive Basic, fresh): Indistinguishable from non-breeding male.

Juvenile (Juvenal): Similar to non-breeding adult but pale fringes on the back and scapulars are paler and complete, thus appearing more scaly than adults. The underparts appear buffy, sometimes showing some dull yellow to the feather bases. The breast 'V' is replaced by dark streaking in the same area. Younger juveniles show a pink bill base that is not seen on adults.

GEOGRAPHIC VARIATION Here we treat 'Lilian's Meadowlark' as a distinctive subspecies of Eastern Meadowlark. However, there is ongoing research that suggests that 'Lilian's Meadowlark' is best treated as a separate species (J. Barlow pers. comm.). If this takes palce, it would need to be decided how many of the current Eastern Meadowlark subspecies would become subspecies of 'Lilian's Meadowlark'. Sibley and Monroe (1990) classify 'Lilian's Meadowlark' as a full species. According to J. Barlow (pers. comm., unpub. data) the Mexican taxa *auropectoralis* is clearly more closely related to 'Lilian's', while the very similar appearing *hoopesi* belongs in Eastern Meadowlark. Saunders (1934) also hypothesized that *lilianae* was most closely related to *auropectoralis*, and that the more southern *alticola* also fit in this group. (See Eastern Meadowlark, Geographic Variation).

HABITAT Desert grasslands, unlike Eastern Meadowlark; 'Lilian's Meadowlark' prefers drier sites than Western Meadowlark. In Arizona, it uses moister sites than in Texas and New Mexico. In winter, desert grasslands are still the preferred habitat, and it is not as commonly found in disturbed areas such as golf courses, urban areas and agricultural fields as are wintering Western Meadowlarks.

BEHAVIOUR Similar to Eastern Meadowlark. No behavioural studies have been made which focus on this form.

DISTRIBUTION AND STATUS Common to uncommon in desert grasslands; where it is found alongside Western Meadowlark, 'Lilian's' is more abundant. Breeds from C and SE Arizona, west to S and C New Mexico, and W Texas west of the Llano Estacado (Rohwer 1972b) and Edwards plateau. Also extends to NE Sonora and N and C Chihuahua in Mexico. The range of 'Lilian's Meadowlark' comes closest to that of Eastern Meadowlark in Texas, but both are separated by 200 km in most of central Texas where only Western Meadowlark is found, but locally as close as 50 km (Lanyon 1995). Resident, but wanders to C Arizona in winter (Lanyon 1995). If the *auropectoralis* subspecies of Eastern Meadowlark is classified as a form of 'Lilian's Meadowlark' then the distribution would be extended farther south in Mexico (see Eastern Meadowlark, Geographic Variation).

MOVEMENTS 'Lilian's Meadowlark' is sedentary, even in its US breeding range.

MOULT Similar to Eastern Meadowlark. Adults undergo one complete moult a year, the definitive pre-basic. This moult occurs after breeding. It appears to take place later than on Eastern and Western Meadowlarks, mainly during mid to late October in *lilianae*. The overall effect is that midwinter 'Lilian's Meadowlarks' appear less worn than Western Meadowlarks wintering in the same areas. The first pre-basic moult is complete, its timing is perhaps earlier than that of adults. Pre-alternate moults are lacking, all plumage change between winter and summer occurs as an effect of wear and bleaching.

MEASUREMENTS ***S. m. lilianae***. Males: (n=39) Wing 119.5; culmen 31.4; tail 73.3; tarsus 39.6. Females: (n=11) Wing 105.7; culmen 28.4; tail 63.6; tarsus 36.7.

NOTES A multivariate analysis of morphology suggested that 'Lilian's' and Western Meadowlarks are less likely to hybridize than Western with Eastern. The status of this form is still in discussion. It is suggested that 'Lilian's Meadowlark' deserves species status based on vocal and molecular (J. Barlow pers. comm.) as well as morphological data (S. Rohwer pers. comm.).

REFERENCES Jon Barlow pers. comm. [voice, systematics, distribution], Rohwer 1972a [distribution], 1972b [identification], 1976 [identification], Saunders 1934 [geographic variation], Zimmer 1984 [identification, habitat].

72 WESTERN MEADOWLARK *Sturnella neglecta* Plate 28
Audubon 1844

A common roadside bird of much of western North America, its distinctive song is a characteristic sound of the west.

IDENTIFICATION Western Meadowlarks are frustratingly similar to Eastern (71) and 'Lilian's' (71X) Meadowlarks; this trio makes for one of the most difficult identification problems in North America. The most reliable way to differentiate between them is by calls and songs. Western Meadowlark gives a *chup* call not heard from the other two species. The song is in two parts, flute-like and lacks the slurred whistles of Eastern and 'Lilian's' Meadowlarks; it is also much lower in frequency and therefore sounds richer (see also Voice). Visually, Western Meadowlark can be differentiated from the other two species by noting the greater extent of yellow on the malar area (Figure 71.1), but note that females can show less yellow than males, and that the yellow is completely obscured in fresh (non-breeding) plumage. The flanks and undertail-coverts of Eastern Meadowlarks are a rich buff, while on Western Meadowlarks they are greyer (J. Dunn pers. comm.). Furthermore, the colour of the wing-coverts and folded secondaries is much greyer in Western Meadowlark. The corresponding parts of Eastern Meadowlark appear more richly coloured (warm brown or rufous-brown) than those of Western Meadowlark. Compared to Eastern and 'Lilian's' Meadowlarks, Western has less white on the tail; there appears to be some overlap in this feature with Eastern Meadowlark, but not with 'Lilian's Meadowlark' (Figure 71.2). Look for the amount of dark on R4 (third from the outside) and the outer two (R5 and R6). Western Meadowlark tends to have a mostly dark fourth rectrix and noticeable dark terminal shaft streaks on the other two. Western Meadowlark's upperparts look paler than those of Eastern Meadowlark, but are comparable to 'Lilian's' which is only slightly darker than Western. The tertials and central tail feathers of Western Meadowlark tend to show free, parallel-sided bars on a pale background. On Eastern Meadowlark, the bars widen toward the rachis and become confluent with adjoining bars, along the mid-vein. Thus, the tail feathers appear dark with deep pale notches along the fringes on Eastern Meadowlark. 'Lilian's Meadowlark' can show an intermediate tail pattern. Western Meadowlarks have significantly longer wings than the other two species, while they are not the largest of the three. Western also has the shortest legs. Where Western and Eastern Meadowlarks winter together, Western favours the shorter grass or bare fields (J. Dunn pers. comm).

VOICE Song: The primary song is loud, liquid and fluty; it appears to be sung only by the male. It has two parts, beginning with several (up to 6) pure whistles on one tone, and then a short jumbled series of notes, the complete performance lasting approximately one to two seconds. Individual males have a repertoire of several slightly different versions of the song; they will repeat one version several times before switching to another. There is considerable geographic variation in the primary song, but all are easily recognized by their two-part structure and the general frequency; call notes do not exhibit geographical variation. Occasionally performs a 'whisper song'. The song of Western Meadowlark is loud, often the only song that can be heard from a car driving at highway speed with the windows open. The song is commonly used in movie soundtracks. Flight songs are more frequent in Western than Eastern Meadowlarks; these songs are a complex and jumbled series of notes that have been compared to Bobolink (103) song. The primary song is learnt and therefore meadowlarks may learn elements or full songs of the 'wrong' species; thus odd songs should not be used as a characteristic of hybrids. See also Eastern Meadowlark.

Calls: The most common call is the *chupp* or *chuck* note which is given by both sexes. Females give a rattle similar to that of Eastern Meadowlark, while males give a rattle very rarely. Males use a similar call that can be best described as a rolling trill. Both sexes may use a sharp, uninflected, whistle when excited or alarmed. Females also give a *tee-tee-tee* call during the early part of the breeding season. Both sexes give an upwardly inflected whistle, the *weet* call most often while in flight and during migration. The *weet* call is by far the most commonly given during winter and is somewhat similar to the *sweet* call of Hooded Oriole (36). If the habitats of these two species were not so different, the two calls would be readily mistaken.

DESCRIPTION A chunky and short-tailed Meadowlark. It is long-billed, much like Eastern Meadowlark. Wing Formula: P9 < P8 < P7 > P6; P9 ≈ P5; P8–P6.

Adult Male Breeding (Definitive Basic, worn): The crown is blackish-brown with a contrasting whitish central stripe; at close range the crown stripes will be seen to be narrowly edged dry-brown, giving them an indistinctly streaked appearance. Western Meadowlarks sport a broad, pale supercilium noticeably yellow in front of the eye. The lores are pale grey as are the auriculars. There is a dark postocular stripe, similar in colour to the crown. The eyering is white, indistinct on the pale face. The whole of the upperparts are patterned in a complex array of streaks and bars. Each back and scapular feather is dark grey-brown with a wide, paler brown tip; they are edged along the full length with pale buff on both sides. These obvious paler edges give the back a streaked appearance. The paler distal ends of the feathers wear as the season progresses, making the back appear darker later in the summer. The marginal coverts are yellow, appearing as a yellow 'bend of the wing' on perched birds. The lesser and median coverts are blackish-brown with dry-brown to pale grey notches along the edges. Greater coverts appear more barred due to the deeper pale notches cutting into the darker blackish-brown centers, leaving neat darker bars along each feather edge. The bars at the tip of each feather are free and cross the full width of the feather. Each covert overlays the next, covering the darker center, and thus the coverts appear neatly barred on perched birds. The tertials appear similarly dry-brown with thin dark bars; the bars are parallel-edged, not widening significantly at the rachis. Tail well-barred, with even, straight dark bars that do not join together at the rachis (see Geographic Variation). Outer three tail feathers white with a dark shaft streak at the tip. The fourth rectrix from the outside sometimes with a small terminal white spot or even a white central stripe. The underparts are mainly bright yellow from the chin to the vent, with the yellow extending up onto the cheeks above the malar area. Most obvious is the black 'V' on the breast. Flanks pale greyish with dark streaks, down to the undertail. The bill is greyish-blue with a dark culmen and tip. Eyes dark brown and legs pale flesh.

Adult Male Non-Breeding (Definitive Basic, fresh): Similar to breeding adult but appears duller due to fringes obscuring bold patterns. The back and scapulars have complete pale fringes, thus making the back appear scaled. The yellow on the underparts appears duller due to buff fringes on the feathers. Similarly the black 'V' on the breast is obscured by pale grey or buff fringes. The yellow on the cheek is completely obscured in this plumage.

Adult Female Breeding (Definitive Basic, worn): Almost identical to male, except paler and with less yellow on the cheek on average. The black 'V' is reduced. A lone female cannot be sexed, but sexing may be done in direct comparison to males, and by using call and behavioural differences.

Adult Female Non-Breeding (Definitive Basic, fresh): Differs from the breeding plumage in the same manner as males (see above).

Juvenile (Juvenal): Similar to non-breeding adult but pale fringes to the back and scapulars paler, thus looks more scaly. Head stripes are duller. The throat is pale and the breast 'V' is replaced by dark streaking in the same area. The yellow on the underparts is not as intense.

GEOGRAPHICAL VARIATION The race *confluenta* is poorly differentiated and is often not recognized as a valid subspecies. It differs from the nominate in its darker plumage, including brighter yellow underparts, and the tendency for bars on the tertials and rectrices to become confluent near the mid-vein, like in Eastern Meadowlark. It occurs in the northwest of the range from S British Columbia south, west of the Cascades, south to central and southern California. There is a slight cline in wing length, with northeastern populations having the longest wings, and southwestern and Mexican birds having the shortest wings. Geographical variation in plumage best regarded as clinal, with *confluenta* the dark extreme; this form is not reliably identifiable in the field.

HABITAT Grasslands of many types, including desert grasslands, savannas, prairie (short and long grass), pastures and abandoned fields. Also found in alfalfa and hay fields, open orchards, roadsides and weedy edges in more heavily industrialized agricultural fields (Lanyon 1994). Shows a preference for native grasslands (Lanyon 1994). In the western end of their sympatric range, Western Meadowlark, prefers drier and sandier grasslands than Eastern Meadowlark (Lanyon 1956a). However in the eastern portion of the area of sympatry (Ontario) there was no consistent difference in the areas chosen by Western and Eastern meadowlarks as breeding sites (Szijj 1966, J.B. Falls pers. comm.). In the southwestern US, Western Meadowlark uses moister grassland sites than 'Lilian's Meadowlark'. These are more likely to be near watercourses and disturbed areas, farmland for example (Lanyon 1962, Rohwer 1976). Winters in open areas, including agricultural areas, golf courses, airports and other open grassy areas similar to those used while breeding.

BEHAVIOUR Males arrive on the breeding grounds at least two weeks earlier than females and begin to proclaim their territory. Males are highly territorial and may sing up until the late summer, after the second brood has been raised. Males will defend their territories against males of the other two species of sympatric meadowlarks. During the breeding season males are conspicuous, singing from an elevated perch, often a fence post or solitary bush or tree. Singing rarely occurs during moult, but resumes before migration. Females are much less conspicuous and quiet, but can be recognized by their rattle calls. Upon the female's arrival at the breeding grounds a pair bond is quickly forged. Usually pairs will perform aerial chases at this point, with males following the flying female; neighbouring males may join in. This is associated with ground displays, but seldom with much vocalization. The mating system is typically polygynous, each male pairing with two or three females. Females are aggressive to other females during egg laying and before, but aggressiveness declines as the female begins to incubate. It is at this point that a male may acquire another mate. Males help the females to feed the offspring, but not to build the nest. In winter, Western Meadowlarks are apt to flock in small to medium-sized groups. They are rather quiet at this season. Both sexes fly in a distinctive bow-winged, fluttery flight which is powered by sudden bursts of wingbeats separated by glides. The white on the tail is usually flashed and obvious in flight and when landing. See Eastern Meadowlark account for more information.

NESTING The nest is a domed structure made of grass and placed on the ground. Often runways lead to the nest, sometimes these are covered tunnels built by the female. In the north, nest building by the female begins in late April and is finished with in approximately one week. Nests appear to be constructed such that the opening points to the direction away from the prevailing winds. Clutche size 3–6 eggs, with incubation taking two weeks. Eggs are white with brown or chestnut markings, concentrated at the broader end. Successful females tend to renest during the same summer. Frequently parasitised by Brown-headed Cowbird (101), but success rates do not appear to be high.

DISTRIBUTION AND STATUS Breeding: A common to abundant bird in the west, much more scarce and local in the easternmost part of its range. This species expanded its range eastward considerably between 1900 and 1950 and may have now stabilized or may even be retreating from its easternmost breeding localities. Breeds from C British Columbia (north to the Chilcotin District) and in the Peace River District where it is rare and breeding has not been confirmed; this is a westernmost extension of the NC Alberta population. Ranges east through NC Alberta, C Saskatchewan, S Manitoba and W and S Ontario; south along the Pacific coast to NW Baja California, NW Sonora; east through S Arizona, C Texas, NW Louisiana; north through W Arkansas, C Missouri, S Illinois and N Indiana to NW Ohio. It has been suspected of breeding once in the lower Hudson valley in New York State, and may also breed sporadically in S Quebec. There is one breeding record for Tennessee (Robinson 1990). An isolated population is found in Mexico in desert grasslands of C Chihuahua south through Durango, Zacatecas to NE Jalisco and S San Luis Potosí. Introduced to Kuai Island, Hawaii.

Non-Breeding: Northernmost breeders retreat to the south. Winters south in Mexico to S Baja California, Michoacán, Nuevo León and Tamaulipas, and in US in Texas, Louisiana and Mississippi; winters north to S British Columbia, S Prairie provinces (very rare), Idaho, Montana, South Dakota, S Minnesota and S Wisconsin. The eastern edge of the wintering range is uncertain due to confusion with the commoner Eastern Meadowlark. However, it does appear to winter east to E Illinois, W Kentucky, E Missouri, W Tennessee, W Mississippi, W and S Louisiana, S Alabama and the westernmost part of the Florida panhandle. In Kentucky, Tennessee, E Arkansas and N Mississippi, it is a rare to uncommon annual wintering species and spring migrant; its abundance increases in the vicinity of the Mississippi river floodplain. There is one breeding record for Tennessee (Robinson 1990).

MOVEMENTS A short to medium distance migrant. Northern populations retreat southward and northwestern populations leave the interior and head to the coast, or southward in winter. Spring arrival is governed by temperatures and does vary, but is typically in mid March in the northeastern part of its range. Arrival dates are earlier further south and in the west. For example, late February in the British Columbia interior. Fall migration spans September–November. Two birds banded in Ontario were recovered wintering in South Carolina and North Carolina, while another was recovered wintering in Mississippi implying that easternmost breeders also winter in the east (BBL data). Of the 82 recoveries, the most interesting include the following: a Wisconsin breeder recovered in Arkansas; a Kansas breeder recovered in Texas; a Montana breeder recovered in Texas; a Nebraska breeder recovered in Arkansas; a bird wintering in Arizona recovered breeding in Wyoming; a bird breeding in Alberta recovered two years later in Texas during August and two Saskatchewan breeders recovered wintering in Oklahoma and Arkansas (BBL Data). Western Meadowlark usually migrates during the day. A bird banded as an adult in Colorado was retrapped in the same area aged six years and 6 months, this is the oldest recorded Western Meadowlark (Klimkiewicz and Futcher 1987). Western Meadowlarks have reached the age of 10 years in captivity (Lanyon 1994).

This species is a vagrant to the north. In Alaska, it has been recorded several times, including three specimen records (Gibson and Kessel 1997), one of these came from the far north from Craig, Alaska on October 22, 1921 (Willett 1923). There are no records for the Yukon, but it has been sighted twice in Fort Simpson, S Mackenzie, Northwest Territories (Lanyon 1994). Also known from Churchill, Manitoba, Moose Factory and Moosonee, Ontario, all on the shores of Hudson's Bay and James Bay (Godfrey 1979). Western Meadowlark is a rare vagrant east of its regular range, bearing in mind that in winter Western Meadowlark is regularly found farther east than during the summer. However, there are records from most eastern states and provinces, except perhaps Newfoundland, Prince Edward Island, Connecticut and Maryland (but see below). Most of the reports occur during the early spring when meadowlarks are most vocal, but there are a few fall and winter records. Due to the problems in identification, many of the reports are of birds vocalizing. However, it appears that a high proportion have included birds singing Western Meadowlark-type songs, and no mention is made of their diagnostic calls. These birds may still be Eastern Meadowlarks in fact, unless the calls are heard (see Identification), therefore some voice-confirmed records may not be valid. Curiously, between the 1950s and 1970s, Western Meadowlarks were more regular in the east than they are now, there appears to have been a range expansion and contraction for some unknown reason.

In New Brunswick, there are approximately four records, while it has been recorded at least once from Nova Scotia. It is a rare vagrant to the New England states. For example, in Massachusetts it has been recorded on over 25 occasions, most of these records between 1957 and 1974; it has been much rarer during the 1980s and 1990s (Veit and Petersen 1993). Farther south in New Jersey, there are 10 reports, all April–July from 1940 to 1965. There are no recent records (Laurie Larson pers. comm.). There are no official records from Maryland, but in the last year or so a bird has been heard singing a Western Meadowlark-type song in Cecil County. In the Carolinas, it is accidental (Potter *et al.* 1980). In Georgia, there are three reports, two of them before 1960, and one specimen. In Florida, it is rare but regular in winter in the Panhandle. There are 15 or so records from the peninsula, none substantiated, but likely some were real; some of these were apparently in song (Stevenson and Anderson 1994). Inland from the coast, it is very rare but somewhat regular in Pennsylvania with over 10 records, several of these in summer, but with no evidence of breeding. There are several records for West Virginia. In Kentucky, Tennessee, E Arkansas and N Mississippi it is a rare to uncommon annual wintering species and spring migrant; its abundance increases in the vicinity of the Mississippi river floodplain. Previously, it has been under estimated how regular Western Meadowlark is this far east in winter, and these birds should not be considered vagrants here.

MOULT Adults undergo a complete definitive pre-basic moult while still on the breeding grounds and after breeding has finished, August–October. The breeding plumage is acquired through wear and fading, there are no pre-alternate moults. Therefore, fall birds are fresh and late summer birds are most worn. Juveniles have a complete first pre-basic moult in August–October while on the breeding grounds. The first pre-basic moult is complete, after this moult adults and immature birds cannot be aged as they appear identical (Pyle *et al.* 1987, Lanyon 1994).

MEASUREMENTS Males are 10–15% larger than females. Males: (n=65) Wing 127.2 (120–134, 2.99); bill (from nostril) 23.1 (19.1–28.8, 1.55); tail 77.6 (71–83, 2.83); tarsus 37.91 (34.0–41.2, 1.46). Females: (n=63) Wing 114.8 (109–119, 2.18); bill (from nostril) 21.08 (18.9–24.0, 1.22); tail 68.7 (62–75, 2.77); tarsus 36.04 (33.0–40.6, 1.52).

NOTES Hybrids are known between Western and Eastern Meadowlark, and are extremely difficult to identify due to the similarity of these two species. A small percentage of birds, usually females, within the zone of sympatry are intermediate in body morphology. These birds appear to be hybrids. The two species will hybridize in captivity, but the young appear to have low fertility. 'Hybrid songs' are not a sign of hybridization, as 'pure' birds can learn songs of other meadowlarks or incorporate vocal elements of the other species. Call notes, on the other hand, appear to be hard wired and therefore species specific. There is evidence that in Ontario some birds can learn the call notes of the other species, but this is rare and probably only happens in special circumstances. Call notes are best thought of as species specific. Hybridization with 'Lilian's Meadowlark' is seemingly rarer than with Eastern Meadowlark. As noted above, Western Meadowlarks defend territories from other meadowlark species. Confrontations between males involve a ritualized display of the underparts. Curiously, male Western Meadowlarks, which are paler on the underparts than Eastern Meadowlarks, are darker below in areas where they coexist with the latter species. This convergence in character state implies that there has been selection for darker underparts in order to better compete with Eastern Meadowlarks. Thus, Western Meadowlarks look more similar where they coexist with Eastern Meadowlarks, something to note when identifying a Western Meadowlark in the eastern fringe of its range.

REFERENCES Bent 1958 [nesting, behaviour], Dickinson and Falls 1989 [behaviour], Klimkiewicz and Futcher 1987 [age], Lanyon 1957, 1958, 1962, 1994 [voice, description, habitat, nesting, behaviour, distribution, measurements], Pyle *et al.* 1987 [moult], Rathbun 1917 [geographical variation], Rohwer 1972, 1973, 1976 [identification, hybridization, morphological convergence], Szijj 1963 [hybridization].

73 YELLOW-RUMPED MARSHBIRD *Pseudoleistes guirahuro* (Vieillot) 1819

Plate 29

The genus *Pseudoleistes* means 'false meadowlark'; however, Yellow-rumped Marshbird is less meadowlark-like than its southern relative.

IDENTIFICATION Yellow-rumped Marshbird is quite similar to its close relative Brown-and-yellow Marshbird (74), but the identification is straightforward. Brown-and-yellow Marshbird always lacks a yellow rump, this is easily observed while in flight but may not be obvious while perched. The epaulet of Yellow-rumped Marshbird is more extensive than on Brown-and-yellow, including the marginal, lesser and median coverts; the epaulet is restricted to the marginal and lesser coverts on the latter species. Note that Yellow-rumped Marshbird has yellow flanks, unlike the brown flanks of Brown-and-yellow. Overall, Yellow-rumped Marshbird is more blackish-brown than the olive-brown of Brown-and-yellow Marshbird. Juveniles of both species are duller and show yellow on the throat, but may be identified by the same criteria as the adults. Note that females and immatures of the rare Saffron-cowled Blackbird (55) may be confused for a marshbird due to the yellow and brown plumage. However, the blackbird differs in plumage pattern. It always shows a yellow or yellowish throat and breast, unlike a marshbird. Saffron-cowled Blackbirds are smaller and slimmer than marshbirds and have a characteristic yellow supercilium. Blackbirds and marshbirds may flock together.

VOICE Song: A rather noisy species. It is almost always found vocalising in groups, making it difficult to differentiate a single song of this species. Belton (1985) describes the song as 'three to five very high, shrill, often slurred notes in variable patterns, one or more of them rapidly trilled'. The song appears to be quite variable. It is composed of several elements strung together, and the variability in songs emerges from the different order in which these elements are given. The individual elements include slurred whistles, trills, buzzes and odd hollow-sounding whistles as well as sharp, tinkling whistles. At times it will utter a few flute-like notes in between these vocalisations. Sick (1993) states that the basic vocalisation is a strident *grooeep-gruit-gruit* which is often repeated. Flocks sing together, in what may be a communal song. Commonly sings while flying, usually in flocks.

Calls: A *chuck-chuck* (Belton 1985) or *tac-tac* (Sick 1993).

DESCRIPTION A chunky and somewhat short-tailed, medium-sized icterid. Its bill is long and pointed, with straight culmen and gonys. Wing formula: P9 < P8 > P7 > P6; P9 ≈ P7; P8–P5 emarginate.

Adult (Definitive Basic): All of the soft part colours are dark, the bill is black and the eyes are dark brown. The head and breast are blackish-brown, this colour continuing on the upperparts as a brown back. The rump, however, is a contrasting bright yellow while the uppertail-coverts are brown. The brown of the hood includes the throat and upperbreast but ends abruptly in a straight line cut-off on the chest. The rest of the underparts, including the flanks, are yellow except for the brown undertail-coverts. The brown wings have obvious yellow epaulets formed by the yellow lesser and median coverts; the wing-linings are also yellow. The tail is brown and unpatterned. The thick legs and feet are black.

Juvenile (Juvenal): Similar to adult, but dull brown above with buff tips to the back feathers. On some, the auriculars may have an obvious yellow wash. The rump is a dull yellow, with brown streaks. In addition, the throat is yellowish, rather than brown, and there is a breast band of brown streaks on a pale brown background. The underparts are paler yellow than on the adult.

GEOGRAPHIC VARIATION Monotypic.

HABITAT Occurs in dry grasslands, interspersed with marshy areas. It is not adverse to foraging in very tall grass, where the birds will be entirely out of view. Typically avoids forested places. On the other hand, it is commonly found in agricultural land particularly cattle pastures. Breeding occurs mainly in marshes, but forages mainly in upland sites. This requirement for both dry and wet grasslands or marshes is a key factor in what makes a particular area suitable for this marshbird. It is not clear if there is a shift in habitat preference during the non-breeding season.

BEHAVIOUR Forages in small flocks of up to 20 birds, sometimes with Brown-and-yellow Marshbird or Saffron-cowled Blackbirds. Forages on the ground, looking very much like a European Starling (*Sturnus vulgaris*), walking and probing deliberately at the ground while gaping. The latter two species are more likely to be found in wetter areas, however. Yellow-rumped Marshbird sings near the breeding site as well as while foraging; often several birds will sing in close proximity to each other or even while the flock flies from one site to another. During the most intense bouts of song, the yellow rump feathers are raised.

NESTING Breeds mid October–mid November in Rio Grande do Sul, Brazil (Belton 1985). Nests are placed in dense marsh vegetation, usually in colonies. The nests themselves are bulky, their bases made of vegetation cemented together by quantities of mud. Clutch 3–4 eggs which have a greenish background colour and brown and purplish-brown markings. Eggs of Shiny Cowbird (102) have been found in its nests, but successful rearing of this parasite is not known. The breeding system of this species has not been studied, but 'nest helping' should be looked for.

DISTRIBUTION AND STATUS Tends to be uncommon or locally common, seldom reaching the ubiquitous nature of Brown-and-yellow Marshbird. Found to 1000 m, mainly in Brazil, but also in the adjoining nations to the south. In Brazil, it is found in the following states: S Matto Grosso, S Goiás, SW Minas Gerais and W Rio de Janeiro, São Paulo, Paraná, Santa Catarina and Rio Grande do Sul. The range extends to eastern Paraguay, including all of the provinces east of the Paraguay river and easternmost Presidente Hayes. In Uruguay, this species may be found anywhere in the country but is distinctly more common and widespread in the north. Restricted to the northeast section of Argentina, in E Formosa, E Chaco, E Santa Fé, Misiones, Corrientes, N and C Entre Ríos. It has not been confirmed for Buenos Aires province (Narosky and Di Giacomo 1993).

MOVEMENTS Resident throughout its range, but apparently wanders during the non-breeding season. It is likely less migratory than Brown-and-yellow Marshbird.

MOULT Largely unknown. Adults go through one complete moult per year, the definitive pre-basic moult. It is timed to occur after the breeding season. A mid February specimen from Uruguay was beginning its definitive pre-basic moult (Wetmore 1926). There is no evidence that pre-alternate moults occur. The extent of the first pre-basic moults is not known.

MEASUREMENTS Males: (n=7) wing 124.4 (121–128); tail 101.9 (90–120); culmen 29.1 (27–31); tarsus 34.4 (31–38). Females: (n=3) wing 120 (118–122); tail 92.3 (91–95); culmen 27.3 (27–28); tarsus 32.7 (32–33).

NOTES The two marshbirds are surely closely related based on their close resemblance, behaviour and odd vocalisations. The relationship between the marshbirds and Saffron-cowled Blackbird has not been directly studied, but sequence data from mitochondrial DNA suggests that Saffron-cowled Blackbird is the 'sister species' to the marshbirds, and that the three are probably closely related to Yellow-hooded (59) and Chestnut-capped (60) Blackbirds (Lanyon 1994).

REFERENCES Belton 1985 [voice, behaviour, nesting], De la Peña 1989 [nesting], Lanyon 1994 [systematics, notes], Orians 1985 [movements], Ridgely and Tudor 1989 [distribution, habitat], Sick 1993 [voice, nesting], Wetmore 1926 [moult].

74 BROWN-AND-YELLOW MARSHBIRD *Pseudoleistes virescens* (Vieillot) 1819 — Plate 29

A common, gregarious icterid of southern South American grasslands and wetlands. Its generic name, *Pseudoleistes*, means false meadowlark quite an appropriate descriptive for this chunky, ground-loving species.

IDENTIFICATION This is a most distinctive bird and is unlikely to be confused with anything other than its close relative Yellow-rumped Marshbird (73). Some plumages of Saffron-cowled Blackbird (55) may also appear similar. Brown-and-yellow Marshbird lacks the yellow rump of its congener, Yellow-rumped Marshbird. In addition, Brown-and-yellow Marshbird is browner, not as blackish, and has brown flanks and thighs as well as a smaller patch of yellow on the epaulets, restricted to the marginal and lesser coverts. In shape and structure, the two species are similar, with Brown-and-yellow Marshbird having a slightly thinner bill. Saffron-cowled Blackbird shows yellow at the bend of the wing and is largely yellow below like the marshbird; however, males of the former have yellow heads and blackish upperparts with a yellow rump. The female blackbird is brownish with yellow underparts, but note that the yellow extends to the chin and throat, and that a yellow supercilium is present as is a yellow rump.

VOICE Noisy and loud, but the vocalisations are almost always given in groups, making it difficult to isolate the different types.

Song: The song appears to be a mix of *chrrr-wee* calls with whistles and other sounds. The other notes are mainly chatters or trills accented by short whistles. For example, *chrr-wee, trr-trr-weetwo, trr-pweo, chrr,* an entire series lasting several seconds in length. Consecutive songs differ in the order of the elements and in the addition of new elements and omission of others.

Calls: A distinctive, chattering *chrrr-wee* given in flight or on the ground. The *chrr-wee* call is composed of a short chatter, usually of three closely spaced notes, followed by a sharp downslurred whistle, the *wee* of the call. Flying flocks are very noisy, the 'corporate sound' made by them is like a bubbling set of screeches and rattles.

DESCRIPTION A chunky blackbird with a long, pointed bill. The tail is of moderate length and square-tipped. Wing Formula: P9 < P8 > P7 > P6; P9 ≈ P7; P8–P5 emarginate.

Adult (Definitive Basic): The bill is black, and the eyes are brown. The head is brown with an olive tint. The lores and auriculars are more blackish, while the nape and upper back are significantly greener. The mantle is brownish with an olive-grey wash; it is paler than the crown. Lower back, rump and uppertail-coverts olive-brown. The breast is olive-brown, stopping abruptly at a straight line cut-off on the breast. The lower breast and belly are a vivid lemon-yellow, while the flanks, thighs and crissum are a contrasting olive-brown. The wings are brown with lemon-yellow lesser coverts, creating a yellow shoulder patch. The other coverts and the primaries are concolourous, brown. The marginal coverts and wing-linings are also yellow, showing up well in flight. The tail is blackish-brown. The legs and feet are blackish.

Juvenile (Juvenal): Similar to adult in overall pattern but has a yellowish throat and shows blurry brown streaks on the yellow breast. The plumage is also duller overall, and the median and greater coverts tend to be tipped with yellow. The lower mandible is largely yellowish in the juvenile.

GEOGRAPHIC VARIATION Monotypic.

HABITAT This is a species of open country, particularly grasslands and marshes. It is common in agricultural areas, especially pastures but prefers moist ones as opposed to drier sites. Where it is found with the related Yellow-rumped Marshbird, the two may flock together, but Brown-and-yellow Marshbird does not venture into the drier upland sites. Plowed fields are often used as foraging sites; it more commonly forages at the edge of marshes. Nests in marshes or wetland edges, but often nests in the vegetation immediately adjoining the marsh particularly in Pampas Grass (*Cortaderia selloana*). As agriculture takes over more and more marshland in Argentina, this species has adapted by nesting in the vegetation offered by wet roadside ditches and by foraging in the agricultural areas.

BEHAVIOUR In the non-breeding season this marshbird may be observed in sizeable flocks. Early in the spring the flocks break up as the breeding season begins. Individuals are not territorial and singing appears to function largely in the attraction of a mate and the formation of the pair bond. Once pairs are established, they travel together continuously likely as a 'mate guarding strategy' by the male. Monogamy appears to be the prevailing mating system. Mating displays have not been described in detail, but there is a report of a 'flight display' with both birds flying together with quick wingbeats, 'towering' and then gradually parachuting down. Another display in the non-breeding season involved birds in flight carrying sticks in their bills, with the heads pointed upwards and the wings bowed as they were flapped. The significance of this display is unknown, but shares elements of a display of Baywing (96). When the female is on the nest the male mainly stays nearby, often perching on a prominent site. Once nesting is underway and the young hatch, adults other than the pair begin to arrive at the nest and feed the young. The relationship of these 'helpers' to the breeding pair is unclear, as is the sex of the helpers. Up to eight birds may actively feed young at any one nest, these birds remaining with the young even after they fledge the nest. Nests with helpers are less likely to have young starve to death (brood reduction). During the breeding season, Brown-and-yellow Marshbirds forage primarily at the edges of marshes or wet meadows. In winter, flooded fields and pastures also suffice. Most or all of the feeding takes place on the ground, by probing and gaping into the earth or at the base of emergent vegetation. In this respect, this species closely resembles European Starling (*Sturnus vulgaris*) in its foraging strategy.

NESTING The marshbird is an early breeder and is on eggs by late September in the pampas of Argentina, with nesting continuing into December. The nests are built entirely by the female, most often in the centre of a dense tuft of Pampas Grass which is by far the preferred nesting site. Other plants used to support the nests include Narrow-leafed Cattail (*Typha latifolia*), sedge (*Cyperus* sp.), rushes (*Juncus acutus*) and thistles (*Cynara* sp., *Carduus* sp.). The nest is constructed of grasses and vegetable fibres, often reinforced with mud towards the base; usually placed less than one meter from the ground. Brown-and-yellow Marshbirds may appear to be colonial, with nests often less than a few meters apart. However, 'clumping' of nest sites (coloniality) appears to be related to distribution of adequate stands of Pampas Grass rather than birds seeking out close associations. Clutch size is 3–4 eggs, usually the latter. The eggs are white or pale pink with chestnut markings throughout, concentrated at the broader end. Incubation is also undertaken by the female, but males may help by feeding her on occasion. This marshbird is commonly, and multipally (more than one egg per nest), parasitised by Shiny Cowbird (102) with rates of parasitism of up to 74% in parts of Buenos Aires province. Shiny Cowbirds lay either a spotted egg or an immaculate white egg. In marshbird nests, white eggs are rarer than expected, only 4.5% as opposed to 50% in other hosts, and experiments have shown that the marshbird rejects white but not spotted eggs (Mermoz and Reboreda 1994). A known host of the Screaming Cowbird (97) (Mermoz and Reboreda 1996).

DISTRIBUTION AND STATUS Common to locally abundant. Found mostly in Argentina and Uruguay, but also in S Brazil. Ranges from NE Argentina (S Chaco) east through Corrientes (and perhaps S Misiones?), to extreme S Brazil (Río Grande do Sul, and southernmost Santa Catarina); south along the Atlantic coast, including all of Uruguay, to S Buenos Aires province (Bahia Blanca); north through E La Pampa, E San Luis, Cordoba, and SE Santiago del Estero. In winter it may be found to Tucumán and E Jujuy, in NW Argentina and perhaps elsewhere, but more information is needed. The Brown-and-Yellow Marshbird has not occurred in Bolivia (Remsen and Traylor 1989) but it should be looked for during the austral winter in the extreme south (Tarija).

MOVEMENTS Wanders in winter to NW Argentina, little else is known about its movements. It is present in the winter throughout the breeding range, thus, the majority of the population appears to be sedentary.

MOULT The complete annual moult of the adults (definitive pre-basic) occurs in February to March. Pre-alternate moults appear to be lacking. The extent of the first pre-basic moult is unknown.

MEASUREMENTS Males: (n=10) wing 122.3 (117–131); tail 92.5 (88–98); culmen 31.9 (28–35); tarsus 31.8 (30.0–34). Females: (n=9) wing 112.4 (104–122); tail 89.3 (80–103); culmen 29.6 (28–31); tarsus 31.2 (29–33).

NOTES In Argentina, Brown-and-yellow Marshbird is known as El Dragón (the Dragon), due to its strange appearance and noisy nature. This name is sometimes also applied to Saffron-cowled Blackbird.

REFERENCES Belton 1985 [behaviour], Mermoz and Reboreda 1994 [behaviour, nesting], Orians 1980 [habitat, behaviour, nesting], Orians *et al.* 1977b [behaviour, nesting], Ridgely and Tudor 1989.

75 SCARLET-HEADED BLACKBIRD *Amblyramphus holosericeus* (Scopoli) 1786

Plate 25

A gorgeous denizen of the marshes of southern South America, however it is local and seldom common.

IDENTIFICATION Adults are unmistakable, their slim black bodies contrasting with brilliant red heads and thighs is diagnostic. Note also the long, slender bill, long tail and long legs. Juveniles are more confusing, they begin life wearing a blackish plumage but gain patches of orange-red on the breast or throat as they age. The orange-red increases in extent, covering (patchily) the head in older birds. It is unlikely that during the short period when juveniles are nearly completely blackish they will be mistaken for another species, particularly since they will be attended by the unmistakable adults. However, note the long, slim shape of the bird. Even juveniles, with partly-grown bills, show rather slim bills. The other marsh-nesting black icterids which overlap with this species are Unicolored, Yellow-winged and Chestnut-capped Blackbirds. Yellow-winged Blackbird has yellow shoulders and underwings, while Chestnut-capped has a chestnut cap and throat. Unicolored Blackbird is all black, but more glossy than juvenile Scarlet-headed Blackbird, and averages smaller in size. Note however, that orange-red feathers are obtained rather early in life so that completely blackish Scarlet-headed Blackbirds are rarely seen. Most have some colour on the throat, even when the tail is not fully grown. Furthermore, the colour of juvenile Scarlet-headed Blackbirds is blackish-brown, not matte black and always lacks iridescence.

VOICE Song: A melodious *ch-wee-wee-wee-wee-wee*, each note given in quick succession such that each song lasts approximately one second and the song is repeated roughly every 10 seconds. Belton (1985) described the bell-like trill (song) as reminiscent of a ringing telephone, yet not as frequent.

Calls: The call is a *check check* (Wetmore 1926). Also gives a clear whistle (Belton 1985).

DESCRIPTION A rather large, slim and long-tailed blackbird. The bill is long, straight and chisel-like with an obviously flattened culmen and gonys, perfectly suited for gaping into stalks of fleshy vegetation. Wing formula: P9 < P8 ≈ P7 > P6 > P5; P9 ≈ P4; P8–P6 emarginate.

Adult (Definitive Basic): The bill is black and the eyes are dark. This species is primarily black with a contrasting scarlet head and neck as well as scarlet thighs. The lores and ocular area are black, creating a small black mask. The black of the body has a dull gloss which is only visible in the hand. The wings and tail are entirely black; the underwings are black as well. The tail is slightly rounded, but not obviously so. The legs and feet are black. Gyldenstolpe (1945b) states that females have paler-tipped bills that males. This needs to be confirmed.

Immatures: **(First Basic).** This plumage is quite variable. Some are similar to adults, but the scarlet head is not as intensely coloured, being orange-red. In addition, immatures often show brown tips to the crown and nape, and brown flecks on the auriculars. The lower portion of the orange-red breast often appears streaked with brown. Other individuals may show a mixture of orange-red and brown feathers on the head, this includes birds which have solidly blackish crowns. Similarly, the thighs are usually reddish mixed with brown. Any patchiness in the red of the head or thighs is a sign of immaturity.

Juvenile (Juvenal): Young juveniles are quite unlike adults, as they lack all red on the body. The entire plumage is brownish-black, the throat and breast being paler and showing a yellowish wash. The upperparts from the crown to the uppertail-coverts are unicolored, brownish-black. As juveniles age they acquire orange-red feathers on the head. This transition is prolonged, such that it is difficult to determine when a bird is actually still in juvenal plumage and when it is in immature (first basic) plumage. Older individuals are similar to adults, but the scarlet of the head is dull and patchy, and largely replaced by dull brown. The body plumage is duller, brownish-black rather than jet black. As the juvenile ages and moults, more red is obtained on the head. It is more appropriate to categorise juveniles as birds which are entirely brown and lacking red, all others are already in first pre-basic moult.

GEOGRAPHIC VARIATION Monotypic. No evidence of significant geographic differences between populations.

HABITAT Restricted to marshes with tall emergent vegetation, particularly those with high proportions of cattail (*Typha* sp.) or bulrush (*Scirpus* sp.). It tends to prefer the more extensive and deeper water marshes, rather than shallow ones. However, particularly in the non-breeding season, Scarlet-headed Blackbird may be found foraging in upland areas, sometimes distant from marshes. Will feed on fields in mixed flocks with other blackbirds (Belton 1985).

BEHAVIOUR In the breeding season they are conspicuous birds, often perching high on emergent vegetation. Pairs begin defending territories as early as September or more commonly in early October. Their territories are quite large and have been estimated to be up to 40 hectares in area (Orians 1980). During the early stages of breeding, the male spends a good deal of time singing from an elevated perch or perching quietly in an exposed situation. At this time of year the female may be quite inconspicuous, keeping low in the marsh vegetation, likely spending most of her time foraging. Once the female is incubating, the male does not feed her, instead she periodically leaves the nest to forage for herself. After the young are hatched, both sexes bring them food. However, one sex typically stays close to the nest while the other is foraging. The mating system is monogamous, with pairs widely scattered in breeding marshes. Both sexes share the majority of nesting duties, but males sing more than females and conduct a greater proportion of the territorial defense and the chasing of predators. When chasing off an intruder, Scarlet-headed Blackbird performs a peculiar display flight where the red thighs are dangled and the thigh feathers are noticeably 'fluffed-up'. They will chase Yellow-winged Blackbirds (58) and Wren-like Rushbirds (*Phleocryptes melanops*) which are too close to the nest site. Only the female incubates, and the nest is built primarily by the female with help from the male (Orians 1980). Non-breeding individuals form small flocks, which forage around the edge of breeding marshes. The majority of these non-breeders appear to be one year-old birds with duller plumage than the adults. After breeding, adults may join these wandering flocks for the winter. Pair formation likely occurs late in the winter in these foraging flocks. Wintering flocks often mix with other blackbirds, including Chestnut-capped Blackbirds (60) and marshbirds (73, 74).

NESTING The nest is a bulky cup composed of dry cattail stem strips. These may be anchored to growing stems of cattails or bulrushes, but the sturdy Duraznillo (*Solanum glaucophyllum*), which has a woody stem, is preferred as a nesting site. The nests tend to be placed 1.5 m from the surface of the water. Eggs are laid late October–late November in Buenos Aires, but can be found in mid December or perhaps later. In Rio Grande do Sul, Brazil, young juveniles have been observed between late January and late July (Belton 1985), suggesting that the breeding season there is more extensive than in the south. Clutch 3–4 eggs which are greenish with dark brown spots and blotches around the broader end. This species is not commonly parasitised by Shiny Cowbird (102).

DISTRIBUTION AND STATUS Uncommon to fairly common, never abundant. The known distribution is somewhat patchy perhaps reflecting that intervening areas are not inhabited by the species or merely that this species has not been noticed due to low densities. A bird of the lowlands, reaching 600 m in Bolivia. One population is present in pampas habitats in the lowlands of NE Bolivia from Beni south to NW Santa Cruz. The main distribution reaches its northern limit in the upper reaches of the Río Paraguay in Mato Grosso, Brazil. South of there the distribution includes easternmost Bolivia, most of Paraguay, except the extreme west; all of Uruguay; extreme S Misiones, and throughout Corrientes and Entre Ríos in Argentina. The western limit of its contiguous range in Argentina crosses through E Formosa, E Chaco, E Santa Fé and NW Buenos Aires south at least to Mar del Plata. In Brazil it is found in SW Matto Grosso, E and S Río Grande do Sul north to extreme SW Santa Catarina. There are scattered records from NW Argentina in Jujuy, Salta and N Córdoba (Ridgely and Tudor 1989). These may refer to isolated populations, but may more likely pertain to wandering non-breeding birds.

MOVEMENTS Appears to be largely sedentary, wandering somewhat in flocks during winter. There is no evidence of migratory movements, but in drought years birds may be forced to disperse to new areas.

MOULT Not well known. Adults undergo a complete pre-basic moult after breeding, during the late austral summer or early fall. There is no evidence of a pre-alternate moult, at least in adults. Juveniles are born with a blackish-brown plumage that changes gradually as they age. This first pre-basic moult appears to be partial, involving only the body plumage and begins almost immediately after fledging. It is unclear if this moult is merely variable in its extent or exceedingly protracted in timing, either one would account for the variable appearance of young Scarlet-headed Blackbirds. Individuals in this patchy, orange plumage have been collected in the middle of the southern hemisphere winter, as well as the fall and spring. More work is needed to determine if further changes in the colour of the head occur due to a first pre-alternate moult.

MEASUREMENTS Unsexed: (n=2) wing 110, 111; tail 98, 100; culmen 27, 28; tarsus 30, 31. Males: (n=10) wing 113.9 (105–124); tail 99.0 (82–110); culmen 30.0 (28–33); tarsus 31.8 (30–34). Females: (n=8) wing 109.9 (104–117); tail 94.4 (84–98); culmen 26.7 (23–29); tarsus 30.3 (28–32).

NOTES The distinctive appearance of this bird has earned it the nickname 'flame-headed chisel-bill'. Sequence data from mitochondrial DNA suggests that the 'sister species' to Scarlet-headed Blackbird is Austral Blackbird (77) (Lanyon 1994).

REFERENCES Belton 1985 [habitat, voice, nesting], De la Peña 1987 [nesting], Gyldenstolpe 1945b [description, measurements], Orians 1980 [behaviour, nesting], Ridgely and Tudor 1989 [distribution], Wetmore 1926 [voice, behaviour].

76 RED-BELLIED GRACKLE *Hypopyrrhus pyrohypogaster* (de Tarragon) 1847 Plate 30

A handsome black and red blackbird of Colombia's montane forests; sadly most of its habitat has been cleared and this species is now rare.

IDENTIFICATION This large, distinctive and restricted range blackbird is unmistakable. Note the bird's large size, stout bill, yellow eyes, black plumage and red underparts. If the red on the underparts is not visible, it may be confused for an all-black icterid species. However, in its subtropical distribution there are no other blackbirds to confuse this species with. Colombian Mountain-Grackle (83) is found in the East Andes of Colombia, north of the Red-bellied Grackle's range, so sympatry is unlikely. However, the Mountain-Grackle has dark eyes, a much slimmer body, a longer tail and a thinner bill. In addition, from the front it lacks red on the belly and may show chestnut axillary tufts given a good view. Scarlet-bellied Mountain-Tanager (*Anisognathus igniventris*) shares the red belly with Red-bellied Grackle, and may be confused given a brief view. The tanager is very much smaller than the grackle, and in addition has blue on the rump and shoulder as well as a red patch on the ear-coverts.

VOICE This species is very vocal when foraging in groups.

Song: Produces a number of different sounds, including gurgles, liquid sounds and wheezy notes, *glok-glok, shleee-o, shleeee* (Ridgely and Tudor 1989).

Calls: The flock contact calls are a loud shrieking *peep*, similar to that of White-capped Tanager (*Sericossypha albocristata*) (P. Salaman pers. comm.).

DESCRIPTION A large and stocky blackbird with a stout, blunt-tipped bill and long tail. The bill is thick at its base, and shorter than the head in length. The culmen is straight for most of its length, curving down at the tip. The gonys shows a slight curve throughout its length. Wing Formula: P9 < P8 < P7 > P6 > p5; P9 ≈ P3; P8–P7 emarginate.

Adult Male (Definitive Basic): The bill is black and the eyes are pale yellow or white. The head, neck, breast, wings, back, rump and uppertail-coverts are black with a dull blue gloss. The feathers of the crown and nape are noticeably pointed and show obvious glossy shafts which continue to the auriculars and are less pointed and glossy on the throat. The underwings are black. There are red breast tufts which can be hidden under the wing or breast feathers, these continue as red on the flanks. There is a scarlet band on the belly, directly in front of the legs, while the thighs and the area behind the legs is black. The crissum is scarlet as well; this creates an alternating series of red and black areas on the underparts – black on the throat and breast, then a scarlet belly, black vent and scarlet undertail-coverts. The tail is black. The sturdy legs and feet are black.

Adult Female (Definitive Basic): Similar to male in plumage, but much smaller in size.

Immature (First Basic): Similar to adults, but shows contrasting browner (retained juvenal) primaries, secondaries and primary coverts. This contrast is most visible when comparing these feathers to the blackish secondary coverts and tertials. The head feathers of immatures are pointed as in adults. The chin is usually brownish.

Juvenile (Juvenal): Similar to adults, but the black areas are replaced by dark brown which lack iridescence. The chin is obviously brown. The scarlet on the body is duller, tending toward orange, than on adults. The thighs and hind flanks are not solidly reddish, but instead the feathers are brown with reddish restricted to their tips. In addition, the feathers of the head lack the glossy shafts and are not obviously pointed.

GEOGRAPHIC VARIATION Monotypic.

HABITAT Lives in humid subtropical forests, particularly near forest borders, in more open situations. It also inhabits scrubby growth and older second growth. Throughout most of its range its appropriate habitat has been cleared, leaving only isolated pockets of habitat. Fragmentation of this type is predicted to lead to extinction for this species.

BEHAVIOUR Travels and forages in small family parties, usually under a dozen in number. Forages from the subcanopy to canopy of the trees. Flocks are invariably observed to be accompanied by Subtropical Caciques (16), and less frequently with Green Jays (*Cyanocorax yncas*), larger tanagers and Masked Tityras (*Tityra semifasciata*). Flocks are noisy and conspicuous. This large icterid is an acrobatic and active forager, hopping along branches, and aggressively gleaning in hanging moss clumps under branches (P. Salaman pers. comm.).

NESTING Breeds January–February in the Central Andes, or March–August in other parts of its range. This grackle is not colonial, but several adults may tend to a nest suggesting that cooperative nesting occurs, as it does in several other neotropical icterids. The bulky nest is constructed from sticks and large dead leaves, and is placed in the fork of a branch. Its eggs are greenish-grey with dark brown and lilac spotting. One nest, placed 9 m up in a *Cupania cinerea* tree held a clutch of two. The young took between 14 and 17 days to fledge (Ochoa and Maya 1998).

DISTRIBUTION AND STATUS Endemic to the mountains of Colombia (800–2400 m), it is now a rare and localised species which has been extirpated from much of its historical range. However, it remains locally common and is observed regularly in some of the more well-known sites (P. Salaman pers. comm.). Early in the 1900s this species was common in some sites. Historically it has been found in the subtropical zone in the West, Central and East Andes. In the West Andes, it is known from the eastern and western slopes from the department of Antioquia (as far north as Peque, in the Serranía de Uramá) south to Risaralda department. In the Central Andes, it has been recorded from both slopes, as far north as Puerto Valdivia (Antioquia) and to Gaitania in the Los Nevados NP (Tolima) in the south. In the Eastern Andes, known only on the east slope, from the area below Andalucia (Huila department) to Cueva de los Guácharos NP and in the department of Caquetá above the town of Florencia. Recent records come from the Central Andes in Antioquia, Ucumarí Regional Park (Risaralda), Cueva de los Guácharos NP (Huila) and north of Florencia (Caquetá).

MOVEMENTS Sedentary, no known seasonal movements.

MOULT Largely unknown. Adults appear to have only one moult a year. This is the complete definitive pre-basic moult which occurs after breeding, the timing of this moult ranges from late September–early November. In Antioquia, moults early June–August. There is no evidence that Red-bellied Grackles undergo a pre-alternate moult. Juvenal plumage is lost through the first pre-basic moult. The extent of this moult is not fully known, the body plumage is replaced but retention of tail and wing feathers is not certain. The first pre-basic moult occurs in October or early November. One specimen from Antioquia was undergoing its first definitive pre-basic (= second pre-basic) moult during late May. The tail replacement is centripetal, ending with the central tail feathers.

MEASUREMENTS Male: (n=10) wing 142.6 (131–152); tail 145.8 (132–162); culmen 32.4 (31–34); tarsus 36.0 (34–38). Females: (n=10) wing 127.3 (120–132); tail 130.3 (125–140); culmen 29.3 (28–38); tarsus 34.0 (31–36).

NOTES The loss of Colombia's subtropical forests is blamed for the decrease in population of this endemic blackbird. Some large pockets of suitable habitat do exist within the species' range, although adequate searches have not been undertaken.

REFERENCES Chapman 1917 [habitat, behaviour], Collar *et al.* 1992 [distribution, conservation, habitat, behaviour, nesting], Ridgely and Tudor 1989 [voice, behaviour], Sclater and Salvin 1879 [nesting].

77 AUSTRAL BLACKBIRD *Curaeus curaeus* (Molina) 1782 — Plate 31

The only all-black blackbird that inhabits the temperate forests of southern South America, this is the southernmost-ranging of all icterids nesting as far south as Cape Horn.

IDENTIFICATION A large and sturdy black icterid with a long and spike-like bill, as well as iridescent pointed head feathers. In its range, it is only likely to be confused with male Shiny Cowbird (102), the only black bird sympatric with it. Austral Blackbird is a larger bird than the cowbird, and differs significantly from it in shape. Austral Blackbirds have long, spike-like bills that are longer than, or nearly the length of the head. Shiny Cowbirds have shorter, more conical bills that are significantly shorter than the head length. From a distance Austral Blackbird can be seen to have a significantly longer tail than the cowbird, a feature that also shows up in flight. In flight, the blackbird shows slower, more deliberate wingbeats than Shiny Cowbird, which has a quick, snappy wingbeat. When seen well, Austral Blackbird shows a weak gloss, and pointed feathers on the throat and crown (particularly the males). Shiny Cowbirds are noticeably more glossy than the blackbirds, when seen in good light, and lack the pointed head feathers. The two species have different voices, cowbirds give sharp whistles and gurgles while Austral Blackbird has lower pitched, more stuttering, calls (see Voice). Chopi Blackbird (79) is not sympatric with Austral Blackbird, but a vagrant of either species could cause confusion. Chopi Blackbird has a shorter bill, not nearly so spike-like, than Austral Blackbird. When observed closely, the lower mandible of Chopi Blackbird has noticeable grooves, which Austral Blackbird lacks. Their voices are superficially similar, but distinctly different in reality and can be used to identify the species (see Voice).

VOICE A rather vocal blackbird, with a varied repertoire. This species is sometimes kept in captivity in Chile and will learn to imitate and even 'talk' (Johnson and Goodall 1967).

Song: Varied, pleasant and musical. The song tends to be slow and methodical, there is always a short period of silence between phrases. The 'stop and start' nature of the song, coupled with hollow notes and striking trills is quite distinctive. Songs of Austral Blackbirds are continuous, not having a set start or ending. They vary in the different notes included in the song as well as the order of repetition of different phrases. It is rare for two consecutive phrases to be identical. The song consists of squeaks, trills and whistles mixed in with the typical *Ku-ra-taaooow* call. Most distinctive are piercing whistles and curious stuttering trills. There are two overall characteristics of the notes which are worth mentioning. First of all, most single whistles and clucks have obvious harmonics which give them a hollow or ringing sound. Secondly, the trills or chatters are commonly followed by a sharp *cheep* note or a whistle.

Calls: A distinctive three syllable call of low frequency, *Ku-ra-taaooow*. Sometimes the first two syllables run together, *Krru-taaoow*. This is the

Figure 77.1 Head and bill shapes of several black icterids.

common flight call as well as the common call given in most circumstances. As is noted above, the call is usually interspersed within the song.

DESCRIPTION A large blackbird with a long and spike-like bill. The bill has a straight culmen and gonys, with a sharp tip. The culmen is distinctly flattened. Its tail is long and square-ended, while the wings are long with pointed primaries. Wing formula: P9 < P8 ≈ P7 ≈ P6 > P5.

Adult Male (Definitive Basic): The bill is black, while the eye colour is dark brown. The entire plumage of this species is black, with a dull blue gloss. The feathers of the head, nape, neck and throat are pointed and can be raised as prominent hackles. The strong legs and feet are black.

Adult Female (Definitive Basic): Similar to adult male, but with a duller gloss on the body. The belly lacks iridescence and is brownish-black rather than black.

Immature (First Basic): Similar to adults, but pointed feathers on the head and throat are both less pointed and not so extensive. The primaries and secondaries are brownish and contrast with the black tertials, coverts and body. On some there may be a contrast between brownish, retained juvenal primaries, and secondaries with blacker, newer feathers. The lowermost scapulars are tipped brown.

Juvenile (Juvenal): It is dull overall, brownish-black rather than black and largely lacking iridescence; the belly is brown. The tail and wings are brownish. The feathers of the head and neck are not pointed as in the adults.

GEOGRAPHIC VARIATION Three subspecies are recognised.

C. c. curaeus is found north of the Straits of Magellan to Coquimbo, Chile. It is described above.

The southern ***C. c. reynoldsi*** occurs in Tierra del Fuego and adjacent islands, including Cape Horn.

The final subspecies, ***C .c. recurvirostris,*** is only found on Riesco Island in Magallanes, Chile. The insular *reynoldsi* is larger than mainland birds with a proportionately longer and slimmer bill. Birds from Wellington Island are intermediate between the nominate and *reynoldsi*. The enigmatic *recurvirostris* is noted to have a peculiar bill. The upper mandible has a slightly spatulate tip and is longer than the lower mandible. The lower mandible is slightly downcurved at the tip.

HABITAT Austral Blackbirds are found in a variety of habitats. In general they avoid open areas without trees or shrubs and do not show an affinity towards livestock, as does the sympatric Shiny Cowbird. They will forage in dense thickets, and appear to need some type of cover, often dense, when foraging in more open agricultural areas. In the north, they frequent matorral woodlands as well as agricultural areas. Further south they are found in openings in forest, open forests, second growth, remnant forest patches, agricultural areas where shrubbery is available and even parks in urban areas. In the far south, they may forage along rocky beaches. In Tierra del Fuego, Austral Blackbird is most closely associated with the forest edge or shrub-covered ravines near forest (Humphrey *et al.* 1970).

BEHAVIOUR A social blackbird that is observed in small flocks or family groups during most of the year. These flocks are quite vocal, often chorusing and producing rather sweet melodies particularly when near the roost site. Rarely forms large flocks, numbering in the low hundreds. When flocking and foraging, one bird typically flies to another foraging spot and one by one the rest of the flock copies the actions of the first, in a 'follow the leader' fashion. Also curious is their habit of all joining in song once one bird begins to sing. During the breeding season, pairs break off from the flocks. When nesting the male sits quietly near the nesting tree, but may sing at times. Austral Blackbird forages both on vegetation and on the ground. Typically, they search for insects in the trunks of large trees especially in Southern Beech (*Nothofagus* sp.) forest, and often keep to the canopy. In more open areas, they will forage by probing in buds and gleaning from leaves and twigs in dense shrubbery. They have a fondness for nectar, and will nectar from species of *Puya* and *Eucalyptus*. Also forages on the ground in forests, commonly in moist sites, as well as in fields, tending to keep to field edges closer to tall vegetation. Austral Blackbird is a known nest robber, eating the eggs or nestlings of other passerines. This species sometimes feeds on wheat and corn, causing a small amount of economic damage (Johnson and Goodall 1967). During the fledgling stage, it is common for more than two adults to come to the defense of the young birds, and also more than two adults have been observed with food, both situations suggesting that 'nest helping' occurs in this species. However, it is also possible that after fledging, neighbouring pairs and their young come together in a small flock, explaining why more than two adults may be involved in caring for the young. Pairs have been observed with a third adult loosely associating with them (Orians *et al.* 1977b).

NESTING Nests during the southern spring and summer (October–January); by January most pairs are feeding well grown young. The timing of the breeding season is similar in the far south in Tierra del Fuego (Humphrey *et al.* 1970). The nest is a large, messy cup made of sticks and coarse grass, which is often reinforced with mud. Nests are placed in thick shrubbery or trees with dense foliage, usually rather close (less than 2 m) to the ground. Clutch 3–6 eggs, which are pale blue with dark spots and streaks which are concentrated at the broader end. Some eggs are nearly unmarked, plain pale blue (Johnson and Goodall 1967). 'Nest helping' may occur in Austral Blackbird (Orians *et al.* 1977b). However, this has not been confirmed and it is not known if it occurs while the young are still in the nest or not.

DISTRIBUTION AND STATUS Common to abundant in most of its range, becoming rarer in northern Chile. Apparently not as abundant in Tierra del Fuego as it is on the mainland. In the south it is found to the treeline, but further north it is seldom found as high as 2000 m. The range of this species is restricted to Chile and Argentina. In Chile, it ranges from Atacama south to Cape Horn, including Tierra del Fuego and the southern islands. In Argentina, from Neuquén and W Río Negro south along the Andes and foothills to Tierra del Fuego. It is rare near the Atlantic coast in Chubut and Santa Cruz (more common in winter?). There is one unconfirmed record from the Falkland Islands (Woods 1988).

MOVEMENTS Partially migratory, but its movements are not well known. Island populations apparently resident. In the non-breeding season, it is found in flocks that may wander north in Chile and east in Argentina. Probably undergoes significant altitudinal migrations, but these are not well documented.

MOULT Not well known. The adults appear to have one moult per year, the complete definitive pre-basic moult. There is no evidence that pre-alternate moults occur. Definitive pre-basic moult January–March. The juvenile plumage is replaced through the partial (sometimes incomplete?) first pre-basic moult. Appropriate data on the extent of this moult is lacking, but a small number of specimens suggest that the entire body, tertials and some of the remiges and rectrices may be changed. On many, the juvenal secondaries and primaries are retained, however. The first basic plumage is slightly different from that of the adults (definitive basic), the full definitive plumage is not achieved until after the second pre-basic moult. The first pre-basic moult is timed similarly to that of the definitive pre-basic moult.

MEASUREMENTS ***C .c. curaeus***. Males: (n=10) wing 128.2 (122–142); tail 108.2 (98–127); culmen 32.3 (30–34); tarsus 33.8 (32–36). Females: (n=10) wing 123.3 (115–135); tail 108.1 (102–115); culmen 31.8 (29–37); tarsus 33.0 (31–35). Both sexes: Wing 128.0 (2.57); culmen 32.6 (0.55); tail 105.0 (2.74).

C. c. reynoldsi. Males: (n=2) Wing 125, 131; (n=1) tail 105; (n=2) culmen 33, 38; (n=1) tarsus 32. Female: (n=1) wing 130; tail 102; culmen 34; tarsus 35.

NOTES Austral Blackbird was originally described as a species of *Turdus* (Thrush). It was also put into the genus *Notiopsar*, for a while a name change proposed to decrease confusion with (*Cureus* a cuckoo genus) (Wetmore 1926). The genus *Curaeus* was consider monospecific until Forbes's Blackbird (78) was moved into *Curaeus* (Short and Parkes 1979). Sequence data from mitochondrial DNA suggests that the 'sister species' to Austral Blackbird is Scarlet-headed Blackbird (75) (Lanyon 1994). The other South American all-black icterids are more distantly related, Forbes's Blackbird was not included in this study.

REFERENCES Araya and Millie 1991 [distribution], De la Peña 1989 [nesting], Johnson and Goodall 1967 [distribution, behaviour, nesting, measurements], Orians *et al.* 1977b [behaviour], Sclater 1939 [geographic variation, measurements].

78 FORBES'S BLACKBIRD *Curaeus forbesi* (Sclater) 1886 — Plate 31

A rare blackbird of northeastern Brazil. It has a history of being confused for the more abundant Chopi Blackbird (79), both in the field and in the museum; indeed the type specimen was originally passed off as a small Chopi Blackbird!

IDENTIFICATION A medium-sized black icterid with a long, sharply-pointed bill. Within its range, the main confusion is between this species and Chopi Blackbird. Both are entirely black and roughly the same size; however, Chopi Blackbird is glossier. If observed in optimal light and close to, Forbes's Blackbird's bill will not show the diagnostic groove, near the base of the lower mandible, of Chopi Blackbird. In addition, the bill shapes of the two species are different (Figure 77.1). Forbes's Blackbird has a relatively longer bill than Chopi, which is shorter than the head length in Chopi Blackbird and equal to the head length in Forbes's Blackbird. In addition, the bill is sharply pointed with a straight culmen and gonys on Forbes's Blackbird. Chopi Blackbird, on the other hand, has a noticeable curve to the culmen, making its bill appear less spike-like. In addition, the culmen of Forbes's Blackbird is flattened in cross-section while that of Chopi is rounded, but this feature is only visible at close quarters. Both Forbes's and Chopi Blackbirds have pointed head feathers with noticeably shiny shafts which can be raised as hackles. In size, Forbes's Blackbird is slightly smaller than Chopi Blackbird. While displaying, Forbes's Blackbirds throw their heads back and open their bills revealing bright red mouth-linings. Chopi Blackbirds have black mouth-linings. The voice may be the single most reliable way to separate these two species. Forbes's Blackbirds have a flight call composed of three quick notes *ti-ti-lit* while Chopi Blackbird's flight call is an onomatopoetic *chopi* or *shop-wee* as well as a standard icterid *chuk*. The song of Forbes's Blackbird is composed of two unmusical buzzes while that of Chopi Blackbird is a rich series of whistles, squeaks and *chopi* notes that sounds somewhat like a *Turdus* thrush song.

The sympatric Shiny Cowbird may also be confused with Forbes's Blackbird. However, the male cowbird is noticeably glossy, much more so than Forbes's Blackbird. The cowbird has a short bill with a noticeably curved culmen, unlike the straight and spike-like bill of Forbes's. Shiny Cowbird is smaller, shorter-tailed and with a long primary extension. Forbes's Blackbirds are noticeably short-winged. Finally, they are vocally quite different (see corresponding Voice sections). The marsh-inhabiting male Unicolored Blackbird is also black, but it is found in a rather different habitat than Forbes's Blackbird. The bill is straight, like that of Forbes's Blackbird, but slightly shorter. Unicolored Blackbird is also slimmer and smaller, and is usually accompanied by the distinctively-plumaged females. Unicolored Blackbird and Shiny Cowbird both lack the pointed, shiny-shafted crown and throat feathers of Forbes's Blackbird.

VOICE Song: The primary vocalisation, which is considered the song by Studer and Vielliard (1988) is a harsh, unpleasant-sounding, nasal buzz. They are given in pairs, the first buzz longer than the second, lasting three-quarters of a second while the subsequent buzz lasts half a second. This vocalisation has a reedy quality, similar to that of Red-winged Blackbird (61) (Studer and Vielliard 1988). Another song-like vocalisation has been described as a loud, rough *check-check-check-check-check* (Willis and Oniki 1991). The *check-check* calls are often given in doubles, but when excited they are delivered quickly and in a series, more than one bird may *check* simmultaneously.

Calls: The flight call is a characteristic *ti-ti-lit ti-ti-lit*. The flight call is comprised of three short, highly frequency-modulated notes, each slightly higher in frequency than the last; the three notes are given quickly each bout lasting half a second or so. Another common call is a dry chatter of less than one second in length. The chatter is an alarm vocalisation. Finally, Forbes's Blackbird gives several different, and variable, contact notes. Each is short in duration, and monosyllabic (Studer and Vielliard 1988). One of these notes may be described as a short, ringing, *wheeo* (Studer and Vielliard 1988), others include a *preck* and a *wop* (Willis and Oniki 1991).

DESCRIPTION A short-winged and long-tailed species, with a moderately graduated tail; the central rectrices are 12–18 mm longer than the outer rectrices. It has a short primary extension, of only 5–8 mm. The wing is quite short and rounded. Wing formula: P9 < P8 < P7 > P6; P9=P2. The flattened culmen is straight and the bill tapers to a sharp tip, the bill is approximately as long as the head (Short and Parkes 1979).

Adult Male (Definitive Basic): The bill is black, and the eyes are dark. The entire plumage is black with a slight brownish tone, particularly on the rump, lower back and belly. It is not obviously glossy, but on close inspection may show the dullest green iridescence on the wings. The feathers of the crown, nape, head and sides of the neck are pointed (hackle-like), with glossy shafts to the individual feathers. The throat feathers are variable, usually they are not pointed except on some specimens (full adults?). The legs and feet are black.

Adult Female (Definitive Basic): Similar to the male, but smaller (Short and Parkes 1979).

Juvenile (Juvenal): Undescribed, but likely quite similar to the adults. It would be logical, based on other icterids, that juveniles should be duller in colour and lacking the pointed feathers of the face. Birds mentioned to lack pointed head feathers by Short and Parkes (1979) may be juveniles or immatures. This needs to be confirmed.

GEOGRAPHIC VARIATION Monotypic.

HABITAT An inhabitant of forest and forest edge as well as marshes. In Quebrângulo, Alagoas, it inhabits and nests primarily at the edge of a relict wet tropical forest, utilising the nearby grasslands and swamps near the forest edge (Studer and Vielliard 1988). At this site, there was a distinct preference for nesting in Mango trees *(Manguifera indica)*. It was previously reported to be common in canefields, but the original report may have been confusing this species with Chopi Blackbird. Destruction of forests appears to be one of the causal factors of this species' rarity.

BEHAVIOUR During the non-breeding season, particularly during the height of the dry season October–January, this species is encountered in small flocks averaging 20–30 individuals (Studer and Vielliard 1988). Originally it was reported to congregate in large flocks (Forbes 1881), but this may have been due to confusion with Chopi Blackbirds. If old records are correct, today's small flocks may be more an effect of rarity rather than a behavioural choice. Forages both on the ground and in trees. Has been observed poking in rolled-up grass leaves while foraging (Willis and Oniki 1991).

Birds may begin displaying several months before nest construction and this peaks during nest construction. Courtship tends to commence in January, but this depends on when rainfall begins that year. Display behaviour declines while the female is incubating, and stops once the chicks hatch, but increases again in intensity once the chicks fledge, preparing the adults for raising a second brood. During the 'ruff-out' display the bird spreads its wings, and throws the head back such that it is resting on its back, and the bill is held open flashing the intensely red mouth-linings. The nearest individual in the flock responds by copying the same display. Sometimes three or four birds may join in the display. The sex of the displaying birds is not known, but it appears to include both males and females. While nesting, a pair is 'helped' by a cohort of two to four helpers. The relationship and sex of these 'helpers' is not known. The 'helpers' aid throughout the breeding period, including during nest construction (Studer and Vielliard 1988).

Forbes's Blackbirds are noisy and attract attention. They often sing and vocalise from the nest site, making it relatively easy to find the nest. The breeding adults, as well as the 'helpers' will vocalise at the nest site. In addition, the nestlings are also noisy and add their shrill begging calls to the cacophony (Studer and Vielliard 1988).

NESTING The breeding season coincides with the beginning of the rainy season, which differs from year to year. Typically, the rainy season falls March–June; the first rain may kick-start an intense bout of breeding activity. Two clutches are commonly raised during one season. The nest is placed at an average height of 7 m off the ground, but may be of any height between 3–12 m, usually in the crown of a small tree. In the Quebrangulo study site, nest trees were usually (38 out of 46 nests) placed in cultivated mangos (*Mangifera indica*) at the forest edge. The remaining nests were found in thorny trees (4 nests), ephyphitic bromeliads (2), lianas (1) and banana tree (1). Nest construction lasts three to four days and up to a week. The breeding adults are aided by the 'helpers' in nest construction; this behaviour is rare in other co-operative breeding species. The nest itself is a bulky cup of dry grasses and herbs woven into the fork of a branch. The bottom of the nests may be reinforced with mud or dung with a lining of soft grasses on top of that. There is a delay of four to ten days before clutch initiation begins, after the nest is completed. Clutch size 1–4 eggs, but averages 2.84. Incubation lasts 13 days, typical for a blackbird of its size. The nestlings are fed insect food by the parents and helpers. Forbes's Blackbird is a known host for Shiny Cowbirds, which have been considered a major threat to the population of this rare blackbird. The blackbird is a preferred host for the cowbird at Quebrângulo, where the cowbird has recently invaded (in the 1950s) as the forest was opened up. The rate of nest parasitism was 100% in one year, but averaged 64% throughout over several years. Shiny Cowbirds may lay between one and five eggs in a Forbes's Blackbird nest, averaging 1.96. Some of these eggs are laid before the blackbird initiates its clutch and the blackbird may respond by abandoning that nest. Successful fledging of Shiny Cowbirds from Forbes's Blackbird nests has been documented(Studer and Vielliard 1988).

DISTRIBUTION AND STATUS An extremely rare Brazilian endemic teetering on the edge of extinction. It is known from several localities, in two disjunct populations, in E Brazil: Pernambuco, and Alagoas as well as SE Minas Gerais. In Pernambuco it has been recorded from Vista Alegre and Macuca, where the type specimen was secured in 1880, as well as near Igarassu which is the northernmost record. In Alagoas, it is known from the Pedra Talhada biological reserve (Quebrângulo), Usina Sinimbu, Matriz de Camarajibe and Pedra Blanca (Murici). Records from Minas Gerais include the Raul Soares area and near the confluence of Rio Piracicaba and Rio Doce, as well as the nearby Rio Doce State Park. Recently, it has only been recorded from the Rio Doce, Quebrângulo, Pedra Blanca and Matriz de Camarajibe. Forbes's Blackbird has not been observed in Pernambuco during the last two decades. A sighting above Brejo at Januária, Minas Gerais, may have been a Forbes's Blackbird (Willis and Oniki 1991). The population at Quebrângulo was estimated at 150 individuals during the 1980s but that number is now lower. The Rio Doce State Park may be quite important for this species, but no population censuses have been carried out in that area, but flocks of up to 40 birds have been observed there. It is unknown if healthy populations persist in other areas where the species has been seen recently. Originally it was found to be abundant in localised areas (Forbes 1881).

MOVEMENTS Sedentary, no evidence of movements.

MOULT The type specimen was finishing its complete pre-basic moult in September. Nothing else is known regarding moult, however it likely fits the general pattern of having one complete pre-basic moult and lacking a pre-alternate moult as an adult.

MEASUREMENTS Males are 10–15% larger than females in linear measurements. Males: (n=6) wing 107.3 (105–110); tail 102.8 (99–108); culmen 26.1 (25.2–28.2); tarsus 30.5 (29.2–32.0). Females: (n=6) wing 97.3 (95–99); tail 94.0 (93–97); culmen 23.0 (21.6–23.5); tarsus 29.9 (29.5–30.5).

NOTES This species is listed as endangered by the red data book (Collar *et al.* 1992).

This species was noted by Forbes, who collected the first specimen, to resemble a small form of *Aphobus* (=*Gnorimopsar*, the genus of Chopi Blackbird). The initial description of the species placed it in *Agelaius* due to its bill shape. Only a few specimens were known until the late 1960s when a good series was found in a museum collection, incorrectly identified as Chopi Blackbirds. This expanded set of specimens made it clear that this species was not an *Agelaius*, but based on bill shape and feathers structure likely a small *Curaeus* – the genus in which it is still placed. Some of the call notes of Forbes's Blackbirds resemble some given by Austral Blackbird, but for the most part they are not vocally similar.

REFERENCES Collar *et al.* 1992 [distribution], Forbes 1881 [behaviour, distribution], Hardy *et al.* 1998 [voice], Short and Parkes 1979 [description, systematics, distribution, measurements, notes], Studer and Vielliard 1988 [voice, habitat, nesting, behaviour], Willis and Oniki 1991 [voice, behaviour].

79 CHOPI BLACKBIRD *Gnorimopsar chopi* (Vieillot) 1819 — Plate 31

This blackbird shares with Bobolink (103) the fact that its name is onomatopoetic. It is a common species in the non-forested lands south of the Amazon.

IDENTIFICATION This black icterid has no obvious field marks, particularly when seen distantly. When seen in detail, the diagonal grove across the base of the lower mandible is diagnostic; however, this is seldom observed in the field. The conversational and melodious song is also a good field mark. Furthermore, Chopi Blackbird does not show a strong gloss in the field and it has pointed feathers on the head which give it a shaggy-headed appearance. The bill is long and pointed, which is a good mark to use when separating it from Shiny (102) and Screaming (97) Cowbirds. Both of the cowbirds have shorter bills, and are noticeably smaller. In addition, both are glossier than Chopi Blackbird but this can be difficult to see under some light conditions, particularly in the more subdued Screaming Cowbird. Giant Cowbird (98) is much larger and glossier than Chopi Blackbird, and also has yellow or red eyes. The entirely black Unicolored Blackbird (57) may be confused with Chopi Blackbird. Chopi Blackbirds are larger, have longer and thicker bills, stouter legs and are not habitually found in marshes with emergent vegetation as is Unicolored Blackbird. In the Caatinga region of Brazil, the blackbird-like Scarlet-throated Tanager (*Compsothraupis loricata*) (Plate 31) may be confused with Chopi Blackbird. The male tanager has a scarlet throat and breast, but the female is entirely black. The female tanager is slightly more glossy than Chopi Blackbird and shows a shorter bill with an obviously curved culmen.

The most delicate identification problem lies in differentiating Chopi Blackbird from the very rare and sympatric Forbes's Blackbird (78). Both are nearly the same size, but Forbes's Blackbird has a thinner, relatively longer and more sharply-pointed bill (Figure 77.1), quite similar in shape to that of *Agelaius* blackbirds. The wings of Forbes's Blackbird are proportionately short showing a primary projection of 5–8 mm; Chopi Blackbirds have longer wings and show a longer primary projection of 25–35 mm. In addition, Forbes's Blackbird has a relatively longer and noticeably graduated tail, while that of Chopi Blackbird is only slightly rounded (Short and Parkes 1979). Due to the rarity of Forbes's Blackbird, the identification of this species should be confirmed by confirming the lack of a grove at the base of the lower mandible. Forbes's Blackbird is slightly less glossy than Chopi Blackbird and differs from it in voice.

The allopatric Bolivian Blackbird's (80) range is not too distant from areas where Chopi Blackbird may be found; these two all-black species would be difficult to separate in the field. The contrasting brownish primaries of Bolivian Blackbird set it apart from Chopi Blackbird which shows no contrast between the colour of the primaries and the wing-coverts. Bolivian Blackbird lacks the bill grooves of Chopi Blackbird. The two species also have slightly different bill shapes (Figure 77.1). The bill of Bolivian Blackbird is both slightly longer and deeper than that of Chopi Blackbird; in addition, Bolivian Blackbird has a noticeably decurved culmen. The culmen of Chopi Blackbird is decurved but this is not all that obvious. Bolivian Blackbirds show smooth contour feathers on the crown and neck, not the hackles of Chopi Blackbird. See Voice section of both species for additional differences which will aid in field identification.

VOICE The name of this species is onomatopoetic. Due to its fine song, Chopi Blackbird is commonly kept as a cagebird.

Song: A complex series of whistles, tweets, chatters, buzzes and similar sounds. The song of a lone Chopi Blackbird may sound somewhat thrush (*Turdus*) like, as it is composed of single notes separated by pauses; however, Chopi Blackbird songs are not nearly as melodious or as fast as *Turdus* songs. A characteristic sound which is always present in Chopi Blackbird songs is the *shop-whee* or *chopi* call; this is composed of two separate notes, the first a low, quick note *shop* and the second a modulated whistle *whee*. The full *shop-whee* lasts just under a second in length. Often adults sing in groups, making it nearly impossible to distinguish the vocalisations of any one individual. It appears that in these group vocalisations, the individuals sing different songs which blend together into a complex, but not unpleasant sound. This blackbird sings while perched in trees, while foraging on the ground and even in flight. Both males and females sing; it is not known if their songs differ consistently.

Calls: The main call is the *shop-wee* or *cho-pi* call, which is given in many circumstances including as a flight call. Also gives a standard blackbird *chuk*. The female gives a sharp *zwirr* (Skutch 1996). A sharp whistle is given when predators are present (Hudson 1920).

DESCRIPTION A medium-sized blackbird with a short, stout bill. In length this is shorter than the length of the head. The lower mandible shows distinctive grooves at its base, which are nearly impossible to see in the field other than in exceptional circumstances. These grooves are comprised of a taller ridge, bounded by a groove on each side; they begin at the bottom of the base of the lower mandible and are directed obliquely towards an area one-third of the way along (from the base) the cutting edge of the mandible. The gonys is straight, but the culmen shows a slight curve and it is longer than the lower mandible, such that it overhangs it at the tip as a small hook. The tail is relatively short and square-ended. Wing Formula: P9 < ≈ P8 > P7 > P6; P9 ≈ P8; P8–P5 emarginate. The feathers of the forehead, crown, nape auriculars and neck sides are pointed and spiky. The feathers of the throat also show a tendency towards being pointed, but this is not as obvious as on the crown. These pointed head feathers have a shiny rachis. The legs are thick and strong.

Adult Male (Definitive Basic): The bill is black and the eyes dark brown. Chopi Blackbirds are entirely black with a dull blue gloss. The gloss is strongest on the breast and back. The feathers of the crown and nape are pointed and can be raised into a shaggy crest. The strong legs and feet are black.

Adult Female (Definitive Basic): Similar to male, but slightly smaller and showing a less prominent iridescence. In addition, the female's distribution of pointed feathers on the head is less extensive.

Immature (First Basic): This plumage is not well known. Like the adult plumages but usually some or all of the juvenal primaries and secondaries are retained, these contrast with the blacker wing-coverts.

Juvenile (Juvenal): Unlike adults, the juveniles are brownish and not iridescent. The belly is a paler brown than the rest of the body. The wings and tail are brown, like the body plumage. The head and nape lack the pointed and shiny feathers typical of adults.

GEOGRAPHIC VARIATION Two subspecies are recognised. The nominate form ranges from SE Bolivia through C and SE Brazil south through Paraguay, Uruguay and NE Argentina south to Buenos Aires. It is described above.

The other race, ***G. c. sulcirostris*** is considered to occur in two oddly disjunct populations: one in E Bolivia and NW Argentina (Salta); and the other in the extreme NE of Brazil (Maranhão, Piauí, Ceará, Bahia). Perhaps it makes sense to divide *sulcirostris* into two subspecies based on geographical considerations. In the past, the Bolivian subspecies was named *G. c. megistus* (Leverkühn 1889, Gyldenstolpe 1945b). Bolivian *megistus* was considered similar in colour to *sulcirostris* of Brazil, but is smaller in size, while being significantly larger than *chopi*. The race *sulcirostris* differs from *chopi* in being larger and having a larger bill with more obvious grooves. The plumage of *sulcirostris* averages glossier.

HABITAT This species may be found in a variety of habitats. It forages both on the ground and in trees, but more commonly on the former. It will use open forests, swamp forest and forest edge, palm stands, and Monkey-puzzle (*Araucaria angustifolia*) groves. It has a preference for roosting and nesting in palm groves. More often it is found in open grasslands, savannas and cultivated regions as long as a few trees are present. Wet grasslands, marshes, and swamp edge are preferred foraging habitat. Its choice of habitats is varied and perhaps the main generalisations that can be made is that this is a species of edge and savanna needing at least some trees to be present but favouring open areas for foraging.

BEHAVIOUR A most sociable blackbird, it is usually observed in flocks which may be sizeable. They invariably roost in large groups either in a dense stand of trees, often palms, or dense canebrakes. Chopi Blackbirds forage arboreally and on the ground. However, most foraging occurs on the ground, while singing and courtship tends to take place in trees. Chopi Blackbird is extremely vocal, usually singing in a conversational manner while in flocks or even while in flight (Belton 1985, Jaramillo pers. obs.). Males display to females by circling in spirals while holding the wings stiffly (Wetmore 1926). Sometimes sings with the body erect and the wings vibrating (Sick 1993). During the nesting season, Chopi Blackbirds are pugnacious defenders of their breeding site against potential predators and will chase hawks and caracaras without hesitation.

NESTING Males are in breeding condition in early November in Rio Grande do Sul, Brazil (Belton 1985). Nesting begins in October and peaks in November in Mato Grosso, Brazil (Naumburg 1930). Nesting behaviour occurs between September and January in Uruguay and Argentina (de la Peña 1989, Wetmore 1926). This species is noted to be semi-colonial by Ridgely and Tudor (1989). Chopi Blackbird nests in tree cavities, which include woodpecker (*Picidae*) holes, hollow palms or other trees as well as in crannies at the base of palm fronds and on fence posts. In addition, it has been observed to nest in old Rufous Hornero (*Furnarius rufus*) nests, within nooks in large Jabiru (*Jabiru mycteria*) stick nests, dense *Araucaria angustifolia* foliage, holes in embankments, holes in terrestrial termite nests as well as old nests of other species (Sick 1993). It appears that any cavity may be used as a potential nest site given that Chopi Blackbird will nest in holes under the eaves or other parts of older houses (Azara 1802) in the Iguazú area of N Argentina this is the nest placement of choice (Fraga 1996). The height of the nest is variable, and depends on the height of the available cavity. Some nests, those in fence posts, may be only a few feet from the ground. When not using a cavity, the open-cup nests built by Chopi Blackbird tend to be placed in a tree with dense foliage, particularly in palms. Clutch size 4–5 eggs which are pale blue with dark spots and scrawls at the broader end (de la Peña 1989), or pale grey with black scrawls and spots mainly on the broader end (Naumburg 1930). It is not known if both sexes assist in the construction of the nest. In addition, this species is highly social, often partaking in many communal behaviours including communal singing. However it is not known if they ever have 'helpers' at the nest. Chopi Blackbird is a known host of Shiny Cowbird (Friedmann and Kiff 1985), and there are records of it actually successfully raising young of that species (Sick 1993). Also known as a host of Screaming Cowbird, a species which usually only parasitises Baywing (96). Sick (1993) suggested that his observations of Screaming Cowbird parasitising Chopi Blackbird in Rolandia, Paraná, Brazil were due to the recent spread of the cowbird to that area and the lack of the regular host (Baywing) in that area. However, at the Iguazú Airport, Misiones, Argentina, Fraga (1996) found that parasitism by Screaming Cowbird on Chopi Blackbirds was frequent and that a careful reading of Azara (1802) suggests that this is not a new behaviour. However more data is needed to confirm if Screaming Cowbirds parasitise Chopi Blackbird where Baywings are also found (Fraga 1996). Chopi Blackbird has successfully fledged young of Screaming Cowbird.

DISTRIBUTION AND STATUS Common to abundant, mainly in lowlands to 1000 m, recently to 3400 m in Bolivia (S. Herzog pers. comm.). Found from extreme SE Peru (Madre de Dios) east through the lowlands of N and E Bolivia in Beni, Santa Cruz, Tarija and also Cochabamba (fide S. Herzog). A flock of 8–10 Chopi Blackbirds has recently been observed in La Paz city (S. Herzog pers. comm.). Found throughout Paraguay in all departments. In Brazil, it is found throughout the country south of Amazonia. It ranges north to S and C Mato Grosso, Goiás, W Maranhão, and NE Pará; south of there the range includes Piauí, Ceará, Rio Grande do Norte, Paraíba, Pernambuco, Alagoas, Sergipe, Bahia, Minas Gerais, Brasilia, Espírito Santo, Rio de Janeiro, São Paulo, Paraná, Santa Catarina and Rio Grande do Sul. In Uruguay, Chopi Blackbird is distributed throughout the country, being most common in the north and west, and rare in the south (coast). The Argentine range is mainly in the north and east from E Salta, Formosa, Chaco, NE Santiago del Estero, N and E Santa Fé, Corrientes, Misiones, Entre Ríos and extreme N Buenos Aires south to the city of Buenos Aires.

MOVEMENTS Not obviously migratory but appears to be nomadic to some extent. In the non-breeding season flocks wander widely.

MOULT Chopi Blackbird undergoes one complete moult per year, the definitive pre-basic moult. Its timing is post-breeding, and therefore varies geographically as breeding seasons vary. In northern Argentina, the definitive pre-basic moult late March–early July. Further north in Brazil, the moult appears to be earlier, January–April, and peaking in February. Pre-alternate moults are lacking. The juvenal plumage is replaced through the incomplete first pre-basic moult. This moult includes the body, tail and some or all of the primaries and secondaries. Its timing is similar to that of adults.

MEASUREMENTS ***G. c. chopi***. Males: (n=12) wing 120.1 (117.3–127); (n=11) tail 89.2 (85.3–100); culmen 22.1 (21.6–24); (n=9) tarsus 31.5 (29–35). Females: (n=6) wing 117.6 (116–123); (n=5) tail 87.4 (78–93); culmen 21.8 (20–23); (n=4) 31.3 (30–32).

G. c. sulcirostris. Males: (n=7) wing 139.6 (132–150); tail 104.1 (92–114); culmen 25.6 (24–28); (n=3) tarsus 34 (33–35). Females: (n=7) wing 134.9 (129–140); tail 99.3 (89–104); culmen 25.1 (24–26); (n=4) tarsus 33.3 (31–35). Unsexed: (n=1) wing 141; tail 112; culmen 24; tarsus 29.

NOTES This accomplished songster is often trapped in Brazil to be kept as a caged songbird. A study of mitochondrial DNA sequence patterns suggests that this species is most closely related to Bolivian Blackbird (80) (Lanyon 1994). Baywing was not included in that study, but vocal similarities between these three species suggest that the three are closely related.

REFERENCES Hellmayr 1929 [geographic variation, measurements], Narosky and Yzurieta 1987 [distribution], Short 1975 [habitat], Short and Parkes 1979 [identification], Ridgely and Tudor 1989 [distribution], Sick 1993 [voice, habitat, behaviour, nesting], Wetmore 1926 [behaviour, nesting, mesurements].

80 BOLIVIAN BLACKBIRD *Oreopsar bolivianus* Sclater 1939 — Plate 33

A restricted range blackbird, endemic to Bolivia, with the curious habit of nesting on cliffs.

IDENTIFICATION A black icterid with browner primaries and secondaries. Its bill is shorter in length than the head. It is unlikely to be mistaken for another species in the dry valleys it inhabits. The sympatric Shiny Cowbird (102) is also black, but it is glossier and smaller than Bolivian Blackbird. Bolivian Blackbird has obviously brown primaries, which contrast with blackish wing-linings. The brownish primaries and secondaries of Bolivian Blackbird show no iridescence. Furthermore, Bolivian Blackbirds are monomorphic, such that flocks are made up of members that look alike. Female Shiny Cowbirds are brownish, and stand out in mixed-sex flocks. Note also the differences in vocalisations. The allopatric Chopi Blackbird's (79) distribution comes quite close to that of Bolivian Blackbird, and any expansion could create a region of sympatry between the two species. Both are black and similar in size, and would be difficult to identify given a poor view. The contrasting brownish primaries of Bolivian Blackbird set it apart from Chopi Blackbird which does not show a contrast between the colour of the primaries and the wing-coverts. Bolivian Blackbird lacks the grooves that Chopi Blackbird shows on its lower mandible. The two species also have slightly different bill shapes (Figure 77.1). The bill of Bolivian Blackbird is both slightly longer and deeper than that of Chopi Blackbird; in addition, Bolivian Blackbird has a noticeably decurved culmen. The culmen of Chopi Blackbird is decurved but this is not all that obvious. Bolivian Blackbirds show smooth contour feathers on the crown and neck, while those of Chopi Blackbird are pointed, giving the head a spiky (hackled) look. See Voice section of both species for additional differences which will aid in field identification.

VOICE Song: Simple series of *chip* notes interspersed with *chu-pit* notes, *chip chip chip chip chu-pit chu-pit chu-pit chip chip.*

Calls: There are several calls given by this species, it is unknown if all are given by both sexes. The common flight call is *chu-pee*, sometimes lengthened to *chu-pee-pit*; however, a series of three or four *cheep* notes is sometimes given. A sharp, loud *cheep-cheep* is sometimes uttered as birds take off. When the female leaves the nest site she gives a long flight call composed of several elements: *chip-chip, churr, chu-pit, chu-pit, churr churr.* A *tew* call may be heard from perching birds. The common alarm call is a *chip* performed as the bird flicks its tail. A harsh churring trill is often given as the bird vibrates its wings (Orians *et al.* 1977a).

DESCRIPTION A long-legged blackbird with a short, stout bill. The culmen is noticeably curved while the gonys is relatively straight. Wing formula: P9 < P8 < P7 < P6 > P5 > P4, or P9 < P8 ≈ P7 < P6 > P5, the latter from an immature bird; P8–P5 emarginate. In flight it may resemble a grackle, but it often glides for short distances, before resuming flapping.

Adult Male (Definitive Basic): The bill is black, and the eyes are dark. The body is entirely black, with an extremely faint blue gloss. There are odd, velvet-like feathers on the chin and forehead. The lesser and median coverts, and the tertials, are black with contrasting brown primaries, secondaries and greater coverts. The wing-linings are black, showing great contrast with the browner primaries and secondaries. The outermost primaries may show some white marbling at their bases, further accentuating their paleness when in flight. It is not known if all Bolivian Blackbirds show this 'marbling'. The tail is black, contrasting with the brown tips of the primaries on a resting bird. The legs and feet are black.

Adult Female (Definitive Basic): Like male, but smaller and slightly duller.

Juvenile (Juvenal): Like adults, but washed brown on the underparts and wing-coverts. In addition, the white marbling of the outer primaries appears to be absent on juveniles, but this needs to be confirmed.

GEOGRAPHIC VARIATION Monotypic.

HABITAT Lives in arid, intermontane valleys. This species is found in dry valley floors or canyons with cliffs. Typically, the canyon walls are sparsely vegetated with small shrubs and grasses. The valley bottoms are vegetated by shrubby forests (*Prosopis* sp., *Schinus molle*, *Acacia* sp. and *Carica* sp.), interspersed with bunch grass and tall *Cereus* cacti. Bolivian Blackbird will not occupy areas similar to these if cliffs are absent; cliffs are an absolute prerequisite (Orians *et al.* 1977).

BEHAVIOUR Invariably it is observed in small flocks, which apparently defend a home range from other flocks. They forage on the ground for grass seeds, and invertebrates, as well as on higher vegetation when gleaning for insects on bushes or for fruit from cacti. They often perch conspicuously at the tops of trees, and will fly long distances to perch elsewhere. The song display is not as complex as in other blackbirds. The singing bird droops the wings and begins to sing, flicking the tail upwards as *chip* notes are given, then flicking open the wings rapidly when the *chu-pit* notes are uttered. When soliciting females, the male takes on a different posture; while flexing the legs and pointing the bill upwards he cocks and spreads the tail as well as flicking open the wings and calling. There is another display of unknown significance which is given by a bird holding nesting material in its bill, it droops the wings then bends down and cocks the tail. When entering and leaving the nest, the female often flew with deep, slow wingbeats, butterfly-like, while the male copied the same flight style. Bolivian Blackbird almost commonly forages on the ground where they eat grass seeds. The fruits of *Cereus* cacti are eaten when available. In addition, they probe and gape along branches and particularly into *Tillandsia* bromeliads, presumably looking for insects. They also feed on insects which they find by gaping into small crevices and tree bark (Orians *et al.* 1977a).

NESTING Curiously, the nests of this species are placed in crevices in rocky cliffs, a behaviour only seen in Yellow-shouldered Blackbird (65) and rarely in Brewer's Blackbirds (95). However, this is the only blackbird that is an obligate cliff nester. The nest itself is moulded to fit within the crevice, and is made up of fine roots and dry grass. The breeding season begins in mid April, the clutch size appears to be 3 eggs. Egg colouration is greyish-green with darker grey spots, streaks and black blotches throughout, which are slightly more concentrated at the broader end. Incubation apparently is performed only by the female. There is evidence that apart from the male, other 'helpers' come to the nest with food for the young. During the incubation period, the female frequently rejoins the flock, begs for food and is fed by flock members.

DISTRIBUTION AND STATUS Locally common. Endemic to Bolivia in the arid intermontane valleys of Cochabamba, Chuquisaca and Potosí (1500–2800 m). Apparently, Bolivian Blackbird's dispersal to similar habitat in La Paz may be impeded by the high ranges north of Cochabamba. It may occur in Tarija, where ornithological work has not been intensive (Vuilleumier 1969).

MOVEMENTS Sedentary, appears to be group territorial throughout the year.

MOULT Largely unknown. Adult specimens are undergoing their complete pre-basic moult during late June to mid August. If this species follows the typical blackbird pattern, pre-alternate moults are missing. The extent and timing of the first pre-basic moult is not known.

MEASUREMENTS Male: (n=5) wing 119.6 (111–128); culmen 22.7 (21–24.0); tail 101 (92–110); tarsus 30.4 (29–31.9). Female: (n=2) wing 112, 114; tail 91, 96; (n=1) culmen 23; tarsus 30.

NOTES Some of the behavioural traits of Bolivian Blackbird are similar to those of Baywing (96) and Chopi Blackbird, suggesting a close relationship between these species. A recent study comparing mitochondrial DNA sequences showed that Bolivian Blackbird may be closely related to Chopi Blackbird, but Baywing was not included in this study (Lanyon 1994).

REFERENCES Fjeldsa and Krabbe 1990 [description], Orians *et al.* 1977a [voice, habitat, behaviour, nesting], Sclater 1939 [description, measurements], Vuilleumier 1969 [distribution, habitat, behaviour].

81 VELVET-FRONTED GRACKLE *Lampropsar tanagrinus* (Spix) 1824 **Plate 30**

A black icterid of wet lowland tropical forests and river edges.

IDENTIFICATION In the field, this species appears all black and featureless. It may be mistaken for Shiny Cowbird (102), Pale-eyed Blackbird (56), Solitary Cacique (21) or Epaulet Oriole (25). Velvet-fronted Grackle is a slim, long-tailed bird with a proportionately thin and small bill. The bill looks noticeably small for the size of the bird (Figure 77.1). Male Shiny Cowbirds have thicker bills and shorter tails as well as a more vivid blue gloss. Pale-eyed Blackbird is easily separated by its yellowish eyes. Solitary Cacique is much larger and has a large and pale bill. Some forms of Epaulet Oriole have dark chestnut epaulets, which appear black at a distance. The oriole however is slimmer, has a longer and more pointed bill, and has sweeter vocalisations than the grackle.

VOICE **Song**: The song begins with several soft *chucks* and ends with a more explosive and liquid *kuk-weeeteew* (*tanagrinus*).

Calls: Subspecies *tanagrinus* give a scolding *ttr-rrrrt,* interspersed with soft *piit* notes. The form *bolivianus* gives a three note call, *chew-chew-chew* as well as a single deep *chuck* note in alarm (V. Remsen pers. comm.).

DESCRIPTION A slim icterid with a long tail. The tail is moderately graduated and appears fan-shaped. The bill is thin, short (less than the length of the head) and somewhat stubby in appearance. The bill shows a slightly curved culmen and largely straight gonys; the culmen is marginally longer than the gonys creating a small hook at the bill tip. The feathers of the crown are short, dense and plush-like velvet. Wing Formula: P9 < P8 < P7 ≈ P6 > P5; P9 ≈ P2; P8–P5 weakly emarginate.

Adult Male (Definitive Basic): The bill is entirely black and the eyes are dark. The entire plumage is black with a slight blue gloss. The iridescence is almost lacking on the belly. The legs and feet are black.

Adult Female (Definitive Basic): Like male but slightly smaller, not noticeable in the field.

Juvenile (Juvenal): This plumage is virtually identical to that of adults, but the feathers have a looser texture.

GEOGRAPHIC VARIATION Five subspecies recognised, which have been separated into three groups: *tanagrinus* [Velvet-fronted Grackle], *violaceus* [Violaceous Grackle] and *boliviensis* [Bolivian Grackle] (English names from Sibley and Monroe 1990). A great deal of the range of this species has not been adequately sampled, leaving many gaps in our understanding of the geographic variation of this species. Even less is known about how the named taxa are related to each other; thus, this grouping may be somewhat premature. On the other hand, perhaps more than one species is involved.

VELVET-FRONTED GRACKLE:

L. t. tanagrinus ranges in the Amazon basin from E Ecuador and NE Peru east to the Río Urubú and Río Madeira, and as far south as the upper Río Purús in Brazil. It is described above.

L. t. guianensis is a form that is disjunct from the rest of the range of the species. It is found along the Orinoco River and in the Orinoco Delta in Venezuela, east to NW Guyana and extreme N Roraima, Brazil. It is similar to *tanagrinus,* but is smaller and has a stronger blue gloss on the underparts. It also shows browner underwings than *tanagrinus* (Gyldenstolpe 1945a).

L. t. macropterus is restricted to the upper Rio Juruá region in W Amazonas, Brazil. It is like *tanagrinus* in colour, but larger, particularly the wings and tail. The body gloss is steel-blue while the rectrices are glossed greenish (Gyldenstolpe 1945a).

BOLIVIAN GRACKLE:

L .t. boliviensis is found in the departments of Beni and Santa Cruz in Bolivia. It is similar to *tanagrinus* in colour, but is much smaller than it, or *violaceus.* Like *tanagrinus, boliviensis* is much less glossy than violaceus and tends towards a blue rather than a violet gloss. It apparently intergrades with *violaceus* in Santa Cruz.

VIOLACEOUS GRACKLE:

L t. violaceus is restricted to NW Mato Grosso in W Brazil. Similar to *tanagrinus* in size, but it is strongly glossed violet.

HABITAT An inhabitant of lowland tropical forest, primarily near water such as rivers, river islands or Amazonian lakes (cochas). It prefers várzea or igapo forests, both of which are seasonally flooded. Igapo is the blackwater version of várzea, which in the strict sense refers to flooded forests adjoining whitewater rivers. Perhaps this species is less common in igapo than várzea. When found in river islands, it typically is found in the older islands foraging in the understory, mid-levels and canopy (Rosenberg 1990). In Bolivia, clumps of trees, particularly Mutucú Palms in flooded savanna are preferred (V. Remsen pers. comm.).

BEHAVIOUR A shy species not allowing close approach. Forages in flocks of variable size; sometimes these include nearly a hundred individuals but more commonly fewer than ten. These flocks are noisy and individuals maintain close to each other while foraging and travelling through the forest. Flocks may join up with those of other species, particularly caciques (*Cacicus*). At other sites, it is more common to see them mainly in pairs (seasonal?). They forage in the upper or middle levels of the forest, primarily by searching for insects in dead leaves, epiphytes and bark. They will also drop down and forage in dense understory vegetation only 1–2 m from the water in flooded forest or river edges. At the edges of lakes, the grackle may drop down onto emergent or floating vegetation in search of food. They are acrobatic when foraging, hanging and reaching for fruits and insects. A.M. Olalla (in Gyldenstolpe 1945a) reported that flocks of this species were quite fearless, allowing him to approach closely.

NESTING A complete nest of this species is not known. However, a nest in the process of construction was observed in Beni, Bolivia (*bolivianus* subspecies). A pair was observed building this nest during mid February in Bolivia approximately 4 m from the ground in a palm. Both birds were present during the nest-building, but it was unclear if both sexes helped in its construction. The nest was in too early a stage of construction to deduce its final form, but apparently it seemed that this nest was taking the form of a hanging basket, as in orioles (*Icterus*). The nest was woven into the leaves of a Mutucú palm, approximately one meter from the tip of the frond. The palm was part of a small forest clump in a dry island surrounded by flooded savanna, where most of the trees were less than 10 m in height (V. Remsen pers.comm.). Specimens of *bolivianus* from February and October were also in breeding condition. Specimens of *tanagrinus* from August were in breeding condition. The eggs of this species are unknown.

DISTRIBUTION AND STATUS Uncommon species of lowland tropical forest to 400 m. The range is disjunct and not well understood. In the north, it is found along the Orinoco river in Venezuela as far upstream as Amazonas and adjoining Colombia (E Vichada) and dowstream to its delta and spreading to parts of the eastern part of Venezuela (E Bolívar and E Monagas), and continuing east to NW Guyana and N Roraima in Brazil. The main part of the distribution lies in the Amazonian basin. It ranges from E Ecuador (E Napo and E Pastaza) and SE. Colombia (S Caqueta, Putumayo and Amazonas) east along the north shore of the Amazon river to the Rio Madeira in Brazil (including C and S Amazonas, Acre, and W Rondônia) south into NE Peru to the Ucayali river valley and south to NE Bolivia (Beni and Santa Cruz).

MOVEMENTS Apparently sedentary, but almost nothing is known.

MOULT Largely unknown. The plumage of this icterid is replaced once annually by the complete definitive pre-basic moult. In other blackbirds this moult occurs after breeding and it can be safely assumed to be similarly timed in Velvet-fronted Grackle. Breeding may occur throughout the year in this species, thus, moult may also be seen at all times of the year. Specimens of *tanagrinus* from August all showed some wing or body moult. Some *bolivianus* from June and 30% of October specimens were undergoing moult. The timing and extent of the first pre-basic moult are not known. No information is available that suggests that pre-alternate moults exist.

MEASUREMENTS Females are only slightly smaller than males, not enough for this to be visible in the field.

L. t. tanagrinus. Males: (n=15) wing (100–107); tail (85–98); culmen (17–18); tarsus (22–25). Females: (n=15) wing (97–102); tail (84–98); culmen (16–18); tarsus (22–24).

L. t. guianensis. Males: (n=3) wing 98.3 (97–103); tail 95 (93–98); (n=1) culmen 18; tarsus 22. (n=?) wing (97–103); tail (93–98) (Gyldenstolpe 1945a). Females: (n=3) wing 93.7 (90–97); tail 91.3 (91–92); (n=2) culmen 17. 17; tarsus 24, 24.

L. t. macropterus. Males,(n=16) wing (108–117); tail (100–110); culmen (19–21); tarsus (24–27). Females, (n=23) wing (100–112); tail (93–103); culmen (18–20); tarsus (23–27).

L. t. boliviensis. Males: (n=3) wing 92.3 (92–93); tail 86.3 (85–89); culmen 16.7 (16–17); tarsus 25.7 (25–26). Females: (n=4) wing 89 (86–91); tail 85 (80–91); culmen 17 (16–18); tarsus 23.5 (23–24).

L. t. violaceus. Male: (n=1) wing 100; tail 92.

NOTES This is a very poorly known species, both in its biology and systematics. It is conceivable that at least one cryptic species is currently lumped within Velvet-fronted Grackle. Molecular work needs to be performed in order to clarify the relationships between the many forms of this species. A study of patterns of mitochondrial DNA sequence patterns shows that this species is closely related to the *Macroagelaius* genus (Mountain-Grackles) as well as Oriole Blackbird (53), the latter relationship is corroborated with restriction site polymorphism data (Lanyon 1994, Freeman and Zink 1995).

REFERENCES M. Cohn-Haft pers. comm. [habitat], Gyldenstolpe 1942 [geographic variation, measurements], 1945a [measurements], 1951 [geographic variation, measurements], Hellmayr 1937 [geographic variation, measurements], Moore 1994 [voice], Remsen *et al.* 1988 [voice, habitat, behaviour, nesting, moult], Ridgely and Tudor 1989 [behaviour, distribution], Rosenberg 1990b [habitat].

82 TEPUI MOUNTAIN-GRACKLE *Macroagelaius imthurni* Plate 30
(Sclater) 1881

A highly social and noisy icterid restricted to the Tepui highlands of northern South America. The relationship of this and the following species to the rest of the family is not clear.

IDENTIFICATION A slim, black icterid with a strikingly long tail. It has a yellow patch at the base of the underwings (axillaries). Unlikely to be misidentified in its limited range. This is the only black bird in the Tepuis which shows a yellow axillary patch. However, the yellow is sometimes entirely hidden. In this case, note the shape of Tepui Mountain-Grackle, a long, slim species with a slim bill. In addition, the tail is quite long, and looks parallel-edged and graduated, such that the corners look rounded rather than squared-off. Finally, the behaviour of this species is distinctive; they travel around in vocal flocks keeping high in the canopy of montane forest. The squeaky, variable, jumbled vocalisations are diagnostic. Male Carib Grackles (93) and Shiny Cowbirds (102) are also black and similar in size. Male Carib Grackle does not overlap with this species, as it is restricted to the lowlands. Similarly, male Shiny Cowbird does not range high into the Tepui highlands and is unlikely to be found sympatrically with Tepui Mountain-Grackle. Note also that both of these species are much more heavily glossed than Tepui Mountain-Grackle, and are both not nearly as slim, long-tailed and slim-billed. They both lack yellow on the axillaries and the former has yellow eyes, unlike Tepui Mountain-Grackle. Colombian Mountain-Grackle (83) is very similar to this species but it is allopatric. However, the two species differ in more than just the colour of the axillary tufts. Tepui Mountain-Grackle is smaller and proportionately shorter-tailed than the Colombian species. In addition, Tepui Mountain-Grackle is thinner-billed and has black epaulets, not chestnut as on Colombian Mountain-Grackle.

VOICE Given the available information it is difficult to categorise the vocalisations as songs or calls. The problem is that most often Tepui Mountain-Grackles vocalise in groups, making it difficult to differentiate the vocalisations of any one individual. In addition, they appear to have a large repertoire creating further problems in categorisation given our limited knowledge of the species.

Song: What appears to be the song is a variable series of jumbled squeaks, *chup* or *chuck* calls, whistles and other notes. Many of the notes given have a squeaky or shrill nature, and are short in duration, often with notes of disparate frequencies given consecutively and in quick succession such that the overall composition sounds as if it 'tinkles' up and down the scale, with a harsher *chuck* note inserted for accent. It is rare to hear a single bird giving this vocalisation, most often several of the individuals within a flock appear to sing, sometimes while in flight, such that it is difficult to differentiate what notes are being performed by any one individual. The songs don't appear to have a beginning or an end and vary wildly in length; it may be proper to consider this species a continuous singer, but more recordings are needed. Both sexes appear to sing.

Calls: Single individuals in flight often give a flat *chup*, typically icterid in quality, as well as a ringing whistle *whee*. These two notes may be given in conjunction, *chup-whee-chup-whee*. The *chup* note varies somewhat, sometimes sounding like *chak* or *chep* in fact they can give both forms consecutively, *chak-chep* for example. These calls may be incorporated into the presumed song of this species.

DESCRIPTION A slim and long-tailed blackbird. The bill is stout, long and slightly hooked. The primaries are somewhat pointed, and the primary extension is long. Wing Formula: P9 < P8 > P7; P9 ≈ P5; P8–P5 emarginate. The long, thin tail is noticeably graduated, the difference between the longest (outermost) and shortest (innermost) feathers nearing 20 mm.

Adult Male (Definitive Basic): The bill is black, and the eyes are dark. The body is entirely black with a blue gloss, particularly on the head, back and breast. The underwings are black, but the pectoral tufts are lemon-yellow, contrasting with the rest of the body colour. Often the tufts are hidden, but these are diagnostic if seen. The bases of these yellow feathers are more chestnut, and there is some variation in the extent of the chestnut or tawny bases to those feathers.

Adult Female (Definitive Basic): Similar to adult male, but shows a duller gloss to the plumage. In most views it is impossible to reliably sex members of this species; however, given good light and a close view it is possible.

Immature (First Basic): Similar to adults, but the body plumage averages less iridescent. Note that the chin is brown, not black as on adults. In addition, brownish primaries, secondaries and tertials contrast with the black wing-coverts.

Juvenile (Juvenal): Like adult and immature, but duller still. The body is blackish-brown rather than black and lacks iridescence. As in the immature, juveniles have brown chins.

GEOGRAPHIC VARIATION Monotypic.

HABITAT Forages largely in the canopy of moist montane subtropical forest. This forest tends to be rather open and low in stature. The high humidity and frequent low clouds and fog create luxuriant growth of epiphytic plants, which this species of blackbird appears to be attracted to. At the edge of the Gran Sabana (Venezuela) in the upper end of its altitudinal range it may be found at the forest edge or along riparian corridors. The species will use stunted, open forest and doesn't appear to avoid roadsides.

BEHAVIOUR Tepui Mountain-Grackle is almost always found in small noisy flocks, and rarely observed singly. The flocks commonly contain several to a dozen birds, but sometimes as many as 40 individuals (Sclater 1881, Chapman 1931a). They forage by probing into bromeliads, other epiphytic vegetation and canopy flowers, but do not forage on the ground as true grackles (*Quiscalus*). While in these foraging flocks, the birds often sing in unison. During singing bouts, they may stay still for a period of time but will also sing while foraging and probing. No special attitude or display is commonly seen while singing, perhaps it is infrequent in occurrence or only given during very specific occasions. Inherently social, this species may be a co-operative nester but this requires confirmation. One nest that was under construction was visited by a female with nesting material as well as at least twice by the local flock. While the flock was at the nest tree, they inspected the nest, foraged and sang. Given this behaviour, they are certainly not territorial. The interest the other members showed in the nest itself suggests that they may have some role to play in the nesting process. Group singing at the nest is also a clue that co-operative breeding may occur in this species. Brood parasitism by Shiny Cowbird is recorded for *Macroagelaius* by Friedmann and Kiff (1985) but it is not clear if this refers to this or the Colombian species.

NESTING A nearly completed nest was found on March 12, 1997, at the top of the Escalera road in Bolívar, Venezuela. It appeared that only one bird, a female was building the nest. It was placed 12 m up in a tree with loose foliage. The nest was placed nearly at the crown of the tree, only 50 cm from the edge of the crown. The nest tree was part of a stand of four trees of the same species situated on a peninsula, immediately beside a larger river, and flanked on the other side by a tributary just before it entered the main river. The nest was a bulky, 30 cm in diameter, and the cup was composed mainly of what looked like dry grasses but what may have been other dry plant fibres. The nest was not cleanly woven, but was messy in its overall appearance (Jaramillo unpub. data). This appears to be the only known nest of this species. The eggs and clutch size are not known.

DISTRIBUTION AND STATUS Uncommon to rare in the subtropical zone and upper reaches of the tropical zone in the tepuis of Venezuela, Guyana and Brazil (500–2000 m). Apparently exists in two disjunct populations. In the west it occurs from northern Amazonas, including Mount Paraque and Mount Yavi, south to Mount Duida, in Venezuela (Phelps and Phelps 1950). The eastern population occurs in Bolívar, Venezuela from the Gran Sabana to Mount Roraima east to Chimanta-tepui, including Uei-tepui, Acopán-tepui, Auyan-tepui, Ptari-tepui, Aprada-tepui, Uaipán-tepui and Mount Sarisariñama (Phelps and Phelps 1950). It is also found in the tepuis of adjacent Roraima, Brazil, and the Merume Mountains of Guyana.

MOVEMENTS No known migratory movements. This species appears to be resident in its montane habitat and there is no evidence that it undergoes altitudinal migrations at any time in the year.

MOULT Adults replace their plumage once a year through the complete definitive pre-basic moult. This takes place after the breeding season. On Mount Duida, Amazonas, Venezuela, moult mid May–mid June. In other sites in Amazonas, Venezuela, the moult timing appears to be similar. The juvenal plumage is replaced through the partial first pre-basic moult. During this moult, the body feathering is replaced as well as the wing-coverts and underwing-coverts. The tertials are usually retained, and the tail and remiges are always retained. The timing of the first pre-basic moult is similar to that of adults. Pre-alternate moults appear to be lacking in adults. However, an immature individual from Mount Duida moulting both the face and primaries during mid December suggests that first pre-alternate moults may occur and that this moult may include the wing feathers. Clarification is needed as to whether this moult pattern is typical or if this was an aberrant individual.

MEASUREMENTS Males: (n=10) wing 125.5 (118–133); tail 123.6 (108–133); (n=9) culmen 27.8 (25–30); (n=10) tarsus 31.7 (30–35). Females: (n=9) wing 118.2 (112–124); tail 118.1 (110–126); culmen 26.3 (24–28); tarsus 29.9 (28–32).

NOTES This species is sometimes treated as conspecific with Colombian Mountain-Grackle. The differences in plumage and structure as well as biogeographic considerations, make retaining these two forms as separate species the most satisfactory decision based on the current available knowledge. Tepui Mountain-Grackle is often referred to as Golden-tufted Grackle. This species was originally described as a member of the true grackle (*Quiscalus*) genus. Later it was moved to the *Agelaius* genus, under the subgenus *Macroagelaius*, with the black and yellow Yellow-winged Blackbird (58) hypothesised to be its closest relative (Sclater 1881). In fact, the *Macroagelaius* genus (Mountain-Grackles) appear to be most closely related to Velvet-fronted Grackle (81) and Oriole Blackbird (53) based on comparisons of mitochondrial DNA sequences (Lanyon 1994). One specimen had a yellow feather on the belly. Isolated yellow feathers in areas where the body is typically black are seen in several black and yellow icterids. This does not appear to be of any taxonomic significance.

REFERENCES Sclater 1881 [description, notes, behaviour], Meyer de Schauensee and Phelps 1978 [distribution], D. Wege in litt. [distribution].

83 COLOMBIAN MOUNTAIN-GRACKLE *Macroagelaius subalaris* **Plate 30**

(Boissonneau) 1840

A rare Colombian endemic, much of this species' habitat has been cleared and it now is teetering on the brink of extinction. The mountain-grackles are birds of the forest, quite unlike the true grackles.

IDENTIFICATION Unlikely to be misidentified in its limited range. Colombian Mountain-Grackle is a slim, medium-sized black icterid with a noticeably long tail. The chestnut axillaries are not easily seen, but are diagnostic. The shape alone should be enough to identify this species. Apart from its slim shape and long tail, it has a slim bill and the tail is parallel-edged, not appearing fanned in the field, and is graduated such that the tip appears rounded, not square or notched. Colombian Mountain-Grackles travel in noisy flocks through the forest, unlike many other black birds sympatric with this species. The rare Red-bellied Grackle (76) is larger, stockier and thicker-billed than Colombian Mountain-Grackle; in addition, it has a scarlet belly and vent. Furthermore, the two species should not be sympatric but given their rarity and the lack of knowledge of distribution of highland birds, this is still uncertain. Colombian Mountain-Grackle may be confused for one of many of the other all-black icterids. However, none of these are expected to be sympatric with it. Giant Cowbird (98) and male Shiny Cowbird (102) are also black, but Giant Cowbird is much larger. Both of these cowbirds do not commonly range into the highlands, and Shiny Cowbird is unlikely to be found in forest habitats. Both of these cowbirds are quite different in shape from Colombian Mountain-Grackle, both being chunkier and shorter-tailed. Carib Grackles (93) are birds of the lowlands and show obviously yellow eyes; Colombian Mountain-Grackles have dark eyes. Carib Grackles have noticeably keeled tails, and in addition are birds of open areas usually foraging on the ground, quite unlike mountain-grackles. Solitary Cacique (21) is also found in the lowlands, but is a forest bird. The cacique is large and stocky, with a pale coloured-bill, which is noticeably thicker and longer than the slim dark bill of Colombian Mountain-Grackle. For differences between this and Tepui Mountain-Grackle (82) see Identification under that species' account.

VOICE Unknown. No recordings or descriptions of this species' song exist.

DESCRIPTION A slim and long-tailed blackbird, its tail is moderately graduated (the difference between the longest and shortest tail feather is approximately 30 mm). The bill is about as long as the head, long and relatively stout. Its culmen is curved at the tip, while the gonys is straight, giving the bill a curved tip and a bit of an 'overbite'. The wings are rounded. Wing formula: P9 < ≈ P8 < ≈ P7 > P6; P9 ≈ P4; P8–P6 emarginate.

Adult Male (Definitive Basic): The bill is largely black, with a grey spot towards the base of the lower mandible; eyes dark brown. The plumage is entirely black with a dull blue gloss, which is not readily visible in the field. The black underparts are also glossed with dull blue, but less so on the belly. The wings are also black, with a blue gloss throughout. The tail is entirely black, also showing a dull blue gloss. The axillaries and most of the wing-linings are chestnut, the darkness of this colour makes it difficult to observe this in the field except under exceptional circumstances. In addition, the lesser and median coverts (the epaulet) is chestnut like the axillaries. The legs are rather thick and strong-looking and entirely black.

Adult Female (Definitive Basic): Similar to adult male, but shows a slightly duller gloss.

Immature (First Basic): Similar to adults, but with browner primaries and duller gloss. The chestnut patch of the underwings is smaller in extent. Some may show a contrast between brownish retained juvenal secondaries and primaries, and newer blackish feathers.

Juvenile (Juvenal): Similar to adults but the body plumage is brownish-black, largely lacking all iridescence. Juveniles are even duller than immatures. The tips of the back feathers are glossed blue, but their bases are brownish and not glossed. Thus, the little iridescence on the upperparts appears patchy if seen well. Similarly, the upper breast feathers have a small amount of iridescence on their tips. The primaries and secondaries are brownish-black. The chestnut epaulets, and wing-linings are present but these are reduced in extent when compared to those of adults. Soft parts as on adults.

GEOGRAPHIC VARIATION Monotypic.

HABITAT Found in the upper subtropical and temperate zone of the eastern cordillera of the Colombian Andes. Lives in forest and forest edge, but primarily a montane forest species.

BEHAVIOUR It is almost always observed in small groups. No other data available.

NESTING Specimens in breeding condition were collected by Carriker during September from Norte de Santander (Hilty and Brown 1989). The nest appears to be unknown.

DISTRIBUTION AND STATUS This Colombian endemic is rare and local (1950–3100 m). Found on both slopes of the East Andes in the north, but only on the west slope south of Norte de Santander; from Ramirez (Norte de Santander) as far south as Fusagasugá (Cundinamarca). It is apparently common at the Reserva Natural Guanentá-Alto Río Fonce near Virolín, Santander (M. Renjifo and D. Uribe, fide D. Wege). This is the only site where sightings of this species have been made in recent decades (P. Salaman pers. comm.). Forest clearance has apparently caused a population decline and range contraction in this species; it is now largely absent from areas where it was once common.

MOVEMENTS Sedentary, no evidence of seasonal movements.

MOULT The plumage is changed through the complete definitive pre-basic moult. The tail feathers moult centripetally, from the outside inwards; the tail is fully grown before the outer primaries have fully moulted. This moult occurs after breeding. Birds from Santander are undergoing their definitive pre-basic moult late August–late September, while in Cundinamarca this moult takes place in June. The juvenal plumage is lost through the partial (perhaps incomplete?) first pre-basic moult. One moulting individual had moulted most of the primaries and secondaries, retaining several secondaries on each wing. The timing is similar to that of adults, August–September in Santander and June in Cundinamarca.

MEASUREMENTS Male: (n=10) wing 136.6 (130–142); tail 153.4 (150–160); culmen 26.5 (25–28); tarsus 32.0 (30–34). Female: (n=10) wing 127.8 (123–131); tail 142.7 (132–155); culmen 25.3 (24–26); tarsus 30.5 (29–31).

NOTES This species is sometimes considered conspecific with Tepui Mountain-Grackle which it resembles in many ways. In the most basic sense the two species differ only in the colour of the underwing 'tuft'. In fact, this is not the case, as there are several plumage and structural differences which were pointed out by Meyer de Schauensee (1951). Firstly, Colombian Mountain-Grackle has a rufous epaulet, and the feather bases to the body plumage are grey, not white as in the Tepui species. The bill shapes of the two species are appreciably different; Colombian Mountain-Grackle has a thicker, shorter bill. Whilst several large montane forest fragments exist within the range of this species, so little is currently known of this species that it warrants immediate attention and threatened species status (P. Salaman pers. comm.).

REFERENCES Chapman 1917 [habitat], Hilty and Brown 1989 [habitat, nesting], Wege in litt [distribution].

84 CUBAN BLACKBIRD *Dives atroviolacea* (d'Orbigny) 1839 — Plate 32

Cuban Blackbird is affectionately known as El Totí in its native Cuba, this name is onomatopoetic.

IDENTIFICATION This species is restricted to Cuba where it is most likely to be confused for another blackbird such as Greater Antillean Grackle (92), Shiny Cowbird (102), Red-shouldered Blackbird (64) or Tawny-shouldered Blackbird (63). The all-black Cuban Bullfinch (*Melopyrrha nigra*) is much smaller and has a thick, short bill. The main identification problems are likely with the grackle and the cowbird, the latter is rare in Cuba. Cuban Blackbirds are medium-sized blackbirds with a purplish body gloss and dark eyes, their bills are noticeably longer than they are deep (Figure 77.1). Greater Antillean Grackle is larger than Cuban Blackbird and like all grackles sports a noticeably wedge-shaped tail, particularly the males. Both sexes of the grackle have pale yellow eyes. However, juvenile Greater Antillean Grackles have dark eyes, but these birds are also dull black without a noticeable body gloss. The bills of Greater Antillean Grackles, of any age or sex, are proportionately longer than those of Cuban Blackbird and tend to show a hook at the tip. Cuban Blackbirds show a purplish body gloss, while the grackle has more bluish iridescence. Shiny Cowbird has dark eyes like Cuban Blackbird, but in all ages it has a much shorter bill. In addition, the cowbird's culmen is nearly straight while the culmen of Cuban Blackbird shows a noticeable decurve. Adult male Shiny Cowbirds have a largely blue or violet-blue body gloss, which is different from the dark purple gloss of Cuban Blackbird. Cuban Blackbird is also larger, longer-legged and longer-tailed than Shiny Cowbird. Female Shiny Cowbirds are brownish-grey and quite unlike Cuban Blackbird in colour. The remaining two blackbirds, Red-shouldered and Tawny-shouldered both show colourful epaulets but these may be hidden at rest. Note that both of these blackbirds are smaller than Cuban Blackbird and they both have spike-like bills with a straight culmen and gonys. In addition, these two blackbirds lack a visible gloss in most lights. Female Red-shouldered Blackbird is entirely black, but shares the male's structural differences from Cuban Blackbird. She lacks any perceivable gloss in most lights and is relatively shorter-tailed than Cuban Blackbird. Finally, the distinctive *to-tee* call of Cuban Blackbird is not given by any of the sympatric blackbirds with which it could be confused.

Note that the allopatric members of the *Dives* genus all look quite alike. However, Cuban Blackbird is the smallest and it is the only one with a vivid purple body gloss. The other species tend towards greenish or blue iridescence.

VOICE Song: The song has been described to be pleasant and composed of short notes (Gundlach 1876). Songs are varied and composed of different phrases separated by short periods of silence. These phrases include a repetition of a single sharp note, such as *twee-twee-twee-twee* or *tiuw-tiuw-tiuw-tiuw*. One phrase is a mellow *quee-ahh-whaaaa,* also it gives a peculiar nasal *wwwwhhhhhhaaaaaaa* which is reminiscent of the sounds of a distant braying sheep. In addition, a more complex and warbled song is given. This song is quiet and faster, the separate notes not being given as different phrases.

Calls: A common call is a two syllable note *To-teee*, with the accent on the second syllable, which has gained it the local name of Toti. A clear *tee-o* which is repeated and sounds reminiscent of a Tufted Titmouse (*Balaenophorus bicolor*) (Bond 1988). In addition, it gives a typical blackbird *chup* or *chuk* note with a slightly nasal quality. Also utters a series of peculiar nasal whistles and sounds.

DESCRIPTION A medium-sized blackbird with a short, thick and blunt-tipped bill. The culmen is evenly curved, typical of the genus, it is not flattened. The culmen is slightly longer than the gonys, creating a small hook. The tail is square ended. Wing formula: P9 < P8 > P7 > P6; P9 = P6.

Adult Male (Definitive Basic): The bill is jet black and the eyes are dark brown. The entire plumage is black with a strong gloss. The iridescence of the upperparts is violet but violet-blue on the crown. The underparts also show a violet iridescence, strongest on the breast and nearly lacking on the belly. The wings, including the primaries, secondaries, tertials and greater coverts are glossed blue-green. These contrast with the vivid violet gloss on the lesser and median coverts. The black wing-linings show a violet iridescence.

The tail is glossed blue-green like the flight feathers. The legs and feet are black.

Adult Female (Definitive Basic): Similar to adult male, but marginally smaller in size and slightly less iridescent.

Immature (First Basic): Similar to adults, but showing a duller iridescence on the body. The bases of the back feathers are brown. On the wings, the tertials and primaries are browner and the wing-linings are obviously brown as they are retained from the juvenal plumage. The contrast between blackish coverts, tertials and body, and brownish primaries and secondaries is typical of immature birds. The belly and thighs are brown.

Juvenile (Juvenal): Similar to adult and immature, but even duller. The body plumage is brownish-black, largely lacking iridescence. The tips of the upper breast and upper back show a limited amount of violet iridescence which would be difficult to detect in the field. The wings and tail are blackish with a very dull gloss. The thighs are brownish.

GEOGRAPHIC VARIATION Monotypic.

HABITAT Varied in its habitat choice. Cuban Blackbird does well in agricultural areas, or any other grassland, open forest or generally open area. In particular, they are fond of stands of palms for nesting, but these are not a prerequisite, and will nest in a variety of situations. Livestock attract this bird which forages on the insects put up by the animals. Particularly in winter, Cuban Blackbird can be found in urban areas and parks.

BEHAVIOUR Flocks in the non-breeding season and may form mixed flocks with other species, particularly Greater Antillean Grackle and sometimes Tawny-shouldered Blackbird. Roosts in large groups, often in traditional sites and usually with Greater Antillean Grackles. Big roosts exist even in downtown Havana. Once pairing begins in the spring, the flocks break up and the birds disperse. They are generalist foragers, and forage both on the ground and by gleaning from palms and flowering trees, including nectaring at flowers. However, it also eats insects and fruits (Gundlach 1876). They may also be observed picking insects from livestock in the manner of cowbirds. Local farmers consider Cuban Blackbird to eat ticks (Acari) which they pull from livestock; however, studies of stomach samples of the blackbird have not found the presence of ticks as food. They certainly eat insects flushed by livestock, however. This blackbird is also known to follow the plough, foraging on the flushed insects (Gundlach 1876). In addition, they eat many different types of seeds including rice, millet and corn as well as seeds of wild plants. They are generally considered somewhat of an agricultural pest. Fruit is also taken, ripe bananas are a favourite. Notably, Cuban Blackbird has also been observed to feed on small lizards (Garcia 1980). Like most blackbirds, Cuban Blackbird performs a stereotyped display while singing. It bows forward, opens the tail and droops the tail (Gundlach 1876) when in song.

NESTING The breeding season commences in April. Nests are placed in the base of a palm frond, in a forked branch of a tree, and sometimes in a cavity found in a tree or even a building. When nesting in trees, it prefers those with dense foliage. Clutch size 3–4 eggs (Skutch 1996). The eggs are white with ashy or brown spots, clustered around the broader end of the egg (Garcia 1980, Gundlach 1876).

DISTRIBUTION AND STATUS Uncommon to locally common. Restricted to Cuba. It is to be found throughout Cuba, from the eastern to the western extreme (Garcia 1980). There are dubious reports from the Isle of Pines (Bond 1988). Apparently this is due to Gundlach who reported Cuban Blackbirds from the Isle of Pines. However, no specimens exist and no one has observed the species there since. It is assumed that Gundlach misidentified a female Greater Antillean Grackle for the shorter-tailed Cuban Blackbird (Garcia 1980).

MOVEMENTS Does not appear to make large scale seasonal movements. However, wintering flocks appear in areas where they are not common breeders such as in urbanised areas, suggesting that winter distribution is not the same as in the summer (Garcia 1980). They also roost in large flocks and will travel a good distance in order to arrive at the roost site. These evening flocks, often associating with Greater Antillean Grackles, are well known and obvious.

MOULT Not well known. The adults undergo their annual complete moult, the definitive pre-basic, between October–November. There is no evidence that pre-alternate moults exist. The first pre-basic moult replaces the juvenal plumage, its timing is similar to that of adults. This moult appears to be partial, involving the body, coverts, tertials and tail. Some individuals may replace some of the flight feathers. Most of the juvenal wing-linings are retained.

MEASUREMENTS Males: (n=12) wing 136.0 (126–142); tail 112.0 (105–120); culmen 25.2 (23–27.7); tarsus 32.7 (31–35.6). Females: (n=10) wing 127.8 (120–133); tail 103.7 (97.3–111); culmen 23.7 (22–25); tarsus 30.7 (29–33).

NOTES A Cuban saying translates as 'All birds eat rice, but the Totí (Cuban Blackbird) gets the blame', another saying is 'The Totí is to blame for everything' (Garcia 1980). See Notes section under Melodious Blackbird (85) for systematic relationships of the genus.

REFERENCES Barbour 1943 [calls, behaviour], Garcia 19?? [habitat, behaviour, nesting, movements], Gundlach 1876 [voice, habitat, nesting], Hardy *et al.* 1998 [voice], Ridgway 1902 [measurements].

85 MELODIOUS BLACKBIRD *Dives dives* (Deppe) 1830 **Plate 32**

A noisy and easy to observe all-black icterid of Mexico and Central America. It is quite similar to its close relative Scrub Blackbird (86) of South America and some consider them to be part of one species.

IDENTIFICATION Likely to be confused with other black icterids like Great-tailed Grackle (87), Bronzed Cowbird (99) and Brewer's Blackbird (95). Melodious Blackbird is entirely black with dark eyes, it has a moderately long tail, a stocky bill with an appreciable curve to the culmen and strong legs. The male grackle is separated by its long, wedge-shaped tail that Melodious Blackbird lacks. The grackle also shows vivid yellow eyes and a bright blue iridescence which is more noticeable than that of Melodious Blackbird. Female grackles are also yellow-eyed but have brown rather than black plumage. Bronzed Cowbirds are black with a bronzy gloss and red eyes, males have a neck ruff which when present is diagnostic. Bronzed Cowbirds have short, thick bills unlike the longer bill of Melodious Blackbird and the former lacks an appreciable curve to the culmen. Melodious Blackbird has a proportionately longer tail than the cowbird. In several ways, Brewer's Blackbird is most similar, but note male Brewer's gleaming whitish-yellow eyes and the strong purple and greenish gloss on the head and body. In addition, the slim Brewer's Blackbird has an elegant walking style lacking in the more thickset Melodious Blackbird. The bill of Melodious Blackbird is longer and more stout than the finer bill of Brewer's Blackbird. Female Brewer's Blackbirds have dark eyes but their greyish-brown plumage easily separate them from Melodious Blackbird. Yellow-billed Cacique (23) inhabits dense tangles, not the more open areas where Melodious Blackbird is found, but like the blackbird, it is entirely black in plumage. The cacique has both a yellow eye and a pale yellow bill, which helps to separate it from Melodious Blackbird. The all-black anis (*Crotophaga*) may be found sympatrically with Melodious Blackbird but these have thick bills, pointed head feathers, long tails and very different behaviour. Melodious Blackbirds perform a series of striking vocalisations which may aid in identification (see Voice). In addition, Melodious Blackbird often duets, a behaviour absent in sympatric all-black blackbirds.

VOICE Most of the calls are given by both sexes; however, females differ in tone, being somewhat more wheezy.

Song: The song is usually performed as a duet between the male and female, usually started by the female. They tend to begin with a *see* call followed by a clear ascending whistle preceded a *whit* note (cardinal call of Orians 1983, which can be described as a *whit-tiiieeeeeeuuuu* and reminiscent of a Northern Cardinal *Cardinalis cardinalis*), then the male may give a *see* note and a series of two syllable calls made up of an initial rising note followed by a falling note *seetup*. Once started, 'cardinal calls' are alternated with *seetup* notes, creating a rich melodious song. The exact structure of the notes and order used appears to vary geographically. Songs not given as a duet tend to mix the *see* and *seetup* elements in variable sequences. A 'gurgle-whistle' is a male song that begins as a pure rising whistle, that then drops in frequency, rises again and is terminated in a high *see*. It is often interspersed within the duet song or given alone.

Calls: The common call is a loud *cheep* often given with a tail flick. The 'cardinal call' is a descending, clear whistle often begun by a short chip note. It may be given in flight and during territorial altercations as well as part of the song. While foraging, females will execute a double note *wheet-wheet*. A *chuck* call, also associated with tail flicks, is given by females and appears to communicate a level of concern rather than alarm.

DESCRIPTION A stocky blackbird with a longish tail. The bill is slightly shorter than the head, the culmen is slightly downcurved and the gonys is nearly straight. The culmen is slightly longer than the gonys, showing up as a small hook at the tip of the bill. The wings are rounded and short (p9 is shorter than p2) and the legs and feet are thick and sturdy. Wing Formula: $P9 < P8 < P7 \approx P6 > P5$; $P9 \approx P2$; P8–P5 emarginate.

Adult Male (Definitive Basic): The bill is black, and the eyes are dark brown. The body is completely black with a slight bluish gloss throughout. This iridescence is dullest on the belly and brightest on the breast and upper back. The legs and feet are black.

Adult Female (Definitive Basic): Identical to male, but marginally smaller.

Immature (First Basic): Duller than, but similar to, adult plumage. There is a brown wash on the throat. The back is browner, showing some feathers lacking iridescence, particularly on the feather bases. The primaries and secondaries are brownish, duller than the blackish body. The tail is black.

Juvenile (Juvenal): Similar to adult, but brownish-black throughout and lacking iridescence. The wings and tail are also brownish. At close range, the loose texture of the flanks, belly and upper back feathers may be noticeable.

GEOGRAPHIC VARIATION Monotypic.

HABITAT Inhabits early successional habitats and edge as well as pine forests. Melodious Blackbird is quite tolerant in its habitat choice and can be found in almost all habitats except for dense forests, it has a definite preference for forest openings. It is common in settled areas and has adapted well to human-produced habitats. Forages mainly on the ground and has a liking for lawns and cleared areas in general. It is not found where there is a thick undergrowth. Roosts in dense cane groves with other blackbirds.

BEHAVIOUR The mating system is monogamous. During the non-breeding season, pairs may associate into small groups, but the pair units are maintained within the flock. Once the breeding season commences, pairs become highly territorial, aggressively defending against other pairs or lone individuals. Both sexes defend the territory and perform the displays and vocalisations associated with territorial defense. Some of these scuffles between pairs can be quite lengthy and involved. Territories are defended against Yellow-backed Orioles (26). Nest predators like the Brown Jay *(Cyanocorax morio)* are violently attacked by nesting Melodious Blackbirds. The 'song spread' is executed by both sexes. While the song is given, the body plumage is ruffled, while the tail is spread and the wings are extended outward, maximally at the shoulders and less so at the tips, the cheek feathers are ruffled while the forehead to the nape is kept flat, creating an odd two-toned appearance on displaying birds. In a less intense 'song spread', the wings are held closed and the tail is not fanned and the plumage is not ruffled to the same extent. During some of the songs, Melodious Blackbird may pump its legs as if it was performing 'push-ups', pumps are accompanied by tail-flicks. Tail-flicking is also observed from agitated lone birds, the tail is flicked rapidly upwards and spread, then allowed to fall more slowly. There are at least three flight displays given. The 'fluttering flight' has been seen in attacks against nest robbers like Brown Jays; it involves shallow, rapid wingbeats as implied by its name. The 'bill-up flight' is used during territorial disputes, by both sexes, and is similar to a display given by Yellow-headed Blackbird (54). It is a slow, undulating flight with the bill held tilted upwards at an obvious angle. 'Bill raising' is a typical blackbird sign of aggression. As in many other blackbird species, sexual chases take place in this species. The male will chase the female in flight, never actually catching her. The display may end with the female going into a solicitation display where she raises her tail, and flutters her wings while uttering *cheep* calls.

NESTING In Belize, nesting begins in mid May and in Guatemala nests May–July. In Costa Rica, nests April–July (Skutch 1954). Nests are an open cup in a bush or tree, both sexes assisting in its construction. It takes approximately 13 days to finish building the nest (Skutch 1954). The nest is a bulky cup made of coarse strips of vegetation, the bottom of the cup is strengthened with mud and cow-dung and it is lined with slender rootlets (Skutch 1954). Nests are placed from three to seven meters up in a densely-foliaged tree. Lemon and orange (*Citrus* sp.) trees were preferred in one Costa Rican location (Skutch 1954). Clutch size 3–4 eggs (Skutch 1954, Orians 1983). The eggs are blue with brown blotches clustered at the broader end. The female performs all of the incubation, but both sexes help to feed the young. The male also brings some food to the incubating female (Skutch 1954). Melodious Blackbird is not known to be parasitised by any cowbird species (Friedmann and Kiff 1985).

DISTRIBUTION AND STATUS Common from sea-level to 2000 m. Distributed from Mexico to Costa Rica. In Mexico along the Atlantic slope from S Tamaulipas south through E San Luis Potosí, E Hildago, Veracruz, NE Puebla, NE Oaxaca, N and NE Chiapas and the entire Yucatán peninsula, Campeche, Yucatán, Quitana Roo in Mexico, throughout Belize, and the Petén and north of Guatemala (Howell and Webb 1995). In easternmost Guatemala the range extends south and reaches the Pacific coast. Distributed throughout El Salvador, W and C Honduras, south along the Pacific slope of Nicaragua to N Costa Rica. The distribution has been steadily pushing southward and its range in Costa Rica is certainly in heavy flux. The first record was in the south in extreme NW Punta Arenas province (Stiles and Skutch 1989) and it is now quite regular in parts of the northern Pacific slope. It has recently been found at the Rancho Naturalista on the Caribbean slope of Costa Rica (M. Holder pers comm.). Melodious Blackbird has not been found in Panama, but it should be looked for near the Costa Rican border.

MOVEMENTS Sedentary. As forests are cleared, the potential habitat for this species has increased. It is expanding its range, particularly to the south. This expansion has been well chronicled. For example, it appears that the species was absent from El Salvador historically. The first records from that country were during the 1950s, by the 1960s the species was well distributed but still local. A noticeable population increase was noted in the 1970s at which time the species became common throughout the country, the last regions to be colonised were mountain tops. Similarly, Melodious Blackbird appears to have been absent from E Guatemala (Lago Izabal) in the early 1900s, but was widespread in the Caribbean slope during the 1960s. By the 1970s it was also found in Guatemala's Pacific slope. It has been suggested that the invasion proceeded first south along the Caribbean Slope in Guatemala then into Honduras. The expansion continued southward in Honduras into El Salvador and once established in El Salvador's Pacific slope it invaded northward to Guatemala's Pacific slope (Thurber *et al.* 1987). The expansion is continuing, with its leading southern edge now in Costa Rica.

MOULT One complete moult a year, the definitive pre-basic. It is timed to begin after breeding is over. In Guatemala, Scrub Blackbirds are in their definitive pre-basic moult during November. Pre-alternate moults are lacking. The juvenal plumage is changed by the first pre-basic moult, which is incomplete. The entire plumage is changed, but some juvenal underwing-coverts and back or scapular feathers are retained. The first basic primaries and secondaries are still not as black as those of the adults.

MEASUREMENTS Males: (n=11) wing 126.3 (123–132.8); tail 119.5 (114.3–127); (n=10) culmen 30.3 (28–32); tarsus 39.2 (38.1–40.4). Females: (n=10) wing 114.5 (112.3–120); tail 109.3 (102.1–114.3); culmen 28.2 (27.4–29.2); tarsus 37.7 (36–38.6).

NOTES Sometimes considered conspecific with its close relative, South American Scrub Blackbird. Griscom (1932) describes that natives of Guatemala used to perform a ritual during crop failures where they trapped a mouse, a grasshopper, and a Melodious Blackbird and these were put through a ceremony where the three animals were told of dire consequences unless they ceased damaging the crops. The three animals were then taken back to the fields and released so that they would spread the news. Griscom does not state if the ritual worked. Based on comparisons of mitochondrial DNA sequences it appears that the *Dives* genus is closely related to *Euphagus* and the *Quiscalus* grackles (Lanyon 1994, Kevin Johnson pers. comm). The centripetal tail moult of these genera is consistent with this arrangement.

REFERENCES Howell and Webb 1995 [description], Orians 1983 [habitat, behaviour, nesting, vocalizations], Ridgway 1902 [measurements], Skutch 1954 [nesting], Thurber *et al.* 1987 [movements, distribution], Williams 1995 [distribution].

86 SCRUB BLACKBIRD *Dives warszewiczi* (Cabanis) 1861 — Plate 32

A common blackbird of the arid coastal region of Ecuador and Peru. It is noisy and does well around human settlements.

IDENTIFICATION Scrub Blackbird is entirely black with a dull iridescence, dark eyes, strong legs, moderate length tail and a moderately long bill with an appreciable curve to the culmen. It may be confused with any of the all-black icterids sympatric with it such as Shiny Cowbird (102), Giant Cowbird (98) and Great-tailed Grackle (87). Male Shiny Cowbird is by far the greatest identification pitfall. However, Scrub Blackbird is larger, has longer legs, a longer bill and less visible gloss than the cowbird. Shiny Cowbird is rather short-legged and short-billed, its culmen appears straight in the field. In addition, Scrub Blackbirds frequently sing a complex song composed of striking notes and whistles. Shiny Cowbirds give a gurgling, liquid song or a quick series of high-pitched whistles as their flight song. Scrub Blackbirds frequently bob as they sing, cowbirds do not do this. Also, duetting between two birds is diagnostic for Melodious Blackbird as no other sympatric all-black, open country species performs this behaviour. Great-tailed Grackle and Giant Cowbird are both considerably larger than Scrub Blackbird. The grackle has pale yellow eyes and the cowbird has yellow or red eyes. The grackle and Giant Cowbird are both noticeably more iridescent than Scrub Blackbird. The grackle has a long wedge-shaped tail and a long and powerful bill; Giant Cowbird also has a long bill but it shows a peculiar small-headed look while Scrub Blackbird has a standard sized head and tail. The *Dives* genus includes some very similar species. In the field, the northern subspecies of Scrub Blackbird would certainly prove to be indistinguishable from Melodious Blackbird, and it is fortunate that the two do not occur together. However, their voices do differ.

VOICE Song: Individual songs are complex mixes of clear whistles, shrill notes, sharp buzzes and short chatters. The notes included in songs may show a wide frequency range, while others are clear and unmodulated. Examples of notes heard in songs include: a low frequency modulated whistle (*tweee* or *whoooeee*), buzzy, shrill sounds (*kzeeet-tzeeeooo, tzeeew-tzeeew*), a sharply descending buzzy whistle (*tzeeeeeeooo*), and a short, high frequency chatter (*ti-ti-ti-ti-ti*). Each note is clearly separated from the next by a pause when a single bird sings. Pairs commonly duet, but group singing (in flocks) is not known. When duetting, each pair member countersings, each performing a different song type simultaneously. One individual will emit a series of sharp, shrill, buzzy notes (including the range of frequencies between 2–10 kHz) that sound somewhat like *wh-tii-tii*. Meanwhile, the other individual intersperses a very different song, which is a series of lower-pitched (usually below 2 kHz) frequency modulated, clear whistles, i.e. *whoooeeep*. The full duet is therefore a complex, but pleasant-sounding series of sounds: *whoeeeep-wh-tii-tii-wh-tii-tii-whoeeeep-wh-tii-tii....* One bout of duetting may last as little as four seconds, but is frequently longer.

Calls: The flight call, given by both sexes, is a series of sharp *teeeew* notes.

DESCRIPTION A medium-sized, stocky blackbird with a short tail and long legs. Its posture is somewhat upright, perhaps accentuated by its longish legs. The bill is slightly shorter than the length of the head. It shows a slight curve to the culmen, but a straight gonys. The shorter gonys and curvature of the culmen create a slight 'overbite', or hook to the bill tip. The wings are rounded and the tail square. The nominate form is described. Wing Formula: P9 < P8 < P7 < P6 > P5; P9 ≈ P2; P8–P6 emarginate in the nominate form. The southern form's (*kalinoswki*) wing formula is slightly different: P9 < P8 < P7 > P5; P9 ≈ P2; P8–P6 emarginate.

Adult Male (Definitive Basic): The bill and legs are black. The body is entirely black, and noticeably glossed at close range. At a distance, or in low light situations, it appears flat black. The iridescence is blue on the head, neck and breast, becoming greenish-blue on the back and rump. The belly is less iridescent than the breast, and shows more of a green gloss as opposed to blue. The wings show a blue iridescence, both on the coverts and the flight feathers. The tail has a blue-green iridescence.

Adult Female (Definitive Basic): Similar to male, but marginally smaller.

Immature (First Basic): Similar to adults but with a dull body gloss. In particular, the iridescence appears to be restricted to the upperparts and largely absent from the underparts. There may be some gloss to the breast, but none whatsoever on the belly, vent and thighs. The thighs appear dull blackish, browner than the rest of the body. On the back and scapulars, the iridescence is patchy, due to the unglossed feather bases which break up the eveness of the iridescence. The primaries are brownish-black, particularly at their tips. On some individuals the black body, tertials and wing-coverts contrast with the browner, retained juvenal primaries and secondaries.

Juvenile (Juvenal): Similar to adults, but duller not iridescent. The body plumage is blackish-brown, darkest on the head, breast and upper back. Only the breast and upper back show any iridescence at all, and here it is restricted to the tips of the feathers thus appearing patchy at close range. The wings and tail are blackish-brown and unglossed. The belly and thighs are brownish. The plumage also has a soft texture which may be evident at close range, particularly in the more flexible and wispy flank feathers.

GEOGRAPHIC VARIATION Two subspecies are recognised.

The northern ***D. w. warszewiczi*** is found from W Ecuador to La Libertad in Peru, and is described above.

The more southern ***D. w. kalinowskii*** occurs from La Libertad south to Ica, Peru. *D. w. kalinowskii* is significantly larger and has a proportionately longer bill than *warszewiczi*. The deeper bill shows a straighter culmen than *warszewiczi* for most of its length, however it is curved near the tip. This form has been described as being more glossy than the nominate (Berlepsch and Stolzmann 1892). However, what is more noticeable is not that *kalinowskii* is more glossy, but that its iridescence is distinctively more violet than the bluish iridescence of *warszewiczi*. These two subspecies have been proposed separate species; however, their voices are similar (Ridgely and Tudor 1989) and there are intermediate specimens (Bond 1953). Nevertheless, there are specimens typical of each form collected only 35 km from each other. The contact zone between these two subspecies is not adequately known, but contrary to earlier reports the ranges of the two subspecies are continuous through N Peru, and there is no gap in the distribution of the species here. The zone of contact appears to lie between Cajamarca and La Libertad in Peru (Schulenberg and Parker 1981).

HABITAT Lives in open habitats with available perches or trees as well as open woodlands and forest edge. It is often found in agricultural areas, including urban zones. Most of this species' foraging is done on the ground, necessitating some grass, lawn, or fields for this purpose. When in open fields, they tend to keep close to a tree or bush which birds retreat to when alarmed. They may feed near livestock, but are not closely associated with them. In desert situations, they keep to irrigated agricultural land, oases or riparian areas where some shrubbery and greenery is to be found. They are also found in organ pipe cactus (*Trichocereus* sp.) stands with agave, as well as wild canebrakes (*Gynerium* sp.) in Peru (Koepcke 1970). The clearing of coastal forest in the north, and irrigation of deserts in the south is surely creating an abundance of habitat for Scrub Blackbird, no doubt explaining the recent range expansion particularly to the north.

BEHAVIOUR The song duet begins by a *chip* note given as the head is thrown back, followed by bobbing as the bird flexes its legs. While displaying, the wings and tail are partly spread, the body feathers are ruffled and the tail is pointed downwards. 'Leg pumps' and 'head tosses' are synchronised with vocalisations given during the song, one body pump per note sung. Scrub Blackbird is not aggressively territorial like its relative the Melodious Blackbird (85), and several pairs will share a nesting area (Orians 1983).

NESTING Not well known. In the high valleys of central Peru, nesting occurs April–May. Nests are placed high in trees, particularly planted *Eucalyptus* (Orians 1983). In Peru, Scrub Blackbird has been observed feeding fledglings of Shiny Cowbird (Friedmann and Kiff 1985).

DISTRIBUTION AND STATUS Common to 1500 m, local at higher elevations, to 3000 m in Peru (Koepcke 1970). Scrub Blackbird is particularly common in SW Ecuador and along the west base of the Andean foothills of N Peru. For the most part this is a species of the coastal lowlands and lower foothills, keeping below the upper tropical zone. It is found on the Pacific slope from Esmeraldas in Ecuador south to Ica, Peru. In Ecuador, it ranges inland to the foothills and lower elevation of the west slope of the Andes, in W Pichincha and valleys of Azuay, Loja and El Oro (Ridgely and Tudor, 1989). It ranges inland as far as the upper Río Marañón, Cajamarca, Peru. In southern Peru it has been found inland in Huancavelica (Fjeldsa and Krabbe 1990) and as far south as the Pampas River Valley in Apurímac (W.P. Vellinga pers. comm.).

MOVEMENTS No known migration or seasonal movements. However, the range of this species is expanding as suitable habitat is being created due to forest cutting or desert irrigation. It is now found in Esmeraldas province in N Ecuador and it may only be a matter of time before it reaches southernmost Colombia.

MOULT The body plumage is renewed annually by the complete definitive pre-basic moult. Specimens from Manabí, Ecuador (nominate form) undergo definitive pre-basic moult in December, however those from Loja, Ecuador, moult during August. Individuals of *kalinowskii* from Peru begin moult in early March. No evidence is available to suggest that pre-alternate moults exist. The juvenal plumage is lost through the partial first pre-basic moult. This moult includes the body, wing-coverts and tertials but retains the primaries, secondaries and tail. The tail moult is centripetal and quick, in some cases the entire tail may be shed leaving birds tail-less while the new feathers grow in.

MEASUREMENTS ***D. w. warszewiczi***. Males: (n=11) wing 114.2 (106–125); (n=10) tail 96.4 (86–107); (n=11) culmen 26.5 (24–29); (n=3) tarsus 34.7 (31–37). Females: (n=10) wing 103.9 (98–110); (n=9) tail 89.4 (85–94); (n=10) culmen 24.8 (24–26.5); (n=5) tarsus 31.2 (30–33). Unsexed: (n=1) wing 111; tail 92; culmen 26; tarsus 35.

D. w. kalinowskii. Males: (n=8) wing 144.4 (136–150); (n=6) tail 131.5 (118–142); culmen 33.1 (27–36); (n=5) tarsus 38.6 (38–41). Females: (n=4) wing 139.3 (137–141); tail 124 (120–126); culmen 31.8 (31–33); tarsus 38 (37.5–39).

NOTES Sometimes considered conspecific with the similar, but allopatric Melodious Blackbird. The southern subspecies of Scrub Blackbird, *D. w. kalinowskii*, is quite divergent in its larger size and longer bill and it has been suggested that it may warrant full species status but intergradation with the nominate form is known (Schulenberg and Parker 1981). More research is needed to fully resolve this question.

See Notes section under Melodious Blackbird (85) for systematic relationships of the genus.

REFERENCES Berlepsch and Stolzmann 1892 (measurements), Chapman 1926 [measurements], Orians 1983 [behaviour, nesting], Ridgely and Tudor 1989 [distribution, systematics], Schulenberg and Parker 1981 [distribution, geographic variation].

87 GREAT-TAILED GRACKLE *Quiscalus mexicanus* (Gmelin) 1788 — Plate 34

This large grackle is an obvious component of the avifauna. It is also expanding its range, colonizing areas where it had not previously been found.

IDENTIFICATION All populations of Great-tailed Grackle have bright yellow eyes, except when young. The primary identification problem is with Boat-tailed Grackle (88) in the US. Comparing the sympatric race, *Q. mexicanus prosopidicola*, to Boat-tailed Grackle, Great-tailed Grackle is generally larger, longer-and wider-tailed and has a more massive bill. Particularly on males, the line the forehead makes with the bill is almost straight, unlike the steeper forehead of Boat-tailed Grackle. The Florida and Gulf coast form of Boat-tailed Grackle has dark eyes, unlike the pale yellow eyes of adult Great-tailed Grackles. Male Great-tailed Grackle has more extensive violet iridescence on the body, extending to the lower back, abdomen and flanks while on Boat-tailed Grackle these areas are greenish-blue. There is some overlap in this feature and lighting conditions can affect the colour of iridescence. Behaviour in the breeding grounds can be useful to identify these grackles. The 'ruff-out (song spread)' display of Great-tailed Grackle lacks the 'wing flip' of Boat-tailed Grackle (see Behaviour). When alert at the colony, Great-tailed Grackle does not appear as bull-necked as Boat-tailed Grackle. Vocally these two grackles differ greatly (see Voice section for differences).

Females are also difficult to identify, but the general differences in shape and size apply. Female Great-tailed Grackles (other than juveniles) show straw yellow eyes. Only the Atlantic coast and some Gulf coast populations of Boat-tailed Grackle show pale yellow eyes, and even in these forms females may have dark speckling on the irises particularly near the pupil. Females are weakly iridescent on the upperparts in both species, but this is much more extensive on Great-tailed than Boat-tailed Grackles. Female Great-tailed Grackle varies in colouration, but tends to be darker than Boat-tailed Grackle. Female Great-tailed Grackles rarely show the vivid buffy or cinnamon tones of Boat-tailed Grackle, particularly on the underparts. Also, Boat-tailed Grackle tends to show a warm brown (or cinnamon) colour to the crown and nape. This colour is rarely achieved by Great-tailed Grackle, which shows either a blackish crown with some iridescence, or a grey-brown crown. The dark malar streak averages bolder and more noticeable on Great-tailed Grackle. Female Great-tailed Grackles do not sing, unlike female Boat-tailed Grackles. Juvenile and immature males may be inseparable on plumage details, one has to rely on structure. However, juvenile and 1st-year female Great-tailed Grackles are darker below than the corresponding ages of Boat-tailed Grackle. In addition, the latter tends to be more noticeably streaked on the underparts.

The wide-ranging Great-tailed Grackle also overlaps with several other grackles including Common (91), Nicaraguan (90) and possibly the Slender-billed (89), in the past. Common and Nicaraguan Grackles are much smaller than Great-tailed Grackle and should not be readily confused. However, several points can be used to clearly differentiate between these species. Males of the Common Grackle form that shares its range with Great-tailed Grackle have a bronze sheen on the back, unlike the blue or violet of the latter species. Female Common Grackles are duller versions of the male and do not show the obviously paler underparts, buffy colours or malar streak of female Great-tailed Grackle. Male Nicaraguan Grackle is similar to Great-tailed Grackle in colour, but it is thinner-billed, apart from being smaller. Female Nicaraguan Grackle is pale greyish on the underparts, unlike the local form of

Great-tailed Grackle which has very dark females. The extinct Slender-billed Grackle of Mexico was a smaller bird than Great-tailed Grackle with a much finer and thinner bill. Female Slender-billed Grackle was bright buff on the underparts and on the head, unlike the darker *mexicanus* Great-tailed Grackle which has recently spread to the Mexican Central Valley.

Giant Cowbird (98) should be mentioned as a potential confusion species due to its large size. However, the cowbird has quite a different shape. Even the shorter-tailed female Great-tailed Grackles are on the whole, long-tailed birds compared to Giant Cowbird. In addition, Giant Cowbird appears small-headed and with a relatively shorter bill than the grackle. At close range, Giant Cowbird often shows reddish eyes, but yellow-eyed individuals are known. In flight, the undulating flight of the cowbird is unlike the steady, level flight style of grackles.

VOICE Note that the repertoire of noises that this species can produce is substantial.

Song: The song, which is given only by the male, can be broken down into four parts. It begins with harsh notes similar to the breaking of twigs, followed by an undulating *chewechewe*, and then an abbreviated version of the twig-breaking notes, and finally several loud, two syllable *cha-wee* calls (Kok 1971). The song is accompanied by the 'ruff-out' display. The song varies geographically, particularly in its general quality (i.e. pitch), but the pattern remains similar. The songs of *obscurus* and *greysoni* are apparently structurally different, a peculiarity that needs to be studied. Full song is only heard during the breeding season, and to a lesser extent in the fall (Kok 1971).

Calls: Both sexes give a generalized *chut* call used in many contexts including as a flight call. Usually, the tail is flicked as the *chut* call is given. Males perform a loud *clack* that functions as a warning note. In extreme excitement, males will produce a striking, ascending whinny whistle, not heard in Boat-tailed Grackle. Sometimes this whistle is reversed so that it descends in frequency. During solicitation, males give a series of *cheat* notes, which are higher pitched in this species than in the Boat-tailed. When extremely excited, the *cheat* notes become truncated to a *che*. Females utter a series of *peep* notes during solicitation. The females give a chatter similar to that of Boat-tailed Grackle. The chatter is only given during the breeding season and appears to have an aggressive function towards other females; however, males will sometimes also use a chatter call. Both sexes produce a strange soft puffing sound, *pft*, of unknown function (Kok 1971).

DESCRIPTION A large, long-tailed grackle with a massive bill. The expression of this bird is invariably aggressive due to the pale eye and the large, raptor-like bill. The flat head, without an obvious forehead adds to this appearance. Wing Formula: P9 < P8 < P7 > P6; P9 ≈ P5; P8–P5 emarginate. The race *prosopidicola* is described.

Adult Male (Definitive Basic): The bill is black and the eyes are lemon-yellow. The body is entirely black, with a violet gloss on the head, back, throat and breast. The gloss turns more bluish on the lower back and rump, as well as the belly. The tail-coverts are glossed greenish-blue. The wings and tail are greenish glossed, becoming bluer on the wing-coverts. The legs and feet are black.

Adult Female (Definitive Basic): Soft part colours as in male. The upperparts are blackish-brown with a dull greenish gloss on the rump and uppertail-coverts. The supercilium is often noticeably paler than the rest of the head. The wings are blackish with a more vivid greenish gloss on the coverts. The underwings are brown. Throat creamy-buff, becoming greyer on the breast. A dark malar streak is usually present. The belly, thighs and vent are brown. The tail is blackish often with a dull greenish gloss.

Immature Male (First Basic/First Alternate): Similar to adult males, but smaller and duller. The iridescence is not as extensive and noticeable, and blues and purples are greener than on adults. Also, the tail is not as long or wide as on adults. After the first pre-alternate moult, an immature spring male obtains an adult-like head and breast, but the belly and back will be duller than in typical adults. Also note that most immature males have brown, as opposed to black thighs. Juvenile males have brown eyes which gradually become more yellow as the bird ages. During the 1st fall and winter, a progressively smaller proportion of immatures retains a brown eye, by mid-winter half will have obtained a yellow eye but a few may show a largely brown eye well into May.

Immature Female (First Basic): Similar to adult female, but averages paler on the underparts and face, particularly the supercilium. The paler underparts makes the dark malar stripe stand out more on young females than adults. The upperparts tend to show less iridescence than adults, and often the feathers have thin brownish or rufous fringes. Adult females are marginally larger than immatures. It appears that the structure of the feathers is different as young females wear and fade faster than adults which can be quite noticeable during the first spring and summer. Eye colour change is as in 1st-year males.

Juvenile (Juvenal): The two sexes are similar in this plumage. They are buffy-brown below and show distinct thin dark streaks, or blotches, on the underparts. The back is dark brown or even blackish. As this species is rather large-billed, juveniles may show obviously shorter bills than adults early in the season before bill growth has finished. The eyes are dull brown in juveniles.

GEOGRAPHIC VARIATION Eight subspecies are recognized. Note that the three northern races, *prosopidicola, monsoni* and *nelsoni* are spreading northward in the US. For the most part, there is little information regarding which forms have spread to which areas, therefore the range descriptions for these races given below are somewhat out of date.

Q. m. prosopidicola is found from SC Texas east to Calcasieu Parrish in Louisiana; south to S

Coahuila, Nuevo León and Tamaulipas, Mexico. It is described above. This form is slightly smaller than *mexicanus* and the male is identical in plumage colour. However, females have light buffy underparts unlike *mexicanus* and a paler head than that form.

Q. m. nelsoni breeds from N Baja California Norte, SW California and SW Arizona to S Sonora, Mexico. Compared to *mexicanus* and *prosopidicola* this form is smaller with a relatively short tail and the female is paler with the throat, underparts and supercilium pale greyish-buff. It is the smallest and palest of the races. In California it now appears clear that this subspecies has been swamped by the form *monsoni* which has moved west. In southern California most females now more closely resemble *monsoni*, but have measurements closer to *nelsoni*, a small percentage of females show the typical pale *nelsoni* plumage (Wehtje and Cooper 1998, P. Unitt pers. comm.).

Q. m. graysoni occurs in the coastal part of Sinaloa, Mexico. It is like *nelsoni* in being small and having pale females. The male has bluish iridescence on the back, breast and flanks rather than violet.

Q. m. obscurus is the coastal bird from Nayarit to Guerrero in Mexico. This form is smaller than *mexicanus*, but the female is darker even than that subspecies, being largely blackish-brown below with a slightly paler throat. Therefore, the female is much darker than the adjacent subspecies *graysoni* as well as being larger. Apparently this race and *graysoni* sing a significantly different song from the other forms.

Q. m. mexicanus is found in Mexico from E Jalisco and San Luis Potosí south to N Nicaragua. It is the largest form and has a relatively longer and deeper bill than the others. Compared to *prosopidicola*, females of this form are also significantly darker. Even though the underparts of *mexicanus* females are quite dark, the darker belly contrasts with the paler, warmer brown throat and breast. Note that the dark belly colour extends up to the breast on the even darker *obscurus*, thus the throat of that form contrasts with the rest of the underparts.

Q. m. monsoni is found from SE Arizona NC New Mexico and W Texas south to Zacatecas, Mexico. It is a large form, but not as large as *mexicanus* and with a less massive bill. The female is dark like *mexicanus*, with a dark brownish-grey breast unlike the buff-brown of *prosopidicola*. Males of this race have a noticeable purplish gloss to the back, more noticeable than on other forms. See *nelsoni* for details of interactions with that subspecies.

Q. m. loweryi is found in the Yucatán and nearby islands south to Turneffe Cay in Belize, and Isla del Carmen, Campeche, on the west coast of the peninsula. This subspecies is much smaller than adjacent *mexicanus*, enough to be noticeable in the field. Females are darker below and above than *prosopidicola* but paler than *mexicanus*, with a warmer tone to the underparts than the latter.

Q. m. peruvianus is the southernmost race, found from Costa Rica through Panama into N South America south to NW Peru on the Pacific coast and NW Venezuela on the Caribbean coast. Males of this form are like those of *mexicanus*, perhaps slightly more blue. Females are much paler than *mexicanus*, being buffy on the throat and underparts, and showing a noticeably paler supercilium.

HABITAT Great-tailed Grackle is a generalist in its habitat preferences, inhabiting largely open areas with some large trees. In general, the main habitat requirement for this species is open areas where it can feed such as pastures, lawns, fields and grassland. Trees or bushes are needed for nesting, but marshes will be used in some circumstances. Marsh-nesting appears to be most common in the western part of the distribution in the US. In the deserts of the SW US, it is usually present in urban situations including gardens, lawns and golf courses in particular. Preferred nesting localities are near water, either streams or ponds. In parts of the Neotropics, particularly coastal Colombia, mangroves are a frequently used habitat. This is not a species adverse to urban living, and it will commonly be found nesting in tall trees in parks and plazas in cities and towns. It is known to nest in heronries in C US. The spread of this species in the US has been aided by two factors, the irrigation of dry areas and the planting of trees in formerly treeless grassland and desert.

BEHAVIOUR Like other blackbirds, Great-tailed Grackle prefers roosting and foraging in large groups, particularly during the non-breeding season. In winter, foraging groups tend to be of a single sex. In Texas, 80% of the food consumed was animal matter, largely grasshoppers. The proportion of insects eaten varies throughout the year. In winter, fewer insects were eaten, while in spring more insects featured in the diet. Age and sex differences in diet have been noted, with young males eating roughly equal amounts of grain and insects, young females concentrating on grain, adult males being nearly granivorous and adult females more strictly insectivorous (Davis and Arnold 1972).

Early in the spring, males establish small territories and begin attempting to attract females. Breeding is highly polygynous. Females nest within the territory of one male, presumably the male she primarily mates with, and defend the nest site from other females. Several females will nest in a male's territory, and mate with him. The male has no role in nesting or care of the young. As is often typical of this type of mating system, 1st-year males do not breed but young females do, and the sex ratio is highly skewed towards females. Males are at attendance at the colony throughout most of the day and will keep their post for several weeks. Males will stay at a colony until there are no more mating possibilities, at which point they will vacate. Males with larger territories who spend a greater portion of their time on their territory appear to sire a greater number of offspring than other males (Kok 1972).

The main display is the 'ruff-out' which is accompanied by song. Its general form is similar to that of other grackles. The tail is fanned, the body feathers are ruffled and the head is arched upwards as it sings. The wings are drooped, but are quivered somewhat or held out towards the sides. Males also perform a courtship display towards females. It is similar to the 'ruff-out' but the feather ruffling and tail fanning are more exaggerated and the bill is pointed downwards, rather than up. The wings are rapidly quivered as the bird gives distinctive *cheat* notes. A female may respond by giving the solicitation display which involves drooping and quivering the wings as the tail is cocked and giving *che* calls. Mating will usually follow. In most forms, females may give low intensity 'ruff-out' displays toward other females but not as frequently as in Boat-tailed Grackle, and not accompanied by singing. However female *nelsoni* is known to sing and give a full 'ruff-out' display.

NESTING Nest placement varies depending on the height of the substrate. This species tends to nest as high as possible, thus, in a marsh a nest may be less than a meter above the ground while in a tall shade tree it will be much higher. The nest is a bulky, cup-shaped structure made of coarse sticks or marsh vegetation such as cattails (*Typha* sp.). It sometimes nests in emergent vegetation, such as cattails, tules (*Scirpus* sp.), in the manner of Boat-tailed Grackles. It will also nest on dead trees or bushes over water.

Clutch size in *prosopidicola* averages around 3.5, at one site in Texas a total of 517 nests showed an average clutch of 3.29 eggs (Tutor 1962). As is typical in birds, clutch size decreases closer to the tropics. In Guatemala, *mexicanus* lays a clutch averaging 2.7 eggs. The eggs are blue with dense blackish or brownish spotting and streaks which are more concentrated at the narrower end of the egg. Incubation begins after the first egg is laid, therefore hatching is asynchronous and there is a size hierarchy of the young. The incubation period lasts 13 days (Tutor 1962).

DISTRIBUTION AND STATUS Common throughout its range. Breeds north to E Oregon (Malheur), C California (San Luis Obispo along the coast, extreme N San Benito county immediately inland from the coast, Placer and Sacramento Cos. in the Central Valley and Mono Co. east of the Sierra), SE Oregon (Malheur NWR), SC Idaho (Burley), N Montana, S South Dakota (Duel Co.), SW Iowa, C Missouri (Columbia), W Arkansas and to SC Louisiana along the Gulf coast. Extends south throughout Mexico, except Baja California, along the entirety of Central America. In South America, the range is restricted to the Pacific and Caribbean coasts, north to extreme NW Venezuela (Zulia) and south to NW Peru. It is present in several near-shore islands in Central America such as Cozumel, Mexico, and the Bay Islands, Honduras. Winters throughout the breeding range, but many birds vacate the northernmost breeding sites.

MOVEMENTS Banding records show that for the most part, Texas birds are sedentary or short-distance migrants. However, further north individuals are more likely to migrate south during the winter, even though some remain within the breeding range. A Texas banding study showed that *prosopodicola* Great-tailed Grackles shifted south in the winter and north in the summer, the pattern of movement suggests that river valleys are important migratory corridors. All age classes and both sexes showed this migratory pattern; in warmer winters the southward shift is not as noticeable (Arnold and Folse 1977). The oldest banded bird that has been recovered was aged 12 years and six months, and was banded in Texas during the spring and retrapped in Rosamorado, Mexico, in winter (Klimkiewicz and Futcher 1987).

During recent times, Great-tailed Grackle has undergone an explosive range expansion that is still continuing. The grackle is expanding along the northern edge of its range, involving three different subspecies. In states where it breeds and is an established component of the avifauna it was first recorded in the following years: California, a female *nelsoni* was collected in 1964; Oregon, the first state record occurred on May 16, 1980, at Malheur NWR, Harney Co., it first nested there in the late 1990s; Idaho, first sighted in 1985 at Downey and possibly nested in 1994 near Burley; Nevada, one observed near Las Vegas in 1970, the first nesting record occurred near Las Vegas in 1974 (Dinsmore and Dinsmore 1993, Holmes *et al.* 1985); Utah, the first record was of a male collected in Provo, March 21, 1957 (Talley 1957), they nested at several sites in the SW in 1986 (Dinsmore and Dinsmore 1993); in Arizona it spread north from Sonora beginning in 1935; Montana, the nesting pair during the summer of 1996 at Bowdoin NWR is the first record for the state (Wright 1996); in Wyoming this species is observed mainly in the southeast, with the first report from 1989 (Dinsmore and Dinsmore 1993) and has likely bred; Colorado, first detected in 1970 at Gunnison, it was nesting in 1973 at Monte Vista (Stepney 1975); New Mexico, the first record occurred in 1913, nesting in 1920s in the SC part of the state at Las Cruces, by the late 1930s it was nesting just south of Albuquerque at Isleta (Compton 1947); South Dakota, possibly nesting at Salt Lake, Duel Co. Apr.–May, 1996, the first state record occurred in 1988 near Yanknon (Dinsmore and Dinsmore 1993); Nebraska, first record in Phelps Co. in 1976, and first nest at Boys Town in 1977 (Faanes and Norling 1981); Kansas, first record in Harvey Co. in 1963, nesting at Great Bend and Wichita in 1969 (Glassel 1993); Oklahoma, first detected in 1953 at Norman (Selander and Giller 1961), nesting by 1958 at Norman (Glassel 1993) all are *prosopidicola*; Texas, historically restricted to the Rio Grande valley it spread north reaching Austin in 1925 and Fort Worth in 1944 (Pruitt 1975); Iowa, first record in Mills Co. in 1983 and first nesting in Fremont Co. in 1983; Missouri, first record at Springfield in 1972, 10 nests at Big Lake State Park in 1979 were the first breeding records (Robbins and Easterla 1986); Arkansas, first record in 1969 in Lafayette Co., and nesting at Ashdown by 1976 (Glassel 1993); Louisiana, present and breeding by 1959 having probably arrived in the 1940s. The pattern in much of the west has been that breeding occurs within 10 years of the first record in a given state, therefore it is not inconceivable that this species may breed in Canada within 10 years at this rate of spread. Dinsmore and Dinsmore (1993) noted that the spread in California was proceeding differently to the rest of the continent. Throughout, the first stray record to an area occurs in the spring or summer, while in California spring stray records are the most common, a significantly greater proportion are first found in the fall and winter. Males make up the majority (65%) of stray records (Dinsmore and Dinsmore 1993).

There are now at least seven records from Canada: two in British Columbia, including the northernmost record from Cape St James, Queen Charlotte Islands in May 1979 and the other from Vernon during the winter of 1993–94; three in Ontario, the first in October of 1987 in Atikokan, a second from November to January 1988–1989 at Port Rowan and the third in the spring of 1995 at Kingsville; one in Nova Scotia photographed on November 17, 1983, in Annapolis Co., this being the easternmost record. In Nova Scotia, four other reports of large grackles, including two from Sable Island, were not conclusively identified to species and may have pertained to Boat-tailed Grackle, which has not been definitively confirmed from Canada. Similarly, there have been observations of large grackles in Quebec which could not be positively identified to species. In Bromhead, Saskatchewan, a male with a female was observed in May 14, 1979, and eventually a total of 5 or 6 were observed in the area where they remained for a week before departing (Bjorklund 1990). The following states have records, but not of breeding birds: Washington, only record is of a male at Union Gap, Yakima Co., 25–26 May 1987; Minnesota, two records, the first in 1982 in Dakota Co., the second at Cannon Lake, Rice Co., for a week in April 1993 (Smith 1993, Glassel 1993); Illinois, one female, Oct. 4–7, 1974, at Jacksonville (Bohlen 1989); Michigan, during the spring of 1997 a female was found near Whitefish Point, Chippewa Co.; Indiana, first record in 1991 in Sullivan Co. (Glassel 1993) and the second in February of 1995 in the same county; Ohio had its first record in Ottowa Co. in 1985 (Glassel 1993). Reports of large grackles have not adequately identified which species they belonged to, but more than likely these pertained to Great-tailed Grackles. This species is overdue in Wisconsin, Alberta and Manitoba.

In Mexico, stray records exist from parts of Baja California Norte and Isla Tres Marías (Nayarit). Great-tailed Grackle has only recently spread into the Central Valley of Mexico. In Hawaii, Great-tailed Grackles have been observed since 1980 in O'ahu, there are now at least four records. The origin of these birds is not known, but are widely thought to be escapees from a local zoo.

MOULT The first pre-basic moult tends to be incomplete, all feathers are replaced except for some underwing-coverts and infrequently some of the tertials. The first pre-basic moult is complete in 8% of males and 12% of females. The tail feathers are moulted simultaneously, so many birds appear tail-less at the end of the summer, which lasts about two weeks. Juveniles begin moult slightly before adults. A limited first pre-alternate moult occurs only in young males and involves some of the head feathers. This moult replaces some of the duller head feathers with brighter ones similar to those of adult males, and lasts from late February–early April. Definitive pre-basic moult is complete and occurs between July–November in US populations. The Mexican race, *obscurus*, moults between November–December. Moult in females of all age groups lasts 15 days less than in males which average 110 days to finish moult. Pre-alternate moults are lacking other than in 1st-year males. The tail moults from the outside inwards, towards the center.

MEASUREMENTS Males are 21.7% larger than females on average in *prosopidicola*. Larger males tend to have proportionately longer tails.

Q. m. prosopidicola. Males: (n=56) Wing 184.9 (172.0–200.0); tail 204.3 (190.0–224.0); culmen 38.8 (36.5–41.5); tarsus 49.2 (44.0–52.0); (n=7) weight 236.0 (216.4–253.7). Females: (n=24) Wing 144.9 (140–150); tail 145.5 (136.0–152.0); culmen 32.0 (30.0–34.0); tarsus 39.9 (36.5–42.1); (n=17) weight 119.3 (109.0–129.5).

Q. m. mexicanus. Males: (n=?) Wing 193.9 (185–210); tail 215.2 (203–229); culmen 41.2 (38.4–46.0); tarsus 50.9 (49.0–52.5); (n=14) weight 264.6 (238.8–317.0). Females: (n=?) Wing 154.7 (146–164); tail 156.3 (143–170); culmen 33.5 (31.3–36.0); tarsus 42.9 (41.0–45.0); (n=18) weight 141.6 (124.5–163.1).

Q. m. monsoni. Males: (n=22) Wing 187.2 (175–196); tail 216.6 (195–235); (n=20) culmen from nostril 31.3 (29.1–34.8); (n=11) weight 203.2 (187.0–222.0). Females: (n=18) Wing 147.6 (143–156); tail 157.0 (143–165.0); (n=17) culmen from nostril 24.6 (22.3–26.7); (n=7) weight 115.4 (106.0–123.7).

Q. m. nelsoni. Males: (n=9) Wing 166.4 (159.5–177.8); tail 168 (154.9–185.0); (n=7) culmen 38.6 (36–40.4); tarsus 44.0 (41.9–45.5). Females: (n=18) Wing 132.4 (124.2–140.0); tail 124.1 (116–134.0); culmen (n=6) 33.2 (32–35.3); tarsus 38.7 (37–40.1).

Q. m. obscurus. Males: (n=10) wing 182.5 (175–192); tail 198.0 (173–224); culmen 40.4 (38–41.9); tarsus 49.1 (47.8–53). Females: (n=10) Wing 147.4 (139.7–159); tail 144.6 (124.5–163.0); culmen 34.0 (32.0–37); tarsus 40.8 (37–45.0).

Q. m. loweryi. Males: (n=10) Wing 185.5 (180–194); (n=9) tail 202.1 (187–216); (n=8) culmen from nostril 31.7 (30.1–34.8); (n=1) culmen 44; tarsus 49. Females: (n=16) Wing 145.8 (142–153); tail (n=13) 144.5 (136–152); (n=22) culmen from nostril 25.8 (23.7–27.3).

Q. m. peruvianus. Males: (n=10) wing 175.1 (166.5–186.4); tail 184.2 (165.4–200.5); culmen 43.2 (40.3–45.7); tarsus 45.8 (43.3–48.8). Females: (n=10) wing 144.1 (139.9–150.2); tail 144.8 (132.5–155.9); culmen 34.6 (31.7–36.8); tarsus 38.4 (36.0–41.6).

***Q. m. graysoni*.** Males: (n=5) wing 165.9 (163.8–168.1); tail 163.6 (157.5–167.6); culmen 38.6 (36.1–40.6); tarsus 45.7 (44.5–47.5). Females: (n=10) wing 133.9 (125–139.7); tail 124.6 (116–134); culmen 32.4 (30.5–33.3); tarsus 38.2 (36.8–39).

NOTES Boat-tailed and Great-tailed Grackles are closely related and were previously lumped as one species. However, they are good biological species and behave as such in zones of sympatry (Selander and Giller 1961). Genetic differentiation is not great between the two species, protein electrophoresis work has detected negligible differences, however restriction site variation of mitochondrial DNA showed clearer differences between the two species but in the low end of the range, compared to genetic differences shown by other closely related species (Avise and Zink 1988). Hybridization is not known between this species and the closely related Boat-tailed Grackle. The isolating mechanisms appear to be behavioural, with females having the ability to discriminate against heterospecifics. However, it had been documented to have hybridized with Red-winged Blackbird (61), accounting for an enigmatic specimen originally considered a completely different species termed the 'nondescript' blackbird! A recent re-analysis of the mitochondrial DNA of this specimen confirms that the female parent is a grackle, but as yet of an undecided species (T. Glenn pers. comm.). An old record of the race *mexicanus* from S Texas probably refers to a large individual of *prosopidicola*.

REFERENCES Arnold and Folse 1977 [movements], Avise and Zink 1988 [notes, genetic divergence], BBL data [movements], Bent 1958 [geographic variation], Bjorklund 1990 [movements], Bohlen 1976 [movements, Illinois record], Dickerman and Phillips 1966 [geographic variation, measurements], Howell and Webb 1995 [geographic variation, distribution, movements], Klimkiewicz and Futcher 1987 [age], Kok 1971 [voice], Lowery 1938 [geographic variation, measurements], McCaskie *et al.* 1966 [movements], Phillips 1950 [geographical variation, measurements], Schuering and Ivey 1995 [distribution, movements], Selander 1958 [moult, description], Selander and Dickerman 1963 [female song, hybrids], Selander and Giller 1961 [identification, voice, behaviour, nesting, movements, distribution, geographical variation, measurements], Tutor 1962 [habitat, nesting], Wetmore *et al.* 1984 [measurements, geographic variation]. Wehtje, W. and D.S. Cooper. 1998. Great-tailed Grackle invasion of California, 1964 to present: patterns of colonization. Poster given at the North American Ornithological Conference, St Louis, Missouri.

88 BOAT-TAILED GRACKLE *Quiscalus major* Vieillot 1819 — Plate 34

The common large grackle of coastal eastern North America; it's an obvious bird which is largely restricted to salt-marshes.

IDENTIFICATION Boat-tailed Grackle is exceedingly difficult to identify from Great-tailed Grackle in many cases. This discussion will be limited to differentiating the present species from the *prosopidicola* race of Great-tailed Grackle, which is the one that is sympatric with it. The most often cited field mark is a difficult one to use without comparative experience. Boat-tailed Grackles have a rounder head than Great-tailed Grackle and an obvious forehead angle, in Great-tailed Grackle the culmen appears to be almost continuous with the forehead. This difference is less obvious on females than males. In addition, Boat-tailed Grackle is smaller and has a relatively shorter, narrower, tail and shorter wings. For its size, Boat-tailed Grackle has long legs and a long bill. In males, the distribution of the violet iridescence differs, as it does not extend to the lower back, abdomen and flanks on Boat-tailed Grackle, while it does on Great-tailed Grackle; on Boat-tailed, these parts are greenish-blue. However, there is some overlap in this character. During the 'ruff-out (song spread)' display of this grackle, it rapidly flutters the wings above the back, this 'wing flip' display is diagnostic for the species and can be used to identify males at great distances. While alert and at attendance at the colony, males 'fluff out' their neck feathers more than Great-tailed and appear correspondingly thicker-necked. See Voice section of both species for other differences. Iris colour varies in this species, but is a useful field mark nonetheless. Gulf coast and Florida Boat-tailed Grackles have brown irides, unlike the vivid yellow of Great-tailed Grackles. Boat-tailed Grackles from the Atlantic coast, and parts of the Gulf coast, have yellow eyes, but even these do not ever achieve the same level of brightness seen on Great-tailed, this is due to a few brownish flecks being present or a greyish ring around the iris. Females also differ in their colouration. On Boat-tailed, the iridescence of the upperparts is less extensive and not as obvious. Ventrally, female Boat-tailed Grackle is paler in colour. The eye colour differences are as in the male. The *kleteet* call of the female is not heard from Great-tailed Grackles, nor does female song exist in that species. Juvenile and immature males may be inseparable on plumage details, one has to rely on structure. Some juvenile Boat-tailed Grackles will show wider pale feather fringes. However, juvenile and 1st-year female Boat-tailed Grackles are warmer buff or cinnamon below than the corresponding ages of Great-tailed Grackles.

VOICE As in Great-tailed Grackle the vocal repertoire of this species is large.

Song: The primary song of the male is entirely different from that of Great-tailed Grackle. It is composed of three parts: an initial series of *tireeet* notes, followed by an ascending rolling trill, that occurs simultaneously with the 'wing-flip' part of the 'ruff-out' display, and ends with a series of *tireet* notes like those opening the song. Harper (1920) described the song as a *kip, kip, kip, kip-kip-kip-kip-kip-kip-kip, chrr, chrr, chrr, chrr, chrr, chrr, pt-pt-pt-pt-pt-pt-pt-pt.* He and other authors, considered that part of the song was created mechanically by the vibrations of the wings, a notion dispelled when he observed some males performing full songs without the full 'ruff-out' display where they vibrated the wings. Unlike Great-tailed Grackle, the female of this species regularly gives a 'ruff-out' display accompanied by the song. It is similar to that of the male.

Calls: During the courtship (solicitation) display a male utters a series of high frequency *cheat* notes. These are similar to those of Great-tailed Grackle, but are flatter in tone in the present species. A commonly given warning note is a two-syllabled *kle-teet,* not heard from Great-taileds; a soft *clak* may also be heard from both sexes. Females may also give a chatter call. High frequency *tee* notes are given during the female solicitation display.

DESCRIPTION A large, long-tailed grackle with a thick, stout bill. Proportions and eye colour vary geographically. Wing Formula: P9 > P8 > P7 > P6; P9 > P8; P8–P5 emarginate. The nominate form is described.

Adult Male (Definitive Basic): The bill is black and the eyes are dark brown (yellow along the Atlantic Coast, see Geographic Variation). The plumage is entirely black and iridescent. The iridescence is violet on the head, becoming greenish-blue on the breast and back, down to the tail. The crissum is blackish and largely lacks iridescence. The wings and tail show a dull greenish gloss. The legs and feet are black.

Immature Male (First Basic): Like adult, but with a more restricted and greener gloss. The tail is shorter on young males. The thighs are brown instead of black. The tail and primary moults of 1st-fall males occur later than those of adults. It is not unusual to see first basic males still carrying juvenal rectrices and outer primaries into September. First basic males may retain a variable number of juvenal tertials.

Adult Female (Definitive Basic): Dark brown on the head, upperparts, wings and tail. The wings and tail may show a slight greenish iridescence. The underparts are paler, a buffy supercilium is obvious and a darker malar streak is sometimes present. The underparts are pale buff to warm cinnamon-buff, extending down to the legs. The crissum is dark brown as are the thighs.

Immature Female (First Basic): Like adult female, but paler above and a richer cinnamon below.

Juvenile (Juvenal): Like adult female, but paler and a warmer cinnamon in tone both above and below, and lacking iridescence on the upperparts. Rather uniform below, but may show some dull streaking. The wing-coverts and tertials may show thin pale rufous fringes. Juveniles always have dark eyes, irrespective of their subspecies. Older nestlings and fledglings may be sexed mainly by noting their size, males are substantially larger than females. Males also show darker-coloured bills and legs. Males begin moulting into first basic plumage early in life and may show some blackish feathers around the eyes, breast sides and wing-coverts. Pale-eyed males will change eye colour during the first pre-basic moult, the eyes become pale from the periphery inwards. Intermediate stages show a pale greyish-yellow ring encircling a brown ring around the dark pupil.

GEOGRAPHIC VARIATION Four subspecies are now recognized (Stevenson 1978), however many recent works only recognize ***major*** and ***torreyi.*** The four races differ in eye colour and proportions, as well as distribution.

The nominate subspecies is found along the coast from SE Texas east along the Louisiana coast to easternmost Louisiana and adjoining W Mississippi. Both sexes of this race have dark eyes. It has the longest tail of all four forms, and the ratio of the tail to wings is high, meaning that the wings are proportionately short compared to the tail length. There is a cline in this proportion, with Texas birds having the proportionately longest tails and Louisiana birds less so. Note that this is the only race that is regularly sympatric with Great-tailed Grackle.

The race ***Q. m. alabamensis*** breeds from the coast of Alabama west to Jackson Co., Mississippi, including Horn Island in Mississippi. Both sexes of *alabamensis* are yellow-eyed, sometimes showing a dark ring around the pupil. The bill is deeper than that of *major* and *westoni.* It has a proportionately long tail in relation to the wing length, but not as much as in *major.*

The subspecies ***Q. m. westoni*** reaches its western range limit at St Vincent Island, Franklin Co., Florida (in the panhandle); inhabits the entire Florida peninsula including interior areas away from the coast, but not on the Florida Keys. The northern breeding limit along Florida's Atlantic coast is St Augustine, formerly as far north as the mouth of the St Johns river. Both sexes of this race have dark eyes, like *major.* Pale-eyed birds observed in the breeding range of this form are likely migratory *torreyi.* This race has a proportionately long and thin (shallow-based) bill.

The race ***Q. m. torreyi*** breeds along the Atlantic coast from Long Island, New York south to Duval Co., northernmost Florida. It is the well-known yellow-eyed form, most of the literature fails to note that Alabama and Mississippi birds also have yellow eyes. This race has proportionately the longest wings and shortest tail of any form. It is also the only subspecies that is regularly migratory.

HABITAT Boat-tailed Grackle nests almost exclusively in marshes, either fresh or salt water, with *torreyi* preferring the latter. This tie to water is strong, and this species is seldom found away from it. Also, their range is largely coastal, birds only venturing inland along major watercourses, other than in Florida where they have established themselves in interior marshes. They prefer to nest in the taller marsh vegetation such as cattail (*Typha latifolia*), sawgrass (*Cladium effusum*) and Spartina grass (*Spartina alterniflora*). However, in part of its range, *torreyi* will nest in trees near water in a manner more similar to Great-tailed Grackle. In winter, they may venture away from their aquatic habitats and can be found foraging in agricultural fields.

BEHAVIOUR A highly social bird at all times, roosting and foraging in groups. During the non-breeding season, males and females usually form sex-specific flocks and forage in different areas. This ecological separation occurs early in life, as juveniles appear to segregate into single sex flocks after reaching independence (McIllhenny 1937). Roosts may be quite large, and include members of both sexes and often other icterids.

In many ways the behaviour is quite similar to that of Great-tailed Grackle. As in that species, breeding is highly polygynous and males defend a territory to which they attract several females. The territories are larger in this species than in Great-tailed Grackle. The sex ratio is highly skewed toward the females, which will breed in their first year. Year-old males do not hold territories, and do not breed. Females are twice as numerous as males, a ratio which has been noted even with nestlings at the nest (McIlhenny 1937). Males do not offer any help to females in the nesting process. Attendance at the colony by males is not as continuous in this species as in Great-tailed Grackle. In large colonies, several males may be at attendance but are largely ignored by the females. Mating does not occur at the colony, but in a nearby open area. Small colonies tend to have one male at attendance, who will drive off other males. Presumably he is able to defend these smaller groups of females from other males and perhaps achieves a large portion of the matings in that colony. Smaller colonies, with one male in attendance are more common later on in the season. Early colonies tend to be large (McIlhenny 1937).

The testicular cycle of adult males is a full two to three weeks ahead of the cycle of 1st-year males. Young males do not achieve the same sized testes as adults, and therefore sperm production is likely lower. After egg-laying, the male's testes begin to shrink, although there is a slight increase in testes size and hormone levels during the fall. A few adult males hold territories for a few days during October. Young males and females remain in foraging flocks, oblivious to the behaviour of the adult males.

The main display is the 'ruff-out (song spread)' which is accompanied by song. Its general form is similar to that of other grackles. The tail is fanned, the body feathers are ruffled and the head is arched upwards as the bird sings. The wings are drooped, but then jerked violently to a vertical position above the back. The wings remain held high, such that it appears that the tips of the primaries are clapped against each other. This is the 'wing-flip' part of the display which is accompanied by the rolling trill part of the song. This display is particular to this species and can be used to identify males, even from a distance. Displaying by males, presumably aimed at other males, begins in February, well before females become receptive. However, early in the season males spend little of their time displaying, the display rate certainly increases later in the season once females are interested in nesting. Males in display groups spend a great deal of time conducting the aggressive 'bill-tilting' display observed in most blackbirds. The body plumage is held stiffly to the body, and the bill is pointed upwards. Male Boat-tailed Grackles may also contract their pupils, highlighting the yellow irides (depending on the population).

Males also perform a courtship display towards females. It is similar to the 'ruff-out' but the feather ruffling and tail fanning are more exaggerated and the bill is pointed downwards, rather than up. The wings are rapidly quivered as the bird gives distinctive *cheat* notes. A female may respond by giving the solicitation display which involves drooping and quivering the wings as the tail is cocked while giving the *che* calls. Mating usually follows. Females may give low intensity 'ruff-out' displays toward other females and sometimes males. These displays are similar to the male display and are often accompanied by a weaker version of the song.

This grackle is often observed in 'sexual chases', where one or more males slowly chase a female. In this ritualized display, males purposely lag behind females, even though they are able to catch up if need be. A female may land and let a male catch her. In these situations, mating tends to occur immediately after the chase.

The diet of Boat-tailed Grackle is varied, ranging from seeds, rice and insects to small birds, carrion and fish! Grackles have a predatory tendency that is very apparent in this species. Boat-tailed Grackles will not hesitate to attack and kill small birds, even injured sandpipers (*Scolupidacea*). In addition, they have been observed feeding on carrion, and foraging on trapped muskrat (*Onadatra zibethica*) flesh. More commonly, they will feast on whichever food is particularly abundant at a given time, eg. rice, dragonfly (*Odonata*) hatches, tadpoles (*Anura*), crayfish (*Crustacea*), spiders, minnows (*Osteichthyes*) and the like. Some males will even take young of their own species (McIlhenny 1937).

NESTING The breeding season begins March–April, lasting into July. This grackle is highly colonial, nesting in groups ranging from a few nests to nearly 200. Larger colonies exist earlier in the season (McIlhenny 1937). Invariably, Boat-tailed Grackle nests in marsh vegetation, often within a meter of the water's surface. In Louisiana, Boat-tailed Grackle nests largely in sawgrass (*Claudium effusum*), particularly early in the season. Later in the season they will secure their nests to the stalks of tules (*Scirpus californicus*) or cattails (*Typha latifolia*). The preference for one species over the other at different times in the nesting season appears to be related to the rate of growth of the plant. Quick, or uneven growth of substrate vegetation often results in the nest being tipped over or damaged. The rich growth of Cut Grass (*Zizaniopsis miliacea*) adjacent to nesting sites of American Alligator (*Alligator mississippiensis*) is a preferred nesting situation for the grackles in Louisiana (McIlhenny 1937).

Clutch size varies geographically with a larger number of eggs being more typical in more temperate areas, as is typical in many other bird species. In the northern part of the range, the clutch size averages 4 eggs, but further south, along the Gulf Coast, it averages 3 eggs. The eggs are bluish, paler than those of Great-tailed Grackle. They are marked with black and brown spots and lines, which are usually clustered around the broader end of the egg. Eggs laid early in the season (April) in Louisiana tend to be more heavily marked than later (June) eggs (McIlhenny 1937). In some breeding marshes, Least Bittern (*Ixobrychus exilis*) shows a significant tendency to nest within Boat-tailed Grackle colonies.

Incubation begins after the second egg has been laid, therefore the third and fourth eggs hatch later than the first two (Asynchronous hatching). The incubation period lasts 14–15 days. Females undertake all of the work involved in nesting, from building the nest to incubation and feeding the young. Males do not help at all at the nest. In fact, cannibalism of young by males has been observed. In addition, Purple Gallinule (*Porphyrula martinica*) appears to be a common nest predator of Boat-tailed Grackle young (McIlhenny 1937).

Nesting has been documented during the fall, but only in urban parks in the Orlando area, Florida. Small numbers partake in this autumnal breeding, and only adult females nest. Their success rates are low, and this behaviour only occurs in some years. This species may begin nest-building as early as late October, but more normally in early November, and fledglings are observed late December–early January. Fall breeding occurs after the adults have finished their pre-basic moult.

DISTRIBUTION AND STATUS Common, coastal and found only near sea-level.

Breeding: Nests as far north as Connecticut (Stratford), and E Long Island (Shinnecock)in New York state. Colonies are patchy from coastal New York, through coastal New Jersey, Delaware and Maryland. It is present in both Delaware Bay and Chesapeake Bay, and becomes more common south of the latter. A common breeder along the coast of Virginia, North Carolina, South Carolina, Georgia and Florida. There is a gap in the distribution between St. Augustine and the mouth of the St Johns river, Florida. It does not breed on the Florida Keys. Only in peninsular Florida does it breed inland, extending north to 50 km south of the Georgia border (Post *et al.* 1996). Along the Gulf coast it is found from SW Florida through the coast of the Florida panhandle, to St Vincent Island. There is a gap in distribution between St Vincent Island and Alabama. Occurs from coastal E Alabama to E Mississippi (Gautier). There is another gap in distribution between Gautier and St Louis Bay, Mississippi. Occurs from St Louis Bay, west and south to NE Texas (to San Antonio Bay).

Non-breeding: Boat-tailed Grackle's winter range is basically the same as that of the breeding range. However, in winter, birds wander south and vacate some of the northernmost sites along the Atlantic coast such that the distribution there becomes more patchy. They may be found in winter in places where they do not breed in summer such as in the distribution gaps in Mississippi and Florida. In addition, birds wander south along the Texas coast and there is a record from near the Mexican border. Recent vagrancy to the north, for example to Rhode Island and Connecticut has included wintering birds.

MOVEMENTS The nominate subspecies is largely sedentary, but a few appear to wander southwest of their breeding areas during the winter in Texas. There is a record from Brownsville. Boat-tailed Grackle has not been found in Mexico, but it may occur along the northeast coast. The form *torreyi* retreats south to some extent and can be classed as a short-distance migrant, perhaps accounting for the longer wings of this race. This is confirmed by the banded nestling from Charleston Co., South Carolina, recovered in Daytona Beach, Florida. This 413 km journey is the most distant recovery of a banded bird (Post 1994a). In addition, pale-eyed (*torreyi*) individuals have been found south of their range in coastal Florida. However, banding records show that many individuals of the Atlantic population are sedentary. In Louisiana, Boat-tailed Grackles move up waterways in winter and the species has reached Baton Rouge, along the Mississippi river on at least three occasions (Lowery 1974). The few records of this species on the Florida Keys should be regarded as stray birds. Recoveries of banded birds have included birds as old as 11 years and 11 months (Klimkewicz 1997); however, most do not achieve such a long life with six years or less being relatively common.

Boat-tailed Grackle historically bred only to North Carolina, but moved north reaching New Jersey in 1892, first breeding in 1952 and breeding in New York in the early 1980s (Leck 1984, Post *et al.* 1996). The distribution continues to expand slowly to the north along the Atlantic coast. Northern records appear to occur after the passage of hurricanes and some of these birds may stay for extended periods of time. There are now records for Massachusetts, Connecticut, Rhode Island and Maine. In Nova Scotia, one sighting of a large grackle from May 3, 1994, was not identified to species and may have pertained to Boat-tailed Grackle, which has not been definitively confirmed from Canada. There are also two records of large grackles from Nova Scotia before the split of Boat-tailed and Great-tailed Grackles was made; these birds occurred on Sable Island, on May 7, 1968 and Glace Bay on August 5, 1969. Due to the uncertainty of the species involved and the fact that Great-tailed Grackle has been confirmed from Nova Scotia, Boat-tailed Grackle was recently dropped from the Canadian list. There is an unusual inland record from upstate New York photographed in Greece, on Nov. 29, 1980. This is an accepted record. Other inland records, outside of peninsular Florida are much closer to the ocean, for example in Assunpink, New Jersey (Leck 1984).

MOULT The first pre-basic moult tends to be incomplete, all feathers are replaced except some underwing-coverts and infrequently some of the tertials. Blackish first basic feathers are obtained as fledglings, beginning with the face and breast sides. The body and coverts are moulted in, often with juvenile feathers remaining in the center of the breast. The secondaries and inner primaries come in earlier than the tail, outer primaries and tertials. Moult into first basic plumage is accompanied by eye colour change in populations with pale-eyed adults. The tail feathers are moulted simultaneously, so many birds appear tailless at the end of the summer, which lasts about two weeks in early September. A limited first pre-alternate moult occurs only in young males and involves some of the head feathers. This moult replaces some of the duller head feathers with brighter ones similar to those of adult males. Definitive pre-basic moult is complete and its timing, in adult males, is approximately a week earlier than in hatch year males. Tail moult is simultaneous in adults, occurring in late August. Pre-alternate moults are lacking other than in 1st-year males. The small number of fall breeding Boat-tailed Grackles in Florida nest after completing the Prebasic moult. The juveniles which are flying by late December have their first pre-basic moult in December or January, much later than young born during the spring.

MEASUREMENTS Males are larger than females by an average of 19.5%. Larger males do not necessarily have longer tails, as in Great-tailed Grackle.

Q. m. major. Males: (10+) Wing 172.3 (167–184); culmen 39.3 (37.0–41.6); tail 172.3 (161–193); tarsus 49.2 (48.0–52.0). Females: (10+) Wing 134.3 (129–138); culmen 31.9 (29.0–34.0); tail 126.1 (120–135); tarsus 40.6 (39.0–43.0).

Q. m. alabamensis. Males: (n=17) wing 177.1 (169–191); tail 167.0 (153–179); bill from nostril 30.5 (28.8–32.3); tarsus 50.0 (48–52). Females: (n=16) wing 137.3 (130–143); tail 121.9 (109–130); bill from nostril 24.2 (23.2–25.3); tarsus 40.9 (40–43)

Q. m. westoni. Males: (n=47) wing 180.4 (177–188); tail 169.6 (151–181); bill from nostril 32.5 (29.4–36.5). Females: (n=40) wing 138.3 (132–147); tail 119.6 (109–135); bill from nostril 25.2 (22.8–27.7).

Q. m. torrey. Males: (n=47) wing 182.4 (147–194), tail 168.6 (147–185), culmen 30.3 (26.2–35.1), (n=25) tarsus 49.2 (44.1–57.8). Females: (n=39) wing 141.7 (131–151); tail 122.0 (109–143), culmen 23.7 (20.4–26.2), (n=20) tarsus 42.1 (36.3–46.2).

NOTES See Notes on Great-tailed Grackle for a summary of genetic differentiation between these two species. Hybridization is not known between this species and the closely related Great-tailed Grackle. The isolating mechanisms appear to be behavioural, with females having the ability to discriminate against heterospecifics. Males will display to females of the wrong species, which achieves no response.

REFERENCES Harper 1920 [voice], 1934 [geographic variation, measurements], Klimkiewicz 1997 [age], McIllhenny 1937 [habitat, behaviour, nesting, moult], Post and Seals 1993 [nesting], Selander and Giller 1961 [identification, voice, behaviour, nesting, movements, distribution, geographical variation], Selander and Hauser 1965 [behaviour], Selander and Nicholson 1962 [nesting, fall, moult], Stevenson 1978 [distribution, geographic variation, measurements].

89 SLENDER-BILLED GRACKLE *Quiscalus palustris* (Swainson) 1827 — Plate 35

This is the only icterid species that can be confirmed to be extinct. It was an endemic to marshes of the Central Valley of Mexico.

IDENTIFICATION Great-tailed Grackle (87) would have been the only grackle that may have been found sympatrically with Slender-billed Grackle. Compared to Great-tailed Grackle, Slender-billed was noticeably smaller and with a much finer and less massive bill. The bill shape was diagnostic, having a largely straight culmen slightly downcurved at the tip, while the gonys was straight and angled up to the bill tip. This made the bill appear tapered and pointed, somewhat like that of an overgrown *Euphagus* blackbird. The bill base was shallow, not at all like the robust bill of other grackles. Male Slender-billed and Great-tailed Grackles are both black, but differed in their iridescence. Slender-billed Grackles showed a dark violet gloss, rather than the bright violet-blue of Great-tailed. In addition, Great-tailed Grackle shows an obvious change from violet-blue on the head, upper back and breast to a purer blue on the rump and belly. Slender-billed Grackles did show a tendency for the lower back and rump iridescence to be bluer than the head and upper back but the difference in colour was slight. On the underparts, this species was unicolored, showing a dark violet gloss that did not become bluer on the vent and belly. Females differed in additional features to Great-tailed Grackle. The general colour of female Slender-billed Grackle was much paler than Great-tailed, particularly of the form *mexicanus*, and had a warmer cinnamon colour. Even compared to the paler *prosopidicola* Great-tailed, the female Slender-billed was more richly coloured with a noticeably chestnut back as well as a paler whitish throat. In fact, the colour of female Slender-billed Grackle is more similar to that of female Boat-tailed Grackle (88). The dark malar streak was lacking in Slender-billed Grackle. Slender-billed Grackles show extensive contrast between the dark vent and pale belly, much more so than in Great-tailed Grackles. In addition, female Slender-billed Grackles appeared to have an orange base to the lower mandible, and the legs appear to have been orange or reddish, rather than black. Juveniles differed from Great-tailed Grackles in having unstreaked underparts.

VOICE Unknown.

DESCRIPTION This was a small, slender grackle, with a proportionately thin bill. Its bill was noticeably longer than the length of the head. The slender bill gave this grackle a noticeable angle at the forehead, producing a much more gentle expression than in Great-tailed Grackle. The culmen was nearly straight, just showing a slight curve at the very tip, while the gonys was nearly straight. The tail of the male was noticeably graduated and keeled. The wings were quite rounded. Wing formula: P9 <= P8 >= P7 > P6 > P5; P9 ≈ P6; P8–P5 emarginate. In addition, this grackle had rather long and thin legs, which possibly gave it an elegant stance.

Adult Males (Definitive Basic): Like other grackles, this species was completely black and iridescent. The gloss was dark violet on the head and neck, continuing to the back and breast. The belly's iridescence was even darker, continuing to the vent. On the lower back and rump, the iridescence is violet-blue. The wings and tail had a violet gloss, lacking blues or greens. The eyes were pale yellow. The legs and bill were completely black.

Adult Females (Definitive Basic): Female colouration was most similar to that of Boat-tailed Grackle. The face and underparts were rich buffy-cinnamon, lightest on the throat (buffy-white) and lower breast, and darkest on the upper breast. The face showed conspicuously darker lores, a slightly paler supercilium and at most, a very thin and indistinct malar streak. In most cases, the malar streak was lacking. The supercilium was pale buff, and the cheeks were washed cinnamon. The back, nape and top of the head were darker than the underparts, but still showing the warm tones as a chestnut-brown. The lower back, rump and uppertail-coverts were dark brown. The belly and hind flanks were blackish-brown contrasting with the paler lower breast, similarly the thighs were dark brown. The wings and tail were dark brown, contrasting with the rest of the body and showed a slight greenish iridescence, particularly on the coverts. On most specimens, the gloss is lacking. The bill was blackish, but the bill showed a paler, reddish base to the lower mandible. The legs appear to have been orange or flesh, but this has been all but lost in the dry specimens.

Immature Male (First Basic/First Alternate?): Similar to adult male, but smaller, particularly with respect to tail length. The black body showed a marginally smaller amount of iridescence and the thighs and primary tips were brownish.

Juvenile (Juvenal): Similar to adult female, but brighter cinnamon on the upper- and underparts. The crown in particular is bright cinnamon on juveniles but warm brown in adult females. Adult females sometimes appeared to show pale mottling on the breast, while juveniles showed solid cinnamon. Juvenile males were similar to juvenile females, but noticeably larger.

GEOGRAPHIC VARIATION Monotypic.

HABITAT Freshwater marshes of the Lerma region in Mexico. This species was apparently restricted to marshes, unlike Great-tailed Grackle. It has been suggested that Slender-billed Grackle inhabited the Tule (*Scirpus* sp.) and Cattail (*Typha* sp.) marshes of this region rather than the sedge marsh which was inhabited by Bicolored Blackbird (61X).

BEHAVIOUR Unknown.

NESTING Unknown, but likely nested in marsh vegetation.

DISTRIBUTION AND STATUS Extinct, formerly found in the headwaters of the Río Lerma in central Mexico. Last recorded in 1910.

MOVEMENTS Unknown, but assumed to be sedentary.

MOULT Largely unknown, but some patterns can be inferred from specimens. The only complete moult was the definitive pre-basic of adults. Its timing is not known, but likely occurred after the breeding season. The juvenal plumage was replaced through the first pre-basic moult which was incomplete. The first pre-basic included the body, wings and tail but some underwing-coverts and thigh feathers were retained from juvenal plumage. The first pre-basic moult began during July. No evidence for pre-alternate moults, but if this species was similar to Great-tailed Grackle, a limited first pre-alternate moult may have occurred.

MEASUREMENTS Males: (n=2) Length (330.2–368.3); (n=8) wing 162.9 (146–170.2); tail 171.5 (141–206); culmen 34.6 (31–39); tarsus 46.5 (45.7–48). Females: (n=9) Length 279.9 (251.5–309.9); wing 132.6 (128.3–135); tail 129.7 (120–139); culmen 29.4 (26–31); tarsus 38.6 (37–40).

NOTES The extinction of this species came about through the extensive draining of the marshes where it lived. Also, the spread of Great-tailed Grackle into the Central Valley may have aided its demise. There is little hope that this species still exists.

REFERENCES Dickerman 1965 [range, juvenile plumage], Hardy 1965 [habitat].

90 NICARAGUAN GRACKLE *Quiscalus nicaraguensis* Salvin and Godman 1891 — Plate 35

This small grackle is one of the least known of the blackbirds, and is restricted to southernmost Nicaragua and northernmost Costa Rica.

IDENTIFICATION Nicaraguan Grackle is a medium-sized, highly sexually dimorphic grackle with a very restricted world range. It is only likely to be confused with the sympatric Great-tailed Grackle (87). Male Nicaraguan Grackle is much smaller than male Great-tailed Grackle. The tail of Nicaraguan Grackle is even more noticeably keeled than that of Great-tailed Grackle (Stiles and Skutch 1989), but is proportionately shorter-tailed than the latter. The bill of Nicaraguan Grackle is approximately the length of the head, and appears somewhat out of proportion, too large for the size of the head, however it does not give the same massive-billed appearance of Great-tailed Grackle. The iridescence on the head, upperparts and wings shows a stronger green colour which is in contrast to the violet and blue of Great-tailed Grackle. Bronzed Cowbird (99) and Melodious Blackbird (85) are also black; however, they both lack the long, keeled tail of the grackle. Bronzed Cowbirds have reddish eyes and Melodious Blackbirds have dark eyes, while Nicaraguan Grackle has yellow eyes which are visible from a distance. The larger Giant Cowbird (98) often shows yellow eyes, but its shape is quite unlike that of the grackle, lacking the grackle's long, graduated and keel-shaped tail. Giant Cowbirds look comically small-headed, and appear hump-backed due to the neck-ruff of males; the same is true for Bronzed Cowbirds.

Female Nicaraguan Grackle is brown above and pale buff below with brown flanks and belly. Females also show the bright yellow eyes of the male. Again, the female is most easily confused with the female of the much larger Great-tailed Grackle. General size and shape differences noted for the males also apply to the females. In addition, the underparts of female Nicaraguan Grackle are strikingly paler than that of the sympatric form of Great-tailed Grackle. Female Nicaraguan Grackles have nearly white throats, the underparts become pale buff on the breast and darker approaching the belly. Female Great-tailed Grackles are nearly solid pale brown on the underparts. Furthermore, female Nicaraguan Grackles have a pale supercilium that becomes wider behind the eye; the supercilium is accented by the dark ear-coverts. Great-tailed Grackles do not show nearly as much contrast on the face, the supercilium not being noticeably paler than the rest of the face. Female Nicaraguan Grackles have a poorly developed dark malar stripe, while that of Great-tailed Grackle is more developed, albeit masked, by the darker colour of the underparts.

VOICE Song: Described as less varied than that of Great-tailed Grackle. It is an accelerating series of staccato whistles, that increase in frequency *kleee klookleekleekleekleee?* (Stiles and Skutch 1989, Hardy *et al.* 1998). When performing the 'song spread' display, a descending whistled *kleeep* is given (F. G. Stiles pers. comm.).

Calls: The common call note is a nasal *jep*. Also utters a dry *chik* and various whistled notes (Stiles and Skutch 1989). Performs a strikingly sharp and resonating whine that rises evenly in frequency, *whoooeeeeeeee*, lasting just over a second in length. Sometimes a series of high staccato notes are added before or after the whistle (Hardy *et al.* 1998).

DESCRIPTION A small grackle with the typical long bill, graduated and keeled tail of the group. However, the bill is stout and somewhat out of proportion for the size of the bird; it is approximately equal to, or slightly longer, than the head length. Although large, the bill does not appear massive in the manner of Great-tailed Grackle. The culmen is curved at the tip, and is often slightly longer than the mandible creating a visible hook. The gonys is nearly straight. Wing formula: $P9 < P8 < P7 > P6 > P5$, $P9 = P5$; P8–P5 emarginate.

Adult Male (Definitive Basic): The male is black with an iridescent gloss throughout. The head, throat, breast, back and rump shows a dark violet iridescence with a noticeable green component. The gloss becomes darkest violet, lacking green, on the belly, vent, thighs and the upper- and undertail-coverts. The wings largely have a blue-green iridescence, except for the violet lesser and median coverts. The graduated tail shows a violet gloss. The eyes are whitish-yellow. The bill, legs and feet are black.

Adult Female (Definitive Basic): The bill is black, while the eyes are pale yellow. The upperparts are brown, evenly coloured from the forehead to the rump. The underparts are paler. On the face, most noticeable is a paler supercilium that becomes wider behind the eyes. Also, the lores and ear-coverts are dark brown, further accentuating the pale supercilium. The malar streak is indistinct. The underparts are buff, brightest and warmest on the upper breast and the edge of the flanks. The palest parts of the underparts are the throat and lower breast; the chin and throat approach white and contrast with the warm buff breast. The thighs, belly and vent are dark brown, with some of this darker colour extending up the flanks; this contrasts with the paler breast and throat. The wings and tail are dark brown, only slightly darker than the back. A dull greenish-blue, or purplish gloss can be seen on the coverts and tertials, and a purple gloss on the dark brown tail. The wing-linings are dark brown. The legs and feet are black. As in many other female icterids, the female plumage is variable and some may show a limited number of black feathers on the upperparts, particularly the crown.

Immature (First Basic) Male: Similar to adult male, but with a duller gloss. The wings in particular are less glossy, appearing brownish-black and dull brown at the tips. The tail is also duller and shows a less prominent gloss. The belly and thighs are brown, not black. Some brown juvenile feathers are retained in the wing-linings, but these are only visible in the hand. The retention of juvenile feathers is variable, with some showing brownish feathers on the crown or scapulars as well as brown tips to the belly feathers.

Juvenile (Juvenal): Not well known. They are grey-brown above and paler below, with dusky mottling (F. G. Stiles pers. comm.).

GEOGRAPHIC VARIATION Monotypic.

HABITAT It is found in thickets and marshes on the shores of Lakes Nicaragua and Lake Managua. In addition, it forages in cattle pastures, and wherever there are cattle near the shores of the lakes, this grackle is found in numbers. Nicaraguan Grackle appears to have increased in numbers due to the presence of cattle (Howell 1970). It is listed as using marshes, lake shores, riverbanks, wet pastures and scrub by Stiles and Skutch (1989). Obviously, this grackle is closely tied to water, probably throughout the year. Nesting is confined to marshes and the species is sensitive to water levels, leaving areas where these are not appropriate.

BEHAVIOUR Forages on the ground, sometimes in the company of other icterids, such as Red-winged Blackbirds (61) and Bronzed Cowbirds. It will forage by picking at small items (seeds and insects) as well as by turning over small stones and debris (Stiles and Skutch 1989). It also forages on and around cattle (Howell 1970, Stiles and Skutch 1989). Usually observed in small groups. Small groups leave the colony to forage and stay together until they return to the colony (F. G. Stiles pers. comm.), in the manner of oropendolas (*Psarocolius* sp.). This grackle is either monogamous or bigamous, with males usually having one mate and less commonly two. Males assist in parental care, foraging for and feeding the young. The pair bond is much stronger than that of the Great-tailed Grackle. Male Nicaraguan Grackles are often observed together with their mates both at the colony and while foraging (F. G. Stiles pers. comm.). The 'ruff-out (bow)' display is similar to that of other grackles. The male raises his head and bill to vertical and continues by throwing the body forward with the breast feathers fully fluffed and the tail held in a deep wedge-shape while he sings his weak song. Females give a similar display, but much less intense in its delivery (F. G. Stiles pers. comm.).

NESTING Breeding takes place mainly mid March–late May as males collected during these dates have active and enlarged testes. A female specimen from late April was in the egg-laying period. A nesting colony was studied in mid April in N Costa Rica (F. G. Stiles pers. comm.). The season however is more extensive, with nesting reported March–September (Stiles and Skutch 1989). Nests colonially in bushes and trees adjacent to or overhanging marshes, or will nest in emergent marsh vegetation, particularly sedges? The colonies tend to be of fewer than 10 nests. The nest is a coarsely-constructed cup, made from marsh vegetation such as grass, sedge leaves and rootlets. Nests are concealed in the vegetation much more effectively than those of Red-winged Blackbirds nesting in the same marshes (F. G. Stiles pers. comm.). Clutch 2–3 eggs, blue with dark brown or black spots and scrawls, clustered around the broader end of the egg (Stiles and Skutch 1989). Only females incubate.

DISTRIBUTION AND STATUS Uncommon to common around the shores of Lakes Managua and Nicaragua, in the Pacific slope of Nicaragua. In Costa Rica, found mainly in the Lago Caño Negro area. In addition, it is distributed throughout the region immediately south of Lago Nicaragua (Las Camelias and Upala) (Stiles and Skutch 1989).

MOVEMENTS Non-migratory, this grackle is likely sedentary throughout the year. However, it is stated to perform local seasonal movements by Stiles and Skutch (1989). The species appears to be sensitive to water level fluctuations and will vacate areas if these are not suitable for nesting. As forests have been cleared, the range of this grackle spread southward from Lake Nicaragua following the recently created pastures. The population in the Río Frio of Costa Rica may in fact be of recent origin and entirely due to human created habitats here.

MOULT The sequence, timing and extent of moults in this species is largely unknown, although some details can be inferred from specimens. Males in their first spring (first basic), show a largely adult-like plumage which must have been attained through an incomplete first pre-basic moult, possibly late in the summer or early fall (July–October). The first basic bird retains a few juvenile feathers on the wing-linings, thighs, and sometimes scapulars and crown. It is not known if there is a first pre-alternate moult or not, or if the first pre-basic moult of females is as extensive as that of males. Adults appear to undergo one complete moult a year, the pre-basic moult which occurs after breeding (July–October). Definitive pre-alternate moults appear to be lacking.

MEASUREMENTS Males: (n=2) Length 292.1 (179.4–304.8); (n=3) wing 127.3 (127–128); tail 136.8 (134.6–139.7); (n=2) culmen 32, 34.3; (n=3) tarsus 37.2 (36.6–38.1). Females: (n=1) Length 241.3; (n=2) wing 104.1, 116; tail 100, 101.6; culmen 25, 29.2; tarsus 31.8, 37.

REFERENCES Howell 1970 [habitat], Ridgway 1902 [measurements], Stiles and Skutch 1989 [identification, voice, habitat, behaviour, nesting, distribution, movements].

91 COMMON GRACKLE *Quiscalus quiscula* (Linnaeus) 1758 — Plate 35

A common to abundant bird throughout North America east of the Rocky Mountains. It has adapted well to human settlements, and is a common city and garden bird.

IDENTIFICATION Common Grackle may be confused with several North American species, particularly Great-tailed Grackle (87) and Boat-tailed Grackle (88) as well as Brewer's Blackbird (95) and Rusty Blackbird (94). Male Rusty and Brewer's Blackbirds are superficially similar in their entirely glossy black plumages. However, Common Grackle has a long and noticeably wedge-shaped tail, even in females. The two *Euphagus* blackbirds have 'standard' tails and much thinner bills, as well as being noticeably smaller, never showing the raptor-like expression of Common Grackle.

The two 'great' grackles share the wedge-shaped tail with Common Grackle, but are much larger. However, observers not familiar with either species may have difficulty in separating these species strictly on structural characteristics. The two grackles have a purplish head and body gloss, which turns more bluish further back on the body (depending on the species) and a bluish or bluish-green iridescence on the wings and tail. Common Grackle shows two patterns in body gloss; the northern and western birds (*versicolor*) have a blue gloss to the head and neck contrasting with a bronze sheen to the rest of the body, while southeastern birds (*stonei, quiscula*) are purplish-blue on the head and neck, with a purple back and underparts with variable amounts of blue tipping or bronze overtones. Any grackle with a bronze iridescence is a Common Grackle. The gloss colouration of 'Purple' Common Grackle (*stonei*) may appear similar to that of Great-tailed or Boat-tailed Grackles; however, there is usually a distinct line of demarcation on the nape where the colour of the iridescence of the head and neck abruptly changes to the colour of the back, unlike the gradual change in the 'great' grackles. In flight, the tail of Common Grackle is noticeably shorter than the body length, it is longer than the body on many Great-tailed Grackles and falls just short of the body length in Boat-tailed Grackles. Gulf coast populations of Boat-tailed Grackles have dark eyes, a feature never seen on Common Grackle. Female Common Grackles are 'male-like' and do not show the paler underparts of the female Great or Boat-tailed Grackles. Note also that Common Grackles are not nearly as noisy as either Boat-tailed or Great-tailed Grackles, lacking all odd cracking or whistling sounds (see Voice).

VOICE Song: The common song of the male is a squeaky or screechy two syllables that has been interpreted as a *readle-eek, re-lick* or *scudle-eek*; the general impression of the song reminds one of a squeaky gate. The song is always accompanied by a 'ruff-out' display. Females rarely give this song, but instead give a less intense version of the 'ruff-out' display accompanied by a note that sounds like *chuga*. Apparently, males are also known to give this *chuga* call, but it is exceedingly rare. Each male and female gives an individually identifiable song (Wiley 1976a).

Calls: The call that is heard most often is the *chack* or *chuck* call, it is louder and deeper than the similar call of Red-winged Blackbird (61); it is given by both sexes. In other contexts, particularly aggressive ones, the female utters a *waa*. The male version of this is a higher-pitched 'snarl', similar to the *waa*, but unlike in females this is not an aggressive call. Peeping calls are given during mating and in pre-copulatory behaviour.

DESCRIPTION A medium-sized blackbird with a long, wedge-shaped tail and stout, strong bill. The tail is graduated and often held in a wedge, the outer tail feathers fanned in an upturned manner. The tail itself is not as long as the length of the body. The bill shows a prominent hook at the tip. Wing Formula: P9 < P8 ≈ P7 > P6; P9 ≈ P6; P8–P5 emarginate. The 'Bronzed Grackle' *Q.q.versicolor* is described.

Adult Male (Definitive Basic): The bill is black and the eyes are lemon-yellow. The plumage is entirely black and iridescent. The head is glossed with a violet-blue which turns abruptly to a bronzy-brass colour on the rest of the body, including the underparts. The colour of the sheen on the wings is variable, often bronze as on the body, but the wings may show a more purplish or bluer tone to them. The same is true for the tail. The legs and feet are black.

Adult Female (Definitive Basic): Similar to adult male, but smaller and duller. The distribution of iridescence and the colours are as in the male; however, the belly and often the lower breast may show almost no gloss, looking blackish-brown. Soft part colours are as in the male.

Immature (First Basic): Largely like adults, but may appear duller. Some tertials and feathers on the wing-linings may be retained from juvenile plumage, these worn, browner feathers contrast strongly with the newer blackish feathers.

Juvenile (Juvenal): Completely dull brown, with paler brown fringes, particularly on the wing-coverts. On some individuals, indistinct streaks are noticeable on the underparts. They largely lack iridescence to the plumage, but a slight purplish gloss may be seen on the tail. The bill tends to appear paler than on adults and it is often not fully grown, giving the bird a less menacing appearance than the adults. The eyes are dark but become pale during the fall or early winter; by the time that the first pre-basic moult takes place, the eye colour change is well underway. The brown eye colour is caused by a layer of dark pigment lying on top of the yellow iris, the dark pigment is slowly absorbed revealing the yellow colour beneath. This occurs from the center of the eye outwards, such that during intermediate stages, the eyes show a yellowish band around the pupil, and a darker band outside of the paler band (Wood 1934).

GEOGRAPHIC VARIATION Three subspecies are recognized.

Bronzed Grackle ***Q. q. versicolor*** breeds in the west and north part of the range from NE British Columbia east through C Saskatchewan, N Manitoba and Ontario to southern Quebec and SW Newfoundland; southeast of the Rockies to C Texas, west of the Appalachians to SW Louisiana and W Mississippi. It is described above.

Purple Grackle ***Q. q. stonei*** is primarily southeastern. The division between Purple and Bronzed Grackles runs as a diagonal from Louisiana to New Jersey, roughly along the Appalachians, from S Louisiana and Mississippi, east through C Alabama, NW Georgia, E Tennessee and W North Carolina along the boundary between the Virginias, E Pennsylvania to S New Jersey south to C Alabama, N Georgia and W South Carolina.

Florida Grackle ***Q. q. quiscalus*** is found in Florida, including the keys, north to coastal SE Louisiana, S Mississippi, S Alabama to coastal Georgia and South Carolina.

There is a general cline in wing length and bill length from north to south, with northern birds having long wings and short bills. The three forms differ primarily in the iridescent colours of the plumage. Bronzed Grackle is described above and tends to have greenish or blue head and body, with bronzed back and rump. Florida Grackle has a dark violet head and breast, with a greenish back with iridescent bluish bars at the base of each back feather which tend to be concealed; the rump is purplish or violet. Purple Grackle has a greenish or purplish head and breast with a purplish back and rump. Intergrades show intermediate characters between these forms; Purple Grackle is particularly variable especially with respect to the colour of the head.

The intergrades between Bronzed and Purple Grackles were formerly given the name *Q. q. ridgwayi*. The extreme variability in *'ridgwayi'* was previously understood to be due to several morphs being present within that form (Oberholser 1919). This is an illustration of an extremely typological view, trying to fit all individuals into some sort of category, in this case a colour morph within a subspecies when the reality is somewhat more complex. Purple Grackle appears to actually reflect a zone of intergradation between Bronzed and Florida Grackles, and thus is not really a 'proper' subspecies. To bring attention to the fact that these birds are different from the two parental forms, I have chosen to use the name *stonei* as was originally designated, to refer to these birds, even though this may be an improper use of the subspecies concept. Curiously, the center of the zone of intergradation has shifted to the north by 32 km between 1930 and the late 1960s, perhaps due to the general trend for warmer winter temperatures. A study of mitochondrial DNA showed no evidence for genetic structure between the two forms, thus these geographic populations are not significantly different genetically.

HABITAT Common Grackle is found in almost any open or semi-open situation throughout its range. This includes bogs, marshes, forest edge and open woodlands such as pine forests, pine and palmetto savanna, to lawns, suburbs, agricultural areas, shopping centers and other highly urbanized places. In urban situations it is most apt to nest where dense shrubbery or ornamental conifers (i.e. Blue Spruce *Picea pungens*) are present; in the east, the grackle is quite fond of cemeteries as these are often planted with conifers. In the prairie region of Canada and the western Great Plains in the US, Common Grackle is restricted to towns, shelterbelts, farmhouses with planted trees and riparian areas, and are absent from sites lacking a suitable number of trees. In these places, Brewer's Blackbird appears to be supplanted by the grackle and relegated to the more open, treeless sites (Jaramillo pers. obs.). Common Grackle is absent from contiguous forest, unless open swamps or marshes are present. It is thought that originally this was a bird of bogs and marshes, and open riparian areas in the Midwest (Peer and Bollinger 1997a), but that it was quick to adapt to new habitats as soon as humans began clearing the forests. In remote areas, it can still be found in marshes, often nesting colonially. In the far north, nesting is almost entirely restricted to wetlands, in particular at beaver (*Castor canadensis*) ponds and their associated swamps in much of the boreal forest. Common Grackle is also known to nest within the large stick nests of Great Blue Herons (*Ardea herodias*). Particularly while in winter flocks, Common Grackles are fond of feedlots, agricultural fields and rice paddies, where they may be found in mixed flocks with other icterids. Fields near wooded situations are preferred (Peer and Bollinger 1997a). Common Grackles roost in a variety of situations, including city parks, cedar (*Thuja* sp.) swamps, conifer groves, deciduous forest edges, and marshes, mainly cattail (*Typha* sp.) but they will also use reed (*Phragmites communis*) (Peer and Bollinger 1997a, Jaramillo pers. obs.).

BEHAVIOUR During the breeding season, the most commonly observed behaviour is the 'ruff-out' display, which is accompanied by the song. This stereotypical behaviour is similar to that of many other blackbirds, the wings and tail are spread and the body plumage is 'ruffed-out'. In the most extreme versions of this display, the bird may bob on its legs and rapidly flash the nictitating membranes (the 'third eyelid') of the eyes. The highest intensity displays are usually given to another male, thus it is not a simple sexual display to attract females. Female 'ruff-out' displays are not as intense as those of the males. Males can also give a beautiful aerial 'ruff-out', often during 'leader flight' displays. As the male flies, the tail is kept in a deep wedge shape and the plumage is 'ruffed out' as the male gives his squeaky song. Early in the season while pairs are forming, pairs may partake in display flights and mutual displays (duets). These flights are commonly seen in the early spring, as soon as the females return to the breeding site. Commonly, a single female will fly off and one or several males will follow her, but without ever actually catching up with her, this is termed the 'leader flight'. It appears that these displays are useful in pair formation. Once pairs are established, the pair will perform 'mutual displays' beginning with a 'ruff-out' and singing by the male and answered by a 'ruff-out' and a *chuga* call by the female. Often they will perform several mutual displays back to back. Mating is also accompanied by a series of specific displays. Males point their bills downwards while opening the wings level with the ground and spreading the tail, while singing and giving 'peep' calls. The female may respond by giving the solicitation display by raising her head and tail, while fluttering the wings. Mating may occur once this ritual has been performed.

The mating system of this species varies from monogamous to at times polygynous, depending on the ecological situation. Males are notorious for guarding their mates against other males. He follows her around for most of the time, during the pre-laying period, aggressively displaying to other males or chasing them away. Pair formation occurs in foraging flocks away from the nesting site. Males will often begin mutual displays with the female to solidify the pair bond. 'Mate guarding' assures the male that he will be the father of her young. Copulations outside of the pair likely occur while the female is on the nest and the male does not guard her as closely.

The tail keel of this grackle, and perhaps other-species appears to have both display and aerodynamic functions. Deep keeling of the tail is only observed during the breeding season; however, shallow keeling occurs throughout the year. Shallow keeling apparently prevents stalling and increases stability when landing and banking. The tail is kept flat, while in flight, when flying into the wind to decrease drag and during entirely level non-manoeuvering flight.

In fall and winter, grackles are quite social, gathering into large flocks, often with other icterids. Males arrive on the breeding grounds in late February to April in the north, with females arriving a week or two later. During most of the year, grackles roost in groups, usually with other blackbirds.

The feeding habits of this species are extremely variable, this bird is a definition of the term omnivorous. The use of grains and cereals as food is widely known as is their insectivorous nature during the summer. More unusual are the Common Grackle's exploits as a fisher. They often eat crayfish (*Crustacea*) and will even execute shallow dives from either a hover or a low perch to catch fish. They will also wade in shallow water to catch fish. Common Grackle is an accomplished nest predator, but will even kill and eat small birds and mammals! Most often birds are killed by pecks to the back of the head. In one documented instance a single Common Grackle killed and ate at least 39 small passerines during a two week period (Davidson 1994). Common Grackle has a well-developed keel on the palate, better developed than in other grackles. This keel is used in opening hard seeds, including acorns, which are rotated with the tongue and pierced by the keel such that the hard outer shell is cut, revealing the edible insides (Wetmore 1919).

NESTING Common Grackles vary from being solitary nesters to colonial. It is thought that the colonial lifestyle is what was originally dominant, before grackles adapted to human-altered habitats. Preferred nesting sites vary, but in many areas dense conifers are selected preferentially. However, there is a great deal of variety in chosen nesting sites including cattails (*Typha* sp.), cavities and even within heron nests. The use of heron nests presumably protects the grackles from nesting predators, including the herons! Usually only one brood is raised per season, but double brooding is known. Nest construction begins as early as late March and few nests are built after mid June (Peer and Bollinger 1997a). The nest is usually constructed by the female alone, but instances of male helping are known (Peterson and Young 1950). The nest is a large and bulky, cup-like structure composed of small sticks, grass stems and leaves; it sometimes includes anthropogenic materials such as fishing line, plastic, string, and paper (Peer and Bollinger 1997a). The nest is reinforced with mud on its interior before a bed of fine grasses is laid down as a lining (Peer and Bollinger 1997a). Clutch typically 3–6 eggs, but ranges from 1–7 (Peer and Bollinger 1997a). The eggs of Common Grackle range from pale greenish to light rusty. They are marked with contrasting dark brown or purplish splotches and scrawls, overall they are quite variable and some eggs may be immaculate (Peer and Bollinger 1997b). Incubation is only performed by the female and lasts 11.5–15 days, averaging 13.5 days (Peer and Bollinger 1997b). The feeding of the young is largely performed by the female, but males commonly help; however, it is also common for males to abandon the female while she is in the incubation stage (Wiley 1976b). Young fledge the nest between 10–17 days old, most between 12–15 days (Peterson and Young 1950, Howe 1976). Brown-headed Cowbird (101) will parasitise this grackle, but unsuccessfully (Friedmann and Kiff 1985). Parasitism is also very rare, with only 20 reports known (Peer and Bollinger 1997a). Grackles reject 17% of eggs experimentally placed in their nests (Peer and Bollinger 1997b).

DISTRIBUTION AND STATUS Common or abundant throughout its wide range.

Breeding: ranges north to NE British Columbia (Peace River District), throughout Alberta (except in the Rockies), northeast to the southern Mackenzie area, Northwest Territories, southeast along N Saskatchewan, NC Manitoba, N Ontario (to latitude of C James Bay), C Quebec (latitude of James Bay and N shore of the Gulf of St Lawrence at least to Mingan and including Anticosti Island) to SW Newfoundland, it is unreported from Labrador; the distribution extends east to the Atlantic coast from Newfoundland and Quebec, including the Gaspé peninsula, the Maritimes and Prince Edward Island, to southernmost Florida and the Florida Keys; ranges west to NE British Columbia, the foothills of the Rocky Mountains in SW Alberta, W Montana, E Idaho (it is a recent arrival here, it is spreading and actually breeds west of the continental divide), NE Utah, W Colorado, recently in WC Nevada (one breeding record at Stillwater)(Alcorn 1988), WC New Mexico (reaching south of Albuquerque and then southeast towards the SE corner of the state); and reaching south to S Texas (remaining north of the Pecos river and Edwards plateau reaching the coast north of Corpus Christi) and the Gulf coast from Texas east to peninsular Florida.

Non-Breeding: Vacates the north and west portion of the range. In winter they are typically absent from British Columbia, Alberta, Saskatchewan, Manitoba, all except southernmost Ontario, Quebec, Newfoundland, Nova Scotia, Prince Edward Island, New Brunswick, Montana, North Dakota, Idaho, Utah, Wyoming, Colorado and New Mexico but isolated individuals or small flocks may attempt to persist in these areas, particularly during warm winters. Wintering birds are regularly found north to the latitude of S Minnesota, S Wisconsin, S Michigan, S Ontario, N New York, C New Hampshire, C Vermont and S Maine, and sometimes S Nova Scotia. They regularly winter west to SE South Dakota, C Nebraska, W Kansas, panhandle of W Oklahoma and W and S Texas.

MOVEMENTS A medium distance migrant, except in the south part of its range where they are year-round residents. All migratory movements are diurnal, often with other blackbirds. The subspecies *versicolor* is the most highly migratory of the three forms. Northward migration tends to take place mainly late February–March and has terminated by mid April (Dolbeer 1982), note that males precede females by at least a week. The fall migration begins in September, peaks in October–November, and finishes by early December (Dolbeer 1982). On average, females migrate 100 km farther than males, and young of the year migrate 100 to 300 km farther than adults (Dolbeer 1982). Banding records show a tendency for southward movements to be in the direction of the Gulf coast, thus birds in the west tend to move to the southeast and those in the northeast move towards the southwest in fall (Burtt and Giltz 1977). Note that many birds avoid crossing the Appalachian mountains and remain to winter on the same side of the mountains that they bred on (BBL data). Nevertheless, some individuals perform significant east-west movements. For example, an individual banded in June 1993 in South Dakota was recovered in June 1997 from coastal Virginia (Tallman 1997). Individuals banded in Saskatchewan during the breeding season were recovered in a very narrow corridor, mainly west of the Mississippi river. Saskatchewan Common Grackles wintered mainly in E Texas, Louisiana, Mississippi and Arkansas; migrants were also recovered from Minnesota and Missouri (Houston 1968). A study (Stewart 1975) of banded Common Grackles recovered wintering in the Carolinas recorded that grackles wintering in South Carolina tend to breed no further north than Maryland, with most coming from the Virginias. A minority were from Ontario, New York and Pennsylvania, with single records from Minnesota and Indiana. Common Grackles wintering in North Carolina tended to come from areas further north, mainly New England. Grackles breeding in North Carolina moved south in the winter, while those south of South Carolina tended to be resident. Populations nearest the Rocky Mountains incorporate a noticeable northwest-southeast tendency in their migration, primarily due to the barrier created by the edge of the Rocky Mountains. Radio tracked fall migrants in Oklahoma moved during or before cold fronts, between 35 and 53 km per day (Bray *et al.* 1979). The oldest banded individual recovered was an amazing 22 years old (Olyphant 1995)! This is the oldest known North American blackbird.

Common Grackle is on the move in the western part of its distribution. It has recently spread to eastern Idaho, where it breeds west of the continental divide, due to the planting of trees in urban areas. In Colorado, the range has shifted westward in the recent past, with most of the expansion occurring in the mid 1980s. In addition, the breeding population in NE Utah is of recent origin, it was considered an accidental as recently as 1976 (Hayward *et al.* 1976). The first Utah record occurred on March 21, 1957 in Provo (Talley 1957). There has been recent successful nesting in WC Nevada in 1987 (Alcorn 1988). The first individual from Nevada was collected in 1932, and other than the recent nesting Common Grackle has remained a rare migrant/vagrant in the state with most records during spring and summer, fewer in fall as well as a few winter occurrences (Alcorn 1988). It was first detected in New Mexico in 1951 and is now a regular breeder with a rapid spread detected in the 1970s (La Rue and Ellis 1992, Peer and Bollinger 1997a). In Texas, it was formerly restricted to the eastern part of the state, but it has moved westward to WC Texas (Peer and Bollinger 1997a).

A rare stray to the Pacific states and coastal British Columbia (all *versicolor*). In the lower mainland and Vancouver Island, British Columbia, there are now at least 7 records. The first lower mainland record occurred on May 6, 1968 in Vancouver, while the first for S Vancouver Island occurred during Sept. 27–30, 1986. Up to 1984, in Washington, there were three E Washington records and two W Washington records. These included one female collected in July in Walla Walla Co. The Oct. 22, 1950 record from Paradise Creek, Washington (within 118 m of the Idaho state line) was the first Pacific Northwest record and the only one prior to the 1960s (Roberson 1980). In Oregon, there are at least 13 records, mostly in spring and summer (Gilligan *et al.* 1994). California has over 30 records mostly from the spring (inland desert oases), with a smaller number in the fall (mostly coastal sites) but there are several winter and midsummer occurrences; most records are from southern California (J. Morlan pers comm., Small 1994). In Arizona, the first record was as recent as 1980 (La Rue and Ellis 1992). Since then there have been several records and reports which have included birds in all seasons (La Rue and Ellis 1992). Nesting may not be far off in Arizona. In the Big Bend region of Texas, Common Grackle has been observed five times in the spring and once in the fall (Wauer 1985). Vagrants have also occurred in the north. There are at least six (Roberson 1980) records for Alaska, from May–late August. The first was a male collected on June 17, 1943 at Wainwright, Alaska, in the tundra (Bailey 1948). This site is southwest of Barrow, at 70° latitude! The remainder of the records are from the west (1), central (2) and panhandle (2) (Roberson 1980). The two specimen records for Alaska are both of the bronzed (*versicolor*) subspecies (Gibson and Kessel 1997). A male was at Dawson City, Yukon Territories, in November 1996. Common Grackle is a vagrant to Rankin Inlet, Keewatin District, Northwest Territories, where one was observed on June 10, 1970 (Helleiner 1972). The nearest regular locality Common Grackle wanders to is Churchill Manitoba, 480 km south. A female Common Grackle was seen in Baja California during January 1996, the first Mexican record (S.N.G. Howell pers. comm.). There is one record for the Western Palearctic, a spring record from Denmark (Alström and Colston 1991). Although, there is an unconfirmed sighting from the Scilly Islands, UK, on June 2, 1986. It has been reported from North Andros, Bahamas, but this is unconfirmed (T. White pers. comm.).

Purple Grackles (*stonei*) have been collected in Quebec during April, there were two in almost 1000 birds processed and *stonei* must be considered a stray this far north. Another Canadian record of a Purple Grackle is from Kent Island, New Brunswick on November 20, 1931 (Prud'Homme-Cyr *et al.* 1976).

MOULT Juveniles have an incomplete first pre-basic moult during the summer to early fall (July–September), retaining some of the wing-lining and tertials. However, this first pre-basic is rarely complete. Adults undergo a complete pre-basic moult on the breeding grounds (late July–mid October). The tail is replaced last, moult is centripetal with the outer rectrices moulted first and the inner ones last (Wood 1945). As far as is known, the pre-alternate moult does not occur, but some of the other grackles do show this moult It may occur in Common Grackles as well. Pre-alternate moults are absent. The pre-basic moult is complete in the adults and incomplete or complete in juveniles. Moulting takes place July–October. First basic (1st- year) birds are duller than the adults and the worn brownish underwing-coverts and perhaps a tertial or two will contrast with the rest of the plumage.

MEASUREMENTS ***Q. q. quiscula***. Males: (n=20) wing 138.5 (129–146.8); tail 128.4 (116.1–139.7); culmen 33.6 (32.5–36.3); tarsus 36.6 (34.3–37.3). Females: (n=15) wing 124.9 (116.6–133.9); tail 110.1 (100.8–122.7); culmen 29.5 (27.9–31.2); tarsus 34.1 (32.8–34.8).

Q. q. stonei. Males: (n=20) wing 138.5 (129–146.8); tail 128.4 (116.1–139.7); culmen 33.6 (31–36.3); tarsus 36.6 (34.3–38.4). Females: (n=15) wing 149.3 (116.6–133.9); tail 110.1 (100.8–122.7); culmen 29.5 (27.9–31.2); tarsus 34.1 (32.8–36.1).

Q. q. versicolor. Males: (n=109) wing 138.9 (135.0–153.2); (n=99) tail 141.4 (124–158); (n=109) culmen 30.3 (28.0–34.0); tarsus 38.8 (31.0–39.0). Females: (n=105) wing 126.5 (118–133); (n=105) tail 123.3 (110.7–132); culmen 26.4 (23–31.2); tarsus 33.9 (29–34.8).

NOTES Previously, Purple and Bronzed Grackles were considered different species.

A Common Grackle, apparently a female, photographed in Toronto, Ontario, Canada during May 1996 showed dark red irides, rather than the usual yellow (C. MacLaughlan, pers. comm. and photos).

REFERENCES Bray *et al.* 1979 [movements], Burtt and Giltz 1977 [movements], Chapman 1935, 1936, 1940 [geographic variation], Davidson 1994 [behaviour, foraging], Ficken 1963 [behaviour, voice], Hickman 1981 [aerodynamic, behaviour], Houston 1968 [movements], Huntington 1952 [habitat, nesting, geographic variation, movements], Olyphant 1995 [age], Prud'Homme-Cyr *et al.* 1976 [vagrant *stonei*], Pyle *et al.* 1987 [moult], Snyder 1937 [measurements], Stewart 1975 [movements], Taylor and Trost 1985 [movements, distribution], Yang and Selander 1968 [geographic variation], Wood 1934 [description, eye colour] Wood 1945 [moult], Zink *et al.* 1991 [geographic variation, genetics].

92 GREATER ANTILLEAN GRACKLE *Quiscalus niger* (Boddaert) 1783 — Plate 33

The common blackbird of the Greater Antilles, a frequent guest at parks and beaches, foraging for leftovers left by tourists.

IDENTIFICATION Greater Antillean Grackle does not overlap in range with the similar Carib Grackle (93) of the Lesser Antilles. However, Carib Grackles are quite similar to Greater Antillean Grackle, however they are smaller with weaker, thinner bills. Also, female Greater Antillean Grackles are black and glossy, while most races of Carib Grackle have more brownish females with pale underparts. In its range there are four other black icterids with which the grackle could be confused. These are Cuban Blackbird (84), Jamaican Blackbird (52), male Shiny Cowbird (102) and female Red-shouldered Blackbird (64). None of these species have either the long and obviously wedge-shaped tail or the pale eyes of the grackle. Cuban Blackbird is most similar in size, the others are noticeably smaller than the grackle. Cuban Blackbirds lack a vivid gloss to the body like the grackle and have a stout bill with a distinctly curved culmen. Shiny Cowbird is glossy, but is small and has a short bill. Female Red-shouldered Blackbirds are shaped like Red-winged Blackbirds (61) but totally lack any iridescence. Finally, Jamaican Blackbird is an arboreal, forest-dwelling blackbird which resembles an oriole in behaviour, they are small and have a fine, sharply pointed bill.

VOICE Song: Complex song made up of a variety of sounds which have been described as harsh and rasping. The song on the Caymans is musical and composed of four syllables. In Jamaica, the song is made up of two or three metallic clangs or clicks, followed by a harsh grating sound. Meanwhile, in Hispaniola they commonly produce a *jui-jui-jui-jui* or *wees-see-ee*, as well as trills and other non-melodious sounds. However, the song is not considered unpleasant (Wetmore and Swales 1931, Stockton de Dod 1978). Bond (1928b) describes songs from Hispaniola (*niger*) as sounding less noisy and squeaky but more musical than Carib Grackles. Female Greater Antillean Grackle are known to sing.

Calls: Vocalizations include a *chak-chak* and *chin-chin-chi-lin*, both of which give the bird many of its onomatopoeic local names. Sometimes a high frequency *whee-see-ee* is given as well as a ringing note. A *click* call is heard on the Cayman Islands, as well as a rasping call performed with the feathers fluffed out.

DESCRIPTION A grackle similar in size to the more well-known Common Grackle (91), and shares with all the grackles the strong bill and wedge-shaped tail. The bill is long, slightly longer than the length of the head. The culmen is straight except for a noticeable downcurve at the extreme tip, which creates a short hook at the bill tip. The long tail is noticeably graduated. The legs are thick and strong. Wing Formula: P9 < P8 < P7 > P6; P9 ≈ P4; P8–P6 emarginate.

Adult Male (Definitive Basic): The plumage is completely black with purplish-violet or bluish iridescence, on the head and body. The wings and tail are also iridescent. The exact distribution and colours of the iridescence varies geographically (see Geographic Variation). In some forms, the wings and rump have a greener gloss. The bill and legs are black while the eye is a vivid whitish-yellow.

Adult Female (Definitive Basic): Like male, but duller with less extensive iridescence and smaller in size. The eyes are yellow. Females have long and graduated tails, but they do not hold them in a distinctive wedge as males do, this difference is particularly noticeable in flight. The bill and legs are black.

Immature Male (First Basic): Similar to adult male, but smaller and with a shorter tail. Immature males are larger than females of any age, but the difference is not as readily obvious and may have to be ascertained through a direct comparison. In addition, they are not as iridescent. On immature males, the belly, vent and thighs lack iridescence altogether and may look brownish-black. The body, wings and tail do have a gloss, but duller than adult males. The primaries are duller, and look brownish, more worn and faded on their tips than those of adult males. Soft parts as on the adults.

Immature Females (First Basic): Similar to adult females but even duller. They are largely indistinguishable from adults but may show more worn, brownish primary tips.

Juvenile (Juvenal): Largely brownish-black, lacking iridescence. The tail wedge is not well developed. The eyes are dark.

GEOGRAPHIC VARIATION There are seven subspecies recognized, most isolated to one island or island group.

Q. n. niger breeds in Hispaniola, including surrounding smaller islands. This form is the smallest of the races. It is a very purple form, with a greenish gloss restricted to the wings and tail feathers, the wing-coverts being purple glossed, not green.

The Jamaican form is ***Q. n. crassirostris***. This race is similar to the Cuban birds, but has a relatively shorter and stouter bill. Males are black with a violet gloss, becoming blue on the abdomen and upper- and undertail-coverts. Wing-coverts have a bronze-green gloss, with purple tips. Overall there is a greenish tone to the wings. Females are similar, but are smaller and slightly duller, with no iridescence on the belly.

The eastern Cuban race is ***Q. n. gundlachii*** which breeds in C and E Cuba. It is like *crassirostris*, but has a more extensive violet gloss, particularly on the underparts. It is smaller and has a longer, more slender bill than *crassirostris*. It is larger and more violet than *caribaeus*. Body completely black, with violet gloss on the body and lesser wing-coverts. The uppertail-coverts have a bluish iridescence; the greater coverts and secondaries are glossed with blue-green. The tail and primaries are largely black, only weakly showing a blue-green gloss. Females are smaller and less glossy than males, much of the violet replaced by blue.

Q. n. caribaeus is found in W Cuba east to San Cristóbal. Its range also includes the Isle of Pines. It is like the eastern Cuban *gundlachii*, but it is smaller and has a blue gloss throughout the body, many do not show any purplish at all. Females of both Cuban races are much more variable in their measurements than the males with much overlap between separate populations; females of *caribaeus* and *gundlachii* may not be able to be separated in all cases.

Q. n. bangsi is isolated to the islands of Little Cayman and formerly Cayman Brac. Like *crassirostris*, but smaller and with a long, stout bill. It is less purple, with the blue extending to the lower breast and the throat is washed with blue. The upperparts are also bluer.

Q. n. caymanensis is found on Grand Cayman Island. This race is smaller than *caribaeus*, but it has a purplish tinge to the blue gloss, but is not as purple as *gundlachii*. The tail-coverts have a greenish-blue iridescence and the secondaries and coverts are glossed bronzy-green. Females are smaller and duller, often showing a brown wash to the underparts.

Q. n. brachypterus is found in Puerto Rico, including Vieques Island, and was introduced to St Croix in 1917, but eventually became extirpated there. It is similar to the Jamaican form (*crassirostris*), but is smaller and has the purple of the upperparts extending down to the uppertail-coverts. Larger than the nominate form, with a stouter bill with an obviously downcurved tip to the culmen.

HABITAT A bird of the lowlands. It may be found in many open areas with trees, including urbanized zones. It is often seen feeding on lawns and in other park-like settings. In addition, it will feed on beaches and tide pools, shores of lakes and other aquatic habitats. Greater Antillean Grackle has a particular affinity for mangroves and swamps, as well as other wet areas. Lack (1976) suggests that sedge marshes, wooded swamps and mangroves probably comprised this grackle's original native habitat before people arrived in Jamaica. Higher up in mountains, it is not as common, but it is becoming more abundant as forests are destroyed. On the Cayman Islands, *caymanensis* occurs in large colonies in mangroves, with smaller colonies in *Conocarpus* sp., and small numbers in urban and littoral areas. The form *bangsi* occurs in small numbers in mangrove and littoral areas, and the extinction of this subspecies on Cayman Brac is apparently due to the destruction of mangroves there (Bradley 1994).

BEHAVIOUR Forms large roosts like many other icterids; often these are in electrical substations in Puerto Rico! They are noisy, noticeable birds like other grackles. They forage mainly on the ground, but are seen to climb on livestock to feed on ectoparasites, in the manner of a cowbird. They appear to feed mainly on insects and seeds, but will hunt and kill small lizards both on Jamaica and the Cayman Islands (Lack 1976, Bradley 1994). Fallen fruit is also a part of their diet. This grackle is a known nest robber (Bradley 1994). The mating system is monogamous (Post 1981a). Like other grackles, this species performs several highly visible displays. The 'ruff-out' display resembles that of other grackles; the plumage is ruffled, the tail is spread somewhat and elevated, the wings are drooped, and the head is thrown back as the bird sings. The throwing back of the head and singing with the bill pointing upward is quite distinctive in this species. Males often fly with the tail conspicuously spread. Females are known to sing and give a full 'ruff-out' display.

NESTING Commonly nests in colonies, usually in large trees and palms. These colonies are small, seldom having more than 25 nests and are usually all placed in a single tree. On the Isle of Pines, nesting colonies are frequently in mangroves, and the same appears to be true on other islands. Like other grackles, it may also nest in reeds (*Scirpus* sp.) or cattails (*Typha* sp.). Nests tend to be bulky and are made of thick stems and grass stalks. The clutch ranges from 3–5 eggs, which are pale blue, green, olive or whitish, with dark spots and scrawls. The young require at least 23 days to fledge from the nest (Stockton de Dod 1978). Nesting in the Cayman Islands takes place March–July. In Haiti, the nesting period ranges from April–July (Wetmore and Swales 1931). In Puerto Rico, the nesting period peaks April–August, but extends from February–September, while in Jamaica the season stretches from April–June. Cuban and Isle of Pine populations begin nesting in April (Gundlach 1876, Todd 1916). In Puerto Rico, this species is a regular, but uncommon, host of Shiny Cowbird (102); nearly 4% of nests are parasitised (Post *et al.* 1990) or as high as 11% (Wiley 1985). It has been observed to successfully fledge cowbird young (Friedmann and Kiff 1985). Interestingly, experiments have discovered that up to 77% of the grackle females may recognize foreign eggs and reject them (Post *et al.* 1990).

DISTRIBUTION AND STATUS Common, usually one of the most abundant lowland birds. Resident on the Greater Antillean islands of Cuba, Isle of Pines (Cuba), Grand Cayman Island, Little Cayman, extirpated on Cayman Brac, Jamaica, Hispaniola (including Île à Vache and Beata Island), Puerto Rico, and Vieques Island (P.R.) which is fact its easternmost outpost. Jamaican Greater Antillean Grackles are common throughout, particularly in lowlands but present in low numbers in the highlands. In Hispaniola, occurs throughout the main island, including highlands, as well as smaller islands, Île à Vache, and Beata Island. Greater Antillean Grackle is absent from Mona Island (Barnes 1946), this is curious as it is found to the east and west of the island. Extirpation of this grackle on Cayman Brac occurred sometime between 1911 and 1970 (Bradley 1994). In the 1970s, it was noted that small flocks of grackles flew to Cayman Brac from Little Cayman and spent the day there but that they did not breed on the former (Johnston 1975).

MOVEMENTS Sedentary, however Lack (1976) states that they move down to coastal forests, from which they are absent at other times, when ripening of *Bursera simaruba* fruits occur. There was an introduction of this grackle to St Croix in 1917, but this population is now thought to be extirpated.

MOULT Adults moult all of their feathers through the complete definitive pre-basic moult. It occurs after breeding has finished, therefore its timing is dependent on the population. In Haiti (*niger*), the definitive pre-basic moult occurs in October (Wetmore and Swales 1931), and as early as August. Birds in noticeably fresh plumage have been collected in May on Beata Island, Dominican Republic (*niger*), confirming local differences in moult timing (Wetmore and Lincoln 1933). Cuban *gundlachi* is in moult during September. Puerto Rican *brachypterus* moults October–November. On Grand Cayman Island, *caymanensis* is in moult during November. The juvenal plumage is replaced through the incomplete first pre-basic moult which replaces most of the plumage except for a few underwing-coverts, and perhaps a scapular or tertial. Pre-alternate moults are lacking.

Immature birds obtain the definitive (adult) plumage after the second pre-basic moult.

MEASUREMENTS ***Q. n. niger***. Males: (n=5) Length 268.2 (248.9–281.9); (n=22) wing 135.4 (122.4–147); tail 119.0 (104.8–132); culmen 32.8 (30–35.1); tarsus 34.4 (32.0–37.7). Females: (n=2) Length 241.3; (n=22) wing 114.4 (108–120); tail 102.5 (91.4–117); culmen 27.9 (26–29.4); (n=12) tarsus 31.6 (29–37).

Q. n. crassirostris. Males: (n=5) Length 286.3 (279.4–194.6); (n=8) wing 150.1 (144–161); tail 129.3 (124.5–141); culmen 33.3 (32–35.6); tarsus 39.5 (38–41.4). Females: (n=4) Length 240.5 (226.1–274.3); (n=8) wing 128.3 (120–135); tail 111.6 (103.9–122); culmen 28.9 (27–33); tarsus 34.7 (32.0–38).

Q. n. gundlachii. Males: (n=22) wing 148.9 (131–156); tail 128.1 (117–137); culmen 34.8 (31–41.4); (n=9) tarsus 39.0 (38.0–41.4). Females: (n=3) Length 252.2 (231.1–264.2); (n=8) wing 128.1 (118.9–134); tail 111.8 (99.1–116.8); culmen 32.1 (30–33); tarsus 34.6 (33.5–36.6).

Q. n. caribaeus. Males: (n=27) wing 141 (137–146); tail 125 (115–136); culmen 33.7 (31–37.0). Females: (n=10) wing 125.1 (122–128); tail 110.5 (90–116.5); culmen 30.9 (29.5–31.5); (n=1) tarsus 36.

Q. n. bangsi. Males: (n=7) wing 145 (141–147); tail 131 (126–135.5); culmen 35 (33–39). Females: (n=8) wing 124 (118–129); tail 113 (107–123); culmen 31.1 (30.5–32);.

Q. n. caymanensis. Males: (n=17) wing 134.7 (128–141); tail 115.8 (105–126); (n=15) culmen 31.4 (28-34); (n=7) tarsus 33.7 (32–36). Females: (n=6) wing 114.2 (111–118); tail 96.8 (92–103); (n=5) culmen 27.7 (27–29); (n=6) tarsus 30.4 (29–32).

Q. n. brachypterus. Males: (n=5) Length 272.8 (261.6–299.7); (n=10) wing 128.9 (125.0–134.6); tail 116.3 (110–121); culmen 32.6 (30–34); tarsus 35.9 (32–37.8). Females: (n=5) Length 242.8 (233.5–259.1); (n=10) wing 112.0 (108.7–116.3); tail 98.6 (92.7–104); culmen 29 (26–31.8); tarsus 32.3 (30–33.8).

NOTES It was formerly split into two species, the subspecies *niger* and *brachypterus* were considered different from the others, mainly based on their small size. Greater Antillean Grackle is clearly related to Carib Grackle based on morphological features. In fact, this species and Carib Grackle were formerly classified in a genus separate from the other grackles, *Holoquiscalus*. Lack (1976) hypothesises that the closest relatives to these two grackles were Common Grackle and Nicaraguan Grackle (90). Based on plumage features and vocalizations, it is our hypothesis that the *Holoquiscalus* grackles are most closely allied to Great-tailed Grackle (87), not Common Grackle.

REFERENCES Biaggi 1983 [habitat, status, behaviour, nesting], Bond 1928b [voice], 1988 [description, voice, nesting], Bradley 1985 [voice, habitat, behaviour], Bryant 1866 [geographic variation], Garcia 1987 [nesting, Cuba], González Sánchez *et al.* 1991 [geographic variation], Johnston 1975 [distribution], Lack 1976 [habitat], Perez-Rivera 1986 [brood parasitism], Peters 1921 [geographic variation, measurements], Post *et al.* 1990 [nesting, brood parasitism], Raffaele 1989 [nesting, habitat], Ridgway 1902 [measurements], Selander and Dickerman 1963 [female song], Stewart 1984 [song, behaviour], Wetmore and Lincoln 1933 [distribution], Wetmore and Swales 1931 [nesting, moult, voice, measurements].

93 CARIB GRACKLE *Quiscalus lugubris* Plate 33
Swainson 1838

The southern counterpart of Greater Antillean Grackle, it is very similar to it in many respects.

IDENTIFICATION In northwest Venezuela, the range of Carib and Great-tailed Grackles (87) comes quite close. Carib Grackle is a noticeably smaller grackle than Great-tailed, but shares with it the 'keel' or wedge-shaped tail. However, the tail wedge of Great-tailed Grackle is more exaggerated and at least as long as the length of the body, while on Carib Grackle the tail is shorter than the body and with a shallower wedge. The bill is proportionately smaller on Carib Grackle and the forehead is more abruptly sloped, while on Great-tailed Grackle the shallowly sloping forehead and culmen are almost in line. All of the other similar species lack the wedge-shaped tail of Carib Grackle. Two species that may be found feeding on the ground in open areas, like the grackle, are Shiny (102) and Giant Cowbirds (98). Shiny Cowbird always lacks a pale eye and has a short bill and tail, it is also smaller than the grackle. Giant Cowbirds are large, bulky birds with a very odd small-headed appearance. Their eyes tend to be red, but may be yellow at times. The size and shape differences should readily differentiate this species from the grackle. Two other black icterids in the grackle's range are Solitary Cacique (21) and Velvet-fronted Grackle (81), which have dark eyes and lack a vivid gloss on the body; the cacique also has a contrasting pale bill. Both are found in forested habitats, not open areas. One should be aware that there are several black thrushes (*Turdus*) and tanagers (*Tachyphonus*) that are largely black, but note their dark eyes and lack of keels on their tails, as well as their different behaviour and habitats; the thrushes often have colourful bills, legs and orbital rings.

VOICE Song: The song has been described as a series of squeaks and clucks, harsh in quality often terminating in a ringing bell-like note. It may sound like *wee-tsi-ke-tsi-ke-tsi-ke* or *queek-queek-queek*, also *etsywee*. There is a noticeable amount of geographic variation in song, with northern birds singing different songs than southern birds. The vocalizations of birds from St Lucia (*inflexirostris*) have been reported to be quite unlike those from the Grenadines (*luminosus*) (Lawrence 1878). Similarly, the vocalisations of *luminosus* have been said to be significantly different from those of *fortirostris* (Clark 1905). Bond (1928) writes that the song of *inflexirostris* from St Lucia is a *weee-tsi-ke-tsi-ke-tsi-ke* which is quite different from the *betse-weee, sicker, sicker, sicker* of *fortirostris* from Barbados; both have a squeaky quality to their songs. The song is harsher and less musical than that of Greater Antillean Grackle (92).

Calls: A *chuck* note and whistles are given. Both of these calls are given in alarm, or in the presence of predators, the whistle reserved mainly for aerial predators or when extremely excited. The *chuck* call is given to ground predators or when moderately disturbed.

DESCRIPTION Typical grackle shape with a wedge-shaped tail, but noticeably short-tailed for a grackle. The tail is graduated and the wedge-shape is more noticeable on adult males than the female or immatures. The males tend to walk with their chests puffed out and heads elevated, giving them less of a horizontal stance than other grackles. The short, but exaggerated wedge-shaped tail, long legs and upright posture gives Carib Grackle an appearance reminiscent of a black bantam rooster. This species is similar in size to Common Grackle (91). The bill is strong, although bill depth varies geographically, and is slightly hooked with the culmen showing a noticeable downcurve at the tip. The maxilla is almost straight. The bill is long, roughly equal to the length of the head. Wing Formula: P9 < P8 < P7 > P6; P9 ≈ P4; P8–P5 emarginate. The nominate form is described.

Adult Male (Definitive Basic): Plumage entirely black with purple gloss; greenish-blue rather than purple on the wings and tail. The bill and legs are black, the eye is a vivid whitish-yellow.

Adult Female (Definitive Basic): Like a dull version of the male without such an exaggerated wedge to the tail. The soft parts are as in the male. Females are brown with darker upperparts with any iridescence being limited to the wings, tail and back. The crown is grey-brown, continuing in this colour to the neck. The mantle, rump and uppertail-coverts are dark brown, darker than the crown. There is a pale (buffy-white) supercilium that contrasts with the darker lores and auriculars. A dull malar stripe is present. Throat off-white, becoming grey-buff on the breast while the flanks and crissum are a darker grey-brown. The wings are brownish with paler grey-brown coverts; the wing-linings are blackish-brown. The tail is blackish-brown with a dull green gloss. In all forms, the throat is paler than the rest of the underparts. The darkness of the plumage of females varies geographically, with *fortirostris* and *construsus* having blackish females. Female *construsus* shows a brownish-black head with a dull blue gloss, and lacks a noticeable pale supercilium. The neck is blackish-brown, paler than the mantle. Its mantle and rump are blackish with a dull blue iridescence. The throat is pale brown with a darker malar stripe, the breast and belly are slate-grey while the flanks, thighs, vent and undertail-coverts are blackish. The wings are blackish with a dull blue gloss; the coverts and wing-linings are blackish as well. The tail is blackish with a dull blue gloss.

Immature Males (First Basic): Similar to adult males, but show the body proportions of females, most notably a shorter tail, and appear smaller than adult males. In addition, the primaries of these individuals have a brown cast and the thighs may be obviously brown (D. Lemmon pers. comm.).

Immature Females (First Basic): Resemble adult females and may not be separable.

Juveniles (Juvenal): During the first few weeks after fledging, juveniles may show some indistinct streaking on the underparts, particularly the males. Juvenile females are brown, like adult females, only differing in their dark eyes, warmer brown colouration (sometimes showing an olive tone), loosely textured plumage and lack of iridescence on the upperparts. Juvenile males are dark brown, not warm brown as females, with dark eyes. Soon after fledging, juvenile males acquire black feathers, appearing as black patches on a brown background. The appearance of black in the plumage is linked to eye colour change, as patchy black males tend to show a pale eye, but not the white-yellow eye of adults at this stage. Juvenile females retain dark eyes much longer (for several months) than juvenile males (D. Lemmon pers. comm.).

GEOGRAPHIC VARIATION Eight subspecies are currently recognized. They fall into two groups, those with brownish females and those with blackish females. Two of the forms *(fortirostris and construsus)* have blackish females. However, the darkness of the females is highly variable in the brown forms as well.

Q. l. lugubris of Trinidad, and the South American mainland from Colombia east to NE Brazil. Males are black with a violet sheen throughout, more blue on the wing-coverts and green on the flight feathers. Females are brown, paler on the underparts, sometimes with a faint greenish gloss to the upperparts and wings. Juveniles appear to be just faintly streaked below, or not at all. Compare this to juvenile *luminosus*.

Q. l. guadelupensis from Montserrat, Guadeloupe, Marie-Galante, Dominica and Martinique. Males are coloured similarly to *inflexirostris*, but this form is smaller and has a shorter, thicker bill. The brownish females are even paler than *inflexirostris*, being pale brown on the head and whitish on the chin and throat. In fact, these are the palest females of the species. The birds from Guadeloupe have slightly longer wings.

Q. l. inflexirostris from St Lucia. Male similar to *fortirostris*. It has a shorter, stouter bill than *luminosus*. However, female most similar to *luminosus* in colour. The upperparts are brown, darker on the rump. Wings blackish with slightly paler, brown fringes. Tail blackish. The underparts are buffy-brown, becoming greyer on the flanks and crissum. The palest part of the underparts is the throat and chin. They show a slightly paler supercilium on the face.

Q. l. luminosus from Grenada and the Grenadines as well as Los Testigos Island, Venezuela. Male is similar to *guadelupensis* and *inflexirostris*, but has more extensive violet iridescence to the head and body. However, the bill is thinner than those forms. The bill is longer and more slender than similar *insularis*. The female is obviously darker than *inflexirostris* and *guadelupensis*, but is slightly paler than *fortirostris* and *insularis*. The upperparts are dark brown, and the face lacks a paler supercilium. The chin and throat are slightly paler than the head, and the breast and belly are brown. Juveniles of this form appear to be more densely streaked than the others.

Q. l. insularis from Margarita and Los Frailes Islands, Venezuela. Similar to nominate form, but larger, yet smaller than *luminosus*.

Q. l. orquillensis from Los Hermanos Islands, Venezuela. Like *insularis*, but with central tail feathers not glossed, the rest with a green iridescence.

Q. l. fortirostris from Barbados has been successfully introduced to Barbuda, Antigua and St Christopher Islands. This is the smallest Carib Grackle and has a short and stout bill, with a very gradually decurved culmen. It is black with a dull violet gloss on the body and head, the iridescence on the wings is greener. The bill, legs and feet are black. Females are similar to males, but smaller and not nearly as glossy. They are usually browner on the head and upper breast.

Q. l. contrusus is found on St Vincent Island. It is similar to *fortirostris*, except it is larger and has a longer bill. The female is similar to *fortirostris*, but is even blacker, making it the blackest female of the species. Bond (1928) collected a bird on St Vincent which was identified as *luminosus*, a form which was introduced to the Leeward district of St Vincent at one time. It is unclear how this introduction may have altered the appearance of *construsus*, if at all.

HABITAT Urban areas, agricultural land or other open habitats, sometimes at the edge of gallery forest. It forages on the ground and has an affinity for lawns and golf courses and other human-shaped landscapes such as hotel grounds and urban areas. It occurs in downtown Caracas, Venezuela and has become rather bold and urbanized, feeding largely on garbage and other discarded food. In addition, Carib Grackle is often found foraging on sandy or rocky beaches. This grackle is absent from contiguous forest. It appears to be expanding in South America as forests are being felled. It roosts in dense stands of trees, palms or mangroves, particularly if near water.

BEHAVIOUR Feeds on the ground, usually in small flocks. A highly gregarious species, it forms large, noisy, evening roosting flocks. Carib Grackles often become very tame and rather bold. They often jump up to forage on table tops of outdoor restaurants throughout their range. In Caracas, Venezuela, a small number of urban Carib Grackles, including around some of the larger shopping malls, have become extremely aggressive against humans often attacking as people approach (R. Palmitesta pers. comm.). While foraging, there are many subtle displays given, including a deliberate walk with drooped wings and a raised head that signals aggression. Males do not help the female in nest-building or incubation but they will feed the young as in Common Grackle. Males perform approximately 40% of the feeding of the young (D. Lemmon pers. comm.). In Venezuela, the mating system is monogamous, young are always raised by pairs. The high level of 'mate guarding' by the male implies that females may sometimes mate outside of the pair (D. Lemmon pers. comm.). However, polygyny has been reported from Trinidad (Ffrench 1991). Males spend most of their time displaying at the nest tree. The 'ruff-out (song spread) display' is like that of other grackles, the male ruffles all his feathers while singing, with the vibrating wings spread, tail cocked and fully spread into a wedge shape. Bond (1928a) described the males from Barbados as '...ruffling their feathers and adopting the most ridiculous attitudes when singing.'

NESTING Another colonial grackle, preferring to nest in palms or tall trees. Sometimes nests in low bushes or palms, particularly those over water. Ten or more nests may be built on one tree. The nest is a bulky structure, often reinforced with mud and placed in forks of branches or where branches meet the trunk of the nest tree. The clutch ranges from 2–4 eggs which are greenish-white, or pale blue, with darker scrawls. Occasionally, eggs are immaculate (D. Lemmon pers. comm.). In Trinidad, nesting occurs twice during the year, May–September (peaking June–July) and November–February (peaking in December). This latter breeding season is the less important of the two, with fewer birds nesting. Surinam Carib Grackles nest April–August (Haverschmidt 1968). On Barbados, these grackles nest during late December (Bond 1928a). St Lucia Carib Grackles begin nesting in April or May, as is the case for most other species on the island (Semper 1872, Bond 1928). Incubation lasts 12 days and the young fledge a further 14 (sometimes only 12) days later. This is a known host for Shiny Cowbird (102), however in St Lucia the parasitism rate is a low 2% of nests. In Venezuela, parasitism rates are low as is the success of the cowbird (D. Lemmon pers. comm.). This species reacts to experimentally placed foreign eggs by rejecting them, with up to 63% of females performing this behaviour the same probably happens to naturally occurring cowbird eggs (Post *et al.* 1990).

DISTRIBUTION AND STATUS Common to abundant. In the Lesser Antilles it inhabits the islands of Barbuda, Antigua, St Christopher, Montserrat, Guadeloupe, Marie-Galante, Dominica, Martinique, St Lucia, St Vincent, the Grenadines, Grenada, Los Testigos (Venezuela), Barbados, Los Hermanos (Venezuela), Margarita and Los Frailes (Venezuela) Islands as well as Trinidad and Tobago. The species was apparently introduced to Aruba, Netherlands Antilles (Voous 1983). In South America, it is found in the Llanos of E Colombia, east through most of northern Venezuela, except the extreme northwest, continuing east along the coastal slope including the Guianas, to the mouth of the Amazon river in Brazil. In Bolivar, Venezuela, the grackle is now found south of the Orinoco river. It appears that it may be slowly colonizing southward as habitat is opened up.

MOVEMENTS Sedentary, but there is a record of a stray female photographed on Bonaire, Netherlands Antilles on March 7, 1980 (Voous 1983). Records from Aruba, Netherlands Antilles have been previously assigned to vagrant birds, but Voous (1983) states that these were introduced to the island but doesn't comment further. Strays to the Netherland Antilles have apparently been *lugubris* of the South American mainland.

MOULT The definitive pre-basic moult is complete, pre-alternate moults appear to be lacking. The timing of the pre-basic moult differs depending on the timing of the breeding season. The form *fortirostris* (Barbados) moults in September. In northern Venezuela, *lugubris* is in moult in mid November (Wetmore 1939), but begins as early as October. In contrast, *lugubris* from Trinidad moults late May–late June. In Martinique (*guadelupensis*), the moult extends from September–October; apparently nesting occurs after the moult on this island (Pichot 1976). The race *inflexirostris* from St Lucia moults November–early December. The first pre-alternate moult is nearly complete, except for some juvenile tertials, and coverts. The primaries, secondaries and tail are almost always replaced entirely. The first pre-basic moult is timed similar to that of the definitive pre-basic. Specimens of *lugubris* are in first pre-basic moult during November.

MEASUREMENTS ***Q. l. lugubris***. Males: (n=25) wing 116.8 (111–122); (n=24) tail 104.4 (97–115); (n=10) culmen 27.7 (26–30); (n=21) tarsus 38.6 (32–41.0). Female: (n=30) wing 101.2 (97.0–107.0); (n=27) tail 87.3 (82.0–93.0); (n=5) culmen 24.6 (24–27); (n=30) tarsus 35.2 (33.8–36.6).

Q. l. guadelupensis. Males: (n=13) Length 247.0 (228.6–261.6); wing 122.4 (114–124.5); tail 100.4 (93.2–112); (n=9) culmen 30.0 (28.0–32.0); (n=13) tarsus 33.6 (30–35.3). Females: (n=13) Length 220.3 (199.7–241.3); wing 105.7 (99–109.2); tail 86.2 (81.0–91.0); (n=9) culmen 26.8 (25.9–27.9); (n=13) tarsus 31.3 (30–33.3).

Q. l. inflexirostris. Males: (n=6) wing 124.0 (119–127.8); tail 107.1 (103.6–110); culmen 30.9 (30–32); tarsus 34.1 (30–36). Females: (n=2) Length 235.0 (228.6–241.3); (n=6) wing 108.5 (105–111); (n=4) tail 99.5 (96–110); (n=6) culmen 26.9 (24–28.5); tarsus 31.7 (30–33.0).

Q. l. luminosus. Males: (n=2) Length 260.4 (246.4–274.3); (n=10) wing 122.2 (116–126); tail 109.2 (94–119); culmen 34.1 (33–35.8); tarsus 32.8 (32–36.3). Females: (n=1) Length 235.0; (n=9) wing 105.3 (103–109); tail 93.3 (88.9–97); culmen 29.5 (28–32); tarsus 30.7 (29.2–32).

Q. l. insularis. Males: (n=1) wing 125; tail 120; culmen 32. Females: (n=2) wing 102, 104; tail 81, 88.5; culmen 26, 27.3; (n=1) tarsus 29.

Q. l. orquillensis. Males: (n=?) wing 118; tail 83; culmen 27. Females: (n=?) wing 102; tail 83; culmen 23.

Q. l. fortirostris. Males: (n=16) wing 105.3 (101–110.0); tail 97.1 (91–107); culmen 27.8 (26–28); (n=10) tarsus 31.0 (28–33). Females: (n=10) Wing 93.8 (88–102); tail 84.6 (75.7–91); culmen 24.58 (23–26); tarsus 26.9 (22–30.0).

Q. l. construsus. Males: (n=10) wing 119.7 (111–129); tail 110.1 (98–121.5); (n=8) culmen 31.8 (30–33); (n=10) tarsus 32.6 (31–34). Females: (n=6) wing 98.3 (93.0–104); (n=5) tail 90.2 (86–93); culmen 27.0 (25–28); (n=4) tarsus 28.7 (28–30).

NOTES The two races with the blackish females (*fortirostris* and *construsus*) were previously classified as a different species, *Holoquiscalus* (= *Quiscalus*) *fortirostris*. Some authors have lumped Carib Grackle with the previous species, Greater Antillean Grackle. Successful introductions have been made to Barbuda, Antigua and St Christopher Islands. The new population in Tobago may be a naturally occurring one, but it has been published that these birds were introduced from Trinidad in 1905. The history of Tobago grackles is uncertain.

REFERENCES Belcher and Smooker 1937 [nesting, notes], Bond 1988 [voice, nesting, geographic variation], Diamond 1973 [habitat], ffrench 1986, 1991 [description, voice, behaviour, nesting], Hilty and Brown 1986 [habitat, voice], Lawrence 1878 [voice], Lemmon unpub. data [description, behaviour, nesting, measurements], Peters 1921 [geographic variation, measurements], Post *et al.* 1990 [nesting, brood parasitism], Ridgley and Tudor 1989 [movements], Ridgway 1902 [measurements], Voous 1983 [movements].

94 RUSTY BLACKBIRD *Euphagus carolinus* (Müller) 1776 **Plate 36**

A blackbird of wooded swamps, breeding in northern North America; most observers see this species only as a migrant, and in winter.

IDENTIFICATION Both sexes of Rusty Blackbird have yellow eyes which differentiate it from cowbirds and Red-winged Blackbird (61), both of which may look all black in some circumstances. Common Grackle (91) shares the black plumage and yellow eyes of Rusty Blackbird, but it is much larger and has a long and wedge-shaped tail. In addition, the grackle always has a blue or purplish iridescence to the head unlike the mainly greenish gloss of Rusty Blackbird.

Brewer's Blackbird (95) poses a more formidable identification challenge. The general shape and structure of Brewer's Blackbird is similar to that of Rusty Blackbird. Also, male Brewer's is black with a yellow eye like Rusty Blackbird. However, the bill of Rusty Blackbird is slightly thinner at the base, imparting a longer-billed look even though the true length is the same for the two species. Rusty Blackbirds show a more obvious terminal hook to the bill, but this is not consistent. In structure, the two are similar, but Rusty has a shorter tail and the tendency to crouch or walk with legs slightly bent. The tail of Rusty Blackbird is also proportionately shorter than that of Brewer's Blackbirds. Unlike Rusty Blackbird, Brewer's Blackbird has an elegant gait, recalling a Common Grackle. In breeding plumage, males of both species are black and iridescent. Male Rusty Blackbird has a dull greenish gloss, quite different from the boldly glossed plumage of Brewer's Blackbird, which has a largely blue body gloss with a contrasting violet head. Breeding females of these two species are also separable. Female Rusty Blackbird is a very grey bird, particularly on the rump which has a plumbeous tone. Brewer's Blackbird is greyish-brown with a paler head. Female Rusty Blackbirds have yellow eyes, while female Brewer's have dark eyes.

During the non-breeding season, the plumages of males of these two species may overlap in appearance, but not generally during the same time of year. Winter plumaged Rusty Blackbird (all ages) is boldly edged rufous, rusty and cinnamon on the body plumage, and the tertials are almost always fringed rufous. Male Brewer's Blackbird in fall and winter resemble summer males. However, some 1st-year males obtain a variable and sometimes extensive amount of paler edging to the upperparts, like a Rusty Blackbird. Note that the tertials are not fringed rufous. A winter *Euphagus* with rufous-fringed tertials is a Rusty Blackbird, but one that lacks these fringes may be either species. The extent of pale fringes is not as great as Rusty Blackbird; however, the pattern may appear superficially similar. On Rusty Blackbird, the edges on the upperpart feathers from the crown to the back are rusty or chestnut, while these are greyish-buff on Brewer's Blackbird. Note that Rusty Blackbird has contrasting black lores on a colourful face, while the lores of Brewer's Blackbird are a similar colour to the auriculars and thus do not contrast greatly with the rest of the face. As winter progresses, these edges are lost and Rusty Blackbirds appear much more like Brewer's Blackbirds. However, note that by this time of year (eg February) a Brewer's Blackbird may rarely still show some pale fringes. As winter progresses, the characteristic gloss pattern of Brewer's Blackbird is revealed. Brewer's Blackbird looks rustiest early in the season (fall) when the edges are widest; however, this is exactly the time when Rusty Blackbird is most colourful and distinctive.

Early juveniles of the Brewer's and Rusty Blackbirds may not be safely identified from each other. However, juvenile Rusty Blackbirds are only observed on the breeding grounds as they moult before migration, so this should only be a problem in areas where both species breed near each other. Early in the fall, immature Rusty Blackbirds may still show dark eyes, like a female Brewer's Blackbird. These dark-eyed immatures are rare and will have the rusty-fringed plumage that female Brewer's Blackbirds never show. These dark-eyed variants may also be confused with a 1st-fall male Red-winged Blackbird, but note the slimmer shape and thinner bill of Rusty Blackbird. Male Red-winged Blackbirds will show some red on the epaulets by this time of year.

VOICE **Song**: The males sings a squeaky but sweet, rising *kush-a-lee* or *chuck-la-weeeee* or *conk-ee*. While the pattern may be somewhat reminiscent of that of Red-winged Blackbird it lacks its harshness, and it is slightly higher in frequency. The other major song type does not rise in pitch. It begins with two or three musical notes followed by a harsher long note. Different song types may be alternated back to back in a long string. Females sing in the breeding grounds, often responding to male songs, however, the female song has not been adequately described.

Calls: The common call is a soft *chuck*, not as loud as Red-winged Blackbirds's call, but slightly rougher in tone. When alarmed a *chip* note is uttered. A downslurred whistle is sometimes given.

DESCRIPTION A slim, medium-sized blackbird with a proportionately short tail. The bill is thin, but does not taper to a sharp point, rather a small hook may be present at the bill tip. The bill is short, obviously shorter than the head. Wing Formula: $P9 < P8 > P7 > P6$; $P9 \approx P6$; P8–P6 emarginate.

Adult Male Breeding (Definitive Basic, worn): The bill is black and the eyes are yellow. The entire body plumage of the male is black with a dull greenish gloss. The gloss is only visible at close range, and in good light. Usually the male

Rusty Blackbird appears as a dull blackish bird. Legs and feet black. This plumage is acquired through wear of the fresh basic plumage.

Adult Male Non-Breeding (Definitive Basic, fresh): Soft parts as in breeding male. The body is entirely black, heavily edged and tipped chestnut or buff, such that the underlying black colour is largely hidden in places. The lores and area below the eye are always black, forming a noticeable black mask. The supercilium is buff, contrasting with the chestnut crown. The upperparts are boldly tipped chestnut, including the crown, nape, back and scapulars, such that these areas appear solidly chestnut or nearly so. The lower back, rump and uppertail-coverts are only lightly tipped, thus, showing a noticeable amount of black that contrasts with the chestnut mantle. The tertials are thinly, but noticeably fringed chestnut as are some of the wing-coverts. The underparts are similarly tipped paler, but the colour is more buffy rather than deep chestnut; this tipping is densest on the chin, throat and breast, and becomes more sparse posteriorly. The buff of the underparts may invade the face. The tail is entirely black.

Immature Male (First Basic): Similar to adult male, but the colourful edging and tipping averages wider and therefore is retained for a longer period of time. Early in the fall some of these young males may retain a dark eye or have honey coloured eyes rather than the gleaming lemon-yellow eye of the adults.

Adult Female Breeding (Definitive Basic, worn): Soft parts as in adult male, including the yellow eyes. A rather grey bird overall. The upperparts are mouse-grey, sometimes showing a dull greenish gloss. The mantle may have a brownish wash. Underparts and head greyish, slightly paler than the upperparts. The wings and tail are blackish grey.

Adult Female Non-Breeding (Definitive Basic, fresh): As in winter male, this plumage is widely veiled with colourful tips and edges throughout. The face is cinnamon to pale chestnut with a paler buff supercilium. The lores are greyish, not black as in male. The back is also cinnamon or chestnut, contrasting with the largely grey rump which is thinly tipped cinnamon. The underparts are pale cinnamon from the chin to the breast, becoming progressively greyer towards the belly and vent. The wings are greyish-black with chestnut or cinnamon tertial fringes and trim on the wing-coverts. The tail is blackish-grey.

Immature Female (First Basic): Similar to adult female. The wing-linings may retain duller juvenile feathers, but this cannot be seen in the field.

Juvenile (Juvenal): Largely dull grey washed with brown on the back and throat. The dark grey wing feathers are thinly edged brown.

GEOGRAPHICAL VARIATION Two subspecies are recognized.

E. c. carolinus is the subspecies found in the major portion of the range and is described above.

E. c. nigrans breeds in Newfoundland, Nova Scotia, the Magdalen Islands and possibly E New Brunswick. This subspecies has been recorded in North Carolina and Georgia during the winter. *E. c. nigrans* has darker rusty feather fringes in fresh basic plumage, and it is not safely identifiable in the field. In breeding plumage it is blacker with a slight blue gloss to the plumage.

HABITAT At all times of year this bird is closely tied to water, particularly non-flowing water; even in migration it searches out wet areas to feed in. Breeds in more open areas in coniferous, or mixed woodland, such openings are almost always wetlands. Rusty Blackbird will breed in forest along the shores of lakes, beaver ponds or wooded swamps, fens, muskeg, bogs, concentrations of conifer saplings, and less commonly on river shores. Occasionally they are known to nest in forest openings in drier sites such as at the edge of pastures (Avery 1995). During migration, Rusty Blackbird is one of the only North American blackbirds that may be found frequenting wooded swamps or lake shores in forest areas. However, they will also feed in open fields, livestock corrals and pastures, often with other blackbirds. Winters largely in or near wet wooded swamps or creek and pond edges, venturing to nearby agricultural fields.

BEHAVIOUR Usually observed in small flocks foraging on the ground; when alarmed these will fly to the nearest tree or bush and begin calling. Unlike many other North American blackbirds, Rusty Blackbird's diet is made up of a greater proportion of insects, perhaps explaining why it does not winter as far north as the other species. Also, unlike other northern blackbirds, Rusty is less inclined to flock with other species. Pure flocks of up to 1000 Rusty Blackbirds are sometimes encountered in the wintering grounds, but numbers are usually much smaller. The 'ruff-out (song-spread)' display is weak and could be classified as lacking in this species, or at best poorly developed. In full display, the song is given as the tail is spread and the bill elevated. The wings are not spread, as in a true 'ruff-out' display, but may be drooped. When giving *chuck* notes the tail is usually flicked upwards. Females may respond to the male song by singing from the nest. Rusty Blackbird is monogamous. While the female is incubating, the male provides her with food. He approaches with the food and lands on a conspicuous perch, to which the female flies and gives the stereotypic begging behaviour. When times are tough, particularly early in the spring if the blackbirds find themselves in the midst of a snowy cold snap, they become aggressive predators. On more than one occasion they have been observed to pursue and kill small songbirds and birds as big as Common Snipe (*Gallinago gallinago*)! After the prey has been quickly dispatched, the brain is eaten, leaving everything else intact. Perhaps the bill and legs of the Rusty Blackbirds are not strong enough to be able to gain access to the breast meat.

NESTING Breeding season April–July; however, eggs are present between early May–late June. Egg dates are slightly earlier in the southern part of the breeding range. After fledging, young Rusty Blackbirds form small flocks that wander around the breeding area, foraging and eventually moulting, before migrating in the fall. Rusty Blackbirds are solitary nesters except in Newfoundland and Labrador, where small colonies are not infrequent. Nests tend to be placed in a conifer usually over the water's surface, and they are particularly fond of the densest foliage of the conifer for the nest site. Nests may be less than one meter off the ground; more commonly they are built at 3 m and may be up to 7 m off the ground. Ground nests at the base of a shrub or tree have been described. The nest is a bulky cup of twigs, grass, lichen and moss, appearing too large for the size of the bird. Old nests of Rusty Blackbird are often chosen by Solitary Sandpiper (*Tringa solitaria*) to raise its brood, this sandpiper being an arboreal nester. Clutches commonly 4–5 eggs, which are light blue-green spotted with brown on the broader end. Incubation lasts 14 days and is the duty of the female. She begins sitting on the eggs after the first egg is laid, which gives rise to staggered hatching of the eggs and a hierarchy of nestling size in the nest. Nestlings need at least 12 days to fledge from the nest. The young are fed by both the female and the male. Brown-headed Cowbird (101) has been known to lay in the nest of Rusty Blackbird, but its a rare occurrence since the blackbird nests north of the main range of the cowbird.

DISTRIBUTION AND STATUS Common, but declining.

Breeding: Nests right up to treeline in the north: N Alaska, N Yukon, SE Northwest Territories, N Manitoba, N Ontario, N Quebec, central Labrador and Newfoundland; south to the Alaska panhandle; NE British Columbia throughout the northeast south to Prince Rupert, Prince George and highlands in the Okanagan; SC Alberta, C Manitoba, C and S Ontario (Algonquin highlands) and N Minnesota, N Wisconsin (very rare, perhaps more common historically, two recent breeding records), N Michigan, NE New York (Adirondack mountains), W Massachusetts (Berkshire Co.), N Vermont, N New Hampshire, C Maine, S New Brunswick and S Nova Scotia.

Non-Breeding: Almost entirely south of the breeding range, from SC British Columbia, S Alberta, S Saskatchewan, S Manitoba, S Ontario, C New York and southern New England south to E Colorado, SE Texas, Gulf coast and N Florida. It is uncommon in C Florida and very rare in the far south of the state (Stevenson and Anderson 1994). Scarce in the west of the wintering range, i.e. Colorado, Wyoming and Montana. The great majority of the population winters east of the Great Plains and south of the Dakotas, Minnesota, Wisconsin, Michigan, Ontario and New England.

MOVEMENTS A medium distance migrant. It migrates diurnally in small flocks. During the spring, it begins leaving wintering sites as early as mid February, arriving at the latitude of southern Ontario, southern Michigan and New Jersey in March with the final birds lingering into early May. In general, spring birds arrive as the snows begin to melt and the larger numbers arrive after the ground has thawed.

Fall migration begins in September and most do not reach the 49th parallel until late in the month, peaking in October. Numbers slow down in November, and only the rare straggler is left in December. Arrival in the southern wintering grounds begins in October and peaks in November. Most will stay in the winter quarters until February. Western breeders migrate in a southeasterly direction to their wintering grounds.

The banding data show a total of over 280 recoveries. The oldest individual was one banded in Arkansas that was recovered in Mississippi at the age of eight years and seven months (Klimkiewicz and Futcher 1987); four other birds are confirmed to have reached seven years old. Most were banded in the winter or in migration, and re-trapped in similar circumstances, as migrants or during winter. The probability of obtaining recoveries from the breeding grounds is small as these are rather remote. For the most part, movements appear to be north-south, lacking an appreciable east-west component. However, a large sample banded in Arkansas were recovered more commonly to the west and north (Oklahoma, Kansas, Nebraska, the Dakotas and Iowa) than to the east (no recoveries from Illinois, Wisconsin, Kentucky or Tennessee). Birds banded east of the Appalachians were re-trapped east of the mountains. Five birds recovered in Labrador and Nova Scotia were originally banded in Maryland (2), New Jersey (1), New York (1) and West Virginia (1). There are no recoveries of birds banded in the far west of the range; however, these birds likely show a more marked east-west component to their migration, purely from geographical grounds. The few recoveries from Saskatchewan and Manitoba show that some move largely north-south but that others do have a significant east-west component to their migration. Five recovered in Manitoba were originally banded in Maryland, Arkansas, Missouri and Illinois (2). One banded in Arkansas was recovered in Saskatchewan and two banded in Saskatchewan were recovered in North Dakota.

West of the Cascades and Sierras, it is a rare but regular migrant with most records occurring in the fall (October–December), peaking in November. The majority of western records are from southern California, particularly in the east, and Arizona. Some stay to winter in California. Unlike most vagrant species from eastern North America, Rusty Blackbird records are not concentrated at the immediate Pacific Coast. In many ways, Rusty Blackbird appears to behave as a regular, albeit rare, migrant in the west rather than as a vagrant. There are records for SW British Columbia where it is a rare but almost annual visitor to Vancouver and Victoria, primarily in the fall and winter. There is at least one winter record from the Queen Charlotte Islands. In Washington and Oregon there are over 20 records, again most are from the fall and winter. Many records are from E Washington, but there are coastal records in both states. In California, it is a rare but regular migrant, usually with less than 10 recorded per year. Most are in the fall, it is extremely rare in the spring. In Idaho, Nevada and Utah, it is a rare fall migrant and winter resident with few spring records. In Utah, there is at least one June record at Zion NP (Hayward 1976). Rusty Blackbird is very rare in fall and winter in Arizona and New Mexico, particularly in the southern parts of those states. There are at least four records from the Pribilof Islands, Alaska. In addition, there are four records from Bermuda and two reports from the Bahamas (Andros and Grand Bahama).

In the Western Palaearctic, there is a British specimen from Cardiff collected on 4th October 1881. This record is not recognized on the official British list; however, the date of occurrence and the fact that it arrived during a time of strong west winds lends credence to it being a genuine vagrant. There are records of this species landing on ships up to 400 km offshore off the New England coast, both during spring and fall. A summer record from 1938 in London is considered an escapee. It has also been recorded from Greenland. Apart from the Pribilof Island records mentioned above, there is a record from Siberia (Indian Point), as well as from St Lawrence Island in the Bering Sea.

MOULT Adults have one annual moult, the definitive pre-basic which occurs July–September, in both subspecies. Pre-alternate moults are lacking, therefore the black breeding plumage is obtained through wear. Juveniles undergo an incomplete or complete first pre-basic moult (July–September); most commonly at least, some juvenile underwing-coverts, and sometimes tertials, are retained. All of the rectrices and remiges are replaced during this moult. As in adults, there is a considerable change in appearance as the plumage wears and fades. All moults occur in the breeding grounds, previous to migration.

MEASUREMENTS Males slightly larger than females.

E. c. carolinus. Male: (n=10) wing 118.2 (114.3–126), tail 92.6 (86.1–100), culmen 20.7 (18.3–24), tarsus 31.0 (30.0–32). Female: (n=10) Wing 110.2 (106.9–116); tail 85.2 (79–95); culmen 19.7 (17.3–22); tarsus 30.3 (29–31.2).

E. c. nigrans. Males: (n=10) wing 117.1; tail 92.0; culmen 19.5 (18.1–24); (n=3) tarsus 29.7 (29–30). Females: (n=6) wing 107.3 (104–108.6); tail 82.5 (80–86); culmen 18.9 (17.5–21); (n=3) tarsus 28 (27–31).

NOTES While Brewer's and Rusty Blackbirds are thought to be quite closely related, they differ substantially in behaviour, distribution and habitat choice. No hybrids between these two species have ever been noted. Rusty Blackbird is showing an alarming population decrease. Both the Christmas bird counts (CBC) and the breeding bird survey (BBS) have detected significant declines throughout their range. The cause of this decline is unclear. Between 1966 and 1993, annual change in population size estimated by the BBS is –5.8%, similarly the change based on CBC data for 1959–1988 is –7.5%. In the southern fringe of the range (Ontario), there is evidence that historically the breeding range was more extensive. Greenberg and Droege (unpub. ms) have detected up to a 90% decline during the last three decades. They propose that the destruction of wooded swamps in the wintering grounds, acid rain in the breeding grounds and the harvesting of boreal forest as possible causes of this precipitous decline. Rusty Blackbirds are in trouble, but it is still early enough to do something!

REFERENCES Avery 1995 [voice, behaviour, nesting, distribution, movements], Bent 1958 [behaviour, nesting, voice], Burke and Jaramillo 1995 [identification], Burleigh and Peters 1948 [geographic variation, measurements], Greenberg and Drouge (pers. comm.) [notes], Flood 1987 [notes], Klimkiewicz and Futcher 1987 [age], McCaskie 1971 [movements], Pyle *et al.* 1987 [moult], Ridgway 1902 [measurements], Small 1994 [movements].

95 BREWER'S BLACKBIRD *Euphagus cyanocephalus* (Wagler) 1829 — Plate 36

A common icterid of the west, which fills a similar ecological niche to Common Grackle in the east. Where they are found together, Brewer's Blackbird tends to be a bird of open farmland.

IDENTIFICATION Brewer's Blackbird most closely resembles its close relative, Rusty Blackbird (94). Males of both species are glossy black with yellow eyes. While both species are quite similar in shape, they do differ. The bill of Brewer's Blackbird is slightly deeper at the base, making it appear shorter than that of Rusty Blackbird. Brewer's Blackbird also has longer legs and a longer tail. The longer legs give the bird a more elegant gait, more like Common Grackle (91) than Rusty Blackbird. Summer male Brewer's Blackbirds are much more glossy than Rusty Blackbirds. Rusty Blackbird has a dull greenish gloss throughout, unlike the blue body and violet head of Brewer's Blackbird. In winter, Rusty Blackbird is heavily tipped with rusty and chestnut on the upperparts, this can be approached by some 1st-winter (first basic) male Brewer's Blackbirds. Note however, that Brewer's Blackbird never has chestnut fringes to the tertials. The pale tips on the nape and back of Brewer's are buffy-grey, not the warm cinnamon or chestnut of Rusty Blackbird. As well, Rusty Blackbirds show a contrasting black mask, unlike Brewer's Blackbirds. For more detail see Rusty Blackbird account. Rusty Blackbird females are paler grey than Brewer's Females and have pale, not dark eyes.

Also similar to Brewer's Blackbird is Common Grackle which is black with pale eyes. The two species are often confused, even though they differ greatly in size and structure. The grackle has a long tail that is wedge-shaped and graduated, unlike the blackbird. The bill of the grackle is large and deep, appearing raptorial rather than the thinner bill of Brewer's Blackbird. Note also that western and northern Grackles have a bronze iridescence to the back which contrasts with a blue head. In the south and east, grackles have purple backs and blue or purple heads, in all plumages note that the blue or purple iridescence of the head stops abruptly in a straight line on the chest, unlike in Brewer's Blackbird. Common Grackles have strong, thick legs unlike the finer, longer-looking, legs of the blackbird.

Other largely black icterids may look all black. For example, the bicolored form of Red-winged Blackbird (61) often hides its epaulets completely. Brown-headed Cowbird (101) may not look to have a brown head from a distance. It is important to look for the distinctive iridescence pattern of the Brewer's Blackbird and its yellow eye. Shiny Cowbird (102), a recent arrival in North America, is very similar to Brewer's Blackbird in shape and pattern of iridescence. The cowbird has a dark eye, and any dark-eyed male Brewer's Blackbird should be carefully scrutinized to separate it from Shiny Cowbird. Adult Brewer's Blackbirds always have yellow eyes.

VOICE Song: The song, often given during the 'ruff-out' display is a faint and rather unattractive; there are two song types the *schlee* and *schrrup,* both sexes sing and display.

Calls: The common call which is also the flight call is a short *chak* or *chuk*, very similar to that of a Red-winged Blackbird. Primarily females also utter a *kit kit kit...* in aggressive situations. When alarmed it gives a whistled *teeeuuuu* or *sweeee.* Males perform a *chug chug chug* while giving the pre-copulatory display.

DESCRIPTION A slim blackbird of medium size with a relatively thin bill. The culmen and gonys are somewhat parallel and straight, curving distally and tapering to a point. The legs are long as is the tail. Wing Formula: P9 < P8 > P7 > P6; P9 ≈ P6; P8–P6 emarginate.

Adult Male (Definitive Basic): The bill is black and the eyes are gleaming yellow or whitish-yellow. The body is entirely black, but conspicuously glossed throughout. The gloss is strong and noticeable even in low light, but not at a distance. The body gloss is blue, or blue-green, commonly greenest on the wings. This blends into a violet-blue gloss on the head, neck and upper breast. The legs and feet are black.

Adult Female (Definitive Basic): The bill is black, and the eyes are dark brown. Note that in extremely rare cases female Brewer's Blackbirds may show a gleaming yellow eye (pers. obs.). This bird is largely brownish-grey and unstreaked. The upperparts are entirely brownish-grey, usually palest on the forehead. A slightly paler supercilium is present. The throat, face and chin are usually paler than the rest of the body, approaching greyish-buff. This becomes progressively darker towards the belly and vent which are solidly dark brown. The wings are blackish-brown, with a noticeable greenish gloss. The tail is blackish-brown. The feet and legs are black.

Immature Males (First Basic): Juveniles moult quite quickly out of their plumage to the immature, or first basic plumage. During the transition stage, while body moult is active, males look very patchy. The black of the immature plumage tends to show up first on the head, scapulars, coverts or breast. As the black feathers come in, the eye begins to lighten. On some birds the eye may start to lighten before any black feathers are visible, revealing the sex of that particular juvenile. On some immature birds, the black is conspicuously fringed brown. These fringes tend to be relatively thin, but may be extensive. On the head these fringes are dull brown, sometimes forming a noticeable supercilium. On the throat and breast, the fringes tend to be slightly greyer and they are a slightly more rusty on the back and coverts. These birds which have been termed 'fall variants' may appear surprisingly like Rusty Blackbirds, see Identification.

Juveniles (Juvenal): Similar to female, but plumage more evenly coloured, lacks supercilium, and lacks the darker belly. In colour, they are browner than adult females, particularly on the chest and lack all iridescence. The coverts and tertials are indistinctly fringed with pale cinnamon which may appear as two obscure wingbars when seen close up. The eyes are dark in both sexes, but lighten relatively quickly on young males. The bill and legs are blackish; the bill can be still growing on fully feathered juveniles, giving them an obviously different jizz to adult females.

GEOGRAPHIC VARIATION Monotypic. It has been divided into several forms by some authors, but these have not gained wide acceptance. For example, Oberholser (1974) resurrected the name *E.c.breweri* for the population found east of the Continental Divide. This form is apparently darker grey, less rufescent, on the upperparts in females, and more blue-green in males. Brewer's Blackbirds are variable throughout their range, and the appearance of *'breweri'* is well within the variation shown by the species elsewhere, and therefore this race has not been considered valid (Browning 1978). The coastal California population averages smaller than those from east of the Sierras. Oberholser (1974) included these birds in a race, *minusculus,* that spanned a breeding range from S Oregon to Baja California, Mexico. The form *minusculus* was described as being significantly smaller. Oberholser (1974) also recognized a form from the Pacific Northwest and the Great Basin, *aliastus,* which was described has having rather blue males and darker grey females than the nominate form. None of these subspecies are recently recognized, but they are listed here since they may describe some clinal differences in size and plumage which may be real.

HABITAT Brewer's Blackbird is a bird of open areas. It is found in agricultural fields, grasslands, pastures, alpine meadows, beaches and is very common in urban areas. In the east, where it is uncommon, it frequents bare, plowed fields more than other blackbirds. In the far northwest of its distribution it will breed in recent clear cuts within forest. Along the Pacific coast, Brewer's Blackbird commonly feeds along sandy beaches, freshwater pools on the beach and other marine environments, much more so than any sympatric blackbird. Brewer's Blackbirds need dense shrubs for nesting and is particularly fond of dense conifers or other thickets. Roosts in marshes or dense vegetation, usually conifers.

BEHAVIOUR At the colonies the male tends to perch in a prominent site near the nest. Both sexes feed nearby, but not necessarily at the colony. Particularly early in the nesting season, there is a surplus of unpaired males. The nest site is chosen by the female and she will defend and build the nest. The male will guard his mate from the advances of other males by keeping a careful watch on her activities. The 'ruff-out' display of the male is similar to that of the grackles, but not as dramatic. During the display, the tail is spread and the wings held slightly open, and while the body plumage is fluffed up the song is given. This display lasts for at most two seconds. Frequently, 'ruff-out' displays are given in a duet between the male and females, as in Common Grackle. The pre-copulatory display of the female is different, with the wings drooped and fluttered, while the tail is cocked as she gives a sharp note. The male may respond by giving a similar display, but with the bill pointing downward, and alternating it with 'ruff-out' displays, prior to copulation. Copulation is most frequent during the last two days of egg laying. Breeding tends to be synchronous in colonies, the degree of synchronization is related to how compact the colony is, rather than the colony size.

Pair formation is quick in migratory populations, but takes a longer time in resident ones. Pairs form in foraging flocks, away from the breeding area (Williams 1952). In California, the first evidence of pair formation may occur in late January, and no nests are constructed until late April! In these resident populations, pairs from the previous year often join up again for a second season. As in Common Grackle, sexual chases are a common element in pair-bonding. The male may dart at the female to begin the display or the female may initiate by using a peculiar butterfly-like flight, with slow, deep wingbeats. The male will chase the female, typically with at least one other male taking part, but will not catch up to her, even though this would be possible. Duets between male and female are also part of the pair formation and do not occur once nesting begins. Throughout the pairing process and once the pair bond is established, the male 'guards' his mate aggressively from other males. Even while the nest is being constructed the male accompanies his mate on trips to look for nesting material. However, a territory is not defended by the male.

Males are monogamous or polygynous, depending on the site and particularly, the sex ratio. Once the primary female begins incubating there is less of a need for the male to 'mate guard' and he begins to pay attention to other females. He may then mate and pair off again, the pair bond initiation is very quick with secondary females. He will then 'mate guard' his second mate until she is on a nest. Males rarely mate with more than two females per season. When the young hatch from the primary nest, the male will aid in feeding them, he is less likely to help a secondary female.

In a series of experiments where a wild flock of Brewer's Blackbirds were shown stuffed models it was clear that females tended to avoid the dummies during the breeding season. In the non-breeding season they sometimes acted aggressively towards a female-plumaged model. Males were always attracted to females and would attempt to mount them, even during the non-breeding season. However, sperm deposition and pre-copulatory displays only occurred during the breeding season (Howell and Bartholomew 1952, 1954).

Brewer's Blackbirds always forage on the ground. Stomach contents show that this species is much more insectivorous than cowbirds or Red-winged Blackbirds (Beecher 1951). They are also resourceful and have learnt new feeding strategies. In California, Brewer's Blackbirds commonly wait in parking lots for cars to arrive and they quickly rush in to feed on the insects encrusted on the front radiator grills of these cars. Groups of Brewer's Blackbirds at Simon Fraser University, Vancouver, Canada use outdoor halls with glass sides as foraging areas, insects collect on the inside of the glass and the blackbirds quickly snap them up throughout the day. Like Rusty Blackbirds and grackles, Brewer's Blackbird has been observed killing and feeding on other birds, Anthony (1923) reports two instances of them feeding on nestlings of other species.

NESTING Brewer's Blackbirds tend to be colonial. Nest 'clumping' appears to improve their foraging efficiency as well as avoidance of predators, particularly when a colony is in a large expanse of suitable nesting habitat. In eastern Washington, nesting commences from mid April–May. In the north of their range, nesting takes place May– July. The nest is a cup-shaped structure made of dry grass and fastened together with mud. It is placed in a variety of situations, preferred is a dense, low bush or conifer. Nests may also be placed on the ground. No territory is defended around the nest, and adjacent nests may be less than a meter from each other. Clutch size 4–7 eggs, but averages around 5. The eggs are variable in appearance, ranging from greyish to pale green with dark spots and blotches. Other eggs are much darker, due to more extensive dark spotting. The nest construction and incubation is the job of the female, incubation lasts 12 days. The young need 13 days to fledge, and both sexes tend to feed the young. Females usually breed only once in a season, but in some areas two broods per summer are common. However, if the first nest is destroyed, the female will re-nest. Brown-headed Cowbird will parasitise nests of this species and they are known to fledge from their nests (Friedmann and Kiff 1985).

DISTRIBUTION AND STATUS Breeding: The northern range of the breeding distribution extends from British Columbia throughout the south and north to the Chilcotin and Peace River Districts; east through C Alberta, C Saskatchewan, SC Manitoba, W Ontario (Rainy river) and along the north shore of Lakes Superior and Huron, and locally in SW Ontario. South from here, Brewer's Blackbird extends its range to the Pacific coast and east from SW Ontario, S Michigan, N Illinois, S Minnesota, C South Dakota; south along the edge of the Great Plains to the western Oklahoma panhandle and NW Texas panhandle. The southern border of the breeding range extends from N Baja California, Mexico west of the Colorado river in California to S Nevada, SW Utah, NC and WC Arizona, NW and N New Mexico to W Oklahoma.

Non-Breeding: The birds of the Pacific coast from SW British Columbia to N Baja California are year-round residents. Other than the coastal populations, the winter range extends north from W Washington, to SW Idaho, S Montana and SW Nebraska east to the Atlantic coast roughly around the latitude of Chesapeake Bay. In Florida, it is most common in the north and panhandle, rarer but regular farther south. The southern extent of the range lies from S Baja California Sur, Mexico and on the mainland from Jalisco and C Michoacán, to NW Oaxaca, Puebla and N Veracruz; in the US to the Gulf coast and S Florida.

MOVEMENTS Sedentary to medium-distance migrant. It is much more regular in the east during winter, but the provenance of these eastern wintering birds is not known. There are no recoveries of birds banded in, or recovered from, those eastern wintering populations. However, Stepney (1975) suggests that eastern wintering populations are most likely composed of eastern breeders, as the abundance in historical sites has not increased as more eastern breeding populations were established. Females appear to migrate longer distances than males, as northern wintering flocks are largely composed of males.

There are over 1030 recoveries of banded Brewer's Blackbirds. In California, west of the Sierras, Brewer's Blackbirds are mostly sedentary. Almost all recoveries come from the general region they were banded in, throughout the year. A few birds banded in California during winter were recovered in Oregon and Washington, thus there is some movement from further north into California during winter. One Brewer's Blackbird banded in spring in the Sacramento valley was recovered several years later in Idaho, thus some birds do cross the Cascade/Sierra mountains. There is also some interchange between Washington and British Columbia. East of the Continental Divide Brewer's Blackbirds are much more migratory. For the most part, birds breeding in the Prairies/Great Plains and Great Basin move southward without a distinct east-west component. There are few recoveries of banded birds from Alberta, Saskatchewan and Manitoba. However, the following recoveries are of interest: a Texas and a Minnesota recovery of Alberta-banded individuals, two Texas recoveries of Saskatchewan-banded birds and a Minnesota recovery of a Manitoba-banded bird. Distant recoveries of birds banded in the Dakotas include two caught later in Saskatchewan, as well as birds that wintered in Louisiana and Texas. Two birds banded in Colorado are listed as being recovered from Mexico, one from Guanajuato and the other from Jalisco; these are the most distant recoveries of any banded Brewer's Blackbirds. Recoveries of birds breeding in the Midwestern states and Ontario again show that wintering sites are mainly to the south, without a significant east-west component. For example, Illinois breeders were recovered in Louisiana and Mississippi, Iowa breeders wintering in Arkansas, Minnesota breeders wintering in Arkansas, Wisconsin birds wintering in Mississippi, and an Ontario breeder recovered in Mississippi (BBL Data). The oldest recorded Brewer's Blackbird was one which reached the age of 12 years and 6 months (Klimkiewicz 1997). As well, a male caught and re-trapped in Carmel, California, was 11 years old.

The distribution of Brewer's Blackbird has undergone a change in the last century, with a significant eastward expansion. Historically, the easternmost breeding area was the Red River valley of western Minnesota (Stepney and Power 1973). During the last 80 years the breeding distribution has crept eastward in the Great Lakes region, taking advantage of suitable habitat created by the felling of forests. In the late 1960s and early 1970s, Brewer's Blackbird reached southern Ontario, east of the Georgian Bay section of Lake Huron (Stepney and Power 1973). Their expansion continues in that region. The winter range has also expanded, both to the east and to the north. In the early 1950s, almost no Brewer's Blackbirds wintered east of the Mississippi river, and now there are concentrations found all the way east to W North and South Carolina, NW Georgia and N Florida (Stepney 1975). In Florida, the species was historically unknown, but now is a regular and locally common wintering resident in the north (Stevenson and Anderson 1994). In addition, the abundance of wintering Brewer's Blackbirds in the northern Great Basin and the Mountain states, for example Idaho, Utah and W Colorado has increased (Stepney 1975).

Brewer's Blackbird has been recorded as a vagrant well north of its range in Alaska, as far north as Barrow. They are now regular in southernmost Alaska (Ketchikan), and have wintered. It is accidental in the Northwest Territories, where it has been recorded from Baker Lake, Keewatin District, Northwest Territories. There is one record from the Yukon Territory, a pair was observed near Upper Liard on May 29, 1996.

It regularly wanders to the Queen Charlotte Islands, British Columbia. In Ontario, there is a vagrant record from Moosonee, May 16, 1992; Moosonee is near the shore of James Bay. South of its regular range it has been recorded from S Oaxaca, Mexico, and once in W Guatemala. The Guatemalan record was collected at Hacienda Chancol, near the border with Chiapas, Mexico

(Griscom 1932). There are records from Isla Guadalupe, off the coast of Baja California Norte.

It is difficult to properly address the question of vagrancy of Brewer's Blackbirds in eastern North America partially due to their recent change in status there and also because this species can be quite local in regular wintering areas. Brewer's Blackbird has occurred at least once in all of the eastern states and provinces except Newfoundland. It is probably correct to say that south of Virginia and North Carolina, this species winters regularly, often in numbers, farther north it is a very rare but regular migrant mainly in fall and sometimes winter. As far north as Massachusetts, Brewer's Blackbird is reported annually in fall. As one would expect, Brewer's Blackbird becomes rarer as one proceeds north. Records exist on the Atlantic Coast as far north as New Brunswick and Nova Scotia. It is a regular winter visitor to Florida, as far as the extreme south of the peninsula. One vagrant record exists from Grand Bahama Island.

MOULT The complete definitive pre-basic moult of adults occurs after breeding during July to August, and takes place near the breeding grounds. The pre-alternate moult is missing in all age and sex classes. Therefore any changes in appearance between winter and summer plumages occurs as an effect of wear and fading. Juveniles lose the juvenal plumage through the first pre-basic moult (July–September); this moult can be complete but usually a few underwing-coverts (or tertials) are retained from the juvenal plumage. The eye colour change of young males takes place at a roughly similar time as the first pre-basic moult. This moult also takes place near the breeding grounds.

MEASUREMENTS Males: (n=10) wing 127.7 (120.1–133.9); tail 97.8 (91–107.2); culmen 22.1 (20–23.6); tarsus 32.2 (30.5–33). Females: (n=10) wing 117.7 (115.8–119.6); tail 88.6 (86–92.7); culmen 20.1 (19.1–20.8); tarsus 30.5 (28.7–32.0).

NOTES Data from the breeding bird survey (BBS) shows that between 1966 and 1994, there was a significant negative trend in populations. The North American population of Brewer's Blackbird appeared to decrease at a rate of 1.9% per year during this time, this decrease occurred largely between 1980 and 1994 rather than prior to 1980. The Christmas bird count (CBC) data did not detect a significant decline in North America between 1959 and 1988, but the direction of the trend was negative. In British Columbia and Colorado, there were significant declines in sightings of Brewer's Blackbirds during Christmas bird counts in this period. The age-related change in eye colour of Brewer's Blackbird occurs by the formation of reflective cells in an iris initially laden with poorly developed pigment cells (Hudon and Muir 1996). The females have a much smaller number of reflective cells, and these are masked by the presence of melanosomes, which are missing in the male eye (Hudon and Muir 1996).

REFERENCES BBS Data 1995 [populations, notes], CBC Data 1995 [populations, notes], Godfrey 1979 [movements], Horn 1968, 1970 [nesting, behaviour], Howell and Bartholomew 1952, 1954 [behaviour], Klimkiewicz and Futcher 1987 [age], Pyle *et al.* 1987 [moult], Ridgway 1902 [measurements], Williams 1952 [voice, behaviour, nesting].

96 BAYWING (COWBIRD) *Agelaioides (Molothrus) badius* (Vieillot) 1819 Plate 37

A common inhabitant in open woodlands of southern South America, often seen in small groups. Also known as Bay-winged Cowbird, this name is misleading since this species is only distantly related to the true cowbirds. In many ways Baywing is behaviourally more similar to Chopi Blackbird (79) and perhaps Bolivian Blackbird (80).

IDENTIFICATION Unmistakable, no other adult icterid is greyish with contrasting black lores and bright rufous edges to the flight feathers. However, juvenile Screaming Cowbird (97) has a mimetic plumage, an exact replica of Baywing's distinctive pattern. In terms of visible plumage features, juvenile Screaming Cowbird is almost indistinguishable from Baywing save for its dark axillaries and wing-linings which are only visible, with great difficulty, in flight. Screaming Cowbird, like all true cowbirds, has rather long primaries which stand out as a long extension of the primaries past the tertials on the folded wings. Baywing is a rather short-winged icterid, thus its primary extension is significantly shorter than that of juvenile Screaming Cowbird, but bear in mind that very young Screaming Cowbirds may still be growing their primaries. The bill of Screaming Cowbird is slightly thicker than that of Baywing. Baywings show a largely straight culmen, but Screaming Cowbird often demonstrates a slight bulge at the base of the culmen; in addition, its lower mandible is thicker than that of Baywing. Finally, male Screaming Cowbirds are substantially larger than Baywings of either sex. If a juvenile Baywing looks oversize for its parents it is almost certainly a male Screaming Cowbird. Female and immature Unicolored Blackbirds (57)show rufous edges to the wing feathers like Baywings. However, they are yellowish on the underparts and lack the blackish lores of Baywing. In addition, Unicolored Blackbirds are usually seen near marshes. Baywings are seldom found far from open woodlands, their preferred habitat. Female Shiny Cowbirds and Chestnut-capped Blackbirds are largely greyish-brown or olive-brown but lack the dark mask and rufous edges on the wings. In flight, the rufous of the wings is usually noticeable. Another feature that identifies Baywing is the blackish tail which contrasts with the paler grey body.

VOICE Song: The song is given by both sexes and is a lazy, but pleasing, series of whistles on different pitches. The song is continuous, rather than having a set length. Argentinians call Baywings the 'musico' or musician, since their songs sound like a group of musicians (an orchestra) tuning up. In fact, they sound like musicians that never really get in tune, due to their reedy and 'off-key' notes. Baywing's whistles are never clear and pure. The individual notes tend to be buzzy whistles with a harmonic structure that makes them sound hollow, usually decreasing slightly in pitch or remaining at one pitch. The notes last about a quarter to a third of a second and are delivered quickly, with the inter-note period being slightly shorter than the whistles. However, consecutive notes are not given at the same pitch, but rise and fall in a seemingly random manner which gives their songs the characteristic sound. Sometimes a harsher buzz is weaved into the song, and commonly sharp, quick, metallic *chip* or *chik* notes are included which disrupt the rhythm of the song. When more than one Baywing is in song, individuals commonly sing in unison and the overall effect is not displeasing, but it is a cacophony.

Calls: The primary call is a deep and low-pitched (2–4 kHz) *chuck* which may be given on many occasions including between phrases; this call is also given while in flight and in general states of concern. Two other calls are given, a higher *chuck* as an alarm, and a clear, whistled *peeeooh* which is rarely heard.

DESCRIPTION Sexually monomorphic, with sexes impossible to tell apart other than by behaviour, as well as the presence of the cloacal protuberance in males or presence of brood patch in females. Baywing is cowbird-like in shape. The bill is short and stout, the body stocky, with a tail of moderate length. Unlike cowbirds, Baywing has short, rounded wings and is a weak flyer. The short wings show a short primary projection on a perching bird. Wing Formula: P9 < P8 < P7 > P6; P9 ≈ P3; P8–P5 emarginate. The form *A.b.badius* is described.

Adult (Definitive Basic): Mostly brownish-grey with contrasting black lores, creating a masked appearance. Baywings often forage in manure which stains the head and underparts an olive colour. The background colour of the wings and tail is blackish-brown with bold rufous fringes, making the wings appear almost completely rufous. The tertials are the only wing feathers where the dark centers are obvious. When fresh, the outermost tail feathers may show thin rufous edges. The bill is charcoal-black, as are the insides of the mouth. The eyes are dark brown and the legs and feet are black.

Juvenile (Juvenal): Similar to adult, but differs in several ways. The lores of the juvenile are not nearly as dark as the adult's, thus they lack the obviously masked look. Juveniles have noticeable rufous edges on the outer tail feathers. Some juveniles are obscurely streaked on the underparts. The eyes of juveniles may look grayish and the bill colour is duller than adults, but both darken as the bird ages. Juveniles quickly become indistinguishable from adults. One of the last immature features to be retained is the pinkish insides of the mouth!

GEOGRAPHIC VARIATION The form *fringillarius*, Pale Baywing (96X), is given its own account below. It is perhaps best treated as a different species from Baywing, but it is generally considered conspecific (Ridgely and Tudor 1989). Otherwise, two subspecies have been named, differing mainly in size and colour saturation.

A. b. badius refers to the subspecies found through most of the range of the species, it is described above.

The other race, ***A. b. bolivianus***, is found in the highlands of Bolivia in La Paz, Cochabamba, Santa Cruz and Chuquisaca to 3350 m. This race probably extends south into the highlands of Salta and Jujuy. The form *bolivianus* is larger and greyer than the nominate subspecies, but is probably not identifiable in the field other than by range.

HABITAT A bird of edge or open woodlands, particularly those with thorny trees and often those which are deciduous in nature. It is primarily a species of dry (xeric) woodlands, or those that experience a marked dry season during part of the year. It nests in open woodlands such as those of the Talares of the Argentine pampas, composed largely of hackberry (*Celtis tala*); chaco woodlands; monte desert in Argentina; and to some extent the northernmost and most chaco-like parts of the Patagonian steppe region. It may also be found in disturbed areas such as urban parks (i.e. parks of Buenos Aires), and agricultural areas particularly in tall trees, often planted, near buildings and estancia or ranch headquarters. In the open grasslands of the pampas, these introduced trees are usually the only available habitat for this species; thus it is likely human influence has locally expanded Baywing's range. Also a species of dry, intermontane valleys in the Andes, being found in brushy sites areas mixed with *Trichocereus* cactus as well as isolated trees in grassy fields. Feeds both in trees, gleaning for insects, and on the ground, often associating with cowbirds or other blackbirds. When feeding in open fields it tends to stay near the periphery, closer to trees than the true cowbirds. The presence of open fields is not a prerequisite for this cowbird, but small openings in woodlands are.

BEHAVIOUR Typically found in small groups, this is an extremely social blackbird. Baywings are observed as single birds or pairs only near the nest, particularly during the early stages of breeding, otherwise they occur in small flocks. Typical flock sizes vary from approximately four to 20 or more, the flocks are smaller in the summer than in winter. In the winter, the average flock numbers approximately 21 birds (Fraga 1991). Since this species is sedentary, flock sizes do not vary substantially between season. In winter Baywings roost communally, and there is some evidence that there is a minor influx of birds from further afield in these winter roosts (Fraga 1991).

Baywings are monogamous, with pairs breeding together for as much as two consecutive seasons (Fraga 1991). However, low levels of polyandry have been detected (Fraga 1972, 1991). The pairs are active around the nest, sometimes choosing the nesting site and guarding it for weeks before they lay an egg. After the eggs are laid and the young hatch, 'helpers' arrive at the nest to help in feeding and guarding of the nest and young. Females may sometimes copulate with 'helpers' during the nestling period, which is after there is any chance of the female laying eggs. Baywings are not double-brooded. In at least two cases during subsequent seasons, females nested with 'helpers' they had previously mated with (Fraga 1991).

Baywings sing often, but do not use a 'ruff out', or 'song spread' display to accompany the song. Singing typically occurs from a perch, often right at the nest. Both sexes sing, but males sing much more often than females. Allopreening behaviour is often observed, where two or more individuals will preen each other. Another common display is 'bill pointing' as an aggressive signal; a display present in most of the blackbirds. In addition, Baywings perform a fascinating communal display that has been suggested to be a group territorial display (Fraga 1991), although other explanations are possible. Up to 25 birds may gather at a site where most will begin to sing. Some of these individuals pick up a stick or leaf and strut around on the ground giving a 'bill pointing' display while holding the leaf or stick. Other birds watch and sing from low bushes, many of them also holding sticks in their bills. Fighting sometimes erupts during these displays, following which the birds will split up into smaller groups (Fraga 1991, Jaramillo pers. obs.).

NESTING Baywings are not brood parasites like the true cowbirds, however they have been known to pirate the nest of other species, driving the rightful owners away and eventually laying their own clutch in the nest (nest parasitism). Most nests used by Baywings are old and there is no confrontation with the species that built the nest. However, there are documented records of Baywings fighting for nest occupation, perhaps when nest sites are limiting. In Cochabamba, Bolivia, Baywing has been reported to sometimes kill Rufous Horneros (*Furnarius rufus*) in order to gain possession of their nests, but these may be exaggerations (Miller 1917). The preferred nests are covered nests: either stick nests including those of Firewood Gathererer (*Anumbius annumbi*), thornbirds (*Phacellodomus* sp.), canasteros (*Asthenes* sp.); grass nests of Great Kiskadees (*Pitangus sulphuratus*); mud nests like those of Rufous Hornero; or woodpecker cavities (*Colaptes* sp.). They will also use nestboxes or may even build their own cup-shaped nest.

Baywings are late nesters throughout their range. In Buenos Aires, the peak breeding period is late November–early January but breeding may begin in late September. Nesting in the northwest of Argentina (Salta) is later, beginning in January and peaking February–March. Clutch size tends to be 4–5 eggs; the eggs are white, cream, grey or pale blue with darker spots. Accurate clutch sizes are difficult to ascertain due to the large number of Screaming Cowbird eggs that are laid in Baywing nests, and because the brood parasite tends to remove eggs from the nest.

Nesting Baywings do not appear to defend an exclusive territory and nests of different pairs may be only a few meters apart. The breeding system is monogamous, but few pairs retain the pair bond between seasons. However, they are cooperative breeders. One or more 'helpers' will assist the pair in feeding and raising the brood. 'Helpers' are at least one year-olds and tend to be males, although some individuals will 'nest help' if their own nest has been unsuccessful. In the early stages of nesting, 'helpers' are not tolerated near the nest, but as the young hatch, 'helpers' begin to be accepted at the nest. During the pre-laying and incubation period, the nests are ferociously guarded against Screaming Cowbirds, the main enemy (nest parasite) of Baywings. Shiny Cowbirds (102) also parasite Baywings, but uncommonly. Nest defense is carried out by the male. As the nests are closed and domed, there is often only one entrance, and the male will perch in front or just above the entrance, to prevent access by Screaming Cowbirds. The male tends to sit tight, only chasing the parasites when they approach within a meter or so. It appears that male Screaming Cowbirds may help females in egg-laying by distracting the male Baywing.

As a juvenile, Screaming Cowbird is a mimic of Baywing. Nestlings of both species are nearly impossible to identify. The similarities between juveniles of these two species are uncanny. Young Screaming Cowbirds only moult to black adult plumage after becoming independent. No Shiny Cowbird has been observed to be raised to independence by Baywings. It appears that once the Shiny Cowbird fledges, it ceases to be fed, probably due to its non Baywing-like appearance. This may be what drives the mimicry by Screaming Cowbird, but this needs to be confirmed. Baywings suffer a great decrease in reproductive success through parasitism by Screaming Cowbirds. Parasitism rates are typically 100%, and many nests have more than one cowbird egg (multiple parasitism). Apart from the nest defense technique mentioned above, Baywings try to 'fool' Screaming Cowbirds by being unpredictable in their egg-laying. behaviour Most bird species will lay a few days after lining the nest; Baywings on the other hand may begin egg-laying immediately, or delay for up to two weeks after the nest lining has been completed. In any case, several behavioural characteristics make Baywings an ideal host for Screaming Cowbirds. These include nesting in safe, domed nests, the use of 'helpers' at the nest, nests are obvious and easy to locate, the aggressive nature of the parents in fledgling defense and the fact that Baywings are adept at removing ectoparasitic botflies (*Philornis*) from their nestlings, including parasitic nestlings; in many other local bird species botflies are a leading cause of nestling mortality.

DISTRIBUTION AND STATUS Baywing is common throughout its range, being more local in the highlands of Bolivia and Salta, Argentina. The Andean population (*bolivianus*) ranges to 2880 m;

lowland *badius* is rarely found to 1000 m. It is found in the Bolivian Andes and chaco, being recorded from all of the Bolivian departments except Oruro, Potosi and Pando, as it is absent from the Altiplano and lowland tropical forest. The Bolivian highland populations are separated from lowland populations by moist yungas forest. In Brazil, it is found in S Mato Grosso as well as in Rio Grande do Sul. There have been recent reports of this species from São Paulo, Brazil (Ridgely and Tudor 1989). Present in Paraguay throughout the country west of the Río Paraguay but restricted to the west half of the region east of the Río Paraguay. Baywing is widespread in Uruguay. Recorded in N and E Argentina, west to the intermontane valleys of Salta and the Andean foothills of Jujuy, Tucumán, Catamarca, La Rioja, San Juan, Mendoza and south to C Rio Negro and N Chubut.

MOVEMENTS Extremely sedentary. A long term study of banded birds found that Baywings disperse to areas mainly within 1 km of their birth place. It appears that females may disperse further than males, many of which remain close to where they were born and become helpers. One retrapped Baywing was found to be at least six years old. A specimen exists from Curicó, Chile, from December 1923, but the original label lists this as an introduced species. It appears that an unsuccessful introduction of this species may have been attempted in central Chile. The history of this specimen and any population introduction is unclear. However, it is highly unlikely that Baywings could have crossed the Andes by natural means.

MOULT The definitive pre-basic moult is complete and occurs after breeding, during March–April, and sometimes finishing in early May. The pre-alternate moult is absent, wear reduces the rufous fringes of the wings during the winter. The first pre-basic is a complete moult. After the first pre-basic, immatures are basically indistinguishable from adults. Young birds may retain a pinkish cast to the mouth lining, it is black in adults.

MEASUREMENTS Females and males are roughly the same size in the nominate form.

M. b. badius. Both sexes: (n=60) wing 89.0 (75–98); culmen 17.8 (14.5–18.6); (n=26) tarsus 25.6 (23.3–28.8). Males: (n=12) tail 78.2 (72–85). Females: (n=3) tail 72.7 (71–74). From Gyldenstolpe 1945b: Males (n=6) wing (86–89); tail (66–69); culmen (17–18); tarsus (22–23). Females, (n=5) wing (85–89); tail (66–73); culmen (16–17); tarsus (23–24).

M. b. bolivianus. Males (n=3): wing 98.7 (96–101); tail 82.5 (79.5–85); (n=2) culmen 19.0, 20; (n=1) tarsus 27. (n=22) wing (97–105); tail (74–80); culmen (18–21); tarsus (23–26) (Gyldenstolpe 1945b). Females (n=3): wing 94 (92–96); tail 78 (76–80); (n=2) culmen 18, 18.5; (n=1) tarsus 25. (n=13) wing (91–101); tail (71–79); culmen (17–21); tarsus (23–26) (Gyldenstolpe 1945b).

NOTES In captivity, Baywing has successfully hybridised with Chestnut-capped Blackbird (60). A female Baywing mated with a male Chestnut-capped Blackbird produced four clutches of eggs over two years. All nests, except one, failed due to predation. The hybrid young were curiously described to be dirty brown with white wings (Shore-Baily 1928).

Recent molecular work (Lanyon 1992) on this species has concluded that Baywing is not a close relative of the other cowbirds, which is consistent with differences in behaviour, vocalizations and colouration. The English name Bay-winged Cowbird thus seems largely inappropriate and we have chosen to use the name Baywing for this peculiar blackbird.

REFERENCES Araya and Millie 1986 [movements], Fraga 1983b [nesting], 1986 [behaviour, nesting], 1991 [behaviour, nesting], Friedmann 1929 [behaviour, moult, measurements], Hoy and Ottow 1964 [nesting], Jaramillo unpub. data [measurements], Miller 1917 [nesting], Ridgely and Tutor 1989 [distribution], Shore-Baily 1928 [hybrids].

96X 'PALE BAYWING' *Agelaioides (Molothrus) badius fringillarius* — Plate 37

This form is usually considered a subspecies of Baywing (96). It differs from Baywing in appearance, and may be best considered a different species.

IDENTIFICATION Resembles Baywing in pattern almost exactly, but is much paler in colour. The rufous feather fringes of Baywing are replaced by washed out buffy-rufous tones. The dark mask is more extensive on 'Pale Baywing' than on Baywing. This species also differs from Baywing in the lack of grey tones on the underparts, and thus appears much buffier. 'Pale Baywing' differs from Baywing in its vocalisations (J. M. Barnett pers. comm.). It is allopatric with Screaming Cowbird (97), thus the similarly plumaged juvenile of that species is not an identification problem (see Baywing Identification). Sympatric species most likely to be confused with 'Pale Baywing' are female Shiny Cowbird (102) and Chestnut-capped Blackbird (60) as both of these are dull and brownish. However, both of these species lack the dark mask of 'Pale Baywing' as well as the buffy-rufous wing edging. Both the cowbird and Chestnut-capped Blackbird are darker than 'Pale Baywing', and the blackbird is obscurely streaked. Note also that Shiny Cowbird tends to show a darker malar stripe which is absent in Pale Baywing.

VOICE Some of its calls are similar to that of typical Baywing, but on the whole the vocalisations are quite different (J.M. Barnett, pers. comm.). The vocalisations are described as being quite expressive in this form (R. Otoch pers. comm.). Recordings of birds from northwest Brazil sound quite different from *badius* Baywing vocalisations. A single individual gave sharp 'tweep' calls interspersed with a short 'trrp' rattle (B. Whitney recordings). The rattle is reminiscent of the begging calls of juvenile *badius* Baywings. A group which was recorded (B. Whitney recordings) gave a variety of vocalisations most of them sounding metallic or squeaky, as opposed to the sweeter vocalisations of *badius* Baywings. The rate of note delivery appears to be substantially faster in Pale Baywing when compared to the slower and mellower delivery of typical Baywings.

Calls: a harsh *chek-chek*, faster when alarmed. Also utters a four note *chek-chek-tchem-tchem*, the two final notes much weaker than the first two (R. Otoch pers. comm.).

DESCRIPTION A stocky, short-tailed and thick-billed icterid. The bill has a slight curve to the culmen, and appears finch-like as it is short. The culmen may be just slightly longer than the gonys, showing up as a slight 'overbite' or hook. The sexes are alike. P9 < P8 ≈ P7 > P6; P9 ≈ P6; P8–P5 emarginate.

Adult (Definitive Basic): The bill, legs and feet are black. The crown and back is sandy-brown, becoming paler on the rump and uppertail-coverts. The lores are black and continue as a blackish mask, darkest on the lores and below the eye, but turning blackish-grey on the auriculars. The throat and underparts are paler than the upperparts. They are sandy-buff, palest on the throat and belly, and darkest on the breast. The wings are brown with wide cinnamon edges to the feathers. The lesser coverts are sandy-brown, like the back, and lack cinnamon. The median and greater coverts are widely edged cinnamon. The primaries, secondaries and tertials are also widely edged cinnamon, such that the wings look largely cinnamon when folded. The tertials have large brown centers, which are visible on the folded wing. The wing-linings are sandy-grey. The tail is brown.

Juvenile (Juvenal): Not known, presumably differs from adult in a similar manner to Baywing.

GEOGRAPHIC VARIATION Monotypic.

HABITAT Open caatinga woodlands, shrubland and agricultural land. Prefers open areas with shrubby vegetation, often secondary in nature. Also fragmented sites, open areas with a few standing trees (R. Otoch pers. comm.). Often forages in livestock corrals.

BEHAVIOUR Observed in flocks of up to 20. However, it is often observed in pairs apart from other Baywings. Sometimes these pairs forage with groups of Shiny Cowbirds. While feeding in animal corrals they forage mainly for insects, but have been noted to be a pest on rice and corn (R. Otoch pers. comm.). Pairs appear to have a large home range. Individuals sit high in a tree, often vocalising, and may fly long distances to settle in another tree, with the mate following closely behind (M. Cohn-Haft pers. comm.). Like nominate Baywing, this form often sings in groups. Up to 20 or 30 birds may be involved, and will sing together often in the hottest hours of the day, each bird hidden in dense foliage at the crown of a shrubby tree (R. Otoch pers. comm.). Very little is known about the behaviour of this taxon, particularly with respect to how similar or different it is from nominate Baywing.

NESTING Unknown, but likely differs from that of Baywing, as this species does not have to contend with the intense parasitism by Screaming Cowbirds. 'Pale Baywings' also use old, covered (domed) nests of other species, particularly furnariids (ovenbirds), especially nests of Rufous Cacholotes (*Pseudoseisura cristata*). This species nests twice during a season and may raise between one and three young. They are rarely, if ever, parasitised by Shiny Cowbirds (102) (R. Otoch pers. comm.). Evidence suggests that this form is not a cooperative breeder (R. Otoch pers. comm.), but this needs to be confirmed.

DISTRIBUTION AND STATUS Rare throughout its range. Occurs in the Caatinga region of NE Brazil in Piauí, Ceará, Rio Grande do Norte, Paraíba, Pernambuco and Bahia. Previously published as extending south to Minas Gerais (the type specimen), but this appears to be in error (Hellmayr 1929).

MOVEMENTS Sedentary.

MOULT Not known. Probably similar to Baywing, but almost certainly differs in its timing and needs further research.

MEASUREMENTS Both sexes similar in size. Males: (n=3) wing 87.3 (85–91); (n=4) tail 68.9 (65.0–71.5); culmen 18.6 (16.5–20); (n=2) tarsus 26, 26. Females: wing 86.0; tail 72.0; culmen 16.7. Unsexed: (n=1) wing 87; tail 68; culmen 18; tarsus 25.

REFERENCES Friedmann 1929 [description, measurements], Hellmayr 1929 [distribution], Lamm 1948 [behaviour].

97 SCREAMING COWBIRD *Molothrus rufoaxillaris* Cassin 1866 — Plate 37

A common cowbird of the pampas and chaco, seldom found away from its primary host, Baywing (96). Screaming Cowbird tends to go unnoticed due to its resemblance to the much more abundant Shiny Cowbird (102).

IDENTIFICATION Visitors to the pampas often overlook this species due to its close resemblance to its commoner relative, Shiny Cowbird. Male and female Screaming Cowbirds are similar to each other in plumage; both are black with a dull blue gloss throughout the body. The gloss is duller than on male Shiny Cowbird, and Shiny Cowbird also has a distinctly purple-blue gloss, unlike the pure blue gloss of Screaming Cowbird. Screaming Cowbirds have chestnut-coloured eyes, the female's slightly browner than the male's, while Shiny Cowbirds always have brown or blackish eyes. In the hand, the rufous underwing-coverts are diagnostic for Screaming Cowbird, but this is a difficult field mark to see, even on birds in flight.

Body shape and proportions are useful features in identifying this species. Screaming Cowbirds have short thick bills, unlike the longer bill of Shiny Cowbird. Screaming Cowbirds are long-winged, the primary extension is noticeably long, and they tend to droop the wings as they feed, further accentuating this difference. Screaming Cowbirds also have long, loose flank feathers which are often 'fluffed out' to the sides of the body, appearing like black flags on either side of the bird's body. In addition, during the breeding season, Screaming Cowbirds are always in pairs. It is exceedingly rare to see a lone Screaming Cowbird during the summer! If you see a pair of black cowbirds that never leave each other's side, you are more than likely looking at Screaming Cowbirds. Shiny Cowbirds are either seen as lone males, groups of males or in pairs with the grey-brown females. See Voice for other differences.

Juvenile Screaming Cowbird is a mimic of juvenile Baywing and can be exceedingly difficult to identify, even the begging and distress calls are nearly identical. Differences to look for are the much longer wings of Screaming Cowbird, which is most obvious on older juveniles which have fully grown primaries. Screaming Cowbird also has blackish wing-coverts unlike the rufous-grey wing-coverts of Baywing. The bill of a young fledgling tends to be paler on a Screaming Cowbird, on older birds the distinctive short, thick shape will become apparent. If seen at close range or through a telescope, Screaming Cowbird shows a large nostril, circular in shape while that of Baywing is partly covered by skin making it slit-like in profile. A male fledgling Screaming Cowbird will look much larger than its 'siblings', and is even larger than its host parents; females are the same size as the host. As Screaming Cowbirds near independence from their Baywing parents, they acquire tell-tale black feathers on the body.

VOICE Screaming Cowbird is misnamed as none of its calls can be called screams, even though the female's rattle is quite harsh. Both sexes have specific calls, as well at least one that is shared. The male's song is perhaps the simplest and most basic of the cowbird genus' songs.

Song: The male's song begins with a couple of low *trruk* or *puk* calls (often missing) followed by a distinctive, high and sharp *shhhleeee* (*pe tzeeee* in Friedmann 1929). The *shhhleee* call is short (less than half a second), quick and explosive; it has a shrill quality due to the high pitch of the terminal portion. At close range, a soft, low *grrrr* or *brrrr* sound can be heard, immediately before the *shhhleee*; therefore the primary component of the song is a *grrrr-shhleee*. As noted above, there are sometimes other introductory notes. Often duets with the female, see below.

Calls: The female's call is a harsh buzzy rattle, much rougher, quicker and sharper than that of female Shiny Cowbird. The female often gives her rattle in a duet with the male, following directly after his *shhhleeee* song. The composite of the calls sounds as if it was emanating from a single individual. The male's screech and the female's rattle are commonly given in flight. In alarm, and to distract male Baywings from their nests, male Screaming Cowbirds will give a harsh, repeated, *tiip* call, while cocking their tails. Friedmann (1929) describes this last call as resembling the *plunk* sound of two pieces of metal hitting each other. The *tiip* call is similar to the call of Tropical Parula (*Parula pitiayumi*). Both sexes appear to share a low *chuck* call similar to that of Red-winged Blackbird (61). The male either lacks a flight whistle or this call is so rarely given that it has not been documented.

DESCRIPTION Sexual dimorphism is slight and not always noticeable in the field, but males are significantly larger. The juveniles are completely unlike adults in colour. This cowbird has longer and more pointed wings than Shiny Cowbird. Wing Formula: P9 < P8 > P7 > P6; P9 ≈ P7; P8–P6 emarginate.

Adult Male (Definitive Basic): The black bill is much thicker and shorter than other sympatric cowbirds. The bright chestnut eyes may appear dark at a distance. The body plumage is almost completely black, with a dark blue gloss. The lesser underwing-coverts are a contrasting rufous, which is difficult to see in the field. The legs and feet are black. In general proportions, this cowbird is long-winged for its size, more obviously so in the male.

Adult Female (Definitive Basic): Like male, but smaller with a duller gloss, smaller rufous underwing patch and browner eyes.

Immature (First Basic): First-year birds often have contrasting worn, older feathers on the wing-coverts or underwing-coverts, and the primaries are much more worn than on adults and appear brownish in the field.

Juvenile (Juvenal): Full juvenile plumage is only retained for a few weeks. Juveniles are grey with a slightly darker mask and tail, the tail edged in rufous. The edges of the primaries, secondaries and tertials are rufous as are all the wing-coverts. The underwing-coverts are blackish.

GEOGRAPHIC VARIATION No recognized subspecies, but northern birds may be slightly smaller than those from the south. The eggs of birds from Salta (northwest Argentina) are smaller and more spherical than those from Buenos Aires province. (Jaramillo. Unpublished data.)

HABITAT Found in fields, pastures and grassland as well as open woodlands, but avoids dense forest. Tends to feed in open areas, often sharing this habitat with Shiny Cowbirds, but breeding activities are focused in areas where Baywings are found, usually open woodlands. Unlike Shiny Cowbirds, this species is not so closely associated with livestock. I have never seen Screaming Cowbird perched on livestock, but Belton (1985) mentions seeing this behaviour. Screaming Cowbirds tend to feed on the ground, directly underneath livestock.

BEHAVIOUR Forms small flocks in the winter, mixing with both Shiny Cowbird and Baywing. Flocks break up into pairs as the breeding season approaches. Pairing and courtship occurs very early in the season and for only a short period of time. The male and female are inseparable during the breeding season, it is very rare to see a lone adult at this time. Even when traveling in a small groups, the pairs maintain their integrity within the flock. Males have a 'song spread' ('ruff-out') like many other blackbirds. The display resembles that of other cowbirds, and is always given while perched in a tree, not on the ground. The male bows the head, fluffs out the plumage, droops the tail and extends the wings horizontally, before quivering the entire body and singing. Sometimes the male will terminate this display by hopping up and down on the branch as he finishes the song. The female may respond to the male's song in a 'mutual display' (see Voice). While perched they often deliberately flick the tail and spread their flank feathers, which is visible from a distance. They are very mobile during the breeding season and are most often seen traveling between Baywing nests.

NESTING A truly extraordinary bird, a masterpiece of evolutionary tinkering. Screaming Cowbird is an obligate brood parasite, like all cowbirds. Unlike other cowbirds, Screaming Cowbird is a host specialist, using Baywing as its only regular host. In Paraná, Brazil, where there are no Baywings, there are records of Chopi Blackbird (79) being used as a host, and it appears that here this is the regular host for Screaming Cowbird. More recently, Fraga (1996) confirmed that Screaming Cowbirds use Chopi Blackbird as a regular host in Iguazú, Misiones, Argentina, and that this may have occurred historically as early as the 1800s in Paraguay (Azara 1802). In Buenos Aires province, Screaming Cowbird also parasitises Brown-and-yellow Marshbird (74), but this is rare (Mermoz and Reboreda 1996).

During the breeding season this species appears to be monogamous. Screaming Cowbirds are always seen in pairs or groups of pairs during this season. Pairs are probably maintained because it may be impossible for the female to achieve access to the host nests by herself (pers. obs.).

Screaming Cowbirds conduct inspections of hosts nests at all times of the day. Arrival at host nests tends to be rapid and noisy. The male Screaming Cowbird usually begins to utter the *tiip* call while flicking the tail. The male Baywing will often chase the male cowbird, allowing the female parasite to deal with the host female. The male's help in distracting the male Baywing may be invaluable to the female Screaming Cowbird, unlike in other cowbirds except perhaps for Giant Cowbird (98).

Eggs begin to be laid in the host nest before the hosts have started laying. Nearly 100% of Baywing nests are parasitised, mostly with more than one parasitic egg. Female Screaming Cowbirds tend to lay one egg per host nest, but many females may lay in the same nest. Screaming Cowbird eggs are variable in colour and shape. In Buenos Aires, the eggs of Screaming Cowbirds appear to be mimetic to those of Baywing in shape and size, while in Salta (NW Argentina) the eggs are more spherical than those of the host. The evidence suggests that egg mimicry is driven by rejection of eggs by other Screaming Cowbirds, while they lay their own eggs (Jaramillo unpub. data).

The nestling and fledgling Screaming Cowbirds are amazing mimics of Baywing young. Their plumage is identical, as are their calls. At five days old, some Screaming Cowbird young may only be differentiated from Baywings by the shape of their nostrils. Juvenile Screaming Cowbirds begin to acquire the black adult plumage as soon as they become independent of their foster parents. There is evidence that Baywings will not feed fledglings that do not look like 'proper' Baywings, a logical driving force for the evolution of plumage mimicry.

DISTRIBUTION AND STATUS Uncommon to common in its range. Mainly a bird of the lowlands, although there is a record from Guancor, Bolivia from 700 m and Ridgely and Tudor (1989) state that it is found to 1000 m. It is found from as far north Tarija, southeast Bolivia, but one record exists from Guanacos, Santa Cruz (Remsen *et al.* 1987); it has not yet been found in Chuquisaca (Remsen and Traylor 1989). Found throughout Paraguay, except for the extreme NW in Chaco, N Nueva Asunción, and N Alto Paraguay. The Brazilian range extends north to C Mato Grosso and it is also found in Rio Grande do Sul. It is spreading in Brazil so its current status in Santa Catarina and Paraná is unclear; it has recently been found in São Paulo. Screaming Cowbird occurs throughout Uruguay. In Argentina, it is found in the northern half of the country west to the foothills of Jujuy, Salta, Tucumán, Catamarca, La Rioja, E Mendoza and south to S San Luis, N La Pampa and S Buenos Aires, and probably found to N Río Negro.

MOVEMENTS Not migratory, appears to spend most of its life near its place of birth. There are now several records from central São Paulo (Willis and Oniki 1985); the species appears to be spreading northeastwards in Brazil as forests are cleared. Recently added to the Chilean list based on a pair that was observed in Arica in September 1986 (Sallaberry *et al.* 1992); however, this non-migratory species is a most unlikely natural vagrant to northern Chile and may have been human assisted. Arica is well north and west of the range of this species, as well as being directly west of the widest part of the Andes, making natural vagrancy even more improbable.

MOULT Complete definitive pre-basic moult takes place after the breeding season, in March–April. Pre-alternate moults appear to be lacking. First pre-basic moult December–February, sometimes later, and is incomplete. Almost all of the feathers are replaced except for a few of the underwing-coverts, and sometimes some wing-coverts and scapulars. The Baywing-like plumage of Screaming Cowbird begins to acquire black (first basic) feathers between 34–44 days after hatching (Fraga 1996). The tail moults from the outside inwards, towards the center.

MEASUREMENTS Males are larger than females. Males: (n=15) wing 117.8 (110–127); (n=12) tail 83.2 (73–90); (n=15) culmen 17.3 (16–19.4); (n=11) tarsus 26.3 (25–30). Females: (n=18) wing 104.6 (96–112); (n=10) tail 77.5 (71–82); (n=18) culmen 16.2 (15.0–18); (n=14) tarsus 25.0 (23–27).

REFERENCES Belton 1985 [behaviour], Fraga 1979,1986 [description, behaviour, nesting], Friedmann 1929 [behaviour, nesting, measurements], Hudson 1920 [nesting], Jaramillo (unpub. data), Mason 1980, 1987 [behaviour, nesting], Ridgely and Tudor 1989 [distribution], Sallaberry *et al.* 1992 [movements], Sick 1993 [nesting].

98 GIANT COWBIRD *Scaphidura (Molothrus) oryzivora* (Gmelin) 1788 — Plate 6

A well-named icterid as it is truly a giant, particularly for a cowbird. A brood parasite which specializes on its relatives, the caciques and oropendolas.

IDENTIFICATION This species should not be difficult to identify given good views. However, observers seeing it for the first time may be taken aback by its large size and may not realise that this bird is a cowbird, particularly since it is most commonly seen in flight. The shape and proportions of this species are curious. It is stocky, and relatively long-tailed and bull-necked, but looks strangely small-headed. In flight, the wings are long and somewhat pointed, and the flight is swift and direct, with a series of flaps interspersed with glides. The males have a particularly noisy flight, due to the emarginate primaries; once this noise is learnt, it helps in the detection of flying birds. There are many all-black icterids within its range, including several cowbirds, but all of these are substantially smaller than Giant Cowbird. Great-tailed Grackle (87) overlaps with the cowbird in size, but is very different in shape. The grackle is extremely long-tailed, with a keel-shaped tail. In addition, the grackle has a longer bill and is very noisy. Flying oropendolas may appear all dark at times, but all have some yellow on the tail. In shape, oropendolas look slimmer and longer-tailed than the cowbird, even though they can be larger in overall size. There are several other large, dark species within the cowbird's range which could be confused with the cowbird, such as Red-ruffed Fruitcrow (*Pyroderus scutatus*), Purple-throated Fruitcrow (*Querula purpurata*) and Violaceous Jay (*Cyanocorax violaceus*). The reddish throat of the Red-ruffed Fruitcrow should be obvious on a clear view. The purple throat of Purple-throated Fruitcrow is difficult to see and is only sported by the male. However, this fruitcrow travels in groups through the canopy, and utters a distinctive mewing call. Both fruitcrows are short-tailed and round-winged. Violaceous Jay also differs in colour but can look black at a distance or in flight. Look for the broad, rounded wings and laboured flight of the jay as well as its squarish rather than rounded head; jays are also noisy!

VOICE Giant Cowbird is particularly silent for an icterid, songs and calls are seldom heard.

Song: The male's song is a rather unpleasant, screechy whistle followed by three or more short notes with a metallic quality. The final notes may be paired and the entire song lasts approximately three seconds. The initial screech may be short and sharp (one third of a second in length) or longer (1 second in length); this unpleasant sound can be likened to that of someone running their nail on a blackboard. Thus the song sounds like *Tchwweeeee, twii-dlee, tic-tic* or *Shhweaa, t-pic-pic*. Published representations of the song include a screechy *fwrreeeeee?* (Stiles and Skutch 1989) and a strident but melodious *jewli chi chi chi chi* (Sick 1993).

Calls: they are sharp and chattery, a *chechk chehk* or *chehk-chick*, a longer *chrrik rrik-rrik-rrik-rrik-rrik-rrik* as well as a mew and a low *dak*, or *kawk*. Juveniles give a high *meeew* or *miieeu* when flocking with their foster parents.

DESCRIPTION A large cowbird with a distinctly small head. The bill is long, roughly the same length as the head or slightly longer. There is a noticeable curve to the culmen, but the gonys is straight. Giant Cowbird looks distinctly long-legged, particularly the tibia, when on the ground. Giant Cowbirds, particularly adult males, are long-winged and the wings are pointed. The long wing length can be seen as a noticeably long primary projection on the perched bird. A careful examination of size and the extent of the primary projection will allow the meticulous observer to sex birds and age males with a good degree of certainty. Wing Formula: P9 > P8 > P7 > P6; P9 > P8; P8–P6 emarginate. The form *impacifica* is described.

Adult Male (Definitive Basic): The black bill is long, pointed, stout, and the base of the slightly downcurved culmen continues level with the forehead. At the top of the bill, the culmen expands to form a small, rounded frontal shield. Males show an erectile 'ruff' of feathers on the neck and nape; these are sometimes partially erected giving the bird a slightly bull-necked appearance. Plumage entirely black with a purplish or violet iridescence on the body and a more blue-coloured sheen on the wings. The primaries are strongly emarginate. The longish tail is distinctly square-ended. The rectrices are black with a violet edge on the outer vanes of each feather. The legs and feet are black.

The eye colour is variable, bright red in Central America and northern South America, but yellow in the west and south of South America. In eastern Ecuador, most birds are yellow-eyed while some are dark-eyed (all young ?), and others look to have dark blood-red eyes. It appears that some individuals have red eyes even in populations where yellow predominates, and orange eyes are also known. It is not certain whether or not this variation is due to age, sex or just individual differences.

Adult Female (Definitive Basic): Like male, but both duller and smaller. Black with some iridescence, but not nearly as noticeable as on the male. The females do not show a well-developed 'ruff', or hood. Soft part colouration as in the male, but the eye colour is duller.

Immature (First Basic): Like the corresponding adult, but duller, less iridescent, and with noticeably more worn (brownish) primaries. The iridescence on the back and scapulars is patchy since the gloss is restricted to the feather tips. Thus, the matt black feather bases show through, breaking up the iridescence on the upperparts. There is no iridescence on the vent, belly or thighs. Sometimes the thighs may show some dull brownish-black colouration. Immature males are smaller than adult males and have distinctly shorter and less pointed primaries.

Juvenile (Juvenal): Blackish but lacks iridescence. Unlike adults, juveniles have brown or greyish eyes, sometimes whitish-yellow, and a pale horn-coloured bill which slowly darkens by the time the bird is several months old.

GEOGRAPHICAL VARIATION Two subspecies are currently recognized.

The nominate subspecies occurs in the entirety of the South American range north to the Canal Zone of Panama and the island of Trinidad.

The more northerly race, ***S. o. impacifica***, ranges from W Panama north to the Caribbean coast of Mexico and north to Veracruz. This form has a violet, rather than bronze-blue, iridescence to the plumage. *Impacifica* has been described as larger than the nominate, but these comparisons were made with the smaller *oryzivora* populations from the south; overall the two races are the same size. Northern populations of the nominate form are larger than those from the south. It appears that *impacifica* adults always have reddish eyes, but that eye colour is variable in *oryzivora* (see Description).

HABITAT This is the only cowbird that is regularly found deep in the forest. Giant Cowbird may forage in the canopy, particularly as juveniles, but often retreats to rivers, cochas (oxbow lakes) and other more open areas to forage. It may also be found at the forest edge, open forest, degraded tropical forest, agricultural land, or wherever caciques and oropendolas nest. Frequents plantations near forest, particularly banana, where it will feed on the fruit. Also forages with cattle in rangeland, perching on them like other cowbirds as well as feeding beneath them.

BEHAVIOUR The 'song spread' is similar to that of Bronzed Cowbird (99). The male raises its short neck ruff and bends forward so that the bill is nestled in the breast feathers. The body feathering is fluffed up and the tail is brought forward. Unlike smaller cowbirds, this species does not bend forward as it displays. Forages for fruit, nectar (*Erythrina* flowers) and insects. While searching for insects it may forage on the ground, turning over debris, including rocks and livestock feces to expose prey. Also forages in trees, with the larger males stripping bark from branches to expose insects while the smaller females poke into dead, rolled leaves or glean from living leaves; may glean from mammals. In Amazonia, Giant Cowbirds regularly forage on the back of Capybaras (*Hydrochoerus hydrochoerus*), the world's largest rodent, and rid them of horse flies (Tabanidae) which pester these animals. This species is not necessarily attracted to livestock, but will sometimes feed around them in the manner of its smaller relatives. Native Americans who farmed corn considered this species a major pest as it was strong enough to feed on and destroy the coarse native corn they planted. In the manner of other cowbirds, this species is known to solicit preening from other bird species, particularly host species. This has been observed in the wild, with the cowbirds soliciting preening from oropendolas. An extreme case of interspecific preening was reported from a captive Giant Cowbird which would solicit preening from passing people (Payne 1969b).

NESTING A brood parasite specializing on caciques and oropendolas. Eggs have been recorded in nests of the following seven species: Yellow-rumped Cacique (13), Red-rumped Cacique (14), Crested Oropendola (2), Russet-backed Oropendola (5), Green Oropendola (3), Chestnut-headed Oropendola (6) and Montezuma Oropendola (7). The timing of breeding of Giant Cowbird is extremely variable and depends entirely on the timing of nesting of the local hosts. Giant Cowbirds nest between December and April in Surinam (Haverschmidt 1968). However, it is expected that Giant Cowbirds will parasitise any of the colonial caciques or oropendolas. In E Peru, Giant Cowbird parasitises Russet-backed Oropendola, but not Casqued Oropendola (1). There is only one report of a Giant Cowbird possibly parasitising a non-icterid host. Lehmann (1960) observed a juvenile Giant Cowbird being fed by a pair of Green Jays (*Cyanocorax yncas*) during August 1957 near Cajibío, Cauca, Colombia. It is unknown if this cowbird actually hatched in the nest of the jays, or if it was adopted after hatching. Given this observation, it is most interesting to note that in Colombia, Green Jays vigorously chase Giant Cowbirds away from nest sites (Alvarez 1975). Another unusual observation is mentioned by Schäfer (1957) who states that he observed a female Giant Cowbird feeding a juvenile Giant Cowbird which was being raised by a Crested Oropendola!

The eggs are oval, with obvious pores. There are two egg types, unspotted (immaculate morph) and spotted, as in Shiny Cowbird (102). Unspotted eggs are white or off-white, the spotted eggs are pale blue, green or white with dark spots and scrawls concentrated at the broader end of the egg. Studies in Surinam have found that Giant Cowbird eggs found in Crested Oropendola nests are different from those found in Yellow-rumped Cacique colonies. The eggs in the cacique colonies have a bluish background colour and are more rounded than the more elongated eggs with a white background found in Crested Oropendola nests. In Panama, the eggs of Giant Cowbird are significantly different in size and colouration from those of the two hosts, Yellow-rumped Cacique and Chestnut-headed Oropendola. The eggs of the cowbird are larger and less spotted than those of the two hosts; four of 31 cowbird eggs were of the immaculate morph at this site. In mixed colonies of caciques and oropendolas, parasitism was higher in Yellow-rumped Caciques than in oropendolas, as was multiple parasitism, when more than one cowbird egg is present in a nest. In nests with two cowbird eggs, some were laid by the same Giant Cowbird female while others were laid by two different females.

In Amazonian Peru, where Yellow-rumped Caciques as well as Russet-backed Oropendolas are parasitised, the situation is different. Here the cacique is rarely parasitised and all fledgling cowbirds observed were being fed by oropendola hosts. The cacique colonies are seldom left unguarded, leaving very little opportunity for the cowbirds to lay. In contrast, Russet-mantled Oropendolas forage in a flock and often leave the colony tree unattended, allowing for easier access by the cowbird. In addition, Peruvian Giant Cowbirds have trouble entering the smaller nests of the cacique. Both host species will chase Giant Cowbirds away from their nesting trees, even when associated with wasp colonies.

At colonies, Giant Cowbirds may be very bold and aggressive. They arrive and try to enter a nest quickly, but are usually mobbed and driven out by members of the host colony. Several females, and sometimes males, have been observed co-operating in nest raids. A member of the group will distract the hosts and allow them to chase it, giving a female the opportunity to lay her eggs. An extreme measure of opportunism by cowbirds can be seen when they take advantage of the confusion caused by predator raids on host colonies to gain access to nests. Giant Cowbird has been observed entering the nests of Crested Oropendolas in Colombia even though the oropendola female was still present in the nest (Lehmann 1960). In Panama, visits to the colony began in January and continued through the nesting season, with the majority of visits taking place in the morning. In Costa Rica, cowbirds at colonies of Montezuma Oropendola tended to concentrate their visits when the oropendolas were lining their nests with leaves, just before egg laying. Also, most cowbird visits took place during the middle of the day; during the midday heat, few oropendolas tend to be at attendance at the colony. Nevertheless, it is relatively rare for a cowbird to actually be able to enter a host nest. Cacique and oropendola colonies tend to be widely spaced and thus female Giant Cowbirds must cover a large home range to be able to access an adequate number of host nests; nothing is know about the fecundity of this species.

Smith (1968) described a complex association between Giant Cowbirds, their hosts, and the presence of bees or ants which guard the host nests. He determined that in some cases, Giant Cowbird parasitism could be beneficial to the hosts. When host nests were not protected by bees, parasitic Bot Flies (*Philornis*) could gain access to nestlings and severely reduce their chances of survival. Giant Cowbird nestlings, according to Smith, preen the hosts and remove Bot Fly eggs and larvae therefore giving an advantage to hosts that allow cowbirds to parasitise them. This amazing story has been questioned recently. Doubts have arisen due to evidence that other aspects of Smith's work (on gulls *Laridae*) may not have been valid (Snell 1989). Part of the criticism of his work was that the methodology described would be nearly impossible to conduct. Smith's cowbird work involved the inspection of 1750 cacique and oropendola nests in four years. He would bring them down using ladders and long telescoping aluminum poles in order to inspect the nest contents and then the nests were replaced 'with sticky tape or contact cement' (Smith 1968). These research protocols are as problematic as those described by Snell (1989). Furthermore, Smith (1968) described that Giant Cowbirds in Panama produced one of five types of eggs, four of which were mimetic to a specific host's eggs. On the other hand, Fleischer and Smith (1992), also working in Panama, did not detect any evidence that Giant Cowbird eggs were mimetic. Robinson (1988a) and Webster (1994c) did not find any evidence that parasitism by the Giant Cowbird was ever beneficial to hosts. Note also that in another brood parasite and host interaction, Baywing (96) host parents have been observed to remove Bot Fly larvae from chicks of the parasitic Screaming Cowbird (97) (Fraga 1984).

DISTRIBUTION AND STATUS Uncommon to locally common, its abundance partially dependent on the number of available hosts. A bird of the lowlands, rarely observed as high as 1500 m. Found in S Mexico from C and E Veracruz, N Oaxaca east along the Caribbean slope of Tabasco, S Campeche, S Quintana Roo and N Chiapas, but absent from the Yucatán peninsula. Continues in the Caribbean lowlands of N Guatemala, Belize, N Honduras and Nicaragua. In Costa Rica, is is most widely distributed in the Caribbean lowlands, but is also present in the central valley and in the Pacific lowlands in the Golfo Dulce (Punta Arenas province) region (Stiles and Skutch 1989). Found along both slopes in Panama, but absent from highlands and scarce on the Azuero peninsula (Ridgely and Gwynne 1989). Found throughout Colombia, but absent from the Guajira peninsula and either absent or local from the Llanos of the NE (Hilty and Brown 1986). Occurs throughout the lowlands of Venezuela, Guyana, Surinam, French Guiana and throughout most of Brazil except the Caatinga region of the northeast

(Piauí to Alagoas southwest to C Minas Gerais). Also known from Trinidad and Tobago, but it may be a recent colonizer on the latter. Present in the lowlands east and west of the Andes in Ecuador. In Peru, it ranges throughout the lowlands east of the Andes from Loreto south to Madre de Dios, and extends from W Ecuador to N Tumbes. Distributed widely in the lowlands of Bolivia in all departments except Chuquisaca, Oruro and Potosi (Remsen and Traylor 1989). In Paraguay, it is found only in the moister half of the country, east of the Paraguay river. Local in Argentina where it is only known from N Misiones. The sight record from Tarija, Bolivia, suggests that Giant Cowbird should be looked for in N Salta, Argentina.

MOVEMENTS There is no evidence to suggest that Giant Cowbirds are not sedentary. However, large areas are covered while birds are nest searching, and it is not inconceivable that there are substantial local movements that track host abundance. The species is historically known from Trinidad, but appears to have only recently arrived in Tobago, so this species may also be spreading as are most of the smaller cowbirds.

MOULT The adults change their plumage once a year. This moult is the complete definitive pre-basic, and it occurs after the breeding season. Therefore, the timing differs depending on the locality, as the breeding season varies geographically. Moult in Guatemala occurs in winter (February). Specimens from Huila, Colombia, are also moulting in February. The juvenal plumage is lost through the first pre-basic moult which replaces all of the body feathers, the coverts and the tertials. We have not been able to ascertain if this moult also involves the primaries and tail, but we believe that it does. Immatures attain adult (definitive) plumage after the second Prebasic moult. Pre-alternate moults appear to be lacking in all age groups. It is not known if immature and adult moults are similar in their timing.

MEASUREMENTS ***M. o. oryzivora***. Males: (n=10) wing 193.1 (182.5–216); tail 147.2 (137–161.5); culmen 36.6 (34.5–39.5); tarsus 44 (40–46). Females: (n=10) wing 153.2 (150–159.5); tail 116.3 (113–124); culmen 32.3 (31–33.5); tarsus 38.8 (37–40.5) .

M. o. impacifica. Males (n=?) wing (183–210); culmen (37–38.5). Males: (n=8) wing 196.4 (187–210); tail 146.9 (138–158); culmen 37.9 (37–39); tarsus 44.8 (42–46). Females: (n=4) wing 156 (151–159); tail 123.8 (117–130); culmen 31.8 930–33); tarsus 37 (35–39).

NOTES The pale bill of the juvenile may be a form of mimicry as several of the host species have pale bills. This species was formerly called Rice Grackle. Its scientific name has also gone through some major changes in the past; two different generic names used for this species during the last century were *Cassidix* (old genus of the large grackles, now in *Quiscalus*) and *Psomocolax*. The name *Scaphidura* was thought to have been an error, basically a misspelling of *Scaphidurus*, a genus that was erected for the description of the Slender-billed Grackle (now *Quiscalus palustris*). However, Parkes (1954) correctly stated that when *Scaphidura* was introduced it was clearly not in error, and it referred to Giant Cowbird. This is now the accepted name for this species, but the recent work of Lanyon (1992) has determined that this cowbird fits neatly into the *Molothrus* clade which is monophyletic, therefore its name should be changed to address this new interpretation of the relationship between the cowbirds.

REFERENCES Belcher and Smooker 1937 [movements], Chapman 1917 [measurements], 1930 [nesting], Fleischer and Smith 1992 [nesting], Friedmann 1929 [behaviour, nesting], Haverschmidt 1967 [eggs], Howell and Webb 1995 [description, moult, voice], Leak and Robinson 1989 [nesting, host choice], Lehmann 1960 [nesting], Parkes 1954 [nomenclature, notes], Peters 1929 [geographic distribution, measurements], Robinson 1988a [behaviour, nesting], Sclater and Salvin 1873 [behaviour], Sick 1993 [voice], Stiles and Skutch 1989 [description, voice], Webster 1994c [nesting].

99 BRONZED COWBIRD *Molothrus aeneus* (Wagler) 1829 — Plate 38

This is the common cowbird throughout most of Mexico and Central America. As forested areas are being degraded, this cowbird has been able to capitalize on the new open habitats and expand its range, like its congeners.

IDENTIFICATION Throughout most of its range, this is the only small cowbird. In structure it is chunky, short-billed and short-tailed. At a distance, the male looks entirely black and may be confused with Brown-headed Cowbird (101) or Melodious Blackbird (85). Grackles (*Quiscalus*) are larger and longer-tailed, the males showing an obviously keeled tail, and should not be confused. Melodious Blackbird is longer-tailed and has a different bill shape – long, stout and with an evenly curved culmen. In addition it has brown, not red eyes. Bronzed Cowbirds are noticeably larger than Brown-headed Cowbirds, possess brilliant red eyes, and males show an erectile ruff on the back of the neck. Bronzed Cowbird lacks the brown head of its smaller relative, and is glossier with a bronze rather than greenish gloss. The thicker bill of Bronzed Cowbird appears almost continuous with the forehead, while Brown-headed Cowbirds show a steeper forehead, which is visible in both sexes. Females are smaller than males, but still larger than female Brown-headed Cowbirds. The colour variation of female Bronzed Cowbirds can cause confusion. In *loyei* and *assimilis*, the females are the same colour as Brown-headed Cowbirds, but have reddish eyes. Throughout the remainder of Bronzed Cowbird's range, females are blackish and should not be confused with female Brown-headed Cowbirds. Female Shiny Cowbird is similar to female Brown-headed Cowbird and can be separated from Bronzed using the same criteria. Juvenile Brown-headed Cowbirds differs in both structure and size, and both Brown-headed and Bronzed Cowbirds will have dark eyes at this time. Note the differences in bill size again, the average larger size of Bronzed Cowbird and Brown-headed Cowbird's more scaled appearance as a juvenile.

VOICE Some vocalisations are shared by the sexes, and others are not.

Song: The male's 'perch song' has been described as a *ugh gub bub tse pss tseeee*, the first three notes may be omitted and the terminal two may run together. The 'perch song' is a commonly used song, and does not carry very far. However, the final note, the *tseee*, does carry over long distances and is perhaps the best way to locate a male Bronzed Cowbird (P. Warren pers. comm.). The 'flight whistle' of the male is given more rarely than the 'perch song' and is especially used in courtship. It is comprised of a series of high pitched whistles and trills, lasting about 4–5 seconds (P. Warren pers. comm.). During copulation, they repeat the final notes of the call, extending it to more like 10 seconds (P. Warren pers. comm.). Flight whistles show substantial geographic variation and in the United States fall into at least three regional dialects, unlike dialects in Brown-headed Cowbird these cover very large areas as opposed to local ones (P. Warren 1998a).

Calls: Females give a chatter call, very much like that of Brown-head Cowbird, it is an infrequently used vocalisation (P. Warren pers. comm.) Roosting flocks are rather talkative, the overlapping 'flight whistles' of many males may sound rather similar to a flock of European Starlings (*Sturnus vulgaris).*

DESCRIPTION Structurally, this is a chunky cowbird with a short tail. The bill is longer, and more massive than that of Brown-headed Cowbird, but still appears short for an icterid. The bill and legs are black. The bright red eye is one of the most obvious features on this species. At times, the male may raise his 'nape ruff', giving it a diagnostic shape, only seen on its close relative Bronze-brown Cowbird (100) and to a lesser extent on Giant Cowbird (98). Wing Formula: P9 < P8 ≈ P7 > P6; P9 ≈ P6; P8–P6 emarginate.

Adult Male (Definitive Basic): The plumage is entirely black with a bronze gloss on the body. The gloss on the wings is a contrasting blue, with greener gloss on the flight feathers. In some lights, the body's iridescence will look green, or even violet. The tail has a bluish-green gloss. In the western subspecies, *loyei* and *assimilis*, the rump of the male has a contrasting violet gloss.

Adult Female (Definitive Basic): Females of *loyei* and *assimilis* are pale. The upperparts are dark grey with darker feather centers. There may be a dull greenish-blue gloss on the feathers of the back, scapulars and wings. Wings and tail are also dark grey. Underparts paler grey, sometimes with very indistinct streaks. The throat is paler than the breast. Female *aeneus* are much darker, being sooty black all over, slightly paler on the underparts. The upperparts and wings are glossed with blue-green.

Immature Male (First Basic): In fall, like adult males, but duller, lacking the strong gloss. Most retain juvenal feathers which are diagnostic for the age, and most noticeable in *loyei*. These birds look intermediate in plumage between adult females and males, but unlike females they have a well-developed ruff. There is a partial pre-alternate moult which replaces some of the body feathers, particularly on the head, with adult-like feathers. The primaries of immature males are brown on the tips and have a green gloss instead of the blue of adults. The body is violet, not bronze-black as in adult males (P. Warren pers. comm.).

Immature Female (First Basic): Most are indistinguishable from adult females. Some dark *aeneus* may retain the slightly paler juvenal feathers which will give them away as a 1st-year female.

Juvenile (Juvenal): Colour varies depending on the subspecies. Arizona *loyei* is the palest, being similar to the females of that form but with streaks on the underparts. The upperparts are grey-brown, the underparts paler with a yellowish wash on the belly. In addition, they show paler fringes to the wing-coverts. The form *assimilis* is similar to *loyei*. The more eastern *aeneus* are dull blackish, with slightly paler underparts. The juvenile females are said to be slightly paler than males. Juveniles of all three forms have dull-coloured (brown) eyes.

GEOGRAPHIC VARIATION Three subspecies have been named; Bronze-brown Cowbird is often considered conspecific with Bronzed Cowbird.

M. a. aeneus is found from SC Texas south through the Atlantic plain of Mexico, including the Yucatán peninsula, continuing to Belize, Nicaragua, Costa Rica and Panama. The new populations in Florida, Louisiana and adjoining states are of this form.

M. a. loyei is found from S California (mainly near the Salton Sea and Colorado river), S Arizona, S New Mexico and W Texas, south through the western Mexican lowlands to Nayarit. This form is now known as far east as Sanderson, Texas (Terrell Co.) and as far north as Carlsbad, New Mexico (P. Warren pers. comm.). Recently *loyei* and *aeneus* have come into secondary contact in W Texas (Terrell Co.) (P. Warren 1998b).

M. a. assimilis ranges from SW Mexico, Jalisco, Colima, Guerrero and Oaxaca to Chiapas.

Traditionally males of *loyei* and *assimilis* are stated to have a violet sheen to the rump, while *aeneus* has an uniform bronze sheen to the rump and back. However, a study of many museum specimens has determined that *aeneus* has a violet sheen on on the rump much as the other two forms (P. Warren pers. comm.). Female *loyei* differs from *assimilis* in being larger and is likely not identifiable in the field. The nominate subspecies is intermediate in size between the other two. Female *aeneus* are blackish, while female *loyei* and *assimilis* are pale brownish-grey.

HABITAT As is typical of the cowbirds, the foraging habitat is slightly different to the breeding habitat. They forage in open pasture, agricultural land, feedlots, plowed fields, lawns, golf courses, as well as the edge of woodlands – all open areas. Breeding habitats tend to be less open, and include deciduous woodlands, semi-deciduous woodlands, edge of moist lowland forest, scrub, thickets, desert oases and riparian forest. In the mountains of Guerrero, Mexico, this species only reaches to the semi-deciduous tropical forest zone, staying below 1000 m (Navarro 1992). Bronzed Cowbirds may be found in urban areas, particularly during the non-breeding season, and often roost in city parks. However, the newly established Louisiana breeding population is decidedly suburban during the breeding season. Other roost sites include cattail (*Typha* sp.) marshes. In winter, they will congregate with other blackbirds around grain elevators and at ranches. Winter habitats are generally open, more similar to the foraging habitats of the breeding season than to the breeding habitat itself.

BEHAVIOUR Bronzed Cowbirds arrive in S Texas in February and join mixed blackbird roosts. In early March they split up into single species flocks and spread out into breeding habitat by April. Yearling males arrive at the breeding grounds a week later than adults. In Louisiana, spring migrants arrive in early April. Breeds in south Texas mid-April–July. Males establish a non-exclusive home range (approximately 5 ha) that is defended on a part-time basis. Females do not defend their home range (approximately 176 ha), and several female home ranges will overlap (Carter 1984). In fact, females may co-operate while nest searching in groups of two to six; however, they lay their eggs alone in the early morning. First-year males do not defend home ranges, and arrive over a week later than the adults. Like other cowbirds, they forage away from the breeding areas, yet females spend a considerable amount of their foraging time in their breeding home range (Carter 1984, 1986). The foraging and breeding areas are at times separated by very short distances.

No pair bonds are maintained, and the mating system is polygamous. The mating system has been regarded as an 'exploded lek' similar to many hummingbirds (*Trochilidae*) and some shorebirds. A lek is a shared 'meeting place' where males display and females visit solely for the purpose of mating; males do not offer any resources other than sperm. In an 'exploded lek', the males are not all clustered at a central site, but are spread out over a larger zone. However, males will court females at feeding areas away from their primary display site. The sex ratio is extremely male-skewed, five males to one female in Texas, and males do not 'mate guard'. Females appear to reproduce in their first year, but yearling males are not successful in reproduction. Young males exhibit delayed maturation, with their testes being less than half the size of those of adult males (Carter 1984, 1986).

The display of the male shares some similarities to other cowbirds, but is much more exaggerated. The primary male courtship display, other than bill-tilting, is the 'bow display'. During the 'bow display', the male raises the neck ruff to full prominence and then puffs out the rest of the feathers of the body, as he partially spreads the wings in the manner of a cape. The tail is brought forward and the head is arched forward so the bill touches the breast as the bird lets out a song. Sometimes the displaying bird will rock or bounce. While on the ground the male will hop with its ruff raised, wings partially spread, making the bird look rather vampire-like! A second display is the 'hover display', or 'aerial bow', this is the primary courtship display. While courting, the 'hover display' always precedes the 'bow display' (P. Warren pers. comm.). During the 'hover display', the male hovers directly above the female, roughly one meter above the ground. 'Hover displays' range from three to 21 seconds, and many 'hover displays' may be given consecutively, not always followed by the 'bow display' (Clotfelter 1995, P. Warren pers. comm.). The 'hover display' is sometimes given from a branch, rather than the ground (E.D. Clotfelter pers. comm.). 'Hover displays' are also observed in Shiny Cowbird (102), but are very rare in Brown-headed Cowbird. Note that a third display type is similar to the 'hover display', this is the 'circle display'. In this display the male flies in a low circle around the female, singing a flight whistle very loudly. This display is always given just prior to copulation (P. Warren pers. comm.).

After breeding they coalesce into flocks, presumably to moult. Post-breeding flocks peak in Louisiana and Texas in the month of August. The flocks decrease in size in the northern parts of the range as birds migrate south. In areas where this species is a permanent resident they usually remain in flocks throughout the winter.

NESTING A generalist brood parasite. A total of 82 species of host have been recorded, of which 32 have successfully raised juvenile Bronzed Cowbirds. Older reports hypothesized that this species specialized on orioles (*Icterus*) as hosts, but recently it has been considered more of a generalist. There is anecdotal evidence that orioles are preferred hosts, but this needs study. Compared to Brown-headed Cowbird it tends to choose larger hosts, but many host species are shared by these two species. Bronzed Cowbird seems to be more specialized than Brown-headed Cowbird as its list of hosts is much smaller. However, part of this discrepancy may be the lack of research work carried out on the breeding habits of Bronzed Cowbird. As is the case in the other cowbirds, Bronzed Cowbird will also spike or puncture eggs, of either the host or of other cowbirds, before laying her own. Multiple parasitism of nests appears to be more common in Bronzed Cowbirds than in Brown-headed Cowbirds.

The eggs are oval or rounded with *loyei* laying much larger eggs than *aeneus*. Unlike the eggs of Brown-headed Cowbird, the eggs of Bronzed Cowbirds are unmarked. In this respect its eggs resemble those of both Giant and Shiny Cowbirds, both of which sometimes lay unmarked eggs. The egg colour tends to be bluish or greenish. Like other cowbirds, a host egg is removed as the female lays the parasitic egg in the host nest. Incubation takes 10–12 days. Bronzed Cowbirds that hatch more than 48 hours after the host young will not survive, regardless of the host species. It takes 11–12 days for the nestling cowbird to fledge.

Nesting takes place March–July in Costa Rica. In El Salvador, the nesting season is more protracted, from April to mid August. In south Texas, breeding occurs May–July.

There is anecdotal evidence that the new population in New Orleans, Louisiana, has caused a local decline in Orchard Orioles (43), the locally preferred host.

DISTRIBUTION AND STATUS Common to uncommon throughout its range. During the non-breeding season they become more local as birds group up in flocks.

Breeding: It has recently expanded to southernmost California, along the lower Colorado river valley (also on the Arizona side) and Imperial valley (largely around Brawley); it has wandered to Los Angeles, San Diego and San Bernardino Cos. (Small 1994). It should be looked for in the Colorado river delta area of Mexico. Occurs from S Arizona east through S New Mexico, as far north as Carlsbad, to W Texas (Terrell Co.), south into Mexico through C Sonora to the Gulf of California coast south along the coast to Central America in the west; from New Mexico the distribution follows the W edge of the Central Mexican Plateau along E Chihuahua, E Durango, C Zacatecas, E Guanajuato, W Queretaro to the Federal District; perhaps present in parts of the plateau. The eastern population has made a recent outpost in the Mississippi river delta in Louisiana, particularly around New Orleans. Found in Texas north to Big Bend (Presidio, Brewster and S Jeff Davis Cos.) straight east to Galveston. This population continues south along the Gulf coast lowlands and Sierra Madre Occidental from Tamaulipas and Nuevo León south, reaching the coast and extending west to C San Luis Potosí, C Hildago; the range is continuous in Mexico south of this point. Thus it is widely distributed in southern Mexico, including the Yucatán peninsula and Cozumel Island. Found throughout Belize including many of the offshore Cays, Guatemala, Honduras, El Salvador and Nicaragua. In Costa Rica, common in both slopes and the central valley but mainly absent from the NW, the Guanacaste area. In Panama, found in Bocas del Toro on the Caribbean slope as well as the Canal Zone in Colón, and from Chiriquí to C Panama province on the Pacific slope.

Non Breeding: Most populations are sedentary, only those breeding in the US and N Mexico show any tendency to withdraw south. In winter it is not found in California. A few remain in SE Arizona, and S New Mexico but most leave those states; similarly they leave N Sonora and Chihuahua states in Mexico (Howell and Webb 1995). The Texas population retreats to the south of the state, numbers being very much smaller than in the summer. A few winter in the New Orleans area of Louisiana, but again numbers are reduced. There are increasing numbers of wintering birds in S Florida, which perhaps are those which breed in Louisiana-this has yet to be determined. South of these areas the distribution is largely unchanged in winter.

MOVEMENTS Partially migratory, some of the US birds winter in southernmost Texas and Arizona (and recently Florida), but it appears that most fly south, probably to northern Mexico. Altitudinal movements have been recorded in Central America, but most of these populations may be sedentary.

In the northern part of the range, this species is currently undergoing a noticeable range expansion. In the early 1900s the species was restricted to the lower Rio Grande valley of Texas within the US. Previous to the 1960s, Bronzed Cowbird bred as far north as San Antonio in Texas. During the late 1950s there was an increased presence in Austin, Texas, and they bred there for the first time in 1960; at that time, the easternmost regular area of occurrence along the coast was Rockport (Selander and Webster 1963). To the west, Bronzed Cowbird was first observed in the Big Bend region in 1969 and confirmed to be breeding in 1970 (Wauer 1973). The northward spread in Texas appears to have slowed, there are no substantiated records in the NC part of the state, but several convincing sight reports (Pulich 1988). The first records for several states where the species is currently regularly encountered are as follows: Arizona (1909), New Mexico (1947), California (1951), Louisiana (1961), Florida (1962) and Mississippi (1979). The most recent successful colonization has been of the Gulf coast states. Bronzed Cowbird is now present in Louisiana, Mississippi, Alabama and NW Florida with the densest population in the New Orleans area. The first record in that area occurred in Cameron Parish, Louisiana, during December, 1961 (Selander and Webster 1963). They were first confirmed to be breeding in 1976 and their numbers have been increasing. The first post-breeding flocks of over 100 birds were seen in New Orleans in 1985 and by 1993 and 1994 these flocks had reached 300 individuals. The cowbirds are primarily migrants and breeding birds, with few remaining in the winter. In Florida, on the other hand, the species is largely a migrant and winter visitor. It is not unlikely that some of the Louisiana population migrates and winters in Florida and perhaps further south in the Caribbean, and Bronzed Cowbird should be looked for in that area. Note that the first record from Louisiana occurred in 1961 and the first in Florida during 1963.

Stray records, away from these areas of recent colonization exist for the following sites: Seal Island, Nova Scotia (May 6–10, 1991), a female specimen *M.a.aneneus* from Squaw Creek, Missouri (Jan. 5, 1979), a winter record during 1994/95 from Fort Meade, Maryland, and a June record from Baja California, Mexico (Howell and Webb 1995). Bronzed Cowbird may be expanding to the south and east in Panama, and populations in Bocas del Toro may be of recent origin (Eisenmann 1957). Of 50 banding recoveries only one came from an area distant from the original banding site, all other recoveries were from the breeding season near the banding locality. This bird was banded in Texas and recovered during the winter three years and seven months later in Veracruz, Mexico (Klimkiewicz and Futcher 1987). The oldest recovered bird was 6 years old (Klimkiewicz 1997). A bird banded as an adult in Arizona reached 5 years old.

MOULT Juveniles undergo an incomplete (rarely complete) moult into first basic plumage July–October. The retained juvenile feathers tend to be on the underwing-coverts and some body feathers; all flight feathers are all replaced. There appears to be a limited first pre-alternate moult in males only, changing some of the head feathers and perhaps the body feathers. More work is needed on the extent and timing of this moult. The definitive pre-basic moult occurs July–October, and is a complete moult. The definitive pre-alternate moult is missing (or may be limited).

MEASUREMENTS ***M. a. aeneus***. Males: (n=?) wing 116.5 (111.8–120.1); culmen 23.0 (22.2–24.0); tail 79.6 (75.7–82.3) weight (n=144) 68.9 (0.3). Females: (n=?) wing 108.7 (105.2–111.8); culmen 20.6 (19.8–21.3); tail 73.0 (64–74.7); weight (n=220) 56.9 (0.3).

M. a. loyei. Males: (n=?) wing 119.0 (117–122); culmen 23.0 (22.5–23.6); tail 85.0 (81.8–88.8). Females: (n=?) wing 104.5 (101–107.5); culmen 19.8 (19.3–20.5); tail 73.5 (70–76.5).

M. a. assimilis. Males: (n=?) Wing 108.7 (105.2–111.8); tail 73.0 (64.0–74.7); culmen 21.3 (20.8–22.6). Females: (n=11) wing 102.6 (94–110); tail 73.5 (66–79); (n=10) culmen 19.1 (17–20); (n=11) tarsus 26.5 (25–29).

NOTES Bronze-brown Cowbird is often lumped as a race of this species, but other authors split this form (Hilty and Brown 1989). The two taxa are different in plumage colour, size and structure. The fact that Bronze-brown Cowbird is rare and declining, unlike all of the other *Molothrus* cowbirds, suggests that there is a fundamental difference in that species' natural history. The genus *Tangavius* was merged into *Molothrus* by the American Ornithologists Union in 1976. *Tangavius* included the Bronzed and Bronze-brown Cowbirds (Auk 93:878). This change caused the subspecific name *milleri* of Bronzed Cowbird to be changed to *loyei*, as *milleri* was preoccupied (under rules of zoological nomenclature) in the genus *Molothrus*. Giant Cowbird, currently classified in the single species genus *Scaphidura* properly belongs in *Molothrus* based on a phylogeny (tree of descent) published by Lanyon (1992). Bronzed Cowbird was until recently called Red-eyed Cowbird, an appropriate name! Note that the National Geographic field guide to North American birds (NGS 1983) calls the subspecies *loyei*, by the old name *milleri*. Outwardly the red eyes of Bronzed Cowbird and Red-eyed Vireo (*Vireo olivaceus*) appear similar. However, the red colouration is obtained in entirely different ways. The cowbird has large blood-filled venous sinuses in the anterior surface of the eye which give the red colour. The vireo, on the other hand, produce the red colour through red-pigmented organelles in the eye (Hudon and Muir 1996).

REFERENCES BBL Data [movements], Carter 1984 [behaviour, vocalizations, measurements], 1986 [behaviour], Clotfelter 1995 [behaviour], Friedmann 1929 [vocalizations, behaviour, measurements] , Klimkiewicz 1997 [age], Parkes and Blake 1965 [nomenclature], Pyle *et al.* 1987 [moult], Robbins and Easterla 1981 [range expansion].

100 BRONZE-BROWN COWBIRD *Molothrus (aeneus) armenti*

Cabanis 1851

Plate 38

This rare Colombian cowbird is often considered conspecific with Bronzed Cowbird (99).

IDENTIFICATION Likely only to be confused with Shiny Cowbird (102) within its restricted range. Bronze-brown Cowbird is slightly smaller and proportionately shorter tailed than the sympatric Shiny Cowbird. Also, both sexes of Bronze-brown Cowbird have obviously red eyes and the males have a conspicuous ruff on the nape. The similar, and perhaps conspecific, Bronzed Cowbird is spreading eastward in Panama and may eventually reach Colombia. These two species would be difficult to identify at a quick glance, but there are some diagnostic details. Bronze-brown Cowbird is not black with a bronzy sheen like Bronzed Cowbird, but rather it is brown with a bronzy iridescence. Also, the present species is much smaller than Bronzed Cowbird, in fact it is smaller than Shiny Cowbird!

VOICE Little known, there are no known recordings.

Song: The primary song is a *eez-eez-dzlee*

accompanied by a 'song spread' behaviour similar to that of Bronzed Cowbird.

Calls: Unknown.

DESCRIPTION This is the smallest cowbird, but it is chunky and short-tailed like Bronzed Cowbird. It has an erectable 'neck ruff' that is less developed than that of Bronzed Cowbird. The outer primary is not emarginate, the next three are; Bronzed Cowbirds have the outer three or four primaries emarginate. Wing formula: P9 < P8 > P7 > P6, P9 = P6.

Adult Male (Definitive Basic): The bill is black and the eyes are brick-red or brownish-red. Head, mantle, back, throat, breast and abdomen silky-brown, or sepia, with a bronze iridescence. There is a dark mask formed by the darker lores, and supralores that contrasts with the rest of the face. The lower back, rump and undertail-coverts have a purple gloss; sometimes the rump is bronze as well. In addition, the belly and undertail-coverts are a darker brown than the rest of the underparts. The flight feathers are black and iridescent. The primaries, secondaries and greater coverts show a blue-green iridescence; the middle and lesser coverts as well as the innermost greater coverts have violet reflections, and often the greater coverts have more of a blue rather than violet gloss. The legs and feet are black.

Adult Female (Definitive Basic): Similar to male, but duller, lacking much of the bronze gloss. The body, therefore, appears largely dark brown. A blue iridescence is restricted to the rump and wings, but is not as vivid as on the male and is not present in all females. The brown wings may show a slight green iridescence on the coverts and tertials, similarly the tail may have a dull green (sometimes blue) gloss. The neck often shows dull buff tips. Like in the nominate subspecies of Bronzed Cowbird, this species shows little sexual dichromatism as the female is largely dark and unicolored. Soft part colouration similar to male, although, the eye colour tends towards a yellowish-brown or dull reddish-brown.

Juvenile (Juvenal): Even duller than female and lacking all iridescence. Juveniles are pale brown with paler fringes on the underparts and lacking noticeable streaking. The head shows more of a pattern than in the adults; the darker lores are present as well as a paler supercilium, a darker eye-line and pale auriculars. The back appears faintly streaked and the wing-coverts are thinly fringed with buff, creating two thin wingbars. The flight feathers are also thinly fringed buff. The legs are black as in adults, although the bill is a paler horn colour with a dark tip and culmen. The juveniles also differ from the adults in having thinner, more pointed tail feathers, and by lacking the emargination of the primaries.

GEOGRAPHIC VARIATION Monotypic.

HABITAT It is found in dry scrub of the dry Caribbean coast of Colombia. At times it may frequent agricultural areas, and has been recorded foraging on rice. At Isla de Salamanca, they may be seen foraging near the road, often in the sandiest areas as well as in grasslands or stands of the shrubby Trupillo (*Prosopis juliflora*) tree.

BEHAVIOUR Forages on the ground, with tail cocked, on native grass seeds and sometimes rice. Often seen alone or in small groups, but forms larger flocks, up to 100–200 recorded in the past, in the post-breeding period (October). It is known to associate with Shiny Cowbirds, Giant Cowbirds (98), Red-breasted Blackbirds (66), Great-tailed Grackles (87) and Smooth-billed Anis (*Crotophaga ani*). There does not appear to be any attraction to livestock as in many of its congeners.

NESTING Unknown, but assumed to be a brood parasite like all of its close relatives. Females collected May–July showed active egg production confirming that nesting occurs during those months. One almost fully-developed egg extracted from the oviduct of a specimen showed an unmarked, white shell colour.

DISTRIBUTION AND STATUS This endangered species is extremely rare and local. It has been found in Atlántico, Bolívar and Magdalena provinces of Colombia. Recently, the only locality where it has been regularly observed is the Isla de Salamanca NP in Magdalena province.

MOVEMENTS Sedentary, but apparently forms larger flocks after breeding.

MOULT Little known, but likely similar to Bronzed Cowbird. Adults appear to undergo only one moult a year, the complete pre-basic moult which occurs after the breeding season. Specimens from late October–early November are in the last stages of this moult. Juveniles are expected to be seen April–July. The first pre-basic moult presumably occurs August–October. In the first pre-basic, the juvenile male feathers are replaced by dull female-like feathers, not glossy ones as in the adult male. Therefore, young males appear to delay plumage maturation, unlike in other cowbirds. The extent of first pre-alternate moults is unknown. It is presumed that pre-alternate moults are lacking in adults.

MEASUREMENTS Males are only slightly larger than the females. Male: (n=8) wing 96.5 (6.3); tail 61.4 (5.7); culmen 16.7 (1.1); tarsus 25.1 (1.9). Female: (n=8) wing 87.9 (1.6); tail 59.4 (2.2); culmen 16.2 (0.7); tarsus 23.3 (1.2).

NOTES This species is extremely endangered. The greater proportion of the world's population is found in the Isla de Salamanca NP which is being partially degraded by a new road and by poor water management practices which are killing its mangroves. An old record from Leticia in Amazonian Colombia refers to a bird on sale in a market which was unlikely to have been caught in the area; thus, Leticia should not be considered as part of this species' normal range.

REFERENCES Collar and Andrew 1988 [conservation], Dugand and Eisenmann 1983 [behaviour], Friedmann 1929 [description], 1957 [distribution], Hernandez-Camacho and Rodriguez-Mahecha 1986 [description, distribution, measurements, habitat, behaviour], Hilty and Brown 1986 [identification, voice, behaviour, habitat].

101 BROWN-HEADED COWBIRD *Molothrus ater* (Boddaert) 1783 — Plate 39

This is the most common cowbird in North America, and therefore the most widespread brood parasite on the continent; the total population has been estimated to be 20 to 40 million birds! Brown-headed Cowbird is most similar to Shiny Cowbird (102), with which it has recently come into sympatry in Florida.

IDENTIFICATION The male is unmistakable within its range. It is the only glossy black bird with a contrasting brown head. Females and juveniles are nondescript, and are often confused with other species. Their short conical bills are similar to those of sparrows or finches, but the cowbird is larger than most of these species. The streaky juveniles may be mistaken for House Finches *Carpodacus mexicanus*, but again that species is much smaller. Fledglings will give away their identity when their foster parents come in to feed them. In flight, Brown-headed Cowbirds are short-tailed for icterids and they have longish, somewhat pointed wings.

The real identification problem is to separate females and juveniles from the closely related Shiny Cowbird (102). These two species have only recently come into contact and identification criteria are still being developed; identification may be impossible at times. In general, Shiny Cowbirds differ from Brown-headed Cowbirds in shape. Shiny Cowbird has a longer, less finch-like bill (Figure 101.1), as well as a longer tail and shorter wings. Thus Shiny Cowbirds look slimmer, less chunky than Brown-headed Cowbirds. The bill of Brown-headed Cowbird nearly always shows a pale base to the lower mandible while that of Shiny Cowbird is nearly always entirely dark. Adult female Shiny Cowbirds tend to be darker in plumage, often with obvious grey tones rather than the browner colours of Brown-headed Cowbird. Some female Shiny Cowbirds are blackish with noticeable iridescence on the upperparts. This dark extreme is not known in Brown-headed Cowbird. Female Brown-headed Cowbird has a contrasting pale throat, unlike on most female Shiny Cowbirds. The pale supercilium is usually more marked in Shiny Cowbird. Brown-headed Cowbird may look 'empty-faced' or beady-eyed due to its less noticeable supercilium and eyeline. In fact this look is strengthened by the diffuse pale crescents above and below the eye on Brown-headed Cowbirds, as well as the paler lores than Shiny Cowbird. On Shiny Cowbird, these areas are darker, with a darker eyeline which does not allow the dark eye to stand out. Female Brown-headed Cowbird may have a contrasting paler head, mirroring the pattern of the males. This is most apparent on worn females from the western subspecies. Additionally, the pale edges on primaries P8–P6 tend to be white, noticeable and crisp on Brown-headed Cowbird, but buffy, diffuse or absent on Shiny Cowbirds. However, the extent of feather wear affects this feature. Finally, there is a difference in the wing formula of these two species; Brown-headed Cowbird has a more pointed wing. The outer primary (P9) on Brown-headed Cowbird is as long or longer than the second primary from the outside (P8), while on Shiny Cowbird P9 is similar in length to P6 or P7, the third and fourth primaries from the outside.

Identification features to differentiate juvenile Shiny and Brown-headed Cowbirds are still being developed. Preliminary observations suggest that juvenile Brown-headed Cowbirds are substantially paler than their tropical congener. This is particularly noticeable on the face, as Brown-headed Cowbirds have a pale and more featureless face, while the supercilium and darker eyeline are noticeable features on juvenile Shiny Cowbirds. Brown-headed Cowbirds tend to be much more scaly-looking due to the more contrasting paler feather fringes. Fresh Brown-headed Cowbirds have crisp pale fringes on the secondaries and primaries, which are not nearly so obvious on juvenile Shiny Cowbird. The shape differences mentioned above also apply. The lower mandible, and often part of the upper mandible, of juvenile Brown-headed Cowbirds is pale, while Shiny Cowbird juveniles usually have darker bills.

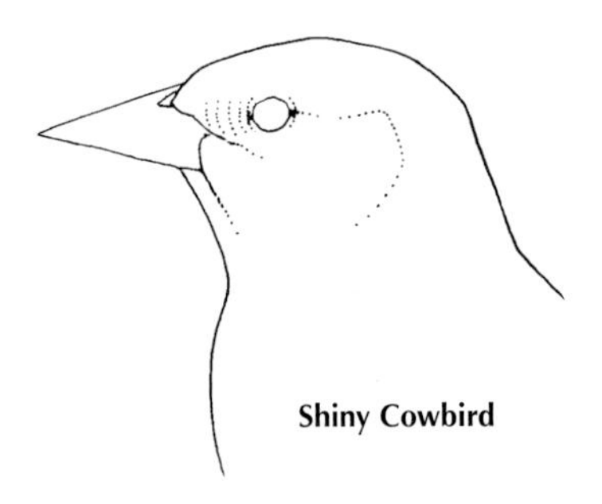

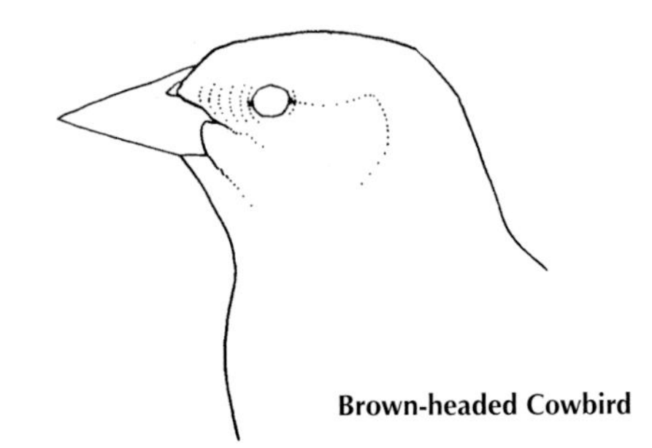

Figure 101.1 Bill shapes of Shiny and Brown-headed Cowbirds.

The larger, stockier Bronzed Cowbird is similar, but males do not have a contrasting browner head and both sexes have bright red eyes. Juvenile Bronzed Cowbirds have dark eyes and could be confused with Brown-headed Cowbird. However, Brown-headed Cowbirds are noticeably smaller, with a much smaller bill. Texas Bronzed Cowbird juveniles are also much darker than Brown-headed Cowbirds.

VOICE The sexes differ in their calls.

Song: The primary song is a series of liquid gurgles followed by a high whistle, *bub ko lum tseeee* or *glug glug glee* and is only performed by the male. *M.a.obscurus* inserts an additional series of quick low frequency bursts at the end of the first phrase, before the whistle.

Calls: The male flight call is commonly a whistled *weee tiuuuu* or *weee tiuu tiuu*; a single note flight whistle is also heard in some populations but it usually is made up of 2–5 whistles. Single note whistles are given more commonly by young males. The variation in the flight whistles is geographical in nature, and there is a tendency for dialects to form, particularly in the Great Basin race *artemisiae*. The common female call is a dry chatter *ch'ch'ch'ch'ch*, given in flight or while perched; males will rarely give a short rattle differing in quality from the female call. Females may give the rattle in an aggressive context to other females or as a response to the male song, often before the male has finished singing. This can be considered a duet and is analogous to that given by Screaming Cowbird (97). Analyses of the female chatter show substantial individual variation in the call, but without a subspecific or geographic pattern unlike male calls. A *kek* or *kuk* call may be given during sexual chases (*kek*) or while feeding (*kuk*) by both sexes.

In captivity, females are more likely to mate with males of their own subspecies. They appear to be able to identify the subspecific identity of males from their songs.

DESCRIPTION Brown-headed Cowbird is a small icterid with a short, finch-like bill. The wings are proportionately long, and the tail is short. In flight, the wings look more pointed than those of sympatric blackbirds.

Adult Male (Definitive Basic): Head and neck brown, becoming paler in midsummer, through wear and bleaching. The lores are noticeably darker than the rest of the head. The remainder of the body is black with a dull greenish gloss, slightly bluer on the upperparts. In some angles, the underparts also look blue glossed. The bill and legs are black and the irides are dark brown.

Adult Female (Definitive Basic): The plumage is brownish-grey throughout. The upperparts sometimes with a dull greenish gloss, and individual feathers with darker centers. Often shows a paler supercilium, but never very striking. The flight feathers are darker, with pale fringes, approaching white on the primaries of worn birds. The underparts are paler than the upperparts being palest on the throat which is usually nearly white; often there are obvious dark malar stripes bordering the pale throat. On some females, particularly when worn, the head is noticeably paler than the upperparts. Soft parts similar to those of male, but bill colour tends to be paler.

Immature Male (First Basic): Similar to adult males but can be identified by the variable number of retained brown juvenile feathers, particularly on the underwing-coverts and tertials. The primaries of these young males are usually worn and appear brown rather than black.

Immature Female (First Basic): Similar to adult females but retain juvenile feathers like in males. Since the juvenile and adult feathers are similar in colour, the two ages are not safely identifiable from each other in the field.

Juvenile (Juvenal): Similar to female but with more obvious pale fringes to the upperpart feathers, ranging from pale greyish to buffy-cinnamon, creating a scaled appearance to the upperparts. Buffy fringes obvious on wing-coverts and tertials. Underparts streaked darker grey. The bill, particularly the lower mandible, and legs, are paler than on adults and individuals usually retain a remnant of the fleshy gape while in this plumage.

GEOGRAPHIC VARIATION Three subspecies are recognized; differing mainly in size and colour saturation of the female. The males differ only in size. However, the geographic variation in flight calls may be useful in identifying birds of different races.

M. a. ater is eastern, breeding south to C Texas and C Arizona, and west to W Ontario, C Minnesota and NW Kansas. It is described above. This form is intermediate in size between the other two and the females average darker than in *artimesiae* and *obscurus*.

M. a. artimisiae is found in the Great Basin, and eastern Great Plains, occurring west of *M. a. ater* and south to central California (east Sierras) and N New Mexico. Note that at least one of the specimens collected from Alaska has been identified as belonging to *artimisiae* (Gibson and Kessel 1997). This race may in fact be much more common further north than currently suspected, in areas where *M. a. ater* has been assumed to be the local subspecies. This is the largest of the three subspecies. The bill shape of *artemisiae* averages both longer, more slender and more narrow than that of *ater*. The culmen of *artemisiae* averages straighter than the curved culmen of *ater*. The males are the same colour as the other two subspecies; however, like *obscurus*, females of *artemisiae* are paler than those of *ater*. Named after Sage-brush (*Artemisia tridentata*) since their ranges are coincident.

M. a. obscurus is the westernmost subspecies, being found along the Pacific coast, west of the Cascades and Sierra Nevada, but it also occupies the range south of the other two species in northern Mexico. Of the three, this is the smallest form. Females are paler than those of the other two, however this difference is most obvious when compared to *ater*. Nestlings of *obscurus* have yellow gapes, while the other two forms have white gapes.

The birds in coastal British Columbia and the Pacific Northwest, where they are relatively recent invaders, have not been adequately identified to subspecies.

HABITAT Breeds in grasslands, open woodlands and particularly forest edge. In the west, riparian corridors are frequented. Brown-headed Cowbirds are also common in residential areas during the breeding season. They avoid continuous forest, and have greatly benefited from forest fragmentation throughout North America. In the far north, Brown-headed Cowbird has an affinity for Aspen (*Populus* sp.) forest, particularly at the forest edge, and tends to avoid coniferous forest. Forages primarily on the ground on lawns, pastures and other open areas, and will not nest too distant from feeding areas. Winters in similar open habitats and forest edge, but in larger flocks. Animal corrals or feed yards are preferred foraging habitat throughout the year.

BEHAVIOUR Brown-headed Cowbirds show a peculiar temporal separation of breeding activities and self-maintenance behaviours. Courtship, nest searching and parasitism are conducted during the morning, but all feeding is done in the afternoon, often away from breeding areas. They have been found to vary both in the adult sex ratio and in the mating system: in Kansas, this species is polygamous but in E California it is monogamous. In male biased populations, monogamy prevails. The strength of the pair bond depends on the mating system, but the tendency is for females to consort with one male and mate with him even though she is courted by other males.

Brown-headed Cowbirds may return to the same breeding areas year after year (Laskey 1950). Males arrive at the breeding grounds before females, typically in early to mid March in the northeast, a little later for birds at the northernmost parts of the range. Males either display singly or in small mobile groups. The most intense display by males is the full 'song-spread', which is accompanied by the song. Males 'fluff up', raising the feathers of the nape and breast, the wings are raised like a cape and the tail is fanned as the bird bows forward and sings, finishing by wiping its bill on the branch. The intensity of the display is dependent on the social context, i.e. whether another cowbird is close by and the sex of that cowbird. The most intense version is given to other males. The most common threat/aggressive display is the 'bill-tilt', as in other blackbirds.

Both sexes appear to maintain a morning home range but these do not appear to be defended as a true territory (Laskey 1950). In the afternoon, they will congregate in flocks at favourite feeding spots. In monogamous situations, males may follow the female around and 'guard' her from other males during the morning. Brown-headed Cowbird is usually a gregarious species. In winter, cowbirds will mix with other icterids in large feeding flocks. These flocks coalesce in the evening in large blackbird roosts, sometimes numbering in the millions. During the non-breeding season, Brown headed Cowbird is mainly a grain eater. However, during the breeding season there is a marked shift towards insect food, a shift more pronounced in females than males (Ankney and Scott 1980).

Groups of male Brown-headed Cowbirds establish a dominance hierarchy. Most of the courtship behaviour is performed by the dominant male. When females are maintained in isolation, the songs of males will elicit pre-copulatory behaviour in the female (West *et al.* 1981b). Females react aggressively towards vocalisations of other females, not males. Males avoid models of other males in experiments, but they actively court female models (Dufty 1982b). Rothstein *et al.* (1988) studied the role of the flight whistle of male Brown-headed Cowbirds. They concluded that 'perched' songs were given in male–male interactions over short distances or during courtship. The monosyllabic and multisyllabic versions of the flight whistle were only given rarely while perched, but commonly while in flight, and were used as communication with male or females over long distances. Males respond aggressively to flight whistles.

Cowbirds participate in unusual behaviour where they will solicit preening from other species by approaching the other bird and freezing with the bill pointed down and the nape feathers raised. This display is not normally given to other cowbirds. A wide variety of birds, including other blackbirds have been seen to respond to this solicitation by preening the cowbird. Often the bird being solicited reacts by pecking at the cowbird. The cowbird is persistent, avoiding the bird's blows but continuing to approach even closer. Eventually the bird calms down and may begin to preen the cowbird, although some species do not react in this way. Both sexes of cowbird partake in this display (Selander and La Rue 1961). The benefits (if any) to the preener are not understood. It has been suggested that this display may aid in decreasing hostilities towards cowbirds in small birds (Selander and La Rue 1961), while others state that it has an aggressive genesis (Rothstein 1977a, Lowther and Rothstein 1980). This behaviour was first observed in Brown-headed Cowbirds, but now appears to be a display given by all of the cowbird species.

NESTING A generalist brood parasite, found to have parasitised more than 220 host species; 144 of these have been seen to fledge young cowbirds. Individual females are not host specialists like some cuckoos. Brown-headed Cowbirds are fecund, holding the record for the highest number of eggs laid in a single breeding season of any non-Galliform bird, 77 eggs were laid by a captive female. The eggs are variable in colour but tend to have a white or off-white background colour overlaid with dark spots which are concentrated at the broader end of the egg. The timing of breeding depends on local conditions, particularly when there are the highest number of available host nests. Experiments have shown that the physiological preparation for breeding and the development of gonads, is triggered by longer day length (Payne 1967). Female cowbirds do not use fat or protein reserves to produce their eggs. The nutrients needed for egg production are obtained directly from the diet (Ankney and Scott 1980).

The hosts used vary depending on the locality. The most commonly used host species recorded are: Yellow Warbler (*Dendroica petechia*), Song Sparrow (*Melospiza melodia*), Red-eyed Vireo (*Vireo olivaceus*), Chipping Sparrow (*Spizella passerina*) and

Eastern Phoebe (*Sayornis phoebe*). There appears to be a preference to choose hosts smaller than the cowbird itself, which allows the cowbird young to have a competitive advantage over host nestlings.

Females do most, if not all, of their egg laying in the morning, typically very early in the day. To find a nest they often perch where they can have a good vantage point and scan for activity, looking for the telltale signs that a nest is present. Otherwise, they 'crash' through the bushes, scaring birds off their nests. Typically, the female cowbird will peck or remove a host egg when it lays its own egg. The timing of egg laying has to be precise. If the cowbird egg is laid in the nest too late in the incubation period it will not be able to hatch. There is one documented case of a nestling Brown-headed Cowbird ejecting a host nestling, in this case an Indigo Bunting (*Passerina cyanea*). This is the only known case of this type of behaviour in any cowbird (Dearborn 1996).

Contrary to common knowledge, many host species have defenses against cowbirds. Hosts can be classified as acceptors, those which accept cowbird eggs, and rejecters, those which reject cowbird eggs from their nests. Rejectors recognize cowbird eggs as 'oddballs' and remove the eggs from their nests. Rejectors include American Robin (*Turdus migratorius*), Grey Catbird (*Dumetella carolina*) and Bullock's Oriole (*Icterus bullockii*). Yellow Warblers may respond to cowbird eggs by making a new nest lining over the parasitised clutch and laying a new set of eggs; some orioles will make a similar second egg chamber.

Finally, all hosts are not equally good. Cowbirds need hosts that feed their young insects. Cardueline finches such as American Goldfinches (*Carduelis tristis*) and House Finches (*Carpodacus mexicaus*) tend to feed their nestlings with seeds; young cowbirds usually die in these nests. An interesting phenomenon has been observed in Ontario, where House Finch is a recent invader. During the early years of the invasion, cowbirds commonly parasitised the nests of House Finch, but never raised any young from them. During subsequent years the parasitism rate decreased, implying that the cowbirds had learnt that this new host, the House Finch was not a good one (D.Kozlovic, pers. comm.). DNA fingerprinting has shown that after the breeding season, four out of 11 pairs of juvenile and female adults caught together were related, suggesting that in some cases female Brown-headed Cowbirds may flock with their progeny in fall (Hahn and Fleischer 1995).

DISTRIBUTION AND STATUS Common to abundant throughout most of its range; the greatest breeding densities are found in the Great Plains.

Breeding. From SE. Alaska (at least two western Alaska records, *Am. Birds* 33: 206 and nine specimens Gibson and Kessel 1997), and the Arctic coast of the Yukon Territory (juvenile being fed by Savannah Sparrow (*Passerculus Sandwichensis*) June 27, 1993, at Komakuk Beach) to N British Columbia (Peace River District), N Alberta, N Saskatchewan, S Manitoba, C Ontario, S Quebec and S Newfoundland south to N Baja California, Guerrero, Oaxaca, Michoacán, Guanajuato, San Luis Potosí

*Figure 101.2 Ovenbird (*Seiurus aurocapillus*) host feeding fledgling Brown-headed Cowbird.*

N Tamaulipas, south Texas along the Gulf coast to central Florida. Expanding southwards in Florida.

Non-Breeding: From SW British Columbia along the Pacific coast to N California east through Arizona, S New Mexico, Kansas, Missouri to the Southern Great Lakes region, New England, S New Brunswick and Nova Scotia south to S Baja California, Oaxaca, C Veracruz, along the Gulf coast to southern Florida. Casual on Cozumel Island.

MOVEMENTS Brown-headed Cowbird is a short distance migrant. Based on banding data for northeastern birds, the distance between the breeding grounds and wintering areas is 800–850 km. Brown-headed Cowbird is often seen migrating diurnally, sometimes in mixed flocks with other icterids.

Stevenson and Anderson (1994) list 53 foreign recoveries in Florida, of birds banded in Delaware, Maine, Massachusetts, New Hampshire, New Jersey, New York, Pennsylvania, Rhode Island, South Carolina, Vermont, Virginia, Alabama, Illinois, Indiana, Louisiana, Mississippi, Ohio, Ontario and Quebec. A similar pattern is found in the number of cowbirds banded in winter in Florida and recovered elsewhere.

Two records exist for the Western Palearctic; a dead female at Telemark, Norway (June 1987) and an adult male found at Islay, Scotland (April 1988). There are several records for the Caribbean: one in Cuba and the rest in the Bahamas (New Providence, Great Inagua, Eleuthera and Grand Bahama). It has also been observed in Bermuda. In Mexico, there are stray records from Guadalupe Island, offshore in Jalisco and San Juanito, Trés Marias Islands (Stager 1957). This species undertakes long migratory flights and has been collected 350 miles out at sea off the coast of northern Oregon (Sanger 1967). This species is regularly encountered 16–32 km offshore during pelagic birding excursions off the California coast. An immature male landed on a ship and died 560 km off the coast of N Oregon, surprisingly this individual was determined to belong the interior subspecies, *M.a.artemisiae* (Sanger 1967).

Earlier this century, Brown-headed Cowbird spread to eastern North America, where it was previously absent. Similar range expansions have occurred in the west. Prior to 1900, cowbirds were absent from California apart from the lower Colorado River valley, and from the Pacific Northwest. This population (*obscurus*) has not crossed the Sierra Nevada and has begun intergrading with *artemisiae*. In the east, the range expansion has stopped, except in Florida where the species has been quickly moving southward in the peninsula.

The oldest Brown-headed Cowbird reached the age of 16 years and 10 months (Klimkiewicz 1997). A male banded during the winter in Pennsylvania was retrapped in Michigan during the spring, aged 15 years and 10 months.

MOULT Adults undergo a complete definitive pre-basic moult in the breeding grounds, July–October. There are no pre-alternate moults in adults. The tail moult is often very rapid, with individuals commonly dropping all of their old rectrices simultaneously and appearing tailless for a short period. Changes in the plumage are caused by fading and wear, thus fall birds are fresh, and late summer birds are most worn. Juveniles have an incomplete first pre-basic moult in August–October. Rarely, the first pre-basic is complete. First basic males usually retain several juvenile underwing-coverts, or rearmost scapulars. Some will also show juvenile feathers retained on the outer edges of the back, mid-line of the belly, and around the eyes. Young males in their first breeding season also show more worn and faded (brownish) primary tips than adults. First basic females apparently retain juvenile feathers, but these are extremely difficult to see as the juvenile and first basic feathers are similarly coloured, making ageing of females very difficult. The first pre-alternate moult appears to be restricted to a few males, particularly of the race *obscurus* (Pyle 1997b).

MEASUREMENTS Males are larger than females.

M. a. ater. Males: (n=23) Wing 110.2; weight 48.9.

M. a. artemisiae. Males: (n=283): Wing 105.9 (2.7); culmen 17.3 (0.9); tarsus 25.0 (0.8); weight 47.5 (3.7). Females: (n=352): Wing 94.9 (2.5); culmen 15.4 (0.7); tarsus 23.3 (0.8); weight 37.6 (3.6).

M. a. obscurus. Males: (n=63): Wing 103.7 (2.3); culmen 6.5 (0.7); tarsus 23.2 (0.8); weight 40.2 (3.4). Females: (n=35): Wing 93.0 (2.2); culmen 14.4 (0.6); tarsus 21.8 (0.8); weight 32.0 (3.8).

NOTES Brown-headed Cowbird has been blamed for a number of decreases in host populations, particularly in rare and restricted range species. However, it is becoming clear that cowbirds are only one, and usually not the most severe, of several factors influencing host population declines. There are many cowbird population control programs, in effect, throughout the continent. The cowbird control program established to help the recovery of Kirtland's Warbler (*Dendroica kirtlandii*) in Michigan successfully decreased cowbird parasitism rates from 59% to 6% in nests of this endangered species. The parasitism rate currently averages 3.4% over a 10 year period. The result was an increase in the number of young Kirtland's Warblers reared; however, this did not give rise to an increase in the population of the warbler, suggesting that they were limited by other factors such as lack of suitable breeding habitat.

REFERENCES Baird 1958 [moult], Bond 1988 [movements], Burnell and Rothstein 1994 [voice], Dearborn 1996 [nesting], Elliott 1977 [behaviour], Fleischer *et al.* 1991 [geographical variation, movements], Friedmann 1929 [Biology, behaviour, geographic variation], Grinnell 1909 [geographic variation, measurements], Holford and Roby 1993 [nesting], Howell and Webb 1995 [movements], Kelly and De Capita 1982 [Kirtland's Warbler, cowbird control], Lowther 1993 [behaviour, vocalizations], McKay 1994 [European records], Pyle *et al.* 1987 [moult], Rothstein 1978 [geographic variation], Rothstein and Fleischer 1987 [voice], Rowley 1984 [distribution], Sanger 1967 [movements], Selander and La Rue 1961 [behaviour], Yokel 1986, 1987, 1989 [behaviour].

102 SHINY COWBIRD *Molothrus bonariensis* (Gmelin) 1789 — Plate 39

The ubiquitous cowbird of South America, and recently the Caribbean. Its northward spread has brought it to Florida and the southern US in the last few years. It may already be breeding in the US, but confirmation is difficult. In many ways, this is a very similar species to Brown-headed Cowbird (101), and like all *Molothrus* it is a brood parasite.

IDENTIFICATION There are two identification pitfalls, firstly, separating this species from the other all-black icterids, and secondly making the identification between it and the similar Brown-headed and Screaming Cowbirds (97). The former problem will be dealt with under the species accounts of each of the look alikes.

Male Shiny Cowbirds are entirely black with a vivid bluish or purplish gloss, particularly on the head and breast. Male Brown-headed Cowbirds are easily separated due to their contrasting brown heads. Both sexes of the all-black Screaming Cowbird are a tougher identification problem. At close range, male Shiny Cowbird has a stronger gloss and brown eyes, rather than the chestnut or reddish-brown colour of Screaming Cowbird. The bill of Shiny Cowbird is relatively longer and shallower at the base, unlike the more finch-like bill of Screaming Cowbird. Relative to its body size, Screaming Cowbird has longer wings than Shiny Cowbird, and this is noticeable in the field, particularly in male Screaming Cowbirds. In addition, Screaming Cowbirds have longer flank feathers that often blow about in a breeze or are flared out voluntarily by the birds, unlike Shiny Cowbirds which looks sleeker. Note that during the breeding season, Screaming Cowbirds travel in pairs or groups of several pairs, while Shiny Cowbirds are in small flocks of both sexes or many males, as well as single individuals. See Voice for other differences. In North America, a potential pitfall is Brewer's Blackbird (95). Male Brewer's Blackbird is similarly coloured to Shiny Cowbird and is a similar size. Unlike the cowbird, Brewer's Blackbird has bright yellow eyes, is slimmer and longer-billed. Nevertheless, an out of range Shiny Cowbird could be overlooked as a dark-eyed Brewer's Blackbird.

Female Shiny Cowbirds are nondescript and easily confused with several species. Particularly similar, in fact it has been considered nearly identical, is female Brown-headed Cowbird which is sympatric with it in parts of Florida and wherever Shiny Cowbird strays in North America. The primary features used to separate the two species are structural. Shiny Cowbird is a slim, long-tailed, longish-billed bird that looks more elegant than the stockier Brown-headed Cowbird. Female Brown-headed Cowbird has a shorter, more finch-like bill (Figure 101.1) as well as a shorter tail than Shiny Cowbird. The bill of female Shiny Cowbird is almost always entirely black, while on Brown-headed Cowbird the lower mandible shows a pale base. In general, female Shiny Cowbirds are darker plumaged and show less contrast on the throat. Both species have pale throats, but Brown-headed Cowbird's is more strikingly pale. Shiny Cowbirds also have paler, more contrasting supercilia than Brown-headed Cowbirds. The eyestripe of Shiny Cowbird is bolder, hiding the shape of the eye, while the area around the eye on Brown-headed Cowbird tends to be pale, making the eye stand out and appear beady. Part of this is due to the paler lores on Brown-headed Cowbird, but is also due to diffuse pale brown eye crescents. Brown-headed Cowbirds from the western US may show a contrasting paler head, particularly when worn. This is not seen on Shiny Cowbird. A feature that may help in the identification of these two difficult species is the edging on the outer primaries. Brown-headed Cowbird shows consistently whiter and crisper edges to primaries P8–P6. The edging is not nearly so obvious, being buff or absent, on female Shiny Cowbirds. Note that Shiny Cowbirds may sometimes be very dark, blackish or nearly so, and show a dull iridescence on the upperparts; this individual variation in darkness is not observed in Brown-headed Cowbird. A female cowbird that looks noticeably darker than all others should be scrutinized for other features of Shiny Cowbird. Vocal differences between female chatters have not been noted, and are therefore not useful in field identification. Finally, the two species differ in their wing formula. On Shiny Cowbird, P9 (the outer primary) is equal in length to P7 or P6, while on Brown-headed Cowbird, the outer primary (P9) is equal to or longer than P8, the second outermost primary. This is due to a more pointed wing on Brown-headed Cowbird. In addition, Brown-headed Cowbird averages longer-winged than Shiny Cowbird, both relatively and absolutely. This may be noticeable as a difference in the primary projection of Brown-headed Cowbird, but needs to be confirmed.

Juveniles differ in similar ways to the females, largely in structure, once soft parts are fully grown. Juvenile Brown-headed Cowbirds appear to be paler and more scaly due to more contrasting fringes on the upperpart feathers. This paleness is particularly noticeable on the face, as Brown-headed Cowbirds have a pale and more featureless face, while the supercilium and darker eyeline are noticeable features on juvenile Shiny Cowbirds. Fresh Brown-headed Cowbirds have crisp pale fringes on the secondaries and primaries, these are not nearly so obvious on Shiny Cowbird juveniles. Bill colour averages darker on juvenile Shiny Cowbirds, commonly being black or showing a restricted pale base to the mandible. On Brown-headed Cowbird, the bill is often largely pale, showing only a dark culmen and tip. However, there is some overlap in bill colour. Identification features for juvenile Shiny and Brown-headed Cowbirds are still being developed.

In parts of South America, female Shiny Cowbird may be confused with female Chestnut-capped Blackbird (60). Female Chestnut-capped Blackbird is slightly smaller and slimmer than the cowbird and has a relatively longer and thinner bill. Like the cowbird, Chestnut-capped Blackbird is dull and lacks strong features, but note that the underparts have a brighter olive or yellowish wash, not present on female Shiny Cowbirds. In addition, Chestnut-cap always has obscured dark streaking on the underparts that is seldom as well expressed on female Shiny Cowbirds. Baywing (96) also shares its range with Shiny Cowbird, but it is easily separated from it by its dark lores and obvious rufous on the wings, and the contrasting black tail.

VOICE The sexes differ in their calls, males whistle and sing and females give a chatter.

Song: The primary song of the male is composed of two parts, first by several liquid purrs followed by a high frequency whistle: *purr purr purr-pe-tssss-teeeee*, lasting about 3 seconds. The song is similar to that of Brown-headed Cowbird. Males also sing a more complex 'twitter song' that begins with three introductory notes followed by a twitter, the social context of this rarer song type is not known. Sometimes a flurry of 'twitter song', including 'flight whistles' among those notes, will follow the proper song extending the length by several seconds.

Calls: The male 'flight whistle' is variable. In Argentina and Chile, it tends to be a 4 note *tee-tee-tee-tiuuu*, or a single *pseeeee* but sometimes it is a longer version of the 4 note whistle. In E Ecuador, the flight whistles were more complex, being made up of many high-pitched notes, typically including at least one frequency modulated whistle. It would not be surprising if 'flight whistles' in this species experience the extreme geographic variation shown in male 'flight whistles' of Brown-headed Cowbird. Females have a harsh chatter or rattle call, similar to that of Brown-headed Cowbird, but a little slower and more metallic, and perhaps averaging longer in duration. The female rattle tends to last 2–3 seconds, depending on the level of excitement. In extremely rare circumstances a male may give a short chatter. Both sexes give a *chuck* note, this is uttered much more commonly than the corresponding *kuk* note of Brown-headed Cowbird.

DESCRIPTION Shiny Cowbird is a small, unadorned, slender icterid with a medium length bill. The wings are long and the tail is of medium length. While feeding, they often cock their tails. Wing Formula: P9 < P8 < P7 > P6; P9 ≈ P5; P8–P5 emarginate. *M.bonariensis bonariensis* is described.

Adult Male (Definitive Basic): An all-black icterid, glossed throughout. The body gloss is bluish, sometimes purplish, particularly on the head, while the wings are glossed bluish. The bill and legs are black, while the eyes are dark brown.

Adult Female (Definitive Basic): Brownish-grey throughout, paler on the underparts. The upperparts may have a slight bluish gloss. The throat is slightly paler than the breast, and often there is obscured streaking on the underparts. Supercilium slightly paler than the rest of the face, but this is often not noticeable. The bill and legs are black while the eyes dark brown. The form *'melanogyna'* is completely black with faint bluish iridescence on the back only.

Immature Male (First Basic): Similar to adult males, but they have a few retained contrasting browner feathers of the juvenal plumage, particularly on the underwing-coverts, tertials or body feathers. The primaries look brown and worn, contrasting with the blackish remainder of the wings.

Immature Female (First Basic): Basically indistinguishable from adult females, except they may have more worn primaries.

Juvenile (Juvenal): This is the only species of cowbird that shows sexual dimorphism in colouration at the juvenile stage. In general, juveniles are like females but are noticeably streaked on the underparts. and the wing and upperparts have paler fringes. Males are darker than females, and can look blackish-grey at times; females are light tawny or buff. The subspecies that has recently invaded the Caribbean and North America, *M. b. minimus*, does not show this dimorphism in the juvenal plumage.

GEOGRAPHIC VARIATION There are seven subspecies of Shiny Cowbird, differing mainly in size and the colouration of the females.

M. b. bonariensis is the most widespread race, found south of Amazonia, S through E and S Brazil, E Bolivia, Paraguay, Uruguay to S Chile (Aysen) and S Argentina (Santa Cruz). This race is described above. Birds from the highlands of Bolivia are larger than those from Argentina (Miller 1917). In the range of *M. b. bonariensis*, a dark female morph exists which was previously given the subspecific epitaph *'M. b. melanogyna'*; these females are most abundant in NE Argentina and SE Brazil, but are observed as far south as Buenos Aires province. The black morph is like the male, but lacks the strong gloss. The slight gloss of *melanogyna* is restricted to the upperparts. In direct comparison they can be identified as females due to their smaller size, as well as their behaviour. Another previously described race, *M. b. milleri*, from Mato Grosso, Brazil, and N Bolivia was described as having dark females similar to *aequatorialis* (Naumburg and Friedmann 1927, Gyldenstolpe 1945b). This race also appears to apply to the dark females previously known as *melanogyna*. It is unclear if there are two elements involved in explaining *melanogyna*, the first being that some individual females are dark while others are pale (a morph), while the second being that this is actually caused by a cline with northern *bonariensis* being darker than the southern ones.

M. b. minimus is found from NE Venezuela (Delta Amacuro) east through the Guianas to NE Brazil (Ceará, Maranhão and Pará); its range extends north through the Lesser and Greater Antilles to SE US and is expanding (see Movements). This is the smallest subspecies, that being its most distinctive feature.

M. b. venezuelensis is found in Venezuela from Mérida and Tabay east to Monagas, including easternmost Colombia, south to the edge of the Amazonian forest. This race has recently invaded the Netherlands Antilles, having been found in the early 1980s in Curaçao. They were relatively common by the early 1990s and breeding has been confirmed (Debrot and Prins 1992). The bill and tail average longer than on *bonariensis*. The female is much like *bonariensis* in its noticeably paler underparts compared to the upperparts, but it has a longer tail. Males are more purple than the other races, as in *occidentalis*. It is unclear if birds from eastern Ecuador are of this form.

M. b. cabanisii is found from E Panama (Darién) and in Colombia along the Andes, the Magdalena valley, and the Cauca valley from Santa Marta south to Cauca. It is the largest form of Shiny Cowbird, females are paler than *bonariensis*. This form is most similar to *venezuelensis*, but it is larger and more purple than that form.

M. b. aequatorialis extends from coastal SW Colombia south along the Ecuadorian Pacific coast to Guayaquil. This is a dark form, darker than *cabanisii* and *occidentalis*, and nearing *bonariensis*. The underparts of the females are almost as dark as the upperparts, but far from being black. It is smaller than *bonariensis*, yet larger than *occidentalis*.

M. b. occidentalis occurs along the Pacific coast of SW Ecuador (Casanga valley) to Peru, south to Arequipa. Females have distinctly paler bellies than the rest of the underparts, which are inconspicuously streaked with olive-brown. These are the palest females of the species. Males are more purplish than the other subspecies, resembling *venezuelensis* in this respect. The subspecific identity of individuals from northern Chile is unknown, but they may be of this form. However in the field, female Shiny Cowbirds from the Lluta valley near Arica, Chile, appear to be much darker below than is described for *occidentalis*. They also appear to be darker than the average *bonariensis* from central Chile (Jaramillo and Burke pers. obs.).

M. b. riparius ranges in Amazonia, west to the Río Ucayali in E Peru and strictly south of the Amazon river. Males are like *bonariensis*, but smaller. Females are much more contrasting than typical *bonariensis*, being blacker above and paler below, with the chin and throat abruptly paler than the chest. A form with black females described as *nigricans* appears to be a black form of *riparius*, similar to *bonariensis*. It is unclear if there is a break in the distribution between *riparius* and *bonariensis*; *riparius* may turn out to be merely a clinal variation of *bonariensis*. The subspecific identity of Shiny Cowbirds in Amazonia, north of the Amazon river, is not known.

As forest destruction proceeds in South America, suitable habitat is opening up for Shiny Cowbird, and many of these previously separate forms are coming into contact.

HABITAT Frequents pastures, grasslands, open forests and clearings in forest as well as open flood plain, oxbow lakes (cochas) and river islands in Amazonia. Forages on the ground like other cowbirds, often well away from trees. It takes well to agriculture and will become a pest on certain crops such as corn and rice (Miller 1917). In parts of its range, this species shows a close association with livestock, both foraging beneath and often perching on them. In the Lesser Antilles, it is rarer on islands with fewer animal feedlots even though host species abundance is equally high, suggesting that agriculture type is an important habitat and limiting factor in this cowbird's population. It winters in similar habitats, but usually in larger concentrations. Large nocturnal roosts often congregate in marshes.

BEHAVIOUR Like Brown-headed Cowbird, Shiny Cowbird divides its time during the breeding season between sexual and breeding activities in the morning, and foraging and maintenance behaviours (e.g. preening) in the afternoon. Most male display thus occurs in the morning, and egg laying is also a morning activity, sometimes beginning before sunrise. In the afternoon feeding areas, most display activity is aggressive in nature ('bill tilting') as birds compete for foraging space. There is no evidence that territories are established by either males or females. Their presence in any one breeding areas is ephemeral, sometimes only lasting for a few days or weeks (Fraga 1985, Mason 1987, Jaramillo unpub. data). During the night, they often congregate in large roosts; these are largest during the non-breeding season and are often in cattail (*Typha* sp.) marshes.

Shiny Cowbirds forage on the ground, and are fond of foraging around livestock. The will perch on the livestock and may clean them of parasites. Primarily they feed on insects disturbed by the livestock, as well as those insects that come to manure piles. They may use their bill and head to 'flip' over manure in order to search for food beneath it. Shiny Cowbirds also forage on caterpillar and beetle larvae, including insects pests. Reed (1913) concluded that the corn eaten by cowbirds was likely evened out by the number of corn insect pests they consumed. In Argentina, Shiny Cowbird also feeds on rice crops and is considered a pest (Miller 1917).

'Song spread', or 'ruff-out' displays are similar to those of Brown-headed Cowbirds, but are more often performed on the ground than in trees. On the ground, the male ruffles the feathers, partially opens the wings, and drops the tail while delivering the 'primary' song. When a female is present, he will hop in front of her and give this display vigorously, flapping and beating his wings as he displays. At times, the male will hover on distinctly 'bowed' wings about 50 cm from the ground and circle or fly in crescents in front of the female in an intricate display. The female may crouch and lift her tail; at this point the male may alight and the pair will mate. Displays while perched in trees are similar to those of Brown-headed Cowbirds, but the male seldom bows as deeply as its North American relative. An extreme display has been observed in Tucumán, Argentina, where a male on a fence bent his tail forward, and opened his wings to a horizontal position then jumped in the air and quickly repeated the jump, while giving a liquid *purrr* call. While the displays of Shiny Cowbird most closely resemble those of its sister species, Brown-headed Cowbird, they also show some elements reminiscent of Bronzed Cowbird (99), particularly during ground displays.

The mating system appears to be polygynous. No evidence exists for stable pair bonds during the breeding season (Mason 1987, Jaramillo unpub. data). Males sometimes consort with a female, even following them while they are looking for nests, but consortships do not appear to last. These consortships may be due to 'mate guarding', where the male guards the female against the advances of other males, securing that her next series of eggs will be fertilized by his sperm. Yearling males display to females and participate in the activities of the breeding season, but it is not known if they are successful at breeding or not. First-year females are sexually mature and lay eggs during their first breeding season.

NESTING Shiny Cowbird is a generalist brood parasite. It has been reported to parasitise a total of 201 species, of which 53 have successfully raised cowbirds. This species is not as well known as Brown-headed Cowbird, but it is likely equally fecund as that species. Unlike Brown-headed Cowbird, this species readily parasitises cavity nesting hosts. In a given region, the egg laying season tends to be synchronized with the nesting period of the 'highest quality' hosts. In Argentina and Chile, this season extends from September–January, while in Trinidad it is timed with the start of the rains, April–August. In Puerto Rico, the cowbird egg-laying season ranges from April–July. In the Magdalena valley of Colombia, egg-laying occurs from June until at least December (Boggs 1961).

Host choice is moderately well understood. There are two factors to consider, the abundance of the host and the 'quality' of the host. In the Caribbean, the most common species of passerines are not necessarily the most commonly parasitised. Here there appears to be a preference for orioles (*Icterus* sp.), Black-whiskered Vireo (*Vireo altiloquus*) and Yellow Warblers (*Dendroica petechia*) as hosts. The factors that affect the quality of a species as a host include: rejection of foreign eggs, nestling diet, nest survivorship, host size, host aggressiveness. In Argentina, one of the most commonly used hosts is Rufous-collared Sparrow (*Zonotrichia capensis*); however, it has a tendency to abandon multiple-parasitised nests, and nests have an extremely low probability of survivorship. The abundance of this host makes up for these losses. Preference for a host appears to be a balance between its quality and abundance. In Puerto Rico, the two most 'high quality' hosts are Yellow-shouldered Blackbirds (65) which fledge an average of 1.5 cowbirds per nest and Black-cowled Orioles (45) which fledge an average of 3.6 cowbirds per nest (Wiley 1985). On Andros Island, Bahamas, Black-cowled Oriole is the only known host of the newly-arrived population of Shiny Cowbirds (Baltz 1995, 1996).

Shiny Cowbird parasitism negatively affects the reproductive success of the hosts. Most of this effect is due to the fact that female Shiny Cowbird (like other cowbirds) removes a host egg when laying her own egg. As the nestlings grow, the cowbird outcompetes the young of smaller species, further decreasing the nesting success of the host. With larger hosts, the cowbird grows up alongside its foster siblings without affecting their chances of survival. Curiously, Shiny Cowbirds that parasitise Yellow-hooded Blackbirds (59) do not remove a host egg as they lay. The reasons for this are unclear, but it may be due to the aggressive nest defense by the male blackbirds, leaving little time for the cowbird to do anything more than lay her egg, which may take as little as 30 seconds.

The eggs are variable in colour. In eastern Argentina and southern Brazil, the eggs come in two distinct morphs. Some females lay completely unmarked immaculate white eggs, while others lay more typical spotted eggs. Immaculate white eggs have been reported from other parts of South America (i.e. Ecuador), so this trait may be more widespread than previously thought. Most host species readily accept both egg morphs, while some will reject both. However, it appears that some species accept one, but not the other morph. Both Chalk-browed Mockingbird (*Mimus saturninus*) and Brown-and-yellow Marshbird (74) accept spotted eggs, but reject immaculate eggs. The one species that is thought to reject spotted eggs but accept immaculate ones is Yellow-winged Blackbird (58), but the evidence is not conclusive. Therefore, the maintenance of this polymorphism may be selected for by only a handful of host species. Similarly, Rufous Horneros (*Furnarius rufus*) reject most Shiny Cowbird eggs but based on the width of the egg. Cowbird eggs which are less than 88% of the width of an hornero egg are rejected from their dark, domed nests. In Uruguay, where the hornero is a frequent host, cowbird eggs are substantially wider than in nearby Buenos Aires, where the cowbird rarely fledges from a hornero nest. Several other species are known to reject Shiny Cowbird eggs, such as some kingbirds (*Tyrannus* sp.), grackles (*Quiscalus* sp.), and mockingbirds (*Mimus* sp.). In the Caribbean, Northern Mockingbirds (*Mimus polyglottos*), Tropical Mockingbirds (*Mimus gilvus*), Grey Kingbirds (*Tyrannus dominicensis*) and Greater Antillean Grackles (92) were noted to reject Shiny Cowbird eggs (Post *et al.* 1990).

DISTRIBUTION AND STATUS Common to abundant throughout its range; however, it is rare and local along rivers in the Amazonian lowlands and is uncommon in some of the Lesser Antilles. In the S US and E Panama where it is a recent arrival it is still rare and local. It is a bird of lowlands, but has been observed to 3000 m in Bolivia (Miller 1917). In South America, it is found throughout the continent, avoiding forested areas (absent in most of Amazonia) and highlands. Ranges throughout Colombia, Venezuela, Trinidad, Tobago, Guyana, Surinam and French Guiana but its distribution in SE Colombia and S Venezuela is unclear. It may be found throughout Brazil, but there are areas where it may be absent (at least for the moment) in the far NW and in the Amazonian basin south of the Amazon river and north of Bolivia. In Ecuador and Peru, it is present in the lowlands east of the Andes as well as the Pacific Coast lowlands, at least as far south as Arequipa, Peru. An isolated population exists in Chile between Atacama and Aisén, but it is also present in the oasis valleys of Arica in the far north. It is not clear if these birds are southern extensions of the Peruvian population or northern extensions of the Chilean populations. Shiny Cowbird is found east of the highlands in Bolivia, and throughout Paraguay and Uruguay. In Argentina, this species is absent from the Andes but common throughout the lowlands in the north and as far south as Santa Cruz in Patagonia. Originally this species was restricted to South America, but it is quickly spreading. It is now found in Panama regularly west to the Canal Zone and perhaps even farther west sporadically (Ridgely and Gwynne 1989). It has 'hopscotched' northward in the Caribbean to Florida and other regions of the US; however, it is currently only regularly encountered in S Florida (see Movements) and there is no unambiguous evidence that the species has bred there even though it is extremely likely. In winter, it is only seen in south Florida. In the Caribbean, it has been found in the Bahamas (Balz 1995), Jamaica (Jeffrey-Smith 1972), Cuba, Hispaniola, Puerto Rico, Mona Island, St John (British Virgin Islands), Marie-Galante (Guadeloupe), Antigua, Martinique, St Croix, St Lucia, St Vincent, Barbados and Grenada, the Grenadines as well as Curaçao (Cruz *et al.* 1985). In most of these islands, Shiny Cowbird is now an established breeding resident.

MOVEMENTS Partially migratory in southern Argentina and apparently in North America. Since 1970 this species has spread into Panama, west to the Canal Zone, and the spread is still continuing. It is a recent arrival in Chile, first reported around 1877. The suggestion has been made that introductions may have caused that expansion of range; however, an unaided expansion is consistent with what is occurring in the rest of the range. Shiny Cowbirds have also been spreading northwards at an explosive rate through the Caribbean islands. The initial dates of first detection in the Caribbean are as follows: Carriacou Island, Grenadines (1899), Grenada (1901), St Vincent (1924), St Lucia (1931), St Croix (1934), Martinique (1948), St John (1955), E Puerto Rico (1955), Marie Galante (1959), Antigua (1959), W Puerto Rico (1965), Mona Island (1971), Dominican Republic, Hispaniola (1972), Jamaica (Rocklands feeding station in Jeffrey-Smith 1972 but no date given), Haiti, Hispaniola (1980), Cárdenas, W Cuba (1982), North Andros Island, Bahamas (1994) (Baltz 1995, Cruz *et al.* 1985) and Barbados (previous to the late 1920s) (Bond 1928a). It has been suggested that the spread of this cowbird has been aided by introductions as Shiny Cowbirds are kept as cagebirds. However, the long dispersal distances observed in the US suggest that long distance movements between islands may not be that unusual. The earliest report from the islands was of a bird on Vieques Island (near Puerto Rico) in 1860, much earlier and further west than the next few records. This anomalous record may have been due to an escaped cagebird. The spread is still continuing on the islands, and many have still not had a record (i.e. Jamaica), but it is probably just a matter of time before it reaches these islands. Curiously, it has been reported that the cowbird has disappeared from some of the Lesser Antilles, such as Marie Galante. A separate invasion of Curaçao (Netherlands Antilles) by the race *venezuelensis* began in the 1980s.

Shiny Cowbird reached the Florida Keys in 1985 and mainland Florida two years later. There is now a resident population in south Florida which is suspected to breed (males collected had enlarged testes), and this is augmented by new arrivals each spring, with the majority leaving in fall. Spring concentrations are observed commonly on the Dry Tortuga Islands, otherwise most occur on the west coast of the peninsula, with fewer on the east coast north of Miami. In Florida, almost all records are from coastal counties (Stevenson and Anderson 1994). During spring, dispersing individuals have ranged far to the north and west. There have now been records north to New Brunswick in Canada and one from Maine. To the west, they have been recorded from Texas twice and once in Oklahoma. Otherwise, most records away from Florida are along the Gulf Coast and coastal southeastern states. The largest flock observed so far in North America totalled 52 birds. During the first several years of the expansion into the US, the increase in numbers was explosive. In recent years, the expansion appears to have slowed, either because less attention has been paid to them by birders, or because spring conditions may not have been conducive to migration to the mainland, or due to a real decrease in numbers.

Outside Florida, where the species may breed, there are North American records for: New Brunswick (adult male); Maine (male); Oklahoma (first basic male, specimen); Texas (two records of male specimens plus four other reports); Louisiana (several); Alabama (many); Mississippi (several); Tennessee (a pair observed in July 1995); Virginia (one record); North Carolina (more than 10 records including a specimen); South Carolina (several, including specimens); Georgia (several beginning in 1989). Most records away from Florida have occurred in the spring.

MOULT Adults undergo a definitive pre-basic moult in February–April (*bonariensis*); this is a complete moult. Pre-alternate moults are missing in adults, and probably first basic individuals. Alternate plumage is acquired through wear; therefore, fall birds are fresh and late summer birds are most worn. The timing of moults depends on the breeding schedule. Pre-basic moult occurs after the breeding season. This varies depending on the breeding schedule of hosts, so moult schedules will differ for each subspecies. An adult male specimen (*minimus*) from the US was finishing its pre-basic moult in October. The form *venezuelensis* is also known to be in moult during October, but may be in moult as early as September. In Argentina, the nominate form is in moult February–April. Juveniles experience an incomplete first pre-basic moult in January–March (*bonariensis*), usually retaining several underwing-coverts or tertials as well as a variable number of body feathers. Retained juvenile feathers are obvious on young males, but difficult to detect in young females. First basic males show browner (more bleached) primaries than adults.

MEASUREMENTS Males are larger than females.

M. b. bonariensis. Males: (n=45) wing 111 (105–117); culmen 19.4 (16.5–20.3); (n=?) tail 83 (73–91); (n=42) tarsus 28.9 (27.3–30.8). Females: (n=64) wing 99 (94–112); culmen 17.6 (14.2–18.9); (n=?) 72 (69–75); (n=50) tarsus 26.8 (1.2).

M. b. minimus. Males: (n=14) wing 98.3 (92–112); tail 76.7 (72–86); (n=13) culmen 18.4 (16–20); (n=11) tarsus 24.2 (22–27). Females: (n=10) wing 85.9 (83–90); tail 66.9 (63–71); culmen 16.3 (15–17); (n=7) tarsus 23 (22–25).

M. b. venezuelensis. Males: (n=9) wing 113.6 (104–126); tail 86.5 (70–99); (n=8) culmen 20 (19–21); (n=7) tarsus 27.3 (25–29). Females: (n=6) wing 97.7 (92–104); tail 75.4 (70–81); culmen 17.8 (17–19); (n=4) tarsus 25.5 (25–26).

M. b. cabanisii. Males: (n=10) wing 127.6 (122.5–132); tail 100.5 (95.5–104); culmen 20.7 (20.0–21.5); tarsus 29.5 (29–30). Females: (n=10) wing 109.1 (102–118); tail 87.8 (80–95); culmen 18.6 (17–20); tarsus 27.2 (26–28).

M. b. aequatorialis. Males: (n=8) wing 118.6 (114–122.5); tail 91.8 (88–97.5); culmen 20.9 (20–22); tarsus 28.0 (23–29.5). Females: (n=6) wing 101.8 (99–105); tail 77.8 (76–79.5); culmen 18.4 (17–20); tarsus 26.4 (25–28).

M. b. occidentalis. Males: (n=?) wing 109.6 (108.5–113.0) ; tail 84 (81–86); culmen 20.5 (20.0–21.0). Females: (n=4) wing 95.5 (92–98); tail 75.0 (72–80); culmen 18.5 (18–19); (n=2) tarsus 25, 27.

M. b. riparius. Male: (n=3) wing 108 (106–111); tail 77.3 (72–88); culmen 19.6 (18.5–21); (n=2) tarsus 25.5, 26.5. Female: (n=1) wing 106; culmen 20; tail 82.

NOTES Fjeldsa and Krabbe (1990) state incorrectly that the first record of this species in Chile was in 1968; old records extend to the 1800s. There is some controversy regarding whether this species was introduced to Chile or whether it arrived here independently from Argentina.

REFERENCES Baltz 1994 [movements], Chapman 1915 [measurements, geographic variation], 1917 [geographic variation], Cruz *et al.* 1985 [movements, Caribbean], Dalmas 1900 [description, measurements], Fraga 1985 [nesting], Friedmann 1929 [voice, behaviour, moult], Griscom and Greenway 1937 [geographic variation], Gyldenstolpe 1945a [measurements], Mason 1986a and 1986b [nesting, host choice],1987 [behaviour], Mason and Rothstein 1986 [nesting], Mermoz and Reboreda 1994 [nesting], Miller 1917 [habitat, behaviour, nesting, measurements], Post, Cruz and McNair 1993 [movements, North America], Post and Wiley 1977b [movements, Caribbean], Ridgely and Gwynne 1989 [movements], Stone 1891 [geographic variation], Traylor 1948 [measurements, geographic variation], Wiley 1988 [nesting, host choice], Wiley and Wiley 1980 [nesting].

103 BOBOLINK *Dolichonyx oryzivorus* (Linnaeus) 1758 — Plate 28

An irresistibly likeable little blackbird of grasslands and meadows of North America, its wonderful flight songs are difficult to miss. Its life in winter is not so well understood.

IDENTIFICATION The breeding male is distinctive in its largely black and white plumage and is difficult to misidentify. However, Lark Bunting (*Calamospiza melanocorys*) is also largely black and white and can live sympatrically with Bobolink in open grasslands. Note that the extent of black on the bunting is greater, lacking the buff nape patch and white rump and lower back. The bunting is also noticeably smaller; both birds perform song flights.

The identification of females and non-breeding (basic) birds is more problematic. Their brown and streaky plumage is similar to that of many sparrows (Emberizinae) or grosbeaks (*Pheucticus* sp.). The *pink* call of Bobolink is characteristic and is helpful in identification. The long primary extension is also useful in identifying Bobolink. Note that North American sparrows that are streaked on the underparts tend to show streaked breasts and flanks, not just streaking on the flanks. Most North American sparrows lack a pale central crown stripe. One species that is similar to Bobolink in both these respects is Le Conte's Sparrow (*Ammodramus leconteii*), which also has pointed tail feathers. Note that the sparrow is very small and has a buff breast that contrasts with a white belly, and shows a purplish-streaked nape. Bobolink is noticeably larger than sparrows, but approaches that of Black-headed (*Pheucticus melanocephalus*) and Rose-breasted Grosbeaks (*Pheucticus ludovicianus*) which are brown and streaked as females and immatures. The grosbeaks have much deeper, stronger bills and have a different face pattern with a dark face contrasting with a pale throat, and lack a dark eyeline. They also have noticeable wingbars. The grosbeaks are birds of open woodlands, not open country. In the wintering grounds there are few brown, streaky birds. The species that are most similar to Bobolink in appearance are female White-browed Blackbird (67) and female Red-breasted Blackbird (66), both of these resemble each other and can be treated together. They are similar in size to Bobolink, have streaked backs, largely unstreaked underparts, a pale central crown stripe and similar face patterns. Note that the blackbirds are duller, lacking the rich buff colour of the underparts and fringes of the upperparts of Boblink. The blackbirds have barred tail feathers and tertials. In addition, they lack the two pale 'suspenders' (braces) of Bobolink, are often streaked on the center of the breast and may show a pinkish wash to the belly. The grass-finches (*Emberizoides* sp.) are similar to Bobolink in being streaked on the upperparts, having a pale face and pointed tail feathers. However, the grass-finches lack streaking on the flanks, are whitish below, lack a dark post-ocular stripe and have much longer tails. Female Yellow-winged Blackbird (58) has yellow shoulders and dark ear-coverts.

VOICE Song: A long joyous bubbling song, usually given while in the air. It is onomatopoeic, and may be interpreted as *bob-o-link bob-o-link link wink bob-o-link...* Each male has two different song types ('alpha' and 'beta') which differ in length and include certain specific notes. The 'alpha' song is approximately 7 seconds long, and begins with several introductory 'alpha' notes (*puck-puck-pi-deedla-eh*), followed by a series of interior notes (*ah-eee-ew-d-t-d-t-dee*) and ends with a sequence of quick gurgled warbled notes and a finishing *ew*. The 'beta' song lasts approximately 4.5 seconds and consists of 'beta' introductory notes (*Pete-n Pete-n*) followed by the latter half of the interior notes and the same warbled notes that close the 'alpha' song (Wittenberger 1983, Capp 1992). Individual males differ somewhat in the introductory notes that they give. The two song types may be given in a mixed order. Both types are given both in flight and from a perch. Different colonies have specific dialects, with colonies that are closer to each other having dialects that are more similar to each other than to those which are farther away. See Behaviour.

Calls: The basic call, used throughout the year and given in flight is a sweet *pink*. Several other calls are only given during the breeding season, most of which are performed by the male. A short buzz is given after the circle-flight display; while chasing other males a *tchek* is heard; during incubation a *chunk* is used as an alarm note; a *tchenk* is the alarm note once the young hatch; and a *see-yew* is an alarm call given while in flight over the intruder. Females use a *quipt* note as an alarm note after the young hatch and a *zeep* is given by females in aggressive interactions with other females.

DESCRIPTION A small blackbird with a short conical bill. Each feather of the short tail tapers to a sharp point, similar to some sparrows of the genus *Ammodramus*. The pointed wings are long, and show a long primary extension. Wing Formula: P9 > P8 > P7 > P6; P9 > P8; P8–P7 emarginate.

Adult Male Breeding (Definitive Alternate): The bill is black with the extreme base of the lower mandible blue-grey. The eyes are dark. The head, other than the nape, and the entirety of the underparts are black. The undertail-coverts may still retain pale olive tips. The nape is buff or cream and becomes whitish late in the summer through wear and bleaching. The back is black with buff edges, making it look somewhat streaked. These edges are widest in the center of the back, forming a wider whitish stripe down the mid-line of the back. This pale stripe connects with the buff nape. The scapulars are white, contrasting with the back and wings. The lower back, rump and uppertail-coverts are also white. The wings are black with white fringes on the tertials and inner greater coverts. The tips of the primaries often show whitish edges. The underwings are black with white marginal coverts. The spiky black tail has thin olive fringes. The legs are brownish-black.

Adult Male 'Pre-breeding' (Definitive Alternate, fresh): Similar to breeding plumage but the black is heavily veiled with pale yellow tips, particularly on the underparts and back. The white rump is tipped olive or greyish-buff. The coverts are fringed pale brown and the white fringes on the tertials are wider. Tail feathers may be obviously tipped and edged olive. These tips wear away by May revealing the largely black and white breeding plumage.

Adult Male Non-Breeding (Definitive Basic): Indistinguishable from non-breeding female other than by its slightly larger size. Some may show a few blackish feathers on the throat.

Adult Female Breeding (Definitive Alternate): Bill pinkish-horn with a black culmen and tip. The eyes are dark. A streaked, sparrow-like (Emberizidae) bird, quite unlike breeding male. Pale buff lores and supercilium that contrasts with the darker crown. The face is pale yellowish-buff with a contrasting blackish eyeline behind the eye. The crown is brownish-black with small olive streaks, and there is an obvious whitish or pale buff central crown stripe. The nape is olive-buff with light dark streaking, forming a paler patch in the same position as the male's buff nape patch. The crown and back are both darker than the nape. The back is blackish-brown with olive edging making it appear widely streaked. Two of the lateral rows of pale back streaks are whitish, creating the appearance of two 'suspenders' on the back. The scapulars are similar to the back feathers but more olive and with darker centers. The rump is olive with darker central streaks on each feather. The underparts are yellowish-buff, darkest on the lower throat and paler on the chin and belly. There are crisp blackish streaks that are confined to the flanks and undertail-coverts. The wings are dark brown with pale fringes on all feathers. The coverts are fringed grey, warmest in tone on the innermost coverts. The tertials are fringed pale yellowish-brown. The fringed rather than tipped pattern of the coverts does not produce noticeable wingbars. The bend of the wing is yellowish as are the wing-linings. The tail is greyish, darkest on the center of each feather and paler around the fringe. Legs and feet pinkish or orange-pink.

Adult Female Non-Breeding (Definitive Basic): Similar to the breeding female breeding plumage but brighter. The buffy underparts are richer in colour, appearing buff-orange on the breast and face. The upperparts are boldly edged warm, bright buff or golden-buff. The 'suspenders' on the back are buffy-orange. The fringes on the coverts and tertials are yellow-buff, darker on the tertials. The fresh tail feathers are not as spiked as those of the breeding plumage. The bill is a brighter pink than in the breeding plumage.

Immature (First Basic): Similar to non-breeding adult female, but brighter on the underparts. The underparts may appear golden in some individuals. The throat is warm buff or yellowish, not whitish as in non-breeding adult. The upperpart fringes tend towards pale buff rather than orange-buff and are thus not as bright as in adults. The tertials in particular are not as contrasting and bright on immature birds as they are in adults. After the first pre-alternate moult, young birds are indistinguishable from adults.

Juvenile (Juvenal): Similar to immature, but lack flank streaking. Instead there are thin streaks on the breast and neck sides. The wings are dull brownish and edged buff, lacking the warm tones of immatures or non-breeding adults. The upperpart feathers are entirely edged buff, giving the back a scaled appearance.

GEOGRAPHIC VARIATION Monotypic. It has been suggested that birds from the west are slightly different from those in the east, but there do not appear to be any consistent differences.

HABITAT Breeding: A bird of grassy prairie that has adapted to old field habitats, particularly those that have been left fallow for at least eight years. It is also fond of grassy meadows, particularly in the west. Bobolinks prefer moist (mesic) meadows to wet or dry ones (Wittenberger 1978). In boreal forest it will nest in old, dry beaver (*Castor canadensis*) meadows. Moults in marshes and sewage ponds with abundant vegetation. **Non-Breeding**: Originally a bird of the northern Pampas and Pantanal, living in grasslands and marshland. Recently it has invaded rice-fields as these natural areas have been converted to agriculture. It is considered an agricultural pest in parts of its non-breeding range.

BEHAVIOUR Bobolinks return to the same nesting area year after year. Almost all males return to the same breeding site, but a smaller number of females return, suggesting that they are not so strongly tied to the breeding area (Wittenberger 1978). Old males arrive on the breeding grounds as a group before all other age or sex classes. Younger males arrive slightly later, followed by females over a more protracted period. Early in the breeding season, singing and song flights are most obvious. Males have two song types, the 'alpha' and 'beta' songs described under Voice. Males in a colony share the same two songs, and different colonies have separate dialects. Recent immigrants from nearby colonies sing a different song, and these males are not as successful at procuring mates as males that share the colony songs. The local 'song-sharing' males arrive earliest in the spring and appear to arrive as a group. The most obvious display is the 'song flight'. The male flies with a fluttery flight, on bowed wings, not raised above the horizontal, while 'ruffling out' the rump, and pointing the tail downwards, raising the white epaulets and elevating the head as it opens its bill wide to sing. The song flight tends to occur less than 10 m from the ground.

It has been proposed that the 'alpha' song functions in mate attraction while 'beta' song is used as communication between males (Wittenberger 1983). However, Capp (1992) did not find that the two songs types conveyed different messages. There was also no evidence that the switching from one song to the other had any function in communicating aggressiveness or sexual motivation. Capp (1992) did find evidence that Bobolinks use the alternate song type when responding to a specific song of an intruder. This is functionally equivalent to 'song matching', where a male answers an intruder with an identical song type in a situation where only two song types exist as in Bobolink.

Males use both song and the song-flights to defend their territory and attract females. Territories are quickly established early in the spring, and violent altercations can occur at territory boundaries. Commonly, males chase each other in flight or may get involved in a somewhat ritualised fight either on the ground or tumbling through the air. The mating system is highly polygynous with successful males mating up to four females. However, females preferentially pair with unmated males and later with those that already have a first mate. 'Secondary' females take longer to finalise pair formation than 'primary' females. The first nesting is highly synchronous at a breeding site (Wittenberger 1978). The degree of polygyny varies geographically and is presumed to be dependent on territory quality. Many males do not breed in a season, these non-breeders are often young males. In one Oregon study, roughly 29% of males were polygynous, 50% were monogamous while 21% were unmated, and most polygynous males had two, rarely three, mates, (Wittenberger 1978). By contrast, in Wisconsin, Bobolink males have three mates and sometimes four (Martin 1974). The sex ratio is roughly even. Note that using genetic fingerprinting, it was found that nearly 17% of 12 nests in New York state were sired by more than one male, therefore females are polyandrous, they mate with more than one male (Gavin and

Bollinger 1985). Females appear to incur a cost by becoming secondary rather than primary females, as the former tend to successfully raise fewer young (Wittenberger 1978). Females raise only one brood per season.

When a female is attracted to a male, the male will 'song flight' and eventually direct some displays at the female. Most obvious is a 'parachuting' flight with the wings held in a stiff 'V' over the body and with the legs dangling, accompanied by the 'buzz' call. He may repeat this several times. The female may respond by initiating a 'sexual chase', where the male chases the female but ritually does not catch up to her; other males may join in this chase. Before mating, the male will perform a pre-copulatory crouch display, with the bill pointed downwards, the wings fluttered and the tail fanned. The female may respond by crouching and vibrating her wings as the tail is cocked; at this point the male mounts and the pair mates. The male flys away from the female after mating, often singing at the same time. Most of the young are sired by the territory-holding male, but up to 15% of a female's clutch may be fathered by another male. It is thought that the female is responsible for soliciting these extra-pair copulations. Once the young hatch, the males abandon territory maintenance and help to feed the offspring, thus territories are rather short-lived.

NESTING The breeding season begins with territory establishment in early May in the south and mid May in the north and west. Females lay their first eggs as early as late May. The nest is placed on the ground, often at the base of a large clump of grass, and is made of fine grasses. It is cup shaped and built in a slight depression, often with an overhanging canopy of dead grass that protects it. The clutch size varies from 3–7 eggs but is typically 5. The eggs are blue-grey, grey or cinnamon with darker brown blotches and spots. Incubation begins after the second last egg is laid and lasts for 13 days. The young require up to 11 days in the nest to fledge. In the southern part of the range, females may raise a second brood in the same season. Males help to feed the young of the first nesting female, but those other than the primary female's are usually raised entirely by the mother. However, in Oregon, it is not unusual for males to feed the young of secondary females (Wittenberger 1980). Curiously, 'helpers' at the nest of both sexes have been reported; this is largely unknown in transequatorial migrants (Beason and Trout 1984). Bobolink is parasitised by Brown-headed Cowbird (101) but it occurs rarely. Cowbird young do manage to fledge from Bobolink nests, so its difficult to understand why they are not more commonly chosen as hosts. One reason for this may be that adult Bobolinks are extremely aggressive towards cowbirds in the breeding grounds.

DISTRIBUTION AND STATUS Common throughout its range, much more local in the south and west. A bird of lowlands but has been found to 3400 m as a transient in Cochabamba, Bolivia (specimen, Field Museum).

Breeding: From the Okanagan valley in SC British Columbia south to E Washington and E Oregon (mainly Malheur area) and possibly extreme NE California in the past; east through N Idaho, Montana, C and SE Alberta, S and C Saskatchewan, S Manitoba, W Ontario (Thunder Bay), S and E Ontario as far north as Moosonee, S Quebec continuing along the north coast of the Gulf of St Lawrence. Found in SW Newfoundland. The southern limit of the distribution is most spotty and disjunct in the west of the range from N, C, and SE Idaho to NE Utah; isolated populations occur in NE Nevada and SW Utah; through W, N, and C Wyoming to SC Colorado; an isolated population occurs in SE Arizona; continuing along the north fringe of Wyoming the range extends south to most of Nebraska except the far SW and southern fringe; several isolated populations occur in Kansas; east through N Missouri, C Illinois, S. Indiana, S Ohio, most of Pennsylvania, south along the Appalachians to S Virginia, W Maryland, N New Jersey and including all of New England and the Maritimes. There is an isolated population in C Kentucky.

Wintering: The non-breeding distribution of this species is improperly known and may be more extensive than is recognised at the present time. The main part of the winter range is in N Argentina, Paraguay, N Bolivia and S Brazil. In Argentina, it is found in the wetter E Chaco (Formosa, Chaco and N Santa Fé) and the marshes of Corrientes and Entre Ríos; occasionally further west to Salta and as far south as NE Buenos Aires, S Santa Fé, S Cordoba and N San Luis. In Paraguay, it is sighted most commonly west of the Paraguay river; winters in Pando, La Paz, Beni, Cochabamba and Santa Cruz departments in Bolivia; and in Brazil in S Mato Grosso. A smaller population may winter west of the Andes in Peru (see Movements). It is accidental in winter in Florida.

MOVEMENTS A long distance migrant. Bobolink undergoes the longest migrations of any icterid, a roughly 20,000 km round trip from its temperate North American breeding grounds to the wintering grounds in N Argentina. This is longer than the migration of most other New World passerines. Both sexes are faithful to their breeding grounds and return year after year; it is unknown if wintering areas are also the same in consecutive years. They appear to be somewhat nomadic in winter, following the ripening of rice crops and native grasses, so perhaps do not display winter site fidelity. A comparison between a population in Oregon with one in Wisconsin determined that the former suffered higher mortality rates during the time it was away from the breeding site. Assuming that the Oregon birds travel east to E South America in the winter they may move an extra 6000 km each year, this extra migratory distance may account for the greater mortality (Wittenberger 1978). However, it is not clear if the Oregon birds are part of the population that winters along the W coast of South America, see below.

Adult males arrive on territory earlier than females and somewhat synchronously (first week of May at the latitude of southern Ontario). Observations of known, banded birds suggest that males that bred at a site during the previous year return together as a group, perhaps both migrating and wintering together! Females arrive at a more leisurely pace, beginning to arrive one week after the males are on territory, and may continue arriving for the next week or two (into late May at latitude of southern Ontario and northern New York state). After breeding, birds congregate in marshes where they undergo the pre-basic moult (July–early August).

Southbound migration is strong by mid and late August. Large flocks gather along the Atlantic Coast from New Jersey south to Florida at this time. Bobolinks move east and use long non-stop flights in autumn. In fall they are uncommon to rare throughout the inland eastern states, from Louisiana and W Florida north to Arkansas, Tennessee and Kentucky, particularly when compared to spring numbers. Their migratory route appears to pass largely through Florida and the Greater Antilles (mainly Cuba and Jamaica, being rare further east in Puerto Rico) as well as over the Atlantic in a direct flight to South America, inferred from good numbers observed flying over Bermuda during the day. In Florida, the fall passage is not as noticeable as in the spring since most Bobolinks fly over the state without stopping. During fall they are common only in the southern part of the Florida peninsula. However, offshore storms in the fall may produce falls of thousands, such as the estimated 250,000 on Aug. 28, 1953, as Hurricane Cleo passed by (Stevenson and Anderson 1994). Good numbers pass in fall through Cuba and Jamaica (Chapman 1890). They are regular in the fall in the Netherlands Antilles between Sept 13–Nov 21 (Voous 1983). They are present in Florida in mid September, peak in the West Indies during October and begin to be seen in N South America in late October. Over Bermuda, they are common mid September–early October; it is possible that these birds may have departed from sites as far north as Nova Scotia and New England and traveled to the islands in one non-stop flight, however this needs confirmation. In Barbados, they are an uncommon passage migrant in small flocks (M. Gawn pers. comm.). Smaller numbers migrate further west crossing through the Yucatán and parts of Central America; perhaps these are birds thrown off course by unfavourable winds. They are rare in fall in Mississippi and Louisiana. In Central America, they are regularly observed in fall, but are very rare in the spring; in contrast, they are more common in spring than fall in the Yucatán Peninsula. In the Santa Marta region of Colombia, Bobolinks are noted in September and October but not in the spring (Todd and Carriker 1922). Passage through this latitude occurs in September and October. Migration occurs east of the Andes through the Amazonian lowlands, but there have been flocks reported from the Andean highlands of Lake Titicaca. Birds arrive in Brazil and Paraguay by November and apparently do not reach the southern extreme of the wintering range until January. The delay in arrival to the southern extreme of the wintering range is apparently due to the later ripening of rice crops as well as the ripening of native grasses.

There is a small number of birds that migrates along the west coast of South America. They are regularly observed on the Galapagos Islands in fall, regularly recorded in coastal Peru and records exist for coastal Ecuador (Marchant 1958). There are at least three reports from northern Chile. It has been hypothesised that these birds are western breeders and that a separate wintering population exists west of the Andes, however this has not been confirmed. These birds may be the same individuals that are observed as migrants in California. However, most fall California migrants do appear to be 1st-year birds rather than adults.

The northbound passage begins in March as birds begin leaving the wintering grounds. Some remain in Argentina into late March and perhaps even early April. Spring migration retraces some of the same routes as those taken during the fall, but the spring routes tend to be further west than the fall routes. They are more common in the spring than the fall in the Yucatán Peninsula and Texas. They are common in Florida, particularly in the Keys and Dry Tortugas, but may be seen throughout the state in spring. They are uncommon to rare in the West Indies compared to the fall. This is true in the Lesser Antilles; for example in Martinique, this species is typically observed in September and October but is rarely observed in spring (Pichot 1976). However, noticeable movements through Jamaica do occur in April (Scott 1893). In Cuba, there is a spring migration but fewer birds are involved and they only remain for a few days, in comparison to the fall (Gundlach 1876). Spring flocks in Cuba are typically sexually segregated (Gundlack 1876). In the Netherlands Antilles, they are observed between February 25 and May 14 (Voous 1983). In addition, they are regular along the Gulf Coast in spring, unlike fall. Very few Bobolinks are detected over Bermuda in spring. Birds begin reaching the S US in April with males arriving before females. Passage continues through the southern US until late May at which time most of the population has settled in the breeding grounds.

Bobolink has been widely used in experiments aimed at understanding migration and thus it seems appropriate to mention some of the major results from these studies. It appears that navigation is achieved by a series of different clues including the position of the stars and the earth's magnetic field. The information from the magnetic field overrides that of the stars, but birds rely on these for finer scale orientation. Bobolinks from North Dakota kept in the lab. during autumn in situations where they could observe the night sky showed a tendency to orient towards the southeast, roughly in the direction of Florida. Fall Bobolinks from New York oriented towards the south (Hamilton 1962).

This species is a regular vagrant to the Pacific coast of North America. In British Columbia, there are several records from Vancouver, and six from Victoria, Vancouver Island; most of these are during September–October, with few in spring. It is rare, but almost annual, in fall on the coasts of Washington and Oregon. In California, it is best considered a rare migrant, rather than a vagrant. In fall almost all records are from the coast, with a lesser number from the interior deserts. It is rarer in spring than in the fall, and a larger proportion of the records are from the interior deserts. There are summer records from the NE part of the state and breeding should be looked for; it may have bred there in the past, but this is unconfirmed. It is a rare migrant in Nevada, Arizona and New Mexico, away from breeding areas. There are two specimen records from Alaska including one of a female on June 23, 1976 at Point Barrow (Gibson and Kessel 1997). Other northern stray records include one on June 10, 1994 from Churchill, Manitoba; there are no records from the Yukon or Northwest Territories. In Newfoundland, the Bobolink is a regular vagrant, particularly in fall.

In South America, a few birds winter in Peru and there are three records from N Chile: one from the Lluta Valley, Arica; another at Pedro de Valdivia in the highlands of Antofagasta; the last in the outskirts of the city of Antofagasta. They may be an overlooked, yet regular component of northern Chile's avifauna rather than overshooting vagrants. There is one record from Punta Rasa, coastal Buenos Aires, Argentina from mid May; this bird was in fresh breeding plumage (M. Rumboll pers. comm.). One was recorded well east of the regular range in Rio de Janeiro, Brazil.

In Europe, most vagrant records are from the fall. The only spring record is of a male sighted at Gibraltar in May. There are over 14 records from Great Britain, all between September and October; one October record from Ouessant, France; one in November in Oslo, Norway; and finally one in mid-September in Tuscany, Italy (Lewington *et al.* 1992). There are also records from Greenland, as well as Labrador and many from Newfoundland.

The banding data shows that the oldest known individual reached the age of 8 years and 1 month old (Klimkiewicz 1997). Other banded birds have reached the age of seven years old (BBL data).

MOULT The moult is peculiar both for an icterid and a passerine. There are two complete moults in a year. In adults the complete pre-basic moult occurs in marshes near the breeding grounds (July–September), but may be suspended during migration and finished in the Neotropics (October). The pre-basic moult is rapid, with many feathers being concurrently moulted, radically altering the flying abilities of the birds. The rectrices are often all dropped and moulted at the same time, rendering many birds tailless for a short period. The complete pre-alternate moult begins in the wintering grounds (February–March) but may be suspended during migration and finished in the breeding grounds (May–June). Fresh alternate plumage is heavily veiled with pale tips which wear off by the summer; this plumage has been termed the pre-breeding plumage by some authorities. The change of male soft part colours occurs after this moult is well underway; the bill and legs turn from pinkish or horn to black. Juveniles moult out of their plumage through the partial first pre-basic moult (July–October), retaining the juvenal wings and tail. A minority (only males?) also moult out the remiges and sometimes the rectrices. The first basic (immature) plumage is replaced by a first pre-alternate moult as in the adults at which time they become largely indistinguishable from adults.

MEASUREMENTS The males are roughly 20% larger than the females. Males: (n=301) wing 97.4; (n=38) tail 68.4 (64.0–74.5); culmen 14.3 (13.0–16.8); tarsus 26.1 (24.0–29.0). Females: (n=261) wing 87.5; (n=23) tail 60.9 (55.0–65.5); culmen 13.4 (12.7–14.4); tarsus 25.2 (24.0–26.0).

NOTES The breeding bird survey data show that during the last 30 years the population has shown a general decline in North America. However, this decline is not significant. By comparison, taking the same data between 1980 and 1989 there was a significant decline for the continent. This decline may be due to a loss of old field habitat as it reverts to second growth forest, at least in the eastern part of the continent. A very rough approximation of the world population of this species during the early 1980s estimated 11 million pairs (Pettingill 1983). Previously, Bobolink was commonly known as the Ricebird or Reedbird. Both of these names were used only for birds in the dull non-breeding plumage.

REFERENCES Araya and Millie 1991 [vagrants, movements], BBS Data 1996 [notes], Beason and Trout 1984 [nesting, nest helpers], Chapman 1893, 1897 [moult], Fjeldsa and Krabbe 1990 [movements], Gavin and Bollinger 1985 [behaviour], Hamilton 1962 [movements], Koepcke 1983 [movements], Lewington *et al.* 1992 [vagrants, movements], Martin and Gavin 1995 [voice, habitat, behaviour, nesting, movements, measurements], Parkes 1952 [moult], Pettingill 1983 [movements, moult], Remsen 1986 [distribution], Remsen and Traylor 1989 [distribution], Pyle *et al.* 1987 [moult], Ridgely and Tudor 1989 [distribution, movements], Small 1994 [movements], Wittenberger 1978 [behaviour, nesting, movements].

BIBLIOGRAPHY

Albers, P.H. 1978. Habitat selection by breeding Red-winged Blackbirds. *Wilson Bull.* 90:619–634.

Alcorn, J.R. 1988. *The Birds of Nevada.* Fairview West Publisher, Fallon.

Allen, J.A. 1890. Description of a new species of *Icterus* from Andros Island, Bahamas. *Auk* 7:343–346.

Alstrom, P., and P. Colston. 1991. *A field guide to rare birds of Britain and Europe.* Harper Collins, London.

Alvarez, H. 1975. The social system of the Green Jay in Colombia. *Living Bird* 14:5–44.

Alvarez del Toro, M. 1971. *Las aves de Chiapas.* Universidad Autonoma de Chiapas, Tuxla Gutierrez, Mexico.

American Ornithologists Union. 1983. *Check-list of North American Birds.* 6th ed. Allen Press, Lawrence, Kansas.

Ankney, C.D., and D.M. Scott. 1980. Changes in nutrient reserves and diet of breeding Brown-headed Cowbirds. *Auk* 97:684–696.

Ankney, C.D., and D.M. Scott. 1982. On the mating system of Brown-headed Cowbirds. *Wilson Bull.* 94:260–268.

Anonymous, 1996. Reflections on the list of Canadian Birds 1996, Black-cowled Oriole. *Birders Journal* 5:208.

Anthony, A.W. 1923. Predatory Brewer's Blackbirds. *Condor* 25:106–106.

Applegate, R.D. 1975. Co-roosting of barred owls and common grackles. *Bird Banding* 46:169–170.

Araya, B., and G. Millie. 1991. *Guia de campo de las aves de Chile.* Editorial Universitaria, Santiago.

Arellano, A.G., J.I. Glendinning, J.B. Anderson, and L.P. Brower. 1993. Interspecific comparisons of the foraging dynamics of Black-backed Orioles and Black-headed Grosbeaks on overwintering monarch butterflies in Mexico. In: Biology and conservation of the monarch butterfly. Eds. S.B. Malcom and M.P. Zalucki.315–322.

Arendt, W.J., and T.A. Vargas Mora. 1984. Range expansion of the Shiny Cowbird in the Dominican Republic. *J. Field Ornithol.* 55:104–107.

Armstrong, E.A. 1963. *A study of bird song.* Oxford University Press, London.

Armstrong, T.A. 1992. Categorization of notes used by female Red-winged Blackbirds in composite vocalizations. *Condor* 94:210–223.

Armstrong, T.A. 1995. Patterns of vocalizations used by female Red-winged Blackbird (Aves: Emberizidae, Icterinae): variation and context. *Ethology* 100:331.

Arnold, K.A., and L.J. Jr. Folse. 1977. Movements of the Great-tailed Grackle in Texas. *Wilson Bull.* 89:602–608.

Ashman, P. 1977. Northern (Bullock's) Oriole eats hummingbirds. *Western Birds* 8:105.

Avery, M., and L.W. Oring. 1977. Song dialects in the bobolink (*Dolichonyx oryzivorus*). *Condor* 79:113–118.

Avery, M.L. 1995. Rusty Blackbird (*Euphagus carolinus*). In The Birds of North America, No. 200 (A. Poole and F. Gill, eds.). The Academy of Natural Sciences, Philadelphia, and the American Ornithologists' Union, Washington, D.C.1–16.

Avise, J.C., and R.M. Zink. 1988. Molecular genetic divergence between avian sibling species: King and Clapper Rails, Long-billed and Short-billed Dowitchers, Boat-tailed and Great-tailed Grackles and Tufted and Black-crested Titmice. *Auk* 105:516–528.

Aweida, M.K. 1995. Repertoires, territory size and mate attraction in Western Meadowlarks. *Condor* 97:1080–1083.

Azara, F. De 1802. *Apuntamientos para la historia natural de los páxaros del Paraguay y Río de la Plata. Vol. 1.* Imprenta de la viuda de Ibarra, Madrid.

Babarskas, M., and B. López Lanús. 1993. Nidos nuevos o poco conocidos para la provincia de Buenos Aires. *Nuestras Aves* 28:27–28.

Babbs, S., S. Buckton, P. Robertson, and P. Wood. 1988. *Report of the 1987 University of East Anglia – ICBP St Lucia expedition.* Cambridge, U.K.: ICBP Study Report 33.

Babbs, S., S. Ling, P. Robertson, and P. Woods. 1987. *Report of the 1986 University of East Anglia Martinique Oriole expedition.* Cambridge, U.K.: ICBP Study Report 23.

Bailey, A.M. 1948. Birds of Arctic Alaska. *Col. Mus. Nat. Hist. Pop. Ser.* 8:1–317.

Bailey, F.M. 1910. The palm-leaf oriole. *Auk* 27:33–35.

Baillarge, W. 1930. A Bronzed Grackle foster parent. *Can. Field Nat.* 44:166–167.

Baird, J. 1958. The postjuvenal molt of the Brown-headed Cowbird. *Bird Banding* 29:224–228.

Balda, R.P., and S. Carothers. 1968. Nest protection by the Brown-headed Cowbird (*Molothrus ater*). *Auk* 85:324–325.

Ball, R.M. jr., S. Freeman, F.C. James, E. Bermingham, and J.C. Avise. 1988. Phylogeographic population structure of Red-winged Blackbirds assessed by mitochondrial DNA. *Proc. Natl. Acad. Sci. USA* 85:1558–1562.

Baltz, M.E. 1994. Shiny Cowbirds (*Molothrus bonariensis*) on North Andros Island, Bahamas. *El Pitirre* 7(3):4–4.

Baltz, M.E. 1995. First Records of the Shiny Cowbird (*Molothrus bonariensis*) in the Bahama Archipelago.

Auk 112:1039–1041.
Baltz, M.E. 1996. The distribution and status of Shiny Cowbirds on Andros Island. *Bahamas Journal of Science* 3:2–6.
Baltz, M.E. 1997. Status of the Black-cowled Oriole (*Icterus dominicensis northropi*) in the Bahamas. Report submitted to the Ministry of Fisheries and Agriculture, Nassau, Bahamas.
Bangs, O. 1899. On a small collection of birds from San Sebastian, Colombia. *Proc. New England Zool. Club* 1:79.
Bangs, O. 1901. A new meadowlark from South America. *Proc. New England Zool. Club* 2:55–56.
Bangs, O. 1913. New birds from Cuba and the Isle of Pines. *Proc. New England Zool. Club* 4:92.
Bangs, O. and T.E. Penard. 1918. Notes on a collection of Surinam birds. *Bull. Mus. Comp. Zool.* 62:85.
Barbour, T. 1923. The Birds of Cuba. *Mem. Nutt. Orn. Club* 6:1–141.
Barbour, T. 1943. Cuban Ornithology. *Nuttall Ornithol. Club* 9:119–124.
Barlow, J., B.W. McGillvray, and T. Dickinson. 1994. Song structure and speciation in meadowlarks. *J. fur Ornithologie Band* 135:25.
Barnes, V. Jr. 1945. A new form of *Agelaius* from Mona Island, Puerto Rico. *Auk* 62:299–300.
Barnes, V. Jr. 1946. The birds of Mona Island, Puerto Rico. *Auk* 63:318–327.
Barros Valenzuela, R. 1956. El Tordo Argentino: Una Plaga de la Avifauna Chilena. *Revista Universitaria* 40–41 (no.1):89–94.
Barrowclough, G.F. 1993. There are 18043 extant species of birds in the world. Paper given during the 111th Stated Meeting of the American Ornithologists Union, Fairbanks, Alaska.
Beason, R.C. 1986. Natural and induced magnetization in the bobolink, *Dolichonyx oryzivorus. J. Exp. Bio.* 125:49–56.
Beason, R.C. 1987. Interaction of visual and non-visual cues during migratory orientation by the Bobolink (*Dolichonyx oryzivorus*). *J. Ornithol* 128:317–324.
Beason, R.C. 1989. Use of an inclination compass during migratory orientation by the Bobolink (*Dolichonyx oryzivorus*). *Ethology* 81:291–299.
Beason, R.C., and L.L. Trout. 1984. Cooperative breeding in the Bobolink. *Wilson Bull.* 96:709–710.
Beebe, W. 1917. *Tropical wild life in British Guiana.* New York Zool Soc.
Beecher, W.J. 1950. Convergent Evolution in the American Orioles. *Wilson Bull.* 62:51–86.
Beecher, W.J. 1951. Adaptations for food-getting in the American Blackbirds. *Auk* 68:411–440.
Beedy, E.C., and W.J. III Hamilton. 1997. Tricolored Blackbird Status Update and Management Guidelines. September. (Jones & Stokes Associates, Inc. 97-099.) Sacramento, CA. Prepared for U.S. Fish and Wildlife Service, Portland, OR and California Department of Fish and Game, Sacramento, CA.
Belcher, C., and G.D. Smooker. 1937. Birds of the colony of Trinidad and Tobago. Part 5. *Ibis* 14:504–550.
Beletsky, L.D. 1982a. Vocalizations of female Northern Orioles. *Condor* 84:445–447.
Beletsky, L.D. 1982b. Vocal behavior of the Northern Oriole *Wilson Bull.* 94:372–381.
Beletsky, L.D. 1983. Aggressive and pair-bond maintenance songs of female Red-winged Blackbirds *(Agelaius phoeniceus*). *Z. Tierpsychol.* 62:47–54.
Beletsky, L.D. 1985. Intersexual song answering in Red-winged Blackbirds. *Can. J. Zool.* 63:735–737.
Beletsky, L.D., and M.G. Corral. 1983. Song response by female Red-winged Blackbirds to male song. *Wilson Bull.* 95:643–647.
Beletsky, L.D., and G.H. Orians. 1987. Territoriality among male Red-winged Blackbirds. I. Site fidelity and movement patterns. *Behav. Ecol. Sociobiol.* 20:21–34.
Beletsky, L.D., and G.H. Orians. 1993. Factors affecting which male Red-winged Blackbirds acquire territories. *Condor* 95:782–791.
Belton, W. 1985. Birds of Rio Grande Do Sul, Brazil. Part 2. Formicariidae through Corvidae. *Bull. Am. Mus. Nat. Hist.* 180:166–184.
Benito-Espinal, E., and P. Hautcastel. 1988. Les oiseaux menac,s de Guadelope et de Martinique. Pp. 37–60. In: J.-C. Thibault and I. Guyot (eds.). Livre rouge de oiseaux menacés des regions francaises d'outre-mer. Saint-Cloud: Conseil Internacional pour la protection des Oiseaux (ICBP). Monograph 5.
Bent, A.C. 1958. Life Histories of North American blackbirds, orioles, tanagers and allies. *U.S. National Mus. Bull.* 211
Berger, A.J. 1951. The cowbird and certain host species in Michigan. *Wilson Bull.* 63:26–34.
Berlepsch, H.V. 1888. Descriptions of new species and subspecies of birds from the Neotropical region. *Auk* 5:454–456.
Berlepsch, M.M. H.Von, and J. Stolzmann. 1892. On birds from Peru. *Proc. Zool. Soc. London* 1892:371–410.
Besser, J.F. 1985. Breeding blackbird populations in Iowa. *Iowa Bird Life* 55:35–42.
Biaggi, V. 1983. Las aves de Puerto Rico. Editorial de la Universidad de Puerto Rico, Rio Piedras, Puerto Rico.306–313.
Bianchi, V.C. 1971. Un ejemplar de color canela del tordo bayo. *Hornero* XI:128–128.
Binford, L.C. 1989. A distributional survey of the birds of the Mexican state of Oaxaca. *Ornithol. Monogr.* 43:1–418.
Bjorklund, C.F. 1990. Bromhead rare bird records. *Blue Jay* 48(4):212–217.
Bjorklund, M. 1991. Evolution, phylogeny, sexual dimorphism and mating system in the grackles

(*Quiscalus* spp., Icterinae). *Evolution* 45:608–621.
Blake, E.R. 1968. In: Paynter, R.A. Jr. (ed). *Check-list of Birds of the World. A Continuation of the Work of James L. Peters.* Museum of Comparative Zoology, Cambridge, Mass.
Bledsoe, A.H. 1988. Nuclear DNA evolution and phylogeny of the New World nine-primaried oscines. *Auk* 105:504–515.
Boardman, R. 1992. Hooded Oriole at Long Point Ontario. *Birders Journal* 1:228–229.
Boggs, G.O. 1961. Notas sobre las aves de 'El Centro' en el valle medio del Rio Magdalena-Colombia. *Novedades Colombianas* 1:401–423.
Bohlen, H.D. 1976. A great tailed grackle from Illinois. *American Birds* 30:917.
Bohlen, H.D. 1989. *The birds of Illinois.* Indiana University Press, Bloomington.
Bollinger, E.K. 1988. Comparison of transects and circular plots for estimating bobolink densities. *J. Wildlife Management* 52:777–786.
Bollinger, E.K. 1988. A longevity record for the Bobolink. *N. Am. Bird Bander* 13:76.
Bollinger, E.K., and T.A. Gavin. 1991. Patterns of extra-pair fertilization in Bobolinks. *Behav. Ecol. Sociobiol.* 29:1–7.
Bond, J. 1928a. On the birds of Dominica, St Lucia, St Vincent, and Barbados, B.W.I. *Proc. Acad. Nat. Sci. Phil* 80:523–545.
Bond, J. 1928b. The distribution and habits of the birds of the republic of Haiti. *Proc. Acad. Nat. Sci. Philadelphia* 80:514–515.
Bond, J. 1939. Some birds from Montserrat, British West Indies. *Auk* 56:193–195.
Bond, J. 1950. Some remarks on West Indian Icteridae. *Wilson Bull.* 62:216–217.
Bond, J. 1951. Taxonomic notes on South American Birds. (Regarding: Tanager *Pyrrota valeryi* is a synonym of *Lampropsar tanagrinus*). *Auk* 68:528.
Bond, J. 1953. Notes on Peruvian Icteridae, Vireonidae and Parulidae. *Not. Nat.* (*Philadelphia*) 255:1–15.
Bond, J. 1988. *Birds of the West Indies: A guide to the species of birds that inhabit the Greater Antilles, Lesser Antilles and Bahama Islands.* Collins, London.
Bond, J., and R. Meyer de Schauensee. 1941. The birds of Bolivia. Part I. *Proc. Acad. Nat. Sci. Philadelphia* 94:307–407.
Bonham, P.F. 1970. Recent reports. *Icterus galbula* third British record at Haverfordwest, Wales. *British Birds* 63:262–264.
Boucard, A. 1883. On a collection of birds from Yucatan. *Proc. Zool. Soc. London* 1883:434–460.
Bradley, P. 1985. *Birds of the Cayman Islands.* Self Published, George Town, Grand Cayman.
Bradley, P.E. 1994. The avifauna of the Cayman Islands: an overview. In: M.A. Brunt and J.E. Davies (eds.). The Cayman Islands: Natural History and Biogeography. 377–406.
Bray, O.E., W.C. Royall, J.L. Guarino, and De Grazio. 1973. Migration and seasonal distribution of Common Grackles banded in North and South Dakota. *Bird Banding* 44:1–12.
Bray, O.E., W.C. Royall, J.L. Guarino, and R.E. Johnson. 1979. Activities of radio-equipped common grackles during fall migration. *Wilson Bull.* 91:78–87.
Brewster, W. 1988. Descriptions of supposed new birds from lower California, Sonora, and Chihuahua, Mexico, and the Bahamas. *Auk* 5:91–92.
Bringham, E.M. 1963. The song of a meadowlark in Oaxaca, Mexico. *Jack-Pine Warbler* 41:90–90.
Briskie, J.V., and S.G. Sealy. 1990. Evolution of short incubation periods in the parasitic cowbirds, *Molothrus* spp. *Auk* 107:790–794.
Brittingham, M.C., and S.A. Temple. 1983. Have Cowbirds Caused Forest Songbirds to Decline? *BioScience* 33:31–35.
Broad, R.A. 1978. Baltimore oriole at Fair Isle. *Scottish Birds* 10:58–59.
Brodkorb, P. 1937. New or noteworthy birds from the Paraguayan chaco. *Occas. Papers Mus. Zool. Univ. Michigan* 345:1–2.
Brookfield, C.M., and O. Griswold. 1956. An exotic new oriole settles in Florida. *Nat. Geographic* 109: 261–264.
Brooks, A. 1942. Additions to the distributional list of the birds of British Columbia. *Condor* 44:33–34.
Brower, L.P., and W.H. Calvert. 1985. Foraging dynamics of bird predators on overwintering monarch butterflies in Mexico. *Evolution* 39:852–868.
Browning, M.R. 1975. First Oregon specimen of *Icterus galbula galbula. Auk* 92:162–163.
Browning, M.R. 1978. An evaluation of the new species and subspecies proposed in Oberholser's Bird Life of Texas. *Proc. Biol. Soc. Wash.* 91:85–122.
Brudenell-Bruce, P.G.C 1975. *The Collins Guide to the Birds of New Providence and the Bahama Islands.* Stephen Green Press
Brush, A.H. 1970. An electrophoretic study of egg whites from three blackbird species. *Univ. Connecticut Occas. Papers* 1:243–264.
Brush, T. 1998. A Closer Look: Altamira Oriole. *Birding* XXX:46–53.
Bryant, H. 1866. A list of birds from Porto Rico. *Proc. Boston Soc. Nat. Hist.* 10:254–255.
Buckingham, B. 1976. Range of immature Common Grackles. Inland *Bird Banding* News 48:171–176.
Bull, J. 1974. *Birds of New York State.* Doubleday, Garden City, NY.
Burke, P., and A. Jaramillo. 1995. Fall and winter plumages of male Rusty and Brewer's blackbirds. *Birders*

Journal 4:97–101.
Burleigh, T., and H.S. Peters. 1948. Geographic Variation in Newfoundland Birds. *Proc. Biol. Soc. Washington* 61:111–124.
Burnell, K., and S.I. Rothstein. 1994. Variation in the structure of female Brown-headed Cowbird vocalizations and its relation to vocal function and development. *Condor* 96:703–715.
Burtt, H.E., and M.L. Giltz. 1967. Population trends for 'blackbirds'. *Redstart* 34:86–90.
Burtt, H.E., and M.L. Giltz. 1976. Sex differences in the tendency for Brown-headed Cowbirds and Red-winged Blackbirds to re-enter a decoy trap. *Ohio J. Sci.* 76:264–267.
Burtt, H.E., and M.L. Giltz. 1977. Seasonal directional patterns of movement and migrations of starlings and blackbirds in North America. *J. Field Ornithol.* 48:259–271.
Butcher, G.S. 1981. Northern Orioles disappear with Mt St Helens ashfall. *Murrelet* 62:15–16.
Butcher, G.S. 1984. Sexual color dimorphism in orioles (the genus *Icterus*): tests of communication hypotheses. *Diss. Abst. Int.* 45:482.
Butcher, G.S. 1984. The predator-deflection hypothesis for sexual colour dimorphism: a test on the northern oriole. *Anim. Behav.* 32:925–926.

Cabanis, J.L. 1861. Uebersicht der im Berliner Museum befindlichen Vögel Cost Rica. *Journ. f. Ornith.* 9:83–97.
Cabanis, 1873. Deutsche ornithologische gesellschaft an Berlin. Protokoll der LII. Monats-Sitzung. *Journ. f. Ornith.* 21:307–310.
Caccamise, D.F. 1977. Breeding success and nest site characteristics of the Red-winged Blackbird. *Wilson Bull.* 89:396–403.
Campbell, R.W., and G.P. Sirk. 1969. Common Grackle sighted at Vancouver, British Columbia. *Murrelet* 50:38.
Camperi, A.R. 1988. Sobre la supuesta presencia de *Agelaius thilius thilius* en la Argentina (Aves, Icteridae). *Hornero* 13:86–88.
Canevari, M., P. Canevari, G.R. Carrizo, G. Harris, J.R. Mata and R.J. Straneck. 1991a. *Nueva Guia de las Aves Argentinas.* Tomo 1. Fundacion Acindar, Buenos Aires.
Canevari, M., P. Canevari, G.R. Carrizo, G. Harris, J.R. Mata and R.J. Straneck. 1991b. *Nueva Guia de las Aves Argentinas.* Tomo II Fundacion Acindar, Buenos Aires.
Cannings, R.A., R.J. Cannings, and S.G. Cannings. 1987. *Birds of the Okanagan Valley, British Columbia.* Royal British Columbia Museum, Victoria.
Capp, M.S. 1992. Tests of the function of the song repertoire in Bobolinks. *Condor* 94:468–479.
Capp, M.S., and W.A. Searcy. 1991. Acoustical communication of aggressive intentions by territorial male Bobolinks. *Behav. Ecol.* 2:319–326.
Cardiff, S.W., and J.V. Jr. Remsen. 1994. Type specimens of birds in the Museum of Natural Sciences, Louisiana State University. *Occ. Pap. Mus. Nat. Sci.* Lousiana State University. 68:1–33.
Carter, M.D. 1984. The social organization of parasitic behavior of the Bronzed Cowbird in south Texas. Ph.D. dissertation , University of Minnesota, Minneapolis.
Carter, M.D. 1986. The Parasitic Behavior of the Bronzed Cowbird in South Texas. *Condor* 88:11–25.
Case, N.A., and H. Hewitt. 1967. Nesting and productivity of the Red-winged Blackbird in relation to habitat. *Living Bird* 2:7–20.
Cassin, J. 1847. Descriptions of three new species of the genus Icterus (Briss.) specimens of which are in the museum of the Academy of Natural Sciences of Philadelphia. *Proc. Acad. Nat. Sci.,Philadelphia.* 3:332–334.
Cassin, J. 1860. Catalogue of birds collected during a survey of a route for a ship canal across the Isthmus of Darien, by order of the government of the United States made by Lieut. Michlen of the U.S. Topographical Engineers with notes and descriptions of new species. *Pr. Acad. Nat. Sci. Phil.* 1860:138–139.
Cassin, J. 1866. A study of the Icteridae. *Proc. Acad. Nat. Sci.* 1866:10–25.
Cassin, J. 1867. A third study of the Icterinae. *Proc. Acad. Nat. Sci. Philadelphia* 19:45–74.
Cavalcanti, R.B., and T.M. Pimentel. 1988. Shiny Cowbird parasitism in Central Brazil. *Condor* 90:40–43.
Chapman, F.M. 1890. On the winter distribution of the Bobolink (*Dolichonyx oryzivorus*) with remarks on its routes of migration. *Auk* 7:39–45.
Chapman, F.M. 1891. The origin of the avifauna of the Bahamas. *Amer. Nat.* 25:528–539.
Chapman, F.M. 1893. On the changes of plumage of the Bobolink (*Dolichonyx oryzivorus*). *Auk* 10:309–311.
Chapman, F.M. 1897. Remarks on the spring moult of the Bobolink. *Auk* 14:149–154.
Chapman, F.M. 1911. Description of a new oriole (*Icterus fuertesi*) from Mexico. *Auk* 28:1–4.
Chapman, F.M. 1914. Diagnoses of apparently new Colombian birds, II. *Bull. Amer. Mus. Nat. Hist.* 33:190.
Chapman, F.M. 1915. Diagnoses of apparently new Colombian birds, IV. *Bull. Amer. Mus. Nat. Hist.* 34:657–662.
Chapman, F.M. 1917. The distribution of bird-life in Colombia; a contribution to a biological survey of South America. *Bull. Amer. Mus. Nat. Hist.* 36:625–636.

Chapman, F.M. 1919. Descriptions of proposed new birds from Peru, Bolivia, Argentina and Chile. *Bull. Amer. Mus. Nat. Hist.* 41:323–333.
Chapman, F.M. 1920. Unusual types of apparent geographic variation in color and of individual variation in size exhibited by *Ostinops decumanus. Proc. Biol. Soc. Washington* 33:26.
Chapman, F.M. 1926. The distribution of bird-life in Ecuador. *Bull. Amer. Mus. Nat. Hist.* 55:692–700.
Chapman, F.M. 1930. The Nesting Habits of Wagler's Oropendola on Barro Colorado Island. *Amer. Mus. Nat. Hist. Bull.* 43:123–166.
Chapman, F.M. 1931a. The upper zonal bird-life of Mts Roraima and Duida. *Bull. Amer. Mus. Nat. Hist.* 63:133–134.
Chapman, F.M. 1931b. The nesting habits of Wagler's Oropendola on Barro Colorado Island. *Bull. Amer. Mus. Nat. Hist.* 58:347–386.
Chapman, F.M. 1935. Further remarks on the relationships of the grackles of the subgenus *Quiscalus. Auk* 52:21–29.
Chapman, F.M. 1936. Further remarks on *Quiscalus* with a report on additional specimens from Louisiana. *Auk* 53:405–417.
Chapman, F.M. 1939a. Nomenclature in the genus *Quiscalus. Auk* 364–365.
Chapman, F.M. 1939b. *Quiscalus* in Mississippi. *Auk* 56:28–31.
Chapman, F.M. 1940. Further studies of the genus *Quiscalus. Auk* 57:225–233.
Cherrie, G.K. 1916. A contribution to the ornithology of the Orinoco Region. *Sci. Bull. Brooklyn Inst. Arts and Sciences* 2:133–374.
Chesser, R.T. 1994. Migration in South America: an overview of the austral system. *Bird Conservation International* 4:91–107.
Chubb, C. 1921. On new forms of South American birds. *Ann. Mag. Nat. Hist.* ser 9,8:444–446.
Clark, 1905. Birds of the Southern Lesser Antilles. *Proc. Boston Soc. Nat. Hist.* 1905:220.
Clark, A.H. 1902. The birds of Margarita Island, Venezuela. *Auk* 19:258–267.
Clewson, S.D. 1980. Comparative ecology of the Northern Oriole (*Icterus galbula*) and the Orchard Oriole (*Icterus spurius*) in Western Nebraska. Master's Thesis, University of Nebraska, Lincoln.
Clotfelter, E.D. 1995. Courtship displaying and intrasexual competition in the Bronzed Cowbird. *Condor* 97:816–818.
Coffey, B.B. Jr., and L.C. Coffey. 1993. Birds songs and calls from southeastern Peru. ARA Records and the Tambopata Nature Reserve. Cassette Tape.
Collar, N.J., and P. Andrew. 1988. *Birds to Watch. The ICBP World Check-list of Threatened Birds.* ICBP Technical Publication No.8. Smithsonian Institution Press, Washington, D.C.
Collar, N.J., L.P. Gozanga, N. Krabbe, A. Madroño Nieto, L.G. Naranjo, T.A. Parker III, and D.C. Wege. 1992. *Threatened Birds of the Americas: The ICBP/IUCN Red Data Book.* Cambridge, U.K.
Compton, L.V. 1947. The Great-tailed Grackle in the Upper Rio Grande Valley. *Condor* 49:35–36.
Contreras, A. 1977. Details: Rusty Blackbird. *Oregon Birds* 3:31–32.
Contreras, J.R. 1980. New data on Shiny Cowbird Parasitism (*Molothrus bonariensis bonariensis*: Aves, Icteridae). *Historia Natural* 1:151–152.
Corbin, K.W., and C.G. Sibley. 1977. Rapid evolution in orioles of the genus *Icterus. Condor* 79:335–342.
Corbin, K.W., C.G. Sibley, and A. Ferguson. 1979. Genic changes associated with the establishment of sympatry in orioles of the genus *Icterus. Evolution* 33:624–633.
Corman, T., and G. Monson. 1995. First United States nesting records of the Streak-backed Oriole. *Western Birds* 26:49–53.
Cox, J. 1982. Chromosomal variation in North American icterids. AOU Chicago, IL
Cracraft, J. 1983. Species concepts and speciation analysis. *Current Ornithology* 1:159–187.
Crandall, L.S. 1937. The giant orioles of the tropics. *Bull. NY Zool. Soc.* 40:85–88.
Crase, F.T., and R.W. DeHaven. 1972. Current breeding status of the Yellow-headed Blackbird in California. *California Birds* 3:39–42.
Crase, F.T., R.W. DeHaven, and P.P. Woronecki. 1972. Movements of Brown-headed Cowbirds banded in the Sacramento Valley, California. *Bird Banding* 43:197–204.
Crawford, R.D. 1977. Breeding biology of year old and older female Red-winged and Yellow-headed Blackbirds. *Wilson Bull.* 89:73–80.
Crawford, R.D. 1978. Temporal patterns of spring migration of Yellow-headed Blackbirds in North Dakota. *Prairie Nat.* 10:120–122.
Crawford, R.D., and W.L. Hohman. 1978. A method for aging female Yellow-headed Blackbirds. *Bird Banding* 49:201–207.
Crawford, R.L. 1973. Wintering Northern Orioles in Thomasville, Georgia. *Oriole* 38:6–9.
Crins, B., and S. O'Donnell. 1993. Breeding habitats of Brewer's Blackbird in central Ontario. *Ontario Birds* 11:113–117.
Cruden, R.W., and S.M. Hermann-Parker. 1977. Defense of feeding sites by orioles and hepatic tanagers in Mexico. *Auk* 94:594–596.
Cruz, A. 1978. Adaptive evolution in the Jamaican Blackbird *Nesopsar nigerrimus. Ornis Scan.* 9:130–137.
Cruz, A., and J.W. Wiley. 1989. The decline of an adaptation in the absence of a presumed selection pres-

sure. *Evolution* 43:55–62.

Cruz, A., T. Manolis, and J. Wiley. 1985. The Shiny Cowbird: a brood parasite expanding its range in the Caribbean region. in P.A.Buckley, M. Foster, E. Morton, R. Ridgely and F. Buckley eds. Neotropical Ornithology. Ornithological Monographs 36:607–620.

Cruz, A., T.D. Manolis, and R.W. Andrews. 1990. Reproductive interactions of the Shiny Cowbird *Molothrus bonariensis* and the Yellow-hooded Blackbird *Agelaius icterocephalus* in Trinidad. *Ibis* 132:436–444.

Cruz, A., J.W. Wiley, T.K. Nakamura, and W. Post. 1989. The Shiny Cowbird *Molothrus bonariensis* in the West Indian region – Biogeographic and ecological implications. p. 519–540. In C.A. Woods [ed.], Biogeography of the West Indies – Past, present, and future. Sandhill Crane Press, Gainesville, FL.519-540.

Curson, J., D. Quinn, and D. Beadle. 1994. *New World Warblers.* A & C Black (Publishers), London/Houghton-Mifflin Company, Boston

Curtis, W.F. 1988. Highlights of a South Atlantic Tour – 1986/87 *Sea Swallow* 37:3–7.

d'Agincourt, L.G., and J.B. Falls. 1983. Variation of repertoire use in the Eastern Meadowlark, *Sturnella magna. Can. J. Zool.* 61:1086–1093.

Daguerre, J.B. 1924. Observaciones sobre la nidificacion de los tordos *Molothrus brevirostris* y *M. badius. Hornero* 3:285–285.

Dalmas, Le C.R. de 1900. Note sur une collection d'oiseaux de l'ile de Tobago (Mer des Antilles). *Mem. Soc. Zool. France* 13:132–144.

Darden, T. 1974. Common grackles preying on fish. *Wilson Bull.* 86:85–86.

Darley, J.A. 1971. Sex ratio and mortality in the Brown-headed Cowbird. *Auk* 88:560–566.

Darley, J.A. 1978. Pairing in captive Brown-headed Cowbirds (*Molothrus ater*). *Can. J. Zool.* 56:2249–2252.

Darley, J.A. 1982. Territoriality and mating behavior of the male Brown-headed Cowbird. *Condor* 84:15–21.

Darley, J.A. 1983. Territorial behavior of the female Brown-headed Cowbird (*Molothrus ater*). *Can. J. Zool.* 61:65–69.

Darlington, P.J. Jr. 1931. Notes on the birds of the Río Frio (near Santa Marta), Magdalena, Colombia. *Bull. Mus. Comp. Zool., Harvard* 71:349–421.

Davidson, A.H. 1994. Common Grackle predation on adult passerines. *Wilson Bull.* 106:174–175.

Davis, D.E. 1942. The number of eggs laid by cowbirds. *Condor* 44(1):10–12.

Davis, L.I. 1972. *A Field Guide to the Birds of Mexico and Central America.* Univ. of Texas Press, Austin, Texas.

Davis, W.R. II, and K.A. Arnold. 1972. Food habits of the Great-tailed Grackle in Brazos Co., Texas. *Condor* 74:439–446.

De la Peña, M.R. 1987. *Nidos y huevos de aves Argentinas.* Self-published, Santa Fe.

De la Peña, M.R. 1989. *Guia de Aves Argentinas.* L.O.L.A., Buenos Aires.

Dearborn, D.C. 1996. Video documentation of a Brown-headed Cowbird nestling ejecting an Indigo Bunting nestling from the nest. *Condor* 98:645–649.

DeBenedictis, P. 1976. Gleanings from the technical literature. *Birding* 8:349–352.

DeBenedictis, P. 1982. Gleanings from the technical literature. *Birding* 14:51–55.

Debrot, A.O., and T.G. Prins. 1992. First record and establishment of the Shiny Cowbird in Curaçao. *Caribbean J. Science* 28:104–105.

DeHaven, R.W., F.T. Crase, and M.R. Miller. 1974. Aging Tricolored Blackbirds by cranial ossification. *Bird Banding* 45:156–159.

DeHaven, R.W., F.T. Crase, and P.P. Woronecki. 1975. Movements of Tricolored Blackbirds banded in the Central Valley of California. *Bird Banding* 46:220–229.

DeHaven, R.W. 1975. Plumages of the Tricolored Blackbird. *Western Bird Bander* 50(4):59–60.

Delaney, D. 1992. Bird Songs of Belize, Guatemala and Mexico. A selection of rarities, regional endemics and distinctive subspecies. Cornell Laboratory of Ornithology. Ithaca, NY.

Demeritt, W.W. 1936. *Agelaius humeralis* a new bird for North America. *Auk* 53:453–453.

Denis, K. 1976. Scott's Oriole near Thunder Bay, Ontario. *Canadian Field-Naturalist* 90:500–501.

Denis, K. 1978. Bullock's Oriole at Thunder Bay – a first for Ontario. *Newsletter Thunder Bay Field Naturalists* 32:17.

Dennis, J.V. 1948. Observations on the Orchard Oriole in lower Mississippi delta. *Bird Banding* 19:12–21.

Devincenzi, G.J. 1925. Notas ornitologicas: observaciones sobre una colecion de nidos. *An. Mus. Montevideo* 2(2):67–102.

Diamond, A.W. 1973. Habitats and feeding stations of St Lucia forest birds. *Ibis* 115:313–329.

Dickerman, R.W. 1960. Red-eyed Cowbird parasitises Song Sparrow and Mexican Cacique. *Auk* 77:472–473.

Dickerman, R.W. 1965a. The nomenclature of the Red-winged Blackbird (*Agelaius phoeniceus*) of south-central Mexico. *Occas. Papers Mus. Zool., Louisiana State University.* 31:1–6.

Dickerman, R.W. 1965. The juvenal plumage and distribution of *Cassidix palustris* (Swainson). *Auk*

82:268–270.
Dickerman, R.W. 1974. Review of Red-winged Blackbirds (*Agelaius phoeniceus*) of eastern, west-central, and southern Mexico and Central America. *Am. Mus. Novit.* 2538:1–18.
Dickerman, R.W. 1981. A taxonomic review of the Spotted-breasted Oriole. *Nemouria* 26:1–10.
Dickerman, R.W. 1987. Type localities of birds described from *Guatemala. Proc. Western Found. Vert. Zool.* 3:51–107.
Dickerman, R.W. 1989. Notes on *Sturnella magna* in South America with a description of a new subspecies. *Bull. BOC.* 109:160–162.
Dickerman, R.W., and A.R. Phillips. 1966. A new subspecies of the Boat-tailed Grackle from Mexico. *Wilson Bull.* 78:129–131.
Dickerman, R.W., and A.R. Phillips. 1970. Taxonomy of the Common Meadowlark (*Sturnella magna*) in Central and southern Mexico and Caribbean Central America. *Condor* 72:305–309.
Dickerman, R.W., and D.W. Warner. 1962. A new Orchard Oriole from Mexico. *Condor* 64:311–314.
Dickey, D.R., and A.J. Van Rossem. 1938. The Birds of El Salvador. *Field Mus. Nat. Hist., Chicago. Zool. Series* 23:1–609.
Dickinson, T.E., and J.B. Falls. 1989. How Western Meadowlarks respond to simulated intrusions by unmated females. *Behav. Ecol. Sociobiol.* 25:217–225.
Dickinson, T.E., J.B. Falls, and J. Kopachena. 1987. Effects of female pairing status and timing of breeding on nesting productivity in Western Meadowlarks (*Sturnella neglecta*). *Can. J. Zool.* 65:3093–3101.
Dickson, J.G., J.H. Williamson, R.N. Conner, and B. Ortego. 1995. Streamside zones and breeding birds in eastern Texas. *Wildlife Society Bulletin.* 23:750–755.
Dinsmore, J.J., and S.J. Dinsmore. 1993. Range expansion of the Great-tailed Grackle in the 1900s. *Journ. of Iowa Academy of Science* 100:54–59.
Doane, B.K. 1971. A Black-cowled Oriole? *Nova Scotia Bird Society Newsletter* 13:79–82.
Dolbeer, R.A. 1976. Reproductive rate and temporal spacing of nesting of Red-winged Blackbirds in upland habitat. *Auk* 93:343–355.
Dolbeer, R.A. 1978. Movements and migration patterns of Red-winged Blackbirds: a continental overview. *J. Field Ornithol.* 49:17–34.
Dolbeer, R.A. 1982. Migration patterns for age and sex classes of blackbirds and starlings. *J. Field Ornithol.* 53:28–46.
Downer, A., and R. Sutton. 1990. *Birds of Jamaica. A photographic guide.* Cambridge Univ. Press, Cambridge.
Drury, W.H. Jr. 1962. Breeding activities, especially nest building, of the Yellowtail (*Ostinops decumanus*) in Trinidad, West Indies. *Zoologica* 47:39–58.
Dubois, A. 1887. Deux nouvelles especes d'oiseaux. *Bull. Mus. Hist. Nat. Belg.* 5:1–4.
Dubs, B. 1992. *Birds of southwestern Brazil.* Betrona-Verlag, Küsnacht, Switzerland.
Dufty, A.M. Jr. 1982. Movements and activities of radio-tracked Brown-headed Cowbirds. *Auk* 99:316–327.
Dufty, A.M. Jr. 1983. Variation in the egg markings of the Brown-headed Cowbird. *Condor* 85:109–111.
Dufty, A.M. Jr. 1982. Response of Brown-headed Cowbird to simmulated conspecific intruders. *Anim. Behav.* 30:1043–1052.
Dufty, A.M. Jr. 1994. Rejection of foreign eggs by Yellow-headed Blackbirds. *Condor* 96:799–801.
Dugand, A., and E. Eisenmann. 1983. Rediscovery of, and new data on, *Molothrus armenti* Cabanis. *Auk* 100:991–992.
Dunham, D.W. 1971. Interspecific cacique colonies in Surinam. *Auk* 88:178.
Dunn, J. 1975. Field Notes – Immature Orioles Western Tanager 41:7.
Dwight, J. 1903. Some new records for Nova Scotia. *Auk* 20:440.

Eastzer, D.H., A.P. King, and M.J. West. 1985. Patterns of Courtship Between Cowbird Subspecies: Evidence for Positive Assortment. *Anim. Behav.* 33:30–39.
Edinger, B.B. 1985. Limited hybridization and behavioral differences among sympatric Baltimore and Bullock's Orioles. M.Sc. Thesis, Univ. of Minnesota, Minneapolis.
Edinger, B.B. 1988. Extra-pair courtship and copulation attempts in Northern Orioles. *Condor* 90:546–554.
Eguiarte, L., C. Martinez del Rio, and H. Arita. 1987. El nectar y el polen como recursos: el papel ecologico de los visitantes a las flores de *Pseudobombax ellipticum. Biotropica* 19:74–82.
Eisenmann, E. 1957. Notes on the birds of the province of Bocas del Toro, Panama. *Condor* 59:247–262.
Eisenmann, E. 1970. Letter to the editor. *Condor* 72:381–381.
Elliot, P.E. 1980. Adaptive significance of cowbird egg distribution. *Auk* 94:590–593.
Emlen, J.T. 1941. An experimental analysis of the breeding cycle of the Tricolored Red-wing. *Condor* 43:209–219.
Emlen, J.T. 1985. Morphological correlates of synchronized nesting in a Tricolored Blackbird colony. *Auk* 102:882–884.
Enstrom, D.A. 1992a. Breeding season communication hypotheses for delayed plumage maturation in passerines: tests in the Orchard Orioles, *Icterus spurius. Anim. Behav.* 43:463–472.

Enstrom, D.A. 1992b. Winter adaptation hypotheses for delayed plumage maturation in passerines: tests in the orchard oriole (*Icterus spurius*). *Behav. Ecol. Sociobiol.* 30:35–42.
Enstrom, D.A. 1993. Female choice for age-specific plumage in the Orchard Oriole: implications for delayed plumage maturation. *Anim. Behav.* 45:435–442.
Erickson, J.E. 1969. Banding studies of wintering Baltimore Orioles in North Carolina, 1963–1966. *Bird Banding* 40:181–198.
Erickson, J.E., and T.L. Quay. 1964. Wintering Baltimore Oriole study. *Chat* 28:90.
Erskine, A.J. 1971. Some new perspectives on the breeding ecology of common grackles. *Wilson Bull.* 83:352–370.
Evans, L.G.R. 1994. *Rare birds in Britain 1800–1990.* LGRE Publications Ltd.
Evens, J., and R. Le Valley. 1981. Middle Pacific Coast Region. *American Birds* 35:224–224.
Ewan, J. 1944. Hooded oriole nesting in banana plant at Beverly Hills, California. *Condor* 46:205–205.

Faanes, C.A. 1976. A nesting study of Common Grackles in southeastern Minnesota. *Loon* 48:149–156.
Faanes, C.A., and W. Norling. 1981. Nesting of the Great-tailed Grackle in Nebraska. *Am. Birds* 35:148–149.
Falls, J.B. 1985. Song matching in Western Meadowlarks. *Can. J. Zool.* 63:2520–2524.
Falls, J.B., and L.G. d'Agincourt. 1981. A comparison of neighbor-stranger discrimination in Eastern and Western Meadowlarks. *Can. J. Zool.* 59:2380–2385.
Falls, J.B., and L.G. D'Agincourt. 1982. Why do meadowlarks switch song types? *Can. J. Zool.* 60:3400–3408.
Falls, J.B., and J.R. Krebs. 1975. Sequences of songs in repertoires of western meadowlarks (*Sturnella neglecta*). *Can. J. Zool.* 53:1165–1178.
Falls, J.B., and L.J. Szijj. 1959. Reactions of Eastern and Western Meadowlarks in Ontario to each others' vocalizations. *Anat. Record.* 134:560.
Fankhauser, D.P. 1968. A comparison of migration between blackbirds and starlings. *Wilson Bull.* 80:225–227.
Fankhauser, D.P. 1971. Annual survival rates of blackbirds and starlings. *Bird Banding* 42:36–42.
Fankhauser, D.P. 1971. Percentage of grackles taken in subsequent breeding seasons in a different breeding area from the area where banded. *Bird Banding* 42:43–45.
Feekes, F. 1977. Colony-specific song in *Cacicus cela* (Icteridae, Aves): The pass-word hypothesis. *Ardea* 65:197–202.
Feekes, F. 1981. Biology and colonial organization of two sympatric caciques, *Cacicus c.cela* and *Cacicus h. haemorrhous* (Icteridae: Aves) in Surinam. *Ardea* 69:83–107.
Feekes, F. 1982. Song mimesis within colonies of *Cacicus cela* (Icteridae:Aves): a colonial password? *Zeitschrift fur Tierpsychologie* 58:119–152.
ffrench, R. 1986. *Birds of Trinidad and Tobago.* Mcmillan, London.
ffrench, R. 1991. *A guide to the birds of Trinidad and Tobago* 2nd ed. Cornell University Press, Ithaca, NY.
Ficken, R.W. 1963. Courtship and agonistic behavior of the common grackle, *Quiscalus quiscula. Auk* 80:52–72.
Findholt, S.L., and S.D. Fitton. 1983. Records of the Scott's Oriole from Wyoming. *Western Birds* 14:109–110.
Fink, L.S., and L.P. Brower. 1981. Birds can overcome the cardenolide defence of monarch butterflies in Mexico. *Nature* 291:67–70.
Fisk, E.J. 1975. On Northern Oriole plumages: questions for banders to answer. *EBBA News* 38:146–147.
Fjeldsa, J., and N. Krabbe. 1990. *Birds of the High Andes.* Zoological Museum, Univ. of Copenhagen and Apollo Books, Svendborg.
Flanigan, A.B. 1977. Northern oriole in perplexing role. *Inland Bird Banding News* 49:141–142.
Fleischer, R.C. 1985. A new technique to identify and assess the dispersion of eggs of individual brood parasites. *Behav. Ecol. Sociobiol.* 17:91–99.
Fleischer, R.C., and S.I. Rothstein. 1988. Known secondary contact and rapid gene flow among subspecies and dialects in the Brown-headed Cowbird. *Evolution* 42:1146–1158.
Fleischer, R.C., S.I. Rothstein, and L.S. Miller. 1991. Mitochondrial DNA variation indicates geneflow aross a zone of known secondary contact between two subspecies of the Brown-headed Cowbird. *Condor* 93:185–189.
Fleischer, R.C., and N.G. Smith. 1992. Giant Cowbird eggs in the nests of two icterid hosts: the use of morphology and electrophoretic variants to identify individuals and species. *Condor* 94:572–578.
Flood, N.J. 1984. Adaptive significance of delayed plumage maturation in male Northern Orioles. *Evolution* 38:267–279.
Flood, N.J. 1985. Incidences of polygyny and extra pair copulation in the northern oriole. *Auk* 102:410–413.
Flood, N.J. 1987. Rusty Blackbird – In Cadman, M.D., P.F.J. Eagles, F.M. Helleiner (eds.). Atlas of the Breeding Birds of Ontario. University of Waterloo Press, Waterloo.476–476.
Flood, N.J. 1989. Coloration in New World orioles. 1. Tests of predation-related hypotheses. *Behav. Ecol. Sociobiol.* 25:49–56.

Flood, N.J. 1990. Aspects of the breeding biology of Audubon's Oriole. *J.Field Ornithol.* 61:290–302.
Forbes, W.A. 1881. Eleven weeks in north-eastern Brazil. *Ibis* 1881:312–362.
Fraga, R.M. 1972. Cooperative Breeding and a case of Successive Polyandry in the Bay-winged Cowbird. *Auk* 89:447–449.
Fraga, R.M. 1978. The Rufous-collared Sparrow as a Host of the Shiny Cowbird. *Wilson Bull.* 90(2):271–284.
Fraga, R.M. 1979. Differences Between Nestlings and Fledglings of Screaming and Bay-winged Cowbirds. *Wilson Bull.* 91(1):151–154.
Fraga, R.M. 1983. Parasitismo de cria del Renegrido (*Molothrus bonariensis*) sobre el Chingolo (*Zonotrichia capensis*): Nuevas observaciones y conclusiones. *Hornero* 12 No. extrao.:245–255.
Fraga, R.M. 1983. The eggs of the parasitic Screaming Cowbird (*Molothrus rufoaxillaris*) and its host, the Baywinged Cowbird (*M.badius*): is there evidence for mimicry? *J. Ornithol.* 124:187–193.
Fraga, R.M. 1984. Bay-winged Cowbirds (*Molothrus badius*) remove ectoparasites from their brood parasites, the Screaming Cowbirds (*M.rufoaxillaris*). *Biotropica* 16:223–226.
Fraga, R.M. 1985. Host-parasite interactions between Chalk-browed Mockingbirds and Shiny Cowbirds. In: Neotropical Ornithology. Buckley, P.A, M.S. Foster, E.S. Morton, R.S.Ridgely and F.G. Buckley (eds.). Ornithological Monographs 36:829–844.
Fraga, R.M. 1986. The Bay-winged Cowbird (*Molothrus badius*) and its brood parasites: Interactions, coevolution and comparative efficiency. Ph.D. Thesis, University of California, Santa Barbara.
Fraga, R.M. 1987. Vocal mimicry in the Epaulet Oriole. *Condor* 89:133–137.
Fraga, R.M. 1988. Nest Sites and Breeding Success of Baywinged Cowbirds (*Molothrus badius*) *J. Ornithol.* 129:175–183.
Fraga, R.M. 1989. Colony sizes and nest trees of Montezuma Oropendolas in Costa Rica. *J.Field Ornithol.* 60:289–295.
Fraga, R.M. 1990. El tordo amarillo al borde de la extinción. *Nuestras Aves* 23:13–15.
Fraga, R.M. 1991. The social system of a communal breeder, the Bay-winged Cowbird *Molothrus badius. Ethology* 89:195–210.
Fraga, R.M. 1996. Further evidence of parasitism of Chopi Blackbirds (*Gnorimopsar chopi*) by the specialized Screaming Cowbird (*Molothrus rufoaxillaris*). *Condor* 98:866–867.
Freeman, S. 1990. Molecular systematics and morphological evolution in the blackbirds. Ph.D. Dissertation, University of Washington, Seattle
Freeman, S., D.F. Gori, and S. Rohwer. 1990. Red-winged Blackbirds and Brown-headed Cowbirds: some aspects of a host-parasite relationship. *Condor* 92:336–340.
Freeman, S., and R.M. Zink. 1995. A phylogenetic study of the blackbirds based on variation in mitochondrial DNA restriction sites. *Syst. Biol.* 44:409–420.
Friedmann H. 1929. *The Cowbirds. A Study in the Biology of Social Parasitism.* Charles C. Thomas, Springfield, Illinois.
Friedmann, H. 1949. Additional data on victims of parasitic cowbirds. *Auk* 66:154–163.
Friedmann, H. 1957. The Rediscovery of *Tangavius armenti* (Cabanis). *Auk* 74:497–498.
Friedmann, H. 1971. Further Information on the Host Relations of the Parasitic Cowbirds. *Auk* 88:239–255.
Friedmann, H., and L.F. Kiff. 1985. The parasitic cowbirds and their hosts. *Proc. West. Found. Vert. Zool.* 2:225–304.
Friedmann, H., L.F. Kiff, and S.I. Rothstein. 1977. A Further Contribution to Knowledge of the Host Relations of the parasitic Cowbirds. *Smithsonian Contrib. Zool.* 235:1–75.
Friedmann, H., and F.D. Jr. Smith. 1950. A contribution to the ornithology of northeastern Venezuela. *Proc. U.S. Nat. Mus.* 100:411–538.
Friedmann, H.L., L. Griscom, and R.T. Moore. 1957. Distributional Check-List of the birds of Mexico, Part II. *Pacific Coast Avifauna* 33:1–435.
Frisch, S., and J.D. Frisch. 1964. Aves Brasileiras Irmaos. Vitale S/A., São Paulo.
Frost, D.R., and D.M. Hillis. 1990. Species in concept and practice: Herpetological applications. *Herpetologica* 46:87–104.
Furrer, R.K. 1975. Breeding success and nest site stereotypy in a population of Brewer's Blackbird (*Euphagus cyanocephalus*). *Oecologia* 20:339–350.

Gabrielson, I.N., and S.G. Jewett. 1940. *Birds of Oregon.* Oregon State College, Corvalis, Oregon.
Garcia, F. 1987. *Las Aves de Cuba. Tomo II. Subespecies endemicas.* Editorial Gente Nueva, Ciudad de la Habana, Cuba 77–90.
Garcia, F. 1980. *Las aves de Cuba. Tomo I. Especies endemicas.* Electron, Cuba.
Garrido, O. 1970. Variacion del genero *Agelaius* (Aves: Icteridae) en Cuba. *Poeyana* 68:1–18.
Garrido, O., and A. Kirkconnell. 1996. Taxonomic status of the Cuban form of the Red-winged Blackbird. *Wilson Bull.* 108:372–374.
Garrido, O.H. 1985. Cuban endangered birds. In: Neotropical Ornithology. Buckley, P.A., M.S. Foster, E.S.Morton, R.S.Ridgely, and F.G.Buckley (eds).992–999.
Gavin, T.A., and E.K. Bollinger. 1985. Multiple paternity in a territorial passerine: the Bobolink. *Auk*

102:550–555.
Gavin, T.A., and E.K. Bollinger. 1988. Reproductive correlates of breeding-site fidelity in Bobolinks (*Dolichonyx oryzivorus*). *Ecology* 69:96–103.
Gavin, T.A., R.A. Howard, and B. May. 1991. Allozyme variation among breeding populations of Red-winged Blackbirds: the California conundrum. *Auk* 108:602–611.
Gibbs, H.L., P.J. Weatherhead, P.T. Boag, B.N. White, L.M. Tabak, and D.J. Haywick. 1990. Realized reproductive success of polygynous Red-winged Blackbirds revealed by DNA markers. *Science* 250:1394–1397.
Gibson, D.D. 1972. Sight records of two birds new to interior Alaska. *Murrelet* 53:31–32.
Gibson, D.D., and B. Kessel. 1997. Inventory of the species and subspecies of Alaska birds. *Western Birds* 28:45–95.
Gibson, E. 1880. Ornithological notes from the neighbourhood of Cape San Antonio, Buenos Aires. *Ibis* 1880:1–39.
Gibson, E. 1918. Further ornithological notes from the neighbourhood of Cape San Antonio, province of Buenos Ayres. *Ibis* 1918 363–415.
Gill, E.L., M.B. Serra, S.B. Canavelli, C.J. Feare, and et al. 1994. Cinnamamide prevents captive Chestnut-capped Blackbirds (*Agelaius ruficapillus*) from eating rice. *Int. J. of Pest Management* 40:195–198.
Gill, G., and M.A. Gill. 1954. Some banding notes on Purple Grackles. *Bird Banding* 25:110–111.
Gilligan, J., and D. Irons. 1987. Oregon's first Orchard Oriole. *Oregon Birds* 13:295.
Gilligan, J.M., M. Smith, D. Rogers, and A. Contreras. 1994. *Birds of Oregon: status and distribution.* Cinclus Publ., McMinnville.
Glassel, R. 1993. A Minnesota Great-tailed Grackle with notes on the species' range expansion in the mid-west. *The Loon* 65:148–150.
Gochfeld, M. 1975. Comparative ecology and behavior of red-breasted meadowlarks and their interactions in sympatry. Ph.D. diss. City University of New York.
Gochfeld, M. 1978. Social facilitation of singing: Group size and flight song rates in the Pampas Meadowlark, *Sturnella defilippii. Ibis* 120:338–339.
Gochfeld, M. 1979a. Interspecific Territoriality in Red-breasted Meadowlarks and a method for estimating the mutuality of their participation. *Behav. Ecol. Sociobiol.* 5:159–170.
Gochfeld, M. 1979b. Begging by Nestling Shiny Cowbirds: Adaptive or maladaptive. *Living Bird* 17:41–50.
Gochfeld, M. 1979c. Brood Parasite and Host Coevolution: Interactions between Shiny Cowbirds and Two Species of Meadowlarks. *Am. Nat.* 113(6):855–870.
Gochfeld, M., D.O. Hill, and G. Tudor. 1973. A second population of the recently described Elfin Woods Warbler and other bird records from the West Indies. *Caribbean J. Sci.* 13:231–235.
Godfrey, W.E. 1979. *The Birds of Canada.* National Museum of Natural Sciences, Ottawa.
Godinez, E., and P. Blanco. 1993. Nido de *Dives atroviolacea* (Icteridae) en condiciones antropizadas. *Orn. Neotrop.* 4:95–96.
Goeldi, E. 1897. On the nesting of *Cassicus persicus, Cassicus oryzivora, Gymnomystax melanicterus* and *Todirostrum maculatus. Ibis* 1897:361–370.
Gonzáles Sánchez, L., M.E. Garcia Romero, and A. Batile Vierra. 1991. Variacion morphologica del genero *Quiscalus* (Aves: Icteridae) en Cuba. *El Pitirre* 4:4–5.
Goodfellow, W. 1901. Results of an ornithological journey through Colombia and Ecuador [part 2]. *Ibis* 1901:458–480.
Graber, R.R., and J.W. Graber. 1954. Comparative notes on Fuertes and Orchard Orioles. *Condor* 56:274–282.
Graham, D.S. 1988. Response of Five Host Species to Cowbird Parasitism. *Condor* 90:588–591.
Grant, P.R. 1965. A systematic study of the birds of the Tres Marias Islands, Mexico. *Postilla* 90:42–44.
Grant, P.R., and I. McT. Cowan. 1964. A review of the avifauna of the Très Marias Islands, Nayarit, Mexico. *Condor* 66:221–228.
Graves, G.R. 1978. Predation on hummingbird by oropendola. *Condor* 80:251–251.
Greenwood, H., and P.J. Weatherhead. 1982. Spring roosting dynamics of the Red-winged Blackbird: biological and management implications. *Can. J. Zool.* 60:750–753.
Greenwood, H., P.J. Weatherhead, and R.D. Titman. 1983. A new age- and sex-specific molt scheme for the Red-winged Blackbird. *Condor* 85:104–105.
Grinnell, H.W. 1944. The Hooded Oriole's choice of nesting sites in the settled portions of southern California. *Condor* 46:298–298.
Grinnell, J. 1909. A new cowbird of the genus *Molothrus*, with a note on the probable genetic relationships of the North American forms. *Univ. Calif. Publ. Zool.* 5:275–281.
Grinnell, J. 1914. A new Red-winged Blackbird from the Great Basin. *Proc. Biol. Soc. Washington* 27:107–108.
Grinnell, J. 1927. Six new subspecies of birds from lower California. *Auk* 44:70–71.
Griscom, L. 1930. Studies from the Dwight collection of Guatemalan birds III. *Amer. Mus. Novit.* 438:13.
Griscom, L. 1932. The distribution of bird-life in Guatemala. *Bull. Amer. Mus. Nat. Hist.* 64:1–439.
Griscom, L. 1934. The ornithology of Guerrero, Mexico. *Bull. Mus. Comp. Zool.* 75:402–409.

Griscom, L., and J.C. Jr. Greenway. 1937. Critical notes on new neotropical birds. *Bull. Mus. Comp. Zool.* 81:434–435.

Grisdale, T. 1882. On the birds of Montserrat. *Ibis* 24:485–493.

Gundlach, J. 1876. *Contribución a la ornitólogia Cubana.* Impresa 'La Antilla' de N. Cachonegrete.

Gyldenstolpe, N. 1942. Preliminary diagnoses of some new birds from Bolivia. *Arkiv for Zoologi* 33B(13):1–10.

Gyldenstolpe, N. 1945a. The bird fauna of Rio Juru in western Brazil. *Kungl. Svenska Vet.-Akad. Handl. Band* 22, No. 3:293–302.

Gyldenstolpe, N. 1945b. A contribution to the ornithology of northern Bolivia. *Kungl. Svenka Vet.-Akad. Handl. Ser.*3, 23(1):1–300.

Gyldenstolpe, N. 1951. The ornithology of the Rio Purús region in western Brazil. *Arkiv. fur Zoologi* (*ser*2) 2:1–320.

Haffer, J. 1969. Speciation in Amazonian forest birds. *Science* 165:131–137.

Haffer, J. 1974. Avian Speciation in Tropical South America. *Publ. Nuttall Ornith. Club* 14:1–390.

Haffer, J. 1975. *Avifauna of Northwest Colombia, South America.* Bonner Zool. Monog. 7, Bonn:

Hahn, D.C., and R.C. Fleischer. 1995. DNA fingerprint similarity between female and juvenile Brown-headed Cowbirds trapped together. *Animal Behaviour* 49:6.

Hallinan, 1924. Notes on some Panama Canal zone birds with special reference to their food. *Auk* XLI: 304–326.

Hamel, P.B. 1974. Age and sex determination of nestling Common Grackles. *Bird Banding* 45:16–23.

Hamilton, W.J. III 1962. Bobolink migratory pathways and their experimental analysis under night skies. *Auk* 79:208–233.

Hammel, P.B. 1974. Age and sex determination of nestling Common Grackles. *Bird Banding* 45:16–23.

Hammer, W.M., and J. Stocking. 1970. Why don't Bobolinks breed in Brazil? *Ecology* 51:743–751.

Hann, H.W. 1941. The cowbird at the nest. *Wilson Bull.* 53:211–221.

Harber, D.D. 1963. Baltimore oriole in Sussex. *British Birds* 56:464–465.

Hardy, J.W. 1965. Evolutionary and ecological relationships between three species of blackbirds (Icteridae) in central Mexico. *Evolution* 21:196–197.

Hardy, J.W. 1970. Duplex nest construction by Hooded Oriole circumvents cowbird parasitism. *Condor* 72:491–491.

Hardy, J.W. 1983. Voices of Neotropical Birds. ARA 1- Cassette tape

Hardy, J.W., and R.W. Dickerman. 1965. Relationships between two forms of the Red-winged Blackbird in Mexico. *Living Bird* 4:107–130.

Hardy, J.W., G. B. Reynard, A. J. Begazo, T. Taylor. 1998. Voices of the Troupials, Blackbirds and Allies. Family Emberizidae, Subfamily Icteridae. ARA Records. Currently in production.

Harms, K.E., L.D. Beletsky, and G.H. Orians. 1991. Conspecific nest parasitism in three species of new world blackbirds. *Condor* 93:967–974.

Harper, C., and K. Taylor. 1986. The Northern Oriole, on the Saanich Peninsula. *Victoria Nat.* 43:4–5.

Harper, F. 1920. The song of the Boat-tailed Grackle. *Auk* 37:295–297.

Harper, F. 1934. The Boat-tailed Grackle of the Atlantic Coast. *Proc. Acad. Sci. Phila.* 86:1–2.

Harris, A. 1989. First Yellow-headed Blackbird nest for Thunder Bay District. *Ontario Birds* 7:29–30.

Harrison, C.J.O. 1963. Interspecific preening display by the Rice Grackle, *Psomocolax oryzivorus. Auk* 80:373–374.

Hartert, E. 1902. Die mit sichenheit festgestellen vogel der inselin Aruba, Curaçao und Bonaire. *Novit. Zool.* 9:299–300.

Hartert, E. 1914. Dr. Ernst Hartert exhibited and described examples of the following new subspecies of birds: *Bull. BOC* 33:76.

Harvey, P.H., and L. Partridge. 1988. Of cuckoo clocks and cowbirds. *Nature Lond.* 335:586–587.

Haverschmidt, F. 1949. The Guianan Meadowlark in Surinam, Dutch Guiana. *Auk* 66:208–208.

Haverschmidt, F. 1951. Notes of *Icterus nigrogularis* and *I. chrysocephalus* in Surinam. *Wilson Bull.* 63:45–47.

Haverschmidt, F. 1967. Additional notes on the eggs of the Giant Cowbird. *Bull. BOC* 87:136–137.

Haverschmidt, F. 1968. *Birds of Surinam.* Oliver & Boyd, Edinburgh and London.

Haverschmidt, F. 1972. More notes on interspecific cacique and oropendola colonies in Surinam. *Auk* 89:676.

Hayes, F.E. 1995. Status, Distribution and Biogeography of the Birds of Paraguay. *ABA Monographs in Field Ornithology No. 1.*

Hayes, F.E., S.M. Goodman, J.A. Fox, T.G. Tamayo, and N.E. Lopez. 1990. North American bird migrants in Paraguay. *Condor* 92:947–960.

Hayes, F.E., P.A. Scharf, and R.S. Ridgely. 1994. Austral bird migrants in Paraguay. *Condor* 96:83–97.

Hayward, C.L., C. Cottam, A.M. Woodbury, and H.H. Frost. 1976. Birds of Utah. *Great Basin Naturalist Memoirs* 1:179–184.

Helleiner, F.M. 1972. Common Grackle at Rankin Inlet, Keewatin District. *Can. Field Nat.* 86:84.

Hellmayr, 1906. Lampropsar tanagrinus violaceus *Abh. Bayern Akad. Wiss. Math.-phys.* Kl. 22:616.

Hellmayr, C.E. 1929. A contribution to the ornithology of northeast Brazil. *Field Mus. Nat. Hist. Publ. Zool. Ser.* 12:272–277.
Hellmayr, C.E. 1937. Catalogue of birds of the Americas and adjacent islands. Part X, Icteridae. *Field Mus. Nat. Hist., Zool. Ser.* 13:1–228.
Hemming, J.E. 1965. Unusual occurrences of birds in northern Alaska. *Murrelet* 46:6.
Hergenrader, G.L. 1962. The incidence of nest parasitism by the brown-headed Cowbird (*Molothrus ater*) on roadside nesting birds in Nebraska. *Auk* 79:85–88.
Herlyn, H.G., S. Jones, and J.L. Simmons. 1994. Oregon's first Streak-backed Oriole. *Oregon Birds.* 20:75–77.
Hernandez-Camacho, J.I., and J.V. Rodriguez-Mahecha. 1986. Status geographico y taxonomico de *Molothrus armenti* Cabanis 1851 (Aves: Icteridae). *Caldasia* XV:655–664.
Hernandez-Prieto, E., and M. González. 1992. The Yellow-shouldered Blackbird on Monito Island, Puerto Rico. *Ornitologia Caribena* 3:54–56.
Hickman, S. 1981. Evidence for aerodynamic advantages of tail keeling in the Common Grackle. *Wilson Bull.* 93:500–505.
Hill, N.P., and J.M. III Hagan. 1991. Population trends of some northeastern North American landbirds: A half century of data. *Wilson Bull.* 103:165–182.
Hill, R.A. 1976. Host-parasite relationships of the Brown-headed Cowbird in a prairie habitat of west-central Kansas. *Wilson Bull.* 88:555–565.
Hill, R.A. 1976. Sex ratio and sex determination of immature Brown-headed Cowbirds. *Bird Banding* 47:112–114.
Holford, K.C., and D.D. Roby. 1993. Factors limiting fecundity of captive Brown-headed Cowbirds. *Condor* 95:536–545.
Holmes, J.A., D.S. Dobkin, and B.A. Wilcox. 1985. Second nesting record and northward advance of the Great-tailed Grackle (*Quiscalus mexicanus*) in Nevada. *Great Basin. Nat.* 45:483–484.
Horn, H.S. 1968. The adaptive significance of colonial nesting in the Brewer's *Blackbird (Euphagus cyanocephalus*). *Ecology* 49:682–694.
Horn, H.S. 1970. Social behavior of nesting Brewer's Blackbirds. *Condor* 72:15–23.
Houston, C.S. 1968. Recoveries of Bronzed Grackles banded in Saskatchewan. *Blue Jay* 26:136–138.
Hovekamp, N.R. 1996. Intersexual vocal communication in the Red-winged Blackbird. *J. Field Ornithol.* 67(3):376–383.
Howe, H.F. 1976. Egg size, hatching asynchrony, sex, and brood reduction in the common grackle. *Ecology* 57:1195–1207.
Howe, H.F. 1977. Sex-ratio adjustment in the Common Grackle. *Science* 198:744–746.
Howe, H.F. 1978. Initial investment, clutch size, and brood reduction in the Common Grackle. *Ecology* 59:1109–1122.
Howe, H.F. 1978. Aspects of parental investment and sex-ratio adjustment in the Common Grackle. *Diss. Abst. Int.* 38:5230–5231.
Howe, H.F. 1979. Evolutionary aspects of parental care in the Common Grackle, *Quiscalus quiscula* L. *Evolution* 33:41–51.
Howell, A.H., and A.J. Van Rossen. 1928. A study of the Red-winged Blackbirds of the southeastern United States. *Auk* 45:155–163.
Howell, S.N.G., and S. Webb. 1995. *A Guide to the Birds of Mexico and Northern Central America.* Oxford Univ. Press, Oxford.
Howell, S.N.G., S. Webb, and B.M. de Montes. 1992. Colonial nesting of the Orange Oriole. *Wilson Bull.* 104:189–190.
Howell, T.R. 1964. Mating behavior of the Montezuma Oropendola. *Condor* 66:511–511.
Howell, T.R. 1970. Avifauna in Nicaragua. In Buechner, H.K. and J.H. Buechner (eds.). The avifauna of northern Latin America. Smithsonian Institution Press, Washington, D.C.
Howell, T.R. 1971. A comparative ecological study of the birds of the lowland pine savanna and adjacent rain forest in northeastern Nicaragua. *Living Bird* 10.
Howell, T.R. 1972. Birds of the lowland pine savanna of northeastern Nicaragua. *Condor* 74:316–340.
Howell, T.R., and G.A. Bartholomew. 1952. Experiments on the mating behavior of the Brewer's Blackbird. *Condor* 54:140–151.
Howell, T.R., and G.A. Bartholomew. 1954. Experiments on the social behaviour in nonbreeding Brewer's Blackbirds. *Condor* 56:33–37.
Hoy, G., and J. Ottow. 1964. Biological and Oological Studies of the Molothrine Cowbirds (Icteridae) of Argentina. *Auk* 81:186–203.
Hubbard, J.P. 1974. Identification of wintering orioles in the northeast. *Delmarva Ornithologist* 7(2):10–12.
Hubbard, J.P. 1983. The tail pattern of meadowlarks in New Mexico. *New Mexico Orn. Soc. Bull.* 11:61–66.
Hudon, J., and A.D. Muir. 1996. Characterization of the reflective materials and organelles in the bright irides of North American blackbirds (Icterinae). *Pigment Cell Research* 9:96–104.
Hudson, W.H. 1874. Notes on the procreant instincts of the three species of *Molothrus* found in Buenos

Ayres. *Proc. Zool. Soc.* 1874 (No.XI):153–174.
Hudson, W.H. 1920. *Birds of La Plata.* J.M. Dent and Sons, London.
Huey, L.M. 1944. Nesting habits of the Hooded Oriole. *Condor* 48:297–297.
Humphrey, P.S., D. Bridge, P.W. Reynolds, and R.T. Peterson. 1970. *Birds of Isla Grande (Tierra del Fuego).* Smithsonian Institution preliminary manual. Washington.
Humphrey, P.S., and K.C. Parkes. 1959. An approach to the study of molts and plumages. *Auk* 76:1–31.
Hunt, L.B. 1977. Brown-headed Cowbirds form a pair bond for three seasons. *Inland Bird Banding News* 49:223–224.
Huntington, C.E. 1952. Hybridization in the Purple Grackle, *Quiscalus quiscula. Syst. Zool.* 1:149–170.
Hurly, T.A., and R.J. Robertson. 1984. Agressive and territorial behavior in female Red-winged Blackbirds. *Can. J. Zool.* 62:148–153.
Hutcheson, W.H., and W. Post. 1990. Shiny Cowbird collected in South Carolina: first North American specimen. *Wilson Bull.* 102:561–561.

Imhof, T.A. 1976. *Alabama Birds.* The Univ. of Alabama Press, Univ. of Alabama, AL.
Irwin, O.F. 1956. Recoveries of Bronzed Grackles banded at Memphis, Tennesee. *Inland Bird Banding News* 28:35–40.
Irwin, R.E. 1989. A comparative study of sexual selection on song repertoire size in the avian subfamily Icterinae (Passeriformes, Emberizidae). PH.D. dissertation, University of Michigan, Ann Arbor.
Irwin, R.E. 1990. Directional sexual selection cannot explain variation in song repertoire size in the New World blackbirds (Icterinae). *Ethology* 85:212–224.
Irwin, R.E. 1994. The evolution of plumage dichromatism in the New World blackbirds: social selection on female brightness? *Am. Nat.* 144:890–907.
Isleib, M.E., and B. Kessel. 1973. Birds of the North Gulf Coast–Prince William Sound region, Alaska. *Biol. Pap. U. Alaska* 14:1–149.

Jackson, J.H., and D.D. Roby. 1992. Fecundity and egg-laying patterns of captive yearling Brown-headed Cowbirds. *Condor* 94:585–589.
James, F.C. 1983. Environmental component of morphological differentiation in birds. *Science* 221:184–186.
James, F.C., R.T. Engstrom, C. Nesmith, and R. Laybourne. 1984. Inferences about population movements of Red-winged Blackbirds from morphological data. *Am. Midl. Nat.* 111:319–331.
Jeffrey-Smith, M. 1972. *Bird-watching in Jamaica.* Bolivar Press, Kingston.
Johnson, A.W., and J.D. Goodall. 1965. *The Birds of Chile and adjacent regions of Argentina, Bolivia and Peru.Vol. II.* Platt Establecimentos Graficos, S.A., Buenos Aires.
Johnson, D.M., G.L. Stewart, M. Corley, R. Ghrist, J. Haguer, A. Ketteres, B. McDonnell, W. Newsom, E. Owen, and P. Samuels. 1980. Brown-headed Cowbird (*Molothrus ater*) mortality in an urban winter roost. *Auk* 97:299–320.
Johnson, R.E., J.D. Reichel, and C.J. Herlugson. 1984. Recent bird records from eastern Washington. *Murrelet* 65:60–61.
Johnson, T.H. 1988. Biodiversity and conservation in the Caribbean: profiles of selected islands. Int. Comm. Bird Preserv 1.
Johnston, D.W. 1975. Ecological analysis of the Cayman Island avifauna. *Bull. Florida State Mus., Biol. Sci.* 19:235–300.
Johnston, R.F. 1960. Behavioral and ecological notes on the Brown-headed Cowbird. *Condor* 62:137–138.
Jonson, 1968. Changing status of the Bronzed Cowbird in Arizona. *Condor* 70:183.

Kale, H.W. II, M.H. Hundley, and J.A. Tucker. 1969. Tower-killed specimens and observations of migrant birds from Grand Bahama Island. *Wilson Bull.* 81:258–263.
Kaufman, K. 1983. Identifying Streak-backed Orioles: a note of caution. *American Birds* 37:140–141.
Kaufman, K. 1987. The practiced eye, notes on female orioles. *American Birds* 41:3–4.
Kaufman, K. 1989. Answers to August Photo Quiz. Scott's Oriole. *Birding* 21:256–258.
Kelley, A.H. 1960. Migration of Baltimore Orioles in late summer. *Jack-Pine Warbler* 38:44–45.
Kelly, S.T., and M.E. DeCapita. 1982. Cowbird control and its effect on Kirtland's Warbler reproductive success. *Wilson Bull.* 94:363–365.
Kennard, J.H. 1975. Longetivity records of North American birds. *Bird Banding* 46:55–73.
Kennard, J.H. 1978. A method of local limitation of brood parasitism by the Brown-headed Cowbird. *North Amer. Bird Bander* 3:100–102.
Kessel, B., and D.D. Gibson. 1978. Status and distribution of Alaska birds. *Studies Avian Biol.* 1.
Keys, G.C., R.C. Fleischer, and S.I. Rothstein. 1986. Relationships between elevation, reproduction and the hematocrit level of Brown-headed Cowbirds. *Comp. Biochem. Physiol.* 83A:765–769.
Kiff, L.F. 1975. Notes on southwestern Costa Rican birds. *Condor* 77:101–102.
King, A.P. 1973. Some factors controlling egg laying in the parasitic cowbird (*Molothrus ater*). *Amer. Zool.* 13:1259–1259.
King, A.P., and M.J. West. 1977. Species Identification in the North American Cowbird: Appropriate

Responses to Abnormal Song. *Science* 195:1002–1004.
King, J.R. 1973. Reproductive Relationships of the Rufous-collared Sparrow and Shiny Cowbird. *Auk* 90:19–34.
Kingery, H.E. 1980. Mountain West Region. *Am. Birds* 34:917–917.
Klimaitis, J.F. 1973. Estudio descriptivo de una colonia de Tordos Varilleros (*Agelaius rufucapillus*). *Hornero* 11:193–202.
Klimkiewicz, M.K. 1997. Longevity Records of North American Birds. Version 97.1 Patuxent Wildlife Research Center. *Bird Banding* Laboratory. Laurel MD.
Klimkiewicz, M.K., and A.G. Futcher. 1987. Longevity records of North American birds: Coerebinae through Estrildidae. *J. Field Ornithol.* 58:318–333.
Klimkiewicz, M.K., and A.G. Futcher. 1989. Longevity records of North American Birds. *J. Field Ornithol.* 60:469–494.
Knapton, R.W. 1988. Nesting success is higher for polygynously mated females than for monogamously mated females in the eastern meadowlark. *Auk* 105:325–329.
Knight, R.L., and S.A. Temple. 1988. Nest-defence behavior in the Red-winged Blackbird. *Condor* 90:193–200.
Knittle, C.E., G.M. Linz, B.E. Johns, J.L. Cummings, and et al. 1987. Dispersal of male Red-winged Blackbirds from two spring roosts in central North America. *J. Field Ornithol.* 59:490–498.
Koepcke, M. 1970. *The Birds of the Department of Lima, Peru.* Livingston Publishing Company, Pennsylvania.
Koepcke, M. 1972. Uber die resistenzformen der vogelnester in einem begrentzten gebiet des tropischen regenwaldes in Peru. *Journal f. Orn.* 113:138–160.
Koepcke, M. 1983. *The birds of the department of Lima, Peru.* Harrowood Books, Newtown Square, Pennsylvania.
Kok, O.B. 1971. Vocal behavior of the Great-tailed Grackle (*Quiscalus mexicanus prosopidicola*). *Condor* 73:348–363.
Kok, O.B. 1972. Breeding success and territorial behavior of male Boat-tailed Grackles. *Auk* 89:528–540.
Kramer, P. 1965. Bobolink and Summer Tanager on Galapagos Islands in late summer. *Condor* 67:90–90.
Kratter, A.W. 1993. Geographic Variation in the Yellow-billed Cacique, *Amblycercus holosericeus*, a partial bamboo specialist. *Condor* 95:641–651.
Kroodsma, D.E., and J.R. Baylis. 1982. Appendix: A world survey of evidence for vocal learning in birds. In D.E. Kroodsma, E.H. Miller and H. Ouellet (eds.) *Acoustic communication in birds.* Academic Press, New York.35–58.
Kroodsma, D.E., and F.C. James. 1994. Song variation within and among populations of Red-winged Blackbirds. *Wilson Bull.* 106:156–162.

La Rue, C.T. 1992. The Common Grackle in Arizona: First specimen record and notes on occurrence. *Western Birds* 23:84.
Labedz, T.E. 1984. Age and reproductive success in Northern Orioles. *Wilson Bull.* 96:303–305.
Lack, D. 1976. *Island Biology Illustrated by the Land Birds of Jamaica.* Univ. of California Press, Berkeley.
Lack, D., and J.T. Emlen. 1939. Observations on breeding behavior in Tricolored Red-wings. *Condor* 41:225–230.
Lamm, D.W. 1948. Notes on the birds of the states of Pernambuco and Paraiba, Brazil. *Auk* 65:261–283.
Land, H.C. 1962. A collection of birds from the arid interior of eastern Guatemala. *Auk* 79:1–11.
Langston, N.E., S. Freeman, S. Rohwer, and D. Gori. 1990. The evolution of female body size in Red-winged Blackbirds: The effect of timing of breeding, social competition, and reproductive energetics. *Evolution* 44:1764–1779.
Lanyon, S.M. 1992. Interspecific brood parasitism in blackbirds (Icterinae): A phylogenetic perspective. *Science* 255:77–79.
Lanyon, S.M. 1994. Polyphyly of the blackbird genus *Agelaius* and the importance of assumptions of monophyly in comparative studies. *Evolution* 48:679–693.
Lanyon, W.E. 1956a. Ecological aspects of the sympatric distribution of meadowlarks in the north-central states. *Ecology* 37:98–108.
Lanyon, W.E. 1956b. Territory in Meadowlarks, genus *Sturnella*. *Ibis* 98:485–489.
Lanyon, W.E. 1957. The comparative biology of the meadowlarks (*Sturnella*) in Wisconsin. *Publ. Nuttall Orn. Club*, no 1. Cambridge, MA.
Lanyon, W.E. 1962. Species limits and distribution of meadowlarks of the desert grasslands. *Auk* 79:183–207.
Lanyon, W.E. 1994. Western Meadowlark (*Sturnella neglecta*). In Poole, A. and Gill, F. (eds)., *The Birds of North America*, No. 104. Philadelphia: Academy of Natural Sciences.
Lanyon, W.E. 1995. Eastern Meadowlark (*Sturnella magna*). In Poole, A. and Gill, F. (eds)., *The Birds of North America*, No. 160. Academy of Natural Sciences, Philadelphia; Amer. Orn. Union, Washington.
Lanyon, W.E., and Fish. 1958. Geographical variation in the vocalizations of the Western Meadowlark. *Condor* 60:339–341.
La Rue, C.T., and D.H. Ellis. 1992. The Common Grackle in Arizona: First Specimen record and notes on

occurrence. *Western Birds* 23:84–86.
Laskey, D. 1950. Cowbird Behavior. *Wilson Bull.* 62:157–174.
Latta, S.L., and J.M. Jr. Wunderle. 1996. The composition and foraging ecology of mixed-species flocks in pine forests of Hispaniola. *Condor* 98:595–607.
Laubmann, A. von 1934. Zur Avifauna Argentiniens Verh. *Ornith. Ges. Bayern* 20:331–333.
Lawrence, G.N. 1864. Descriptions of new species of birds of the families Caerebidae, Tanagridae, Icteridae and Scolopacidae. *Proc. Acad. Nat. Sci. Philadelphia* 106–108.
Lawrence, G.N. 1878. Cataloge of the birds collected in Martinique by Mr. Fred A. Ober for the Smithsonian Institution. *Proc. U.S. Nat. Mus.* 1:349–360.
Lawrence, G.N. 1880. Descriptions of New Species of *Icterus* from the West Indies. *Proc. United States National Museum* 3:351–351.
Lawrence, R.E., and H. Brackbill. 1957. Winter returns of Baltimore Orioles in the Washington-Baltimore area. *Auk* 57–59.
Leak, J., and S.K. Robinson. 1989. Notes on the social behaviour and mating system of the Casqued Oropendola. *Wilson Bull.* 101:134–137.
Leck, C. 1974. Further observations of nectar feeding by orioles. *Auk* 91:162–163.
Leck, C. F. 1984. *The status and distribution of New Jersey's Birds.* Rutgers University Press, New Brusnswick, New Jersey.
Leck, C. F. 1985. *Range expansions in North American Agelaiini* (*blackbirds, cowbirds, meadowlarks, and grackles*). AOU Tempe, AZ.
Lee, C. and A. Birch. 1998. Field Identification of Bullock's and Baltimore Orioles. Birding. In Press.
Lehman, P. 1988. Orchard and immature Hooded Orioles: a field identification nightmare. *Birding* 20:98–100.
Lehmann, V. 1957. Contribuciónes al estudio de la fauna de Colombia XII. *Novedades Colombianas* 3:101–156.
Lehmann, V. 1960. Contribuciónes al estudio de la fauna de Colombia XV. *Novedades Colombianas* 1:256–276.
Lever, C. 1987. *Naturalized Birds of the World.* Longman, Bath.
Leverkuhn, P 1889. Sudamerikanische nova aus dem kieler museum. *Journ. fur Orn.* 1889:104–107.
Lewington, I., P. Alström, and P. Coston. 1992. *A Field Guide to the Rare Birds of Britain and Europe.* HarperCollins. London.
Lewis, J.B. 1931. Behavior of Rusty Blackbird. *Auk* 48:125–126.
Lightbody, J.P., and P.J. Weatherhead. 1987. Interactions among females in polygynous Yellow-headed Blackbirds. *Behav. Ecol. Sociobiol.* 21:23–30.
Ligon, J.S. 1926. Nesting of the Great-tailed Grackle in New Mexico. *Condor* 28:93–94.
Lincoln, F.C. 1940. The Arizona Hooded Oriole in Kansas. *Auk* 57:420.
Lindell, C. 1996. Benefits and costs to Plain-fronted Thornbirds (*Phacellodomus rufifrons*) of interactions with avian nest associates. *Auk* 113:565–577.
Linz, G.M. 1986. Temporal, sex, and population characteristics of the first prebasic molt of Red-winged Blackbirds. *J. Field Ornithol.* 57:91–98.
Linz, G.M., S.B. Bolen, and J.F. Cassel. 1983. Postnuptual and postjuvenal molts of Red-winged Blackbirds in Cass County, North Dakota. *Auk* 100:206–209.
Loat, W.L.S. 1898. Field notes on the birds of British Guiana. *Ibis* 4(7):558–567.
Loftin, H., D.T. Jr. Rogers, and D.L. Hicks. 1966. Repeats, returns and recoveries of North American migrant birds banded in Panama. *Bird Banding* 37:35–44.
Long, C.F. 1971. Common grackles prey on big brown bat. *Wilson Bull.* 83:196.
Lowery, G.H. 1938. A new grackle of the *Cassidix mexicanus* group. *Occas. Papers Mus. Zool. Lousiana State Univ.* 1:1–11.
Lowery, G.H., and J.P. O'Neill. 1965. A new species of Cacicus (Aves: Icteridae) from Peru. *Occ. Pap. Mus. Zool. Louisiana State Mus.* 33:1–5.
Lowery, G.H. Jr. 1974. *Louisiana Birds.* Louisiana State Univ. Press, Baton Rouge.
Lown, B.A. 1980. Reproductive sucess of the Brown-headed Cowbird: a prognosis based on breeding bird census data. *Am. Birds* 34:15–17.
Lowther, P.E. 1975. Geographic and Ecological Variation in the Family Icteridae. *Wilson Bull.* 87(4):481–495.
Lowther, P.E. 1983. Brown-headed Cowbird (*Molothrus ater*). In *The Birds of North America,* No. 47 (A. Poole and F. Gill Eds.). Philadelphia: The Academy of Natural Sciences; Washington, D.C.: The American Ornithologists' Union.
Lowther, P.E. 1995. Bronzed Cowbird (*Molothrus aeneus*). In *The Birds of North America,* No. 144 (A. Poole and F. Gill, Eds.). The Academy of Natural Sciences, Philadelphia, and the American Ornithologists' Union, Washington, D.C.
Lowther, P.E., and R.F. Johnson. 1977. Influence of habitat on cowbird host selection. *Kans. Ornithol. Soc. Bull.* 28:36–40.
Lowther, P.E., and S.I. Rothstein. 1980. Head-down or 'Preening Invitation' Displays Involving Juvenile Brown-headed Cowbirds. *Condor* 82:459–460.

Mack, A.L., and C.D. Fisher. 1988. Notes on the birds from the llanos of Meta, Colombia. *Gerfaut* 78:397–408.
Madroño, A., R. Clay, and H. Hostettler. 1997. Sites to Save: San Rafael National Park, Paraguay. *World Birdwatch* 19:6–7.
Mailliard, J. 1910. The status of the California Bi-colored Blackbird. *Condor* 12:63–70.
Marchant, S. 1958. The birds of the Santa Elena Peninsula, S.W. Ecuador. *Ibis* 100:349–387.
Marchant, S. 1959. The breeding season in S.W. Ecuador. *Ibis* 101:137–152.
Marchant, S. 1960. The breeding of some S.W. Ecuadorian birds. *Ibis* 102:584–599.
Markham, 1971. Catalogo de los amphibios, reptiles, aves y mamiferos de la provincia de Magallanes, Chile. Publ. del Instituto de la Patagonia. Serie Monografias No. 3
Marshall, J.T. Jr. 1957. Birds of Pine-oak woodland in southern Arizona and adjacent Mexico. *Pacific Coast Avifauna* 32:112–113.
Martin, S.G. 1974. Adaptations for polygynous breeding in the bobolink, *Dolichonyx oryzivorus. Am. Zool.* 14:109–119.
Martin, S.G. and T.A. Gavin. 1995. Bobolink (*Dolichonyx oryzivorus*). In *The Birds of North America*, No. 176 (A. Poole and F. Gill, eds.). The Academy of Natural Sciences, Philadelphia, and The American Ornithologists' Union, Washington, D.C.
Marvil, R.E., and A. Cruz. 1989. Impact of Brown-headed Cowbird parasitism on the reproductive success of the Solitary Vireo. *Auk* 106:476–480.
Mason, E.A. 1942. Recoveries from migrating Bronzed Grackles. *Bird Banding* 13:105–107.
Mason, P. 1980. Ecological and evolutionary aspects of host selection in cowbirds. Ph.D. thesis, University of California, Santa Barbara.
Mason, P. 1985. The nesting biology of some passerines of Buenos Aires, Argentina. *Neotropical Ornithology*. Ornithol. Monogr. 36:954–972.
Mason, P. 1986a. Brood Parasitism in a Host Generalist, The Shiny Cowbird:I. The Quality of Different Species as Hosts. *Auk* 103:52–60.
Mason, P. 1986b. Brood Parasitism in a Host Generalist, The Shiny Cowbird: II. Host Selection. *Auk* 103:61–69.
Mason, P. 1987. Pair Formation in Cowbirds: Evidence Found for Screaming but not Shiny Cowbirds. *Condor* 89:349–356.
Mason, P., and S.I. Rothstein. 1986. Coevolution and Avian Brood Parasitism: Cowbird eggs show evolutionary response to host discrimination. *Evolution* 40(6):1207–1214.
Mason, P., and S.I. Rothstein. 1987. Crypsis Versus Mimicry and the Color of Shiny Cowbird Eggs. *Am. Nat.* 130(2):161–167.
Maxwell, G.R. II, J.M. Nocilly, and R.I. Shearer. 1976. Observations at a cavity nest of the Common Grackle and an analysis of grackle nest sites. *Wilson Bull.* 88:505–507.
Mayer, S. 1996. Bird Sounds of Bolivia. CD-ROM, Version 1.0. Bird Songs International B.V.
Mayfield, H. 1961. Vestiges of a proprietary interest in nests by the Brown-headed Cowbird parasitizing the Kirtland's Warbler. *Auk* 78:162–167.
Mayfield, H. 1965. Chance distribution of cowbird eggs. *Condor* 67:257–263.
Mayfield, H.F. 1965. The Brown-headed Cowbird with old and new hosts. *Living Bird* 4:13–28.
Mayr, E. 1969. *Principles of Systematic Zoology*. McGraw-Hill Book Company, New York.
Mayr, E. 1982. *The Growth of Biological Thought*. Belknap Press, Harvard Univ., Cambridge, Mass.
McCaskie, G. 1971. Rusty Blackbirds in California and western North America. *Calif. Birds*. 2:55–68.
McCaskie, G., R. Stallcup, and P. De Benedictis. 1966. Notes on the distribution of certain icterids and tanagers in California. *Condor* 68:595–597.
McCowan, M. 1980. Common grackle winters at Brandon, Manitoba. *Blue Jay* 83:123.
McGeen, D.S. 1971. Factors affecting cowbird sucess with Yellow Warbler and Song Sparrow Hosts. *Jack-Pine Warbler* 49:53–57.
McGeen, D.S. 1972. Cowbird-Host Relationships. *Auk* 89:360–380.
McGeen, D.S., and J.J. McGeen. 1968. The cowbirds of Otter Lake. *Wilson Bull.* 80:84–93.
McIlhenny, E.A. 1937. Life history of the Boat-tailed Grackle in Louisiana. *Auk* 54:274–295.
McKay, C.R. 1994. Brown-headed Cowbird in Strathclyde: new to Britain and Ireland. *British Birds* 87:284–289.
McNicholl, M.K. 1968. Cowbird egg in Mourning Dove nest. *Blue Jay* 26:22–23.
McNicholl, M.K. 1979. Passerines feeding from floating vegetation. *Blue Jay* 37:223–223.
McNicholl, M.K. 1981. Interactions between Forster's Terns and Yellow-headed Blackbirds. *Colonial Waterbirds* 4:150–154.
McNicholl, M.K. 1981. Fly-catching by male Red-winged Blackbirds. *Blue Jay* 39:206–207.
McNicholl, M.K. 1983. Early fledging record of Western Meadowlark in Manitoba. *Blue Jay* 41:201–204.
McNicholl, M.K. 1987. Red-winged Blackbirds nesting in urban downtown Toronto. *Ontario Birds* 5:74–75.
Meanley, B. 1967. Aging and sexing blackbirds, Bobolinks and starlings. Special report to Patuxent Wildlife Research Center; work unit F-24.1.
Meanley, B., and G.M. Bond. 1970. Molts and plumages of the Red-winged Blackbird with particular ref-

erence to fall migration. *Bird Banding* 41(1):22–27.
Meanley, B., and W.C. Royall Jr. 1976. *Nationwide estimates of blackbirds and starlings.* Pp. 39–40 in Proceedings seventh bird control seminar, Bowling Green State Univ., Bowling Green, OH.
Mermoz, M.E., and J.C. Reboreda. 1994. Brood parasitism of the Shiny Cowbird, *Molothrus bonariensis,* on the Brown-and-yellow Marshbird, *Pseudoleistes vivescens. Condor* 96:716–721.
Mermoz, M.E., and J.C. Reboreda. 1996. New effective host for the Screaming Cowbird. *Condor* 98:630–632.
Merrill, J.C. 1877. Notes on *Molothrus aeneus,* Wagl. *Bull. Nutt. Orn.Club* 2:289–299.
Meyer de Schauensee, R. 1945. On the type of *Cassicus melanurus* Cassin. *Auk* 62:456–457.
Meyer de Schauensee, R. 1946. Colombian Zoological Survey. Part IV. Further notes on Colombian birds, with the description of new forms. *Notulae Naturae (Philadelphia)* 167:10–13.
Meyer de Schauensee, R. 1951. The birds of the Republic of Colombia. Caldasia pts. 1–5, number 25251-1212.
Meyer de Schauensee, R., and W.H. Phelphs Jr. 1978. *A Guide to the Birds of Venezuela.* Princeton University Press, Princeton, New Jersey.
Middleton, A.L.A. 1991. Failure of Brown-headed Cowbird parasitism in nests of the American Goldfinch. *J. Field Ornithol.* 62 (2):200–203.
Miller, 1978. Notes on birds of San Salvador Island. *Auk* 95:281–263.
Miller, A.H. 1931. Notes on the song and territorial habits of Bullock's Oriole. *Wilson Bull.* 43: 102–108.
Miller, A.H. 1947. The tropical avifauna of the upper Magdalena valley, Colombia. *Auk* 64:351–381.
Miller, A.H. 1960. Additional data on the distribution of some Colombian birds. *Novedades Colombianas* 1:235–237.
Miller, L.E. 1917. Field notes on *Molothrus bonariensis* and *M.badius. Bull. Amer. Mus. Nat. Hist.* 37:579–592.
Miller, R.S. 1968. Conditions of competition between redwings and yellow-headed blackbirds. *J. Anim. Ecol.* 37:43–61.
Miller, W. de W. 1924. Notes on the genera of Caciques. I. *Auk* 41:463–467.
Miller, W. de W., and L. Griscom. 1925. Further notes on Central American birds, with descriptions of new forms. *Amer. Mus. Novit.* 184(4):4–5.
Miller, W.de W. 1906. List of birds collected in northwest Durango, Mexico, by F. H. Batty, during 1903. *Bull. Amer. Mus. Nat. Hist.* 22:161–183.
Millie, W.R. 1938. Las aves del valle del Huasco y sus alrededores (Provincia de Atacama). *Rev. Chilena. Hist. Nat.* 42:181–205.
Miskimen, M. 1980. Red-winged Blackbirds: Age-related epaulet color changes in captive females. *Ohio J. Sci.* 80:232–235.
Misra, R.K., and L.L. Short. 1974. A biometric analysis of oriole hybridization. *Condor* 76:137–146.
Monroe, B.L. 1963. Three new subspecies of birds from Honduras. *Occ. Pap. Mus. Zool. Lousiana State Univ.* 26:1–7.
Monroe, B.L. Jr. 1968. A distributional survey of the birds of Honduras. *Ornithol. Monograph* 7:1–458.
Moore, J.V. 1993. Sounds of La Selva, Ecuador. Self-produced cassette.
Moore, J.V. 1994a. Ecuador: More bird vocalizations from the lowland rainforest. Volume I. Self-produced cassette.
Moore, J.V. 1994b. A bird walk at Chan Chich. Self-produced cassette tape.
Moore, J.V. 1996. Ecuador. More bird vocalizations from the lowland rainforest. Volume 2. Cassette tape.
Moore, W.S., and R.A. Dobeer. 1989. The use of banding recovery data to estimate dispersal rates and gene flow in avian species: case studies in the Red-winged Blackbird and Common Grackle. *Condor* 91:242–253.
Muller, P. 1968. Nachweis von *Leistes militaris superciliaris* fur den Osten des Staates von Sao Paulo (Brasilien). *Ornithologische Mitteilungen* 20:107–108.
Munro, J.A. 1955. A sight record of Baltimore Oriole in British Columbia. *Murrelet* 36:43.

Narosky, T. 1985. *Aves Argentinas. Guia para el reconocimiento de la avifauna Bonaerense.* Editorial Albatros, Buenos Aires.
Narosky, T., and A.G. Di Giacomo. 1993. *Las aves de la Provincia de Buenos Aires: Distribucíon y estatus.* L.O.L.A., Buenos Aires.
Narosky, T., and D. Yzurieta. 1987. *Guia Para la Identificación de las Aves de Argentina y Uruguay.* Asociación Ornitológica del Plata, Buenos Aires.
National Geographic Society 1983. *Field Guide to the Birds of North America.* National Geographic Society, Washington, D.C.
Nauman, E.D. 1930. The nesting habits of the Baltimore Oriole. *Wilson Bull.* 42:295–296.
Naumburg, E.M.B. 1930. The birds of Mato Grosso, Brazil. *Bull. Am. Mus. Nat. Hist.* 60:1–432.
Naumburg, E.M.B., and H. Friedmann. 1927. A new race of *Molothrus bonariensis* from Brazil. *Auk* 44:494–494.
Navarro S., A.G. 1992. Altitudinal distribution of birds in the Sierra Madre del Sur, Guerrero, Mexico. *Condor* 94:29–39.

Neff, J.A. 1933. The Tricolored Red-wing in Oregon. *Condor* 35:234–235.
Neff, J.A. 1937. Nesting distribution of the Tricolored Red-wing. *Condor* 39:61.
Neff, J.A. 1942. Migration of the Tricolored Red-wing in Central California. *Condor* 44:45–53.
Nelson, E.W. 1897. Preliminary descriptions of new birds from Mexico and Guatemala in the collection of the United States department of agriculture. *Auk* 14:42–76.
Nelson, E.W. 1900. Descriptions of thirty new North American birds in the biological survey collection. *Auk* 17:266–267.
Nero, R.W. 1954. Plumage aberrations of the Red-winged Blackbird (*Agelaius phoeniceus*). *Auk* 71:137–155.
Nero, R.W. 1956. A behavior study of the Red-winged Blackbird. I. Mating and nesting activities. *Wilson Bull.* 68:5–37.
Nero, R.W. 1957. Vestigial claws on the wings of a Red-winged Blackbird. *Auk* 74:262–262.
Nero, R.W. 1960. Additonal notes on the plumage of the Redwinged Blackbird. *Auk* 77:298–305.
Nero, R.W. 1964. Comparative behavior of the Yellow-headed Blackbird, Red-winged Blackbird, and other icterids. *Wilson Bull.* 75:376–413.
Newman, G.A. 1970. Cowbird parasitism and nesting success of lark Sparrows in southern Oklahoma. *Wilson Bull.* 82:304–309.
Nice, M.M. 1939. Observations on the behavior of a young cowbird. *Wilson Bull.* 51:233–239.
Nice, M.M. 1949. The laying rhythm of cowbirds. *Wilson Bull.* 61:231–233.
Nice, M.M. 1952. Song in hand-raised meadowlarks. *Condor* 54:362–363.
Nickell, W.P. 1955. Notes on cowbird parasitism on four species. *Auk* 72:88–92.
Nickell, W.P. 1958. Brown-headed Cowbird fledged in nest of Catbird. *Wilson Bull.* 70:286–287.
Niles, D.M. 1970. A record of clutch size and breeding in New Mexico for the Bronzed Cowbird. *Condor* 72:500–501.
Nores, M., and D. Yzurieta. 1979. Una nueva especie y dos nuevas subespecies de aves (Passeriformes). *Acad. Nac. Cienc. Cordoba, Arg. misc* 61:4–7.
Norman, R.F., and R.J. Roberstson. 1975. Nest searching behavior in the Brown-headed Cowbird. *Auk* 92:610–611.
Norris, R.T. 1947. The cowbirds of Preston Frith. *Wilson Bull.* 59:83–103.
Norton, R.L. 1979. New records of birds for the Virgin Islands. *American Birds* 33:145–146.
Norton, R.L. 1981. Additional records and notes of birds in the Virgin Islands. *American Birds* 35:144–147.

O'Loghlen, A.L. 1995. Delayed Access to Local Songs Prolongs Vocal Development in Dialect Populations of Brown-headed Cowbirds. *Condor* 97:402–414.
O'Loghlen, A.L., and S.I. Rothstein. 1993. An extreme example of delayed vocal development: song learning in a population of wild Brown-headed Cowbirds. *Anim. Behav.* 46:293–304.
Oberholser, H.C. 1902. Some new South American birds. *Proc. U.S. Nat. Mus.* XXV:68.
Oberholser, H.C. 1919a. Description of a new Red-winged Blackbird from Texas. *Wilson Bull.* 31:20–23.
Oberholser, H.C. 1919b. Notes on the races of *Quiscalus quiscula* (Linnaeus). *Auk* 36:549–555.
Oberholser, H.C. 1974. *The Bird Life of Texas.* Univ. of Texas press, Austin, Texas.
Ochoa, J.M., and A.C. Maya. 1998. Apuntes sobre la anidación del cacique candela (*Hypopyrrhus pyrohypogaster*) en el municipio de Barbosa (Antioquia). *Boletin SAO.*
Olivares, A. 1969. *Aves de Cundinamarca.* Universidad Nacional de Colombia, Bogota.
Olson, S. 1985. Weights of some Cuban birds. *Bull. BOC* 105:68–69.
Olson, S. L. 1983. A hybrid between *Icterus chrysater* and *I. mesomelas. Auk* 100:733–735.
Olson, S.L. 1981. Systematic notes on certain oscines from Panama and adjacent areas (Aves: Passeriformes). *Proc. Biol. Soc. Wash.* 94:363–373.
Olyphant, J.C. 1995. An incredible recovery. *N. Am. Bird Bander* 20:28–28.
Oniki, Y., and E.O. Willis. 1983. Breeding records of birds from Manaus, Brazil: V. Icteridae to Fringillidae. *Rev. Bras. Biol.* 43:55–64.
Orians, G.H. 1960. Autumnal breeding in Tricolored Blackbird. *Auk* 77:379–398.
Orians, G.H. 1961a. Social stimulation within blackbird colonies. *Condor* 63:330–337.
Orians, G.H. 1961b. Ecology of blackbird (*Agelaius*) social systems. *Ecological Monographs* 31:285–312.
Orians, G.H. 1963. Notes on fall-hatched Tricolored Blackbirds. *Auk* 80:552–553.
Orians, G.H. 1969. On the evolution of mating systems in birds and mammals. *Amer. Nat.* 103:589–603.
Orians, G.H. 1972. The adaptive significance of mating systems in the Icteridae. Proc. XVth Int. Ornithol. Congress 1972:389–398.
Orians, G.H. 1973. The Red-winged Blackbird in tropical marshes. *Condor* 75:28–42.
Orians, G.H. 1980. Some adaptations of marsh-nesting blackbirds. *Monographs in Population Biology* 14:1–295.
Orians, G.H. 1983. Notes on the behavior of the Melodius Blackbird. *Condor* 85:453–460.
Orians, G.H. 1985. Allocation of reproductive effort by breeding blackbirds, family Icteridae. *Rev. Chil. Hist. Nat.* 58:19–29.
Orians, G.H., and G.M. Christman. 1968. A comparative study of the behavior of Red-winged, Tricolored

and Yellow-headed Blackbirds. *Univ. California Publ. Zool.* 84:
Orians, G.H., and G. Collier. 1963. Competition and Blackbird social systems. *Evolution* 17:449–459.
Orians, G.H., M.L. Erckmann, and J.C. Schulz. 1977a. Nesting and other habits of the Bolivian Blackbird *(Oreopsar bolivianus) Condor* 79:250–255.
Orians, G.H., and H.S. Horn. 1969. Overlap in foods and foraging of four species of blackbirds in the potholes of central Washington. *Ecology* 50:930–938.
Orians, G.H., C.E. Orians, and K.J. Orians. 1977b. Helpers at the nest in some Argentine blackbirds. *In* Stonehouse, B. and Perrins, C. (eds)., *Evolutionary Ecology*, Macmillan, London.
Orians, G.H., E. Roskaft, and L.D. Beletsky. 1989. Do Brown-headed Cowbirds lay their eggs at random in the nests of Red-winged Blackbirds? *Wilson Bull.* 101:599–605.
Ortega, C., and A. Cruz. 1992. Gene flow of the obscurus race into the north-central Colorado population of Brown-headed Cowbird. *J. Field Ornithol.* 63:311–317.
Ortega, C.P. 1991. A comparative study of Cowbird Parasitism in Yellow-headed and Red-winged Blackbirds. *Auk* 108(1):16–24.
Ortega, C.P., and A. Cruz. 1988. Mechanisms of Egg Acceptance by Marsh-Dwelling Blackbirds. *Condor* 90:349–358.
Ortega, C.P., and A. Cruz. 1992. Differential growth patterns of nestling Brown-headed Cowbirds and Yellow-headed Blackbirds. *Auk* 109:368–376.
Ortega, J.P., C.P. Ortega, and A. Cruz. 1993. Does Brown-headed Cowbird egg coloration influence Red-winged Blackbird responses towards nest contents. *Condor* 95:782–791.
Ortiz, F., P. Greenfield, and J.C. Matheus. 1990. *Aves del Ecuador.* FREPOTUR and CECIA, Quito, Ecuador.

Packard, F.M. 1936. Notes on the plumages of the Eastern Red-wing. *Bird Banding* 7:77–80.
Palmer, T. 1992. Population changes in a long-term Northern Oriole winter roost in central Florida. *Florida Field Natur.* 20:18–20.
Parker, T.A. III 1985. Voices of the Peruvian Rainforest. Library of Natural Sounds, Cornell Laboratory of Ornithology. Cassette tape.
Parker, T.A. III 1989. An avifaunal survey of the Chimanes Ecosystem Program of Northern Bolivia, 17–26 June, 1989. Unpublished report, Conservation International.
Parker, T.A. III, S.A. Parker, and M.A. Plenge. 1982. *An Annotated Checklist of Peruvian Birds.* Buteo Books, Vermillion, South Dakota.
Parker, T.A. III, and J.V. Jr. Remsen. 1987. Fifty-two Amazonian bird species new to Bolivia. *Bull. BOC* 107:94–107.
Parker, T.A., III 1982. Observations of some unusual rainforest and marsh birds of southeastern Peru. *Wilson Bull.* 94:477–493.
Parkes, K.C. 1952. Post-juvenal molt in the Bobolink. *Wilson Bull.* 64:161–162.
Parkes, K.C. 1954. The generic name of the rice grackle. *Condor* 56:229–229.
Parkes, K.C. 1966. Geographic variation in Azara's Marsh Blackbird, *Agelius cyanopus. Proc. Biol. Soc. Washington* 79:1–12.
Parkes, K.C. 1970. A revision of the Red-rumped Cacique, *Cacicus haemorrhous* (Aves, Icteridae). *Proc. Biol. Soc. Washington* 83:203–214.
Parkes, K.C. 1972. Tail molt in the family Icteridae. *Proc. XVth Int. Orn. Congress*:674.
Parkes, K.C. 1997. Tail molt in the family Icteridae. Poster presented at the 115th Stated Meeting of the American Ornithologists' Union.
Parkes, K.C., and E.R. Blake. 1965. Taxonomy and nomenclature of the Bronzed Cowbird. *Fieldiana-Zoology* 44:207–216.
Paterson, H.E.H. 1985. The recognition concept of species. *Transvaal Mus. Monogr.* 4:21–29.
Paulson, D.R., G.H. Orians, and C.F. Leck. 1969. Notes on birds of Isla San Andrés. *Auk* 86:755–758.
Payne, R.B. 1965. Clutch size and number of eggs laid by Brown-headed Cowbirds. *Condor* 67:44–60.
Payne, R.B. 1967. Gonadal responses of Brown-headed Cowbirds to long daylengths. *Condor* 69:289–297.
Payne, R.B. 1969a. Breeding seasons and reproductive physiology of Tricolored Blackbirds and Red-winged Blackbirds. *Univ. Calif. Pubs. Zool.* 90:1–137.
Payne, R.B. 1969b. Giant Cowbird solicits preening from human. *Auk* 86:751–752.
Payne, R.B. 1973. The breeeding season of a parasitic bird, the Brown-headed Cowbird in Central California. *Condor* 75:80–99.
Paynter, R.A. 1954. Two new species to the Mexican avifauna. *Auk* 71:204–204.
Paynter, R.A. Jr. 1952. Birds from Popocatepetl and Ixtaccihuatl, Mexico. *Auk* 69:293–301.
Pearman, M. 1994. Neotropical Notebook. *Cotinga* 2:26–31.
Pearson, D.L. 1974. Used of abandoned cacique nests by nesting Troupials (*Icterus icterus*): precursor to parasitism? *Wilson Bull.* 86:290–291.
Peek, F.W. 1972. An experimental study of the territorial function of vocal and visual display in the male Red-winged Blackbird (*Agelaius phoeniceus*). *Anim. Behav.* 20:112–118.
Peer, B.D., and E.K. Bollinger. 1997a. Explanations for the infrequent cowbird parasitism on Common Grackles. *Condor* 99:151–161.

Peer, B.D., and E.K. Bollinger. 1997b. Common Grackle (*Quiscalus quiscula*). In The Birds of North America, No. 271 (A. Poole and F. Gill eds.). The Academy of Natural Sciences, Philadelphia, PA, and The American Ornithologists' Union, Washington, D.C.
Pereyra, J.A. 1933. Miscelanea ornitologica. *Hornero* 5:215–219.
Perez-Rivera, R. 1978. Notas sobre los Icteridos vulnerables o en peligro de extincion en Puerto Rico. *Science-Ciencia* 5.2:70–74.
Perez-Rivera, R. 1980. Algunas notas sobre la biologia y status de la mariquita (*Agelaius xanthomus*), con enfasis en la subespecie de Mona. Memorias del Segundo Coloquio sobre la Fauna de Puerto Rico. Univ. de P.R. Colegio Universitario de Humacao.54–63.
Perez-Rivera, R.A. 1986. Parasitism by the Shiny Cowbird in the interior parts of Puerto Rico. *J. Field Ornithol.* 57:99–104.
Peters, H.S., and T.D. Burleigh. 1951. *The Birds of Newfoundland.* Department of Natural Resources, Province of Newfoundland, St John's.
Peters, J.L. 1921. A review of the grackles of the genus *Holoquiscalus. Auk* 38:435–453.
Peters, J.L. 1929. The identity of *Corvus mexicanus* Gmelin. *Proc. Biol. Soc. Washington* 42:123–123.
Petersen, 1985. New bird species for the Icelandic checklist. *Bliki* 4:57–67.
Peterson, A., and H. Young. 1950. A nesting study of the Bronzed Grackle. *Auk* 67:466–476.
Peterson, R.T., and E.L. Chalif. 1973. *A Field Guide to Mexican Birds.* Houghton Mifflin Co., Boston.
Pettingill, O.S. Jr. 1983. Winter of the Bobolink. *Audubon* 85:102–109.
Phelps, W.H., and W.H. Jr. Phelps. 1950. Lista de las aves de Venezuela con su distribución. *Bol. Soc. Venezolana Cienc. Nat.* 12 (75):1–351.
Phelps, W.H. Jr., and R. Aveledo. 1966. A new subspecies of *Icterus icterus* and other notes on the birds of northern South America. *Amer. Mus. Novit.* 2274:1–14.
Philippi, R., and Landbeck. 1861. *Anal. Univ. Chile* 19:616–616.
Phillips, A.R. 1950. The Great-tailed Grackles of the Southwest. *Condor* 52:78–81.
Phillips, A.R. 1966. Further systematic notes on Mexican birds. *Bull. BOC* 86:125–131.
Phillips, A.R. 1975. Why neglect the difficult? *Western Birds* 6:69–86.
Phillips, A.R., and R.W. Dickerman. 1965. A new subspecies of *Icterus prosthemelas* from Panama and Costa Rica. *Wilson Bull.* 77:298–299.
Phillips, A.R., J. Marshall, and G. Monson. 1964. *The birds of Arizona.* U. of Arizona Press, Tucson.
Pichot, P.P. 1976. *Faune des Antilles Francaises: Les oiseaux.* 2nd Edition. Fort – de – France.
Picman, J. 1981. The adaptive value of polygyny in marsh-nesting red-winged blackbirds: renesting, territory tenacity, and mate fidelity of females. *Can. J. Zool.* 59:2284–2296.
Piers, H. 1894. Notes on Nova Sciotian Zoology: No.3. *Proc. Trans. Nova Scotian Inst. Sci.* 8:395–410.
Pinto, O.M. de O. 1935. Aves da Bahia. Notas criticas e orservacoes dobra uma collecao feita no Reconcavo e na parte meridonal do Estado. *Rev. Mus. Paulista* 19:1–326.
Pinto, O.M. de O. 1944. *Catalogo das aves do Brasil. Part 2.* Dpt. Zool. Sicr. Agr. Sao Paulo.
Pinto, O.M.O. 1967. Do parasitismo provavel de *Icterus jamacaii* (Gmelin) em *Pseudoseisura cristata* (Gmelin). *Hornero* 10:447–449.
Pinto, O.M.O. 1975. Icterus nest parasitism *Pap. Avulsos Zool. S. Paulo* 29:35–36.
Pleasants, B.Y. 1979. Adaptive significance of the variable dispersion pattern of breeding Northern Orioles. *Condor* 81:28–34.
Pleasants, B.Y. 1981. Aspects of the breeding biology of a subtropical oriole, *Icterus gularis. Wilson Bull.* 93:531–537.
Pleasants, B.Y. 1993. Altamira Oriole (*Icterus gularis*). In The Birds of North America No. 56:1–8.
Post, P.W. 1968. Photographs of New York State rarities. 14. Bullock's Oriole. *Kingbird* 18:122–123.
Post, W. 1976. Population ecology of the Yellow-winged Blackbird. Final Report to Dept. Int. U.S. Fish & Wildlife Serv., Office of Endangered Species.
Post, W. 1981a. Biology of the Yellow-shouldered Blackbird – *Agelaius xanthomus* on a tropical island. *Bull. Fl. State Mus. Biol. Sci.* 26:125–202.
Post, W. 1981b. The prevalance of some ectoparasites, diseases, and abnormalities in the Yellow-shouldered Blackbird. *J. Field Ornithol.* 52:16–22.
Post, W. 1986. Reproductive success in New World marsh-nesting passerines. Acta XIX Congr. Int. Orn.2645–2657.
Post, W. 1986. Reproductive success in New World marsh-nesting passerines. In: Ouellet (ed.) Acta XIX Congressus Internationalis Ornithologici. Ottawa, Canada.
Post, W. 1987. Boat-tailed Grackle lays eggs in abandoned nest containing eggs. *Wilson Bull.* 99:724–724.
Post, W. 1988. Boat-tailed Grackles nest in freshwater habitat in interior South Carolina. *Wilson Bull.* 100:325–326.
Post, W. 1992. Dominance and mating success in male Boat-tailed Grackles. *Anim. Behav.* 44:917–929.
Post, W. 1992. First Florida specimens of the Shiny Cowbird. *Fla. Field Nat.* 20:17–18.
Post, W. 1993. First specimen of the Shiny Cowbird, *Molothrus bonariensis* (Aves: Emberizidae), in North Carolina. *Brimleyana* 19:205–206.
Post, W. 1994. Are female Boat-tailed Grackle colonies neutral assemblages? *Behav. Ecol. Sociobiol.* 35:401–407.

Post, W. 1994. Redirected copulation by male Boat-tailed Grackles. *Wilson Bull.* 106:770–771.
Post, W. 1994a. Banding confirmation that some Middle Atlantic Coast Boat-tailed Grackles visit Florida in the winter. *Florida Field Nat.* 22:51.
Post, W. 1995. Reproduction of female Boat-tailed Grackles: comparisons between South Carolina and Florida. *J. Field Ornithol.* 66:221–230.
Post, W., and M.M. Browne. 1982. Active anting by the Yellow-shouldered Blackbird. *Wilson Bull.* 94:89–90.
Post, W., A. Cruz, and D.B. McNair. 1993. The North American invasion pattern of the Shiny Cowbird. J. *Field Ornithol.* 64:32–41.
Post, W., T.K. Nakamura, and A. Cruz. 1990. Patterns of shiny cowbird parasitism in St Lucia and Southwestern Puerto Rico. *Condor* 92:461–469.
Post, W., and K.W. Post. 1987. Roosting behavior of the Yellow-shouldered Blackbird. *Fla. Field Nat.* 15:93–105.
Post, W., J.P. Poston, and G.T. Bancroft. 1996. Boat-tailed Grackle (*Quiscalus major*) In The Birds of North America, No. 207 (A. Poole and F. Gill, eds.). The Academy of Natural Sciences, Philadelphia and the American Ornithologists' Union, Washington, D.C.
Post, W., and C. Seals. 1989. Common Moorhen parasitizes a Boat-tailed Grackle nest. *Wilson Bull.* 101:508–509.
Post, W., and C.A. Seals. 1991. Bird density and productivity in an impounded cattail marsh. *J. Field Ornithol.* 62:195–199.
Post, W., and C.A. Seals. 1993. Nesting Associations of Least Bitterns and Boat-tailed Grackles. *Condor* 95:139–144.
Post, W., and J.W. Wiley. 1976. The Yellow-shouldered Blackbird-present and future. *Am. Birds* 30:13–20.
Post, W., and J.W. Wiley. 1977. Reproductive Interactions of Shiny Cowbird and the Yellow-shouldered Blackbird. *Condor* 79:176–184.
Post, W., and J.W. Wiley. 1977. The Shiny Cowbird in the West Indies. *Condor* 79:119–121.
Post, W., and J.W. Wiley. 1992. The head-down display in Shiny Cowbirds and its relation to dominance behavior. *Condor* 94:999–1002.
Power, D.M. 1970a. Geographic variation of Red-winged Blackbirds in Central North America. *Univ. Kansas Publ. Mus. Nat. Hist.* 19:1–83.
Power, D.M. 1970b. Geographic variation in the surface/volume ratio of the bill of Red-winged Blackbirds in relation to certain geographic and climatic factors. *Condor* 72:299–304.
Pratt, H.D. 1974. Field Identification of Great-tailed and Boat-tailed grackles in their zone of overlap. *Birding* 6:217–223.
Pratt, H.D. 1991. Hybridization of Great-tailed and Boat-tailed Grackles (*Quiscalus*) in Louisiana. *J. La. Ornith.* 2:2–14.
Pratt, H.D., B. Ortego, and H.D. Guillory. 1977. Spread of the Great-tailed Grackle in southwestern Louisiana. *Wilson Bull.* 89:483–485.
Preston, F.W. 1948. The cowbird (*M.ater*) and the cuckoo (*C.canorus*). *Ecology* 29:115–116.
Prud'homme-Cyr, J., R. McNeil, and A. Cyr. 1976. First Quebec record of *Quiscalus quiscula stonei*, the 'purple' race of the Common Grackle. *Can. Field Nat.* 90:172.
Pruitt, J. 1975. The return of the Great-tailed Grackle. *Am. Birds* 29:985–992.
Pulich, W.M. 1988. *The birds of north-central Texas.* Texas A & M University Press.
Pyle, P. 1997a. Molt limits in North American Passerines. *North American Bird Bander* 22: 49–89.
Pyle, P. 1997b. *Identification Guide to North American Birds. Part 1.* Slate Creek Press, Bolinas.
Pyle, P., K. Hanni, and D. Smith. 1994. Bird notes from Isla Guadalupe, including three new records. *Euphonia* 3(1):1–4.
Pyle, P., S.N.G. Howell, R. P. Yunick, and D. F. De Sante. 1987. *Identification Guide to North American Passerines.* Slate Creek Press, Bolinas.

Raffaele, H.A. 1989. *A guide to the birds of Puerto Rico and the Virgin Islands.* Princeton University Press, Princeton, NJ.
Rahn, H.L., Currant-Everett, and D.T. Booth. 1988. Eggshell differences between parasitic and nonparasitic Icteridae. *Condor* 90:962–964.
Ramsden, C.T. 1912. *Xanthocephalus xanthocephalus* in eastern Cuba. *Auk* 29:103–103.
Rand, A.L., and M.A. Traylor. 1954. *Manual de las aves de el Salvador.* Universidad de el Salvador.
Rasmussen, J.F., C. Rahbek, E. Horstman, M.K. Poulsen, and H. Bloch. 1994. Aves del parque nacional podocarpus, una lista anotada. CECIA, Quito, Ecuador.
Rathburn, 1917. A new subspecies of the Western Meadowlark. *Auk* 34:68–70.
Reed, C.W. 1913. Datos para la biologia del *Molothrus bonariensis. Rev. Chil. Hist. Nat.* 17:172–178.
Remsen, J.V. Jr. 1977. Five bird species new to Colombia. *Auk* 94:363–363.
Remsen, J.V. Jr. 1986. Aves de una localidad en la sabana humeda del norte de Bolivia. *Ecología en Bolivia* 8:21–35.
Remsen, J.V. Jr., and S.K. Robinson. 1990. A classification scheme for foraging behaviour of birds in terrestrial habitats. *Studies in Avian Biology* 13:144–160.

Remsen, J.V. Jr., C.G. Schmitt, and D.C. Schmitt. 1988. Natural history notes on some poorly known Bolivian birds. Part 3. *Gerfaut* 78:363–381.
Remsen, J.V. Jr., M.M. Swan, S.W. Cardiff, and K.V. Rosenberg. 1991. The importance of the rice growing region of south-central Louisiana to winter populations of shorebirds, raptors, waders, and other birds. *J. La. Ornith.* 1:36–47.
Remsen, J.V. Jr., and M.A. Jr. Traylor. 1989. *An Annotated List of the Birds of Bolivia.* Buteo Books, Vermillion, South Dakota.
Remsen, J.V. Jr., M.A. Jr. Traylor, and K.C. Parkes. 1987. Range extensions for some Bolivian Birds, 3 (Tyrannidae to Passeridae). *Bull. BOC* 107:6–16.
Richmond, C.W. 1893. Notes on a collection of birds from eastern Nicaragua and the Río Frio, Costa Rica, with a description of a supposed new trogon. *Proc. U.S. Nat. Mus.* 16:479–532.
Richmond, C.W. 1898. Description of a new species of Gymnostinops. *Auk* 15:326–327.
Ridgway, R. 1884. On a collection of birds made by Messrs. J.E. Benedict and W. Nye, of the United States Fish Commission steamer 'Albatross'. *Proc. U.S. Nat. Mus.* 7:172–176.
Ridgway, R. 1885. *Icterus cucullatus*, Swainson, and its geographical variations. *Proc. U.S. Nat. Mus.* 8:18–19.
Ridgway, R. 1887. *Agelaius phoeniceus bryanti* Manual of North American Birds 370. J.P. Lippincott Co., Philadelphia.
Ridgway, R. 1901. New birds of the families Tanagridae and Icteridae. *Proc. Washington Acad. Sci.* 3:149–155.
Ridgway, R. 1902. The birds of North and Middle America. *Bull. U.S. Natl. Mus.*,50, pt.2, i–xx1–834.
Rising, J.D. 1969. A comparison of metabolism and evaporative water loss of Baltimore and Bullock's Orioles. *Comp. Biochem. Physiol.* 31:915–925.
Rising, J.D. 1970. Morphological variation and evolution in some North American orioles. *Syst. Zool.* 19:315–351.
Rising, J.D. 1973. Morphological variation and status of the orioles, *Icterus galbula, I.bullockii* and *I. abeillei* in the northern Great Plains and in Durango, Mexico. *Can. J. Zool.* 51:1267–1273.
Rising, J.D. 1983. The progress of oriole hybridization in Kansas. *Auk* 100:885–897.
Rising, J.D. 1983. The Great Plains hybrid zones. In, R. F. Johnston (ed.). *Current Ornithology* 1:131–157.
Rising, J.D. 1996. The stability of the oriole hybrid zone in western Kansas. *Condor* 98:658–663.
Robbins, M.B., and D.A. Easterla. 1986. Range Expansion of the Great-tailed Grackle into Missouri, with Details of the First Nesting Colony. *Bluebird* 53:24–27.
Robbins, M.B., and D.A. Easterlea. 1981. Range expansion of the Bronzed Cowbird with the first Missouri record. *Condor* 83:270–272.
Roberts, L.B., and W.A. Searcy. 1988. Dominance relationships in harems of female Red-winged Blackbirds. *Auk* 105:89–96.
Robertson, W.B. Jr. 1962. Observations on the birds of St. John, Virgin Islands. *Auk* 79:44–76.
Robinson, S.K. 1984. Social behavior and sexual selection in a neotropical oriole. Ph.D. diss., Princeton University.
Robinson, S.K. 1985. Fighting and assessment in the Yellow-rumped Cacique (*Cacicus cela*). *Behav. Ecol. Sociobiol.* 18:39–44.
Robinson, S.K. 1985a. Coloniality in the Yellow-rumped Cacique as a defence against nest predators. *Auk* 102:506–519.
Robinson, S.K. 1985b. The Yellow-rumped Cacique and its associated nest pirates. In Neotropical Ornithology (P.A. Buckley, M.S. Foster, E.S. Morton, R.S. Ridgely and F.G. Buckley eds.). *Ornithol. Monograph* 36.
Robinson, S.K. 1986. The evolution of social behaviour and mating systems in the blackbirds (Icterinae). In:Ecological Aspects of Social Evolution. P.I. Rubenstein and R.A. Wrangham (eds). Princeton University Press, Princeton. New Jersey.
Robinson, S.K. 1986a. Benefits, costs, and determinants of dominance in a polygynous oriole. *Anim. Behav.* 34:241–255.
Robinson, S.K. 1986b. The evolution of social behavior and mating systems in the blackbirds (Icterinae) In D.I. Rubenstein and R.W. Wrangham (eds.), Ecological aspects of social evolution. Princeton University Press.
Robinson, S.K. 1986c. Social Security for Birds. *Natural History* 95(3):38–47.
Robinson, S.K. 1986d. Three-speed foraging during the breeding cycle of Yellow-rumped Caciques (Icterinae: *Cacicus cela*). *Ecology* 67:394–405.
Robinson, S.K. 1986e. Competitive and mutualistic interactions among females in a neotropical oriole. *Anim. Behav.* 34:113–122.
Robinson, S.K. 1988a. Foraging ecology and host relationships of Giant Cowbirds in Southeastern Peru. *Wilson Bull.* 100:224–235.
Robinson, S.K. 1988b. Anti-social and social behaviour of adolescent Yellow-rumped Caciques (Icterinae: *Cacicus cela*). Animal Behav. 36:1482–1495.
Rodriguez, J.V. 1982. *Aves del parque nacional natural los Katios, Choco, Colombia.* Proyecto ICA, INDERENEA, USDA, Bogota.

Rodríguez-Ferraro, A. 1996. Notes on the nesting behavior of the Olive Oropendola (*Gymnostinops yuracares*) in southern Venezuela. Unpublished Manuscript.
Rogers, D.T. Jr., D.L. Hicks, E.W. Wischusen, and J.R. Parrish. 1982. Repeats, returns, and estimated flight ranges of some North American migrants in Guatemala. *J. Field Ornithol.* 53:133–138.
Rogers, M.J. 1995. A possible Common Grackle. *British Birds.* 88:156.
Rohwer, S.A. 1972a. A multivariate assessment of interbreeding between the meadowlarks, *Sturnella. Syst. Zool.* 21:313–338.
Rohwer, S.A. 1972b. Distribution of meadowlarks in the central and southern Great Plains and the Desert Grasslands of eastern New Mexico and west Texas. Trans. *Kansas Acad. Sci.* 75:1–19.
Rohwer, S.A. 1973. Significance of sympatry to behavior and evolution of Great Plains meadowlarks. *Evolution* 27:44–57.
Rohwer, S.A. 1973. Distribution of meadowlarks in the central and southern Great Plains and the desert grasslands of eastern New Mexico and west Texas. Trans. *Kansas Academy of Sciences* 75:1–19.
Rohwer, S.A., and M.S. Johnson. 1992. Scheduling differences of molt and migration for Baltimore and Bullock's Orioles persist in a common environment. *Condor* 94:992–994.
Rohwer, S.A., and J. Manning. 1990. Differences in timing and number of molts for Baltimore and Bullock's Orioles: implications to hybrid fitness and theories of delayed plumage maturation. *Condor* 92:125–140.
Root, T. 1988. *Atlas of wintering North American birds.* The University of Chicago Press, Chicago.
Rosenberg, G.H. 1990a. Arizona birding pitfalls, Part I: species we take for granted. *Birding* 22:120–129.
Rosenberg, G.H. 1990b. Habitat specialization and foraging behavior by birds of Amazonian river islands in northeastern Peru. *Condor* 92:427–443.
Roskaft, E., and S. Rohwer. 1987. An experimental study of the function of the red epaulettes and the black body colour of male Red-winged Blackbirds. *Anim. Behav.* 35:1070–1077.
Roskaft, E., S. Rohwer, and C.D. Spaw. 1993. Cost of puncture ejection compared with costs of rearing cowbird chicks for Northern Orioles. *Ornis Scandinavica* 24:28–32.
Rothstein, S.I. 1972. Territoriality and mating system in the parasitic Brown-headed Cowbird (*Molothrus ater*) as determined from captive birds. *Am. Zool.* 12:659–659.
Rothstein, S.I. 1976. Experiments on defences Cedar Waxwings use against cowbird parasitism. *Auk* 93:675–691.
Rothstein, S.I. 1977a. The Preening Invitation or Head-Down Display of Parasitic Cowbirds:I. Evidence for Intraspecific Occurrence. *Condor* 79:13–23.
Rothstein, S.I. 1977b. An Experimental and Teleonomic Investigation of Avian Brood Parasitism. *Condor* 77:250–271.
Rothstein, S.I. 1977c. Cowbird parasitism and egg recognition of the Northern Oriole. *Wilson Bull.* 89:21–32.
Rothstein, S.I. 1978. Geographical Variation in the Nestling Coloration of Parasitic Cowbirds. *Auk* 95:152–160.
Rothstein, S.I. 1982. Successes and failures in avian egg and nestling recognition with comments on the utility of optimality reasoning. *Amer. Zool.* 22:547–560.
Rothstein, S.I. 1990. A Model System for Coevolution: Avian Brood Parasitism. *Annu. Rev. Ecol. Syst.* 21:481–50.
Rothstein, S.I. 1993. An Experimental test of the Hamilton-Orians hypothesis for the origin of avian brood parasitism. *Condor* 95:1000–1005.
Rothstein, S.I. 1994. The cowbird's invasion of the far west: history, causes, and consequences experienced by host species. *Studies in Avian Biology* 15:301–315.
Rothstein, S.I., and R.C. Fleischer. 1987. Vocal dialects and their possible relation to honest status signalling in the Brown-headed Cowbird. *Condor* 89:1–23.
Rothstein, S.I., J. Verner, and E. Stevens. 1980. Range expansion and diurnal changes in dispersion of the Brown-headed Cowbird in the Sierra Nevada. *Auk* 97:253–267.
Rothstein, S.I., D.A. Yokel, and R.C. Fleischer. 1986. Social dominance, mating and spacing systems, female fecundity, and vocal dialects in captive and free-ranging Brown-headed Cowbirds. *Current Ornithology* 3:127–185.
Rothstein, S.I., D.A. Yokel, and R.C. Fleischer. 1988. The agonistic and sexual function of vocalizations of male Brown-headed Cowbirds, *Molothrus ater. Anim. Behav.* 36:73–86.
Rowher, S. 1976. Species distinctness and adaptive differences in southwestern meadowlarks. *Occ. papers Mus. Nat. Hist. U. of Kansas* 44:1–14.
Rowley, J.S. 1984. Breeding records of land birds in Oaxaca, Mexico. Proc. Western Found. Vert. Zool. 3:76–221.
Royall, W.C. Jr. 1973. The Common Grackle in Texas–a review of fifty years of band recovery data. *Bull. Texas. Ornithol. Soc.* 6:20–22.
Royall, W.C. Jr., J.L. Guarino, J.W. De Grazio, and A. Gammell. 1971. Migration of banded Yellow-headed Blackbirds. *Condor* 73:100–106.
Russell, S.M. 1964. A distributional study of the birds of British Honduras. *Ornithol. Monograph* 1:1–194.
Russell, S.M., J.C. Barlow, and D.W. Lamm. 1979. Status of some birds on Isla San Andres and Isla

Providencia, Colombia. *Condor* 81:98–100.

Sallaberry, M., J. Aguirre, and J. Yañez. 1992. Adiciones a la lista de aves de Chile: descripción de especies nuevas para el pa¡s y otros datos ornitológicos. *Noticiario Mensual, Museo Nacional de Historia Natural.* 321:3–9.
Salvador, S.A. 1983. Parasitismo de cria del Renegrido (*Molothrus bonariensis*) en Villa Maria, Cordoba, Argentina (Aves: Icteridae). *Historia Natural* 3:149–158.
Salvador, S.A. 1984. Estudio de parasitismo de cria del renegrido (*Molothrus bonariensis*) en Calandria (*Mimus saturninus*), en Villa Maria, Cordoba. *Hornero* 12(3):141–149.
Salvador, S.A., and L.A. Salvador. 1984. Notas Sobre Hospendantes del Renegrido (*Molothrus bonariensis*) (Aves: Icteridae). *Historia Natural* 4:121–130.
Salvin, O. 1859. Letter on Guatemalan birds. Ibis 1:468.
Salvin, O., and F.D. Godman. 1891. Descriptions of five new species of birds discovered in Central America by W.B. Richardson. *Ibis* Ser VI,Vol.III:612.
Sanger, G.A. 1967. Brown-headed Cowbird collected far at sea. *Condor* 69:89.
Sauer, J.R., J.E. Hines, G. Gough, I. Thomas, and B.G. Peterjohn. 1997. The North American Breeding Bird Survey results and analysis. Version 96.3 Internet- Patuxent Wildlife Research Center, Laurel, MD. http://www.mbr.nbs.gov/bbs/htm96/htmra/all.html
Saunders, G.B. 1934. Description of a new meadowlark from southwestern Mexico. *Auk* 51:42–45.
Saunders, W.E. 1902. Birds of Sable Island. N.S. *Ottawa Nat.* 16:15–31.
Sawyer, M., and M.I. Dyer. 1968. Yellow-headed Blackbird Nesting in Southern Ontario. *Wilson Bull.* 80:236–237.
Schaefer, V.H. 1976. Geographic variation in the placement and structure of oriole nests. *Condor* 78:443–448.
Schaefer, V.H. 1980. Geographic variation in the insulative qualities of nests of the northern oriole. *Wilson Bull.* 92:466–474.
Schäfer, E. 1953. Resultados parciales de una investigación comparativa de biologi de incubación de *Psarocolius decumanus* (Conoto negro) y *Psarocolius angustifrons* (Conoto verde). *Tirada especial del Boletin de la Academia de Ciencias Fisicas, Matematicas y Naturales.* 51:5–17.
Schäfer, E. 1957. Les conotos: etude comparative de *Psarocolius angustifrons* et *Psarocolius decumanus. Bonner Zoologische Beitrage* 1957:1–149.
Schaldach, W.J. Jr. 1963. The avifauna of Colima and adjacent Jalisco, Mexico. *Proc. Western Found. Vert. Zool.* 1:1–100.
Scharf, W.C., and J. Kren. 1996. Orchard Oriole (*Icterus spurius*). In The Birds of North America, No. 255 (A. Poole and F. Gill eds.). The Academy of Natural Sciences, Philadephia, PA, and the American Ornithologists's Union, Washington, D.C.
Schemske, D.W. 1975. Territoriality in a nectar feeding northern oriole in Costa Rica. *Auk* 92:594–595.
Scheuering, E.J., and G.L. Ivey. 1995. First nesting record of the Great-tailed Grackle in Oregon. *Wilson Bull.* 107:562–563.
Schorger, A. W. 1941. The Bronzed Grackle's method of opening acorns. *Wilson Bull.* 53:238–240.
Schulenberg, T.S., and T.A. Parker. 1981. Status and distribution of some northwest Peruvian birds. *Condor* 83:213–214.
Sclater, P.L. 1857. Notes on the birds in the museum of the Academy of Natural Sciences of Philadelphia, and other collections in the United States of America. *Proc. Zool. Soc. London* 27:6–7.
Sclater, P.L. 1860. List of birds collected by Mr. Fraser at Babahoyo in Ecuador, with descriptions of new species. *Proc. Zool. Soc. London* 1860:272–277.
Sclater, P.L. 1861a. Notice of the occurrence of the American Meadow-Starling (*Sturnella ludoviciana*) in England. *Ibis* 1861:179–179.
Sclater, P.L. 1861. List of a collection of birds made by the late Mr. W. Osburn in Jamaica, with notes. *Proc. Zool. Soc. London* 1861:69–82.
Sclater, P.L. 1881. On some birds collected by Mr. E.F. im Thurn in British Guiana. *Proc. Zool. Soc. London* 1881:213–214.
Sclater, P.L. 1883. Review of the species of the family Icteridae, Part 1. *Ibis* 1883:145–163.
Sclater, P.L. 1884. A review of the species of the family Icteridae. Part IV. *Ibis* 1884:156–157.
Sclater, P.L., and O. Salvin. 1864. Notes on a collection of birds from the Isthmus of Panama. *Proc. Zool. Soc. London* 1864:342–372.
Sclater, P.L., and O. Salvin. 1873. On the birds of eastern Peru. *Proc. Zool. Soc. London* 1873:252–311.
Sclater, P.L., and O. Salvin. 1879. On the birds collected by T.K. Salmon in the state of Antioquia, United States of Colombia. *Proc. Zool. Soc. London* 1879:486–550.
Sclater, W.L. 1939. A note on some American orioles of the family Icteridae. *Ibis* 3:140–145.
Scott, D.M. 1977. Cowbird parasitism on the Gray Catbird at London, Ontario. *Auk* 94:18–27.
Scott, D.M. 1978. Using sizes of unovulated follicles to estimate the laying rate of the Brown-headed Cowbird. *Can.J.Zool.* 56:2230–2234.
Scott, D.M., and C.D. Ankney. 1979. Evaluation of a method for estimating the laying rate of Brown-headed Cowbirds. *Auk* 96:483–488.

Scott, D.M., and C.D. Ankney. 1980. Fecundity of the Brown-headed Cowbird in southern Ontario. *Auk* 97:677–683.
Scott, D.M., and D.W. Ankney. 1983. The laying cycle of Brown-headed Cowbirds: passerine chickens? *Auk* 100:583–592.
Scott, D.M., and A.L.A. Middleton. 1968. The annual testicular cycle of the Brown-headed Cowbird (*Molothrus ater*). *Can.J.Zool.* 46:77–87.
Scott, D.M., P.J. Weatherhead, and C.D. Ankney. 1992. Egg-eating by female Brown-headed Cowbirds. *Condor* 94:579–584.
Scott, W.E.D. 1893. Observations on the birds of Jamaica, West Indies. *Auk* 10:177–181.
Scott, W.E.D. 1901. Data on songbirds: observations on the song of Baltimore Orioles in captivity. *Science* 14:522–526.
Sealy, S.G. 1979. Prebasic molt of the Northern Oriole. *Can. J. Zool.* 57:1473–1478.
Sealy, S.G. 1980a. Breeding biology of Orchard Orioles in a new population in Manitoba. *Can. Field. Nat.* 94:154–158.
Sealy, S.G. 1980b. Reproductive responses of Northern Orioles to a changing food supply. *Can. J. Zool.* 58:221–227.
Sealy, S.G. 1985. Where do Northern ('Baltimore') Orioles spend the winter? *North American Bird Bander* 10:12–17.
Sealy, S.G. 1986. Fall migration of Northern Orioles: an analysis of tower-killed individuals. *North American Bird Bander* 11:43–45.
Sealy, S.G., and D.L. Neudorf. 1995. Male Northern Orioles Eject Cowbird Eggs: Implications for the Evolution of Rejection Behavior. *Condor* 97:369–375.
Searcy, W.A. 1979. Female choice of mates: a general model for birds and its application to red-winged blackbirds (*Agelaius phoeniceus*). *Am. Nat.* 114:77–100.
Searcy, W.A. 1979. Male characteristics and pairing success in red-winged blackbirds. *Auk* 96:353–363.
Searcy, W.A. 1979. Sexual selection and body size in male red-winged blackbirds. *Evolution* 33:649–661.
Searcy, W.A. 1979. Size and mortality in male Yellow-headed Blackbirds. *Condor* 81:304–305.
Searcy, W.A. 1986. Are female Red-winged Blackbirds territorial? *Animal Behav.* 34:1381–1391.
Searcy, W.A. 1989. Function of male courtship vocalizations in Red-winged Blackbirds. *Behav. Ecol. Sociobiol.* 24:325–331.
Searcy, W.A. 1990. Species recognition of song by female Red-winged Blackbirds. *Animal Behav.* 40:1119–1127.
Searcy, W.A., and K. Yasukawa. 1981. Sexual size dimorphism and survival of male and female blackbirds (Icteridae). *Auk* 98:457–465.
Searcy, W.A., and K. Yasukawa. 1990. Use of the song repertoire in intersexual and intrasexual contexts by male Red-winged Blackbirds. *Behav. Ecol. Sociobiol.* 27:123–128.
Selander, R.K., and C.J. La Rue Jr. 1961. Interspecific preening invitation display of Parasitic Cowbirds *Auk* 78:473–504.
Selander, R.K. 1958. Age determination and molt in the Boat-tailed Grackle. *Condor* 60:355–375.
Selander, R.K. 1964. Behavior of Captive South American Cowbirds. *Auk* 81:394–402.
Selander, R.K. 1970. Parental feeding in a male Great-tailed Grackle. *Condor* 72:238–238.
Selander, R.K., and R.W. Dickerman. 1963. The 'nondescript' blackbird from Arizona: an intergenetic hybrid. *Evolution* 17:440–448.
Selander, R.K., and D.R. Giller. 1960. First-year plumage of the Brown-headed Cowbird and Red-winged Blackbird. *Condor* 62:202–214.
Selander, R.K., and D.R. Giller. 1961. Analysis of sympatry of Great-tailed and Boat-tailed Grackles. *Condor* 63:29–86.
Selander, R.K., and R.J. Hauser. 1965. Gonadal and behavioral cycles in the Great-tailed Grackle. *Condor* 67:157–182.
Selander, R.K., and L.L. Kuick. 1963. Hormonal control and development of the incubation patch in Icterids, with notes on the behaviour of cowbirds. *Condor* 65:73–90.
Selander, R.K., and D.J. Nicholson. 1962. Autumnal breeding of Boat-tailed Grakles in Florida. *Condor* 64:81–91.
Selander, R.K., and F.S. Jr. Webster. 1963. Distribution of the Bronzed Cowbird in Texas. *Condor* 65:245–246.
Semper, J.E. 1872. Observations on the birds of St Lucia. *Proc. Zool. Soc. London* 1872:647–653.
Servat, G., and D.L. Pearson. 1991. Natural history notes and records for seven poorly-known bird species from Amazonian Peru. *Bull. BOC.* 111:92–95.
Shake, W.F., and J.P. Mattson. 1975. Three years of cowbird control: An effort to save the Kirtland's Warbler. *Jack Pine Warbler* 53(2):48–53.
Shore-Baily, W. 1928. Breeding hybrids between *Agelaius frontalis* and *Molothrus badius. Avicult. Mag.* 6:291–292.
Short, L.L. 1968. Sympatry of Red-breasted Meadowlarks in Argentina, and the Taxonomy of Meadowlarks (Aves: *Leistes, Pezites,* and *Sturnella*). *Am. Mus. Nov.* 2349:1–30.
Short, L.L. 1969. A new species of blackbird (*Agelaius*) from Peru. *Occ. Pap. Mus. Zool. Louisiana State*

Univ. 36:1–8.
Short, L.L. 1975. A zoogeographic analysis of the South American Chaco avifauna. *Bull. Am. Mus. Nat. Hist.* 154:165–352.
Short, L.L., and K.C. Parkes. 1979. The status of *Agelaius forbesi. Auk* 96:179–183.
Sibley, C.G., and B.L. Jr. Monroe. 1990. *Distribution and Taxonomy of Birds of the World.* Yale University Press, New Haven.
Sibley, C.G., and L.L. Jr. Short. 1964. Hybridization in the orioles of the Great Plains. *Condor* 66:130–150.
Sick, H. 1957. Robhaarpilze als nestbau material brasilianischer vogel. *Journal f. Orn.* 98:421–431.
Sick, H. 1979. Notes on some Brazilian birds. *Bull. BOC.* 99:115–120.
Sick, H. 1993. *Birds in Brazil.* Princeton University Press, Princeton, New Jersey.
Siderius, J.A. 1984. Behavioural, mensural and ecological correlates of polygyny in the Eastern Meadowlark (*Sturnella magna*). M.Sc. Thesis, Brock University, St. Catharines.
Siegel, A. 1983. *Birds of Montserrat.* Montserrat National Trust, Montserrat, West Indies.
Skorupa, J.P., R.L. Hothem, and R.W. De Haven. 1980. Foods of breeding tricolored blackbirds in agricultural areas of Merced County, California. *Condor* 82:465–467.
Skutch, A. F. 1996. *Orioles, blackbirds and their kin. A natural history.* The University of Arizona Press, Tucson.
Skutch, A.F. 1954. Life histories of Central American birds. *Pac. Coast Avif.* 31:1–448.
Skutch, A.F. 1967. Life History notes on the Oriole-Blackbird (*Gymnomystax mexicanus*) in Venzuela. *Hornero* 10:379–388.
Skutch, A.F. 1969. A study of the Rufous-fronted Thornbird and associated birds. Part II: Birds which breed in thornbird's nests. *Wilson Bull.* 81:123–139.
Skutch, A.F. 1971. *A Naturalist in Costa Rica.* Univ. of Florida Press, Gainesville.
Skutch, A.F. 1972. Studies of tropical American Birds. *Publ. Nuttall Orn. Club* 10:173–181.
Slud, P. 1964. The birds of Costa Rica. *Bull. Amer. Mus. Natur. Hist.* 128:336–344.
Small, A. 1994. *California Birds: Their Status and Distribution.* Ibis Publishing Company, Vista, CA.
Smith, D.G. 1972. The role of the epaulets in the red-winged blackbird (*Agelaius phoeniceus*) social system. *Behaviour* 41:251–268.
Smith, D.G., and D.O. Norman. 1979. 'Leader-follower' singing in Red-winged Blackbirds. *Condor* 81:83–84.
Smith, D.G., and F.A. Reid. 1979. Roles of the song repertoire in red-winged blackbirds. *Behav. Ecol. Sociobiol.* 5:279–290.
Smith, F.R. 1967. Report on rare birds in Great Britain in 1966. *Brit. Birds.* 60:309–338.
Smith, J.K., and E.G. Zimmerman. 1976. Biochemical genetics and evolution of North American blackbirds, family Icteridae. *Comp. Biochem. Physiol.* 538:317–324.
Smith, J.N.M. 1981. Cowbird parasitism, host fitness, and age of host female in an island song sparrow population. *Condor* 83:153–161.
Smith, J.N.M., and P. Arcese. 1994. Brown-headed Cowbirds and an Island Population of Song Sparrows: A 16-Year Study. *Condor* 96:916–934.
Smith, K. 1993. A Minnesota Great-tailed Grackle. *The Loon* 65:148–148.
Smith, N.G. 1968. The advantage of being parasitized. *Nature* 219:690–694.
Smith, N.G. 1978. Alternate responses by hosts to parasites which may be helpful or harmful. In encounter: The interface between populations. Ed. B. Nikol. Academic Press, New York.
Smith, N.G. 1979. Alternate responses by hosts to parasites which may be helpful or harmful. In Host-parasite interfaces. Academic Press, New York.
Smith, N.G. 1980. Some evolutionary, ecological, and behavioural correlates of communal nesting by birds whith wasps of bees. Proc. Intern. Ornithol. Congr. XVII:1199–1205.
Smith, N.G. 1982. Some evolutionary, ecological, and behavioral correlates of communal nesting by birds with wasps or bees. Proc. 13th Int. Orn. Congress, Berlin.
Smith, N.G. 1983. *Zarynchus wagleri* In Janzen, D.H. (ed.), Costa Rican Natural History, University of Chicago Press.614–616.
Smith, P.W., and A., IV Sprunt. 1987. The Shiny Cowbird reaches the United States. *Am. Birds.* 41:370–371.
Smith, T.S. 1972. Cowbird parasitism of Western Kingbird and Baltimore Oriole nests. *Wilson Bull.* 84:497–497.
Smithe, F.B. 1966. *Birds of Tikal.* American Museum of Natural History, New York.
Smyth, C.A. 1928. Descripción de una collection de huevos de aves argentinas. *Hornero* 4:1–125.
Snell, R.R. 1989. Status of *Larus* Gulls at Home Bay, Baffin Island. *Colonial Waterbirds* 12:12–23.
Snethlage, E. 1925. Neue vogelarten aus Nord-Brasilen. *Journ. f. Ornithol.* 73:264–265.
Snyder, D.E. 1966. *The Birds of Guyana.* Peabody Museum, Salem.
Snyder, L.L. 1937. Some measurements and observations from Bronzed Grackles. *Can. Field Nat.* 51:37–39.
Southern, W.E., and L.K. Southern. 1980. A summary of the incidence of cowbird parasitism in northern Michigan from 1911–1978. *Jack Pine Warbler* 578:77–84.
Stager, K.E. 1957. The avifauna of the Trés Marias Islands, Mexico. *Auk* 74:413–432.

Stepney, P.H.R. 1975. Wintering distribution of Brewer's Blackbird: historical aspects, recent changes and fluctuations. *Bird Banding* 46:106–125.
Stepney, P.H.R. 1975. First recorded breeding of the Great-tailed Grackle in Colorado. *Condor* 77:298–210.
Stepney, P.H.R., and D.M. Power. 1973. Analysis of eastern breeding expansion of Brewer's Blackbird plus general aspects of avian expansions. *Wilson Bull.* 85:452–464.
Stevenson, H.M. 1978. The populations of Boat-tailed Grackles in the southeastern United States. *Proc. Biol. Soc. Wash.* 91:27–51.
Stevenson, H.M., and B.H. Anderson. 1994. *The Birdlife of Florida.* Univ. of Florida Press, Gainesville.
Stewart, D.B. 1984. *Gosse's Jamaica 1844–45.* Institute of Jamaica Publications, Kingston, Jamaica.
Stewart, P.A. 1975. Breeding localities of Common Grackles wintering in the Carolinas. *Chat* 39:32–34.
Stiles, F.G. 1994. Migration of landbirds in central Costa Rica. *Bird Cons. Int.* 4:71–89.
Stockton de Dod, A. 1978. *Aves de la República Dominicana.* Museo Nacional de Historia Natural, Santo Domingo.
Stoddard, H.L. 1951. Bullock's and Baltimore Orioles, *Icterus bullockii* and *I. galbula,* in southwest Georgia. *Auk* 68:108–110.
Stone, W. and Roberts. 1934. 'regarding *Icterus cayanensis' Proc. Acad. Nat. Sci. Philadelphia* 86:394.
Stone, W. 1891. A revision of the species of *Molothrus* allied to *M.bonariensis* (Gm.). *Auk* 8:344–347.
Stone, W. 1897. The Genus *Sturnella Proc. Acad. Nat. Sci. Philadelphia*146–152.
Stone, W. 1928. On a collection of birds from the Par region, eastern Brazil (with field notes by James Bond and Rodolphe M. de Schauensee). *Proc. Acad. Nat. Sc.. Philadelphia* 80:149–176.
Stoppkotte, G.W. 1975. Scott's Oriole Reported. *Nebraska Bird Review* 43:64–66.
Straneck, R. 1990. Canto de las aves de Misiones I. Cassette Tape, published by PESPIR, Argentina.
Straube, F.C., and M.R. Bornschein. 1995. New or noteworthy records of birds from northwest Paraná and adjacent areas (Brazil). *Bull. BOC.* 115:224–225.
Stresemann, 1954. Ferdinand Deppe's Travels in Mexico, 1824–1829. *Condor* 56:90.
Studer, A. 1982. La redecuventre de l'Icteride Curaeus forbesi ao Bresil. Ex. D.E.S. Univ. Nancy
Studer, A., and J. Vielliard. 1988. Premières données étho-écologiques sur l'Ictéridé brésilien *Curaeus forbesi* (Sclater 1886) (Aves, Passerifomes). *Rev. Suisse Zool.* 95(4):1063–1077.
Summers, S. 1977. A Common Grackle record for Oregon. *Western Birds* 8:156.
Summers, S. 1977. Details: Common Grackle. *Oregon Birds* 3:81–82.
Sutton, G.M. 1938. Oddly plumaged orioles from western Oklahoma. *Auk* 55:1–6.
Sutton, G.M. 1942. Winter range of Oklahoma's hybrid oriole. *Condor* 44:79–79.
Sutton, G.M. 1948. Comments on *Icterus cucullatus cucullatus* Swainson in the United States. *Condor* 50:257–258.
Sutton, G.M., and O.S. Jr. Pettingill. 1943. The Alta Mira Oriole and its nest. *Condor* 45:125–132.
Szijj, J.J. 1966. Hybridization and the nature of the isolating mechanism in sympatric populations of meadowlarks (*Sturnella*) in Ontario. *Z. Tierpsychol.* 6:677–690.
Szijj, L.L. 1963. Morphological analysis of the sympatric populations of meadowlarks in Ontario. *Proc. XIII Int. Ornithol. Congr.* 1963:176–188.

Taczanowski, L. 1874. Liste des oiseaux recuillis pour M. Constantin Jelski dans la partie centraleski du Perou occidental. *Proc. Zool. Soc. London* 1874:501–565.
Talley, G.M. 1957. Common Grackle in Utah. *Condor* 59:400.
Tallman, D. 1997. Common Grackle Banded in South Dakota Recovered in Virginia. *South Dakota Bird Notes.* 49:96–96.
Tashian, R.E. 1957. Nesting behavior of the Crested Oropendola (*Psarocolius decumanus*) in northern Trinidad, B.W.I. *Zoologica* 42:87–98.
Taylor, D.M., and C.H. Trost. 1985. The Common Grackle in Idaho. *Am. Birds* 39 (2):217–218.
Taylor, E.C. 1864. Five months in the West Indies. *Ibis* 6:157–173.
Templeton, A.R. 1989. The meaning of species and speciation: A genetic perspective. In: Speciation and its consequences (D. Otte and J. A. Endler, Eds.) Sinauer Associates, Sunderland, Mass.3–27.
Terres, J.K. 1980. *The Audubon Society Encyclopedia of North American Birds.* Alfred A. Knopf, New York.
Thomas, B.T. 1979. The birds of a ranch in the Venezuelan llanos. pp 213–232 In. Vertebrate ecology in the northern Neotropics. J.F. Eisenberg (ed.). Smithsonian Institution Press, Washington, DC.
Thomas, B.T. 1983. The Plain-fronted Thornbird: nest construction, material choice and nest defence behaviour. *Wilson Bull.* 95:106–117.
Thomas, R.H. 1946. An Orchard Oriole colony in Arkansas. *Bird Banding* 17:161–167.
Thompson, C.F., and B.M. Gottfried. 1976. How do cowbirds find and select nests to parasitize? *Wilson Bull.* 88:673–675.
Thompson, C.F., and B.M. Gottfried. 1981. Nest discovery and selection by Brown-headed Cowbirds. *Condor* 83:268–269.
Thurber, W.A., J.F. Serrano, A. Sermeño, and M. Benitez. 1987. Status of uncommon and previously unreported birds of El Salvador. *Proc. Western Found. Vert. Zool.* 3:111–293.
Thurber, W.A., and A. Vidella. 1980. Notes on parasitism by Bronzed Cowbirds in El Salvador. *Wilson*

Bull. 92:112–113.
Timken, R.L. 1970. Food habits and feeding behavior of the Baltimore Oriole in Costa Rica. *Wilson Bull.* 82:184–188.
Todd, W.E.C, and W.W. Worthington. 1911. A contribution to the ornithology of the Bahama Islands. *Ann. Carnegie Mus.* 7:439–440.
Todd, W.E.C. 1913. Preliminary diagnoses of apparently new birds from tropical America. *Proc. Biol. Soc. Washington.* 26:169–174.
Todd, W.E.C. 1916. The birds of the Isle of Pines. *Ann. Carnegie Mus.* 10:276.
Todd, W.E.C. 1917. Preliminary diagnoses of apparently new birds from Colombia and Bolivia. *Proc. Biol. Soc. Washington* 30:3–6.
Todd, W.E.C. 1924a. Remarks on the genus *Amblycercus* and its allies. *Proc. Biol. Soc. Washington* 37:113–117.
Todd, W.E.C. 1924b. Descriptions of eight new neotropical birds. *Proc. Biol. Soc. Washington* 37:121–124.
Todd, W.E.C. 1932. Seven apparently new South American birds. *Proc. Biol. Soc. Washington* 45:219.
Todd, W.E.C., and M.A. Carriker. 1922. The birds of the Santa Marta region of Colombia: a study in altitudinal distribution. *Annals of the Carnegie Museum* 14:1–611.
Townsend, C.W. 1927. Notes on the courtship of the Lesser Scaup, Everglade Kite, Crow, and Boat-tailed and Great-tailed Grackles. *Auk* 44:59–554.
Trainer, J.M. 1987. Behavioral associations of song types during aggressive interactions among male Yellow-rumped Caciques. *Condor* 89:731–738.
Trainer, J.M. 1988. Singing organization during aggressive interactions among male Yellow-rumped Caciques. *Condor* 90:681–688.
Trainer, J.M. 1989. Cultural evolution in song dialects of Yellow-rumped Cacique in Panama. *Ethology* 80:190.
Traylor, M.E. 1948. New birds from Peru and Ecuador. *Fieldiana Zool.* 31:198.
Tufts, R.W. 1986. *Birds of Nova Scotia.* (3rd Edition). Nimbus publishing Ltd., Nova Scotia Museum, Halifax.
Tutor, B.M. 1962. Nesting studies of the Boat-tailed Grackle. *Auk* 79:77–84.
Twedt, D.J. 1982. The Yellow-headed Blackbird in Kentucky: First reported capture and previous sight records. *Kentucky Warbler* 58:59–60.
Twedt, D.J., W.J. Bleier, and G.M. Linz. 1994. Geographic Variation in Yellow-headed Blackbirds from the Northern Great Plains. *Condor* 96:1030–1036.
Twedt, D.J., W.J. Bleier, and G.M. Linz. 1994. Genetic variation in male Yellow-headed Blackbirds from the northern Great Plains. *Can. J. Zool.* 70:2280–2282.
Twedt, D.J., and R.D. Crawford. 1995. Yellow-headed Blackbird (*Xanthocephalus xanthocephalus*). In The Birds of North America, No. 192 (A. Poole and F. Gill, eds). The Academy of Natural Sciences, Philadelphia, and The American Ornithologists' Union, Washington, D.C.

Van Rossem, A.J. 1926. The California forms of *Agelaius phoeniceus* (Linnaeus). *Condor* 28:215–230.
Van Rossem, A.J. 1927. A new race of the Sclater Oriole. *Condor* 29:76–77.
Van Rossem, A.J. 1934. Critical notes on Middle American Birds. *Bull. Mus. Comp. Zool.* 77:404–405.
Van Rossem, A.J. 1938. Descriptions of twenty-one new races of Fringillidae and Icteridae from Mexico and Guatemala. *Bull. BOC* 58:124–138.
Van Rossem, A.J. 1945. A distributional survey of the birds of Sonora, Mexico. *Occas. Papers Mus. Zool. Lousiana State Univ.* 21:1–379.
Van Velzen, W.T. 1972. Distribution and abundance of the Brown-headed Cowbird. *Jack-Pine Warbler* 50:110–113.
Veit, R.R., and W.R. Petersen. 1993. *The birds of Massachusetts.* Massachusetts Audubon Society, Lincoln, Massachusetts.
Verner, J. 1975. Interspecific aggression between Yellow-headed Blackbirds and Long-billed Marsh Wrens. *Condor* 77:329–331.
Verner, J., and M.F. Willson. 1966. The influence of habitats on mating systems of North American passerine birds. *Ecology* 47:143–147.
von Sneidern, K. 1954. Notas sobre algunas aves del Museo de Historia Natural de la Universidad del Cauca, Popayán, Colombia. *Novedades Colombianas* 1:3–13.
Voous, K.H. 1983. *Birds of the Netherlands Antilles.* De Walburg Pers, Utrecht.
Voous, K.H. 1985. Additions to the avifauna of Aruba, Curacao, and Bonaire, South Caribbean. In Neotropical Ornithology. Buckley, P.A., M.S. Foster, E.S.Morton, R.S.Ridgely, and F.G.Buckley (eds.). *Ornithological Monographs* 36:247–254.
Vuilleumier, F. 1969. Field notes on some birds from the Bolivian Andes. *Ibis* 111:559–608.

Walkingshaw, L.H. 1949. Twenty-five eggs apparently laid by a cowbird. *Wilson Bull.* 61:82–85.
Walkingshaw, L.H. 1961. The effect of parasitism by the Brown-headed Cowbird on *Empidonax* flycatchers in Michigan. *Auk* 78:266–268.

Warren, P. 1998a. Song Dialects in the Bronzed Cowbird. Abstract, North American Ornithological Conference. St Louis, Missouri.

Warren, P. 1998b. Male Bronzed Cowbirds Show Stronger Responses to Foreign Songs in Paraptry than in Allopatry. Abstract, North American Orthithological Conference. St Louis, Missouri.

Wauer, R.H. 1973. Bronzed Cowbird extends range into the Texas Big Bend country. *Wilson Bull.* 85:343–344.

Wauer, R.H. 1985. *A field guide to birds of the Big Bend.* Texas Monthly Press, Austin.

Weatherhead, P.J. 1981. The dynamics of Red-winged Blackbird populations at four late summer roosts in Quebec. *J. Field Ornithol.* 52:222–227.

Weatherhead, P.J. 1989. Sex ratios, host-specific reproductive success, and impact of Brown-headed Cowbirds. *Auk* 106:358–366.

Weatherhead, P.J. 1991. The adaptive value of thick-shelled eggs for Brown-headed Cowbirds. *Auk* 108:196–198.

Weatherhead, P.J., and R.J. Robertson. 1977. Harem size, territory quality, and reproductive success in the Red-winged Blackbird (*Agelaius phoeniceus*). *Can. J. Zool.* 55:1261–1267.

Weatherhead, P.J., and R.J. Robertson. 1977. Male behavior and female recruitment in the Red-winged Blackbird. *Wilson Bull.* 89:583–592.

Weatherhead, P.J., and K.L. Teather. 1994. Sexual size dimorphism and egg-size allometry in birds. *Evolution* 48:671–678.

Weber, W.C. 1976. Mourning Warbler and Northern Oriole in northeastern British Columbia. *Murrelet* 57:68–69.

Webster, M.S. 1991. The dynamics and consequences of intrasexual competition in the Montezuma Oropendola: harem-polygyny in a Neotropical bird. Ph.D. diss., Cornell University, Ithaca, NY.

Webster, M.S. 1992. Sexual dimorphism, mating system and body size in new world blackbirds (Icterinae). *Evolution* 46:1621–1641.

Webster, M.S. 1994a. Female-defence polygyny in a Neotropical bird, the Montezuma Oropendola. *Anim. Behav.* 48:779–794.

Webster, M.S. 1994b. Spatial and Temporal Distribution of Breeding Female Montezuma Oropendolas: Effects on Mating Strategies. *Condor* 96:722–733.

Webster, M.S. 1994c. Interspecific Brood Parasitism of Montezuma Oropendolas by Giant Cowbirds: Parasitism or Mutualism? *Condor* 96:794–798.

Webster, M.S. 1995. Effects of female choice and copulations away from colony on fertilization success of male Montezuma Oropendolas (*Psarocolius montezuma*). *Auk* 112:659–671.

Weller, M.W. 1967. Notes on some marsh birds of Cape San Antonio, Argentina. *Ibis* 109:391–411.

West, M.J., A.P. King, and D.H. Eastzer. 1981. The cowbird: reflections on development from an unlikely source. *Am. Sci.* 69:56–66.

West, M.J., A.P. King, and D.H. Eastzer. 1981. Validating the female bioassay of cowbird song: relating differences in song potency to mating sucess. *Anim. Behav.* 29:490–501.

Wetmore, A. 1919. Notes on the structure of the palate in the Icteridae. *Auk* 36:190–197.

Wetmore, A. 1926. Observations of the birds of Argentina, Paraguay, Uruguay and Chile. *Bull. U.S. Nat. Mus.* 133:372–391.

Wetmore, A. 1939. Observations on the birds of northern Venezuela. *Proc. U.S. Nat. Mus.* 87:173–260.

Wetmore, A., and F.C. Lincoln. 1933. Additional notes on the birds of Haiti and the Dominican Republic. *Proc. U.S. Nat. Mus.* 82:1–88.

Wetmore, A., R.F. Pasquier, and S.L. Olson. 1984. The birds of the Republic of Panama. Part 4. *Smithsonian Misc. Collection* 150:338–382.

Wetmore, A., and B.H. Swales. 1931. The birds of Haiti and the Dominican Republic. *Bull. U.S. Nat. Mus.* 155:1–483.

Whittingham, L.A., A. Kirkconnell, and L.M. Ratcliffe. 1992. Differences in song and sexual dimorphism between Cuban and North American Red-winged Blackbirds (*Agelaius phoeniceus*). *Auk* 109:928–933.

Whittingham, L.A., A. Kirkconnell, and L.M. Ratcliffe. 1996. Breeding behavior, social organization and morphology of Red-shouldered (*Agelaius assimilis*) and Tawny-shouldered (*A.humeralis*) Blackbirds. *Condor* 98:832–836.

Wiens, J.A., and J.T. Rottenberry. 1981. Habitat associations and community structure of birds in shrub-steppe environments. Ecol. Monog. 51:21–41.

Wiley, E. 1981. *Phylogenetics.* John Wiley and Sons, New York.

Wiley, J.M., and A. Cruz. 1991. Endangered passerine hosts of the Shiny Cowbird in the Greater Antilles – A case study of the Yellow-shouldered Blackbird. *El Pitirre* 4:4–4.

Wiley, J.W. 1982. Ecology of avian brood parasitism at an early interfacing of host and parasite populations. Ph.D. Thesis. University of Miami, Coral gables, Florida.

Wiley, J.W. 1985. Shiny cowbird parasitism in two avian communities in Puerto Rico. *Condor* 87:165–176.

Wiley, J.W. 1986. Growth of Shiny Cowbirds and host chicks. *Wilson Bull.* 98:126–131.

Wiley, J.W. 1988. Host Selection by the Shiny Cowbird. *Condor* 90:289–303.

Wiley, J.W., W. Post, and A. Cruz. 1991. Conservation of the Yellow-shouldered Blackbird *Agelaius xan-*

thomus, an endangered West Indian species. *Biological Conservation* 55:119–138.
Wiley, R.H. 1976a. Communication and spacial relationships in a colony of Common Grackles. *Anim. Behav.* 24:570–584.
Wiley, R.H. 1976b. Affiliation between the sexes in Common Grackles, I: Specificity and seasonal progression. *Z. Tierpsychol.* 40:244–264.
Wiley, R.H. 1976c. Affiliation between the sexes in Common Grackles. II. Spatial and vocal coordination. *Z. f. Tierpsychol.* 40:244–264.
Wiley, R.H., and A. Cruz. 1980. The Jamaican Blackbird: A 'natural experiment' for hypotheses in socioecology. *Evol. Biol.* 13:261–293.
Wiley, R.H., and M.S. Wiley. 1980. Spacing and timing in the nesting ecology of a tropical blackbird: Comparison of populations in different environments. *Ecol. Monogr.* 50:153–178.
Willett, G. 1923. Bird records from Craig, Alaska. *Condor* 25:105–106.
Williams, L. 1952. Breeding behavior of the Brewer's Blackbird. *Condor* 54:3–47.
Williams, P.L. 1982. A comparison of colonial and non-colonial nesting by Northern Orioles in central coastal California. Unpublished M.Sc. Thesis, Univ. of California, Berkeley.
Williams, R. 1995. Neotropical Notebook. *Cotinga* 4:65–69.
Willis, E.O., and Y. Oniki. 1985. Bird specimens new for the state of Sao Paulo, Brazil. *Rev. Brasil. Biol.* 45:105–108.
Willis, E.O., and Y. Oniki. 1991. Avifaunal transects across the open zones of northern Minas Gerais, Brazil. *Ararajuba* 2:41–58.
Willson, M.F. 1966. The breeding ecology of the Yellow-headed Blackbird. *Animal Behaviour* 34: 113–122.
Willson, M.F. 1966. Breeding ecology of the Yellow-headed Blackbird. *Ecol. Monographs* 36:51–77.
Willson, M.F., and G.H. Orians. 1963. Comparative ecology of Red-winged and Yellow-headed Blackbirds during the breeding season. Proc. XVI Int. Congr. Zool. 3:342–346.
Wilson, D.B. 1979. Notas sobre casos de parasitismo del Renegrido (*M.bonariensis*) sobre varios passeriformes observados en la Pcia. de Corrientes. *Hornero* 12 No.extraord:69–71.
Wittenberger, J.F. 1978. The breeding biology of an isolated Bobolink population in Oregon. *Condor* 80:355–371.
Wittenberger, J.F. 1980. Vegetation structure, food supply, and polygyny in Bobolinks (*Dolichonyx oryzivorus*). *Ecology* 61:140–150.
Wittenberger, J.F. 1980. Feeding of secondary nestlings by polygynous male Bobolinks in Oregon. *Wilson Bull.* 92:330–340.
Wittenberger, J.F. 1982. Factors affecting how male and female Bobolinks apportion parental investments. *Condor* 84:22–39.
Wittenberger, J.F. 1983. A contextual analysis of two song variants in the Bobolink. *Condor* 85:172–184.
Wolf, L.L. 1971. Predatory Behavior in Montezuma Oropendola. *Wilson Bull.* 83:113–212.
Wood, H.B. 1934. Color of the iris of the Purple Grackle. *Auk* 51:527–528.
Wood, H.B. 1945. The sequence of molt in Purple Grackles. *Auk* 62:455–456.
Wood, P. 1987. Report of the 1986 University of East Anglia Martinique Oriole Expedition. Int. Coun. Bird. Preserv. Study Report 23.
Woods, R.W. 1988. *Guide to Birds of the Falkland Islands.* Anthony Nelson, Shropshire, England.
Woods, R.W., and A. Woods. 1997. *Atlas of Breeding Birds of the Falkland Islands.* Anthony Nelson, Oswestry, U.K.
Woodward, P.W. 1983. Behavioral ecology of fledgeling Brown-headed Cowbirds and their hosts. *Condor* 85:151–163.
Woodward, P.W., and J.C. Woodward. 1979. Survival of fledgeling Brown-headed Cowbirds. *Bird Banding* 50:66–68.
Workman, W.B. 1963. Baltimore Oriole on Lundy, Devon (1958). *Brit. Birds* 56:52–53.
Wormington, A., and W. Lamond. 1987. Orchard Oriole: new to northern Ontario. *Ontario Birds* 5:32–34.
Worthen, G.L. 1973. First Utah record of the Baltimore Oriole. *Auk* 90:677–678.
Wright, B.S. 1962. Baltimore Oriole kills hummingbird. *Auk* 79:112–112.
Wright, P.L. 1996. Status of rare birds in Montana, with comments on known hybrids. *Northwest Naturalist* 77:57–85.
Wright, P.L., and M.H. Wright. 1944. The reproductive cycle of the male Red-winged Blackbird. *Condor* 46:46–59.

Yakusawa, K., and et al. 1987. Seasonal change in the vocal behaviour of female Red-winged Blackbirds. *Animal Behaviour* 35:1416–1423.
Yang, S.Y., and R.K. Selander. 1968. Hybridization in the Grackle, *Quiscalus quiscula* in Louisiana. *Syst. Zool.* 17:107–143.
Yasukawa, K. 1979. Territory establishment in Red-winged Blackbirds: Importance of aggressive behavior and experience. *Condor* 81:258–264.
Yasukawa, K. 1981. Male quality and female choice of mate in the Red-winged Blackbird (*Agelaius phoeniceus*). *Ecology* 62:922–929.

Yasukawa, K., J.L. Blank, and C.B. Patterson. 1980. Song repertoires and sexual selection in the Red-winged Blackbird. *Behav. Ecol. Sociobiol.* 7:233–238.
Yasukawa, K., R.A. Boley, J.L. McClure, and J. Zanocco. 1992. Nest dispersion in the Red-winged Blackbird. *Condor* 94:775–777.
Yasukawa, K., and W.A. Searcy. 1982. Aggression in female Red-winged Blackbirds: a strategy to ensure male parental investment. *Behav. Ecol. Sociobiol.* 11:13–17.
Yasukawa, K., and W.A. Searcy. 1995. Red-winged Blackbird (*Agelaius phoeniceus*). In The Birds of North America, No. 184 (A.Poole and F. Gill, eds.). The Academy of Natural Sciences, Philadelphia and the American Ornithologist's Union.
Yokel, D.A. 1986. Monogamy and brood parasitism: an unlikely pair. *Anim. Behav.* 34:1348–1358.
Yokel, D.A. 1987. Sexual selection and the mating system of the Brown-headed Cowbird (*Molothrus ater*) in Eastern California. PhD Dissertation. University of California, Santa Barbara.
Yokel, D.A. 1989. Payoff asymmetries in contests among male Brown-headed Cowbirds. *Behav. Ecol. Sociobiol.* 24:209–216.
Yokel, D.A. 1989. Intrasexual aggression and the mating behaviour so Brown-headed Cowbirds: Their relation to population densities and sex ratios. *Condor* 91:43–51.
Yokel, D.A., and S.I. Rothstein. 1991. The basis for female choice in an avian brood parasite. *Behav. Ecol. Sociobiol.* 29(1):39–45.
Young, H. 1963. Breeding success of the cowbird. *Wilson Bull.* 75:115–122.

Zimmer, J.T. 1924. New Birds from Central Peru. *Field Mus. Nat. Hist. Publ., Zool. Ser.* 12:51–67.
Zimmer, J.T. 1930. Birds of the Marshall Field Peruvian expedition, 1922–1923. *Field Mus. Nat. Hist. Zoology* 17:428–435.
Zimmer, K. 1985. *The Western Bird Watcher.* Prentice-Hall, Inc., Engleward Cliffs, New Jersey.
Zimmer, K.J. 1984. ID Point: Eastern vs. Western Meadowlarks. *Birding* 16:155–156.
Zink, R.M., and M.C. McKitrick. 1995. The debate over species concepts and its implications for ornithology. *Auk* 112:701–719.
Zink, R.M., and J.V. Remsen. 1986. Evolutionary processes and patterns of geographic variation in birds. *Current Ornithol.* 4:1–69.
Zink, R.M., W.L. Rootes, and D.L. Dittman. 1991. Mitochondrial DNA variation, population structure, and evolution of the Common Grackle (*Quiscalus quiscula*). *Condor* 93:318–329.
Zotta, A.R. 1937. Una nueva subespecie de pecho colorado *Pezites militaris catamarcanus,* subsp. nov. *Hornero* 6:449–454.

INDEX

Figures in bold refer to plate numbers

ENGLISH NAMES

Altamira Oriole **12**, 13, 181
Audubon's Oriole **17**, 13, 231
Austral Blackbird **31**, 312

Baltimore Oriole **10**, 12, 13, 21, 199
Band-tailed Oropendola, **1** 134
Bar-winged Oriole **17**, 233
Baudo Oropendola **3**, 128
Baywing (Cowbird) **37**, 12, 21, 364
'Bicolored Blackbird' **23**, 266
Black Oropendola **3**, 132
Black-backed Oriole **10**, 12, 211
Black-cowled Oriole **16**, 12, 221
Black-vented Oriole **16**, 12, 218
Boat-tailed Grackle **34**, 338
Bobolink **28**, 12, 18, 19, 390
Bolivian Blackbird **33**, 260
Brewer's Blackbird **36**, 16, 360
Bronze-brown Cowbird **38**, 378
Bronzed Cowbird **38**, 374
Brown-and-yellow Marshbird **29**, 307
Brown-headed Cowbird **39**, 9, 23, 380
Bullock's Oriole **10**, 12, 13, 21, 206

Campo Troupial **18**, 196
Carib Grackle **33**, 21, 353
Casqued Oropendola **2**, 107
Chestnut-capped Blackbird **20**, 256
Chestnut-headed Oropendola **1**, 123
Chopi Blackbird **31**, 317
Colombian Mountain-Grackle **30**, 325
Common Grackle **35**, 23, 346
Crested Oropendola **1**, 19, 108
Cuban Blackbird **32**, 327

Dusky-green Oropendola **4**, 114

Eastern Meadowlark **27**, 292
Ecuadorian Cacique **8**, 150
Epaulet Oriole **9**, 12, 159

Forbes's Blackbird **31**, 315
'Fuerte's Oriole' **9**, 217

Giant Cowbird **6**, 12, 371
Golden-winged Cacique **7**, 145
Great-tailed Grackle **34**, 333
Greater Antillean Grackle **33**, 350
Green Oropendola **2**, 19, 21, 112

Hooded Oriole **11**, 12, 189
Hybrid Baltimore X Bullock's Orioles 205

Jamaican Blackbird **33**, 12, 21, 238
Jamaican Oriole **15**, 13, 167

'Lilian's Meadowlark' **27**, 300
Long-tailed Meadowlark **26**, 290

Martinique Oriole **15**, 12, 227
Melodious Blackbird **32**, 329
Montezuma Oropendola **3**, 19, 125
Montserrat Oriole **15**, 226
Moriche Oriole **9**, 157
Mountain Cacique **7**, 147

Nicaraguan Grackle **35**,, 344

Olive Oropendola **2**, 129
Orange Oriole **11**, 12, 169
Orange-backed Troupial **18**, 197
Orange-crowned Oriole **13**, 174
Orchard Oriole **9**, 12, 213
Oriole Blackbird **18**, 12, 240

'Pale Baywing' **37**, 367
Pale-eyed Blackbird **19**, 12, 248
Pampas Meadowlark **26**, 12, 287
Para Oropendola **3**, 131
Peruvian Meadowlark **26**, 12, 285

Red-bellied Grackle **30**, 311
Red-breasted Blackbird **25**, 281
Red-rumped Cacique **5**, 140
Red-shouldered Blackbird **21**, 275
Red-winged Blackbird **22**, 9, 14, 15, 19, 23, 258
Russet-backed Oropendola **4**, 117

Rusty Blackbird **36**, 17, 18, 357
Saffron-cowled Blackbird **29**, 245
Scarlet-headed Blackbird **25**, 12, 309
Scarlet-rumped Cacique **5**, 142
Scarlet-throated Tanager **31**, 318
Scott's Oriole **17**, 13, 235
Screaming Cowbird **37**, 368
Scrub Blackbird **32**, 331
Selva Cacique **7**, 23, 149
Shiny Cowbird **39**, 385
Slender-billed Grackle **35**, 342
Solitary Cacique **8**, 151
Spot-breasted Oriole **12**, 178
St Lucia Oriole **15**, 12, 229
Streak-backed Oriole **12**, 13, 21, 83
Subtropical Cacique **5**, 144

Tawny-shouldered Blackbird **21**, 273
Tepui Mountain-Grackle **30**, 323
Tricolored Blackbird **23**, 269
Troupial **18**, 12, 193

Unicolored Blackbird **19**, 249

Velvet-fronted Grackle **30**, 12, 322

Western Meadowlark **28**, 302
White-browed Blackbird **25**, 284
White-edged Oriole **13**, 176

Yellow Oriole **14**, 13, 164
Yellow-backed Oriole **14**, 13, 162
Yellow-billed Cacique **8**, 155
'Yellow-billed Oropendola' **4**, 119
Yellow-headed Blackbird **24**, 12, 19, 242
Yellow-hooded Blackbird **19**, 253
Yellow-rumped Cacique **6**, 136
Yellow-rumped Marshbird **29**, 306
Yellow-shouldered Blackbird **21**, 277
Yellow-tailed Oriole **13**, 172
Yellow-winged Blackbird **20**, 251
Yellow-winged Cacique **6**, 153

SCIENTIFIC NAMES

abeillei, Icterus **10**, 12, 211
aciculatus, Agelaius phoeniceus 261
aeneus, Molothrus **38**, 374
aeneus, Molothrus aeneus 375
aequatorialis, Molothrus bonariensis 387
affinis, Cacicus haemorrhus 141
Agelaioides (Molothrus) badius **37**, 364
Agelaius assimilis **21**, 275
Agelaius cyanopus **19**, 249
Agelaius humeralis **21**, 273
Agelaius icterocephalus **19**, 253
Agelaius phoeniceus **22**, 9, 14, 15, 19, 23, 258
Agelaius phoeniceus gubernator **23**, 266
Agelaius ruficapillus **20**, 256
Agelaius thilius **20**, 251
Agelaius tricolor **23**, 269
Agelaius xanthomus **21**, 277
Agelaius xanthophthalmus **19**, 12, 248
albipes, Sturnella bellicosa 286
alfredi, Psarocolius (angustifrons?) alfredi **4**, 120
alticola, Agelaius thilius 252
alticola, Icterus pustulatus 187
alticola, Sturnella magna 296
Amblycercus holosericeus **8**, 155
Amblyramphus holosericeus **25**, 12, 309
angustifrons, Psarocolius **4**, 117
argutula, Sturnella magna 296
armenti, Molothrus (aeneus) **38**, 378
arthuralleni, Agelaius phoeniceus 261
artimisiae, Molothrus ater 381
assimilis, Agelaius **21**, 275
assimilis, Molothrus aeneus 375
ater, Molothrus **39**, 9, 23, 380
ater, Molothrus ater 381
atrocastaneus, Psarocolius (angustifrons?) alfredi 120
atroolivaceus, Agelaius cyanopus 249–250
atroviolacea, Dives **32**, 327
atrovirens, Psarocolius **4**, 114
audubonii, Icterus graduacauda 232
auratus, Icterus **11**, 12, 169
auricapillus, Icterus **13**, 174
auropectoralis, Sturnella magna 296
australis, Amblycercus holosericeus 155

badius, Agelaioides (Molothrus) **37**, 12, 364
badius, Agelaioides (Molothrus) badius 365
bairdi, Icterus leucopteryx 168
bangsi, Quiscalus niger 351
bellicosa, Sturnella **26**, 12, 285
bellicosa, Sturnella bellicosa 286
beniensis, Agelaius cyanopus 250
bifasciatus, Psarocolius **3**, 131
bogotensis, Agelaius icterocephalus 254
bolivianus, Agelaioides (Molothrus) badius 364
bolivianus, Oreopsar **33**, 320
boliviensis, Lampropsar tanagrinus 322
bonana, Icterus **15**, 12, 227
bonariensis, Molothrus **39**, 385
bonariensis, Molothrus bonariensis 386
brachypterus, Quiscalus niger 352
bryanti, Agelaius phoeniceus 261
bullockii, Icterus **10**, 12, 13, 21, 206

cabanisii, Molothrus bonariensis 387
Cacicus cela **6**, 136
Cacicus chrysonotus **7**, 147
Cacicus chrysopterus **7**, 145
Cacicus haemorrhous **5**, 140
Cacicus koepckeae **7**, 23, 149
Cacicus melanicterus **6**, 153
Cacicus microrhynchus **5**, 142
Cacicus sclateri **8**, 150
Cacicus solitarius **8**, 151
Cacicus uropygialis **5**, 144
californicus, Agelaius gubernator 268
caribaeus, Quiscalus niger 351
carolinus, Euphagus **36**, 17, 18, 357
carolinus, Euphagus carolinus 358
carolynae, Icterus pectoralis 179
carrikeri, Icterus mesomelas 173
cassini, Psarocolius **3**, 128
castaneopectus, Icterus wagleri 219
catamarcana, Sturnella loyca 291
caurinus, Agelaius phoeniceus 261
cayanensis, Icterus **9**, 12, 159
cayanensis, Icterus cayanensis 160
caymanensis, Quiscalus niger 351

cela, Cacicus **6**, 136
cela, Cacicus cela 137
chopi, Gnorimopsar **31**, 317
chopi, Gnorimposar chopi 317–318
chrysater, Icterus **14**, 13, 162
chrysater, Icterus chrysater 162
chrysocephalus, Icterus **9**, 157
chrysonotus, Cacicus **7**, 147
chrysonotus, Cacicus chrysonotus 148
chrysopterus, Cacicus **7**, 145
Compsothraupis, loricata **31**, 318
contrusus, Quiscalus lugubris 355
crassirostris, Quiscalus niger 351
croconotus, Icterus **18**, 197
croconotus, Icterus croconotus 198
cucullatus, Icterus **11**, 12, 189
cucullatus, Icterus cucullatus 191
Curaeus curaeus **31**, 312
curaeus, Curaeus **31**, 312
curaeus, Curaeus curaeus 314
Curaeus forbesi **31**, 315
curasoensis, Icterus nigrogularis 165
cyanocephalus, Euphagus **36**, 16, 360
cyanopus, Agelaius **19**, 249
cyanopus, Agelaius cyanopus 249

decumanus, Psarocolius **1**, 19, 108
decumanus, Psarocolius decumanus 109
defillippi, Sturnella **26**, 12, 287
dickeyae, Icterus graduacauda **17**, 232
Dives atroviolacea **32**, 327
Dives dives **32**, 329
dives, Dives **32**, 329
Dives warszewiczi **32**, 331
Dolichonyx oryzivorus **28**, 12, 18, 19, 390
dominicensis, Icterus **16**, 12, 221
dominicensis, Icterus dominicensis 222

eleutherus, Icterus bullockii 208
espinachi, Icterus pectoralis 179
Euphagus carolinus **36**, 17, 18, 357
Euphagus cyanocephalus **36**, 16, 360

falklandica, Sturnella loyca 291
flavescens, Icterus gularis 182
flavicrissus, Cacicus cela 138
flavirostris, Amblycercus holosericeus 156
flavus, Xanthopsar **29**, 245
floridanus, Agelaius phoeniceus 260
forbesi, Curaeus **31**, 315
formosus, Icterus pustulatus 187
fortirostris, Quiscalus lugubris 355
fringillarius, Agelaioides (Molothrus) badius **37**, 367
frontalis, Agelaius ruficapillus 257
fuertesi, Icterus spurius **9**, 214

galbula, Icterus **10**, 12, 13, 21, 199
gigas, Icterus gularis 182
Giraudii, Icterus chrysater 163
Gnorimopsar chopi **31**, 317
graceannae, Icterus **13**, 176
graduacauda, Icterus **17**, 13, 231
graduacauda, Icterus graduacauda 232
graysonii, Icterus gularis 187
graysoni, Quiscalus mexicanus 335
grinnelli, Agelaius phoeniceus 262
griscom, Sturnella magna 296
guadelupensis, Quiscalus lugubris 354
guatimozinus, Psarocolius **3**, 132
gubernator, Agelaius gubernator 268
gubernator, Agelaius phoeniceus **23**, 266
guianensis, Lampropsar tanagrinus 322
guirahuro, Pseudoleistes **29**, 306
gularis, Icterus **12**, 13, 181
gularis, Icterus gularis 182
gundlachii, Quiscalus niger 351
guttulatus, Icterus pectoralis 179
Gymnomystax mexicanus **18**, 12, 240

haemorrhous, Cacicus **5**, 140
haemorrhous, Cacicus haemorrhous 141
helioeides, Icterus nigrogularis 165
hippocrepis, Sturnella magna 296
holosericeus, Amblycercus **8**, 155
holosericeus, Amblycercus holosericeus 156
holosericeus, Amblyramphus **25**, 12, 309
hondae, Icterus chrysater 163
hoopesi, Sturnella magna 296
humeralis, Agelaius **21**, 273
humeralis, Agelaius humeralis 274
Hypopyrrhus pyrohypogaster **30**, 311

icterocephalus, Agelaius **19** 253
icterocephalus, Agelaius icterocephalus 254
Icterus abeillei **10**, 12, 211
Icterus auratus **11**, 12, 169
Icterus auricapillus **13**, 174
Icterus bonana **15**, 12, 227
Icterus bullockii **10**, 12, 13, 21, 206
Icterus cayanensis **9**, 12, 159
Icterus chrysater **14**, 13, 162
Icterus chrysocephalus **9**, 157
Icterus croconotus **18**, 197
Icterus cucullatus **11**, 12, 189
Icterus dominicensis **16**, 12, 221
Icterus galbula **10**, 12, 13, 21, 199
Icterus galbula x bullockii 205
Icterus graceannae **13**, 176
Icterus graduacauda **17**, 13, 231
Icterus gularis **12**, 13, 181
Icterus icterus **18**, 12, 193
icterus, Icterus **18**, 12, 193
icterus, Icterus icterus 194
Icterus jamacaii **18**, 196
Icterus laudabilis **15**, 12, 229
Icterus leucopteryx **15**,13, 167
Icterus maculialatus **17**, 233
Icterus mesomelas **13**, 172
Icterus nigrogularis **14**, 13, 164
Icterus oberi **15**, 12, 226
Icterus parisorum **17**, 13, 235
Icterus pectoralis **12**, 178
Icterus pustulatus **12**, 13, 21, 183
Icterus spurius **9**, 12, 213
Icterus spurius fuertesi **9**, 214
Icterus wagleri **16**, 12, 218
igneus, Icterus cucullatus 191
impacifica, Scaphidura (Molothrus) oryzivora 372
imthurni, Macroagelaius **30**, 323
inexpectata, Sturnella magna 296
inflexirostris, Quiscalus lugubris 355
insularis, Psarocolius decumanus 109
insularis, Quiscalus lugubris 355

jamacaii, Icterus **18**, 196

kalinowskii, Dives warszewiczi 332
koepckeae, Cacicus **7**, 23, 149

Lampropsar tanagrinus **30**, 12, 322
latirostris, Ocyalus **1**, 134
laudabilis, Icterus **15**, 12, 229
lawrencii, Icterus leucopteryx 168
leucopteryx, Icterus **15**, 13, 167
leucopteryx, Icterus leucopteryx 168
leucoramphus, Cacicus chrysonotus 147
lilianae, Sturnella magna **27**, 300
littoralis, Agelaius phoeniceus 261
loricata, Compsothraupis **31**, 318
loweryi, Quiscalus mexicanus 335
loyca, Sturnella **26**, 290
loyca, Sturnella loyca 291
loyei, Molothrus aeneus 374
lugubris, Quiscalus **33**, 353
lugubris, Quiscalus lugubris 354
luminosus, Quiscalus lugubris 355

Macroagelaius simthurni **30**, 323
Macroagelaius subularis **30**, 12, 325
macropterus, Lampropsar tanagrinus 322
maculialatus, Icterus **17**, 233
maculosus, Psarocolius decumanus 109
magna, Sturnella **27**, 292
magna, Sturnella magna 296
mailliardorum, Agelaius gubernator 268
major, Quiscalus **34**, 338
major, Quiscalus major 339
maximus, Icterus pustulatus 87
mayensis, Icterus chrysater 163
mearnsi, Agelaius phoeniceus 260
megapotamus, Agelaius phoeniceus 261
melanicterus, Cacicus **6**, 153
melanopsis, Icterus dominicensis 222
melanterus, Psarocolius decumanus 109
meridionalis, Sturnella magna 297
mesomelas, Icterus **13**, 172
mesomelas, Icterus mesomelas 173

metae, Icterus icterus 194
mexicana, Sturnella magna 296
mexicanus, Gymnomystax **18**, 12, 206
mexicanus, Quiscalus **34**, 333
mexicanus, Quiscalus mexicanus 335
microrhynchus, Cacicus **5**, 142
microrhynchus, Cacicus microrhynchus 143
microstictus, Icterus pustulatus 187
militaris, Sturnella **25**, 281
minimus, Molothrus bonariensis 386
Molothrus (aeneus) armenti **38**, 378
Molothrus aeneus **38**, 374
Molothrus (Ageloioides) badius **37**, 12, 364
Molothrus (Agelaioides) badius fringillarius **37**, 367
Molothrus ater **39**, 9, 23, 380
Molothrus bonariensis **39**, 385
Molothrus rufoaxillaris **37**, 368
monensis, Agelaius xanthomus 278
monsoni, Quiscalus mexicanus 335
montezuma, Psarocolius **3**, 19, 125

nayaritensis, Agelaius phoeniceus 262
nayaritensis, Icterus graduacauda 232
neglecta, Sturnella **28**, 302
neglectus, Psarocolius (angustifrons?) alfredi 121
neivae, Psaracolius yuracares 130
nelsoni, Agelaius phoeniceus 262
nelsoni, Icterus cucullatus 191
nelsoni, Quiscalus mexicanus 335
Nesopsar nigerrimus **33** 12, 21, 238
neutralis, Agelaius phoeniceus 261
nevadensis, Agelaius phoeniceus 261
nicaraguensis, Quiscalus **35**, 344
niger, Quiscalus **33**, 350
niger, Quiscalus niger 351
nigerrimus, Nesopsar **33**, 12, 21, 238
nigrans, Euphagus carolinus 358
nigrogularis, Icterus **14**, 13, 164
nigrogularis, Icterus nigrogularis 165
northropi, Icterus dominicensis 222

oberi, Icterus **15**, 12, 226
obscura, Sturnella loyca 291
obscurus, Molothrus ater 381
obscurus, Quiscalus mexicanus 335
occidentalis, Molothrus bonariensis 187
Ocyalus latirostris **1**, 134
oleagineus, Psaroclius (angustifrons?) alfredi 121
Oreopsar bolivianus **33**, 320
orquillensis, Quiscalus lugubris 355
oryzivora, Scaphidura (Molothrus) **6**, 12, 371
oryzivora, Scaphidura (Molothrus) oryzivora 371
oryzivorus, Dolichonyx **28**, 12, 18, 19, 390
oseryi, Psarocolius **2**, 107

pachyrhynchus, Cacicus haemorrhous 141
pacificus, Cacicus microrhynchus 143
pallidulus, Agelaius phoeniceus 261
palustris, Quiscalus **35**, 342
paralios, Sturnella magna 297
parisorum, Icterus **17**, 13, 235
parvus, Icterus bullockii 207
pectoralis, Icterus **12**, 178
pectoralis, Icterus pectoralis 179
periporphyrus, Icterus cayanensis 160
peruvianus, Cacicus chrysonotus 148
peruvianus, Quiscalus mexicanus 335
petersi, Agelaius thilius 252
phillipsi, Icterus spurius 214
phoeniceus, Agelaius **22**, 9, 14, 15, 19, 23, 258
phoeniceus, Agelaius phoeniceus 260
phoeniceus, gubernator Agelaius **23**, 266
portoricensis, Icterus dominicensis 224
practicola, Sturnella magna 297
praecox, Icterus dominicensis 222
prosopidicola, Quiscalus mexicanus 334
prosthemelas, Icterus dominicensis 222
Psarocolius angustifrons **4**, 117
Psarocolius (angustifrons) alfredi **4**, 119
Psarocolius atrovirens **4**, 114
Psarocolius bifasciatus **3**, 131
Psarocolius (bifasciatus?) yuracares **2**, 129
Psarocolius cassini **3**, 128
Psarocolius decumanus **1**, 19, 108
Psarocolius guatimozinus **3**, 132
Psarocolius montezuma **3**, 19, 125
Psarocolius oseryi **2**, 107
Psarocolius viridis **2**, 19, 21, 112
Psarocolius wagleri **1**, 123
Pseudoleistes guirahuro **29**, 306
Pseudoleistes virescens **29**, 307
pustulatus, Icterus **12**, 13, 21, 183
pustulatus, Icterus pustulatus 187
pustuloides, Icterus pustulatus 187
pyrohypogaster, Hypopyrrhus **30**, 311
pyrrhopterus, Icterus cayanensis 160

quinta, Sturnella magna 297
Quiscalus lugubris **3**, 21, 353
Quiscalus major **34**, 338
Quiscalus mexicanus **34**, 333
Quiscalus nicaraguensis **35**, 344
Quiscalus niger **33**, 350
quiscalus, niger Quiscalus 351
Quiscalus palustris **35**, 342
Quiscalus quiscula **35**, 23, 346
quiscula, Quiscalus **35**, 346
quiscula, Quiscalus quiscula 347

recurvirostris, Curaeus curaeus 314
reynoldsi, Curaeus curaeus 314
richmondi, Agelaius phoeniceus 261
ridgwayi, Icterus icterus 194
ridgwayi, Psarocolius wagleri 124
riparius, Molothrus bonariensis 387
ruficapillus, Agelaius **20**, 256
ruficapillus, Agelaius ruficapillus 257
rufoaxillaris, Molothrus **37**, 368

salmoni, Psarocolius (angustifrons?) alfredi 120
salvinii, Icterus mesomelas 173
saundersi, Sturnella magna 296
Scaphidura (Molothrus) oryzivora **6**, 12, 371
sclateri, Cacicus **8**, 150
sclateri, Icterus pustulatus 187

scopulus, Agelaius humeralis 274
sennetti, Icterus cucullatus 191
sincipitalis, Psarocolius (angustifrons) alfredi 121
solitarius, Cacicus **8**, 151
sonoriensis, Agelaius phoeniceus 261
spurius, Icterus **9**, 12, 213
spurius, Icterus fuertesi **9**, 213
spurius, Icterus spurius 214
stonei, Quiscalus quiscula 347
strictifrons, Icterus croconotus 198
Sturnella bellicosa **26**, 12, 285
Sturnella defillippi **26**, 12, 287
Sturnella loyca **26**, 290
Sturnella magna **27**, 292
Sturnella magna lilianae **27**, 300
Sturnella militaris **25**, 281
Sturnella neglecta **28**, 302
Sturnella superciliaris **25**, 284
subniger, Agelaius assimilis 276
subularis, Macroagelaius **30**, 12, 325
subulata, Sturnella magna 296
sulcirostris, Gnorimopsar chopi 319
superciliaris, Sturnella **25**, 284

taczanowskii, Icterus mesomelas 173
tamaulipensis, Icterus gularis 182
tanagrinus, Lampropsar **30**, 12, 322
tanagrinus, Lampropsar tanagrinus 322
thilius, Agelaius **20**, 251
thilius, Agelaius thilius 252
tibialis, Icterus cayanensis 159
torreyi, Quiscalus major 339
tricolor, Agelaius **23**, 269
trinitatis, Icterus nigrogularis 165
trochiloides, Icterus cucullatus 191
troglodytes, Icterus gularis 182

uropygialis, Cacicus **5**, 144

valenciobuenoi, Icterus cayanensis 159
venezuelensis, Molothrus bonariensis 387
versicolor, Quiscalus quiscula 347
violaceus, Lampropsar tanagrinus 322
virescens, Pseudoleistes **29**, 307
viridis, Psarocolius **2**, 19, 21, 112
vitellinus, Cacicus cela 137

wagleri, Icterus **16**, 12, 218
wagleri, Icterus wagleri 219
wagleri, Psarocolius **1**, 123
wagleri, Psarocolius wagleri 124
warszewiczi, Dives **32**, 331
warszewiczi, Dives warszewiczi 332

Xanthocephalus xanthocephalus **24**, 12, 19, 242
xanthocephalus, Xanthocephalus **24**, 12, 19, 242
xanthomus, Agelaius **21**, 277
xanthomus, Agelaius xanthomus 278
xanthomus, Agelaius xanthomus 277–278
xanthophthalmus, Agelaius **19**, 12, 248
Xanthopsar flavus **29**, 245
xenicus, Agelaius cyanopus 242

yucatanensis, Icterus gularis 182
yuracares, Psarocolius (bifasciatus?) yuracares **2**, 129